Automotive Heating and Air Conditioning

Third Edition

Tom Birch

Upper Saddle River, New Jersey
Columbus, Ohio

Library of Congress Cataloging-in-Publication Data

Birch, Thomas W.
Automotive heating and air conditioning / Tom Birch.—3rd ed.
p. cm.
Includes index.
ISBN 0-13-099366-2
1. Automobiles—Heating and ventilation. 2. Automobiles—Air conditioning. I. Title.

TL271 .B57 2003
629.2'772—dc21 2001058821

Editor in Chief: Stephen Helba
Executive Editor: Ed Francis
Production Editor: Christine M. Buckendahl
Production Coordination: Carlisle Publishers Services
Design Coordinator: Diane Y. Ernsberger
Cover Designer: Jason Moore
Cover Art: Visteon Corporation
Production Manager: Brian Fox
Marketing Manager: Mark Marsden

This book was set in Century Book by Carlisle Communications, Ltd., and was printed and bound by Banta Book Group. The cover was printed by The Lehigh Press, Inc.

Pearson Education Ltd.
Pearson Education Australia Pty. Limited
Pearson Education Singapore Pte. Ltd.
Pearson Education North Asia Ltd.
Pearson Education Canada, Ltd.
Pearson Educación de Mexico, S.A. de C.V.
Pearson Education—Japan
Pearson Education Malaysia Pte. Ltd
Pearson Education, *Upper Saddle River, New Jersey*

10 9 8 7 6 5 4 3
ISBN: 0-13-099366-2

PREFACE

The automobile has undergone a continuous evolution since its introduction in the late 1800s. Today's vehicles still incorporate most of the major features of the very early vehicles, yet are vastly different in many aspects. Numerous comfort features, such as heating, air conditioning, and ventilation (HVAC) systems, have been added to make driving safer and more enjoyable. Durability and longevity have increased continuously, although vehicles still break and wear out. Drivability and fuel mileage are slowly but steadily improving, while exhaust emissions are steadily decreasing.

The HVAC system is becoming one of the more durable areas of a vehicle while also becoming very complex with electronic controls, such as dual- and three-zone temperature controls. Also, the lingering effects of environmental regulations for chlorofluorocarbon (CFC) and in the near future hydrochlorofluorocarbon (HCFC) reduction are causing refinements in HVAC system service procedures. These changes are affecting millions of older vehicles that are still on the road. As the automobile becomes more sophisticated and complex, more sophisticated technicians with better training are required to keep up with the changes in regulations, to determine the causes of any malfunction, and to repair problems.

Automotive Heating and Air Conditioning, Third Edition, was written to help the student understand the HVAC system used in cars, sport utility vehicles, pickups, light trucks, and vans. It describes the theory of operation for the major power transfer and control systems, as well as the methods to diagnose and repair common problems. The text covers traditional and modern HVAC systems and engine cooling systems. This edition has been expanded to include new regulations and service methods. It is arranged so that the major components of these systems are described completely with regard to theory of basic operation and methods used to diagnose, adjust, and repair problems.

The theory chapters (1 through 8 and 15) provide the student technician with an understanding of how the various systems work, the terminology used, and the variety of systems that can be encountered. For example, there are three major ways of transmitting motion and force from the HVAC control head to the airflow control doors. Modern HVAC systems use electronics for the same purpose that earlier systems used mechanical cables. Even though these newer systems are based on principles of the past, the modern technician needs to keep abreast of the technical advancements as vehicles evolve. If he or she understands how the systems operate, successful problem diagnosis and repair become much easier.

The service and repair chapters (Chapters 9 through 14 and 16) complete the coverage of each system by describing the procedure used to maintain a system for proper operation; diagnosing the cause of a problem if it does occur; and then repairing or adjusting the unit to correct the root cause of the problem. These topics are covered in a very generic fashion, to apply to as many makes and types of vehicles as possible. Service information in a printed or electronic form is commonly used in automotive repair industry. It takes more than just the ability to read for the student-technician to understand and use this information effectively. This author intends not to replace these manuals, but rather to supplement them so the student-technician can gain full benefit from these valuable sources of information. This book describes how to perform service operations, and the service information tells what operations are necessary for a particular service operation.

The service chapters have been made more realistic by the addition of *Service Tips*, which describe procedures that can bring a job to completion faster or ensure a more thorough repair. *Real World Fixes* have also been added; these are case studies of how technicians have solved and repaired problems.

This book covers all the areas contained in the Automotive Service Excellence (ASE) test for Automotive Heating and Air Conditioning, and the ASE Task List for this area is included as Appendix A. With class instruction and shop experience, the student-technician should have no trouble passing this test.

For the instructor, a complete instructor's guide to accompany this text is available from the publisher. The contents include:

- Sources of teaching aids
- A listing of web sites related to this subject

- A list correlating the ATTS Certification test skills and NATEF tasks to the text
- An outline for each chapter of the text with a set of reading questions, and additional test questions in three different styles
- Student exercises and job report sheets
- The answer key for the chapter review and quiz questions

This book has the support of much of the HVAC and cooling system repair industry. I am grateful to the following companies and individuals for their contributions:

ACDelco
Acme Radiator & Air Conditioning
Airsept, Inc.
American Lokring Corporation
Appollo America Corporation, Lorie Homolish
BLR Enterprises
Bright Solutions
Castrol North America
Cliplight Manufacturing Company
DaimlerChrysler Corporation
Dayco Products, Inc.
Environmental Test Systems, Inc.
Everco Industries
Fedco Automotive Components, Patrick L. O'Conner
Four Seasons
The Gates Rubber Company
General Motors Corporation
Goodyear Tire & Rubber Company
International Mobile Air Conditioning Association (IMACA), Executive Director, Frank Allison
James Halderman
John Fluke Mfg. Co.
Kent-Moore
Mobile Air Conditioning Society (MACS), Paul De Guiseppi
Mobile Air Conditioning Society (MACS), Simon Oulouhojian, Past President
Mastercool
Modine Manufacturing
Nartron Corp./Smart Power Products
Neutronics Inc.
Nissan Motor Corporation
Purdue University, Frederick Peacock
Raytek Corp.
Red Dot Corp.
Robinaire Division/SPX Corp.
Sanden International (USA)
Santech Industries
Saturn Corporation
Selective Technology, Seltec
Sercon Spectronics Corporation
Society of Automotive Engineers (SAE)
Stant Manufacturing
System Guard
TDR Stabilizer Clamp Company
Technical Chemical Company (TCC)
TIF Instruments
Toyota Motor Corporation
Tracer Products
UView Ultraviolet Systems
Visteon Corporation
Waekon Industries
Warner Electric
White Industries
Wynn Oil Company
Yokogawa Corporation of America
Yuba College, Bill Steen
Zexel Illinois

Portions of materials contained herein have been reprinted with permission of General Motors Corporation, Service Operations

Finally, I would like to thank the following reviewers for their helpful suggestions: Roger Donovan, Illinois Central College; Paul Rossiter, Central Florida Community College; Don Schinker, Madison Area Technical College; and Mitchell Walker, St. Louis Community College.

Tom Birch

CONTENTS

PRENTICE HALL MULTIMEDIA SERIES IN AUTOMOTIVE TECHNOLOGY

Automotive Electrical and Electronic Systems: A Worktext, First Edition
Halderman | 0134768132 | © 1999

Automotive Heating and Air Conditioning, Third Edition
Birch | 0130993662 | © 2003

Manual Drive Trains and Axles, Third Edition
Birch | 0130339660 | © 2002

Automatic Transmissions and Transaxles, First Edition
Birch & Rockwood | 013846945 | © 2002

Automotive Heating and Air Conditioning, Second Edition
Birch | 0130135755 | © 2000

Guide to the ASE Exam—Engine Assembly Specialist, First Edition
Halderman | 0130191884 | © 2003

Guide to the ASE Exam—Cylinder Block Specialist, First Edition
Halderman | 013019185X | © 2003

Automotive Chassis Systems, Third Edition
Halderman & Mitchell | 0130484253 | © 2004

Automotive Brake Systems, Third Edition
Halderman & Mitchell | 0130475076 | © 2004

Automotive Steering, Suspension, and Alignment, Third Edition
Halderman & Mitchell | 0130488569 | © 2004

Undervehicle Service, First Edition
Mitchell & Halderman | 0130409685 | © 2004

Guide to the ASE Exam—Brakes, First Edition
Halderman | 0130191841 | © 2003

Guide to the ASE Exam—Automatic Transmissions and Transaxles, First Edition
Halderman | 0130191825 | © 2003

Guide to the ASE Exam—Electrical and Electronic Systems, First Edition
Halderman | 0130191876 | © 2003

Automotive Engine Performance, First Edition
Halderman | 0130488518 | © 2003

Guide to the ASE Exam—Engine Performance, First Edition
Halderman | 0130191892 | © 2003

Guide to the ASE Exam—Manual Drive Trains and Axles, First Edition
Halderman | © 2003

Guide to the ASE Exam—Heating and Air Conditioning, First Edition
Halderman | 0130191914 | © 2003

Guide to the ASE Exam—Steering and Suspension, First Edition
Halderman | 0130191949 | © 2003

Guide to the ASE Exam—Engine Repair, First Edition
Halderman | 0130191809 | © 2003

Automotive Electricity and Electronics, First Edition
Halderman & Mitchell | 0130842249 | © 2003

Guide to the ASE Exam—Advanced Engine Performance, First Edition
Halderman & Mitchell | 0130310956 | © 2003

Introduction to Automotive Service, First Edition
Halderman & Mitchell | 0139090940 | © 2004

Automotive Technology: Principles, Diagnosis, and Service, Second Edition
Halderman & Mitchell | 0130994537 | © 2003

Advanced Engine Performance and Worktext and CD Package, Second Edition
Halderman & Mitchell | 013096784X | © 2002

Diagnosis and Troubleshooting of Automotive Electrical, Electronic, and Computer Systems, Third Edition
Halderman & Mitchell | 0130799793 | © 2001

Automotive Engines: Theory and Servicing, Fourth Edition
Halderman & Mitchell | 013799019 | © 2001

CHAPTER

Basics of Heating and Air Conditioning

LEARNING OBJECTIVES

After completing this chapter, you should:

- Have a basic understanding of the purpose of the heating, ventilation, and air conditioning (HVAC) system in automobiles and other vehicles.
- Have a basic understanding of temperature and humidity comfort zones.
- Know how heat is measured in both temperature and quantity, using either metric or U.S./British units.

TERMS TO LEARN

air distribution	Fahrenheit
air management	heat
British Thermal Unit (BTU)	humidity
	intensity
cabin filter	quantity
calorie	relative humidity (RH)
Celsius	thermal
comfort zone	watt

1.1 INTRODUCTION

For the most part, the HVAC system of an automobile is designed to provide comfort for the driver and passengers. It is intended to maintain in-car temperature and humidity within a range that is comfortable for the people inside and provide fresh, clean air for ventilation. This temperature range also helps keep the driver alert and attentive.

Heating and air conditioning (A/C) systems control several aspects of an in-vehicle environment. This is often called climate control. The most noticeable of these aspects is temperature. Two other major aspects are humidity and air cleanliness. We are all familiar with the temperature aspect—if it is too hot or too cold. Many of us realize the effects of humidity and how unpleasant a hot day can be if the humidity is too high. We all know that dusty, foul, or smelly conditions are unpleasant (Figure 1–1).

The HVAC system in a vehicle (Figure 1–2) can be divided into three closely related subsystems:

- **Air distribution**, also called **air management**, with the control system
- **Heating**
- **Refrigeration, A/C**

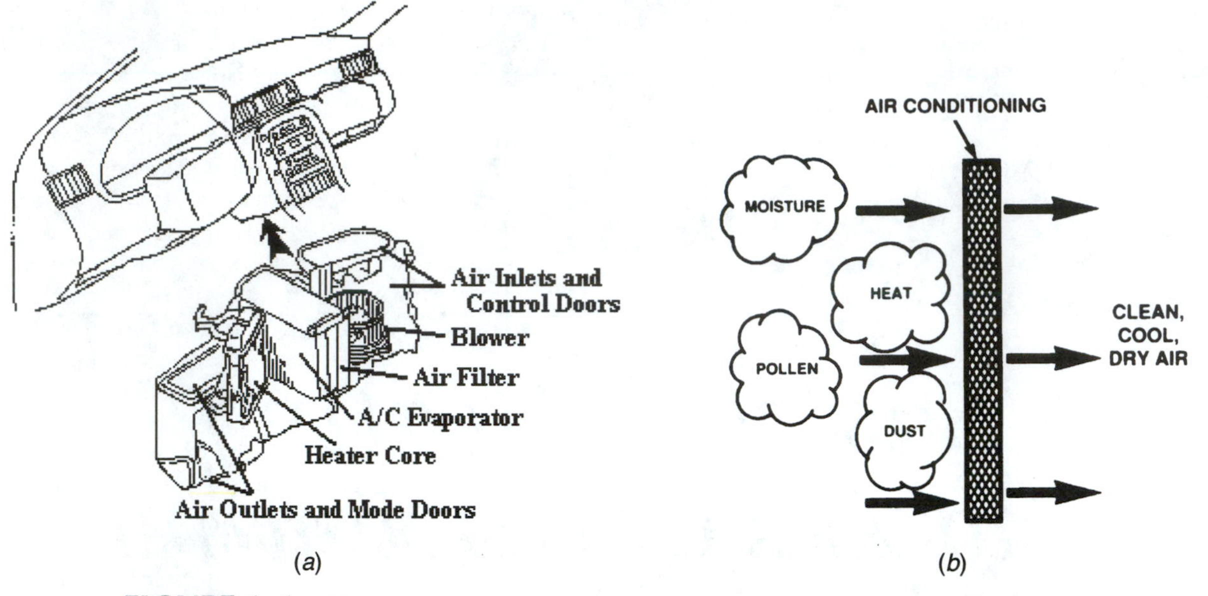

FIGURE 1–1 The air moves through the HVAC system's ducts, evaporator, and cabin filter in modern systems (a). Excess heat, moisture, dust, and pollen are removed to condition the air and make it more comfortable. (b, *reprinted with permission of General Motors Corporation*)

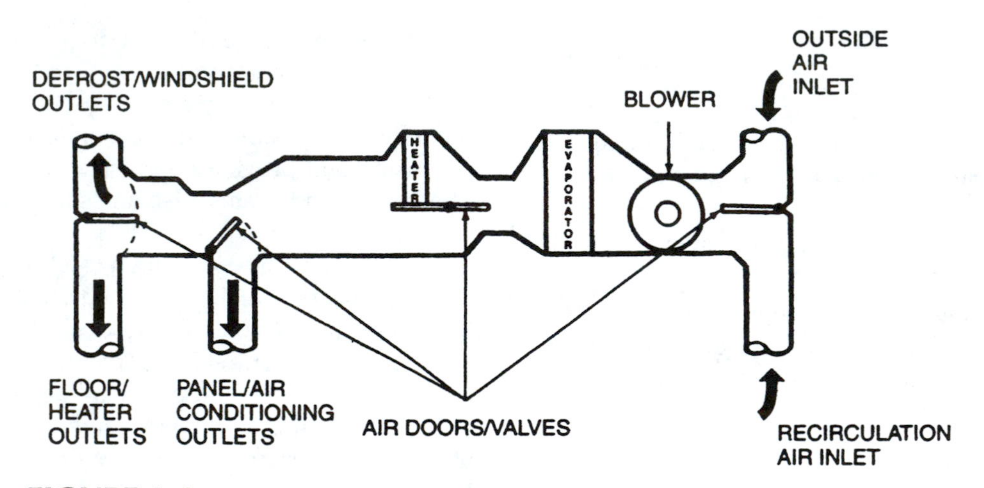

FIGURE 1–2 A typical HVAC system has a choice of two inlets (right), a blower to move the air, an A/C evaporator core and heater core, and several air discharge outlets. *(Reprinted with permission of General Motors Corporation)*

In this text, we will study these factors, concentrating for the most part on the ways we can control air temperature. Because much of this study has to do with heat and how we move it, we need to cover the basic principles of heat. Air conditioning becomes relatively easy to understand when we are familiar with concepts such as latent heat and the pressure–temperature relationships of saturated vapors; it is almost impossible to understand A/C operation otherwise.

1.2 HEAT

Heat is a form of basic energy, and, like other forms of energy, heat cannot be created or destroyed. It can, however, be converted to or from other forms of energy (Figure 1–3). Most automotive students realize that the engine converts the potential energy found in a fuel such as gasoline into heat. This **thermal** (heat) energy increases the air pressure in the cylinders, which in turn forces the crankshaft to revolve. Crankshaft rotation is mechanical energy, and this mechanical energy drives the car. The engine does not make the energy, but it transfers the energy from the gasoline to the rotating crankshaft (Figure 1–4).

An engine is not capable of converting all of the heat from the gasoline into mechanical energy in the short period of a combustion cycle. Because energy cannot be destroyed, this leftover heat energy is sent out the exhaust pipe or to the cooling system. We can feel the waste heat at both of these places (Figure 1–5).

FIGURE 1–3 Heat, like light and electricity, is one of the basic forms of energy. *(Courtesy of Daimler Chrysler Corporation)*

Along with the concept of heat comes the concept of cold. Cold is merely the absence of heat. It is what is left if we remove all heat energy. Heat and cold are much like light and dark. Light is a form of energy, and dark is the absence of light. Light travels toward dark much like heat travels toward cold (Figure 1–6). The action of light is easy to see when we turn on a light in a dark room; the action of heat traveling toward cold can be felt as we come close to a hot stove. Heat is not affected by gravity, and its direction of travel can be upward as easily as downward, depending on where something is colder (Figure 1–7).

As we study heat further, we will find that it is relatively easy to measure and its methods of movement are very predictable and controllable.

1.3 HEAT MEASUREMENT

A heating and air conditioning technician is concerned with measuring two different aspects of heat: **intensity** and **quantity**. Intensity is what we feel; it is measured in degrees, on either a **Celsius** or **Fahrenheit** scale. Quantity is the actual amount of heat; it is measured in either **calories** or **British thermal units (Btu)**.

1.3.1 Heat Intensity

Intensity of heat is important to us because it is what we feel. If it is too cold, we are uncomfortable. Extremely cold temperatures can cause frostbite and hypothermia.

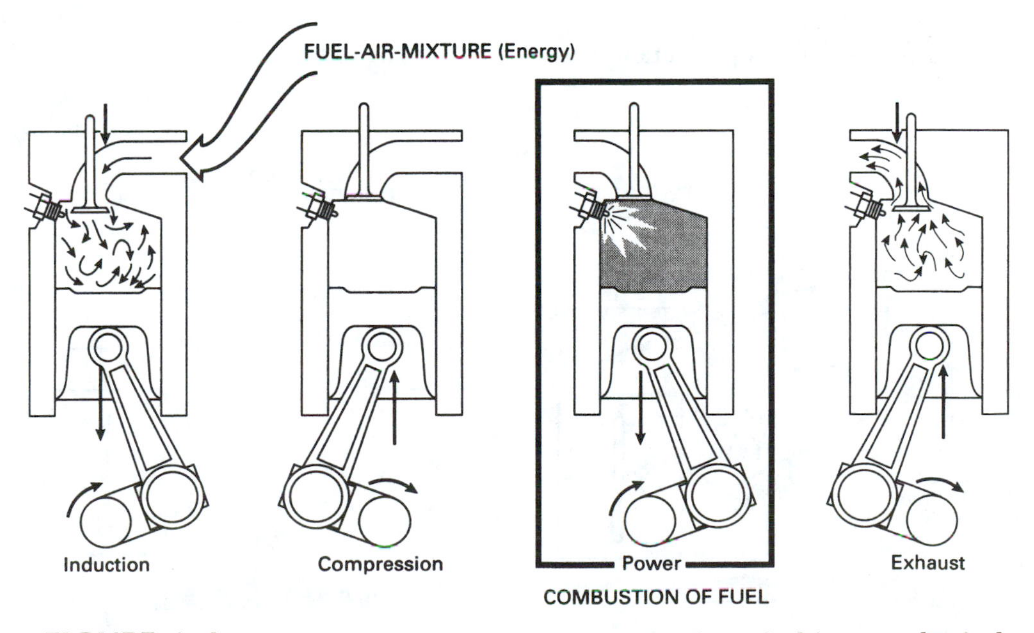

FIGURE 1–4 An engine converts potential energy from fuel into mechanical energy.

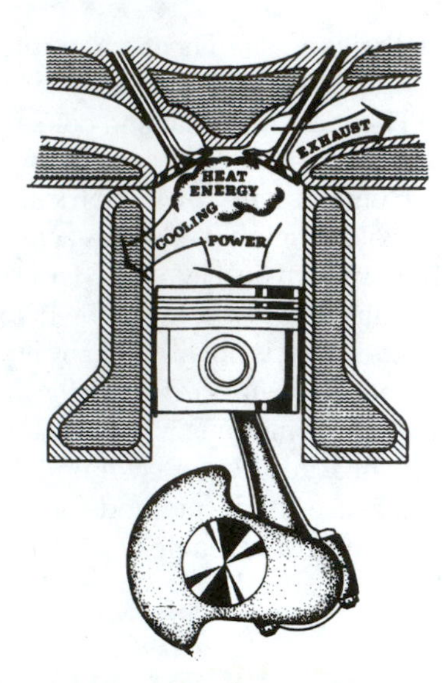

FIGURE 1–5 About one-third of the heat in an engine is converted into useful power. The other two-thirds are lost to the cooling and exhaust systems.

The other end of the scale can also be uncomfortable and may cause heat stress and dehydration. Humans have a temperature **comfort zone** somewhere between 65 and 80°F (21 and 27°C). This comfort zone varies among individuals. Women tend to prefer slightly warmer temperatures than men, and older people tend to have a narrower comfort zone and prefer warmer temperatures than younger people (Figure 1–8).

Critical temperatures are also very important to us. A good example of this is the boiling point of many automotive cooling systems, about 260°F (126°C). If this temperature is exceeded, coolant will boil away, leaving the moving parts of the engine without protection. The next critical temperature is the breakdown point of the oil, a loss of lubrication, and probably bearing burnout or piston seizure to the cylinders.

Temperature or heat intensity scales allow us to discuss the amount of heat in an area or object and let us define the range of operation for people and objects.

1.3.1.1 Measuring Intensity Heat intensity is easily measured using a thermometer; as just mentioned, two different scales are in common usage. The Celsius scale, sometimes called the centigrade scale, is a metric measurement. The base for this scale, 0°C, is the freezing point of water; 100°C is the boiling point of water. The Fahrenheit scale has commonly been used in the United States and England. The freezing point of water on this scale is 32°F, and the boiling point is 212°F. We often use water when we discuss heat because we are familiar with it, and we can easily get an idea of its temperature by observing what state it is in: The temperature of ice must be below 32°F (0°C), and water as a liquid is usually between 32 and 212°F (0 and 100°C) (Figure 1–9). If we take away all heat, we are left with absolute cold or absolute zero. This temperature, which we will never encounter in the normal world, is –459.67°F (–273.15°C).

There are several ways to convert temperature between the Celsius and Fahrenheit scales. This conversion becomes easier if we remember these major points:

- 0°C = 32°F
- 100°C = 212°F
- The Celsius scale is 100° long from freezing to boiling
- The Fahrenheit scale is 212° long from freezing to boiling

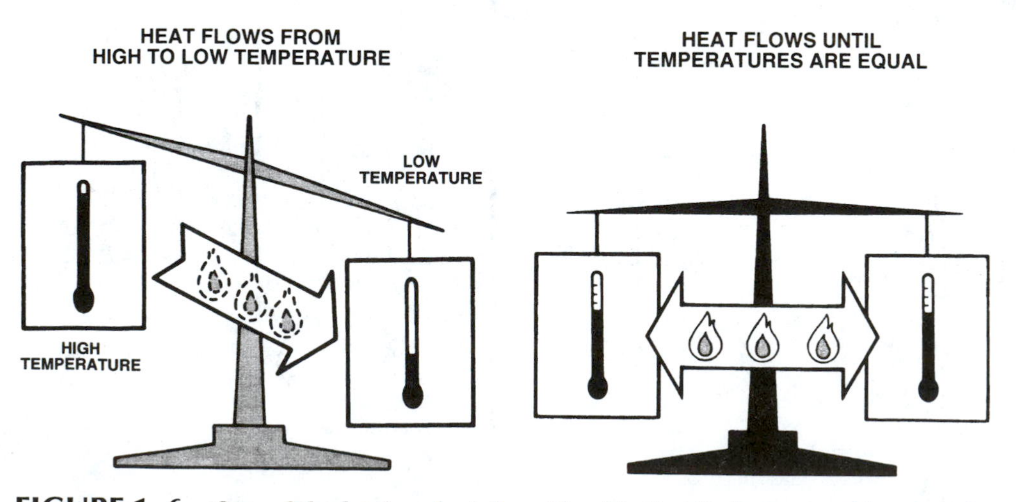

FIGURE 1–6 One of the basic principles of heat is that it always travels toward a colder area. *(Reprinted with permission of General Motors Corporation)*

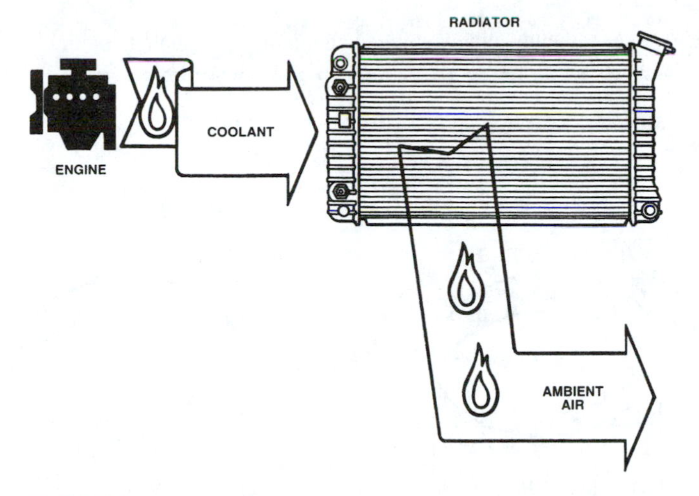

FIGURE 1–7 In a cooling system, heat flows from the hot engine parts to the colder coolant, then to the colder radiator, and then to the colder ambient air. *(Reprinted with permission of General Motors Corporation)*

TEMPERATURE COMPARISON

	Degrees C	Degrees F	
	140°	284°	
	130°	266°	
	20°	248°	
	110°	230°	
NORMAL BOILING POINT OF WATER	100°	212°	
	90°	194°	
	80°	176°	COMFORT ZONE OF MOST HUMANS
	70°	158°	
	60°	140°	
	50°	122°	
	40°	104°	
	30°	86°	
	20°	68°	
	10°	50°	
NORMAL FREEZING POINT OF WATER	0°	32°	
	-10°	14°	R-134a BOILS (–15°F)
	-20°	-4°	
	-30°	-22°	R-12 BOILS (–22°F)
	-40°	-40°	
	-50°	-58°	

FIGURE 1–8 Heat intensity is measured using a thermometer. The two common measuring scales, Celsius and Fahrenheit, are shown here. This thermometer is also marked with important temperatures.

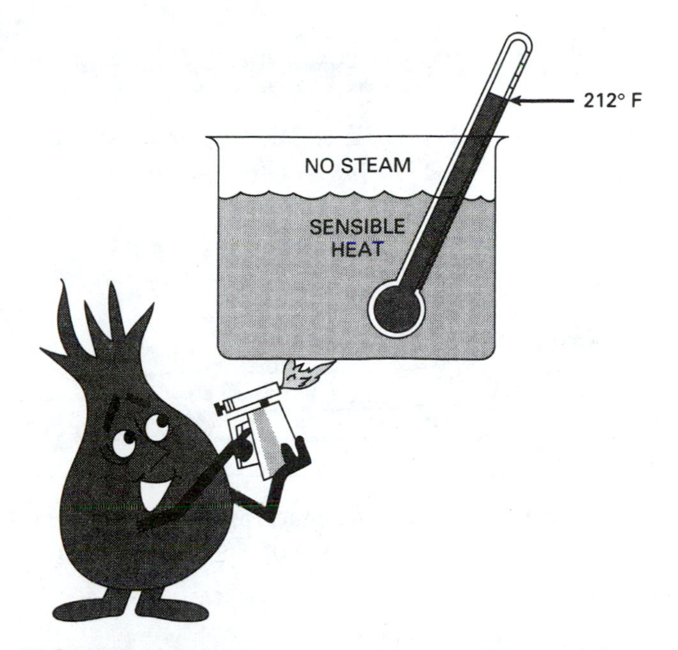

FIGURE 1–9 If water is heated, its temperature can increase to 212°F before boiling. The added heat will cause the increase in temperature. *(Courtesy of DaimlerChrysler Corporation)*

The commonly used conversion formulas are as follows:

$$\text{Temperature in } ^\circ\text{C} \times 1.8 + 32 = \text{Temperature in } ^\circ\text{F } \textit{or}$$

$$\frac{\text{Temperature in } ^\circ\text{C} \times 9 + 32}{5} = \text{Temperature in } ^\circ\text{F } \textit{or}$$

$$\text{Temperature in } ^\circ\text{F} - 32 \times 0.556 = \text{Temperature in } ^\circ\text{C } \textit{or}$$

$$\text{Temperature in } ^\circ\text{F} - 32 \times 5 \div 9 = \text{Temperature in } ^\circ\text{C}$$

The simplest way to convert between the Fahrenheit and Celsius scales is to use a conversion chart (Figure 1–10).

1.3.2 Heat Quantity

We use the concept of heat quantity to illustrate heat movement and transfer and to discuss efficiency. To heat or cool a person or object, we must move a certain amount of heat either to it or away from it. For example, when you buy a gallon of gasoline, you are really purchasing about 115,000 Btu of thermal energy, and your plan is to move that amount of heat into the engine to produce power (Figure 1–11).

Engineers rate heating and air conditioning units in calories—actually kilocalories (1,000 calorie units) (1 kilocalorie per second = 4.1868 kW)—or watts or Btu so that heating or air conditioning units can be sized to fit the load. Once they determine the expected heating or

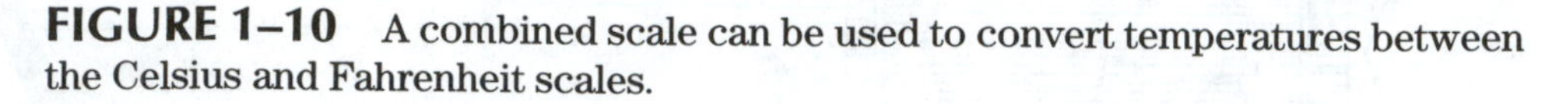

FIGURE 1–10 A combined scale can be used to convert temperatures between the Celsius and Fahrenheit scales.

FIGURE 1–11 We put about 115,000 Btu of heat energy in the car with each gallon of gas we buy.

cooling load, they can select a unit with the correct size specification.

1.3.2.1 Measuring Quantity When we need to cool the interior of a vehicle that is at 122°F (50°C) down to 75°F (24°C), we become involved with removing a certain amount of heat energy from the vehicle. This quantity depends on several factors: The size of the vehicle and the sun load are the most important. As mentioned earlier, we measure these quantities of heat in Btu, calories, or kilowatts.

A Btu is the amount of heat it takes to raise the temperature of 1 lb of water 1°F. Very similarly, a calorie is defined as the amount of heat it takes to increase the temperature of 1 g of water 1°C. A Btu is much larger: 1 Btu equals 252 calories. The burning of a kitchen-size wooden match gives off about 1 Btu when it burns completely (Figure 1–12).

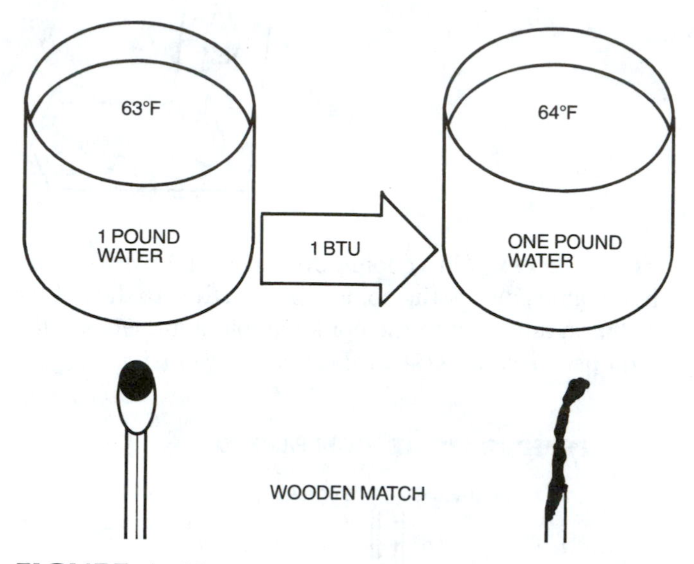

FIGURE 1–12 A wooden match produces about 1 Btu of heat when it burns. This amount of heat will increase the temperature of 1 lb of water 1°F. *(Courtesy of Fluke; reproduced with permission)*

1.4 COMFORT

Our goal in heating and air conditioning is to maintain a comfortable in-vehicle temperature and humidity. This is affected by the size of the vehicle, the number of passengers, and the amount of glass area, to name only a few variables. The internal body temperature of humans is about 98.6°F (37°C), which seems odd when our most comfortable temperature is 65 to 80°F (21 to 27°C). This means that in the summer we must continuously give off heat to be comfortable, but in the winter suitable clothing can keep heat next to us to maintain warmth. Body comfort is also affected by radiant heat: When the sun shines on us, we feel much warmer. Depending on the season of the year, this can be good or bad. Solar engineers are working on ways to control this heat flow as the amount of glass area of a vehicle increases.

The velocity of air past our bodies is another factor in human comfort. In the summer, this helps us feel cooler; in the winter, a chill factor can be created if the warm air that surrounds our bodies is blown away. Air movement is an important part of heating and A/C systems. (The blower motor and air registers are discussed more completely in Chapter 8.)

1.4.1 Humidity

A factor that greatly affects the heat flow to or from our bodies is humidity, the amount of water in the air around us. The amount of water vapor suspended in air can vary from a perfectly dry 0% to a foggy 100%, where drizzle or rain forms. This is called **humidity.** The amount of water vapor air can hold varies with temperature and is referred to as **relative humidity**

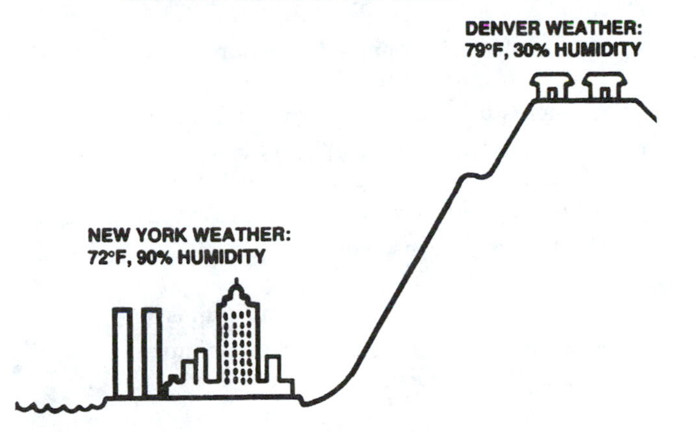

FIGURE 1–13 Humidity greatly affects our comfort. *(Reprinted with permission of General Motors Corporation)*

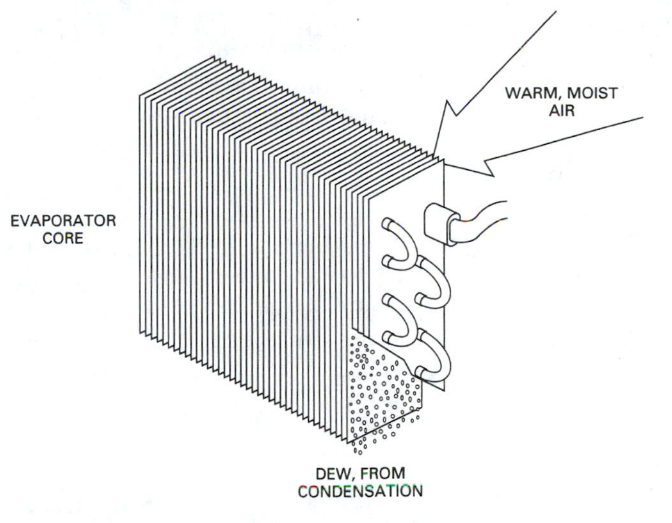

FIGURE 1–14 When air comes into contact with the cold evaporator, excess moisture forms dew. This condensed moisture leaves the car through the evaporator drain.

(RH). Air at 50°F (10°C) can hold a maximum of 0.33 oz (9.41 g) in a cubic meter (100% RH), but it can hold 0.61 oz (17.3 g) at 68°F (20°C). Warm air can hold much more water vapor than cold air. This is why dew forms as air cools and fog burns off as a day becomes warmer.

Humid cold air feels much colder than dry air at the same temperature. Humid hot air slows down our natural body cooling system (evaporation of perspiration), so it can make a day feel much hotter. Air that is too dry also tends to make us uncomfortable. As with temperature, we have a range of humidity in which we feel most comfortable, about 45 to 50% for most people (Figure 1–13).

As it operates, a vehicle's air conditioner dehumidifies, or removes moisture, from air. Water vapor condenses on the cold evaporator fins just as on the glass holding a cold drink. This condensed water then drops off the evaporator and runs out the drain at the bottom of the evaporator case. In-vehicle humidity becomes about 40 to 45% on even the most humid days if the A/C is operated long enough. This level, of course, will be higher if any wet, outside air enters. A good example of this dehumidification process occurs when a vehicle's A/C is operated on cold days when the windows are fogged up; it usually takes only a short time to dry the air and remove the fog from the windows (Figure 1–14).

1.4.2 Cleanliness

A side effect of air conditioning is the cleaning of the air coming into the car as it passes through the cooling ductwork. The act of cooling and dehumidifying air at the A/C evaporator causes water droplets to form on the evaporator fins. Dust and other contaminants in the air that come into contact with these droplets become trapped and are flushed out of the system as the water drops drain from the evaporator (Figure 1–15). Some new vehicles incorporate cabin filters into their A/C and heating systems to clean the air even more thoroughly by trapping dust and pollen particles before they enter the passenger compartment.

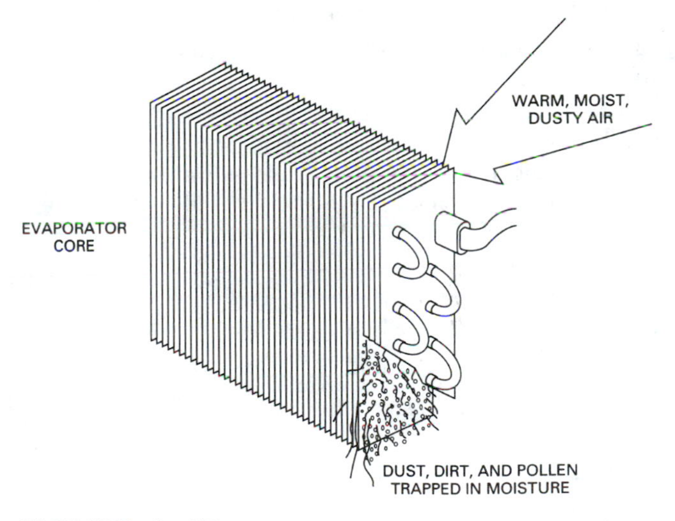

FIGURE 1–15 The dew on the evaporator traps dust and pollen that comes in contact with it; the dust and pollen drain out with the water.

CHAPTER QUIZ

These questions help you study this chapter. Enter the proper word(s) in the blanks to complete each statement.

1. HVAC is short for __________, __________, and __________ __________.
2. A typical HVAC system uses a(n) __________ to cool the air, a(n) __________ to warm the air, and __________ or __________ to control the air flow.
3. __________ is a form of energy that makes us feel warmer, and __________ is the absence of heat.
4. Heat intensity is commonly measured using either the __________ or __________ scale.
5. A temperature of 100° C is equal to __________° F.
6. Heat quantity is commonly measured using units of __________ or __________.
7. Heat always moves from __________ to __________.
8. The A/C system can make air __________, __________, and __________.
9. RH is generally __________ as air warms up.
10. Moisture that collects on the evaporator reduces the __________ in the vehicle.

REVIEW QUESTIONS

These questions allow you to check what you have learned. Select the answer that correctly completes each statement.

1. A vehicle's HVAC system contains a(n)__________ system.
   - **a.** A/C
   - **b.** heater
   - **c.** air distribution
   - **d.** All of these
2. Air conditioning systems control which of the following in a vehicle?
   - **a.** Temperature
   - **b.** Humidity
   - **c.** Air cleanliness
   - **d.** All of these
3. Two students are discussing heat movement. Student A says that heat is a form of energy. Student B says heat travels from hot to cold. Who is correct?
   - **a.** Student A
   - **b.** Student B
   - **c.** Both A and B
   - **d.** Neither A nor B
4. Heat intensity refers to which of the following?
   - **a.** The amount of heat in an area
   - **b.** The number of Btu in an area
   - **c.** The heat individuals feel
   - **d.** None of these
5. The freezing point of water is __________ on the Celsius scale.
   - **a.** 100°
   - **b.** 0°
   - **c.** 32°
   - **d.** 212°
6. The boiling point of water is __________ on the Fahrenheit scale.
   - **a.** 0°
   - **b.** 32°
   - **c.** 100°
   - **d.** 212°
7. Student A says that 1 c of heat can increase the temperature of 1 g of water by 1°F. Student B says the temperature will increase 1°C if 1 c is added. Who is correct?
   - **a.** Student A
   - **b.** Student B
   - **c.** Both A and B
   - **d.** Neither A nor B
8. The normal temperature for the average human is
   - **a.** 80°F
   - **b.** 100°F
   - **c.** 60°F
   - **d.** 98.6°F
9. The amount of moisture contained by air is called
   - **a.** temperature.
   - **b.** ambient.
   - **c.** humidity.
   - **d.** None of these
10. In addition to cooling the passenger compartment, an air conditioning system
    - **a.** heats the passengers.
    - **b.** dehumidifies the air.
    - **c.** puts moisture in the air.
    - **d.** All of these

CHAPTER

Heat Movement Theory

LEARNING OBJECTIVES

After completing this chapter, you should:

- Understand how heat can be transferred from one location to another.
- Be familiar with the three states of matter and the effect heat has on them.
- Understand what latent heat is and why it is important to A/C.
- Understand how pressure is measured.
- Understand the effect pressure has on boiling points.

TERMS TO LEARN

boiling point
chlorofluorocarbon (CFC)
conduction
convection
critical temperature
gas
heat exchanger
hydrochlorofluorocarbon (HCFC)
hydrofluorocarbon (HFC)
insulator
latent heat
latent heat of condensation
latent heat of evaporation
latent heat of fusion
liquid
kilopascal (kPa)
pounds per square inch absolute (PSIA)
pounds per square inch gauge (PSIG)
pressure
pressure-temperature (PT)
radiation
refrigerant
retrofit
saturated vapor
sensible heat
solid
subcool
superheat
vacuum

2.1 INTRODUCTION

In practical terms, an HVAC system moves heat. The A/C system simply transfers the heat out of a place where it is not wanted, and the heating system moves the heat into a place where it is wanted. Several physical principles are involved. If we understand these basic principles, the operation of the A/C system becomes easily understandable. This understanding, in turn, makes diagnosing and servicing an A/C system relatively easy and quite interesting.

Heat always moves from hot to cold. Remember that heat is energy, and cold is lack of energy. The rate or speed that heat moves, then, is simply a factor of the difference in the temperature between the hot and cold areas. A large temperature difference moves heat much faster than if the two areas are almost the same temperature. The heat flow tends to make the hot item cooler and the cool item warmer. If left alone, the two areas become the same temperature, and the heat flow stops (Figure 2–1).

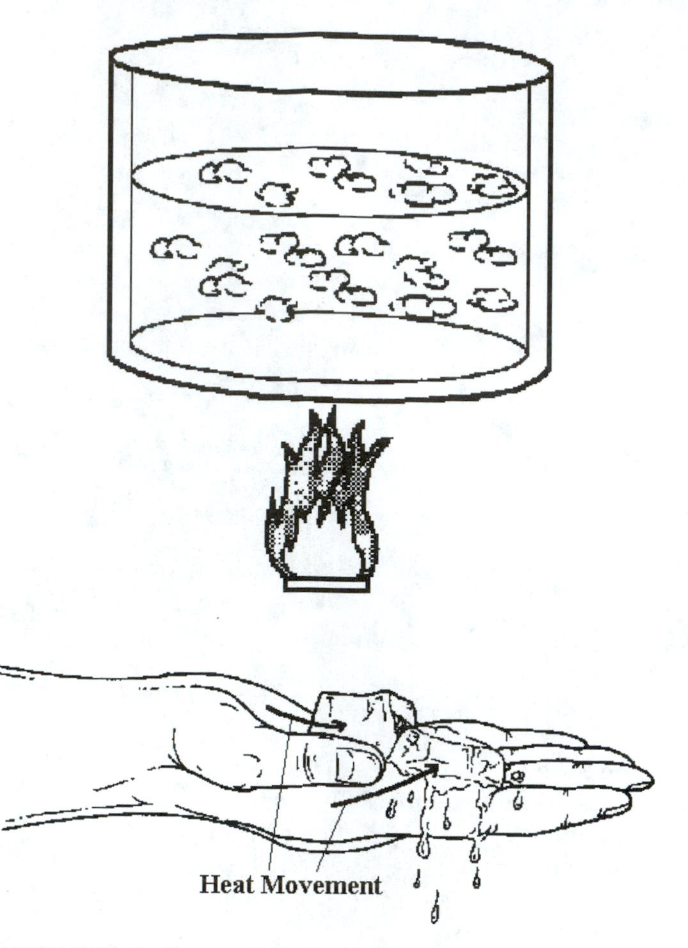

FIGURE 2–1 Heat always moves from hot to cold. In this example, it will travel from the fire to the cooler water and from the hand to the cooler ice.

Understanding heat flow shows us that if we want to cool a hot car, we have to create a place inside the car for the heat to go to; this place must be colder than our desired in-car temperature. Because cooling the car makes the hot place cooler, we have to move the heat elsewhere. The only other place to move it is outside the car, and the ambient air temperature around the car is often hotter than inside. This means we have to take the heat from inside the car and make it hotter than the air outside the car. As we shall see, this takes a little energy from the engine, but it is actually fairly easy to do once we learn a little about heat (Figure 2–2).

2.2 HEAT MOVEMENT

Heat can travel on one or more of three paths as it moves from hot to cold: *conduction*, *convection*, and

FIGURE 2–2 Heat can enter a vehicle's passenger compartment from several sources. The A/C system allows us to move excess heat out of the vehicle.

radiation. Heat travel can be beneficial and intentional, or it can be detrimental. Understanding it allows us to either control its flow or avoid unwanted heat movement as we promote the desired movement; for example, to make living with radiant heat easier, many of us park our cars in the warm sunshine in the winter and the cooler shade in the summer.

2.2.1 Conduction

The simplest heat movement is **conduction**, by which heat travels through a medium such as a solid or liquid, moving from one molecule of the material to the next. If we heat one end of a wire, heat will go in one end and be conducted through the material, and the wire will become hot at the other end (Figure 2–3).

Some materials (most of the metals) are good heat conductors. Copper and aluminum are among the best of the commonly used metals, so we make most **heat exchangers** (radiators, evaporators, and condensers) from them. Some materials (wood, for example) are poor heat conductors, and some (styrofoam, for example) conduct heat so poorly that they are called **insulators**. Most good insulators incorporate a lot of air or gaseous material in their structure because air is a poor heat conductor.

2.2.2 Convection

Convection is a process of transferring heat by moving the heated medium. The medium is fluid, in either a liquid or gaseous state, so we can heat it in one location and move it to another location where the heat is released. A convection current is a continuous flow of the medium and heat.

An example of convection is a vehicle's cooling system. Coolant (the medium) is heated in the water jackets next to the cylinders and combustion chambers. Then it is pumped to the radiator, where the heat is transferred to the air traveling through the radiator. Convection also occurs in the interior of the car, where we circulate air

FIGURE 2–3 The transfer of heat directly through a material is called conduction. (*Courtesy of Daimler-Chrysler Corporation*)

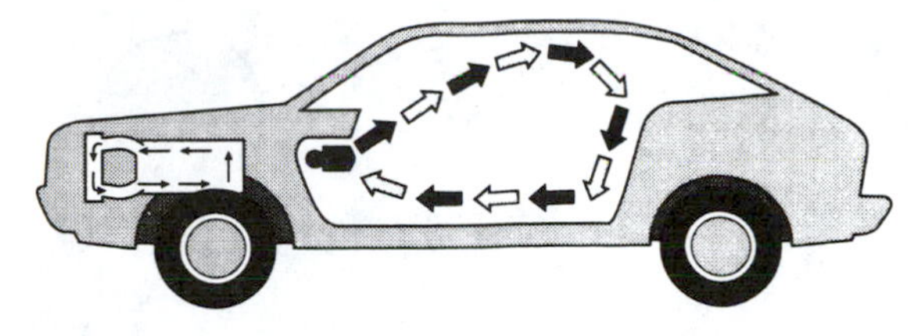

FIGURE 2–4 We move heat from the heater to the passengers or from the passengers to the A/C system by convection. Heat movement from the engine to the radiator using circulating coolant is also an example of convection.

past the driver and passengers to pick up heat and move the air to the evaporator, where the heat is transferred to the cooler evaporator fins (Figure 2–4).

Convection can be forced by a fan or pump, or it can occur naturally. Most materials expand as they are heated, which decreases their weight per volume. Hotter liquids and gases tend to rise, and cooler ones tend to drop. When heating occurs in a liquid or gas, the warmer, lighter parts naturally rise and the cooler, heavier parts fall. Natural convection currents occur in every volume of liquid or gas that is not heated evenly. Sometimes we can feel this movement: If we were to stand next to a hot stove in a cold room, we would feel the warmed air rising above the stove and the cooler air moving horizontally, down low, toward the stove (Figure 2–5).

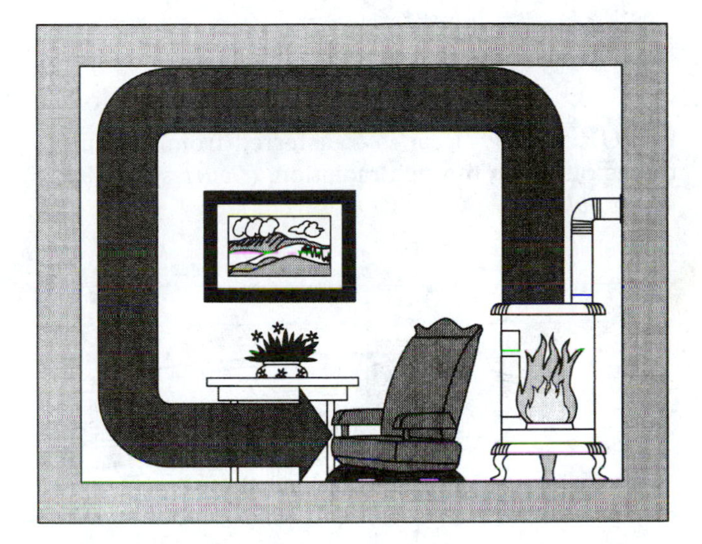

FIGURE 2–5 Heated air next to the stove will rise and cooler air will move in to replace it. This creates a convection current to move the air and heat. (*Courtesy of DaimlerChrysler Corporation*)

2.2.3 Radiation

Heat can travel through heat rays and pass from one location to another without warming the air through which it passes. The best example of this is the heat from the sun, which passes through cold space and warms our planet and everything it shines on. **Radiant heat** can pass from any warmer object through air to any cooler object. It is affected by the color and texture of both the heat emitter, where the heat leaves, and the collector, where the heat is absorbed. Dark, rough surfaces make better heat emitters and collectors than light-colored, smooth surfaces (Figure 2–6).

Radiant heat can be beneficial (for example, passing through the windows and warming the interior of a car or building in winter), but this same heat can be very detrimental in the summer. It is not unusual for sunlight to pass through a window, where it is absorbed by dark dash pads or upholstery, and produce in-car temperatures of 150°F (66°C) or higher (Figure 2–7). (At one time, California Highway Patrol cars were painted all black. Painting the tops white benefited the patrol officers by lowering the in-car temperature significantly.) The engine, radiator, and exhaust system all radiate heat, which can find its way into the passenger compartment.

2.3 STATES OF MATTER

The air conditioning process works through a fluid, called a **refrigerant,** that continuously changes state from liquid to gas and back to liquid. These changes of state are where the movement of heat needed for cooling occurs. All basic materials exist in one of the states of matter—*solid, liquid,* or *gas*—and most of them can

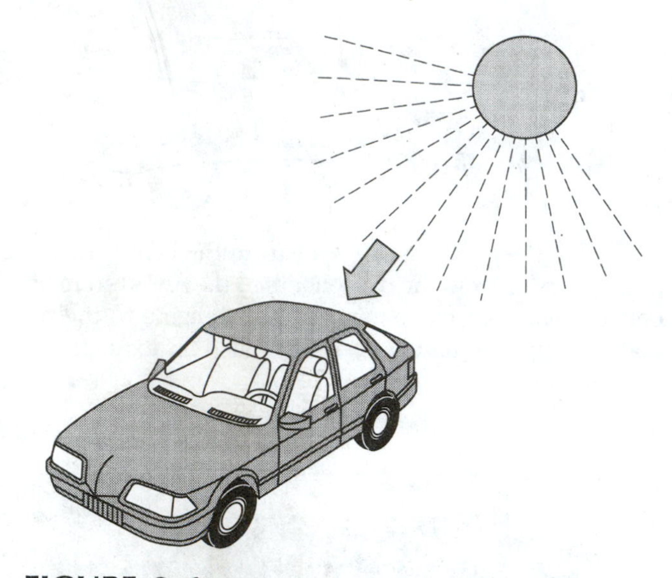

FIGURE 2–6 Heat is transferred from the sun to things on Earth through radiation. (*Courtesy of Toyota Motor Sales*)

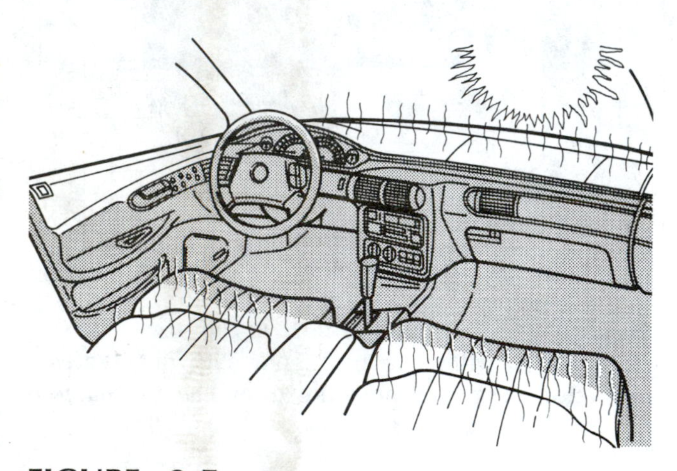

FIGURE 2–7 Radiant heat entering a vehicle through the windows can add a lot of heat to a car's interior. (*Courtesy of DaimlerChrysler Corporation*)

be changed from one state to another by adding or removing heat (Figure 2–8).

Molecules are the building blocks for all things we can see or feel. Molecules are combinations of atoms, which are in turn made up of electrons and protons. The protons are in the center, or nucleus, of the atom; the electrons travel in an orbit around them. There are about 100 basic elements or atoms, each having a different atomic number, that combine with other elements to make the many, varied molecules. The atomic number of an element is based on the number of electrons and protons in that element. The periodic table of elements seen in most chemistry laboratories shows the relationship of these elements.

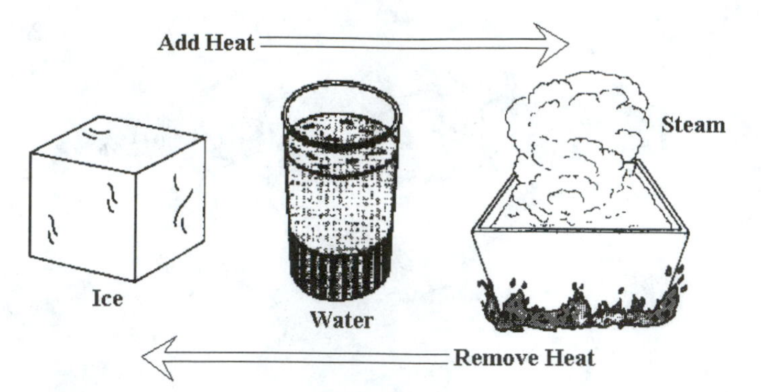

FIGURE 2–8 We can cause matter to change state by adding or removing heat.

In A/C we are primarily concerned with four different molecules: water, refrigerant 12 (R-12), refrigerant 22 (R-22), and refrigerant 134a (R-134a). Water molecules are H_2O; this is a combination of a single oxygen atom and two hydrogen atoms. Hydrogen has an atomic number of 1 (1 proton and 1 electron), and oxygen has an atomic number of 8 (8 electrons and 8 protons). Again, we are describing water because we are familiar with it in all three states: solid ice, liquid water, and gaseous steam. R-12, R-22, and R-134a are described more completely at a later point in this chapter (Figure 2–9).

2.3.1 Solid

Solid matter is familiar; it has a definite shape and substance. Solids exert pressure in only one direction, and that is downward because of gravity.

Most of us know that ice is the solid form of water, will hold its shape, and is cold. Water is normally a solid at temperatures below 32°F (0°C), which is the normal freezing point. The electrons in the molecule's atoms are still orbiting around the protons, but the movement has been slowed because much of the heat energy has been removed (Figure 2–10).

2.3.2 Liquid

Adding heat to most solids produces a liquid as the solid material melts. It is the same material, but heat energy has broken the molecular bond and has made the matter fluid. A fluid has no shape and needs a container to hold it; a fluid takes the shape of its container. Liquid is affected by gravity but also exerts pressure sideways. Another point important to A/C is that liquids flow through a pipe or hose and can be pumped.

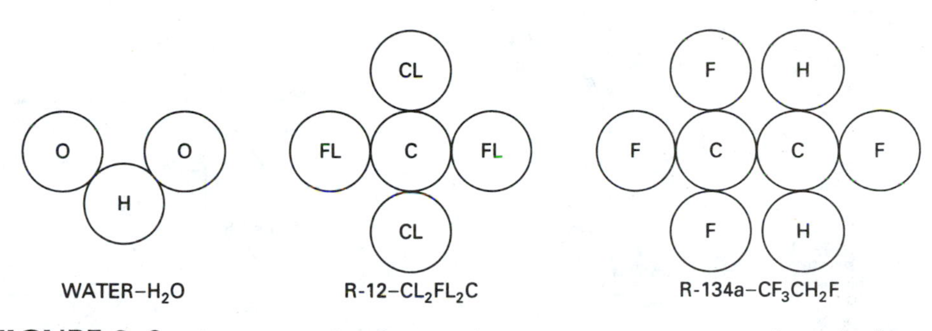

FIGURE 2–9 A water molecule contains two oxygen atoms and one hydrogen atom; R-12 is a combination of one carbon, two chlorine, and two fluorine atoms; and R-134a is a combination of two carbon, four fluorine, and two hydrogen atoms.

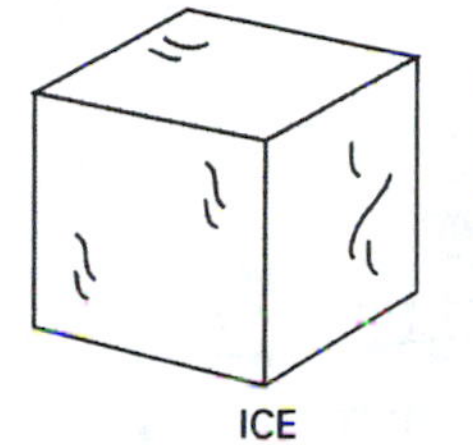

FIGURE 2–10 Ice is a solid form of water with a low temperature and slow molecular action.

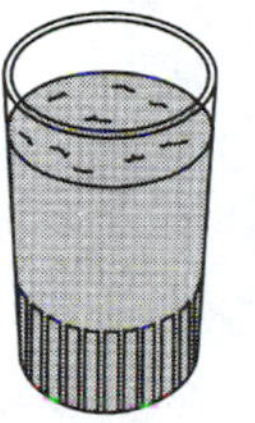

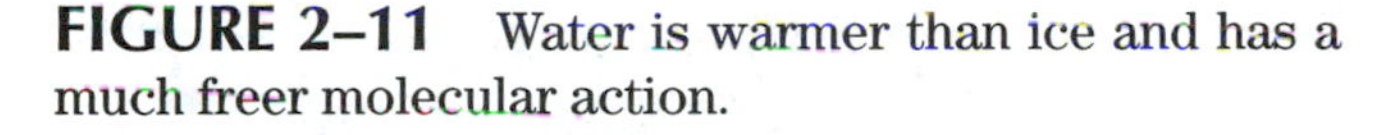

FIGURE 2–11 Water is warmer than ice and has a much freer molecular action.

Water is normally a liquid between 32 and 212°F (0 and 100°C). It is the same molecule as ice, but heat energy has increased the movement of the electrons (Figure 2–11).

2.3.3 Gas

Adding heat to most liquids produces gas as the liquids boil. It is the same material, but the heat energy has freed the molecular bonds still further so that the molecule has no shape at all and has expanded so much that it has very little weight. A gas molecule exerts pressure in every direction. Gases can also be pumped through hoses and pipes, making them easy to move through an A/C system.

At temperatures above 212°F (100°C) water normally boils to become a gas, called steam. Again, this is the same molecule as water or ice, but heat energy has greatly increased molecular movement (Figure 2–12).

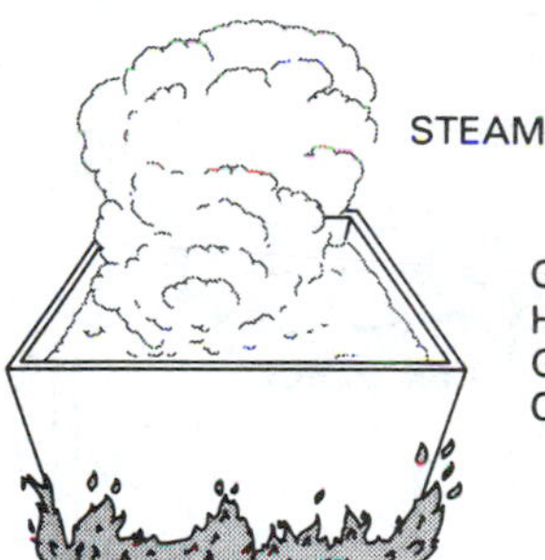

FIGURE 2–12 Adding heat to water produces steam, the gas state, with a much freer molecular action.

2.4 LATENT AND SENSIBLE HEAT

Sensible heat makes sense; it can be felt and measured on a thermometer. If we have 1 lb of water at 40°F and add 1 Btu of heat to it, the temperature will increase to 41°F; adding another Btu of heat will increase the temperature to 42°F; and adding another 170 Btu (212 – 42) will increase the temperature to 212°F, the boiling point.

Sensible heat is fairly easy to understand, but if we add more heat, an odd thing occurs (Figure 2–13). If we add another Btu of heat to water at 212°F, some of the water will boil, but the temperature of both the water and the steam produced will remain at 212°F (Figure 2–14). The added heat has caused some of the water to change state, but it has not changed temperature. This is an example of *latent,* or *hidden, heat.* We can watch this happen each time we boil water. The water boils a little bit at a time. It takes a large amount of heat to get the water to change into steam, and we end up with

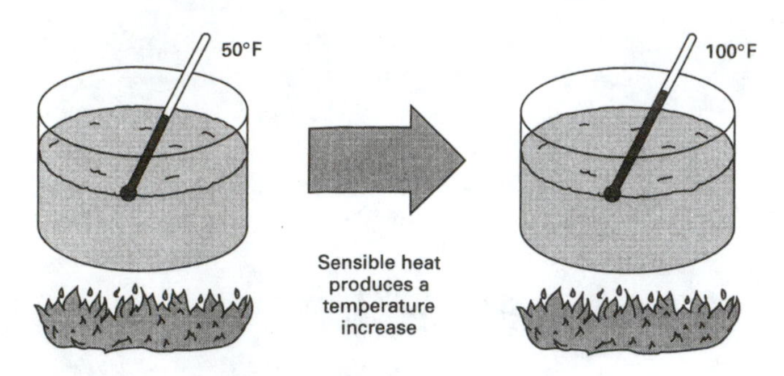

FIGURE 2–13 Heat that causes a temperature increase is called *sensible heat.*

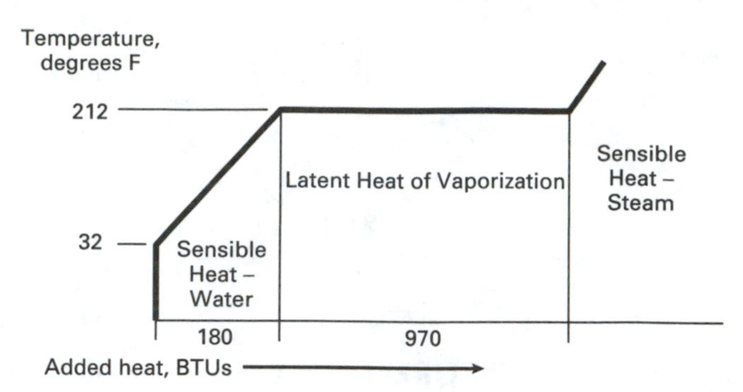

FIGURE 2–16 If we start with 1 lb of water at 32°F, adding 180 Btu will increase the temperature to 212°F. It will take another 970 Btu (the latent heat of evaporation) to boil that pound of water.

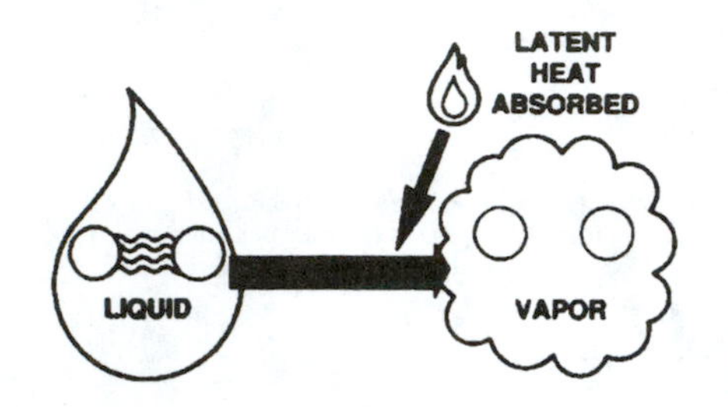

FIGURE 2–14 Latent heat causes a change of state without a temperature change. (*Reprinted with permission of General Motors Corporation*)

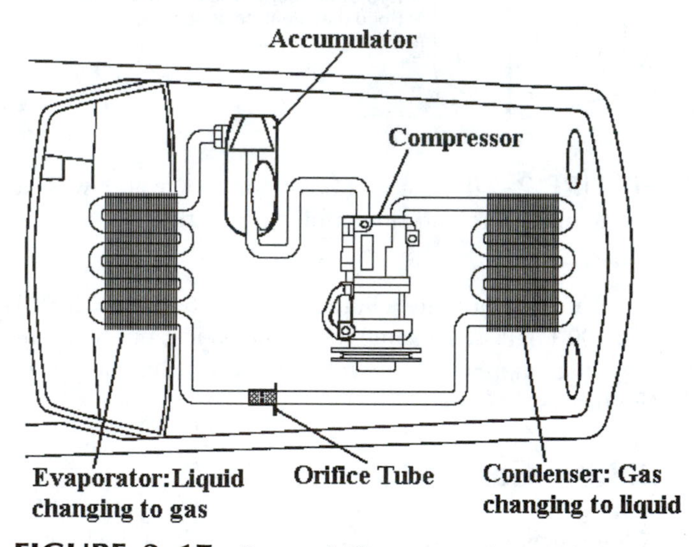

FIGURE 2–17 In an A/C system, the refrigerant changes state and absorbs heat in the evaporator and releases heat as it changes state again in the condenser.

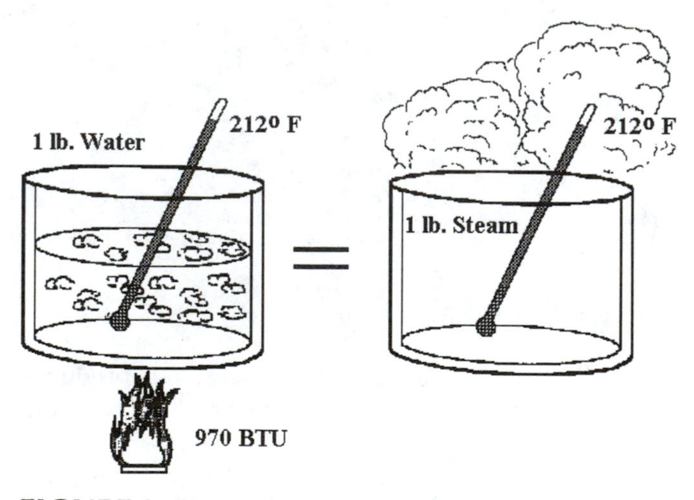

FIGURE 2–15 If we add 970 Btu of heat to one lb of water at 212°F, we will have one lb of steam at the same temperature.

steam that is the same temperature as the water. Latent heat causes a change in state but no change in temperature (Figure 2–15). Once water reaches 212°F, it cannot get any hotter; it simply boils.

To change 1 lb of water at 212°F into 1 lb of steam at 212°F, we must add 970 Btu of heat. Using metric terms, to change 1 g of water at 100°C into 1 g of steam at 100°C, we must add 540 calories of heat. This is called the *latent heat of evaporation.* To reverse this and change 1 lb of steam back into 1 lb of water, we have to remove this same 970 Btu of heat; this is called the **latent heat of condensation**. It should be noted that the latent heat of evaporation requires much more heat than the amount of heat required to raise the temperature from the freezing point to the boiling point (Figure 2–16).

This is the major principle through which A/C works. Heat added in the car's evaporator causes the liquid refrigerant to change state, and it absorbs the latent heat of evaporation. This heat is released from the system in the condenser as the gas changes back into a liquid and latent heat of condensation is removed (Figure 2–17).

The change of state between a solid and liquid also requires a large amount of latent heat. This is called the **latent heat of fusion**. For water, it takes the removal of 144 Btu of heat to change 1 lb of water at 32°F into ice at 32°F (Figure 2–18). Using metric terms, it takes the removal of 79.7 calories of heat to change 1 g of water at 0°C into ice at 0°C. We need to add the same amount of heat to change the ice back into water. The latent heat of fusion is what makes ice chests effective in cooling things; ice can absorb a fairly large amount of heat as it melts (Figure 2–19).

All commercial forms of refrigeration or air conditioning utilize the principle of latent heat and the change of state that it causes. The liquid-to-gas-to-liquid change is used because both media are fluid and can be easily moved by pumping. In addition, the greatest amount of heat energy is absorbed and released between these two states (Figure 2–20). With the solid-to-liquid change of state, the liquid can be pumped but solids must be moved using more cumbersome methods.

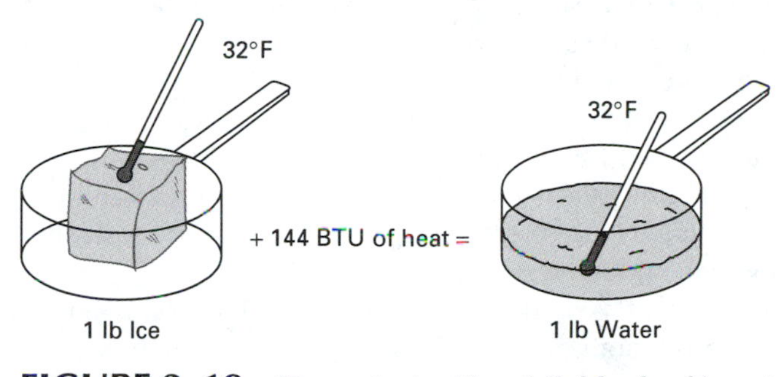

FIGURE 2–18 If we start with a 1-lb block of ice at 32°F, it will take 144 Btu (the latent heat of fusion) to melt all of the ice.

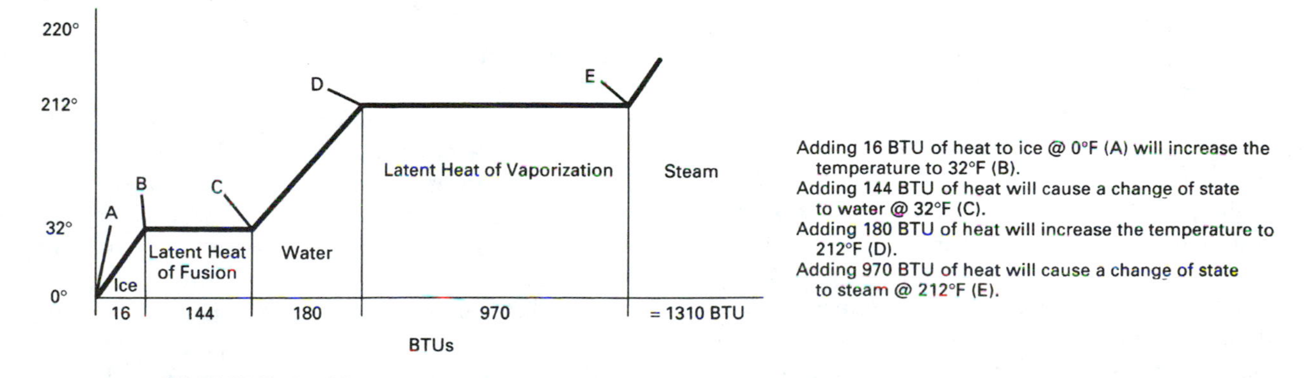

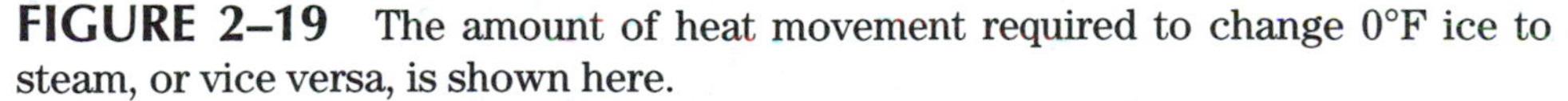

FIGURE 2–19 The amount of heat movement required to change 0°F ice to steam, or vice versa, is shown here.

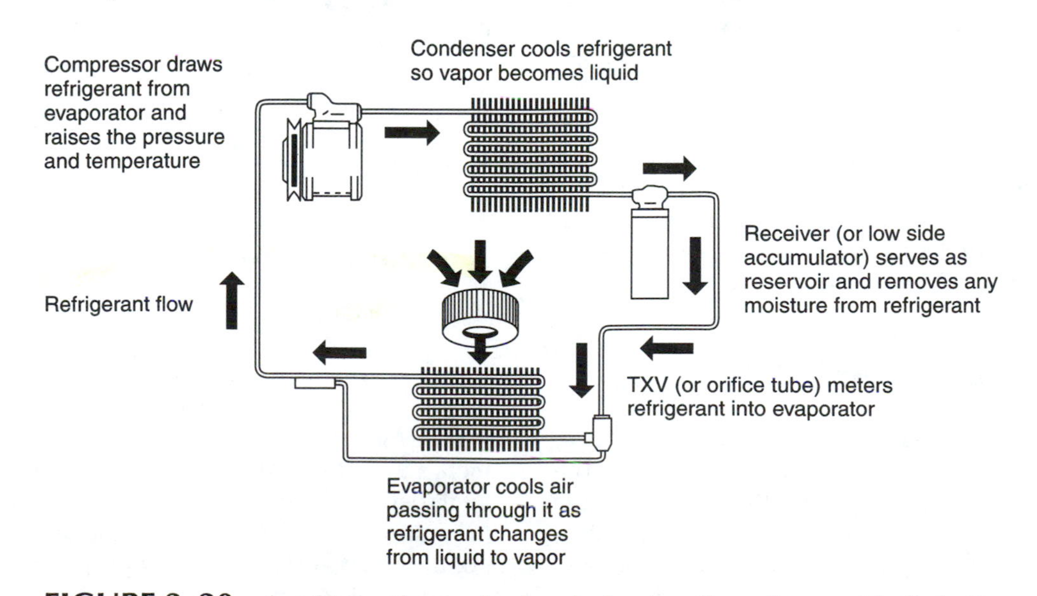

FIGURE 2–20 A refrigeration cycle absorbs heat as the refrigerant boils in the evaporator and removes heat as it changes state back to a liquid in the condenser.

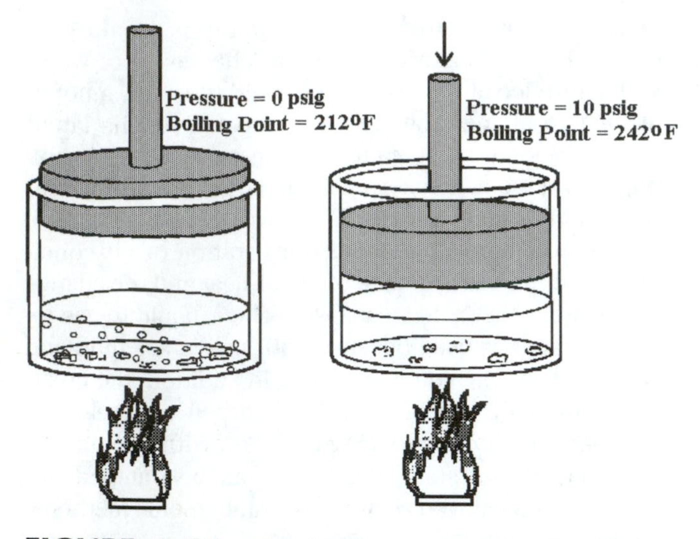

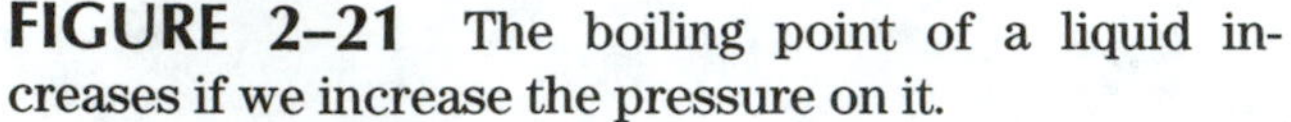

FIGURE 2–21 The boiling point of a liquid increases if we increase the pressure on it.

2.5 BOILING POINTS

Boiling points can be increased or decreased by raising or lowering the pressure on the liquid. With water, the boiling point will rise about 2 1/2°F for each pound per square inch (psi) of pressure, or about 1°C for each 5 kPa.

Note that raising the pressure increases the boiling point and lowering the pressure reduces it. A standard A/C service step is to evacuate the system before recharging it with new or reclaimed refrigerant. This is done to remove any water, which can cause rust or corrosion or mix with the refrigerant to form acids. At a vacuum of 30 inches of mercury ("Hg) (–103 kPa), the boiling point of water is lowered to 0°F (−18°C).

The condensing point of a gas is the same as the boiling point; the only difference is that we add heat to a liquid to make it boil, and we remove heat from a gas to make it condense. Raising the pressure of a gas allows the gas to condense at temperatures above the normal boiling point (Figure 2–21).

2.5.1 Critical Temperature

The **critical temperature** is the maximum point at which a gas can be liquefied or condensed by raising the pressure. The **critical pressure** is the pressure that is necessary to liquefy a gas at that temperature. Looking at Figure 2-28 shows us that R-134a has a critical temperature of 101.15°C (214°F) and a critical pressure of 4.065 MPa (589 psi). This tells us what the upper limits are for the high side of an A/C system.

2.6 SATURATED VAPORS AND THE PRESSURE–TEMPERATURE RELATIONSHIP

Saturated vapor is the term used to describe a liquid and gas inside a closed chamber, which is the condition in an A/C system. When discussing saturated vapors, we need to learn two additional terms: **subcool** and **superheat**. Subcool refers to a liquid whose temperature is well below its boiling point. Superheat refers to the temperature increases of a vapor after all of the liquid has boiled.

If heat is added to the saturated vapor inside a closed container, some liquid will boil, which will increase the pressure within the container. This added pressure will, in turn, increase the boiling point. If heat is removed, the pressure will drop as some of the gas condenses back into a liquid because the boiling point will be lowered. The pressure increase is directly proportional to the temperature increase and is caused by the great increase in volume (about 1,000 times) as the liquid boils (Figure 2–22).

A pressure–temperature (PT) relationship chart (such as that shown in Figure 2–31) shows us the relative temperature–pressure relationship of a refrigerant when it is in a system. Once a technician has measured the pressure in a system, that part of the system should be the same temperature as indicated on the chart. The temperature of a saturated vapor for a particular liquid will always be at a constant point relative to the pressure, and the pressure will always be relative to the temperature. If they do not match, something is wrong. Lower-than-normal pressure for a particular temperature can indicate starvation because there is no saturated vapor. Higher-than-normal pressure for a particular temperature usually indicates contamination from another chemical, air, or the wrong refrigerant.

2.7 PRESSURE: GAUGE AND ABSOLUTE

Pressure is defined as a certain amount of force exerted on a unit area. Traditionally in the United States, pressure is given in pounds and the unit area in square inches, so pressure has been given in pounds per square inch (psi). Most pressure gauges disregard atmospheric pressure and are calibrated to read 0 at their starting point, which is the normal atmospheric pressure surrounding us. The pressure created by the weight of the air in our atmosphere generates a pressure of 14.7 psi at sea level, which is often rounded off to 15 psi (Figure 2–23).

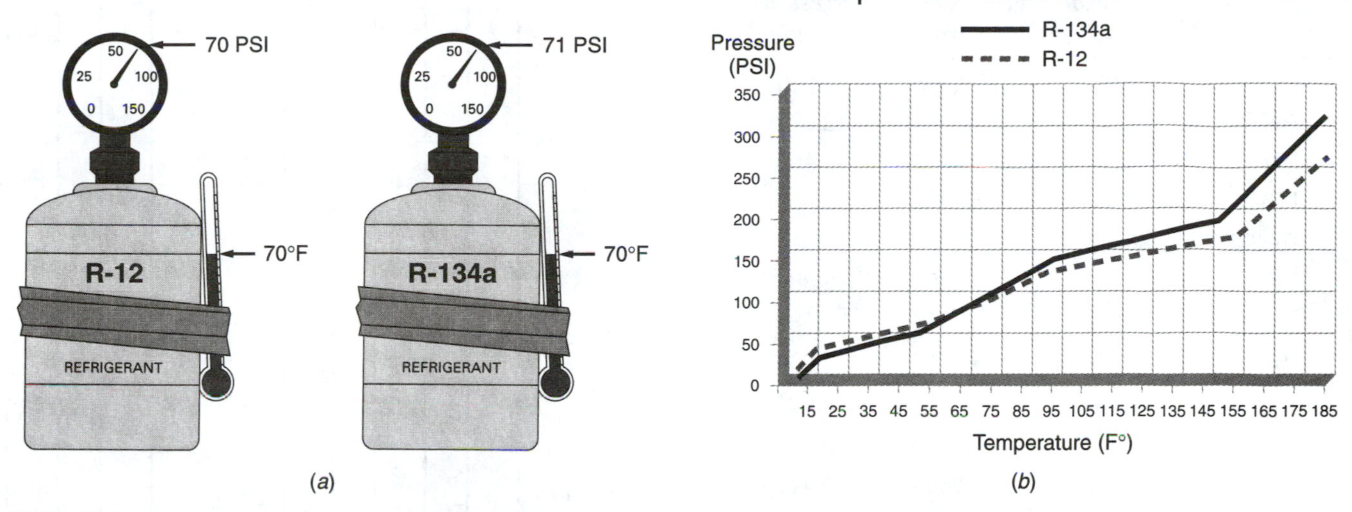

FIGURE 2–22 R-12 in a container is a saturated vapor with gas in contact with a liquid. The pressure in the container is in direct relation to the temperature (*a*); a chart can be used to determine the temperature if we know the pressure or vice versa (*b*). (a *courtesy of DaimlerChrysler Corporation*; b *courtesy of Four Seasons*)

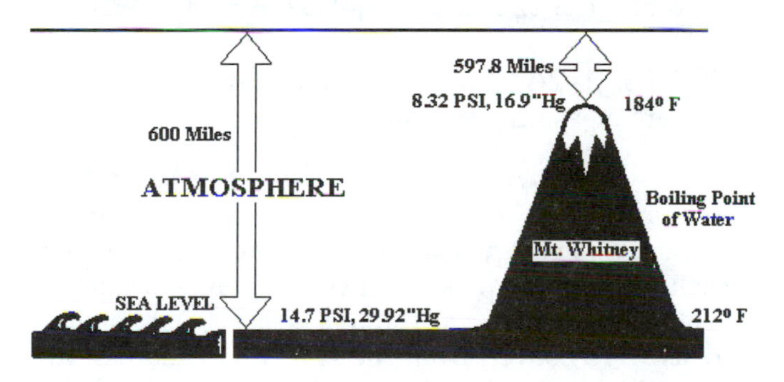

FIGURE 2–23 The weight of the air in our atmosphere generates a pressure of about 15 psi at sea level. Atmospheric pressure and the boiling point of water are lower at higher altitudes (*Courtesy of DaimlerChrysler Corporation*)

Pressures below atmospheric are usually called a **vacuum** and are measured using a gauge calibrated in inches of mercury ("Hg) or millimeters of mercury (mmHg). Most of these gauges start from 0 at atmospheric pressure and read downward to 29.92 "Hg (usually rounded off to 30 "Hg); 29.92 "Hg is a perfect vacuum where there is no pressure at all. A perfect vacuum is really zero pressure (Figure 2–24).

Deep or high vacuum is close to perfect vacuum and is measured in microns of mercury. A micron is one-millionth of a meter; atmospheric pressure is equal to 759,968 microns. Figure 2–25 illustrates the relative pressure in inches and microns, and how the boiling point of water relates to these pressures.

The pressures just discussed are often called **gauge pressures** because their zero points are atmospheric

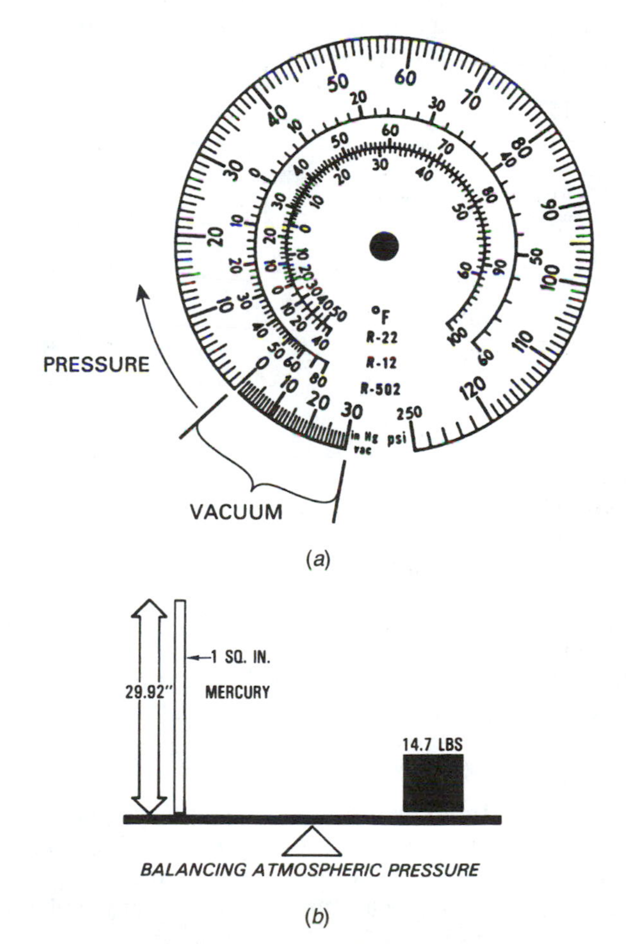

FIGURE 2–24 As shown on this compound gauge, pressures below atmospheric are commonly called a *vacuum* (*a*). A perfect vacuum is 29.92 inches of mercury (*b*). (a *courtesy of TIF Instruments*; b *courtesy of Robinair, SPX Corporation*)

Temperature (°F)	*Vacuum (inches)*	*Pressure* (microns)*	*Pressure (psi)*
212°	0.00	759.968	14.696
205°	4.92	535,000	12.279
194°	9.23	525,526	10.162
176°	15.94	355,092	6.866
158°	20.72	233,680	4.519
140°	24.04	149,352	2.888
122°	26.28	92,456	1.788
104°	27.75	55,118	1.066
86°	28.67	31,750	0.614
80°	28.92	25,400	0.491
76°	29.02	22,860	0.442
72°	29.12	20,320	0.393
69°	29.22	17,780	0.344
64°	29.32	15,240	0.295
59°	29.42	12,700	0.246
53°	29.52	10,160	0.196
45°	29.62	7,620	0.147
32°	29.74	4,572	0.088
21°	29.82	2,540	0.049
6°	29.87	1,270	0.0245
−24°	29.91	254	0.0049
−35°	29.915	127	0.00245
−60°	29.919	25.4	0.00049
−70°	29.9195	12.7	0.00024
−90°	29.9199	2.54	0.000049

*Remaining pressure in system in microns:
1.000 inch = 25,400 microns = 2.540 cm = 25.40 mm
0.100 inch = 2,540 microns = 0.254 cm = 2.54 mm
0.039 inch = 1,000 microns = 0.100 cm = 1.00 mm

FIGURE 2–25 The boiling point of water drops as pressure is reduced. At a near perfect vacuum of 29.9199 "Hg or 2.54 microns, the boiling point is –90°F. *(Courtesy of Robinair, SPX Corporation)*

pressure, and zero on these gauges can be called 0 psig. Some gauges are calibrated so their zero reading is at an absolute vacuum, and they read upward from this point. These gauges read *absolute pressures;* atmospheric pressure shows up on them as 15 psia.

At one time, the metric system used kilograms per square centimeter (kg/cm^2) and **bar** as the pressure standard. Bar is still commonly used, and they have similar values. One bar is equal to 14.5 psi, atmospheric

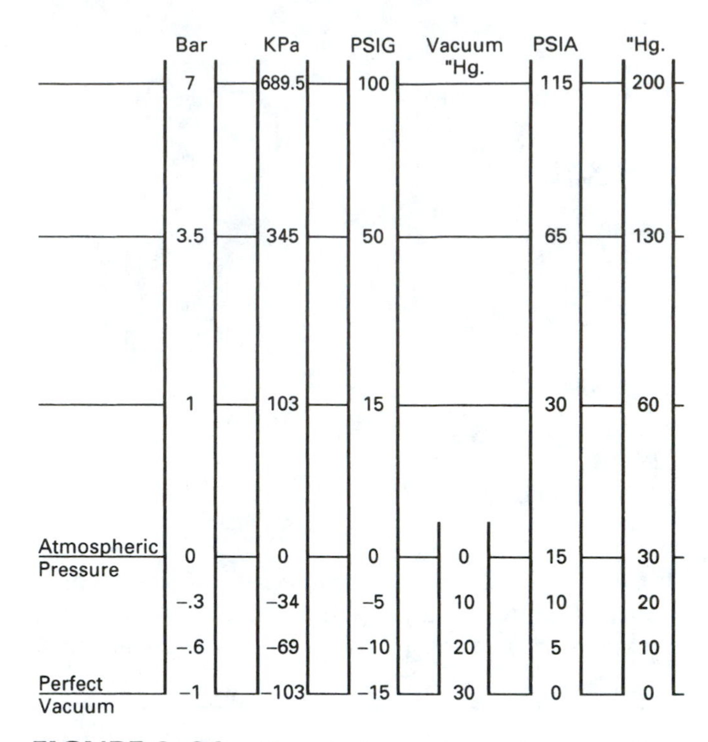

FIGURE 2–26 The relative pressures between five different measuring systems.

pressure. Today, the international unit for pressure is pascal (Pa): 1 psi is equal to 0.006895 Pa. A pascal is very small, so kilopascal (kPa) (1,000 Pa) (1 thousand) or sometimes megapascal (MPa) (1,000,000) (1 million) are used. To convert from one standard to another, you can multiply a number in psi by 6.895 to get kilopascals or a number in kPa by 0.145 to get the reading in psi. An A/C system high-side pressure of 200 psi is equal to 1,379 kPa and 1.38 MPa. A conversion table is given in Appendix F (Figure 2-26).

2.8 REFRIGERANTS

The working fluid of an A/C system is refrigerant. Refrigerants were first developed by the Du Pont Corporation using the name **Freon**. This term is used improperly by many people to mean refrigerant. A new term, *SUVA*, refers to Du Pont's newer refrigerants: Suva MP52 (a blend) and Suva Trans A/C (134a). There are many refrigerants, but the two main ones used in automotive and other mobile systems are R-12 and R-134a. R-22 is commonly used in refrigerators and stationary A/C units for buildings. A refrigerant compound can be a combination of the following:

- Chlorine, fluorine, and carbon, called a *chlorofluorocarbon (CFC)*

- Hydrogen, fluorine, and carbon, called a *hydrofluorocarbon (HFC)*
- Hydrogen, chlorine, fluorine, and carbon, called a *hydrochlorofluorocarbon (HCFC)* (Figure 2–27)

Technically, a refrigerant should be referred to as CFC-12, HFC-134a, or HCFC-22 to show the chemical prefix.

Refrigerants must have a low boiling point—below 32°F (0°C)—to boil and absorb latent heat at low temperatures. The lowest temperature we can use to cool a car's passenger compartment is 32°F. Temperatures below 32°F cause ice to form and block airflow through the fins of the evaporator. When a refrigerant is in a gaseous state, its vapor pressure must be high enough so that a sufficient quantity can be pumped efficiently, but the pressure should not be so high that containing it becomes a problem (Figure 2–28).

HFC:	R-125, R-134a, other halogenated compounds
CFC:	R-12, R-11, R-13, R-113, R-114, R-500, R-503
HCFC:	R-22, R-123, R-124, R-502

FIGURE 2–27 Depending on the combination of carbon, chlorine, fluorine, or hydrogen, a refrigerant is classed as an HFC, CFC, or HCFC. (*Courtesy of TIF Instruments*)

Even though the refrigerant's boiling point is well below ambient temperature, it can be contained in a liquid form in metal cans, canisters, and drums if the container is strong enough to hold the required pressure (Figure 2–29). Remember that this is a saturated vapor, so some of the liquid will boil and generate pressure in the container; the container must be able to contain the pressure generated by that particular chemical at normally encountered temperatures. Some additional potential hazards in handling refrigerants are discussed in Chapter 3.

A refrigerant should mix and become chemically stable with oil, so that the oil is moved through the system to keep the compressor and the expansion valve in some systems lubricated. The oil and refrigerant mixture must be compatible with the various metal, rubber, and plastic materials that make up the system. A refrigerant should also be safe to work with; flammability and toxicity are important concerns (Figure 2–30).

Item	*R-134a*	*R-12*
Molecular formula	CH_2FCF_3	CCL_2F_2
Molecular weight	102.03	120.91
Boiling point	−26.8°C	−29.79°C
Critical temperature	101.15°C	111.80°C
Critical pressure	4.065 MPa {41.452 kgf/cm^2}	4.125 MPa {42.063 kgf/cm^2}
Critical density	511 kg/cm^3	558 kg/cm^3
Saturated liquid density	1206.0 kg/cm^3	1310.9 kg/cm^3
Specific volume (saturated vapor)	0.031009 m^3/kg	0.027085 m^3/kg
Specific heat (saturated liquid at constant pressure)	1.4287 kJ/kg·K {0.3413 kcal/kgf·K}	0.9682 kJ/kg·K {0.2313 kcal/kgf·K}
Specific heat (saturated vapor at constant pressure)	0.8519 kJ/kg·K {0.2035 kcal/kgf·K}	0.6116 kJ/kg·K {0.1461 kcal/kgf·K}
Latent heat of vaporization	216.50 kJ/kg {51.72 kcal/kgf}	166.56 kJ/kg {39.79 kcal/kgf}
Thermal conductivity (saturated liquid)	0.0815 W/m·K {0.0701 kcal/m·h·K}	0.0702 W/m·K {0.0604kcal/m·h·K}
Combustibility	Incombustible	Incombustible
Ozone depletion index (CFC-12 = 1.0)	0	1.0
Global warming index	0.24–0.29	2.8–3.4

FIGURE 2–28 Comparison of the physical characteristics of R-12 and R-134a. (*Courtesy of Zexel USA Corporation*)

FIGURE 2–29 R-134a refrigerant is commonly available in small (about 12-oz) or larger (30- or 50-lb) containers. R-12 is no longer available in small containers. KLEA is a registered trademark for R-134a. *(Courtesy of Sercon)*

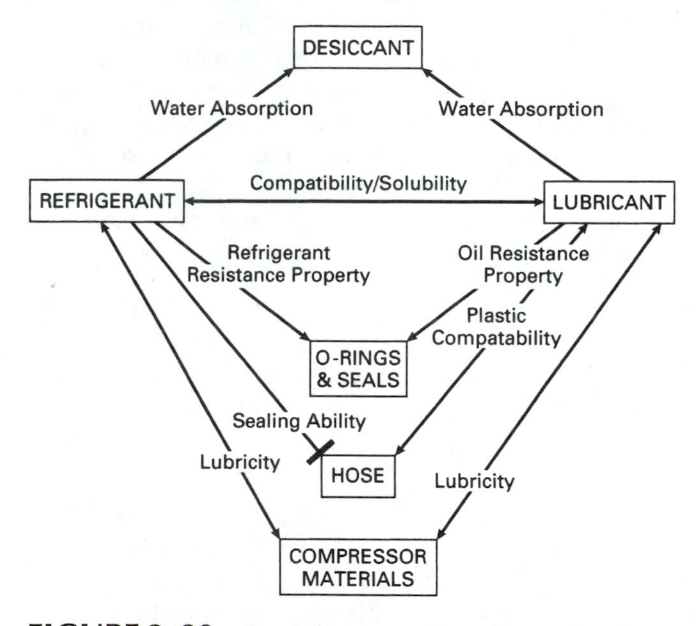

FIGURE 2–30 A refrigerant and its oil must be completely compatible with all of the materials and chemicals in the system.

Temperature		Pressure		
°F	*°C*	*R-12*	*R-134a*	*R-22*
0	−22	9	6	24
5	−15	12	9	28
10	−12	15	12	33
15	−9	18	15	38
20	−7	21	18	43
25	−4	25	22	49
30	−1	28.5	26	55
35	2	33	30.5	61
40	4	37	35	68
45	7	42	40	76
50	10	47	45	84
55	13	52	51	93
60	16	47	57	102
65	18	64	64	111
70	21	70	71	121
75	24	77	79	132
80	27	84	87	144
85	29	92	95	156
90	32	100	104	168
95	35	108	114	182
100	38	117	124	196
105	41	127	158	211
110	43	136	147	226
115	47	147	163	243
120	49	158	171	260
125	52	169	185	278
130	54	181	199	297
135	57	193	214	317
140	60	207	229	337
145	63	220	246	
150	65	235	263	

FIGURE 2–31 The boiling points for R-12, R-22, and R-134a vary depending on the pressure.

2.8.1 R-12

R-12, or more properly CFC-12, is a CFC, a compound of chlorine, fluorine, and carbon. The actual compound is Cl_2Fl_2C, two chlorine atoms combined with two fluorine atoms and one carbon atom. Its chemical name is dichlorodifluoromethane.

R-12 has a boiling point of −21.7°F (−29.8°C). This low temperature allows R-12 to boil easily in the evaporator and absorb the needed latent heat. Its vapor pressure in the evaporator will be about 30 psi and in the condenser about 150 to 300 psi (Figure 2–31). As we will learn, these pressures vary somewhat depending on the temperature. R-12 has a latent heat of vaporization of 70 Btu per lb.

R-12 is soluble in mineral oil and does not react to the metals, hoses, and gaskets used in the system. At one time, R-12 was very inexpensive. R-12 was considered an ideal refrigerant until it was discovered that its molecule can travel into the upper atmosphere before breaking down. Here, the chlorine atoms can react with ozone, O_3, in the ozone layer. This is discussed further in Chapter 3.

2.8.2 R-22

R-22, or more properly HCFC-22, is an HCFC; its chemical compound is monochlorodifluoromethane, $CHClF_2$. R-22 is closely related to R-12, and like R-12, R-22 was once fairly inexpensive.

R-22 has a boiling point of −40°F (−40°C) and a vapor pressure that is about double that of R-12, which is why R-22 is normally used in freezers, which require colder temperatures—about 0°F (−19°C)—and in stationary installations, which use hermetically sealed compressors and metal tubing. A hermetically sealed compressor has both the motor and the compressor mounted in a well-sealed container, eliminating most possibilities for leakage.

R-22 should never be used in an R-12 system or mixed with R-12 or R-134a: It would have a greater tendency to leak from the compressor shaft seal and flexible hoses used in an automotive A/C system. R-22 is not compatible with the rubber hose and seal materials used in most R-12 systems. A mixture of R-12 and R-22 would cause higher pressure and damage to the desiccant, hoses, and O-rings. If a blend that contains R-22 is used to retrofit an R-12 system, the rubber hoses must be replaced with barrier hoses. If mixed, R-12 and R-22 cannot be separated by normal methods.

2.8.3 R-134a

R-134a, or more properly HFC-134a, is an HFC. Its full name is tetrafluoroethane, and its chemical compound is CF_3CH_2F. There is no chlorine in this compound, so it has no effect on Earth's ozone layer. This is one of the major reasons why the automotive industry converted to R-134a. Its cost was once about three times that of R-12, but this has changed. R-12 is becoming much more expensive, and R-134a is becoming much cheaper. The characteristics of R-134a are very similar to those of R-12, and, with some changes, an R-12 system can be converted or **retrofitted** to use R-134a.

R-134a has a boiling point of −15.2°F (−26.8°C) and a latent heat of vaporization of 77.74 Btu per lb. The pressure–temperature relationship is also similar to that of R-12.

R-134a is lighter in weight than R-12. A comparison of the sizes of equal-weight bottles of R-12 and R-134a illustrates the weight differences. When we retrofit an R-12 system to R-134a, we usually decrease the amount of refrigerant charged into the system to compensate for this weight difference.

R-134a is not compatible with the mineral oil used to lubricate an R-12 system. Synthetic **polyalkaline glycol (PAG)** or **polyolester (POE),** commonly called **ester**, lubricants are used with R-134a; these two lubricants do not mix well with R-12. R-134a is also not compatible with the desiccant used in R-12 systems. The process of retrofitting an R-12 system to use R-134a is described in Chapter 13. This process requires changing both the refrigerant and the oil. Other parts of the system need to be replaced if they are not compatible with R-134a or its oil.

During early tests, it was thought that mineral oil and PAG oils would not mix properly in a retrofitted R-12 system. Now it is believed that these early tests were adversely affected by residue left over from flushing the system with R-11; several vehicle manufacturers believe that PAG and mineral oil will mix with no chemical incompatibility.

2.8.4 Blends

Several refrigerant blends (a mixture of several different chemical compounds) are currently being marketed to replace R-12. The chemicals most commonly used are mixtures of HCFC-22, HCFC-124, HFC-134a, HCFC-142B, HCFC-152a, butane (R-600), isobutane (R-600a), and propane (R-290).

Three of these blend chemicals (butane, isobutane, and propane) are very flammable. At least one technician has been seriously burned when the escaping refrigerant has contacted flame. Many states ban the use of flammable refrigerants. The HCFC compounds will be phased out of production in the near future because they add to environmental global warming problems. It is recommended that no new type of refrigerant be used unless it has been approved by the Environmental Protection Agency (EPA) and the Society of Automotive Engineers (SAE). The SNAP rule and the acceptable blends are described in Chapter 3. A list of SNAP approved blends is provided in Appendix D.

Blends offer the advantage of being supposedly "drop-in" replacements for R-12 in that they can be used in an R-12 system with little or no change. *However, there is no legal drop-in refrigerant for an R-12 system.* Blends also offer rather serious drawbacks: fractionation, glide, and contamination. Instead of adding more refrigerant, the system has to be emptied and completely recharged. Normally, a blend cannot be recycled in a shop; it has to be sent off for disposal or recovery. Blend refrigerants are discussed more thoroughly in Chapters 3 and 13.

Contamination is becoming the focus of every A/C technician who works on automotive systems. Abnormally high pressures in a system may be caused by several factors, one of which is the addition of a blend refrigerant. As we will learn, the recovery equipment used to remove the refrigerant from a system pulls the refrigerant out of the system and puts it into a container with other recovered refrigerant. Recycling equipment cannot separate out the blend compounds or even one type of refrigerant from another. Contaminated refrigerants must be disposed of through a rather expensive process. The recovery and recycling equipment may also become contaminated; this, in turn, can contaminate the systems in other vehicles.

2.8.5 Refrigerant Oils

Refrigerant oil is highly refined, with all wax particles and water removed. An R-12 system uses a mineral-based oil with a viscosity of 500–525 Saybolt universal seconds (SUS). (SAE 30 automotive oil has a viscosity of between 410 and 630 SUS.) An original equipment manufacturer (OEM) R-134a system probably uses 1 of about 29 different PAG oils; the particular oil variety is selected by the vehicle or compressor manufacturer after many hours of very expensive testing to ensure that it works perfectly in that system. An oil that is not recommended by a particular manufacturer either has not been tested or has failed these tests. Almost every compressor and vehicle manufacturer recommends using a particular PAG oil.

Because they mix better with mineral oil and have a higher tolerance for any chlorine remaining in the system from R-12, ester oils are commonly used when an R-12 system is retrofitted to become an R-134a system. One expert recommends against using ester oils because he believes that if they are introduced into a system that is contaminated with moisture or R-12, they will break down into acid and alcohol, which will cause long-term system damage.

Aftermarket sources market two or three types of PAG oils and one or two types of ester oils. Both PAG and ester oils include high and low viscosities to suit the requirements of certain compressors.

CHAPTER QUIZ

These questions help you study this chapter. Enter the proper word(s) in the blanks to complete each statement.

1. We must move __________ to the outside of the vehicle in order to cool it.
2. Heat transfer from one molecule to the one next to it is called __________, and moving heat by circulating hot air or water is called __________.
3. The three states of matter are __________, __________, and __________.
4. If we add enough heat to solid matter, it will change to a(n) __________.
5. Heat that is added to change water to gas is called __________ __________ of __________.
6. Adding pressure to a hot liquid will increase the __________ __________.
7. A liquid gas mixture in a closed container is called a(n) __________ __________, and heating this container will __________ the internal pressure.
8. Ten psi is equal to __________ kPa.
9. A pressure lower than atmospheric is called a(n) __________, and this pressure is measured in __________ or __________.
10. Most vehicle and compressor manufacturers recommend using __________ oil in R-134a systems.

REVIEW QUESTIONS

These questions allow you to check what you have learned. Select the answer that correctly completes each statement.

1. Which of the following is true about heat?
   a. Heat always travels from something warm to something cold.
   b. Heat is a form of energy.
   c. Cold is the lack of heat.
   d. All of these.
2. _____ occurs when heat travels through a material, from one molecule to the one next to it.
   a. Radiation
   b. Convection
   c. Conduction
   d. None of these
3. The process of transferring heat by circulating the heated media is called
   a. radiation.
   b. convection.
   c. conduction.
   d. None of these
4. The movement of heat through heat rays is called
   a. radiation.
   b. convection.
   c. conduction.
   d. None of these

5. Molecules are composed of
   a. atoms. c. electrons.
   b. protons. d. compounds.
6. Which of the following is a form of matter?
   a. Solid c. Gas
   b. Liquid d. All of these
7. Which of the following describes sensible heat?
   a. It can be felt. c. Water at 200°F
   b. It can be measured. d. All of these
8. __________ causes a change of state in matter.
   A. Sensible heat B. Latent heat
   Which is correct?
   a. A only c. Both A and B
   b. B only d. Neither A nor B
9. It takes __________ Btu of heat to cause a change of state from 1 lb of water to 1 lb of steam.
   a. 100 c. 970
   b. 212 d. 32
10. The boiling point of a liquid can be increased by raising the
    a. temperature. c. pressure.
    b. latent heat. d. All of these
11. Superheat refers to temperature increases in a vapor after all of the liquid has boiled.
    a. True b. False
12. Zero psig is
    a. the zero point on all gauges.
    b. the lowest point on a vacuum gauge.
    c. equal to about 15 psi on the absolute pressure scale.
    d. All of these
13. Which of the following is not a CFC or HCFC?
    a. R-12 c. R-134a
    b. R-22 d. None of these
14. Two students are discussing heat. Student A says that heat always travels from hot to cold. Student B says that heat is a form of energy. Who is correct?
    a. Student A c. Both A and B
    b. Student B d. Neither A nor B
15. R-12 is normally used with __________ oil.
    a. mineral c. POE
    b. PAG d. Any of these

CHAPTER

Refrigerants and the Environment

LEARNING OBJECTIVES

After completing this chapter, you should:

- Be familiar with the ozone layer and the adverse effects of a CFC on it.
- Be familiar with the effects of the Montreal Protocol on automotive A/C servicing.
- Understand what refrigerants are and the important safety concerns related to them.
- Be aware of the possible effects of the Kyoto Protocol.
- Be aware of what the next generation of HVAC systems might be.

TERMS TO LEARN

blend	recover
bubble point	recycle
Clean Air Act	SAE
dew point	Section 609
drop in	small can
EPA	SNAP
fractionizing	stratosphere
glide	topping off
global warming	vent
greenhouse effect	zeotrope
ozone	

3.1 INTRODUCTION

Planet Earth is unique in many ways. It has an atmosphere that contains a percentage of oxygen high enough to allow mammals, including humans, to live in it. This atmosphere extends outward from Earth for about 31 miles (50 km). The upper layer of the atmosphere is called the **stratosphere**, and it begins about 7 to 10 miles (11 to 16 km) up and extends to the outer limits. A layer of ozone (O_3) extends around the earth in the stratosphere (Figure 3–1).

The ozone layer is important to us because it blocks out ultraviolet wavelengths of light generated by the sun. Ultraviolet rays can be very harmful to our way of life. In humans, an excess of these rays can cause an increase in skin cancer and cataracts of the eyes, as well as damage to our immune systems. These same problems can affect many animals. Ultraviolet rays can also damage plants and vegetables. This damage probably also extends to plankton and larvae in the sea, the base of the food chain for sea animals.

In the late 1900s, it was determined that the ozone layer is getting much thinner, and large holes are being created in it (mostly near the South Pole). The ozone layer is not providing the same protection from ultraviolet (UV) rays as it once did. It has been determined that (1) the breakup or depletion of the ozone layer is caused by human-made chemical pollution, (2) one of the major ozone-depleting chemicals is chlorine, and (3) one of the major sources of chlorine in the atmosphere is R-12.

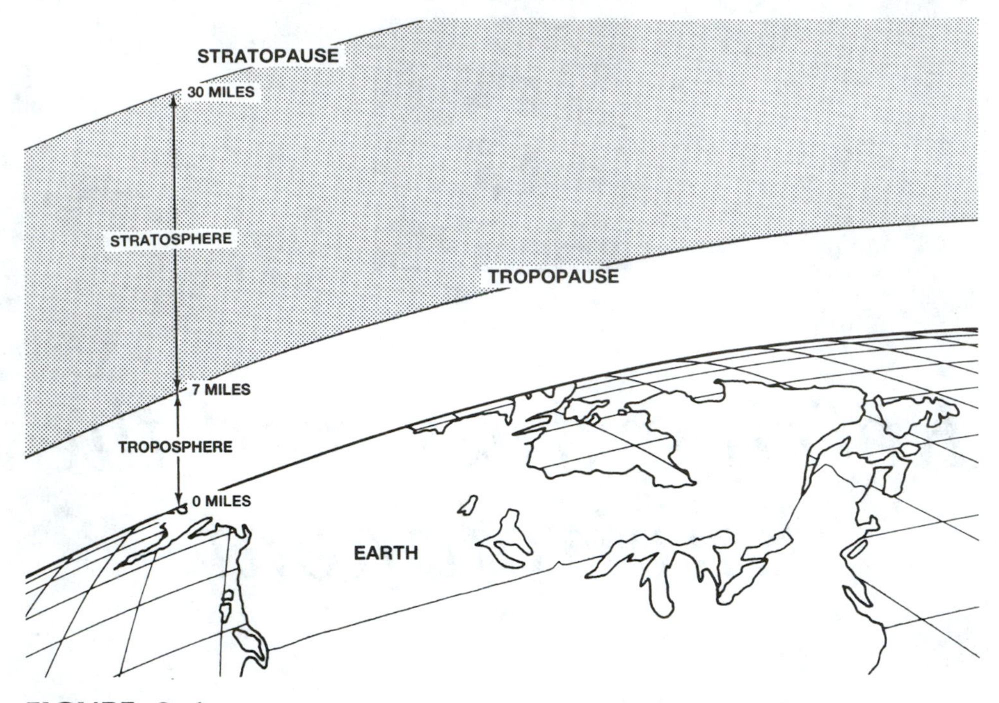

FIGURE 3–1 The ozone layer is in the stratosphere, miles above Earth. *(Reprinted with permission of General Motors)*

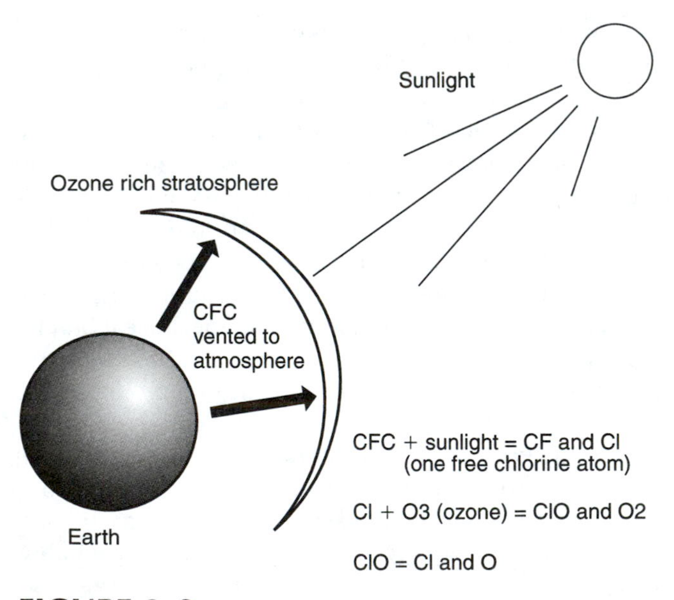

FIGURE 3–2 The ozone-rich stratosphere protects us from harmful rays coming from the sun. CFC that is vented can break down this layer.

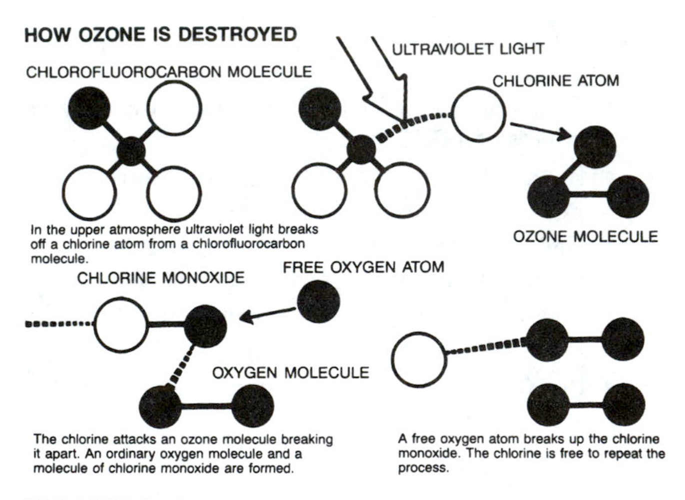

FIGURE 3–3 When CFC molecules break down in the stratosphere, a chlorine atom breaks free to attack ozone molecules. *(Courtesy of DaimlerChrysler Corporation)*

A chlorine atom from a chlorinated fluorocarbon (CFC) such as R-12 can travel into the stratosphere if it escapes or is released. There, under the effects of ice clouds and sunlight, it can combine with one of the oxygen atoms of an ozone molecule to form chlorine monoxide and an ordinary oxygen molecule, O_2. This destroys that ozone molecule (Figure 3–2). The effect does not end there; the chlorine can break away and attack other ozone molecules. It is believed that 1 chlorine atom can destroy 10,000 to 100,000 ozone molecules (Figure 3–3).

Another area of concern is a layer of gases that are causing a **greenhouse effect**. This gas layer traps

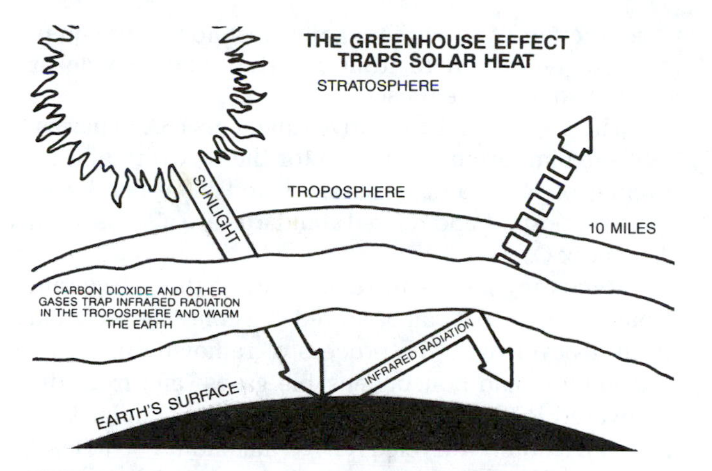

FIGURE 3–4 Greenhouse gases in the lower atmosphere reflect heat back onto Earth and increase temperatures. *(Courtesy of DaimlerChrysler Corporation)*

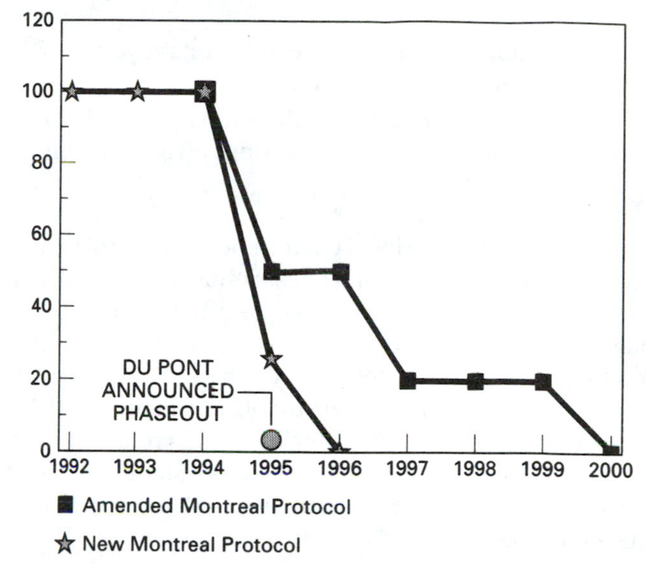

FIGURE 3–5 The time scale adopted as the Montreal Protocol was amended to speed up the phaseout of CFCs. *(Courtesy of Castrol North America)*

heat at the Earth's surface and lower atmosphere, and it is increasing the temperature of our living area. This is called **global warming**. CFC and HC gases are considered greenhouse gases (Figure 3–4). Although very low, R-134a does contribute to global warming. To become more environmentally friendly, vehicle manufacturers are currently testing two "natural" refrigerants: hydrocarbon (HC) and carbon dioxide (CO_2). Both of these are very difficult to use, and substantial development is necessary before they can be used in motor vehicles. HC is flammable, and CO_2 requires system pressures of 1,500 to 2,000 psi (10,342 to 13,790 kPa).

The automotive refrigeration industry went through a major search for a non-CFC refrigerant that was as good a refrigerant as R-12. Another goal was to use as few different chemicals in the systems as possible to reduce mixing and cross-contamination of chemicals. At this time, every major vehicle manufacturer has chosen R-134a; few other compounds are being considered or tested.

3.2 LEGISLATION

At a conference in Montreal, Canada, in 1987, the United States, along with 22 other countries, agreed to limit the production of ozone-depleting chemicals. This agreement is referred to as the Montreal Protocol. In 1990, President Bush signed the Clean Air Act, which phased out the production of CFCs in the United States by the year 2000. In 1993, the phaseout of CFC-12 was moved forward; R-12 production in the United States ceased at the end of 1995 (Figure 3–5).

Section 609 is a portion of the Clean Air Act that places certain requirements on the mobile vehicle air conditioning (MVAC) service field. Important portions of this section require the following:

Effective January 1, 1992

- Technicians who repair or service automotive A/C systems shall be properly trained and certified and use approved refrigerant recovery and recycling equipment.
- Recovery and recycling equipment must be properly approved.

Effective November 15, 1992

- Sales of small containers of R-12 (less than 20 lb) are restricted to certified technicians.

Effective January 1, 1992

- Each shop that performs A/C service on motor vehicles shall certify to the EPA that it is using approved recycling equipment and that only properly trained and certified technicians are using this equipment.

Effective November 14, 1994

- MVAC technicians must be certified to purchase EPA-acceptable blend refrigerants.

Effective January 29, 1998

- Equipment used to recover or recycle R-134a must meet SAE standards.
- MVAC service technicians must be certified to handle non-ozone-depleting refrigerants (including R-134a).

Some areas have local environmental control regulations that include even more restrictions. The European Community set 1995 for a complete phaseout of R-12, and two Canadian provinces have banned R-12. R-12 is no longer produced in the United States.

Under the Montreal Protocol, phaseout of HCFC will begin in 2004 (a 35% reduction). This reduction will begin in 2003 in the United States. Manufacture of HCFC-22 will stop in 2010 in the United States and other developed countries.

3.2 Additional Concerns

There is still a major concern about the global-warming effect from escaping refrigerants. Representatives from many nations met in Kyoto, Japan, in 1997 and established the Kyoto Protocol to address these concerns. The Total Environmental Warming Impact (TEWI) index was developed; it rates the impact of various refrigerants along with the energy required to perform the cooling operation. In the future, it is hoped that each nation will reduce its negative impact on the environment to zero in order to reduce global change to zero. All factors concerning energy use and emissions will be considered. The United States has not signed the Kyoto Protocol.

3.3 RECOVERY AND RECYCLING

A major thrust of the Clean Air Act is to recycle R-12 instead of releasing it to enter the atmosphere. This makes economic sense. At one time, new (also called virgin) R-12 was inexpensive, well under $1.00 per pound. Now added taxes and increased demand due to limited production have raised the price significantly. When R-12 was inexpensive, it was standard practice to simply **vent** the contents of a system to the atmosphere when service was needed. Also, many people kept adding R-12 to a system rather than going to the trouble and expense of repairing a leak. This was called **topping off** a system. With the increased cost of R-12, it pays to repair small leaks and possibly retrofit the vehicle to use a less expensive refrigerant. It is also socially irresponsible not to repair any fixable leaks. Small containers of R-12 are no longer available to do-it-yourselfers. And service shops must now have equipment to recover R-12 from a system and recycle or clean it so it can be reused.

The Society of Automotive Engineers (SAE) has established important standards for the recovery and recycling of R-12, and these are listed in Appendix C. Other important and related standards are also listed in Appendix C.

Recovery means to remove all of the refrigerant from a system so it can be stored in a container in liquid form. **Recycling** is the process of removing moisture (water), oil, and noncondensable gases (air) from the recovered R-12 or R-134a so it meets the standards of new refrigerant. Recycled refrigerant should be at least 98% pure. Limits for these standards are as follows:

- Moisture: 15 parts per million (ppm) by weight
- Refrigerant oil: 4,000 ppm by weight
- Noncondensable gases: 330 ppm by weight

R-134a can also be recovered and recycled, but separate equipment, dedicated to R-134a, is required. Recycled R-12 or R-134a can be used in the same way as new refrigerant.

Blend refrigerants can also be recovered and recycled using a dedicated machine for that blend. However, this refrigerant can only be used in the vehicle from which the recovered mix came. In cases of fleet vehicles, recycled blend refrigerants can be used in other vehicles within that fleet.

Recovery equipment is available as a single unit or in combination with a recycling unit. Recovery-only units are simpler and less expensive than equipment that handles both recovery and recycling. The procedure used to recover and recycle R-12 and R-134a is described in Chapter 13 (Figure 3–6).

3.3.1 Recovering Contaminated Refrigerant

A service problem facing the modern MVAC technician concerns what to do with a system that contains the wrong or a contaminated refrigerant. If it is recovered using R-12 or R-134a recovery equipment, the equipment and the recovery container become contaminated. Recycling equipment is designed to remove only water, oil, and air; other refrigerant compounds cannot be removed. The equipment must be decontaminated, and any refrigerant in the recovery container must be sent off-site, for reclaiming or disposal. Another fear is that if the recovery unit is electric powered, as most are, and if a flammable refrigerant is

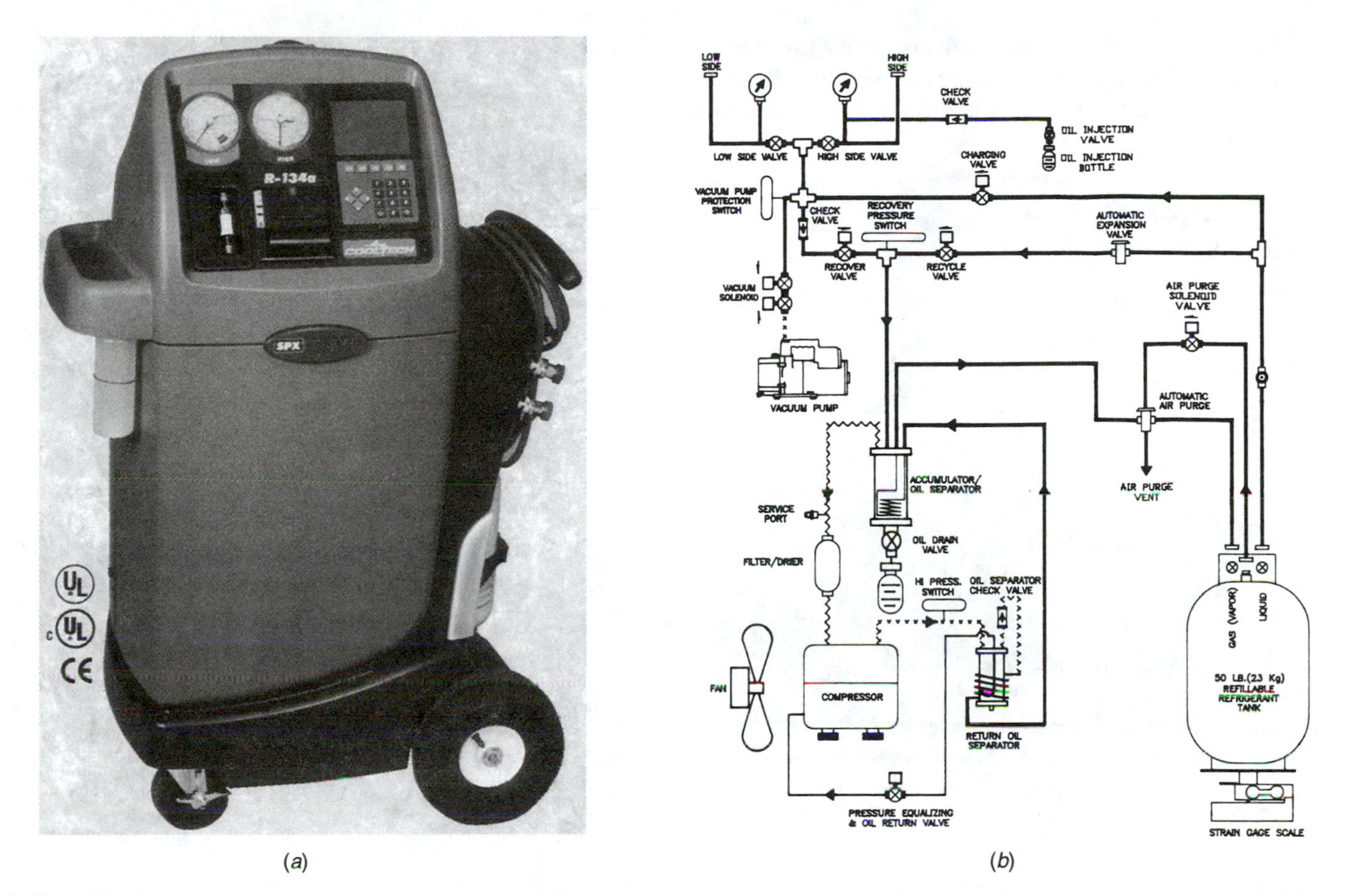

(*a*) (*b*)

FIGURE 3–6 Recovery systems remove the refrigerant from a system so it can be recycled. Some are small enough to be carried to the job; most are on a cart for easy movement to the vehicle (*a*). A schematic of the internal portion is shown in (*b*). (*Courtesy of Robinair Division, SPX Corporation*)

recovered, an explosion or other damage to the equipment is possible. Air-powered recovery machines or a special procedure can be used to recover flammable refrigerants.

An important step in any future recovery process is to identify what refrigerant is in the system: R-12, R-134a, a blend, or a contaminated mixture. Contaminated mixtures are called **unknown, junk**, or **Brand X**. Most larger shops have a separate recovery or recovery–recycling machine for R-12, another one for R-134a, and still another one for blends and unknowns. The technician has two ways of identifying the refrigerant: the label and service fittings on the vehicle and a refrigerant identifier. These labels and fittings are described in Appendix D. The label is the easiest method but is not always accurate. The identifier is the most reliable, but it must be capable of identifying whether the refrigerant contains hydrocarbon (flammable refrigerant) and the type of refrigerant (Figure 3–7). If a contaminated or unknown refrigerant is recovered, it should be recovered into a container that is gray with a yellow top that is labeled as a **mixed, junk, dirty R-12, dirty R-134a, or unknown refrigerant**. The mixture should be sent off to one of the several firms that specialize in disposal or recovery of the mixtures.

3.3.2 Recovery Container Certification

The tank used to recover refrigerant on a recovery–recycling machine must be safety checked every five years. This legal requirement is part of the refrigerant equipment certification requirement in the Clean Air Act. Tank testing is based on the Title 49 U.S. Department of Transportation (DOT) requirements, essentially an external and internal visual inspection and a hydrostatic pressure test. Failure to test the cylinders can put the owners at risk of a $25,000 fine. SAE standard J2296, Retest of Refrigerant Container, identifies the DOT standards that are required by law and also discusses other concerns for storage containers. The five-year time period begins at the date of tank manufacture, which is stamped on the tank's collar (Figure 3–8).

FIGURE 3–7 A refrigerant identifier can be used to determine whether the refrigerant in a system is pure or contaminated. *(Courtesy of Neutronics Inc.)*

3.4 REFRIGERANTS

As discussed in Chapter 2, before the early 1990s, R-12 was the refrigerant used in mobile A/C systems. R-134a is now being used (beginning in model year 1994). R-22, commonly used in stationary units, is not for automotive use (Figure 3–9). R-22 has a lower ozone-depletion potential (ODP) of 0.05 than R-12, which is 0.9; R-134a has an ODP of 0 (Figure 3–10).

Refrigerants are colorless and odorless compounds. Usually the only way we know that they are present is how the container feels when we pick it up or shake it. We can also feel the temperature drop of

FIGURE 3–8 The circled portion of this recovery tank reads "Retest by 02."

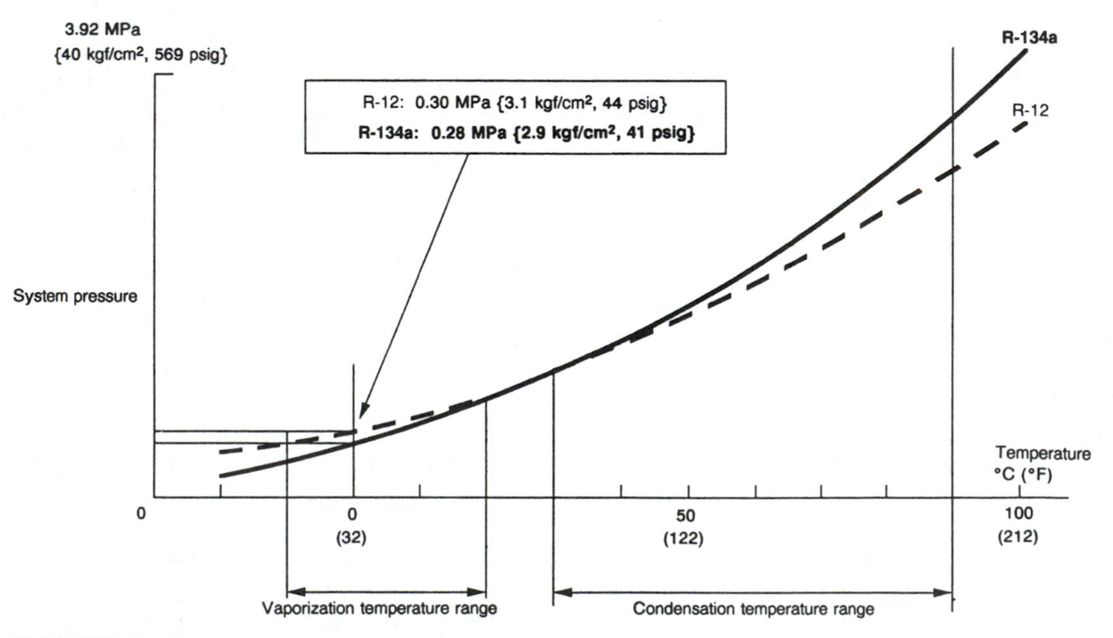

FIGURE 3–9 R-12 and R-134a have similar operating pressures. *(Courtesy of Zexel USA Corporation)*

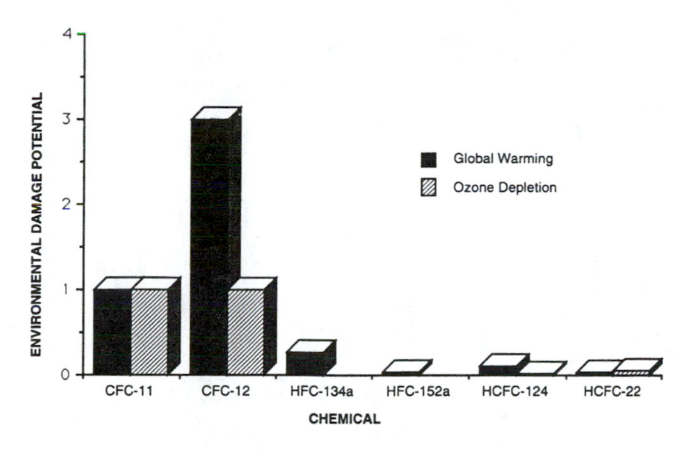

FIGURE 3–10 A comparison of potential atmospheric damage of refrigerants and flushing agents. (*Reprinted with permission from SAE Document M-106, © 1992, Society of Automotive Engineers, Inc.*)

the container as we release the refrigerant into a system. On gauge sets equipped with a sight glass, we can see bubbles in the clear liquid as it passes by the glass.

Two other refrigerants have been used in the automotive field as flushing agents. These refrigerants are R-11 and R-113, and they both have fairly high boiling points, 75°F (24°C) for R-11 and 118°F (48°C) for R-113. Their usage, basically to flush debris out of a system, has stopped because of their possible toxicity and their effect on the environment. Because they are CFCs, they should not be released into the atmosphere.

Refrigerants are commonly available in several sizes of containers with a small can of 12 to 14 oz (400 g—at one time, this was 15 oz of R-12 and 1 oz of can, for a total of 1 lb) and larger drums or canisters of 15 or 30 lb (6.8 or 13.6 kg). As mentioned earlier, small containers of R-12 can only be purchased by certified technicians. Refrigerant containers are color coded: R-12 containers are white, R-22 containers are green, and R-134a containers are light blue (Figure 3–11).

The containers that refrigerants are purchased in are usually disposable (Figure 3–12). These containers should be evacuated into a recovery unit, marked empty, and properly disposed of when they are emptied. The storage containers for recycled refrigerant must be approved by the DOT and carry the proper marking to show this (Figure 3–13).

3.4.1 Alternate Refrigerants for R-12

Both R-12 and R-134a are good single-compound refrigerants. But R-12 adversely affects our environment, and some systems and compressors are not compatible with R-134a. Approved alternate refrigerants are blend refrigerants that are combinations of molecules. They are an *azeotrope* or a *zeotrope*, both are mixtures or blends that cannot be easily separated. Azeotrope chemicals act like a single chemical with a single boiling point. Zeotrope chemicals are also blends, but they behave differently. They have a range of boiling points (called *bubble point*) that *glides* upward until the heavier compounds boil off. The amount of glide indicates the boiling point range. When a blend condenses, the lowest boiling point ingredient will condense first (called the *dew point*), and condensation will continue to the point where all of the ingredients are liquid. If a refrigerant leak should occur, the lighter elements will vaporize and escape first. This is called *fractionizing*, and it changes the characteristics of the refrigerant. Most of the problems with R-134a are caused by compressors that are not capable of handling the slightly higher pressures or by reactions with viton sealing materials. These problems can be solved by replacing the compressor, but this increases the cost of the retrofit. Retrofit procedure and potential problems are discussed more completely in Chapter 13.

It should be noted that R-134a is the only alternate product for an R-12 system that has addressed the SAE retrofit documents (SAE J1657, SAE J1658, SAE J1659, and SAE J1662). These standards are designed to ensure long-term, trouble-free operation of the A/C system.

Section 612 of the 1990 Clean Air Act established the **Significant New Alternatives Policy (SNAP)** program to determine acceptable replacements for Class I and Class II chemicals. Class I chemicals include CFCs, and Class II chemicals are all the HCFCs. SNAP is administered by the EPA and identifies refrigerants that are acceptable from their ozone-depleting potential, global warming potential, flammability, and toxicity characteristics. Alternate refrigerants are not tested on their refrigeration quality, only on their human health and environmental risks. Several alternate refrigerants have been declared acceptable by SNAP, and these are listed in Appendix D. All of these are blends of two or more compounds, and they have characteristics of zeotropic compounds: fractionation and glide. Some contain R-134a, which means they could cause some of the same problems to the system that R-134a does, and others contain R-22. R-22 causes serious seal swell (as great as 40%) with some of the sealing compounds used with R-12, and barrier hoses are required (Figure 3–14). Most of these compounds contain an HCFC, one of the interim refrigerants that

R-12	R-134a
REFRIGERANT	
CONTAINER COLOR White	**CONTAINER COLOR** Light blue
CONTAINER MARKING R-12, Freon®	**CONTAINER MARKING** R-134a, Suva® Trans A/C
PART NUMBER 999MP-A4001	**PART NUMBER** 999MP-R134A
CONTAINER FITTING SIZE 7/16" - 20, also known as "1/4 - flare"	**CONTAINER FITTING SIZE** 1/2" - 16 ACME
CHEMICAL NAME Dichlorodifluoromethane	**CHEMICAL NAME** Tetrafluoroethane
ODP (OZONE DEPLETION POTENTIAL) (R-11 = 1) 1	**ODP (OZONE DEPLETION POTENTIAL) (R-11 = 1)** 0
HGWP (HALOCARBON GLOBAL WARMING POTENTIAL) 3.0	**HGWP (HALOCARBON GLOBAL WARMING POTENTIAL)** Less than 0.3
BOILING POINT -21.62°F (-29.79°C)	**BOILING POINT** -15.07°F (-26.15°C)
LATENT HEAT OF VAPORIZATION (The amount of energy required to change state from vapor to liquid) 36.43 (Kcal/Kg @ 0°C)	**LATENT HEAT OF VAPORIZATION** (The amount of energy required to change state from vapor to liquid) 47.19 (Kcal/Kg @ 0°C)
CHEMICAL STRUCTURE CCl_2F_2 (F - C(F)(Cl) - Cl)	**CHEMICAL STRUCTURE** CH_2F-CF_3 (H - C(H)(F) - C(F)(F) - F)
MOLECULAR DIAMETER 4.4 angstroms	**MOLECULAR DIAMETER** 4.2 angstroms

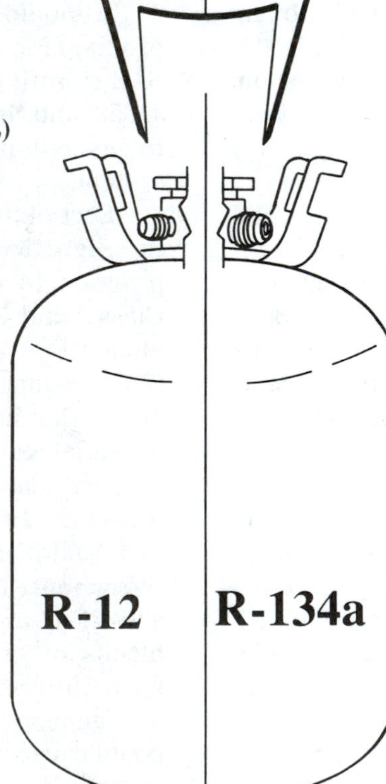

FIGURE 3–11 A comparison of R-12 and R-134a; note differences in container and fittings. (*Courtesy of Nissan Motor Corporation in USA*)

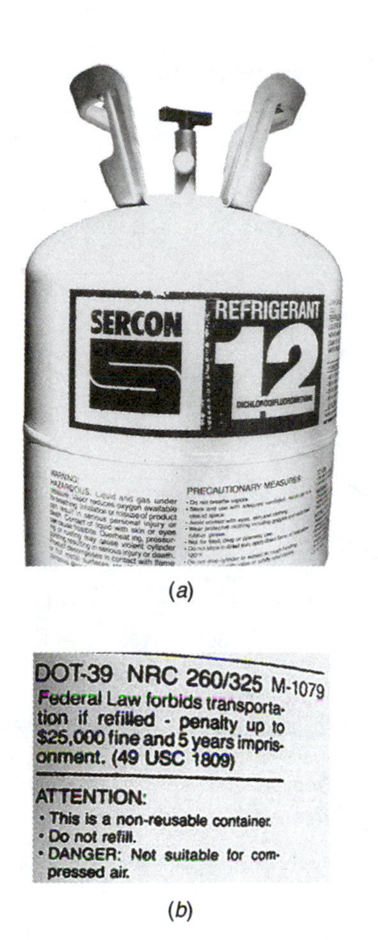

FIGURE 3–12 A disposable container of R-12 (*a*). As mentioned on the printed portion (*b*), this container should not be reused.

will be banned in the near future because of their effect on global warming.

An alternate refrigerant can only be used under the following conditions:

- Each refrigerant must have its own unique set of fittings, and all ports not converted must be permanently disabled. (These fittings are described in Appendix D.)
- Each refrigerant must have a label with a unique color that specifies pertinent information.
- All original refrigerant must be removed before charging with the new refrigerant.
- With blends that contain HCFC-22, hoses must be replaced with less permeable barrier hoses.
- With systems that include a high-pressure release device, a high-pressure shutoff switch must be installed.

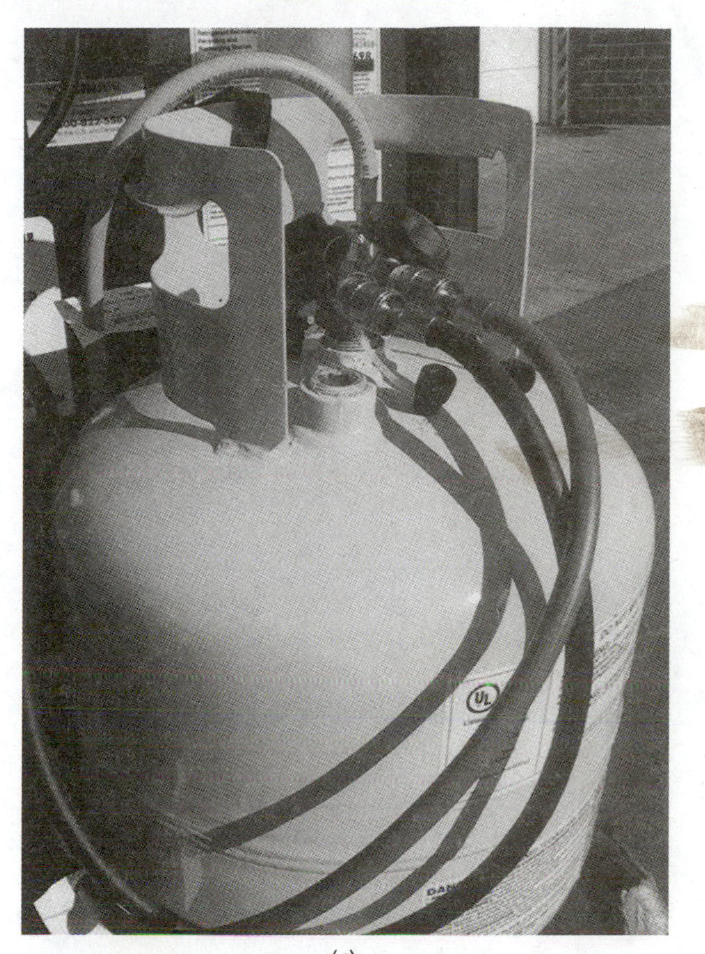

(*a*)

(*b*)

FIGURE 3–13 A DOT-approved refrigerant container has two valves, one for gas and one for liquid (*a*). A portion of the upper band reads "DOT-4BA400 (*b*).

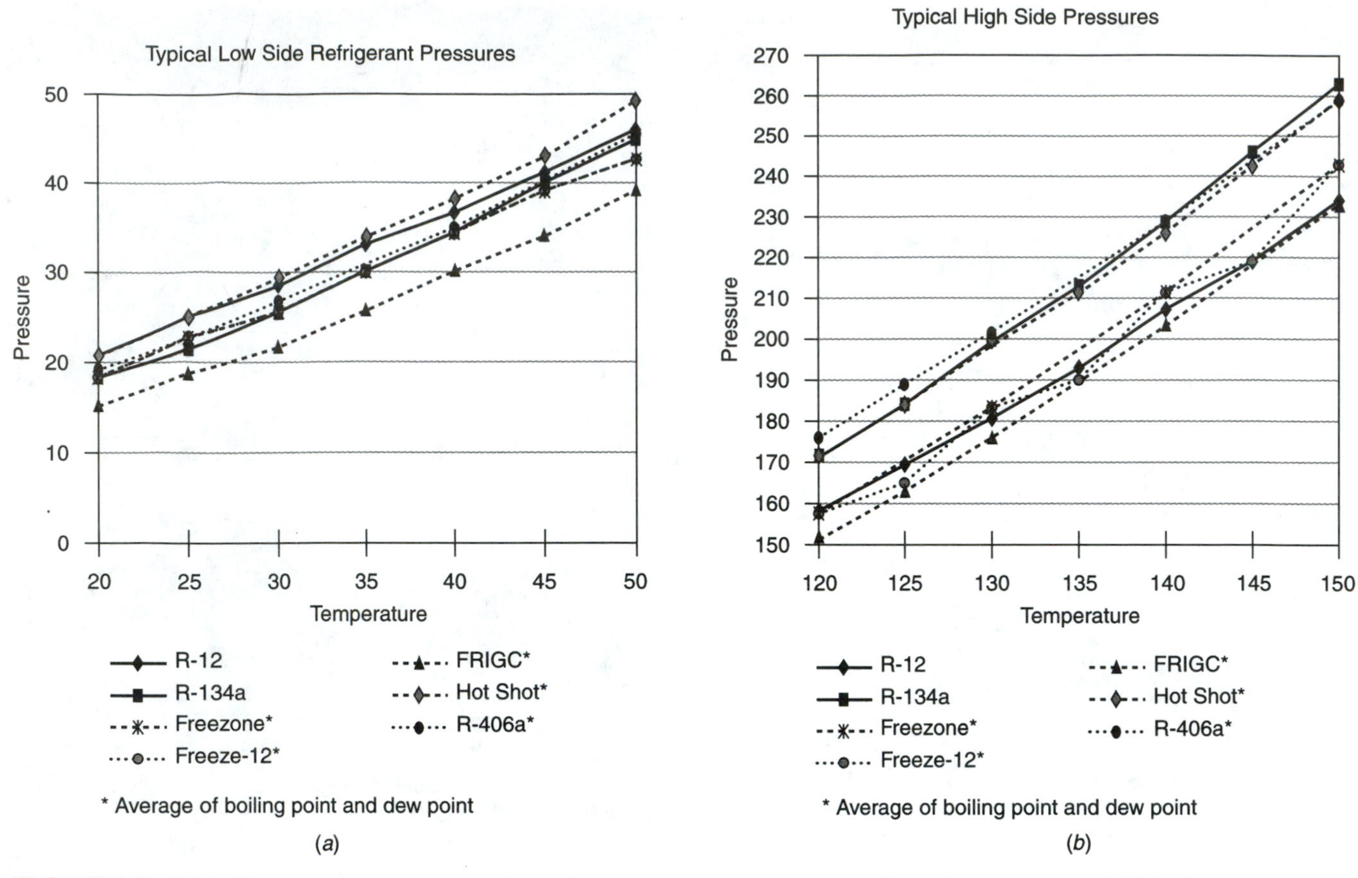

FIGURE 3–14 The pressure–temperature relationship varies between these EPA-approved SNAP refrigerants.

- Blends containing HCFC-22 will be phased out in the near future.

These requirements apply to every vehicle in the United States. It is fairly easy to see that there can be no drop-in refrigerant that will simply replace R-12 or that can be added to an R-12 system.

If you are considering retrofitting a vehicle to an alternate refrigerant, consider the following:

- Cost of refrigerant, fittings, and equipment to service it
- Cost of any components that might need replacement for the particular system
- Long-term effect on components, seals, and hoses in the system
- If a refrigerant that contains R-22 is considered, barrier hoses must be installed (if not already used on the system).
- Cost of future disposal of refrigerant used (R-134a can be recovered and recycled; a blend can be recycled but only for reuse in the same vehicle or for reuse in the same fleet)
- Warranty for most vehicle and replacement parts; manufacturers only warranty parts and equipment used with R-12 or R-134a.
- Operating characteristics relative to R-12; whether the system controls need to be recalibrated (Figure 3–15)
- Availability of refrigerant in the future or in other parts of the country where the vehicle might end up.

The approved retrofit procedure is described in Section 13.4.

3.4.2 Alternate Refrigerants, R-134a

At least one manufacturer has developed an alternate refrigerant for R-134a; this refrigerant is a zeotropic blend containing R-152a, R-134a, and R-125 and is claimed to

Refrigerant	ST4470 Nitrile O-ring	ST7470 HNBR O-ring	ST9670 Neoprene O-ring	HNBR Lathe Cut	Nitrile Lathe Cut
R-12 Mineral Oil	Good	Good	Good	Good	Good
R-134a Polyalkylene Glycol Oil	Good	Good	Good	Good	Good
Freez-12 Mineral Oil	Good	Marginal	Good	Marginal	Good
Freez-12 Polyol Ester Oil	Poor	Poor	Marginal	Poor	Marginal
RB-276 (Freezone) Mineral Oil	Marginal	Marginal	Good	Marginal	Good
FR-12 (Frig-c) Mineral Oil	Marginal	Poor	Good	Poor	Marginal
FR-12 (Frig-c) Polyol Ester Oil	Good	Marginal	Good	Poor	Good
GHG-X4 (Chill-it) Mineral Oil	Poor	Poor	Marginal	Poor	Poor
Hot Shot/Kar Kool Mineral Oil	Marginal Note 1	Marginal Note 1	Good Note 1	Poor Note 1	Good Note 1
R-406a (McCool) Mineral Oil	Good	Poor	Marginal	Poor	Good

Good = Volume Change±15% Marginal = Volume Change +16 to +40 Poor = Volume Change >40%
Note 1: The Hot Shot container received through retail channels contained only R-22 & R-142b.

FIGURE 3–15 This table shows how various refrigerants affect the elastomer seal materials commonly used in an A/C system. *(Courtesy of Santech Industries, Fort Worth, Texas)*

improve the performance of R-134a systems. The manufacturer states that SNAP approval and other retrofit procedure requirements do not apply because SNAP only applies to alternate refrigerants for R-12. (R-134a is neither a Class I or II chemical.) It is very important whenever an alternate refrigerant is used in a system that proper labeling be applied so future service technicians will be aware of the change. Many technicians prefer not to use an alternate for R-134a because of the possible fractionation and contamination problems.

3.4.3 Counterfeit and Bootleg Refrigerants

With the high cost and limited availability of R-12, a refrigerant buyer has to be more careful of where R-12 is purchased. Refrigerant is being illegally imported, and some of it is badly contaminated. There are stories of counterfeit refrigerant being sold in containers marked the same as those of a reputable, domestic manufacturer. Other stories tell of containers that contain nothing but water with air pressure. Refrigerant should be purchased only from a known source that handles reputable products; buyers must be certified.

3.4.4 Future Refrigerants and Systems

In order to meet the requirements of Total Enivironment Warming Impact (TEWI), several possible changes are being considered: improved R-134a efficiency and one of the new natural refrigerants, CO_2 and HC. There are potential problems with both of the natural refrigerants.

A carbon dioxide (CO_2) system is similar to the present-day systems, but it requires extremely high pressures, 7 to 10 times that of an R-134a system.

A butane/propane (HC) system uses a very flammable gas that will function in present-day A/C systems, but the potential danger is too great to use it. A leak or rupture in the evaporator could easily result in a vehicle explosion. An HC system will probably use a secondary

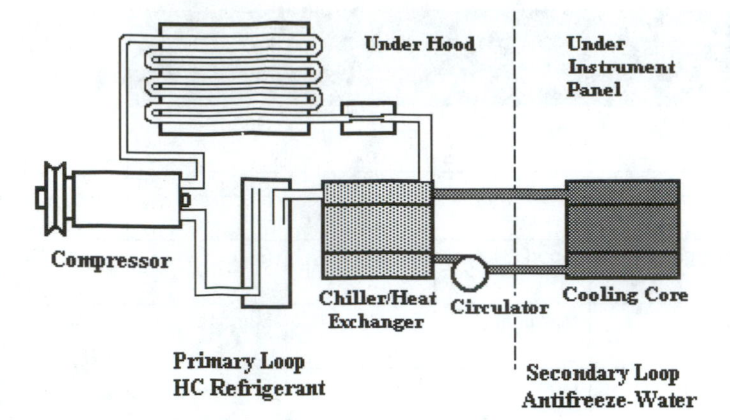

FIGURE 3–16 A secondary loop A/C system keeps the potentially dangerous HC refrigerant out of the passenger compartment by using a chiller/heat exchanger to cool an antifreeze and water mixture. This fluid then transfers heat from the cooling core in the air distribution section to the chiller.

loop. The HC portion of the system will be entirely under the hood; a heat exchanger will provide a very cold fluid to connect with a liquid-to-air heat exchanger that will replace the evaporator (Figure 3–16).

On the whole, R-134a systems are not too bad, and they can get better if the efficiency of the present system can be improved and leakage can be reduced. At this time, a CO_2 system is up to 40% less efficient than an R-134a system and can cost about 30% more to produce. The efficiency of an HC, secondary loop system is about 20% lower than an R-134a system and will be about 20% more expensive to build. System efficiency is important in that less efficient systems require more power to do the same job, and this power comes from the engine. A less efficient system will cause the engine to work harder and burn more fuel, increasing its emissions.

3.5 REFRIGERANT SAFETY PRECAUTIONS

Refrigerants should be handled only by trained and certified technicians because of potential safety hazards:

- Physiological reaction
- Asphyxiation
- Frostbite and blindness
- Poisoning
- Combustion
- Explosion of storage containers

CAUTION

The wise technician is always prudent when working with modern chemicals and avoids breathing fumes and vapors, as well as skin contact, with refrigerants, oils and lubricants, cleaning agents, and other chemicals.

Human bodies react to chemicals in different ways, so exposure to chemicals is not completely predictable. In laboratory tests, a human subject lost consciousness and both his blood pressure and pulse dropped to zero after $4^1/_2$ minutes of exposure to a concentration of 4,000 ppm of R-134a; other subjects experienced a rise in blood pressure and pulse rate as well as noticeable discomfort. The condition of these subjects returned to normal after the tests were stopped. With modern A/C service methods and recovery equipment, it should be easy to avoid breathing in refrigerants.

Refrigerants do not contain oxygen and are heavier than air. If they are released into a confined area, they fill the lower space, forcing air and its oxygen upward. Any humans or animals that breathe refrigerants can be asphyxiated, and lack of oxygen can cause loss of consciousness or death. Current regulations for recycling refrigerants should eliminate any intentional dumping of refrigerants, but in case of accidental release into a confined area, you should immediately move to an area with adequate ventilation.

If liquid refrigerant is splashed onto the skin or into the eye of a human or animal, it immediately boils and absorbs heat from the body part it is in direct contact with. The temperature of the area is reduced to the low boiling point of the refrigerant, which is cold enough to freeze that body part.

SAFETY NOTE

A wise technician wears safety goggles or a clear face shield and protective clothing (gloves) when working with refrigerants. If refrigerant splashes into your eyes, blindness can occur. If refrigerant splashes into your eye or onto your skin, **do not** rub that body part; **flush it with cool, clean water** to restore the temperature. Place sterile gauze over the eye to keep it clean and get professional medical attention immediately.

If a CFC such as R-12 or R-22 comes into contact with flame or heated metal, *a poisonous gas similar to phosgene is formed.* This can occur while using a flame-type leak detector, if refrigerant is drawn into a running engine, or even if it is drawn through burning tobacco. An indication that a poisonous gas is forming is a bitter taste in your throat. Flame-type leak detectors should be used only in well-ventilated areas; if used, they should be held away from your face at arm's length.

Several flammable refrigerants have been marketed, and even though they have been banned and are illegal, they still show up. A mixture of more than 2% hydrocarbon (butane, isobutane, or propane) is considered flammable, and about 4 oz in a car interior can become an explosive mixture. R-134a can become combustible at higher pressures if mixed with air. Air should not be used to flush an R-134a system because of the remote chance of a fire or explosion.

When refrigerant containers are filled, reserve room is left for expansion and the container is marked with its critical temperature, the maximum that it should be subjected to. Refrigerant containers are designed to contain the refrigerant under pressures encountered under normal working and storage conditions. Container pressure is about the same as the vapor pressure for that refrigerant up to a certain temperature point, which is where the liquid has expanded to fill the entire container. Beyond this point, any further expansion of the liquid generates very high hydraulic pressures that will rupture the container. The chance of container rupture is generally low unless the container is overfilled or overheated. You should never apply direct heat to a storage container or place the container where it will be exposed to high temperatures. Containers of R-12 have exploded from the radiant heat of direct sunlight (Figure 3–17).

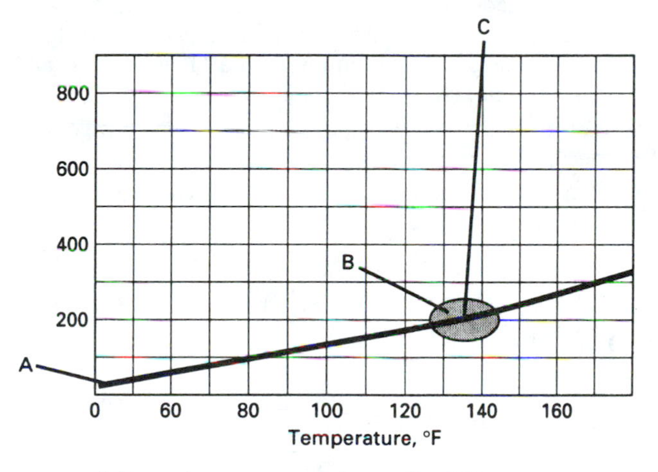

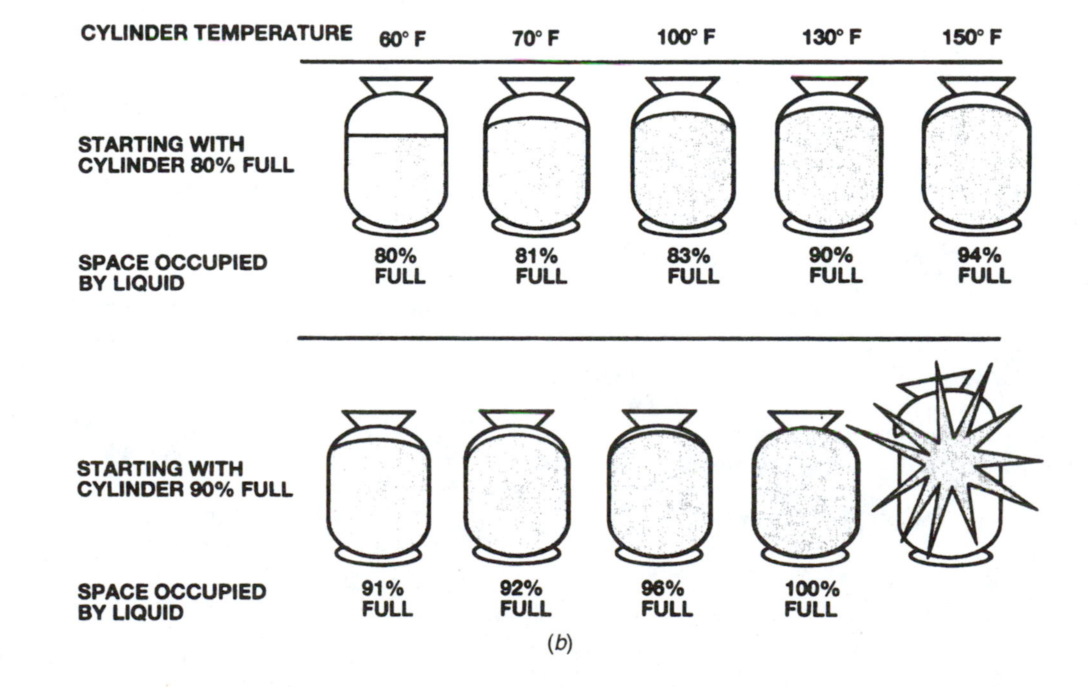

FIGURE 3–17 Pressure in a refrigerant container follows the pressure–temperature relationship until the contents expand to fill the container (*a*). When recovering refrigerant, the container should be filled to a maximum of about 80% (*b*). (b *is reprinted with permission of General Motors Corporation*)

When working with PAG and ester oils, you should take care not to breathe the vapor that can escape from an open container or the mist discharging with refrigerant from a system. Avoid contact between these oils and your skin: Nonpermeable gloves should be worn to prevent skin contact with PAG oil. Another problem encountered with R-134a and PAG oil is damage to paint (possibly removing any paint that it is spilled onto), plastic parts, drive belts, and hoses.

CHAPTER QUIZ

These questions help you study this chapter. Enter the proper word(s) in the blanks to complete each statement.

1. A chlorine atom in the stratosphere will combine with a(n) ________ molecule and ________ it.
2. A layer of gases that can trap heat in the Earth's lower atmosphere is called the ________ effect, and this can produce ________ ________.
3. One of the provisions of Section 609 of the Clean Air Act requires the technicians who ________ or ________ automotive A/C systems to be properly ________ and ________.
4. Removing refrigerant from a system is called ________; ________ removes water, oil, and air from that refrigerant so it can be reused.
5. A mixture of R-12 and R-22 is considered ________ and ________ be recycled.
6. The tank into which the refrigerant is recovered must be ________ every ________ years.
7. Both R-12 and R-134a are ________ ________ refrigerants; a mixture of two or more refrigerants is commonly called a(n) ________.
8. SNAP stands for ________ ________ ________ ________, and SNAP ________ refrigerants have acceptable global warming potential, flammability, and toxicity characteristics.
9. Care should be exercised while working with refrigerants because they present potential ________ ________.
10. If a liquid expands to the point where a container is filled with liquid, adding heat will ________ the ________ very rapidly.

REVIEW QUESTIONS

These questions allow you to check what you have learned. Select the answer that correctly completes each statement.

1. ________ is thought to be the chemical chiefly responsible for depletion of the ozone layer.
   a. Hydrogen c. Nitrogen
   b. Chlorine d. Carbon
2. In 1987, 23 major countries agreed to phase out ________ at a conference in Montreal.
   a. R-22 c. R-11
   b. R-12 d. All of these
3. Two students are discussing Section 609 of the Clean Air Act. Student A says that it requires that A/C technicians be properly trained and certified. Student B says Section 609 requires that recovery and recycling equipment be of an approved type. Who is correct?
   a. Student A c. Both A and B
   b. Student B d. Neither A nor B
4. Student A says that the EPA ensures that all SNAP refrigerants are as good or better than R-12. Student B says that the EPA checks possible new refrigerants to make sure they are safe to work with. Who is correct?
   a. Student A c. Both A and B
   b. Student B d. Neither A nor B
5. Standards for A/C service procedures and equipment have been developed by the
   a. EPA. c. SAE.
   b. Clean Air Act. d. DOT.
6. Two students are discussing the retrofit procedure. Student A says that the service ports must be changed to match the new refrigerant. Student B says a new label must be installed over the old one to identify the new refrigerant and oil. Who is correct?
   a. Student A c. Both A and B
   b. Student B d. Neither A nor B
7. Mixing two different refrigerants
   a. is illegal.
   b. makes a contaminated mixture that is expensive to dispose of.
   c. creates a zeotrope.
   d. All of these.
8. Two students are discussing the practice of topping off an R-12 system. Student A says that this practice is no longer profitable. Student B says that it is not recommended because of ozone depletion. Who is correct?
   a. Student A c. Both A and B
   b. Student B d. Neither A nor B
9. A container of R-134a is color-coded ________.
   a. yellow c. green
   b. white d. light blue

10. A potential safety hazard when working with refrigerants is
    a. frostbite and blindness.
    b. asphyxiation.
    c. container explosion.
    d. All of these.
11. When R-12 comes in contact with flame, a toxic gas called ________ is formed.
    a. chlorine
    b. phosgene
    c. fluorine
    d. All of these
12. When working with refrigerants and refrigerant oils, you should
    A. avoid breathing the vapors.
    B. avoid contact with your skin.

    Which is correct?
    a. A only
    b. B only
    c. Both A and B
    d. Neither A nor B
13. Student A says that butane and propane will work as refrigerants. Student B says that these gases present a fire hazard during service procedures. Who is correct?
    a. Student A
    b. Student B
    c. Both A and B
    d. Neither A nor B
14. Student A says that the recovery tank used in a recovery system must be checked and certified every three years. Student B says that a disposable refrigerant tank should be properly evacuated before disposal. Who is correct?
    a. Student A
    b. Student B
    c. Both A and B
    d. Neither A nor B

Moving Heat: Heating and Air Conditioning Principles

LEARNING OBJECTIVES

After completing this chapter, you should:

- Have a basic understanding of heating and cooling loads.
- Be familiar with the ways to handle a heating load.
- Be familiar with the ways to handle a cooling load.
- Understand the effect that compression and expansion have on the temperature of a gas.

TERMS TO LEARN

cooling load
evaporative cooling
fossil fuel
heater core
heat exchanger
heat load
mechanical refrigeration

4.1 INTRODUCTION

Heating and air conditioning must follow the basic rules of heat transfer. An understanding of these rules helps greatly in understanding the systems:

- Heat always flows toward cold.
- To warm a person or item, heat must be added.
- To cool a person or item, heat must be removed.
- Fuels can be burned to generate heat.
- A large amount of heat is absorbed when a liquid changes state to a vapor.
- A large amount of heat is released when a vapor changes state to a liquid.
- Compressing a gas concentrates the heat and increases the temperature.

4.2 HEATING LOAD

Heating load is the term used when we need to add heat. The actual load is the number of Btu or calories of heat energy that must be added. In a home or office, burning fuel is the usual way to generate heat; the fuel is usually a **fossil fuel** such as coal, gas, or oil. In most cars, the heat is provided by the heated coolant from the engine's cooling system. This coolant is at a temperature of 180 to 205°F (82 to 98°C) when the engine reaches its normal operating temperature (Figure 4–1).

In a vehicle, it is a fairly simple process to circulate heated coolant through a **heat exchanger**, called a **heater core**. Air is circulated through the heater core, where it absorbs heat. Then it is blown into the lower part of the passenger compartment, where the heat travels on to warm the car interior and occupants. Convection is used with the air as the medium to move heat from the heater core to the passengers

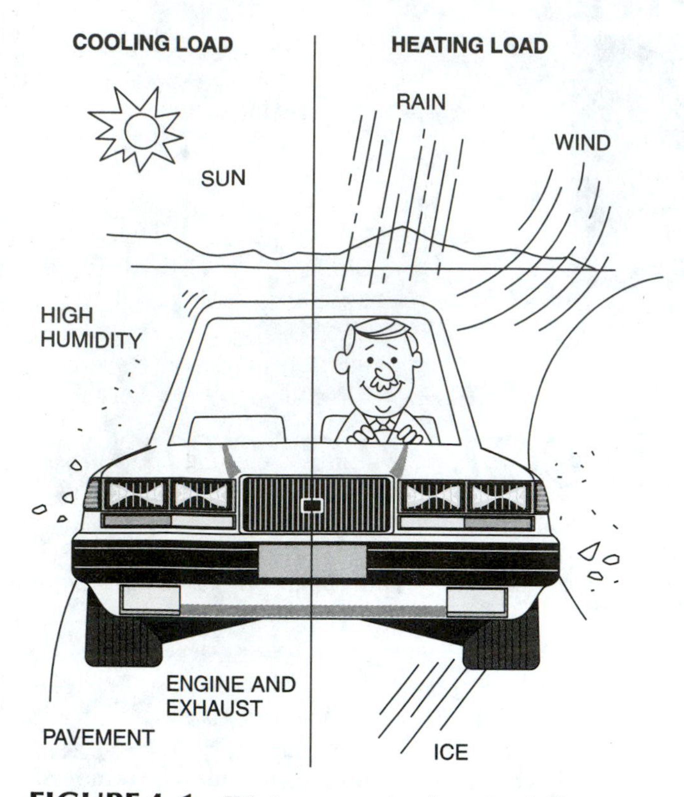

FIGURE 4–1 Winter presents a heat load: Heat must be added for comfort (right). Summer presents a cooling load.

(Figure 4–2). This system is described in more detail in Chapter 7.

Another method of generating heat is through a thermoelectric (TE) device using the Peltier effect. This is a simple device, essentially two alloys of bismuth or tellurium that have dissimilar free electron densities that are fused together. If an electric current is sent in one connector and out the other, the junction will heat up. If the current flow is reversed, the junction will become cold. This is the operating principle of picnic food chests that plug into a cigarette lighter socket and either cool or heat the inside. Some vehicles use this device plus a fan to circulate air inside of the driver and passenger seats. It can heat or cool the seat, circulating the air out through the seat cover. In the future, small electric vehicles will probably use TE HVAC systems to heat and cool vehicles.

4.3 COOLING LOAD

Cooling load describes the removal of heat, which is the purpose of air conditioners and evaporative coolers. We need to move heat to a cooler location in order to handle a cooling load.

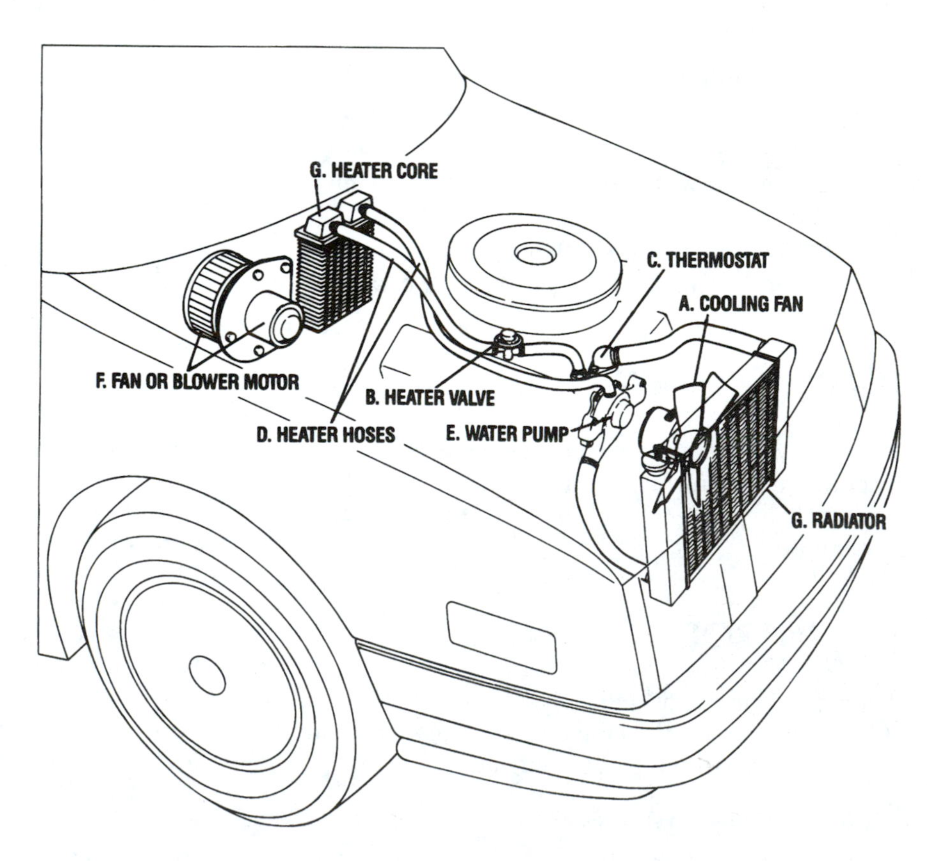

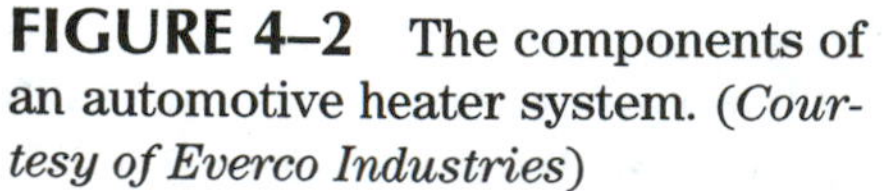

FIGURE 4–2 The components of an automotive heater system. (*Courtesy of Everco Industries*)

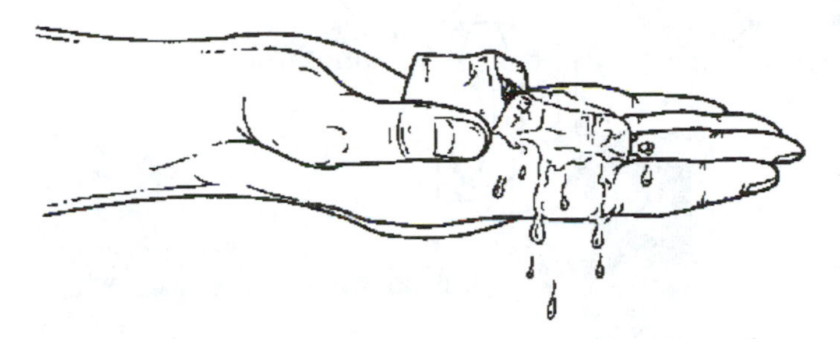

FIGURE 4–3 Ice has a cooling effect because of latent heat of fusion: It absorbs heat as it melts.

One way to move heat is with a block of ice. A substantial amount of latent heat is required to change the state of the solid ice into a liquid: 144 Btu per lb (79.7 calories per gram). A 50-lb block of ice represents 50 × 144, or 7,200 Btu of cooling power, when it changes from 50 lb of solid at 32°F to 50 lb of liquid at 32°F. In the early days of air conditioning, the term **ton** was commonly used. A ton of air conditioning was the amount of heat it took to melt a ton of ice: 2,000 × 144, or 288,000, Btu (Figure 4–3).

A method of cooling that works quite well in areas of low humidity is evaporation of water. This process is commonly called **evaporative cooling**. If we spread water thinly over the extremely large area of a mesh cooler pad and blow air across it, the water evaporates (Figure 4–4). For every pound of water that evaporates, 970 Btu (540 calories per gram) are absorbed. This is the **latent heat of evaporation**, just as when it is boiled. This is a natural process and uses only the energy required by the blower to circulate the air through the cooler pads and on to the space to be cooled. A possible drawback to evaporative coolers is that they increase the humidity; they are often called "swamp" coolers. They are not effective in areas of very high humidity because the water does not evaporate rapidly enough. At one time, window-mounted evaporative coolers were used in cars. They were not very popular because they were unattractive and only worked well in dry areas (Figure 4–5).

A third way to handle a cooling load is by **mechanical refrigeration:** air conditioning. This system also uses evaporation of a liquid and the large amount of heat required for the latent heat of evaporation. We boil the refrigerant so that it changes from a liquid to a gas, but we recycle the gas (Figure 4–6). An evaporative cooler must be continuously supplied with water, and new ice must be added to replace the ice in an ice chest. Mechanical refrigeration uses the gas to absorb and move heat from the evaporator to the condenser and sends the liquid back to the evaporator to boil again and absorb more heat. It requires energy to drive the compressor for this process to occur. As with a heating system, air convection is used to move the heat from the area to be cooled to the evaporator (Figure 4–7). Mechanical refrigeration is used in home and mobile A/C and is described fully in Chapter 5.

FIGURE 4–4 (*a*) Both boiling and evaporating liquids absorb the latent heat of vaporization as they change state. (*b*) This principle is used to cool water in a cloth Lister bag. (*Courtesy of DaimlerChrysler Corporation*)

FIGURE 4–5 At one time evaporative coolers were used to cool car interiors. Air forced through a water-wetted mesh produces evaporation and a cooling effect. *(Courtesy of International Mobile Air Conditioning Association, IMACA)*

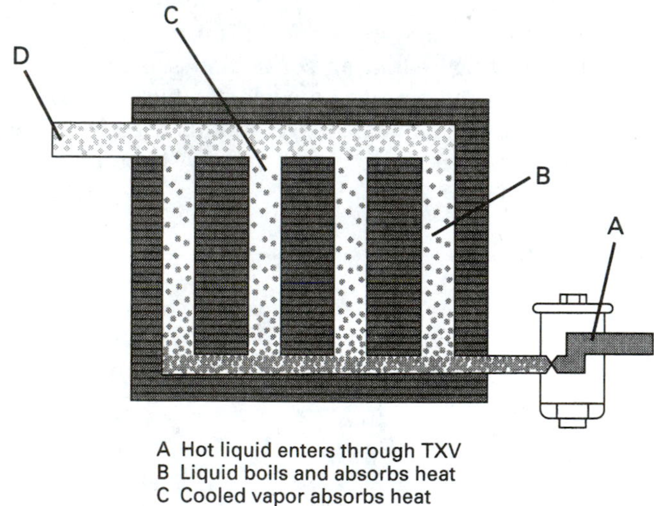

FIGURE 4–6 Liquid refrigerant boils and absorbs heat as it enters the lower pressure of the evaporator. The absorbed heat comes from the air passing through the evaporator fins.

4.4 COMPRESSION HEATING

When we compress a gas into a higher pressure, we also increase the temperature of the gas. A given volume of gas has a certain temperature that, as we know, represents a certain amount of energy. Because

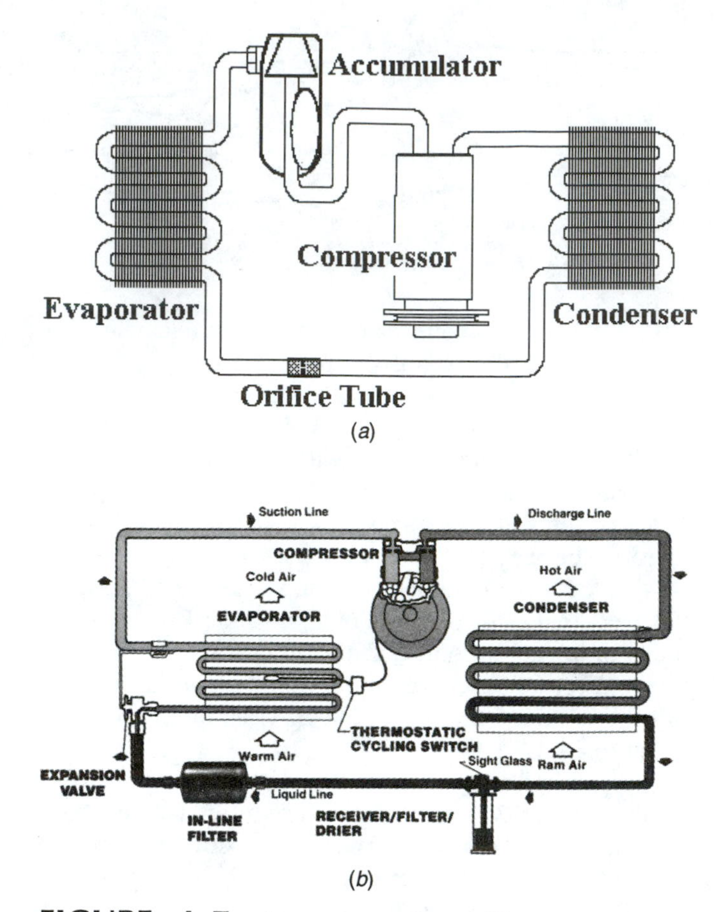

FIGURE 4–7 An automotive A/C system in a schematic (*a*) and somewhat realistic view (*b*). These views show the relationship of the components and the circulation of the refrigerant.

this heat energy becomes more concentrated as the gas is compressed, heat intensity is increased. A good example of this process is the diesel engine. The operating principle of this engine is to compress air a relatively large amount, about 20:1, so that the air becomes hot enough to ignite the fuel as it is injected into the cylinder (Figure 4–8). Another example is the discharge line of an operating air compressor, which is substantially warmer than the inlet line because of the heat of the compressed air. Some of these lines can be very hot.

Mechanical refrigeration uses a compressor to raise the pressure of the refrigerant that has boiled in the evaporator. R-12 is at a temperature of about 32°F in the evaporator, with a pressure of about 30 psi; R-134a has a slightly lower pressure of 27 psi at 32°F. Raising the pressure increases the gas temperature to a point above the ambient temperature. With R-12 and an ambient temperature of 100°F, we have to raise the pressure

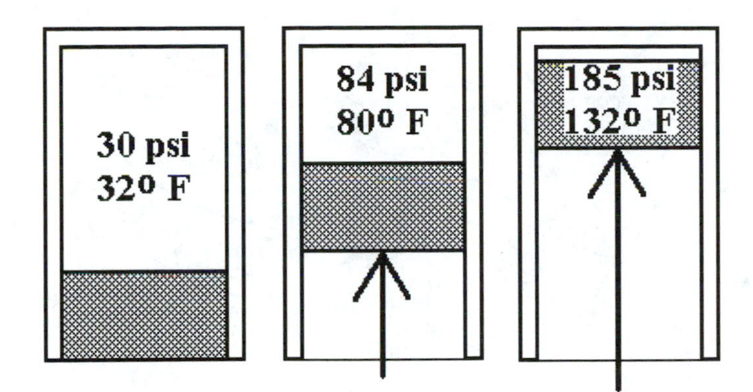

FIGURE 4–8 When a gas is compressed the heat energy is more concentrated, and this causes a temperature increase.

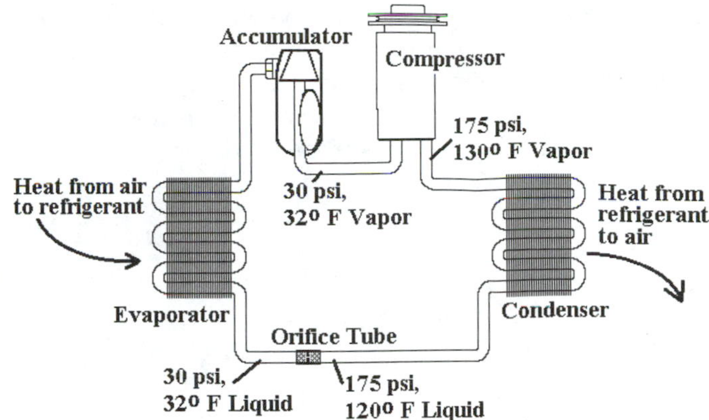

FIGURE 4–9 In a mechanical refrigeration system, the compressor increases the pressure and causes the refrigerant to circulate through the cycle as shown here.

to about 205 psi. This pressure is slightly higher with R-134a and depends to a great extent on the design of the condenser and the airflow through it, about 230 psi (Figure 4–9).

Raising the temperature of the refrigerant allows the required heat flow from the now warmer refrigerant to ambient air; this removal of heat causes the change of state back to liquid. We accomplish two things at the condenser: We remove all of the heat that was absorbed in the evaporator, and we recycle the gas back into a liquid (Figure 4–10).

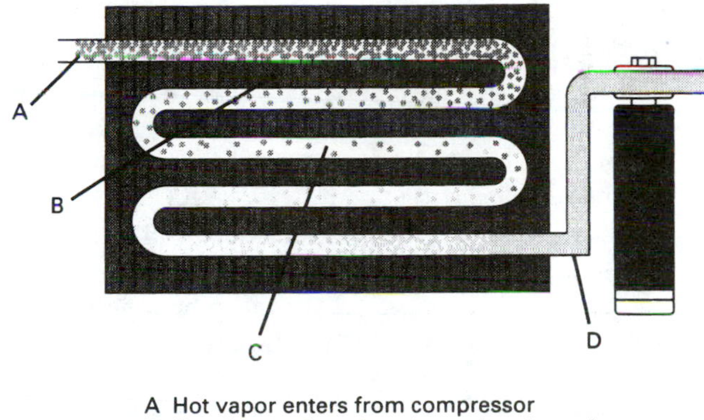

A Hot vapor enters from compressor
B Heat is transferred to air passing through
C Vapor condenses to liquid
D Liquid leaves through liquid line

FIGURE 4–10 Refrigerant enters the condenser as a hot, high-pressure gas and leaves as a liquid. Removing heat to ambient air causes the refrigerant to condense.

4.5 EXPANSION COOLING

If we can raise the temperature of a gas by compressing it, we can lower the temperature if we allow it to expand. Expanding a volume of gas spreads out the heat energy over a larger area and lowers its temperature. An example can be felt with the coolness of an air nozzle as we use it to blow parts clean and dry (Figure 4–11).

In the past, a closed system using air compression and expansion was under development for use in cooling car interiors. Air was compressed and then cooled in a heat exchanger mounted in front of the car. Then this high-pressure, fairly cool air was allowed to expand in a heat exchanger mounted in the passenger compartment. The cooler air absorbed heat from the inside of the car as it expanded. This heat was then concentrated so that it could be released at the outer heat exchanger. This system worked, but it required much larger lines and a bigger heat exchanger to handle the volume of air. Furthermore, it was not as efficient as the mechanical refrigeration system currently in use (Figure 4–12).

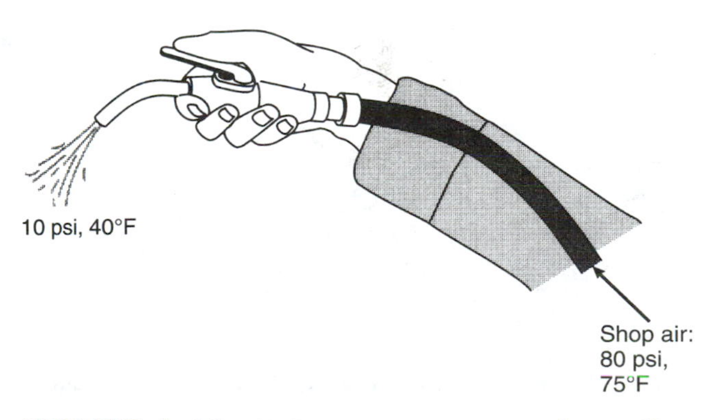

FIGURE 4–11 Releasing a compressed gas allows the air to expand; the air will be cooled because the heat energy is spread out.

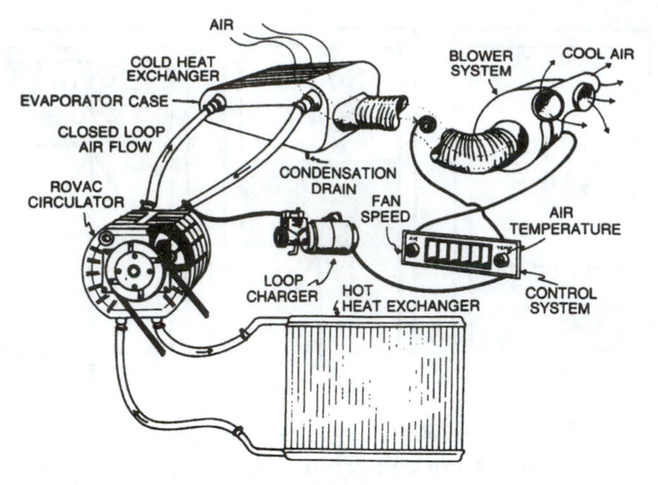

FIGURE 4–12 The Rovac system used a circulator to compress air and to allow the air to expand again. Heat from the compressed air was transferred to ambient air in the hot heat exchanger. Heat from the passenger compartment was absorbed by the expanded air in the cold heat exchanger.

CHAPTER QUIZ

These questions help you study this chapter. Enter the proper word(s) in the blanks to complete each statement.

1. *Heating load* is the term used to describe the amount of __________ that must be __________ to make people comfortable.
2. The object where heat is transferred from one medium to another is called a(n) __________ __________.
3. It takes __________ Btu of heat to melt 1 lb of ice.
4. Evaporative coolers use the principle of __________ __________ of __________, the amount of heat that is required to evaporate water.
5. Mechanical refrigeration cools because of the amount of __________ __________ that is required to boil the __________ __________ in the evaporator.
6. An A/C system changes the refrigerant from a(n) __________ to a(n) __________ in the evaporator.
7. An A/C system changes the refrigerant from a(n) __________ to a(n) __________ in the condenser.
8. An A/C compressor increases the gas pressure in order to increase the __________.
9. As a gas is compressed, the temperature will __________.
10. There will be a(n) __________ effect when we allow compressed air to expand.

REVIEW QUESTIONS

These questions allow you to check what you have learned. Select the answer that correctly completes each statement.

1. In the automobile, the heat load comes from
   a. the engine cooling system.
   b. the sun load on the car.
   c. a cold winter day.
   d. None of these.
2. Which of the following is a method of handling a cooling load?
   a. A block of ice **c.** Mechanical refrigeration
   b. Evaporating water **d.** All of these
3. Two students are discussing evaporative cooling for a car. Student A says that it would increase the RH inside the car. Student B says that an A/C system lowers the RH. Who is correct?
   a. Student A **c.** Both A and B
   b. Student B **d.** Neither A nor B
4. Which of the following describes a drawback of mechanical refrigeration?
   a. Oil must be added continuously.
   b. R-12 must be added continuously.
   c. The passenger compartment does not get cool enough.
   d. It takes energy to drive the compressor.
5. Two students are discussing MVAC systems. Student A says the evaporator moves heat from the passenger compartment to the refrigerant. Student B says the refrigerant boils inside the evaporator. Who is correct?
   a. Student A **c.** Both A and B
   b. Student B **d.** Neither A nor B
6. In an A/C system, the refrigerant changes state from a __________ in the evaporator.
   A. liquid to a gas **B.** gas to a liquid
   Which is correct?
   a. A only
   b. B only **c.** Both A and B
   d. Neither A nor B
7. When an A/C system is operating, the compressor __________ runs hot.
   A. inlet **B.** outlet
   Which is correct?
   a. A only **c.** Both A and B
   b. B only **d.** Neither A nor B
8. Student A says that some devices generate heat by compressing a gas to a higher pressure. Student B says some devices generate cool air by allowing a compressed gas to expand. Who is correct?
   a. Student A **c.** Both A and B
   b. Student B **d.** Neither A nor B

CHAPTER

Air Conditioning Systems

LEARNING OBJECTIVES

After completing this chapter, you should:

- Understand the relationship between the A/C cycle and the components on the low side and high side of a system.
- Understand the functions for the low side and high side of a system.
- Be familiar with the role of each A/C component.

TERMS TO LEARN

accumulator
barrier hose
condenser
cut-in
cutout
cycling clutch
discharge line
evaporator
evaporator pressure regulator (EPR)
fin and tube
flat tube
flooded
high side
liquid line
low side
orifice tube (OT)
overcharge
parallel flow
pilot operated absolute (POA)
pressure switch
receiver–drier
reciprocating piston
rotary vane
Scotch yoke
scroll
serpentine flow
slugging
starved
sub-cooling
suction line
suction throttling valve (STV)
swash plate
thermal expansion valve (TXV)
undercharge
variable displacement
variable orifice valve (VOV)
wobble plate

5.1 INTRODUCTION

The automotive A/C system uses the physical principles described in earlier chapters to move heat from the passenger compartment to the condenser and then to the ambient air moving through the condenser. The heating system uses some of the same principles to move heat from the engine's cooling system to the passenger compartment (Figure 5–1).

Automotive A/C systems are either **orifice tube systems** or **thermal expansion valve systems,** depending on which type of **flow control** or **expansion device** is used (Figure 5–2). Orifice tube systems are also called **cycling clutch orifice tube (CCOT)** and **fixed orifice tube (FOT)** systems. A/C systems can be easily divided into two parts: the **low side,** with its **low pressure and temperature,** and the **high side,** with its **higher pressures and temperatures** (Figure 5–3). The low side begins at the expansion device—the orifice tube (OT) or the thermal expansion valve

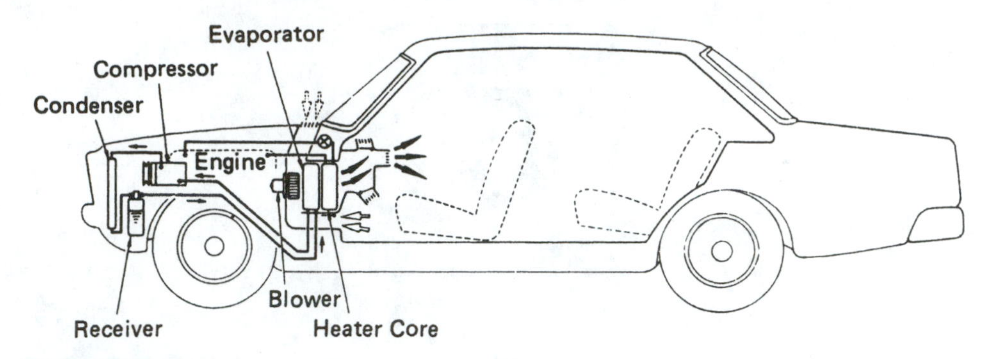

FIGURE 5–1 Air is circulated through the A/C and heating system and the car to either add or remove heat. (*Courtesy of Toyota Motor Sales*)

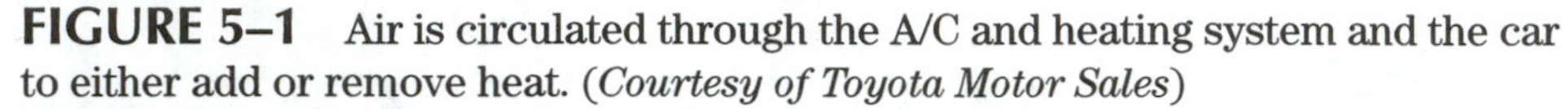

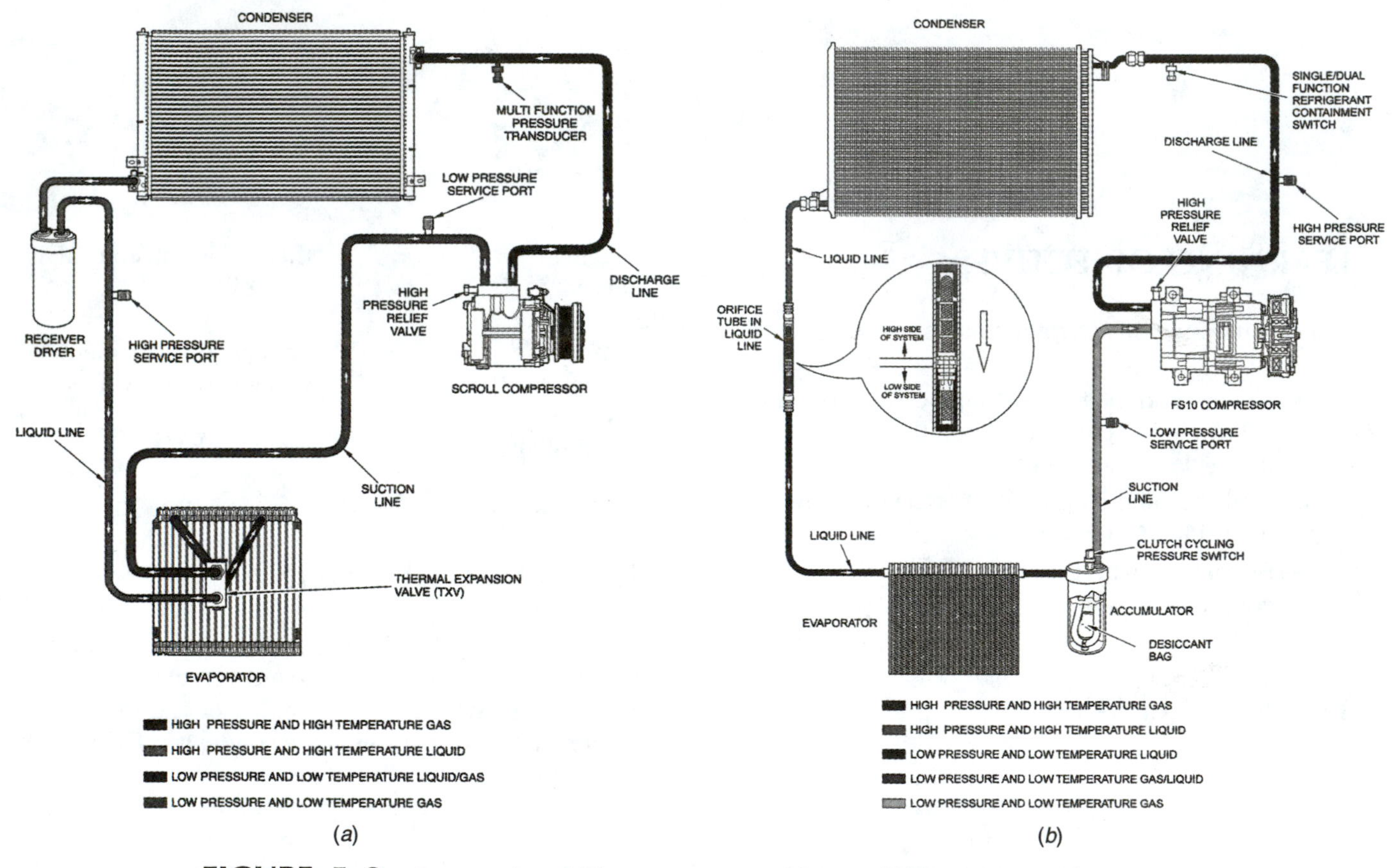

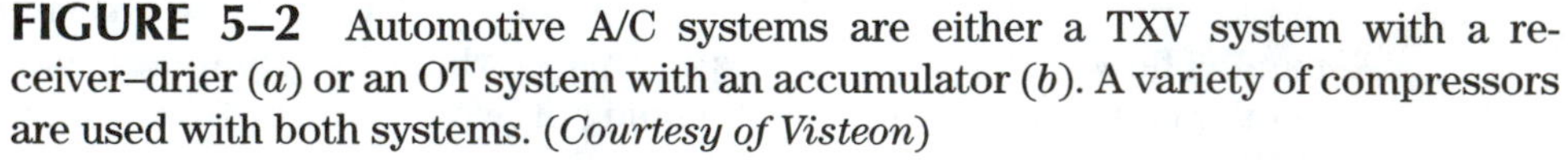

FIGURE 5–2 Automotive A/C systems are either a TXV system with a receiver–drier (*a*) or an OT system with an accumulator (*b*). A variety of compressors are used with both systems. (*Courtesy of Visteon*)

(TXV)—and ends at the compressor; the high side begins at the compressor and ends at the OT or TXV.

Refrigerant **boils** or **evaporates** in the low side and it **condenses** in the high side (Figure 5–4). In an operating system, you can identify the low and high sides by:

- **Pressure:** A gauge set shows low pressure in the low side and high pressure in the high side.
- **Sight:** On high-humidity days, the cold low side tubing often collects water droplets and may even frost.
- **Temperature:** The low side is cool to cold, and the high side is hot.
- **Tubing size:** Low side tubes and hoses are larger (vapor), and high side tubes and hoses are smaller (liquid).

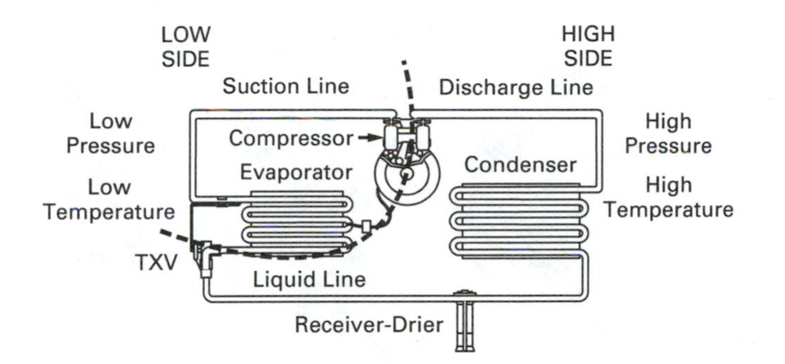

FIGURE 5–3 The high and low sides of an A/C system are divided by the compressor (where the pressure is increased) and either a TXV or OT (where the pressure drops).

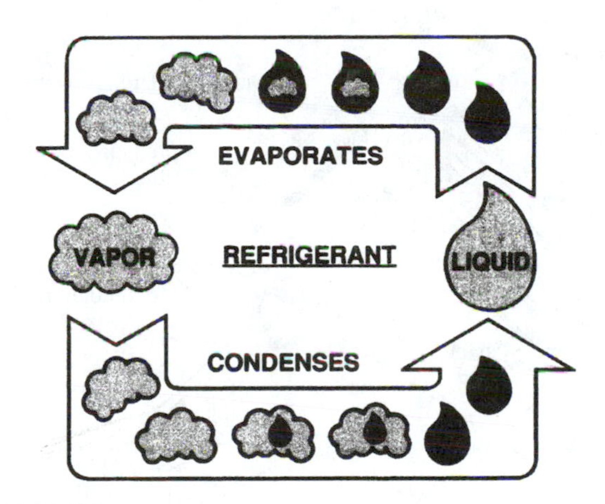

FIGURE 5–4 Refrigerant changes state to a vapor in the low side and into a liquid in the high side. *(Reprinted with permission of General Motors Corporation)*

5.2 LOW SIDE OPERATION

When the A/C system is in full operation, the goal of most systems is to maintain an evaporator temperature just above the freezing point of water, 32°F (0°C). This temperature produces the greatest heat exchange without ice formation on the evaporator fins (evaporator icing significantly reduces the heat transfer).

The cold temperature in the evaporator is produced by boiling the refrigerant. Remember that R-12 and R-134a have very low boiling points, well below 0°F, and that when a liquid boils, it absorbs a large amount of heat, the latent heat of vaporization. To produce cooling, liquid refrigerant must enter the evaporator, and it must boil inside the evaporator. The amount of heat an evaporator absorbs is directly related to the amount of liquid refrigerant that boils inside it (Figure 5–5).

A properly operating evaporator has a temperature just above 32°F (0°C), and refrigerant pressure is

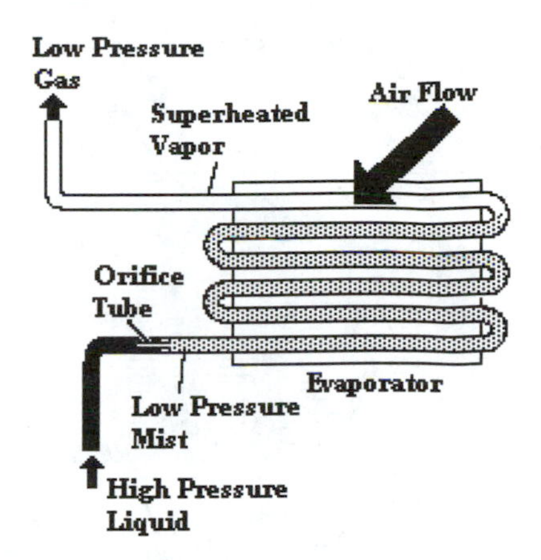

FIGURE 5–5 As liquid refrigerant enters the evaporator, the boiling point will try to drop as low as 32°F because of the drop in pressure. The cold temperature causes the refrigerant to absorb heat from the air circulated through the evaporator.

directly related to temperature because the refrigerant is a saturated vapor. An A/C technician becomes familiar with normal low side temperatures and pressures. Abnormal temperatures and pressures indicate that something is wrong (e.g., a **starved** or **flooded** evaporator). An evaporator that has a low pressure but a temperature that is too warm is called starved; not enough refrigerant is entering to produce the desired cooling effect. A starved evaporator is usually caused by a restriction at or before the expansion device or an undercharge of refrigerant. If more refrigerant enters the evaporator than can boil, the evaporator floods. In this case the pressure is then higher than normal (Figure 5–6).

Major components in the low side are the evaporator and the expansion device. The evaporator is the heat exchanger and absorbs heat from the passenger compartment. The low side begins at the refrigerant expansion or flow metering device, either a TXV or an OT, which produces a pressure drop. The low side ends at the compressor, which causes the pressure to increase (Figure 5–7).

5.2.1 Expansion Devices

The expansion device separates the high-pressure liquid from the low-pressure evaporator. The liquid refrigerant becomes a mist as it passes through, and this allows it to absorb heat easily. The restriction of the expansion device slows the refrigerant flow so the compressor must generate the high pressure.

5.2.1.1 Thermal Expansion Valves A TXV is a variable valve that changes the size of the valve opening

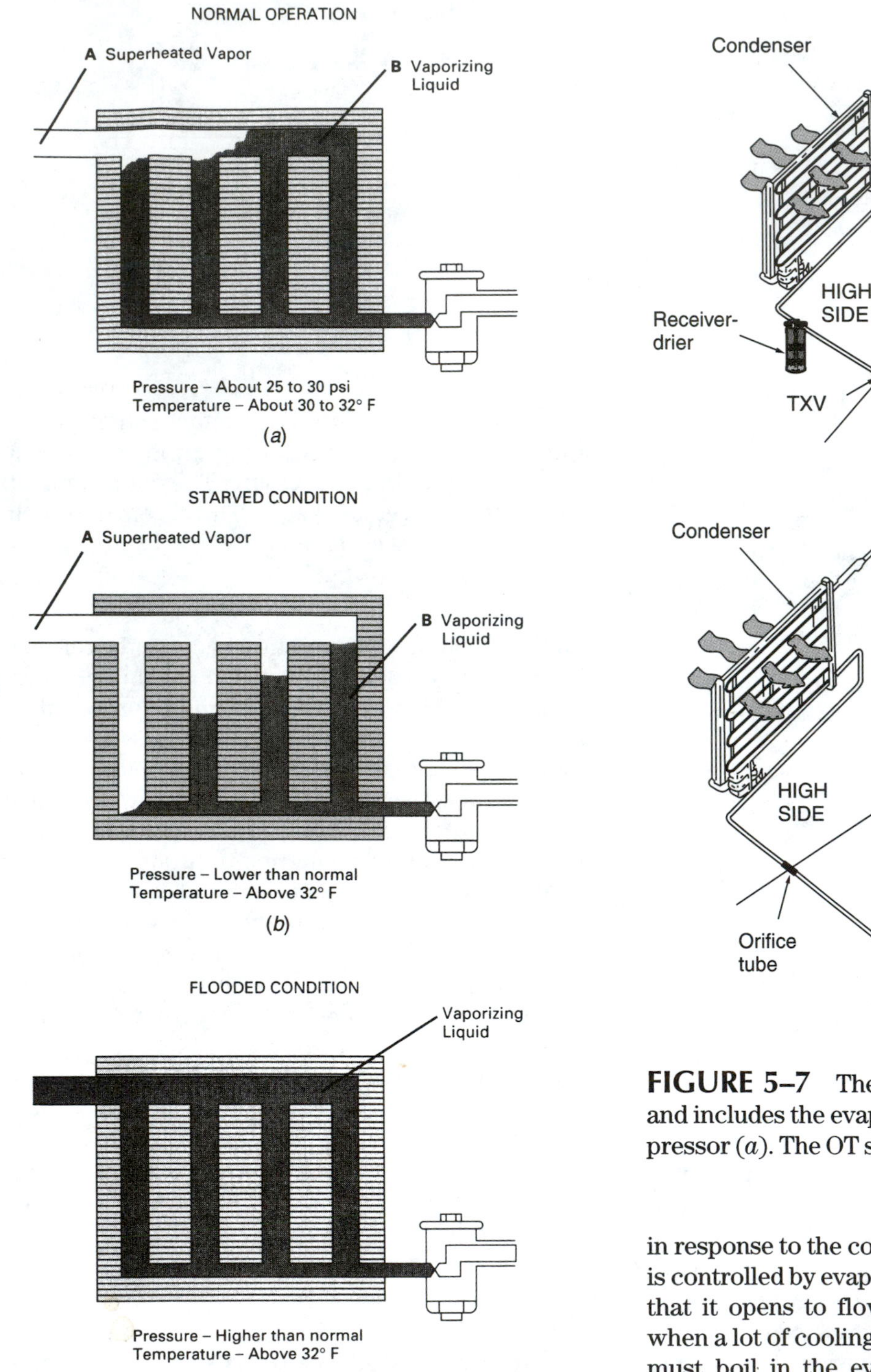

FIGURE 5–6 If the proper amount of refrigerant enters the evaporator, it has a slight superheat as it leaves (*a*). A starved condition, in which not enough refrigerant enters the evaporator, does not produce as much cooling (*b*). If too much refrigerant enters, the evaporator floods because the refrigerant will not all boil (*c*).

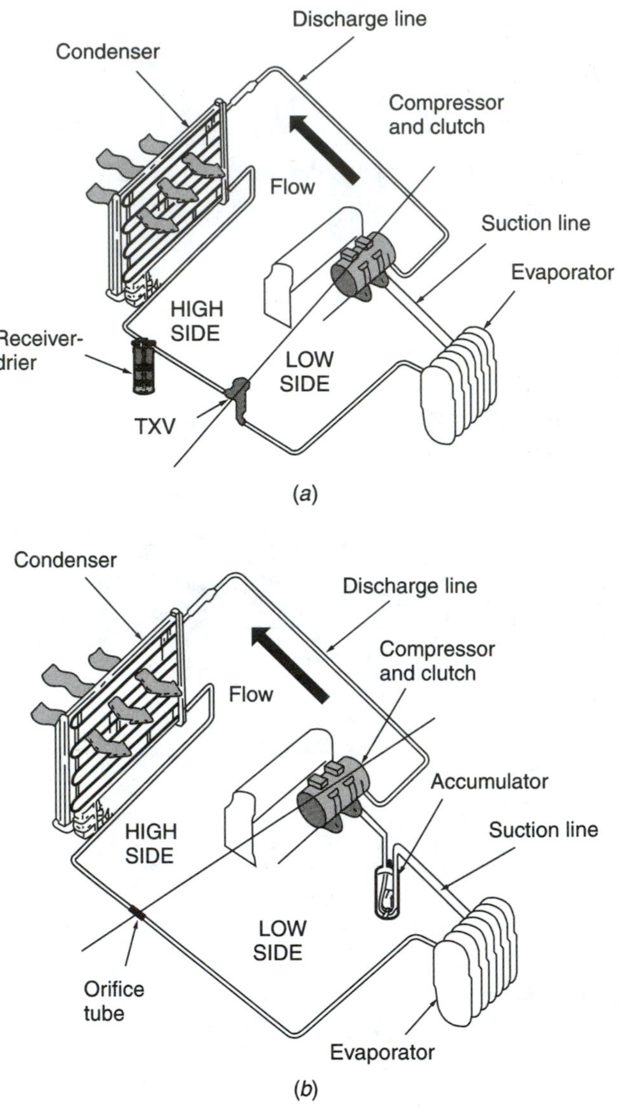

FIGURE 5–7 The low side begins at the TXV or OT and includes the evaporator and suction line to the compressor (*a*). The OT system includes an accumulator (*b*).

in response to the cooling load of the evaporator. A TXV is controlled by evaporator temperature and pressure so that it opens to flow as much refrigerant as possible when a lot of cooling is needed. But all of the refrigerant must boil in the evaporator, and the vapor must be slightly superheated when it reaches the evaporator outlet. Most TXVs are calibrated so that the outlet temperature is a few degrees above the inlet pressure and temperature: The refrigerant has a few degrees of superheat. When there is a lower cooling load, the TXV must reduce the flow. The various types of TXVs and other system components are described more completely in Chapter 6 (Figure 5–8).

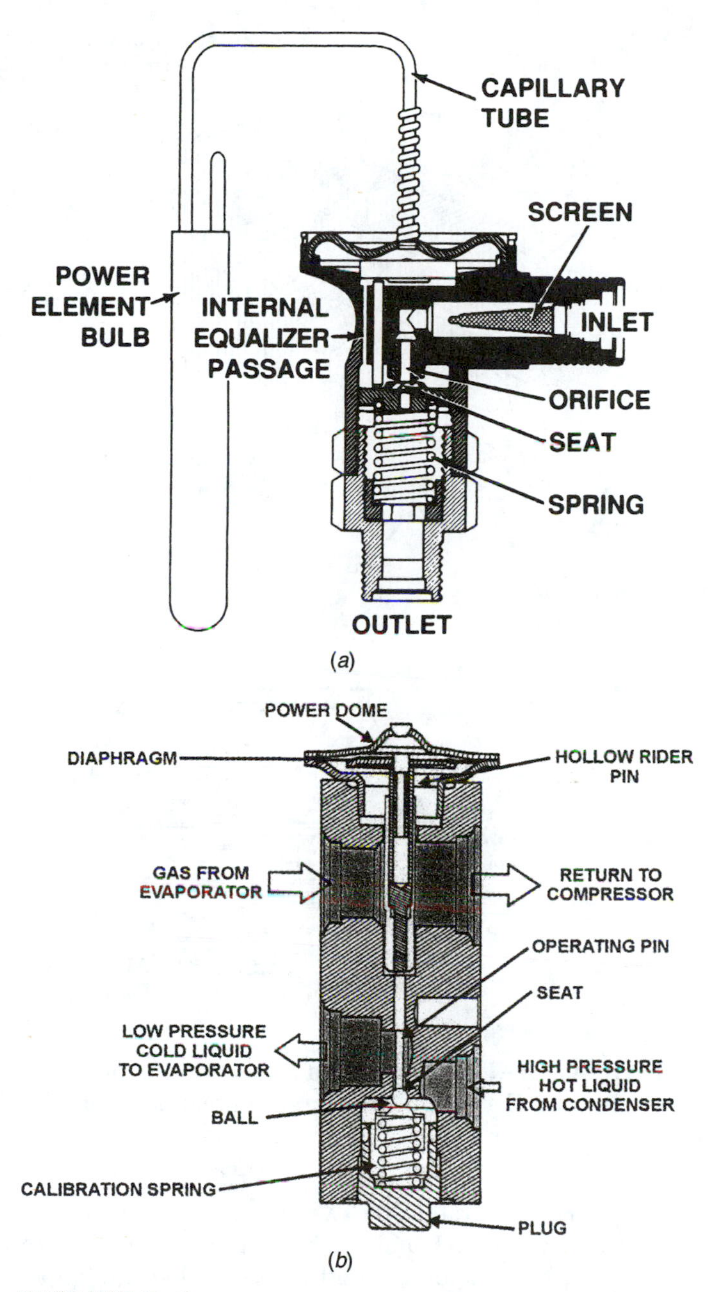

FIGURE 5–8 A TXV is controlled by the diaphragm in the power element that senses the gas temperature leaving the evaporator, a calibration spring, and evaporator pressure (through the internal equalizer passage in *a*). An H-type valve is shown at *b*. (a *is reprinted with permission of General Motors Corporation;* b *is courtesy of DaimlerChrysler Corporation)*

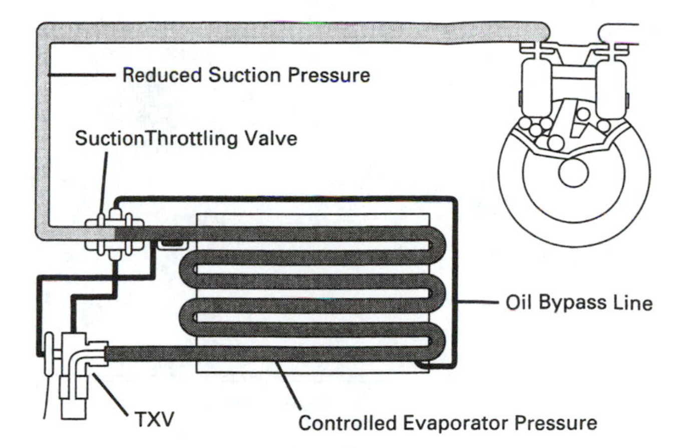

FIGURE 5–9 Some systems use a suction throttling valve to keep evaporator pressure from dropping to the point at which icing can occur.

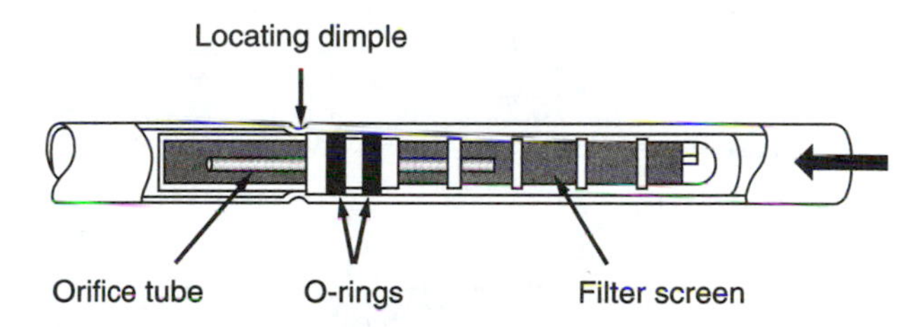

FIGURE 5–10 An OT is a simple restriction that limits the flow of refrigerant into the evaporator.

In most systems that use a TXV, the evaporator outlet is connected to the compressor inlet by a hose with an internal diameter (ID) of about 5/8 or 3/4 inch (16 or 19 mm). Some TXV systems use a suction throttling valve to prevent evaporator pressure from dropping below a certain pressure, about 30 psi (207 kPa). Lower pressure with its lower temperature can cause evaporator icing (Figure 5–9).

5.2.1.2 Orifice Tubes An **orifice tube** (OT) is a fixed-diameter orifice that the refrigerant must flow through; it is also called an **expansion tube** or **fixed orifice tube** (Figure 5–10). The diameter varies between systems and is about 1/16 inch (0.065 inch, 1.588 mm). It is much simpler and cheaper to produce than a TXV, but it cannot respond to evaporator temperature. At times of low cooling loads, it flows too much refrigerant, which floods the evaporator with liquid. An OT system must include a **low side accumulator** between the evaporator and compressor to catch and store liquid refrigerant. The accumulator is constructed so it retains the liquid and allows only gas to flow back to the compressor. The accumulator is usually attached to the evaporator outlet and connected to the compressor by a hose (Figure 5–11). A **variable orifice valve (VOV)** has recently been developed. The VOV includes a valve that is sensitive to flow and pressure or temperature; it is explained more completely in Section 6.4.3.

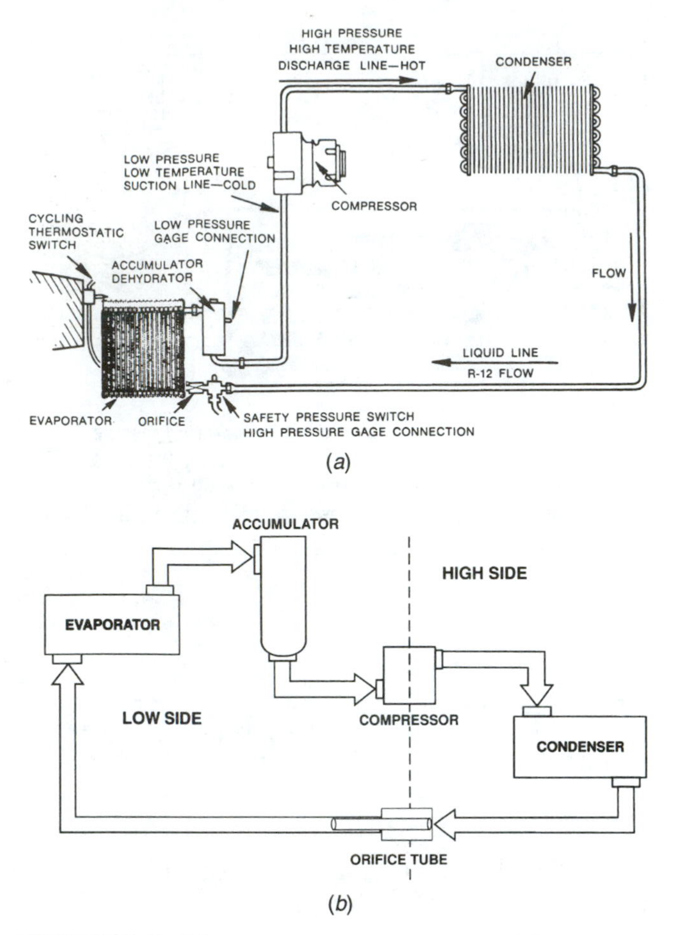

FIGURE 5–11 Two views of a typical OT system; *a* is somewhat realistic and *b* is schematic. Both show the arrangement of the components and the refrigerant flow. (*Reprinted with permission of General Motors Corporation*)

In some modern systems, the OT is mounted upstream, close to the condenser inside the liquid line. When the engine and the A/C system are shut off, high side pressure bleeds off through the OT and makes a hissing sound, which might alarm some motorists; moving the OT upstream reduces this noise.

5.2.2 Evaporators

The refrigerant enters the evaporator as a spray or mist, leaving an area of a few hundred pounds per square inch (psi) and passing through a small orifice into an area of about 30 psi. Like most heat exchangers, a well-designed evaporator has a large amount of surface area in contact with the refrigerant and the air from the passenger compartment. The heat from the air causes the refrigerant to boil and turn into a vapor; the cooler air is returned to the passenger compartment (Figure 5–12).

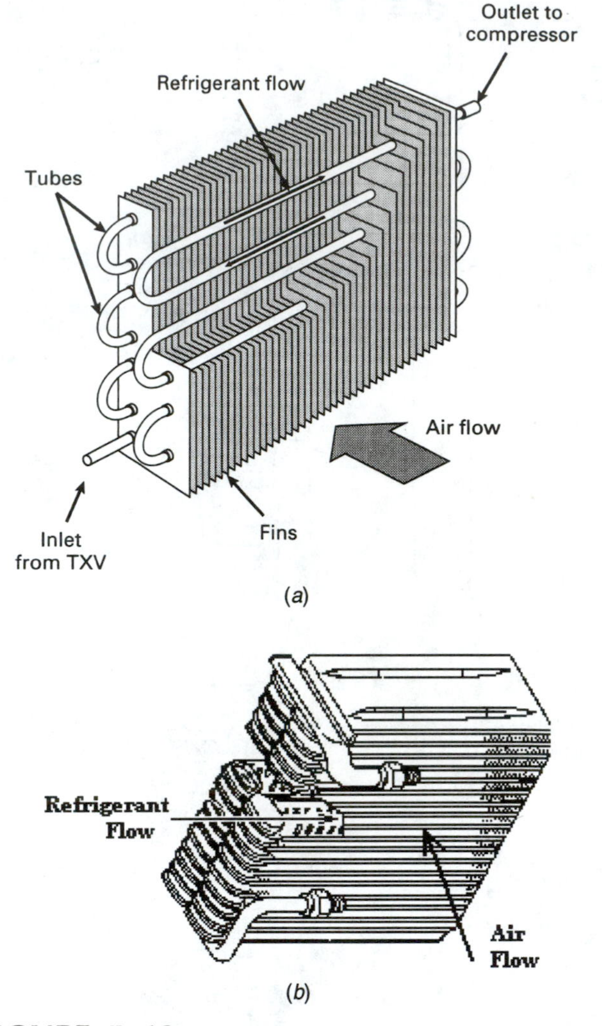

FIGURE 5–12 A tube-and-fin (*a*) and a plate (*b*) evaporator. Each type has a large contact area for heat to leave the air and enter the refrigerant.

In the evaporator of a TXV system, all of the refrigerant is vaporized and slightly superheated before reaching the evaporator outlet, also called a tailpipe. This is also the case in an OT system when there is a hot car, but as the car cools down some of the refrigerant is still in mist form at the tailpipe.

If liquid refrigerant enters the compressor, it can cause compressor damage. This is called **slugging.** The design of a low side accumulator prevents compressor slugging.

5.2.3 Accumulator

The accumulator serves three major functions: (1) to prevent liquid refrigerant from passing to the compressor; (2) to hold the desiccant, which helps remove mois-

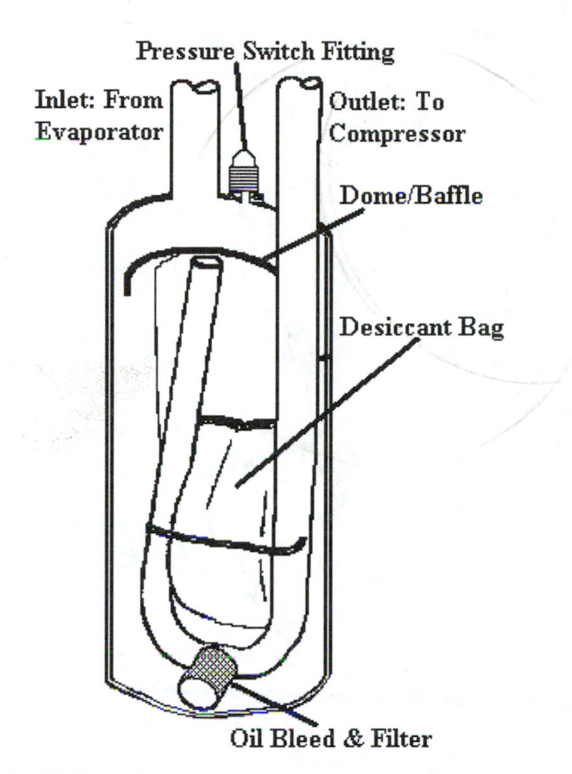

FIGURE 5–13 Accumulators are designed so that vapor from the top leaves to the compressor. They contain desiccant to absorb water from the refrigerant and many include a fitting for low side pressure and the clutch cycling switch.

ture from the system; and (3) to hold a reserve of refrigerant (Figure 5–13). The receiver–drier used in a TXV system serves the second two purposes and is mounted in the high side.

An accumulator is a container that holds about 1 quart (0.95 L) in volume. The inlet line enters near the top. The outlet line begins near the top, passes downward to the bottom, and then goes either out the bottom or back up to its exit. A small opening for **oil bleed** is at the lowermost point; this opening usually has a filter so debris will not block it. This routing of the outlet tube separates the refrigerant vapor at the top from the liquid at the bottom so that only vapor will leave the accumulator. A small amount of liquid refrigerant and oil also leaves through the bleed hole; the oil ensures that the compressor is lubricated .

The desiccant is a chemical drying agent called **molecular sieve** that removes all traces of water vapor from a system. Water can mix with refrigerant to form acids, which cause rust and corrosion of metal parts. Water can also freeze and form ice at the TXV or OT, which can block the flow of refrigerant into the evaporator. Desiccant is usually contained in a cloth bag inside the accumulator or receiver–drier. (Figure 5–14)

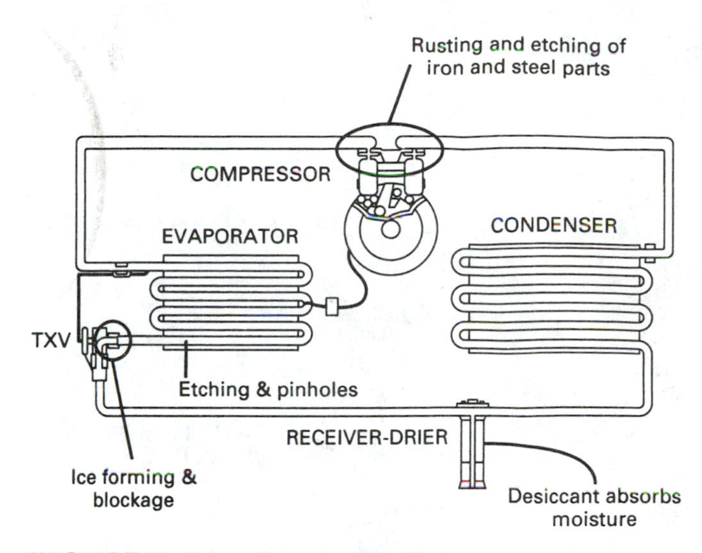

FIGURE 5–14 Water in an A/C system can combine with refrigerant to form acids. These acids can etch and dissolve components, cause rusting of metal parts, and cause ice blockage at the expansion device.

A variety of desiccant referred to as XH-5 has commonly been used with R-12 in automotive systems. XH-5 has the ability to absorb about 1% of its weight in water, but this chemical suffers damage when it absorbs fluorine from R-134a, and it begins to decompose when it comes into contact with R-134a and PAG oil. Other desiccant types (XH-7 and XH-9) are compatible with R-134a and PAG. The receiver–drier or accumulator used in an R-134a system should use XH-7 or XH-9 desiccant; all new accumulators and receiver–driers contain one of these. This desiccant is a little more expensive and requires a little more volume than XH-5. R-134a driers must have greater water absorption capacity because R-134a is more water soluble than R-12, so they are often slightly larger to hold the additional desiccant.

When off the vehicle, a new accumulator or receiver–drier must be kept closed as much as possible to keep it dry. Accumulators and receiver–driers are normally replaced when the system is serviced if the desiccant is suspected to contain water.

A refrigerant reserve is necessary because automotive A/C systems are subject to a wide variety of temperatures. This temperature variation causes the liquid refrigerant to change volume as it expands and contracts. An automotive A/C system also leaks some refrigerant through flexible hoses and the compressor shaft seal. Driving the compressor with an engine-driven belt requires a seal at the compressor drive shaft. Mounting the compressor on the engine requires flexible

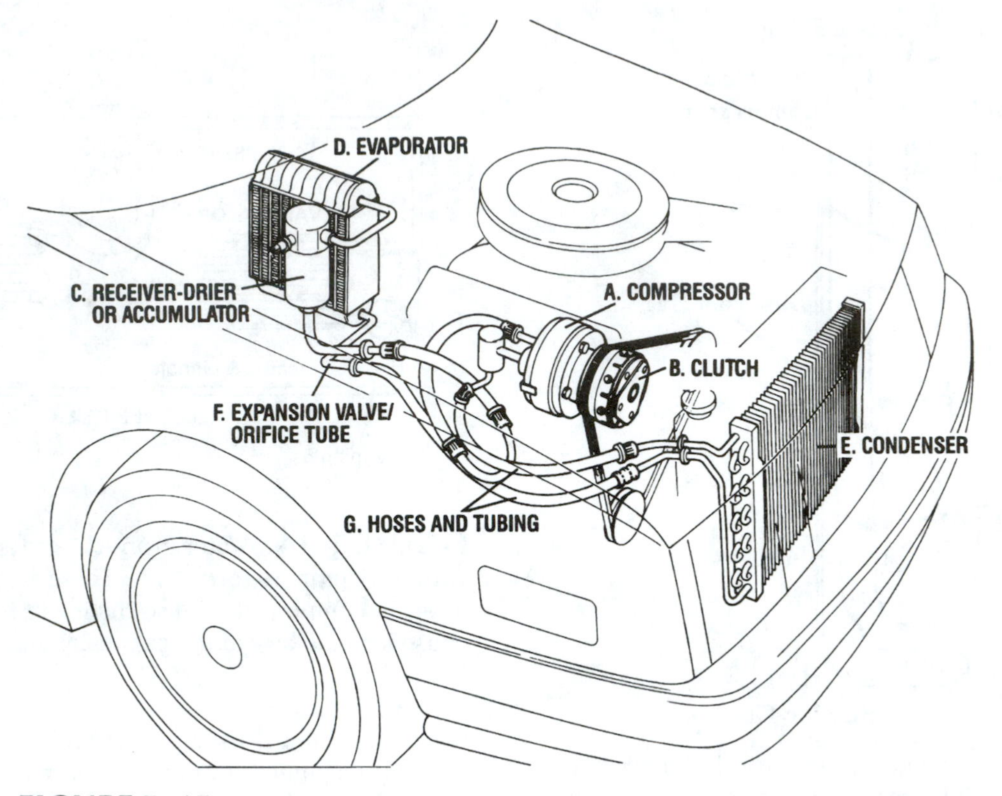

FIGURE 5–15 An automotive A/C system has the potential to lose refrigerant through hoses, the compressor shaft seal, and line fittings. (*Courtesy of Everco Industries*)

hoses because the engine moves on its mounts. The volume in an accumulator (or receiver–drier) allows us to charge excess refrigerant into a system as a stored liquid that can change its volume to compensate for loss or volume change. Early A/C systems used rather large receiver–driers because refrigerant was cheap and the system leaked. Newer receiver–driers and accumulators are much smaller because barrier-type hoses have reduced the leak rate and refrigerant is more valuable (Figure 5–15).

✓PROBLEM SOLVING

An A/C system does not put out cold air like it did when it was new, and you agree to take a look at it. With the system operating and the hood open, you see that the compressor is operating, but the temperature of the discharge hose to the condenser is only a little warmer than the underhood temperature and the accumulator is only slightly cool. Is the system working? How cold should the accumulator be? If it is not that cold, what might be wrong?

5.2.4 Refrigerant Charge Level

For an A/C system to work properly, a constant flow of liquid refrigerant must pass through the TXV or OT. While operating, the evaporator contains a refrigerant mist in the first two-thirds to three-fourths of its volume, with vapor in the remaining portion; the condenser contains a condensing vapor in the upper portion, with liquid in the bottom passages; and the line connecting the condenser to the expansion device is filled with liquid. The accumulator or receiver–drier is about half full of liquid (Figure 5–16).

Most newer A/C systems have reduced the size of some components so they can operate with smaller charge volumes than in the past. At one time, the refrigerant capacity of many domestic systems was in the 3- to 4-lb range; today, most systems hold $1^1/2$ to $2^1/2$ lb of refrigerant. With this reduced volume, the actual charge amount is more critical.

If the volume of liquid drops so that vapor bubbles pass through the TXV or OT, the system is **undercharged** and its cooling effectiveness is reduced. If an excessive amount of refrigerant is put into a system, the excess volume partially fills the con-

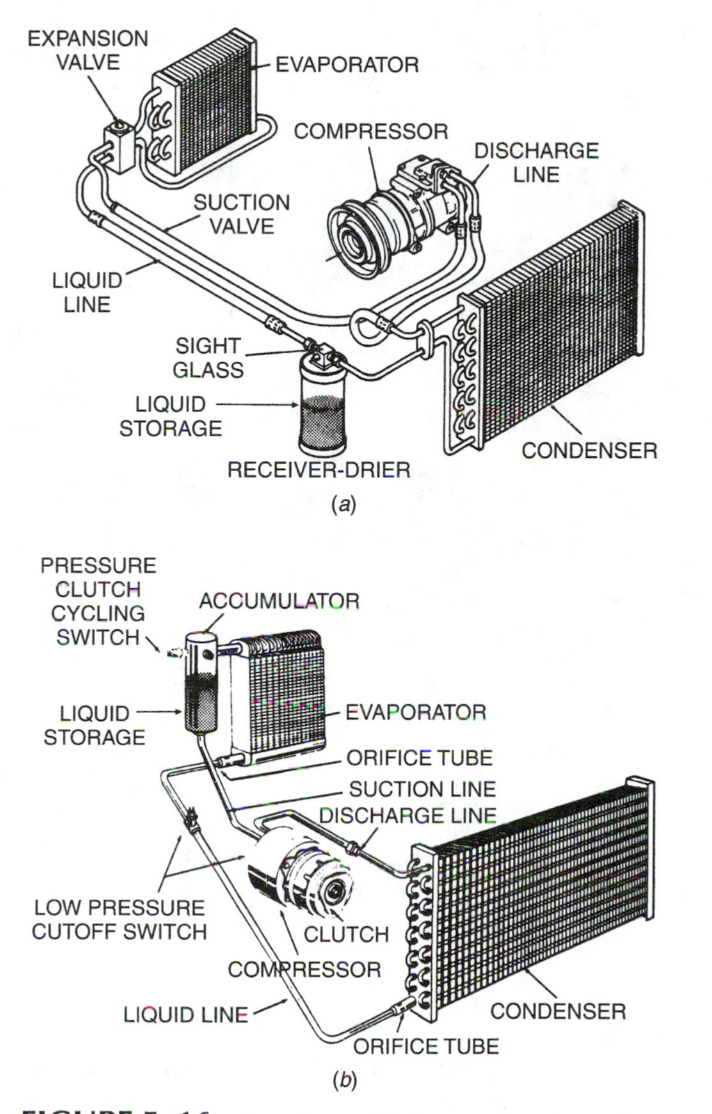

FIGURE 5–16 A system with the proper charge has the receiver–drier (*a*) or the accumulator (*b*) about half full of liquid. (*Courtesy of Everco Industries*)

denser as a liquid and reduces its effective volume. This is called an **overcharge** and causes abnormally high pressures, especially in the high side, and poor cooling at the evaporator (Figure 5–17).

5.2.5 Evaporator Icing Controls

Most A/C systems operate at maximum capacity when it is necessary to cool the car. Compressor size (displacement) and the sizes of the evaporator and condenser determine cooling power; these parts are normally designed to cool a particular car and its passengers at normal speeds on a hot day. Vehicle size and glass area, compressor displacement and operating

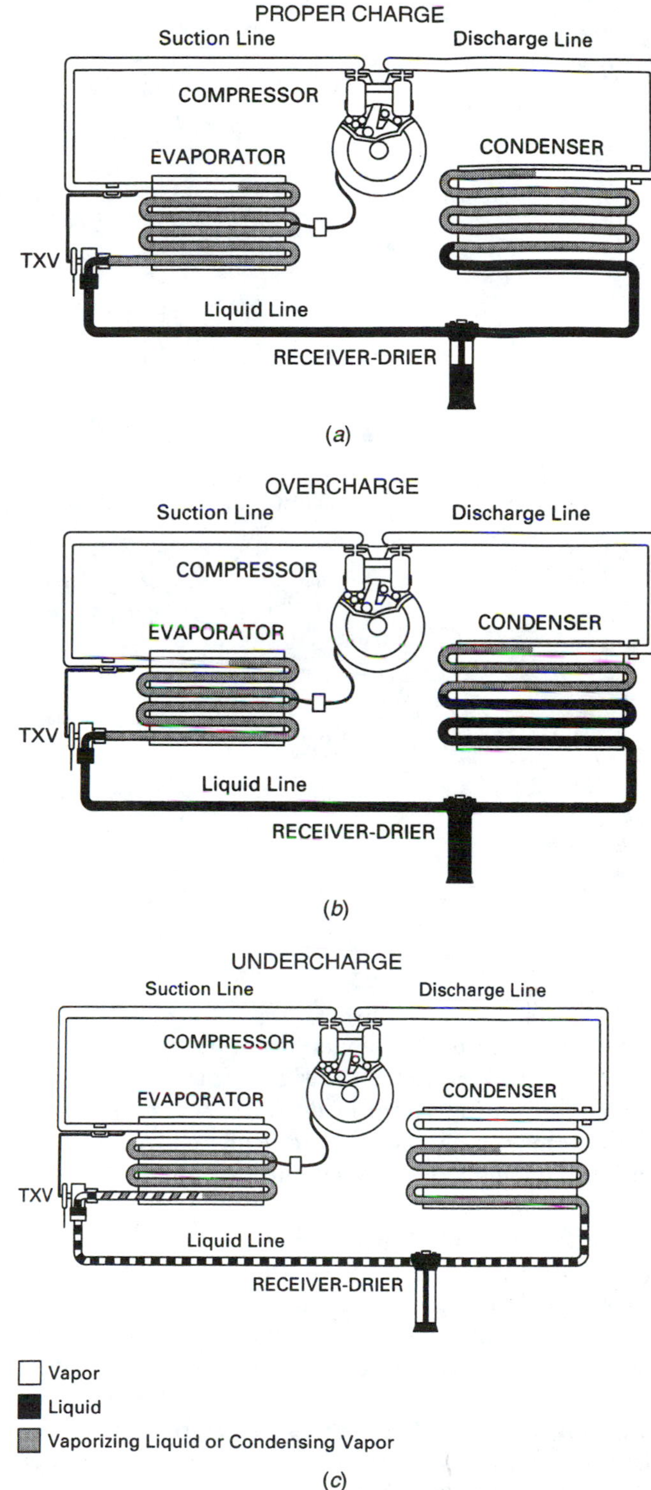

FIGURE 5–17 A properly charged system has the condenser filled with condensing vapor and some liquid, a liquid line filled with liquid, a receiver–drier about half full of liquid, and an evaporator with vaporizing liquid (*a*). An overcharge with too much liquid causes liquid to partially fill the condenser (*b*). An undercharge has vapor in the liquid line and a starved evaporator (*c*).

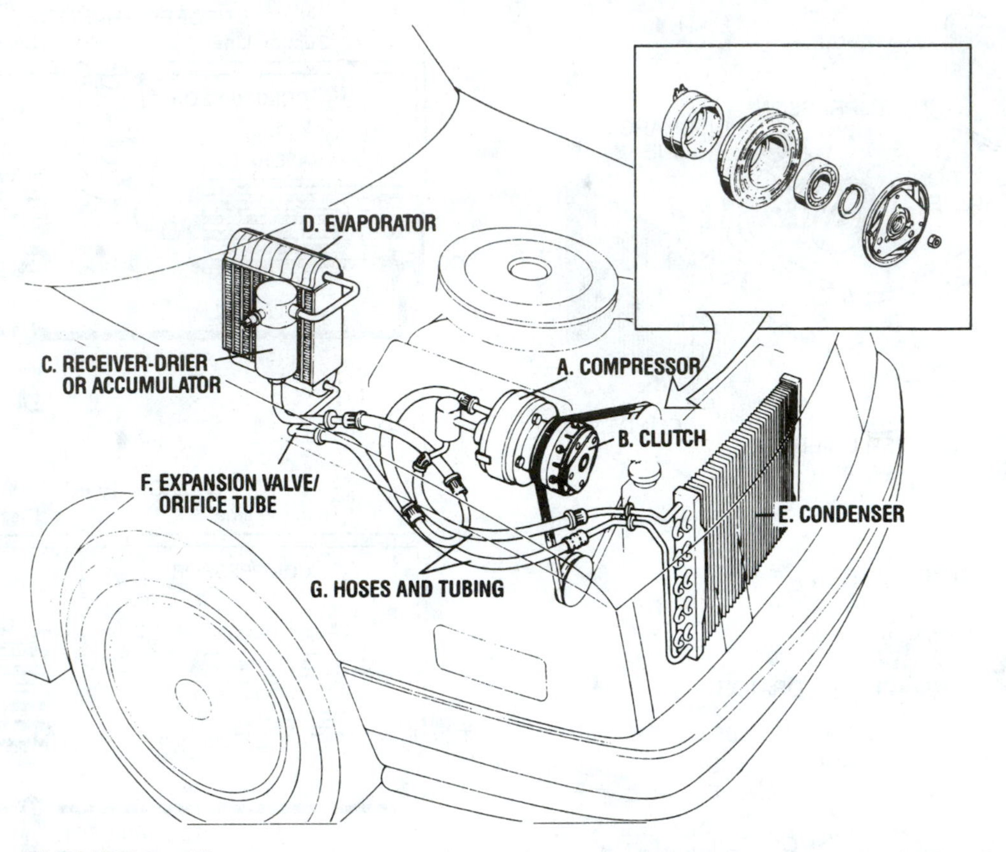

FIGURE 5–18 The compressor clutch allows us to cycle the compressor off and on to control evaporator temperature and to shut the system off. (*Courtesy of Everco Industries*)

speed, number of passengers, ambient temperature, and vehicle speed are all design parameters that are considered during the initial design of the A/C and heating systems. Some systems are designed to cool a car with the engine at idle speed and the compressor running at its slowest speed.

As the car cools down, the cooling load on the evaporator drops, and its temperature also drops. As mentioned earlier, the minimum temperature for an evaporator is 32°F, the point at which water freezes and ice and frost form. There are several ways to prevent evaporator icing. These include **cycling the compressor clutch, controlling evaporator pressure** so it does not drop below 30 psi, and **reducing the displacement of the compressor.**

5.2.5.1 Cycling Clutch Systems A **cycling clutch (CC) system** disengages the compressor and shuts the system off when the evaporator temperature or pressure starts to drop below freezing. Mobile A/C compressors are driven by a belt from the engine through a magnetic clutch (Figure 5–18). When the system is turned on, electricity is sent to energize the clutch. The current flow to the clutch is controlled by one or more switches; one of these is the on–off control at the heater–A/C control head.

Early A/C systems used a **temperature-controlled switch** mounted in the airstream from the evaporator. This thermal switch, also called an **icing or defrost switch,** is set to open and stop the current flow to the clutch when the temperature drops below 32°F and reclose when there is a temperature increase of about 5 to 15°F. This causes a pressure increase of about 10 to 20 psi that, in turn, produces at temperature rise to melt any frost or ice on the fins. Some newer systems use a **temperature sensing thermistor.** A thermistor is a solid-state device that changes its electrical resistance in direct inverse relationship to its temperature; as the temperature goes up, the resistance is reduced and vice versa. It is used as an input to an electronic control module (ECM) to provide the actual evaporator temperature control.

Many OT systems use a **pressure switch** mounted in the accumulator or the suction line to the compressor.

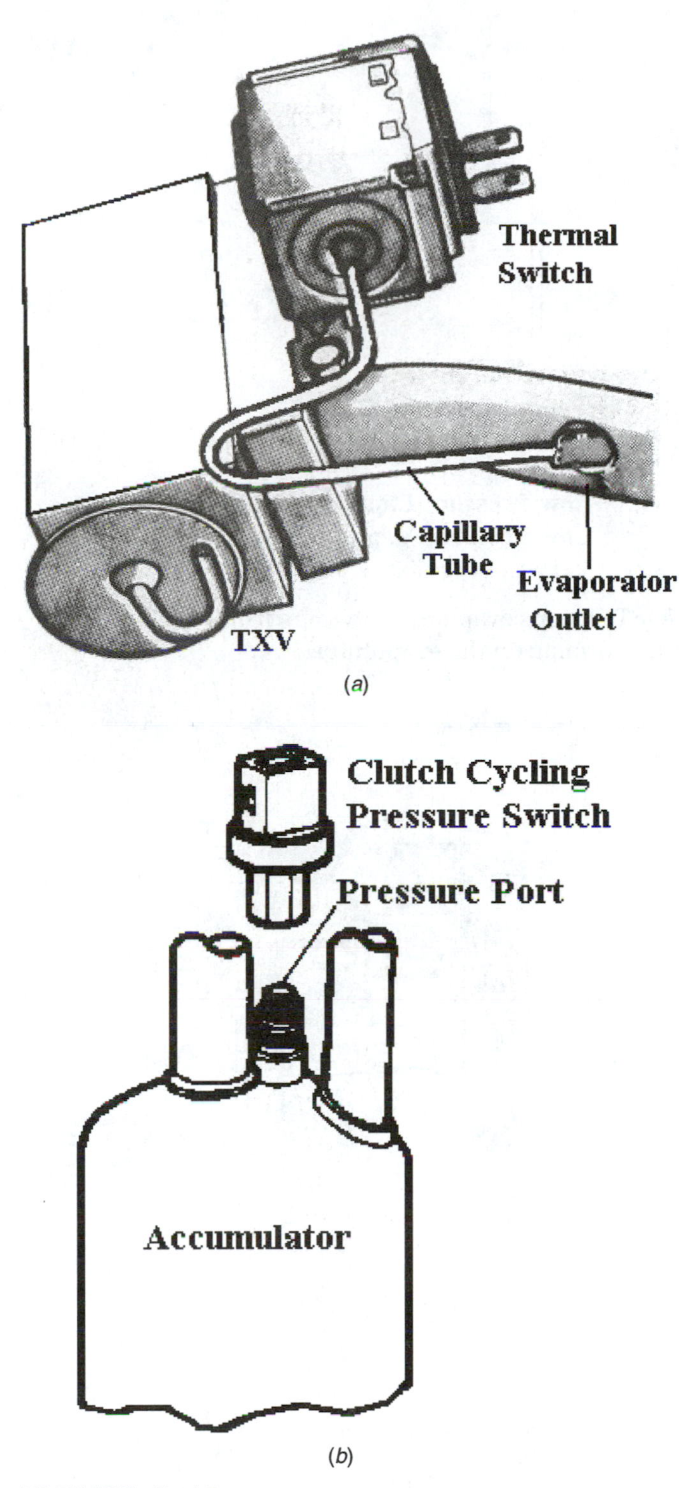

FIGURE 5–19 Most TXV systems use a thermal switch to cycle the compressor out when the evaporator gets too cold (*a*). Most OT systems use a pressure switch to cycle the compressor out when the low side pressure drops too low (*b*).

Remember that evaporator temperature and pressure are closely linked; they drop together. When the pressure switch senses the pressure dropping below a certain point (about 30 psi for R-12, slightly less for R-134a), this switch opens to stop the compressor. This is called the **cutout pressure;** depending on the particular switch, this pressure is between 22 and 28 psi. Like the thermal switch, a pressure switch recloses when the pressure increases; the **cut-in pressure** is about 42 to 49 psi, again depending on the particular switch. With either of these systems, if ice and frost start to form because the evaporator gets too cold, they melt during the off part of the cycle (Figure 5–19). Some vehicles use a **pressure sensor** in place of a pressure switch. The resistance of the sensor changes in direct relation to the pressure. It is an input to an ECM used for compressor clutch, cooling fan, and idle speed control as well as low-pressure and high-pressure protection.

5.2.5.2 Evaporator Pressure Controls If we can keep the evaporator pressure at 30 psi in an R-12 system or 28 psi in an R-134a system, the evaporator temperature will stay at 32°F. Lower pressures cause icing, and higher pressures cause a temperature increase. Two methods have been used in automotive systems to maintain the ideal evaporator pressure: **suction throttling** and **hot gas bypass.**

Many domestic A/C systems in the 1960s and 1970s used suction throttling valves—pilot-operated absolutes (POAs) and evaporator pressure regulators (EPRs) were the most common. These valves are mounted at the evaporator outlet, the compressor inlet, or somewhere between. Most of them sense evaporator pressure; when the pressure starts to drop below a certain point, the valve closes down to restrict refrigerant flow to the compressor. When this occurs, the system has three basic pressures: low side evaporator pressure controlled to 30 psi (R-12), low side compressor inlet pressure below 30 psi, and high side pressure. A system that uses a suction throttling valve maintains an almost constant evaporator temperature of 32°F without cycling the clutch. It should be noted that the compressor drive load drops when the valve restricts the flow (Figure 5–20).

A few very early automotive systems (before 1960) used a valve that would sense evaporator pressure. When the pressure began to drop below 30 psi, the valve would allow hot gas from the high side to enter the evaporator. This valve was called a **hot gas bypass valve.** It worked, but now this system is considered crude (Figure 5–21).

5.2.5.3 Variable Displacement Compressors A **variable displacement compressor** provides the smoothest operation (no clutch cycling), a constant

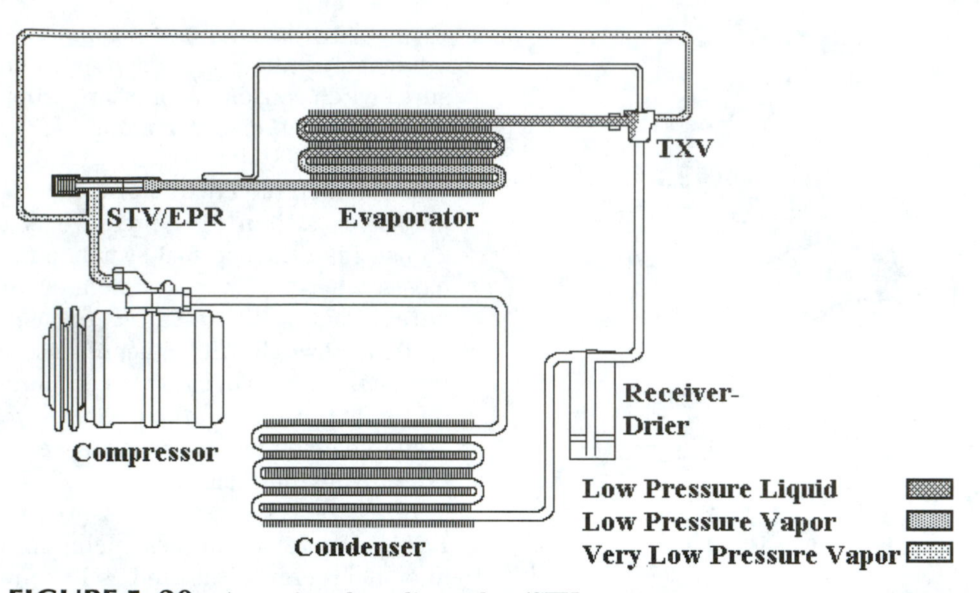

FIGURE 5–20 A suction throttling valve (STV) stops evaporator pressure from dropping below 30 psi, and this keeps ice from forming on the evaporator.

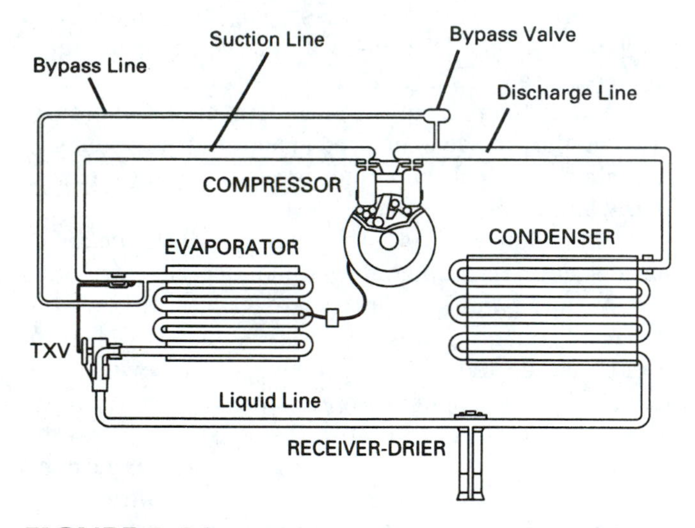

FIGURE 5–21 A hot gas bypass system diverts high side pressure into the evaporator to keep the pressure from dropping to the point at which icing can occur.

32°F evaporator, and the most efficiency. This design includes a large compressor that can pump enough refrigerant to meet high cooling loads, and it reduces the displacement and pumping capacity of the compressor to match the needs of the evaporator as the evaporator cools. At this time, variable displacement compressors are of the wobble plate type and vane type, with development work being done on a variable displacement scroll-type compressor. These compressors are described in more detail in Chapter 6 (Figure 5–22).

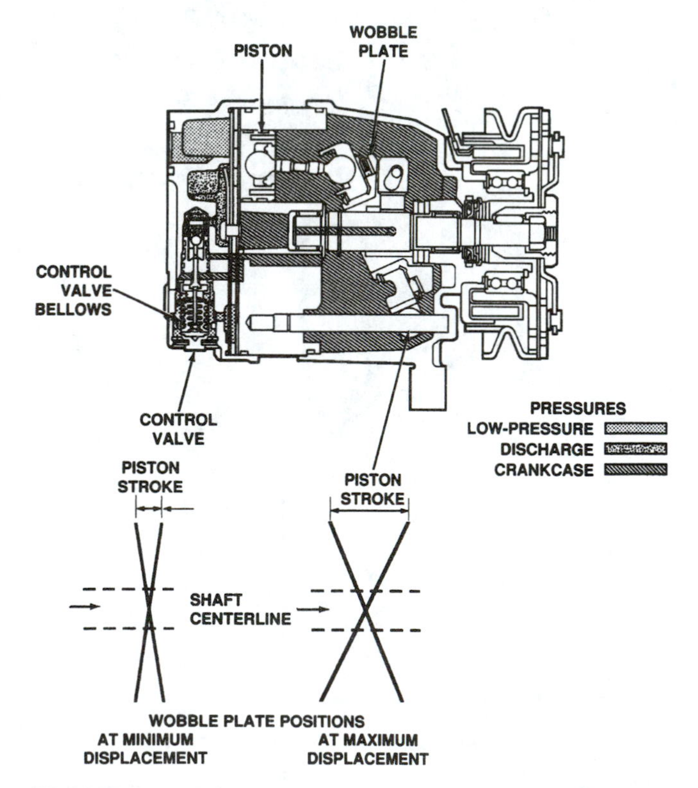

FIGURE 5–22 When the evaporator cools and low side pressure drops, the piston stroke of a variable displacement compressor is reduced so that compressor output matches the cooling load. (*Reprinted with permission of General Motors Corporation*)

5.3 HIGH SIDE OPERATION

The high side of an A/C system takes the low-pressure vapor from the evaporator and returns high-pressure liquid to the expansion device. To do this, the compressor must raise the pressure and concentrate the heat so that the vapor temperature is above ambient. This causes heat to flow from the refrigerant to the air passing through the condenser. Removing the heat from the saturated vapor causes it to change state, to a liquid.

Major components in the high side are the compressor and the condenser. The condenser, like the evaporator, is a heat exchanger. The high side begins at the compressor and ends at the expansion device. The high side of a TXV system includes a receiver–drier.

5.3.1 Compressor

The compressor can be thought of as a pump that circulates refrigerant. It has to work against the restriction of the TXV or OT, and in doing this, it can increase the pressure about 10 times. The compression ratio normally varies from about 5:1 to 8:1, depending on ambient air temperature and refrigerant type. The pressure must be increased to the point where refrigerant temperature is above ambient air temperature and there is enough heat transfer at the condenser to get rid of all the heat absorbed in the evaporator.

At one time, most compressors used two pistons and a crankshaft. Because the pistons moved up and down in a cylinder, they are called **reciprocating piston** compressors. Other compressor designs use reciprocating pistons mounted in an **axial** position and operated by a **wobble plate** or a **swash plate, radial pistons** mounted on a **Scotch yoke.** There are two types of compressors that don't use pistons; these are **rotary vanes** and **scroll** designs (Figure 5–23).

A piston compressor uses reed valves to control refrigerant flow in and out of the cylinder. A reed is a thin, flexible strip of metal that closes off one side of a hole in the metal reed plate or piston. Pressure from the reed side of the plate forces the reed tightly against the plate and keeps the hole closed. Pressure from the other side forces the reed open and causes refrigerant flow; it is a one-way check valve. When the piston moves downward, evaporator pressure forces refrigerant through the suction reed and fills the cylinder; this is called the **suction stroke.** The discharge reed is positioned so it blocks a back flow from the high side. During the upward or **discharge stroke,** piston action and cylinder pressure force the refrigerant through the discharge reeds and into the high side as cylinder pressure forces the suction reeds tightly closed (Figure 5–24).

5.3.1.1 Rotary Compressors Almost any style of air pump can be used as a refrigerant compressor. Several of these styles are used for stationary A/C units, but only two rotary styles are currently used in automotive systems.

Vane- and scroll-type compressors form chambers that enlarge to draw refrigerant into them at one location. This area is open to low side pressure. These chambers rotate to another location; then they contract to force the refrigerant into the high side (Figure 5–25).

5.3.1.1.1 Vane Compressors The vanes of these compressors are mounted in a rotor that runs inside a round and eccentric, or a somewhat elliptical, chamber. The vanes slide in and out of the rotor as their outer end follows the shape of the chamber. Compressors with a round, eccentric chamber have one pumping action per vane per revolution (Figure 5–26). Compressors with elliptical housing have two pumping actions per vane per revolution; this type is sometimes called **balanced** because there is a pressure chamber on each side of the rotor. Early compressors used two vanes, and most newer compressors use five.

As the rotor turns, in one (or two) areas, the chamber behind the vane increases in size. This area has a port connected to the suction cavity. The following vane traps the refrigerant and forms a chamber as it passes by the suction port. The trapped refrigerant is carried around to the discharge port. In this location, the chamber size gets smaller; this increases gas pressure and forces it into the high side.

Vane compressors have been used on aftermarket systems and as OEM compressors. They have the advantage of being very compact and vibration free.

Variable capacity vane compressors are under development. These compressors move the suction port location to change the starting time of the compression process. Moving the suction port in a direction opposite the rotor causes the suction port to be closed earlier and reduces its capacity (Figure 5–27).

5.3.1.1.2 Scroll Compressors Scroll compressors use two major components: a fixed and a movable scroll; both scrolls have a spiral shape. Each of these forms one side of the pumping chamber. The fixed scroll is attached to the compressor housing; the movable scroll is mounted over an eccentric bushing and counterweight on the crankshaft. It does not rotate,

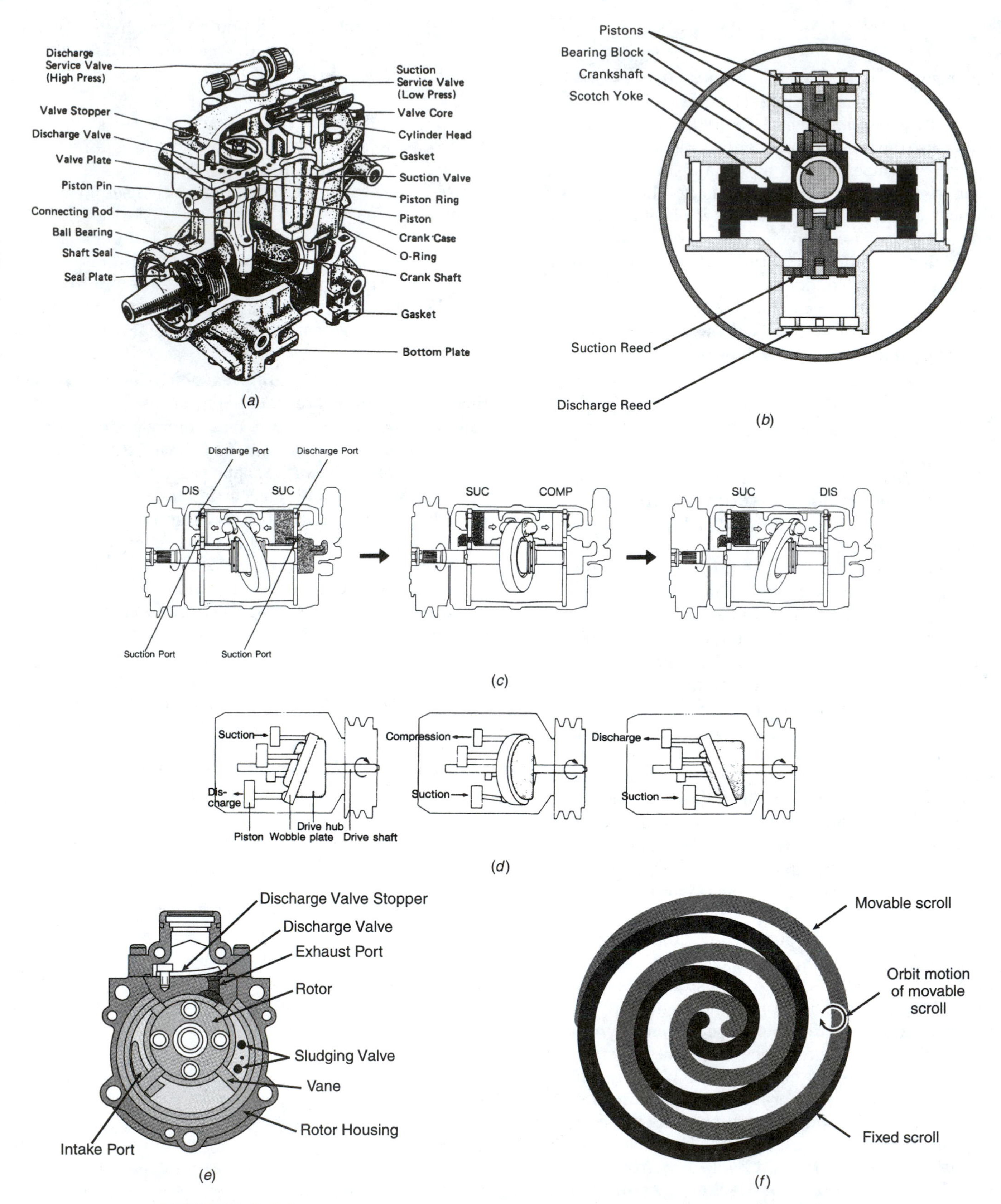

FIGURE 5–23 Piston compressors can drive the piston through a crankshaft (*a*), Scotch yoke (*b*), swash plate (*c*), or wobble plate (*d*). A rotary compressor can use vanes (*e*) or a pair of scrolls (*f*). (a *and* e *are courtesy of Toyota Motor Sales;* c *and* d *are courtesy of Zexel USA Corporation*)

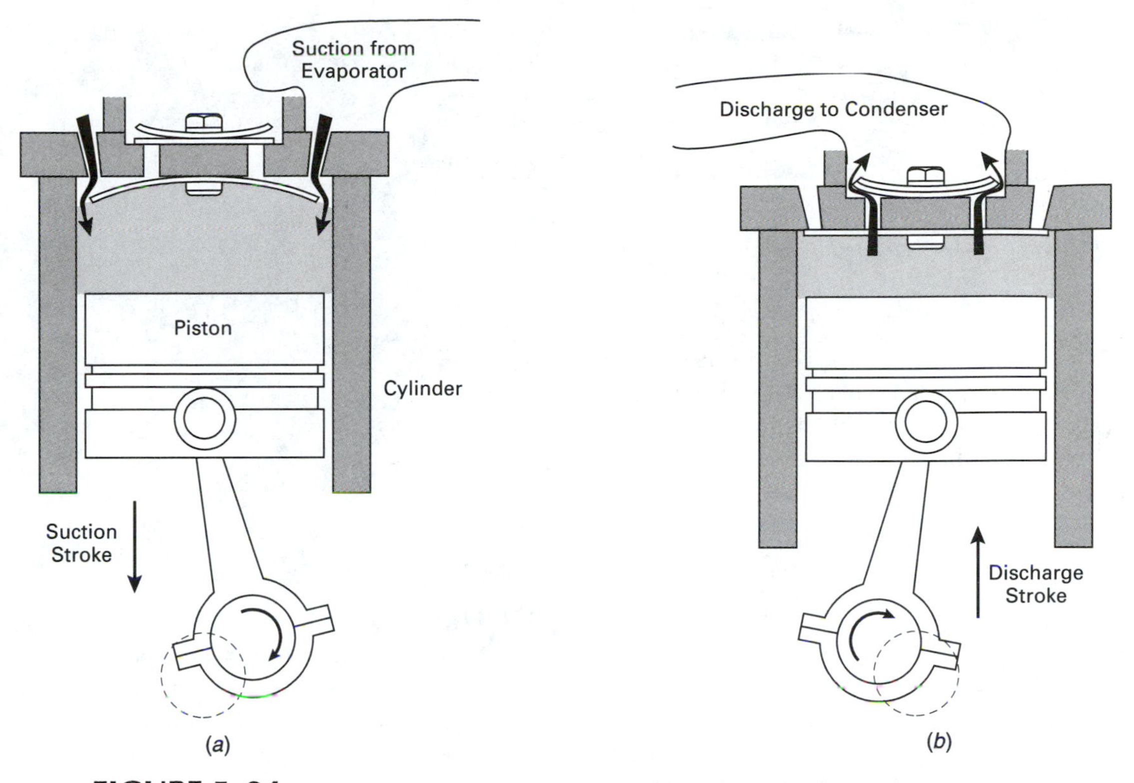

FIGURE 5–24 As the piston moves downward in the cylinder, evaporator pressure opens the suction reed and fills the cylinder with refrigerant (*a*). As the piston moves upward, piston pressure forces the discharge reed open and the refrigerant into the high side (*b*).

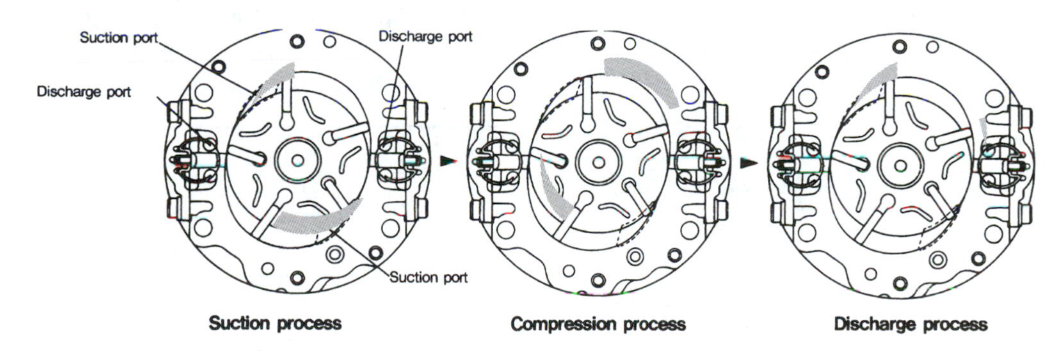

FIGURE 5–25 As the rotor turns, in a clockwise direction, the vanes move in and out to follow the contour of the housing. This action forms chambers that get larger at the suction ports and smaller at the discharge ports. Evaporator pressure fills the chambers as they get bigger, and the reducing size forces the refrigerant into the high side. (*Courtesy of Zexel USA Corporation*)

but it moves in an orbit relative to the stationary scroll—some class the scroll compressor as an **orbiting piston compressor** (Figure 5–28).

As the scroll orbits, it forms a pumping chamber that is open at the outer end. This chamber is moved to the center by the scroll's action. Two or three chambers are present at the same time. The outer ends of the scrolls are open to the suction port, and the inner ends connect to the discharge port (Figure 5–29).

A scroll compressor has the advantage of having very smooth operation and low engagement torque that allows the use of a small clutch. A scroll compressor can also be driven at higher revolutions per minute (rpm) than other designs, so that a smaller drive pulley is used.

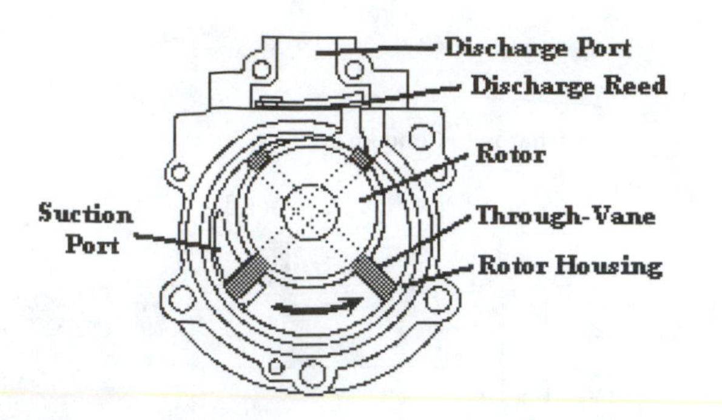

FIGURE 5–26 This through-vane compressor has vanes that contact the rotor housing at each end, and they slide to make a seal at each end as the rotor turns. The vanes form a pumping chamber that gets larger at the suction port and smaller at the discharge port.

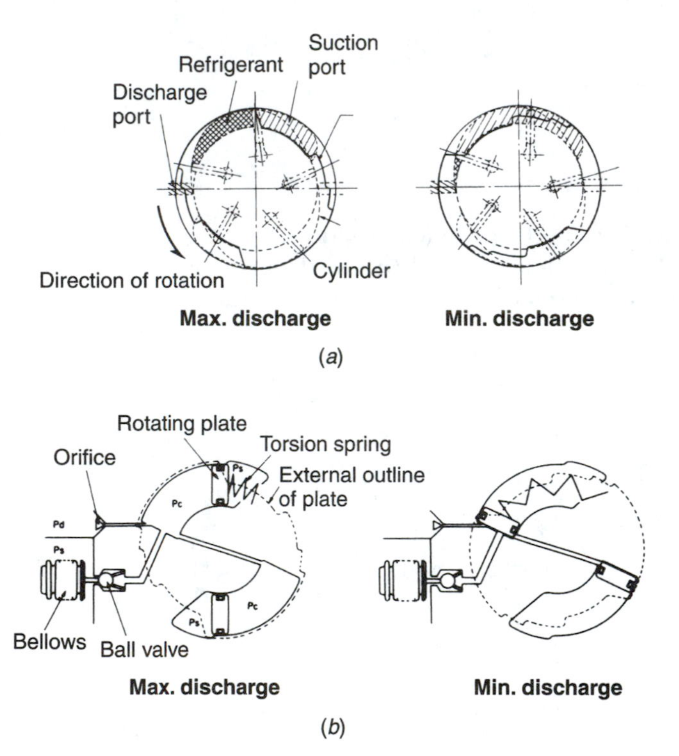

FIGURE 5–27 A variable displacement vane compressor can alter the location of the suction and discharge ports to reduce the amount of refrigerant that is pumped during each revolution (*a*). The position of the control plate is controlled by a bellows that senses evaporator pressure (*b*). (*Courtesy of Zexel USA Corporation*)

This compressor design is also much more efficient than the other compressor styles when it is operated at the design speed; this is a definite advantage for vehicles that tend to run most of the time at cruising speed. A scroll compressor is more expensive to manufacture.

FIGURE 5–28 A cutaway view of a scroll compressor. Note that one scroll is secured to the housing and the other can be moved through its orbit by the driveshaft. (*Courtesy of Sanden International*)

✓PROBLEM SOLVING

An A/C system does not put out cold air like it did when the car was purchased, and you agree to take a look at it. With the system turned on and the hood open, you see that the compressor is not running. What could cause this? What should you do next?

5.3.1.2 Electric Compressors Electric vehicles do not have an engine, and hybrid vehicles do not usually run the engine continuously during vehicle operation.

An electric vehicle can use an electric compressor that operates from the battery pack. An electric compressor uses a scroll compressor driven by a DC electric motor dedicated to the compressor (Figure 5–30). The motor is cycled on and off to produce the desired cooling. This unit can be mounted at any convenient place on the vehicle since it does not need to be connected to an engine. Some electric vehicles use a heat pump for A/C and heat; this unit is a smaller version of heat pumps used in homes. Heat pumps, much like an A/C system, move heat from one location to another. The heat pump operates off the vehicle battery pack.

Hybrid vehicles use an engine to charge the battery pack, and it is also used to drive the vehicle when needed. Most hybrid vehicles shut the engine off when the vehicle

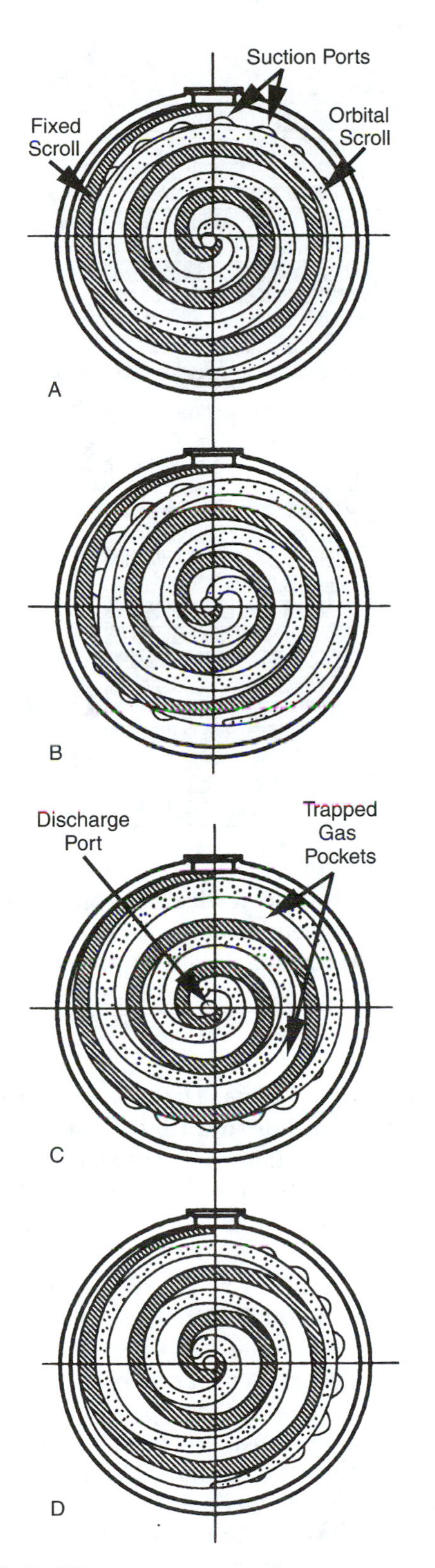

FIGURE 5–29 As the orbital scroll moves, it forms pumping chambers/gas pockets that start at the suction ports and force the refrigerant to the discharge port at the center. (*Courtesy of DaimlerChrysler Corporation*)

(*a*)

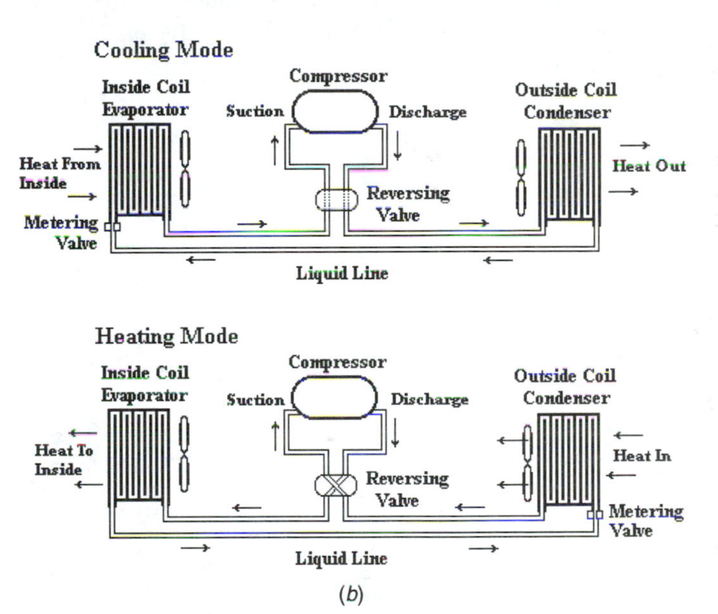

(*b*)

FIGURE 5–30 This electric scroll compressor (*a*) is operated by a DC electric motor operating off batteries. A similar compressor can be used in a heat pump (*b*). Note that the heat pump is very similar to an A/C system that includes a reversing valve that can swap the high and low sides. (a *is courtesy of Sanden International*)

is cruising, operating at slow speeds, or stopped. Some hybrid vehicles use an A/C system much like other vehicles but run the engine constantly when the A/C is turned on.

5.3.2 Condensers

The condenser is the heat exchanger that passes the heat from the refrigerant to the air passing through it. Since this is where the heat leaves the A/C system, the condenser becomes the major component in determining

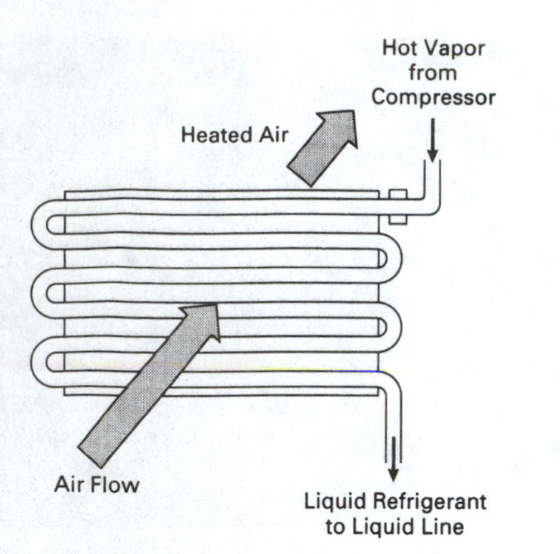

FIGURE 5–31 A condenser is a heat exchanger that transfers heat from the refrigerant to the air flowing through it.

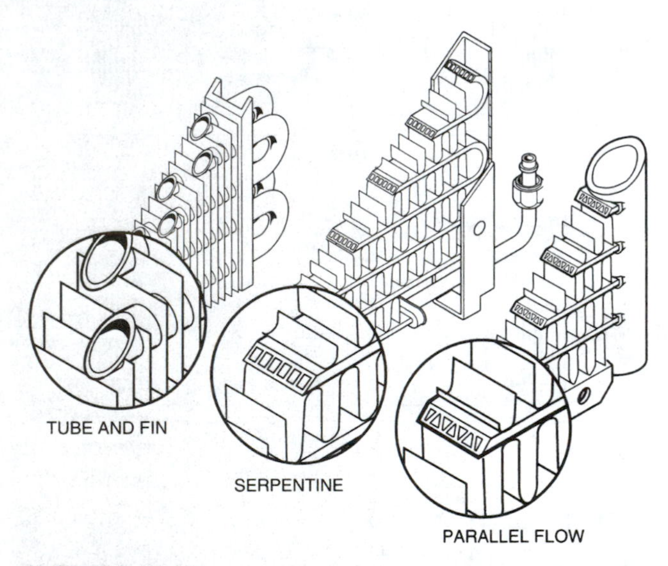

FIGURE 5–32 A tube-and-fin condenser is made up of a series of fins with the tubes passing through them. An extruded tube condenser uses flat tubes with the fins attached between them. Flat tube condensers can use either parallel or serpentine flow. (*Courtesy of Four Seasons*)

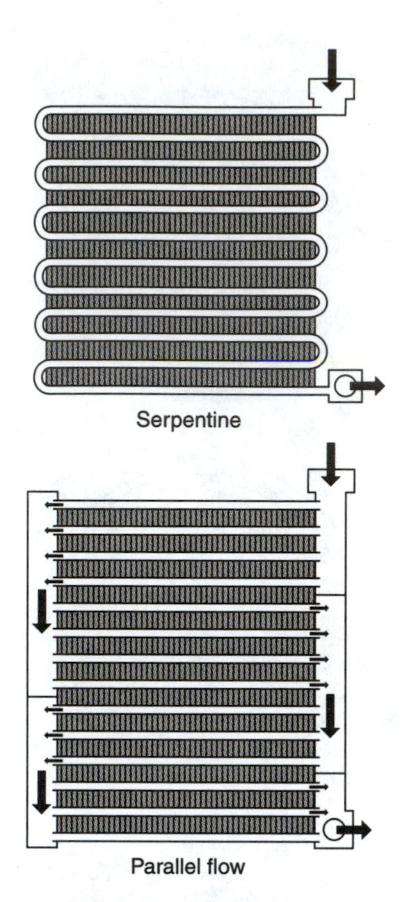

FIGURE 5–33 The refrigerant follows a winding path through a serpentine condenser; it follows a back-and-forth path through a parallel flow condenser.

the overall efficiency of a system. Refrigerant enters the top of the condenser as a hot vapor and leaves from the bottom as a cooler liquid (Figure 5–31).

Many automotive condensers are of the **fin and tube** type and use a round tube that passes through the fins. Some condensers use a **flattened** tube with fins between the tubes (Figure 5–32). Many newer condensers use a flattened, extruded aluminum tube that is divided into several small refrigerant passages. The numerous fins provide the large amount of contact area needed with the airstream. The tubing is formed in either a **serpentine** or **parallel flow** arrangement (Figure 5–33). Newer flat tube, parallel-flow condensers are more efficient and transfer heat much better, and this allows for a smaller and lighter component.

One way to think of condenser action is to consider the size of the inlet fitting and the hose leading to it and compare it with the outlet fitting and hose. The size of the suction hose for the low-pressure vapor (about 0.750-inch inside diameter) can also be compared with the size of the orifice in the expansion device (about 0.060 inch). The gas must condense into a liquid or its volume will not fit through the TXV or OT. Remember that a gas has about 1,000 times the volume of the same liquid. As the latent heat of condensation is transferred to the airstream, the refrigerant vapor makes the necessary change into a liquid (Figure 5–34).

Some vehicles use a **dual condenser,** also called a **secondary condenser** or **subcondenser,** to provide additional capacity (Figure 5–35). In some cases, the receiver, called a modulator by one manufacturer, is positioned between the two condensers.

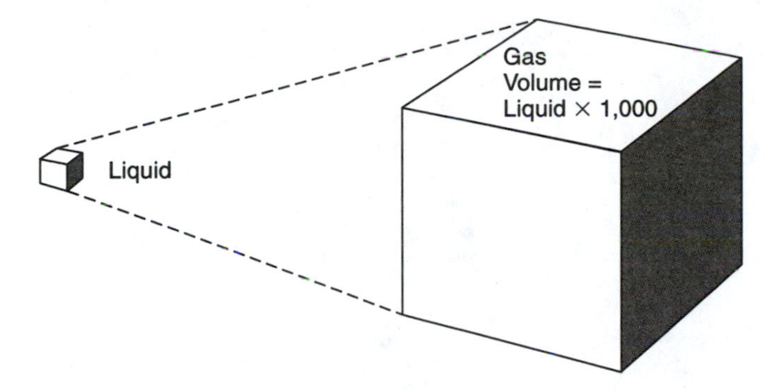

FIGURE 5–34 The volume of gas that enters a condenser is about 1,000 times the volume of liquid leaving it.

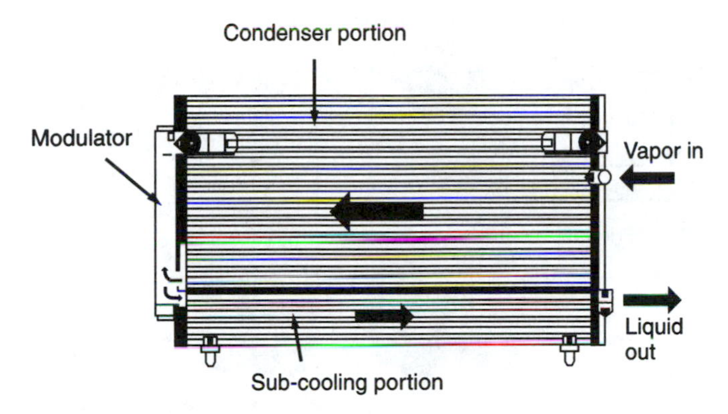

FIGURE 5–35 A dual condenser: the refrigerant flows from the condenser portion through the modulator/receiver–drier portion and then through the subcooling portion.

✓PROBLEM SOLVING

An A/C system does not put out cold air like it did when the car was purchased, and you agree to take a look at it. With the system operating and the hood open, you see that the compressor is operating, but the temperature of the discharge hose to the condenser is very hot and the receiver–drier is quite warm. Are these indications that the system is working correctly or incorrectly? Do you have enough information to determine what might be wrong?

5.3.3 Receiver–Drier

A receiver–drier is used in the high side of a TXV system and, like an accumulator used in OT systems, contains a desiccant to remove moisture and provides a storage chamber for liquid refrigerant. Most receiver–driers also contain a filter to trap debris that might plug the TXV.

A receiver–drier has different internal line routing than an accumulator. Because it must pass liquid on to the TXV, the outlet line begins near the bottom; a filter screen is usually placed over the line opening. Many receiver–driers have a sight glass in the outlet line so the refrigerant flow can be checked to see if it is all liquid or contains bubbles. A receiver–drier should be about half full of liquid, so vapor bubbles are an indication of an undercharge (Figure 5–36). A sight glass is not used in most R-134a systems because the refrigerant has a cloudy appearance in a properly charged system.

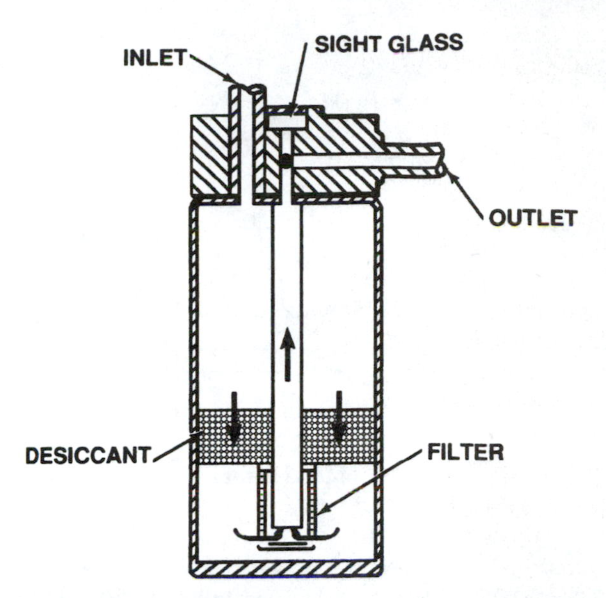

FIGURE 5–36 The outlet of a receiver–drier is close to the bottom so liquid flows on to the TXV. Many units include a sight glass so we can look at this flow. *(Reprinted with permission of General Motors Corporation)*

5.3.4 High Pressure Control

Excessive high side pressure can produce compressor damage and a potential safety hazard if the system should rupture. Many systems contain a **high pressure relief** or **release valve,** and this is mounted on the compressor or at some location in the high side (Figure 5–37). R-12 relief valves are set to release pressure 440 to 550 psi (3,034 to 3,792 kPa), and R-134a are a little higher at 500 to 600 psi (3,448 to 4,137 kPa). A few systems include a **low pressure relief valve** that is mounted in the low side. A relief valve is spring-loaded so excessive pressure will open the valve, and as soon as the excess pressure is released, the valve will reclose.

Many older and some modern systems include a **fusible plug,** also called a **fusible bolt** or **melting plug.** This small plug has a center of low-melting-point solder. Excessive high side pressure or temperature (about 400 psi/2,758 kPa or 220° F/110° C) will melt the center of the plug, and this will allow all of the refrigerant to escape from the system.

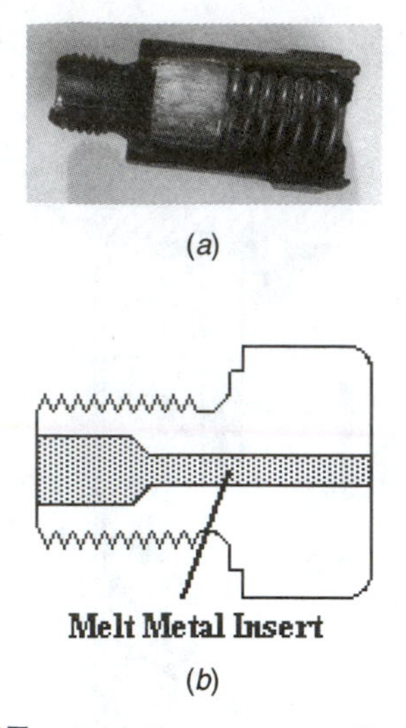

FIGURE 5–37 A high pressure relief valve (*a*) contains a strong spring that keeps the valve closed unless high side pressure (from the left) forces it open; the valve can close when the pressure drops. The fusible plug (*b*) contains a melt metal insert that will blow out if pressure gets too high.

Releasing refrigerant from a system is not good. Modern systems include a **high pressure cutoff (HPCO) switch,** and this will cycle the compressor off before the pressure gets high enough to cause release.

5.4 LINES AND HOSES

The various system components must be interconnected so that refrigerant can circulate through the system. Both flexible rubber and rigid metal hoses are used to link the components. The connections to the compressor must be flexible to allow for engine and compressor movement. Early R-12 hoses were solid rubber or rubber with one or two layers of reinforcing material, and refrigerants could permeate most of these flexible hose materials and escape from the system. Modern refrigerant hose is made from one or two nonpermeable inner layers with internal reinforcement and an outer layer for protection. The nonpermeable nylon layer forms a leakproof barrier; these hoses are commonly called **barrier hoses.** The materials for the various layers are developed to hold refrigerant loss to a minimum (Figure 5–38).

Metal tubing is used in many systems to connect stationary components such as a condenser to the receiver–drier or OT. Although metal tubing does not have permeation problems, corrosion caused by battery spillage or water can make holes in the tubing and produce leakage.

The lines in a system are named for their function or what they contain. Starting at the compressor, the

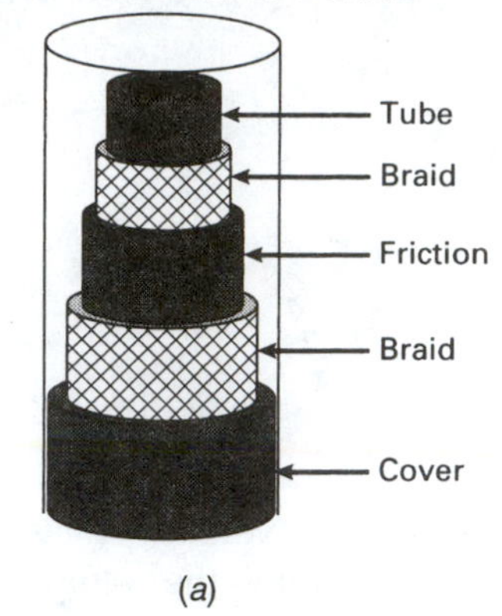

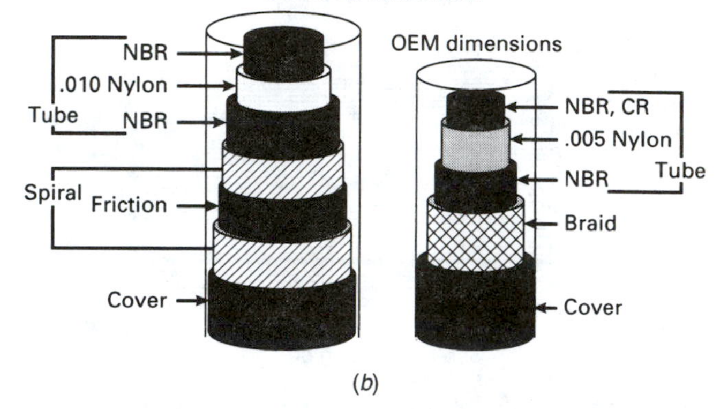

FIGURE 5–38 A refrigerant hose contains one or two reinforcing braid layers around the rubber tube (*a*). A barrier hose includes an impervious nylon layer to reduce leakage (*b*). (*Courtesy of Goodyear Tire & Rubber Company*)

discharge line, sometimes called the **hot gas line,** connects the compressor to the condenser. The **liquid line** connects the condenser to the receiver–drier and TXV or OT. A TXV system can have two liquid lines, one on each side of the receiver–drier. The **suction line** connects the evaporator to the compressor; it has the largest diameter because it transfers a low-pressure vapor (Figure 5–39). Theoretically, when the OT is mounted upstream in the liquid line, the line between the condenser and the evaporator should have two names for the segments above and below the OT. In the A/C trade, it is simply called the liquid line, and a **liquid line repair kit** is required if you have to cut the line to replace the OT.

Four sizes of refrigerant hose are commonly used: #6, #8, #10, and #12. At one time, the numbers indicated the approximate size in 1/16 inch; a #6 hose has an inside diameter (ID) of about 6/16 or 3/8 inch. In modern practice, the sizes have changed, as shown in Figure 5–40. The suction line has an ID of 1/2 or 5/8 inch (12.7 to 15.9 mm) (a #10 or #12 hose). The liquid line has the smallest diameter, usually an ID of 5/16 inch (7.9 mm)

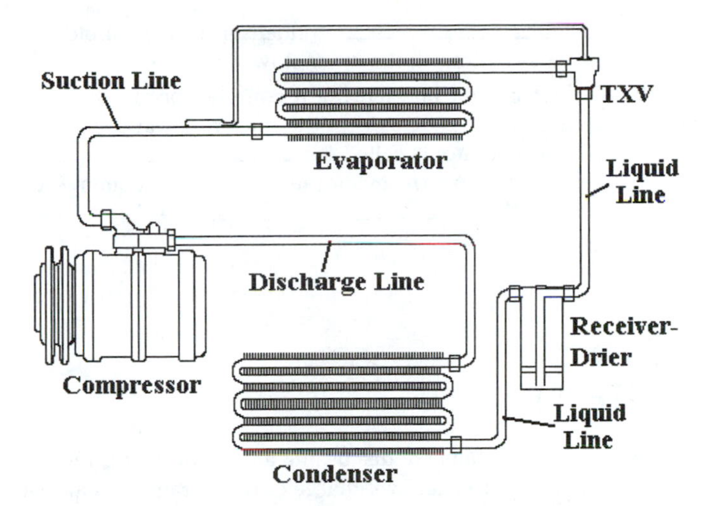

FIGURE 5–39 The three major hoses/lines are the discharge, liquid, and suction lines. Many systems have two liquid lines.

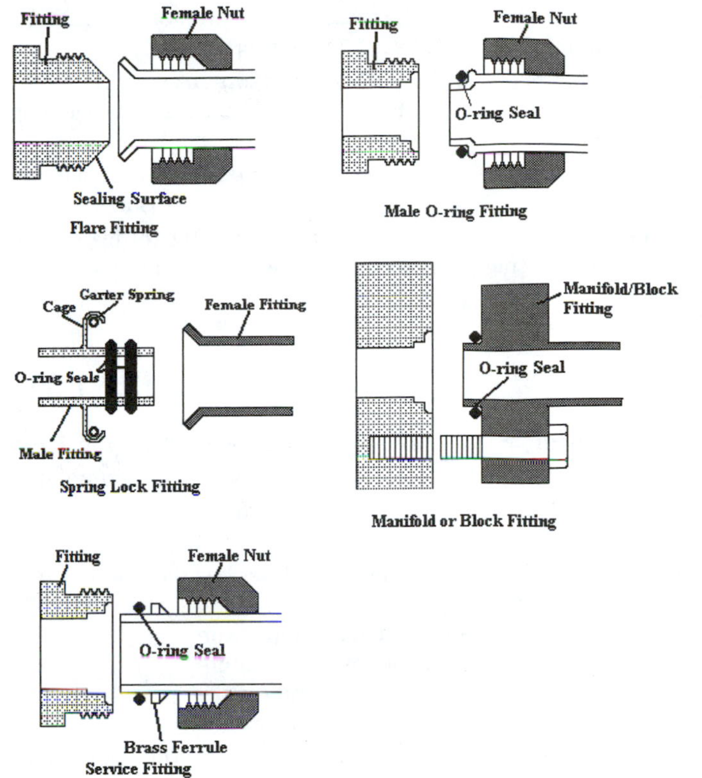

FIGURE 5–41 A variety of fittings are used to seal the refrigerant line connections. The service fitting is used for metal line repairs or to insert an inline filter.

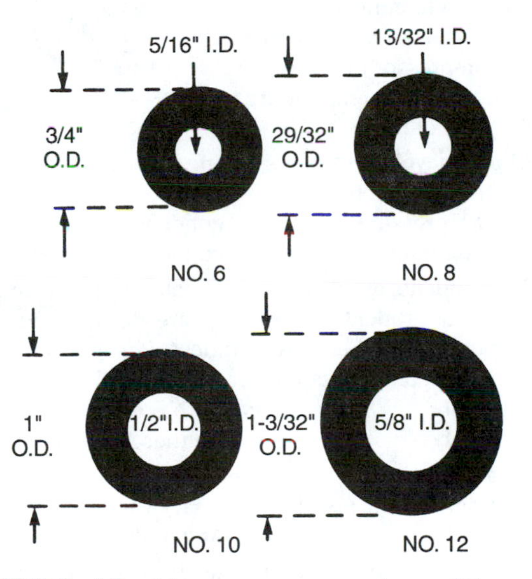

FIGURE 5–40 Most systems use three of these four refrigerant hose sizes. (*Courtesy of Four Seasons*)

(#6 hose). The discharge line has an ID of 13/32 or 1/2 inch (10.3 or 12.7 mm) (#8 hose).

The lines and hoses are connected to the major components using fittings of several different styles (Figure 5–41). These fittings allow the lines to be disconnected and are designed to keep refrigerant leakage to a minimum. Early systems used flare fittings that were very difficult to repair if they did leak. Most newer fittings use an O-ring seal that can be replaced to repair them. The spring-lock fitting uses a garter spring to hold the two portions together.

CHAPTER QUIZ

These questions help you study this chapter. Enter the proper word(s) in the blanks to complete each statement.

1. Automotive A/C systems use either a(n) __________ __________ or a(n) __________ __________ __________ for the expansion device.
2. Refrigerant __________ or __________ in the low side and __________ in the high side.
3. When it is operating, the high side will have a high __________ and __________ .
4. An accumulator is used with on OT system because of the probability of a(n) __________ evaporator.
5. In a TXV system, the refrigerant will have a few degrees of __________ as it leaves the evaporator.
6. The job of the __________ in the __________ or __________ __________ is to remove all traces of water from the refrigerant.
7. A(n) __________ is used with OT systems and a __________ __________ is used with TXV to hold a small refrigerant reserve.

8. A CC system __________ the compressor __________ to prevent __________ of the evaporator.
9. The major styles of compressors use __________, __________, or a pair of __________ for pumping members.
10. Reed valves control the refrigerant flow in and out of the __________ of a(n) __________ compressor.
11. In a(n) __________ condenser, the refrigerant follows a single, winding path from the top of the bottom; a __________ __________ condenser has manifolds at the sides to control the cross-flow.
12. The refrigerant line leaving a compressor is called the __________ __________; the line from the condenser to the evaporator is called the __________ __________; and the line from the evaporator back to the compressor is called the __________ __________.

REVIEW QUESTIONS

These questions allow you to check what you have learned. Select the answer that correctly completes each statement.

1. Which of the following best describes the low side of a system?
   a. High pressure and high temperature
   b. High pressure and low temperature
   c. Low pressure and high temperature
   d. Low pressure and low temperature
2. Which of the following best describes the high side of a system?
   a. High pressure and high temperature
   b. High pressure and low temperature
   c. Low pressure and high temperature
   d. Low pressure and low temperature
3. In a properly operating system, the evaporator temperature is about
   a. 25°F.
   b. 30°F.
   c. 32°F.
   d. 35°F.
4. The TXV is located at the
   a. evaporator outlet.
   b. evaporator inlet.
   c. condenser outlet.
   d. condenser inlet.
5. The refrigerant inside an evaporator is a
   a. gas.
   b. liquid.
   c. liquid changing to a gas.
   d. gas changing to a liquid.
6. An accumulator is very similar to a(n)
   a. receiver–drier.
   b. evaporator.
   c. condenser.
   d. orifice tube.
7. Two students are discussing orifice tubes. Student A says the OT is always placed close to the evaporator inlet. Student B says the OT is placed close to the condenser in some vehicles. Who is correct?
   a. A only
   b. B only
   c. Both A and B
   d. Neither A nor B
8. Two technicians are discussing an A/C problem. Technician A says that an overcharge of refrigerant will cause a higher-than-normal high side pressure. Technician B says that a refrigerant overcharge will cause poor cooling. Who is correct?
   a. A only
   b. B only
   c. Both A and B
   d. Neither A nor B
9. Most A/C systems cycle the compressor clutch in regular intervals to control
   a. condenser temperature.
   b. compressor pressure.
   c. evaporator temperature.
   d. None of these
10. In a noncycling clutch system, evaporator freeze-up is prevented by controlling
    a. TXV pressure.
    b. evaporator pressure.
    c. condenser pressure.
    d. condenser temperature.
11. Two students are discussing variable displacement compressors. Student A says they are designed to control evaporator temperature. Student B says they are designed to control evaporator pressure. Who is correct?
    a. A only
    b. B only
    c. Both A and B
    d. Neither A nor B
12. An automotive A/C compressor is of the __________ type.
    a. reciprocating piston
    b. scroll
    c. rotary vane
    d. Any of these
13. Two students are discussing the lines used to connect the A/C components. Student A says they are usually rubber hoses with metal ends. Student B says that metal tubing is often used. Who is correct?
    a. A only
    b. B only
    c. Both A and B
    d. Neither A nor B

CHAPTER

Air Conditioning System Components

LEARNING OBJECTIVES

After completing this chapter, you should:

- Know how the components used in automotive A/C systems operate.
- Be familiar with the variety of components used in today's A/C systems.
- Be familiar with the variety of controls used in an automotive A/C system.

TERMS TO LEARN

barb fitting
barrier hose
beadlock fitting
captive O-ring
coaxial
condenser seals
discharge stroke
displacement
drive plate
ester
extruded tube
filter
hang-on
hygroscopic
lip seal
mineral oil
muffler
outgassing
polyalkylene glycol (PAG)
polyol ester (POE)
pressure relief
reciprocating piston
reed valves
roof pack
rotary compressor
rotating field
rotor
Scotch yoke
scroll compressor
seal cartridge
seal seat
sight glass
stationary field
STV
suction stroke
swash plate
thermal fuse
valves in receiver (VIR)
vane compressor
variable orifice valve (VOV)
wobble plate

6.1 INTRODUCTION

Automotive A/C components have been evolving steadily since the introduction of A/C in vehicles in 1940. From the early days, when A/C was a very expensive option in luxury cars, to today, when it is standard equipment in many models, many different types of systems have been used. This chapter describes this variety of components, concentrating on the newest and most commonly encountered (Figure 6-1).

6.2 COMPRESSORS

Most automotive compressors are of the **reciprocating piston** type. These components are steadily being downsized, made smaller to fit into tighter locations to reduce the overall vehicle weight. These designs have disadvantages, chief among them the high inertial loads that result from moving a piston at a rather high speed,

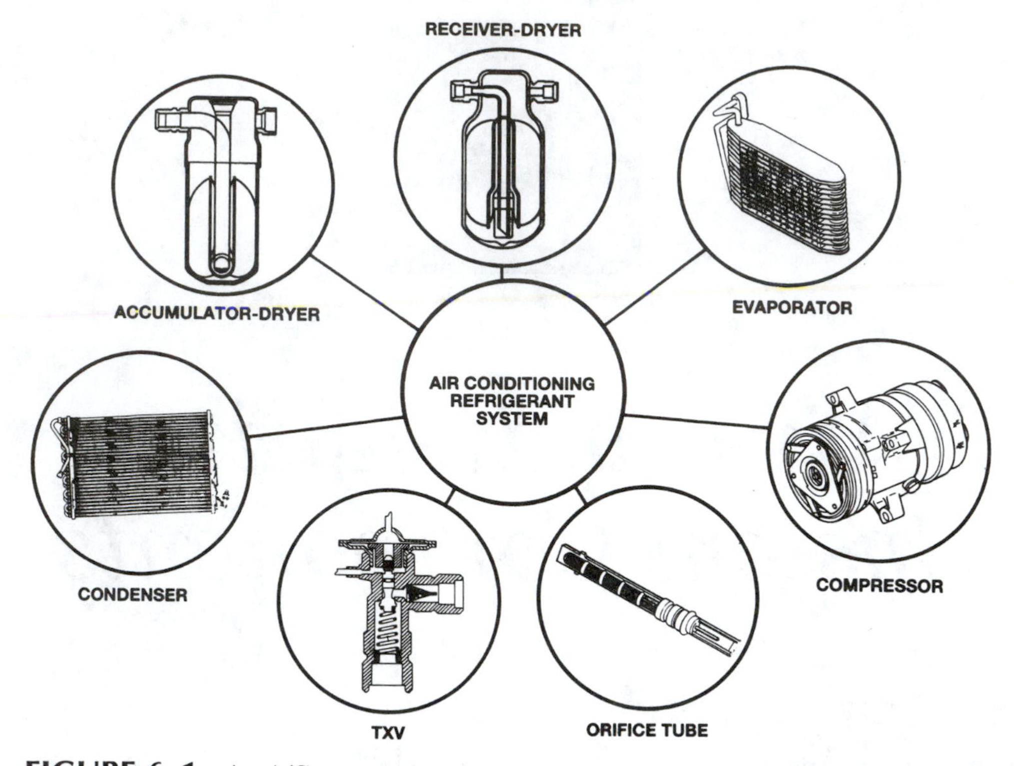

FIGURE 6–1 An A/C system is made up of the components shown. *(Reprinted with permission of General Motors Corporation)*

bringing it to a stop, moving it at a high speed in the opposite direction, bringing it to a stop, and so on. This movement produces vibrations and severe stress on moving parts. **Rotary compressors** merely spin the pumping member (in the case of vane compressors) or cause a member to move in an eccentric orbit (in the case of scroll compressors) (Figure 6-2).

Vane and **scroll** compressors also have problems, chiefly wear of the vane tips and outer chamber (which the vanes have to seal against) and wear between the vanes and the rotor (as the vanes slide in and out). Scroll compressors require rather complex machining to achieve constant sealing between the fixed and movable scrolls.

Currently, about 60 different makes and models of automotive compressors are manufactured. Many of these are versions of a particular model and use different pulleys, switches, mounting flanges, and port configurations (Figure 6-3).

6.2.1 Piston Compressors

With one exception, all of the automotive compressors were of the piston type until the early 1980s (a rotary compressor was used on some General Motors [GM] models in the early 1950s). A piston compressor moves the pistons up and down in a cylinder to produce pumping action and controls the refrigerant flow with two sets of reed valves. The downward, or suction, stroke of the piston causes refrigerant to flow from the compressor suction cavity to push the suction reed open and fill the cylinder. The suction cavity is connected to the evaporator so it contains refrigerant vapor at evaporator pressure. An upward, or discharge, stroke of the piston generates pressure to force the refrigerant through the discharge reed into the discharge chamber and on to the condenser (Figure 6-4).

Pistons can be driven by a **crankshaft, swash plate, wobble plate**, or **Scotch yoke** and can be arranged **in line**, in a **V shape, coaxially**, or **radially**. Most early compressors were of the rather bulky in-line design. Most modern compressors use the more compact coaxial design.

Piston and cylinder diameters and length of stroke determine the internal size or **displacement** of the compressor in **cubic inches (cu. in.)** or **cubic centimeters (cc)**. This is the volume of gas pumped with each revolution of the shaft. The displacement of the compressor is sized to meet the cooling load of the

COMPRESSOR IDENTIFICATION

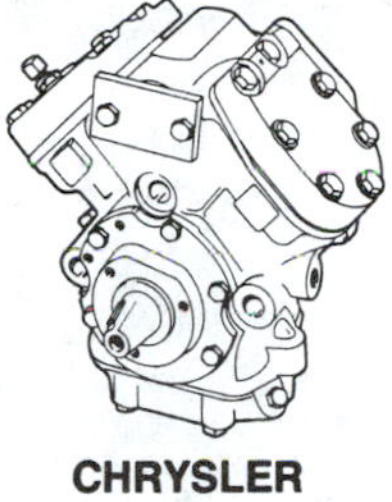

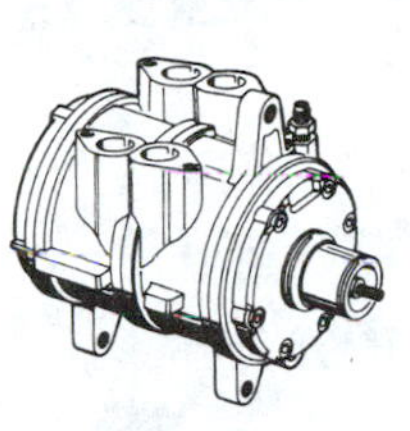

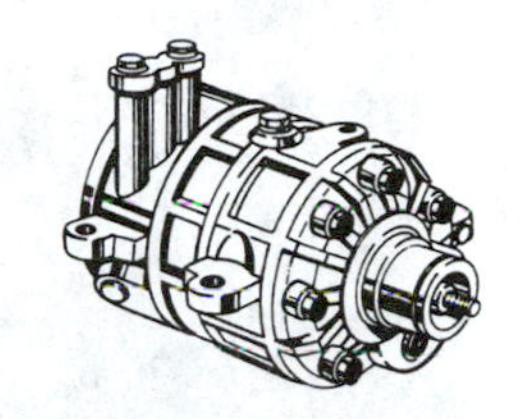

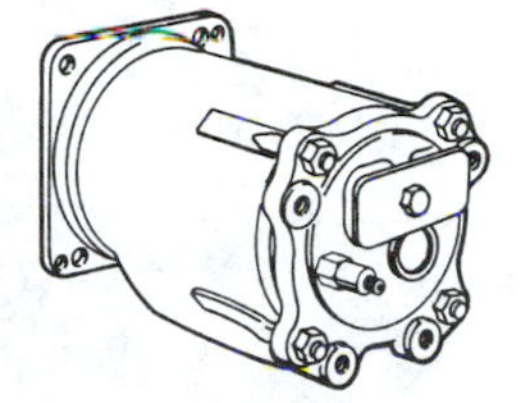

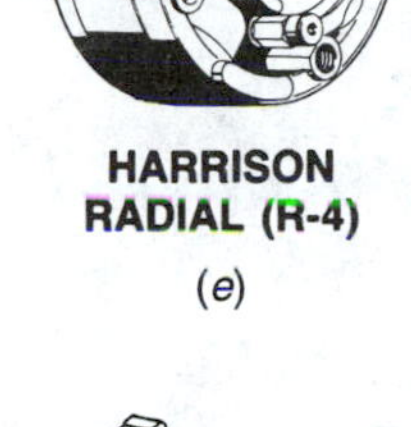

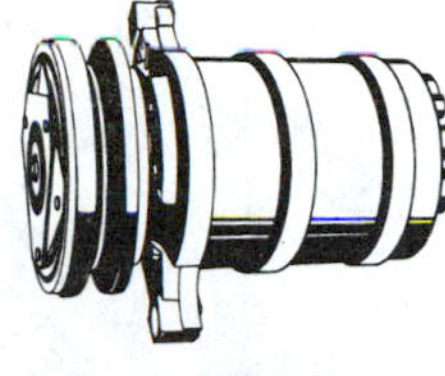

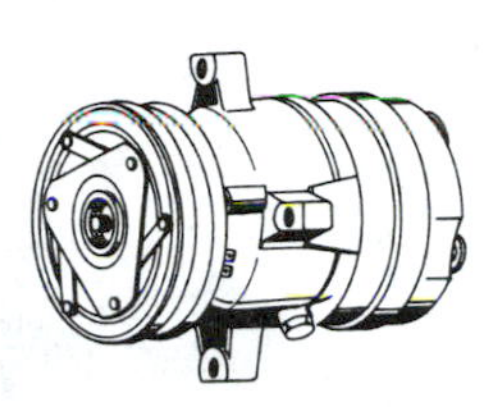

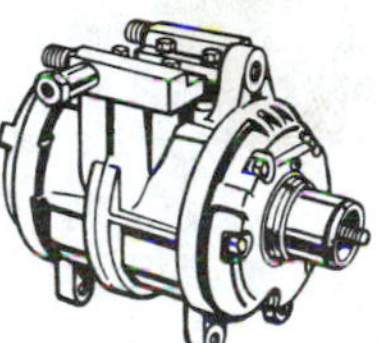

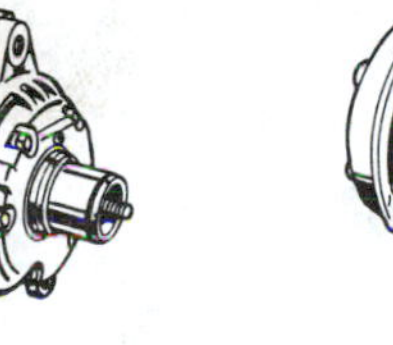

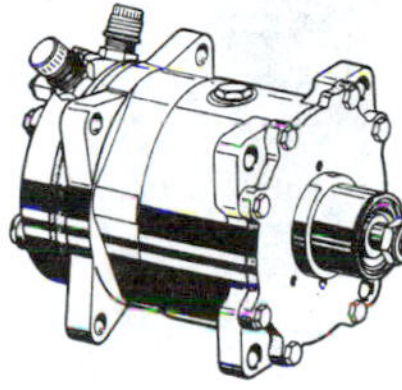

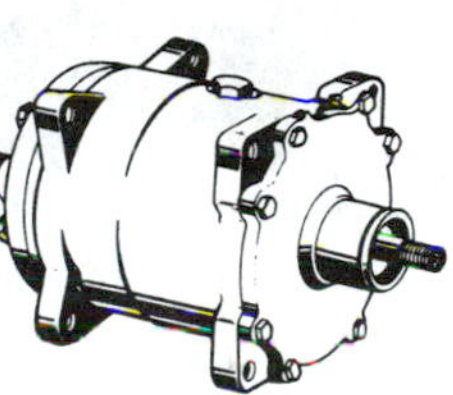

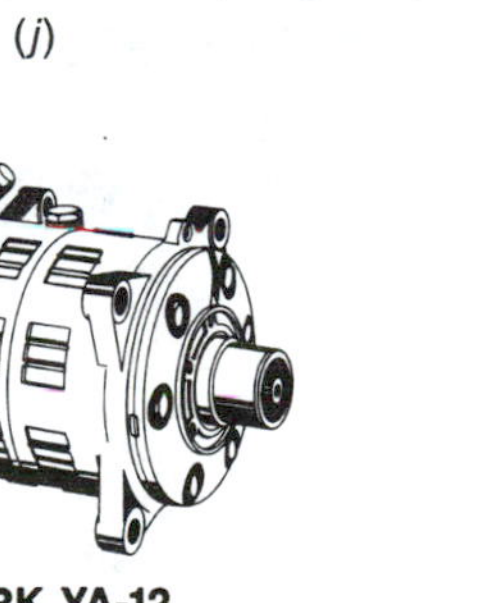

FIGURE 6–2 Some of the variety of compressors. *a* and *l* use a crankshaft; *b*, *d*, *e*, *f*, *h*, and *i* use a swash plate; *c* and *g* use a wobble plate with a variable stroke; *e* and *m* use a Scotch yoke; *j* and *k* use a plain wobble plate; and *n* is a vane compressor. (*Courtesy of Everco Industries*)

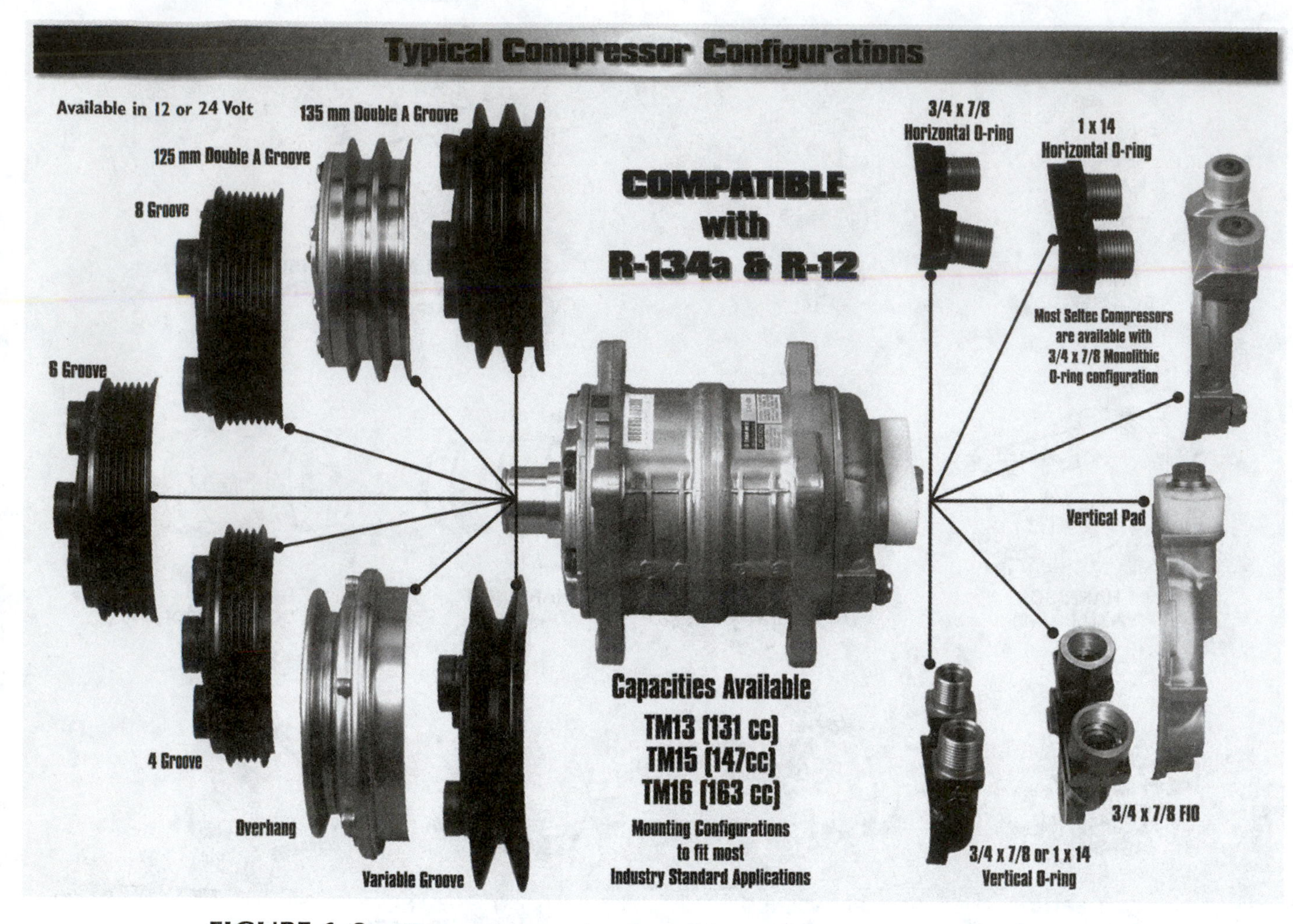

FIGURE 6–3 This compressor is compatible with both R-134a and R-12 and is available in three different capacities or displacements, with one of seven different clutch configurations of either 12 or 24 volts, and with one of seven different rear head and line port configurations. (*Courtesy of Seltec*)

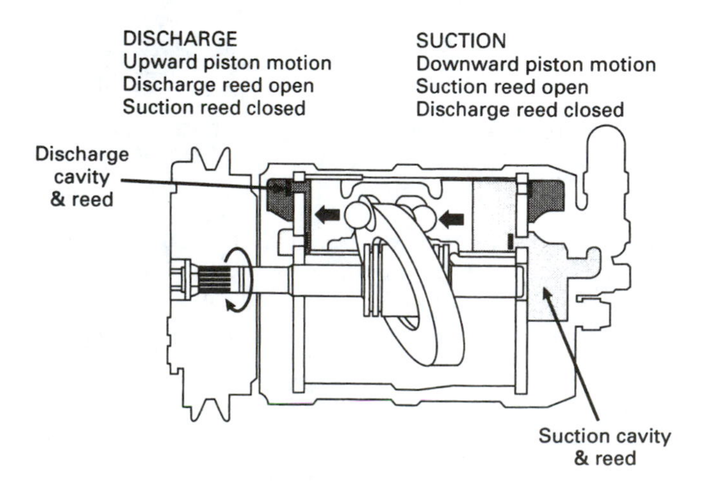

FIGURE 6–4 Rotation of the swash plate causes these double pistons to move through the suction and discharge strokes. Evaporator pressure fills the cylinders with refrigerant during the suction stroke. This refrigerant is pumped into the high side during the discharge stroke.

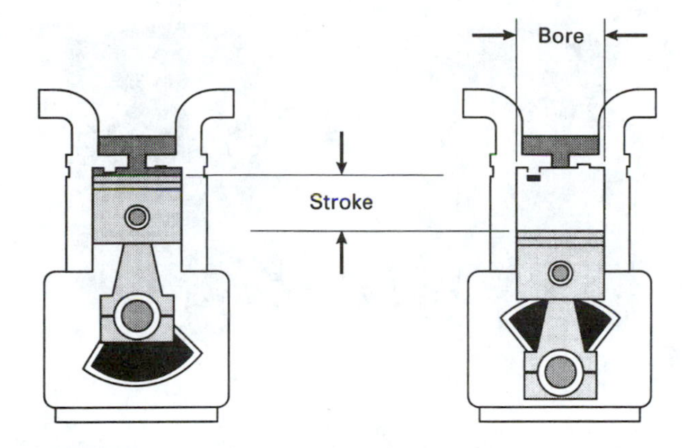

FIGURE 6–5 The displacement of a compressor is determined by the length of the stroke, diameter of the cylinders (bore), and number of cylinders.

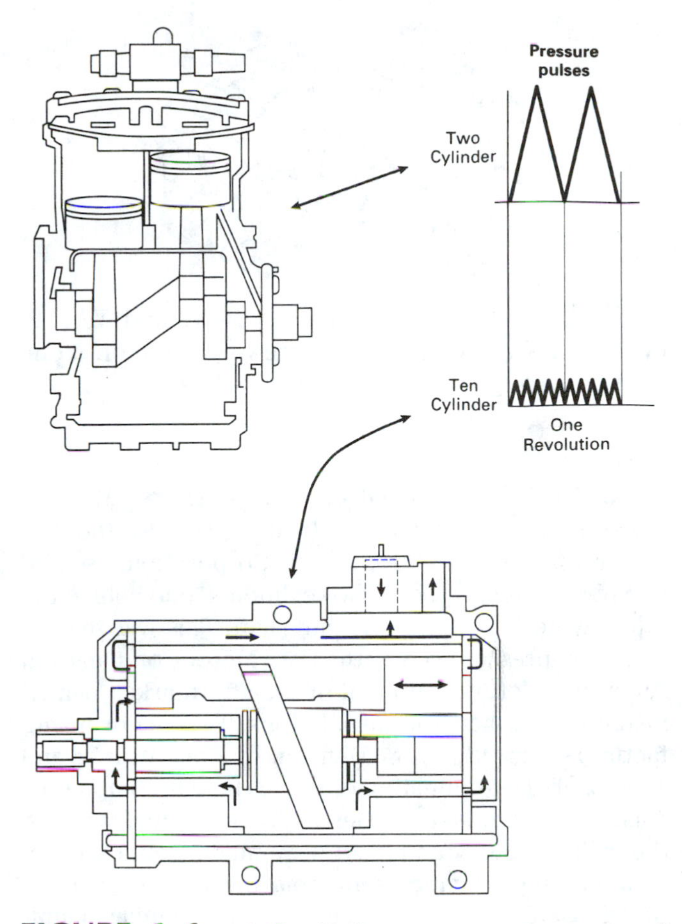

FIGURE 6–6 A 2-cylinder compressor creates 2 large pumping pulses per revolution. A 10-cylinder compressor (same displacement) creates 10 smaller pulses; its operation is much smoother.

vehicle and the size of the engine. A larger compressor has the ability to cool a larger vehicle faster, but it also places a greater power draw on the engine. With some compact vehicles, compressor load places a significant power draw on the engine, and the cycling of the compressor clutch is quite noticeable and can be annoying (Figure 6-5). Many vehicles with small engines either disconnect the compressor clutch at idle speed or increase the idle speed when the compressor is operating to compensate for the load.

Most compressors have the ability to significantly increase gas pressure, depending on the needs of the system. With an engine, compression ratio is determined by the size of the cylinder and combustion chamber. With an A/C compressor, compression ratio is determined by the system's ability to condense refrigerant in the high side. In a system that is working properly, the **compression ratio** is about 5:1 to 7:1. Pressure ratios above 8:1 place added loads on the pistons and bearings in the compressor, as well as very high temperatures that can cause oil breakdown and lacquer and varnish deposits. The compression ratio of a system can be determined by simply dividing the absolute high side pressure by the absolute low side pressure. This formula can be used:

$$\text{Compression ratio} = \frac{\text{High side pressure} + 15}{\text{Low side pressure} + 15}$$

Each piston stroke causes a pressure pulsation in the high- and low-pressure lines at the compressor, resulting in a slight vibration and drumming noise. A compressor with more cylinders has the advantage of running smoother and quieter. A 1-cylinder compressor causes 1 rather large pulsation per revolution; a 10-cylinder compressor causes 10 smaller pulsations (Figure 6-6).

Early compressors were built very much like small engines. Pistons with cast-iron rings operated in cast-iron cylinders; they included oil pumps to circulate oil to lubricate metal parts and prevent wear. Modern compressors use a Teflon band around the pistons, which serves to seal the bore and support the piston. This almost frictionless material allows the piston to run in an aluminum bore with very little wear.

Most compressors developed since the 1960s have not included an oil pump. They use the refrigerant to bring the oil to the cylinders and route the oil scraped off the cylinders to lubricate the rotating parts in the crankcase. Such developments have greatly reduced the size and weight of the compressor and have also changed some of the service procedures.

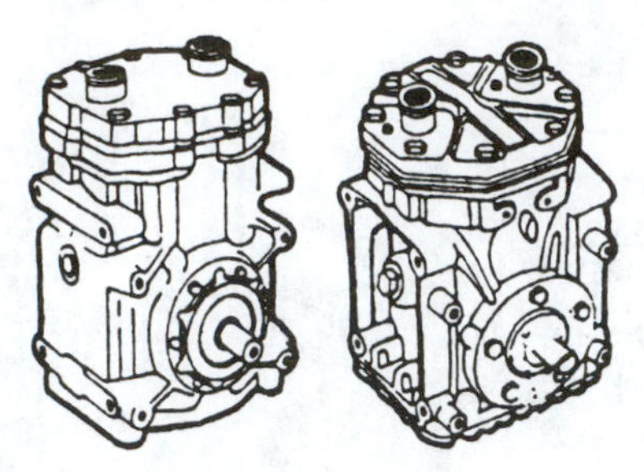

FIGURE 6–7 The Tecumseh (left) and York 2-cylinder, in-line compressors were very common at one time. *(Courtesy of Four Seasons)*

6.2.1.1 In-Line Piston Compressors Most of the systems in the 1960s and 1970s, except for those of General Motors and the Chrysler Corporation, used in-line piston compressors. Two cylinders, one behind the other, were operated by a crankshaft. This was the primary compressor used by the Ford Motor Company and American Motors and in all of the aftermarket, dealer-installed, and hang-on installations. Two major manufacturers were the York Division of Borg Warner and Tecumseh. Determination of the particular compressor was made by model number, appearance, and whether the cylinder block and head were made from cast iron or aluminum. Compressors from these two manufacturers could be interchanged. They had similar mounting holes on each side and the base, and they could be mounted on the engine in an upright, left-hand, or right-hand position (Figure 6-7).

Most York compressors were made from aluminum with reinforcing webs on the cylinder head. The most common model numbers were in the DA 2 (die-cast aluminum, 2-cylinder) series. DA 2 is followed by two numbers that indicate displacement (for example, a DA 209 indicates a 9 cu. in. [147 cc] model). Displacement could also be determined by the shape of the end of the crankshaft. A compact version of this compressor with a smaller overall size was manufactured (Figure 6-8). This compressor is currently used on heavy trucks, mobile industrial equipment, and farm equipment, and it is produced by Climate Control, Inc.

Tecumseh compressors were made from cast iron and were usually painted black. The most common model numbers were in the HG series. These letters were followed by three or four numbers that indicated displacement. Most of these compressors were HG 850 (for an 8.5 cu. in. [139 cc]) and HG 1000 (for a 10 cu. in. [164 cc]). An HG 500 model was also offered; it was a 5 cu. in. (82 cc) single-cylinder unit (Figure 6-9).

(*a*)

(*b*)

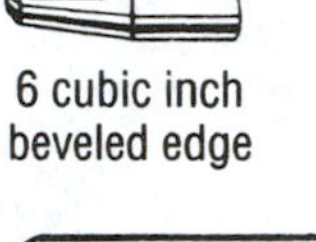

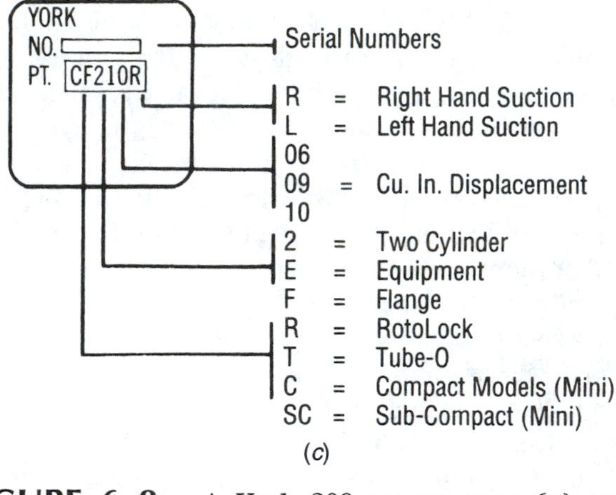

(*c*)

FIGURE 6–8 A York 209 compressor *(a)* and a more compact mini series (*b*). York compressors can be identified by part number and crankshaft appearance (*c*). *(Courtesy of Four Seasons)*

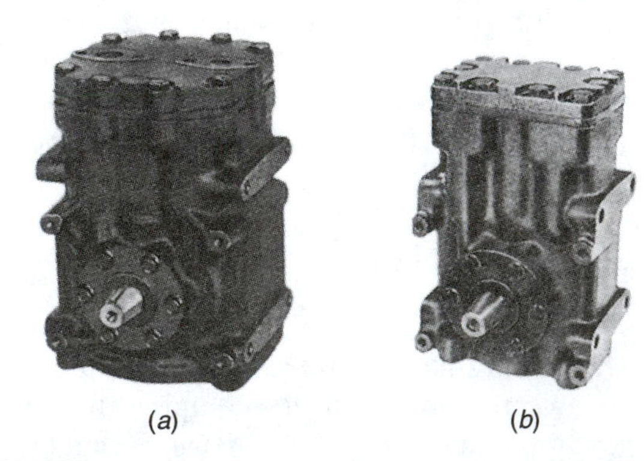

(*a*) (*b*)

FIGURE 6–9 A Tecumseh HG 850 or HG 1000 (*a*) is identified by part number. The HG 500 (*b*) is a single-cylinder compressor. *(Courtesy of Four Seasons)*

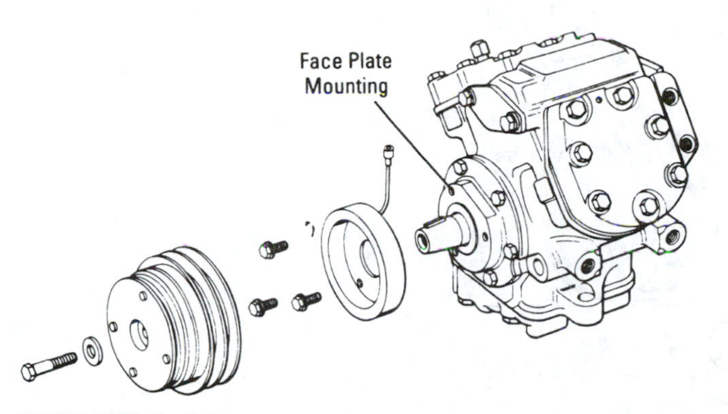

FIGURE 6–10 The two pistons in a Chrysler V compressor are arranged in a V shape. *(Reprinted with permission of General Motors Corporation)*

Several Japanese manufacturers produced 2-cylinder in-line compressors of various sizes. These compressors were used primarily on Japanese vehicles.

6.2.1.2 Two-Cylinder V Compressor Through the 1960s and 1970s, Chrysler Corporation manufactured a 2-cylinder compressor with the cylinders in a V configuration. These compressors were made from cast iron and were used only as OEM installation on Chrysler vehicles (Figure 6-10).

6.2.1.3 Radial Compressors Most radial compressors were made by GM for use on GM vehicles; Tecumseh produced a version for OEM use on Ford vehicles. Radial compressors use two double-ended pistons that cross at a rotating block and Scotch yoke crankshaft. Radial compressors are fairly compact and are much shorter than in-line compressors (Figure 6-11).

The GM version of this compressor was produced from the mid 1970s to the late 1980s and is called the R-4 (radial 4-cylinder). It has a displacement of 10 cu. in. (164 cc) (Figure 6-12).

The Tecumseh version was produced from the mid to late 1980s and is called the HR 980. It has a displacement of 9.8 cu. in. (160 cc).

A version of the radial compressor was also produced by Keihin in Japan for use on Honda automobiles.

6.2.1.4 Coaxial Swash-Plate Compressors Coaxial swash-plate compressors drive the pistons through a swash plate, which is attached to the drive shaft. The swash plate is mounted at an angle so it will wobble and cause the reciprocating action of the pistons. The swash plate revolves with the shaft; each piston has a pair of bearings that can pivot as the swash

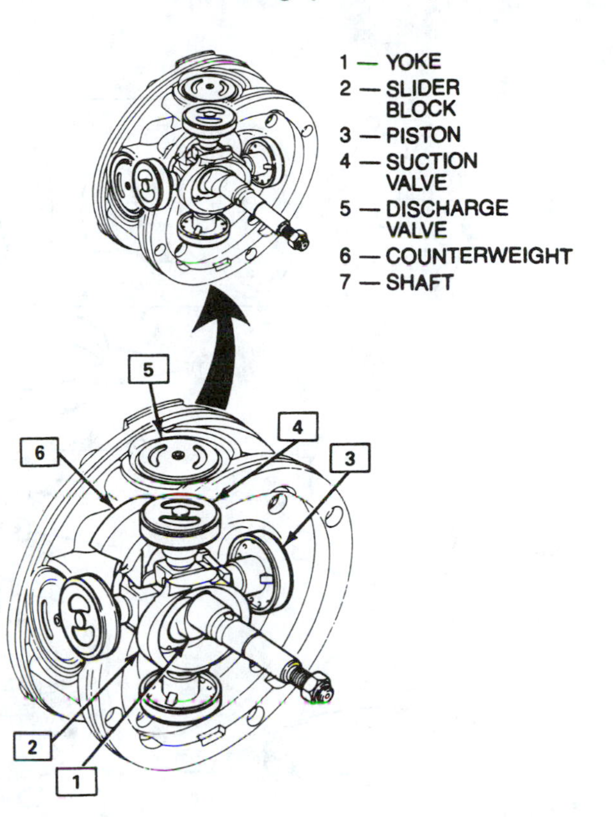

FIGURE 6–11 As the yoke rotates, the two pairs of pistons are forced to slide in their cylinders. Note the suction reeds on the piston tops; incoming refrigerant is routed through the crankcase. *(Reprinted with permission of General Motors Corporation)*

plate slides through them. Each piston is double ended so that each end can pump, and the pistons are arranged parallel to and around the driveshaft. This is called a **coaxial** arrangement. One driveshaft revolution causes each piston end to move through a complete pumping cycle. The most common arrangement is three double pistons making a 6-cylinder compressor and a 10-cylinder using five pistons (Figure 6-13).

A swash-plate compressor must have passages to transfer refrigerant between the suction and discharge chambers at each end of the compressor. The suction crossover passage is usually designed so that it can provide lubrication to internal moving parts (Figure 6-14).

General Motors produced a 6-cylinder compressor called the A-6 from 1962 through the mid 1980s. Most A-6 compressors had a displacement of 12.6 cu. in. (206 cc), but some used dished pistons, which reduced the displacement to 10.8 cu. in. (177 cc). A-6 compressors used a cast-iron cylinder block, cast-iron piston rings, and an oil pump. Most were built with a clockwise (when viewed from the front) crankshaft rotation.

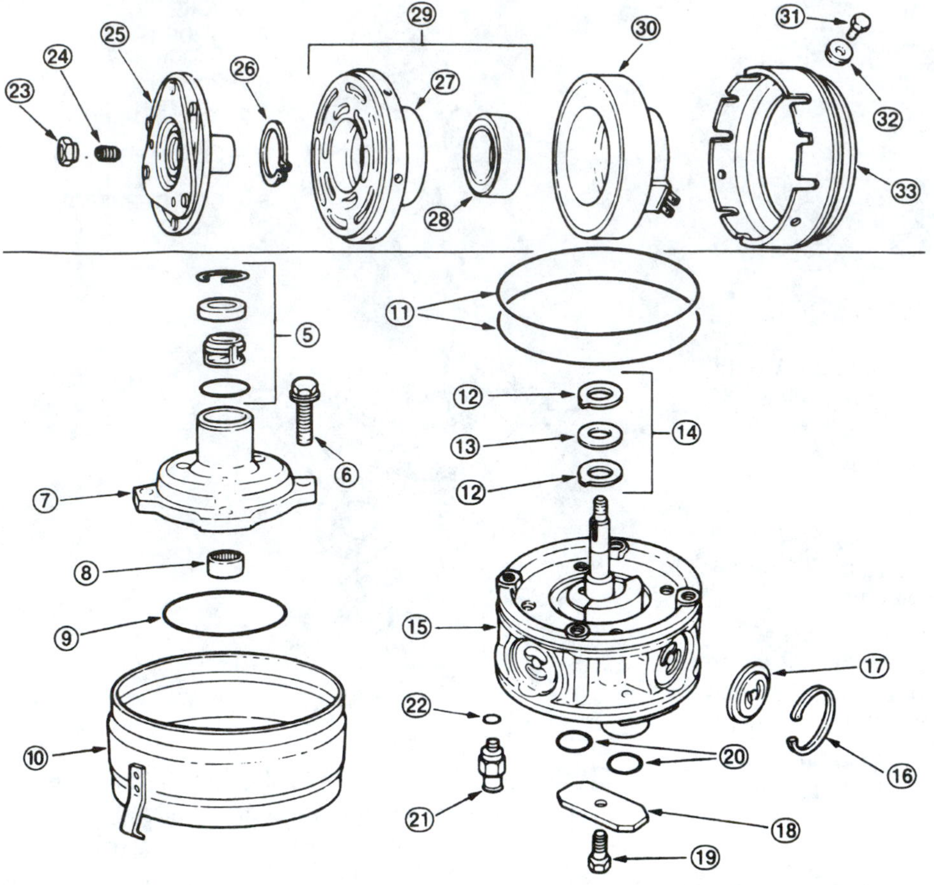

R4 GM COMPRESSOR (4 CYL.) (4 POLE TYPE)

⑤ SHAFT SEAL KIT
⑥ SCREW AND WASHER ASSEMBLY FRONT HEAD MOUNTING
⑦ FRONT HEAD
⑧ MAIN BEARING
⑨ RING SEAL, FRONT HEAD TO CYLINDER
⑩ SHELL
⑪ O-RING, CYLINDER TO SHELL
⑫ THRUST WASHER
⑬ BELLEVILLE WASHER
⑭ THRUST WASHER KIT
⑮ CYLINDER AND SHAFT ASSEMBLY
⑯ RETAINER RING
⑰ VALVE PLATE
⑱ SHIPPING PLATE
⑲ SCREW
⑳ O-RING, SUCTION-DISCHARGE PORTS
㉑ PRESSURE RELIEF VALVE
㉒ O-RING, PRESSURE RELIEF VALVE
㉓ SHAFT NUT
㉔ CLUTCH HUB KEY
㉕ CLUTCH DRIVE ASSEMBLY
㉖ RETAINER RING
㉗ ROTOR
㉘ ROTOR BEARING
㉙ ROTOR AND BEARING ASSEMBLY
㉚ COIL AND HOUSING ASSEMBLY
㉛ PULLEY RIM MOUNTING SCREW
㉜ SPECIAL WASHER-PULLEY RIM MOUNTING SCREW LOCKING
㉝ PULLEY RIM

(a)

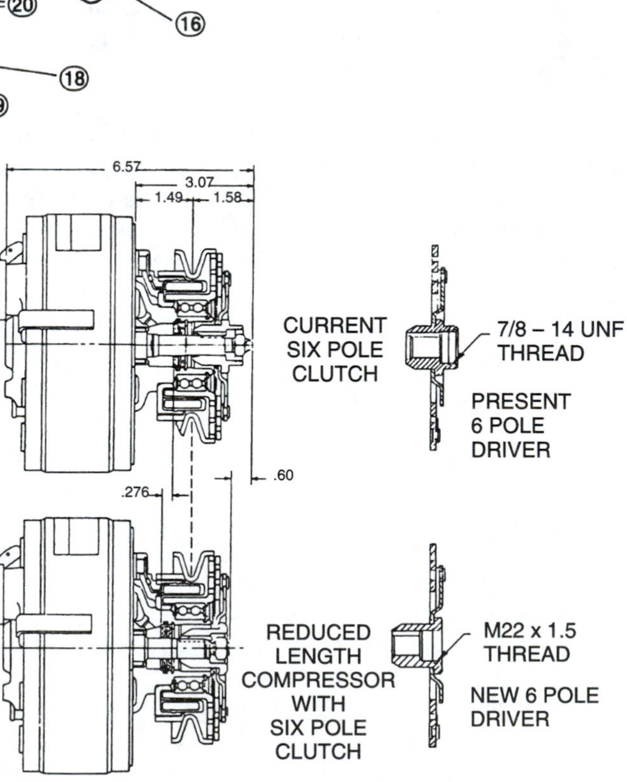

(b)

FIGURE 6–12 A disassembled General Motors R-4 compressor (*a*) and two versions of this same compressor (*b*); note the different clutches and shaft lengths. (*Reprinted with permission of General Motors Corporation*)

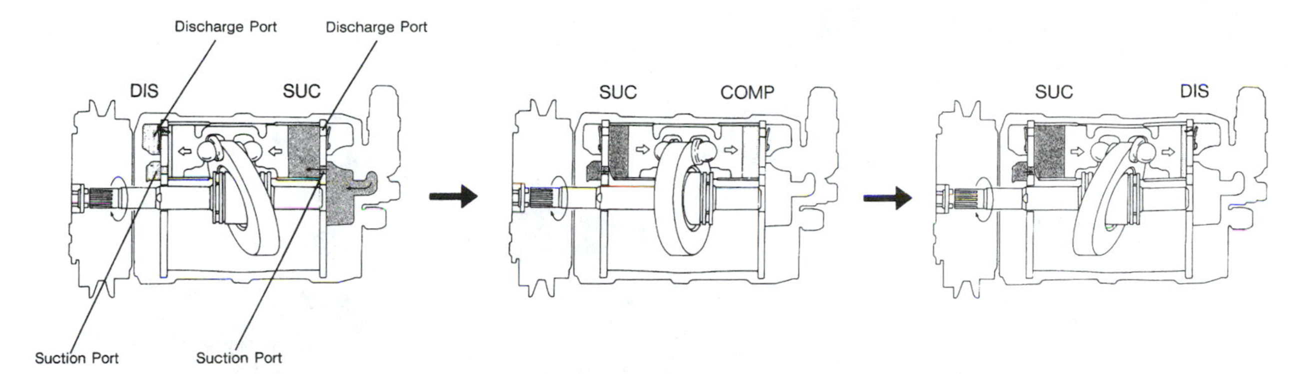

FIGURE 6–13 Note how rotation of the swash plate causes the pistons to slide through their strokes. Balls and shoes act as bearings between the swash plate and pistons. *(Courtesy of Zexel USA Corporation)*

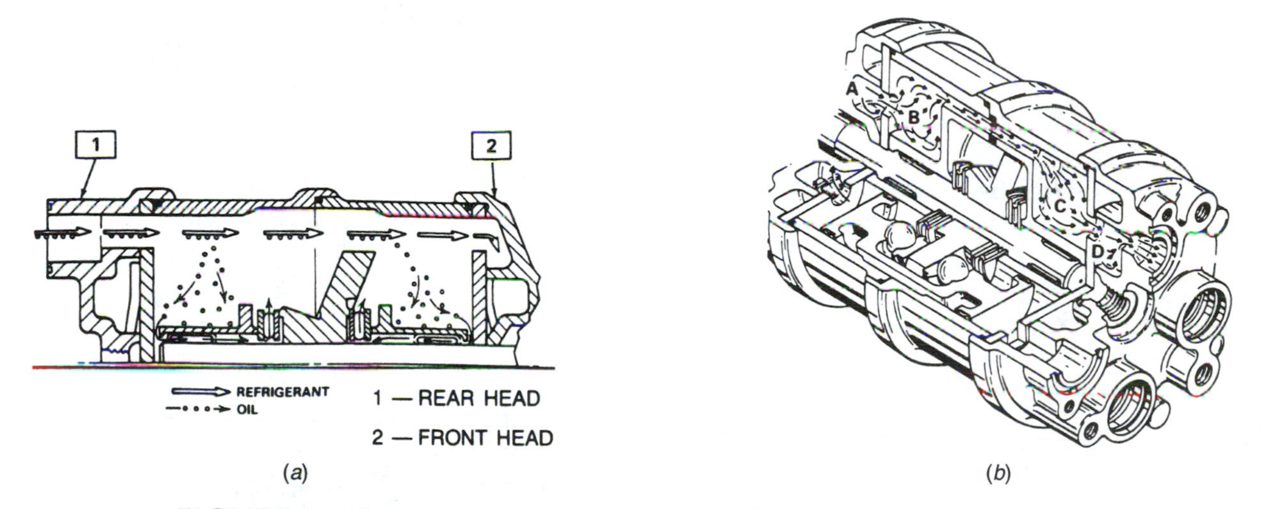

FIGURE 6–14 The suction (*a*) and discharge (*b*) crossover circuits of a swash-plate compressor transfer refrigerant to and from the cylinders at the other end. Note how oil from the suction crossover lubricates the internal parts. *(Reprinted with permission of General Motors Corporation)*

Some were built to operate in the opposite direction for use in the Corvair (Figure 6-15).

In 1982, GM started production of a more modern version of this compressor, called the DA-6. The DA-6 cylinder block is made from die-cast aluminum and is slightly smaller and much lighter than the A-6. Its weight has been reduced from 33.25 lb (15.1 kg) to 12.5 lb (5.7 kg). The DA-6 uses Teflon piston seals, a lip-type shaft seal, and no oil pump. The DA-6 was superseded by the HR-6 (Harrison Redesigned) and then by the HR6HE high-efficiency version (Figure 6-16).

Nippondenso also manufactures 6- and 10-cylinder coaxial compressors. This design is used as OEM equipment by the Chrysler Corporation, the Ford Motor Company, and other vehicle manufacturers around the world. Nippondenso coaxial compressors, like those of other manufacturers, use a four-part aluminum body with two cylinder assemblies, rear head, and front head sealed by O-rings. Either a single-key drive or splined drive is used between the clutch drive plate and the compressor shaft (Figure 6-17).

Several other Japanese manufacturers have produced 6- and 10-cylinder swash-plate compressors. These manufacturers include Calsonic, Hitachi, Mitsubishi, Nihon Radiator, Seltec, and Zexel (formerly Diesel Kiki) (Figure 6-18).

6.2.1.5 Coaxial Wobble-Plate Compressors

Wobble-plate compressors drive the pistons through an angle plate that looks somewhat like a swash plate,

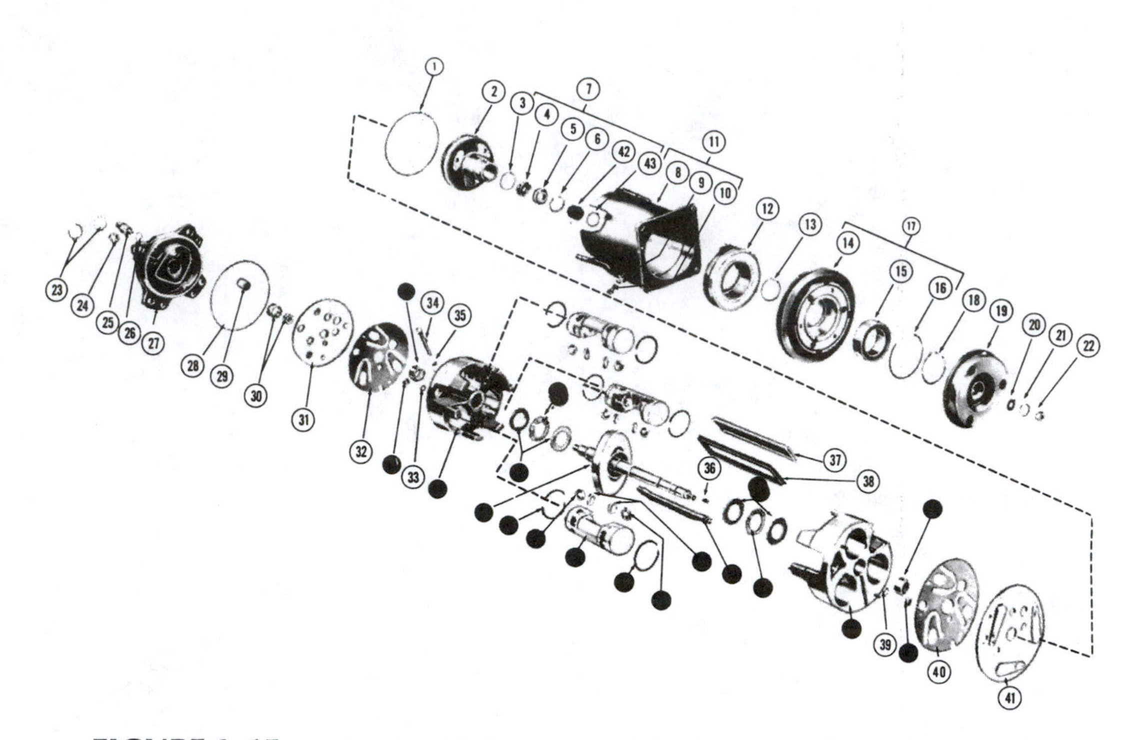

FIGURE 6–15 A disassembled view of an A-6 compressor; this compressor was very popular during the 1960s, 1970s, and early 1980s. (*Reprinted with permission of General Motors Corporation*)

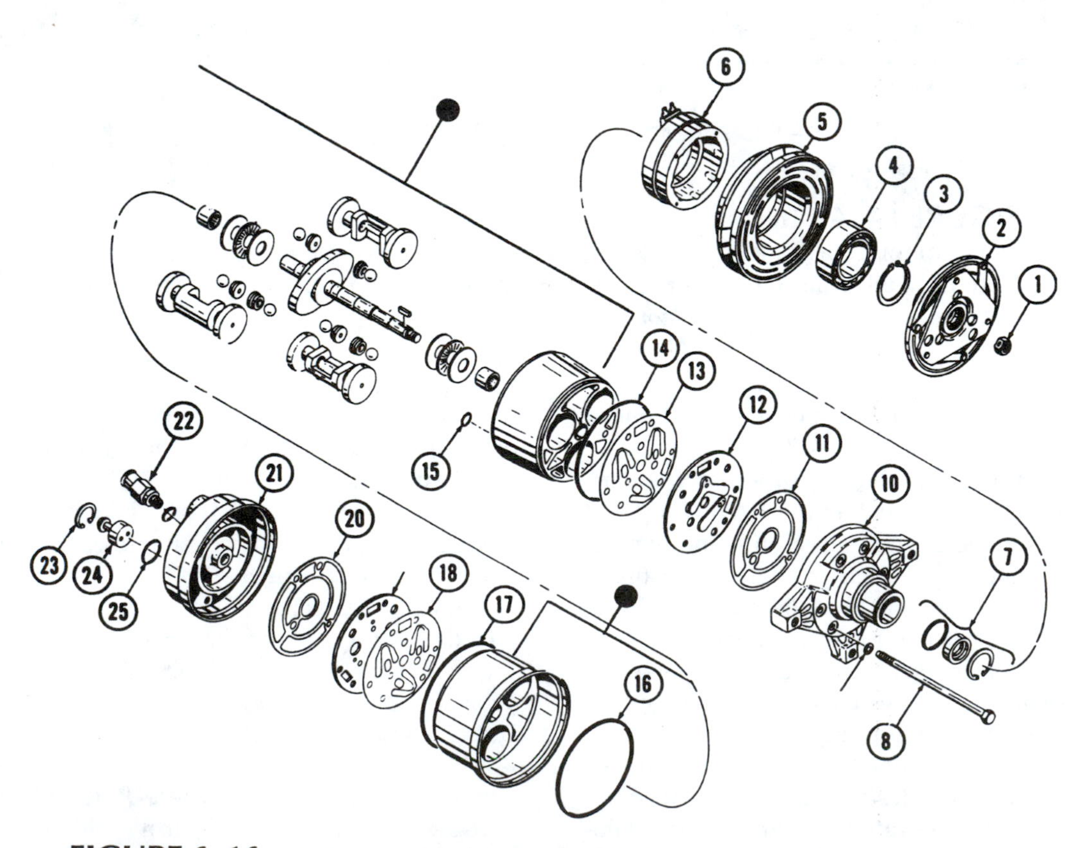

FIGURE 6–16 A disassembled view of a DA-6 compressor; this compressor replaced the A-6. (*Reprinted with permission of General Motors Corporation*)

FIGURE 6–17 Three versions of Nippondenso compressors; note the different mounting bosses. (*Courtesy of Four Seasons*)

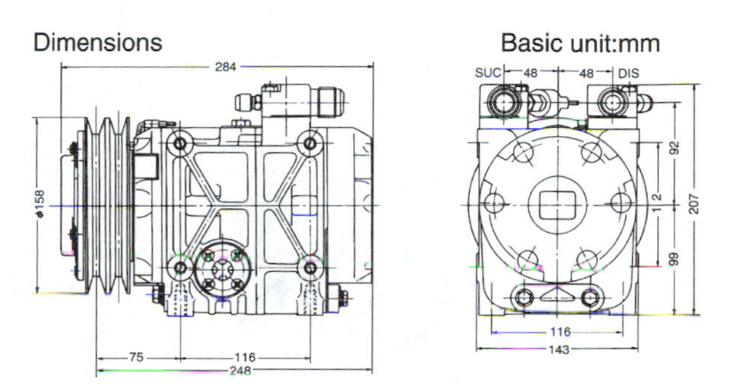

FIGURE 6–18 This Zexel compressor uses a swashplate design. (*Courtesy of Zexel USA Corporation*)

but the wobble plate does not rotate and drives single pistons through piston rods. Bearings are used between the rotating drive plate and the wobble plate. Wobble-plate compressors commonly use five or seven cylinders (Figure 6-19).

General Motors used a 5-cylinder wobble-plate compressor from 1955 to 1961. This compressor was rather large and quite heavy and was replaced by the A-6. Compressors have evolved steadily through the years, becoming smaller, lighter units (Figure 6-20). GM is currently producing a variable displacement, 5-cylinder wobble-plate compressor, the V-5.

Sanden Corporation has produced compressors of this type since the mid 1970s, also under the names Abacus and Sankyo. The SD-505, 5.3 cu. in. (87 cc); SD-507, 6.59 cu. in. (108 cc); and SD-508, 8.4 cu. in. (138 cc) are all 5-cylinder units designed for automotive use. The SD 508-HD, 8.4 cu. in. (138 cc), and SD 510-HD, 9.8 cu. in. (161 cc), are 5-cylinder units designed for trucks and agricultural and industrial equipment. The SD-708, 7.8 cu. in. (129 cc), and SD-709, 9.4 cu. in. (155 cc), are 7-cylinder units designed for automotive use. Sanden compressors, like those of other manufacturers, are manufactured with a variety of mounting points, clutches, and cylinder heads to suit particular installations. Varieties of these

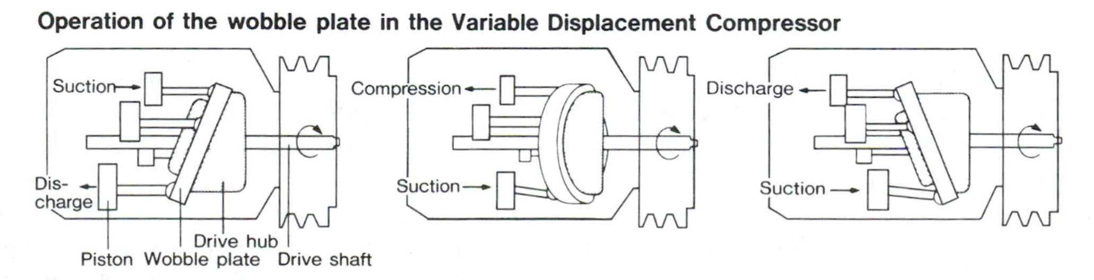

FIGURE 6–19 Rotation of the drive hub causes the wobble action of the wobble plate and forces the single pistons to move through their strokes. (*Courtesy of Zexel USA Corporation*)

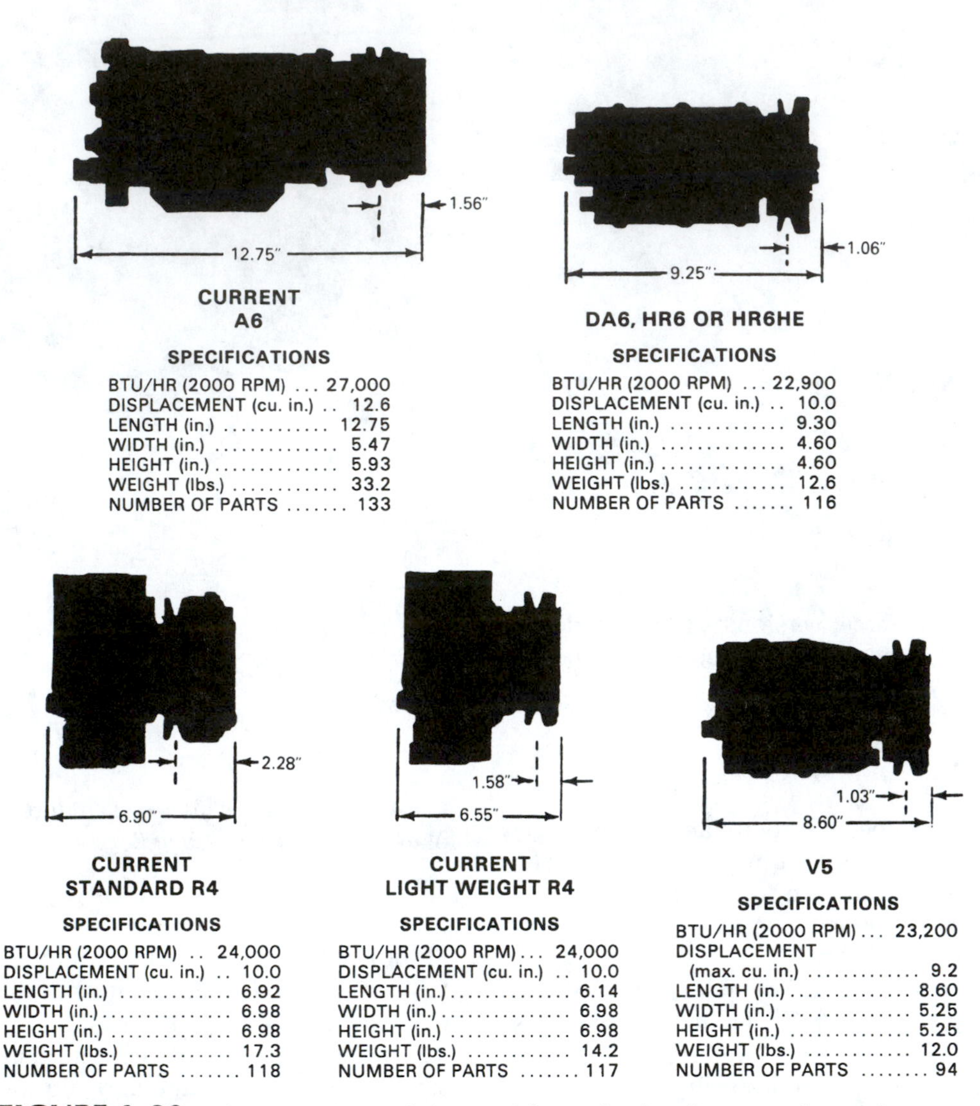

FIGURE 6–20 A comparison of size, weight, and other features shows the evolution of General Motors compressors. (*Reprinted with permission of General Motors Corporation*)

compressors have been used for OEM installation by domestic and import vehicle manufacturers and in aftermarket installations.

6.2.1.6 Variable Displacement Wobble-Plate Compressors Modern wobble-plate designs can be variable displacement compressors. When there is a low cooling load at the evaporator, the wobble plate is moved to a less angled position. This feature makes the compressor more efficient by reducing the drive load when it is not needed; it also eliminates the need to cycle the compressor off and on. Compressor displacement is the only control for preventing evaporator icing (Figure 6-21).

Wobble-plate angle is determined by the relative pressures at each end of the pistons; the exact angle is controlled by changing the pressure in the crankcase. When cooling load calls for high output and maximum displacement, crankcase pressure is kept low, and the wobble plate is at its maximum angle. The control valve bleeds crankcase pressure into the compressor suction cavity to lower the pressure. When cooling demand lessens, the control valve closes the bleed to the suction cavity and opens a passage between the dis-

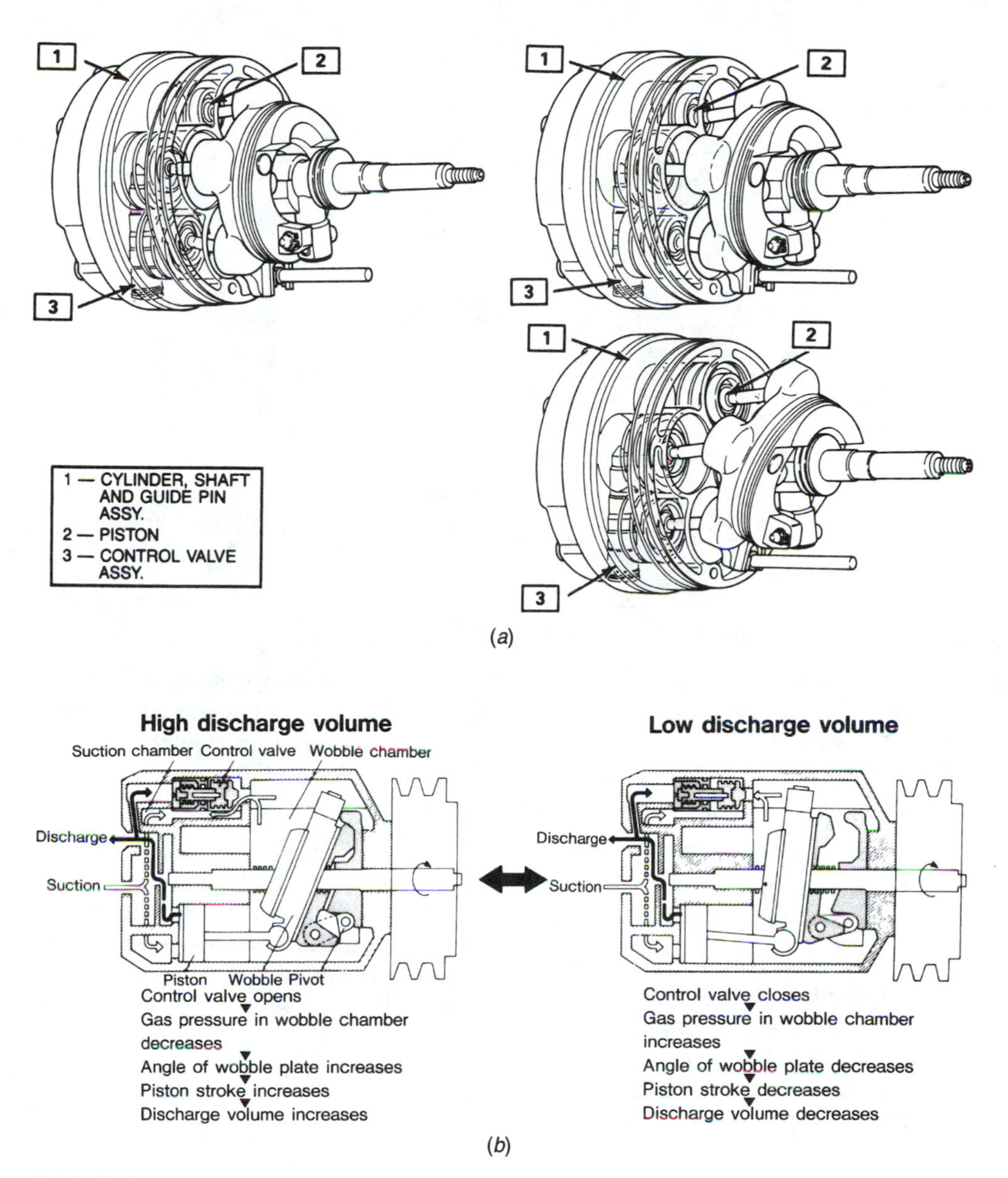

FIGURE 6–21 A variable displacement compressor can change the angle of the wobble plate and piston stroke (*a*). This angle is changed by a control valve that senses evaporator pressure, which, in turn, changes wobble chamber pressure (*b*). (a *is reprinted with permission of General Motors Corporation;* b *is courtesy of Zexel USA Corporation*)

charge cavity and the crankcase, raising the pressure. Increasing crankcase pressure raises the pressure on the bottom side of the pistons and causes the wobble plate to move to low angle, reducing displacement (Figure 6-22).

In 1985, GM began installing a variable displacement compressor called the V-5 on certain vehicle models. This compressor is lightweight, at 12 lb (5.4 kg) (the 5-cylinder of the 1950s weighed 58 lb), and has a displacement of 0.6 to 9.2 cu. in. (10 to 151 cc) (Figure 6-23). Compact variable compressors (CVC) are currently available in 6- and 7-cylinder versions in maximum displacement sizes from 7.6 to 11.2 cu. in. (125 to 185 cc). The GM compact variable compressor, CVC, uses a variable-angle swash plate and single-ended pistons. Some versions use electronic control of compressor output.

Other manufacturers, including Calsonic, Sanden, and Zexel, are currently producing variable displacement, coaxial wobble-plate compressors.

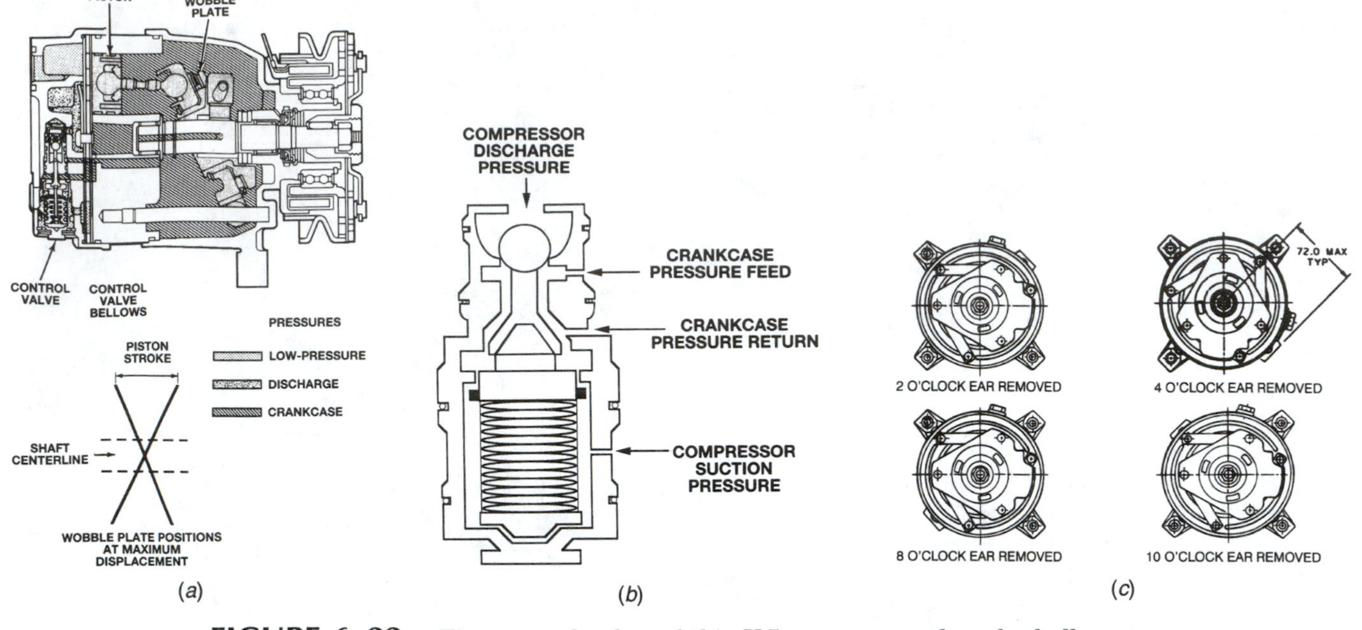

FIGURE 6–22 The control valve of this V-5 compressor has the bellows exposed to evaporator pressure (*a*). It will increase crankcase pressure when evaporator pressure drops, which, in turn, will reduce compressor displacement (*b*). Some of the model varieties are shown in c (a *and* b *are reprinted with permission of General Motors Corporation; c is courtesy of ACDelco*)

✓SERVICE TIP

A technician becomes concerned with the compressor make and model when he needs to obtain a replacement for a failed compressor. Replacement compressors are available as new or rebuilt units, and proper identification is made from the vehicle make, model, and engine size. Then, if needed, proper identification is made by the old compressor make and model (Figure 6-24). At times, a failed compressor is replaced with a different compressor make and model if the mounting points, clutch diameter and belt position, and line fittings are the same.

6.2.3 Compressor Clutches

An important part of every automotive compressor is the clutch, which allows us to easily turn it on or off. Magnetic clutches are used; they have a clutch coil where a magnetic field is generated when electricity is sent through it. The magnetic field pulls the drive plate against the rotating pulley to drive the compressor (Figure 6-25).

The clutch on most 2-cylinder in-line and V compressors is a unit with both the pulley and the drive plate mounted on the compressor shaft. On a few early clutches, called **rotating field**, the field coil was built into the pulley, and brushes were used to conduct electricity to the coil. Most units are **stationary field** clutches, with the pulley and field coil mounted on the front of the compressor. A cavity in the pulley fits over the coil closely so it attracts the magnetic lines of flux (Figure 6-26).

Most modern clutches are three-piece units. The clutch coil and pulley are both mounted on an extension from the front of the compressor housing; the **drive plate** is attached to the compressor shaft. The drive plate is also called an **armature** or **disc**, and the **pulley** is also called a **rotor**. The pulley is mounted on a bearing on an extension of the front head. This placement allows the side load of the drive belt to be absorbed by the pulley bearing and compressor housing. It also allows easier servicing of individual clutch parts (Figure 6-27).

A variety of pulleys can be used on a compressor model. Important details to note when replacing a clutch assembly, pulley, or compressor are the type of drive belt (V or V ribbed), pulley diameter, number of V grooves, and position of V grooves.

Clutches used in some of the more modern systems have been redesigned to develop greater holding power, or torque capacity, to help prevent slipping. Modern clutches can transfer about 100 ft lb of torque, double that of early clutches. This is more necessary with R-134a because of the higher head pressure encountered.

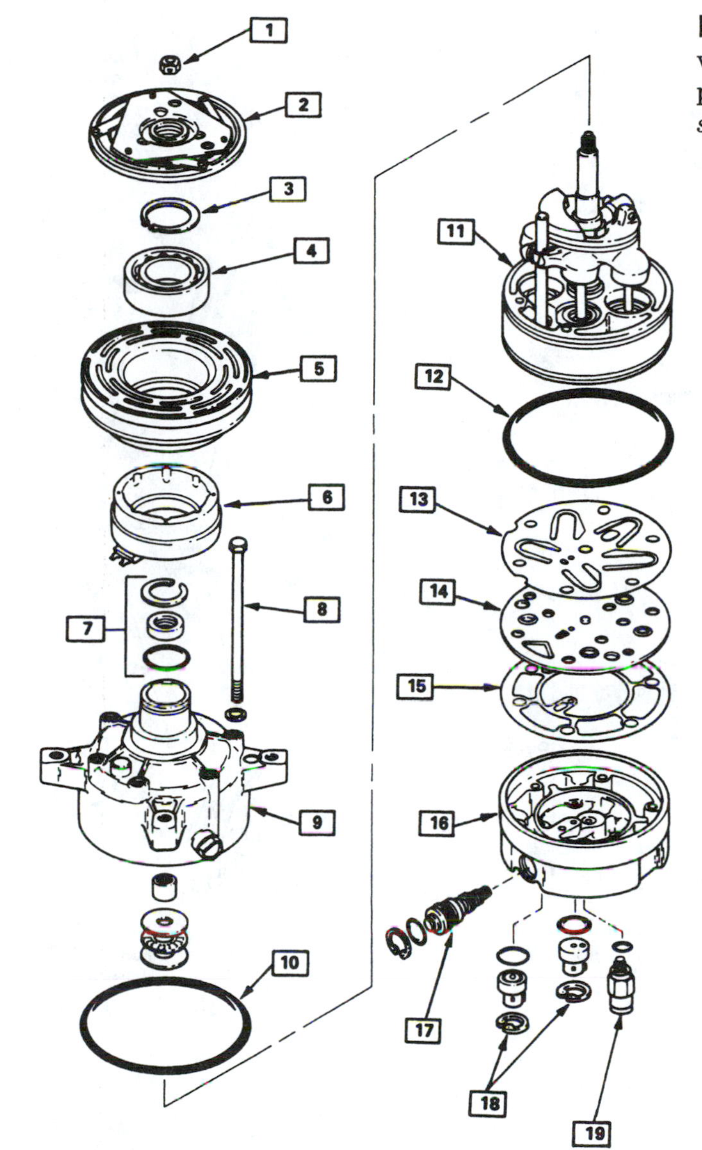

FIGURE 6–23 An exploded view of a General Motors V-5 compressor. (*Reprinted with permission of General Motors Corporation*)

FIGURE 6–24 The decal on this compressor identifies the type (SDB709) and the serial number. Note also that it uses a seven-groove, multi-V clutch, four mounting bolts, and vertical-pad service ports at the side.

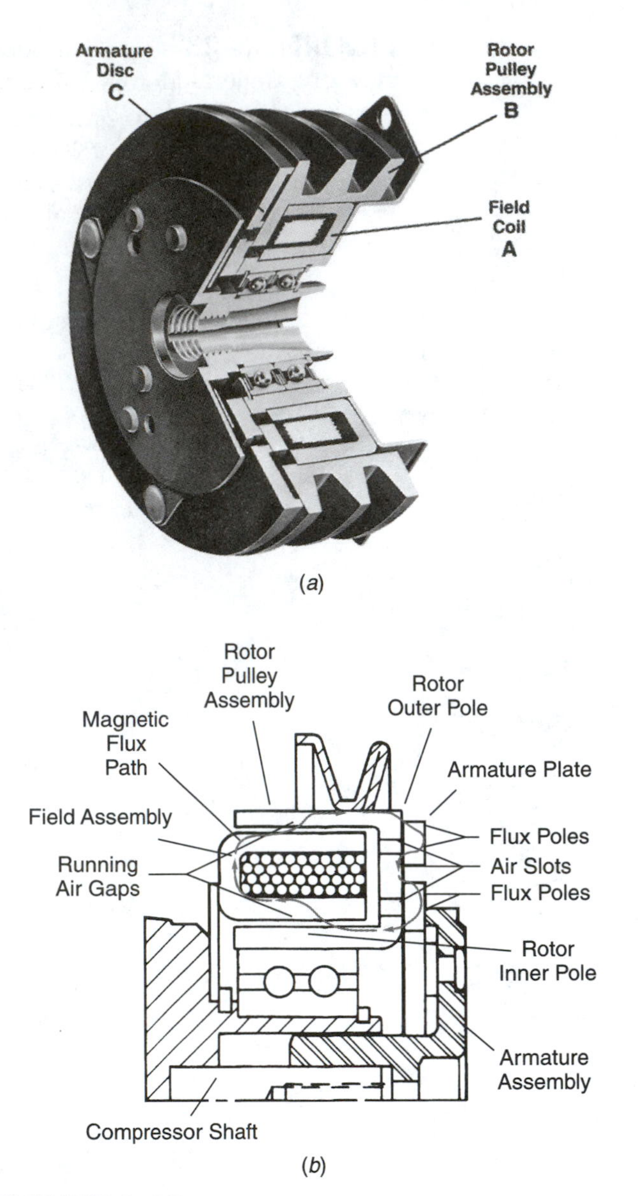

FIGURE 6–25 Cutaway views of a clutch assembly showing its parts (a) and the magnetic flux path (b). (a *is courtesy of Warner Electric;* b *is reprinted with permission of General Motors Corporation*)

Some design factors used to increase holding power are the number of **flux poles** (slots in the face of the clutch armature), the diameter of the rotor and armature, the use of copper or aluminum in the clutch coil (copper produces about 20% greater torque capacity), and the current draw of the coil (Figure 6-28).

Some modern clutches are designed to act as a sort of fuse to protect the compressor. When system problems cause increased compressor pressure loads to the point where damage might occur, the clutch slips for a longer period of time during engagement. This slippage generates heat, and in some designs, the added heat

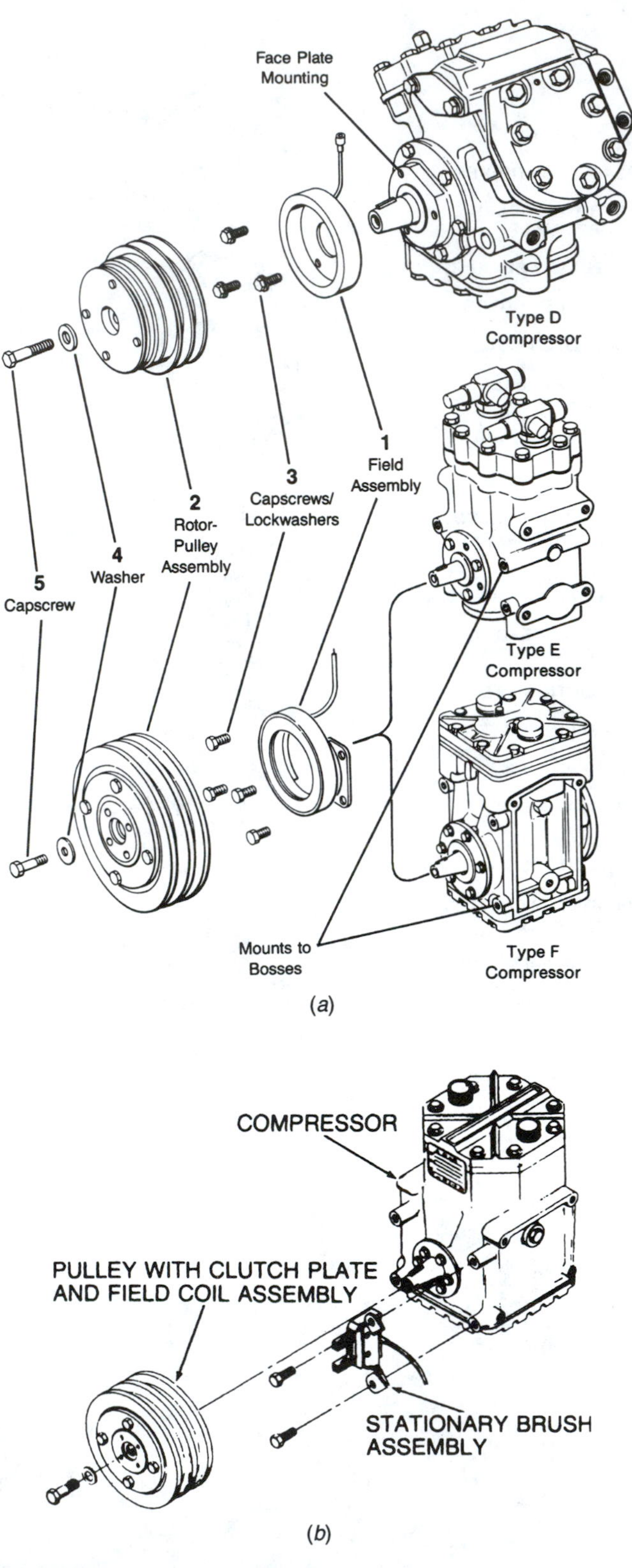

FIGURE 6–26 In most clutches, the coil is stationary, secured to the compressor (a). In a rotating field clutch, the coil is built into the rotor, and a brush assembly is used to conduct electricity into and out of the coil (b). (*Courtesy of Warner Electric*)

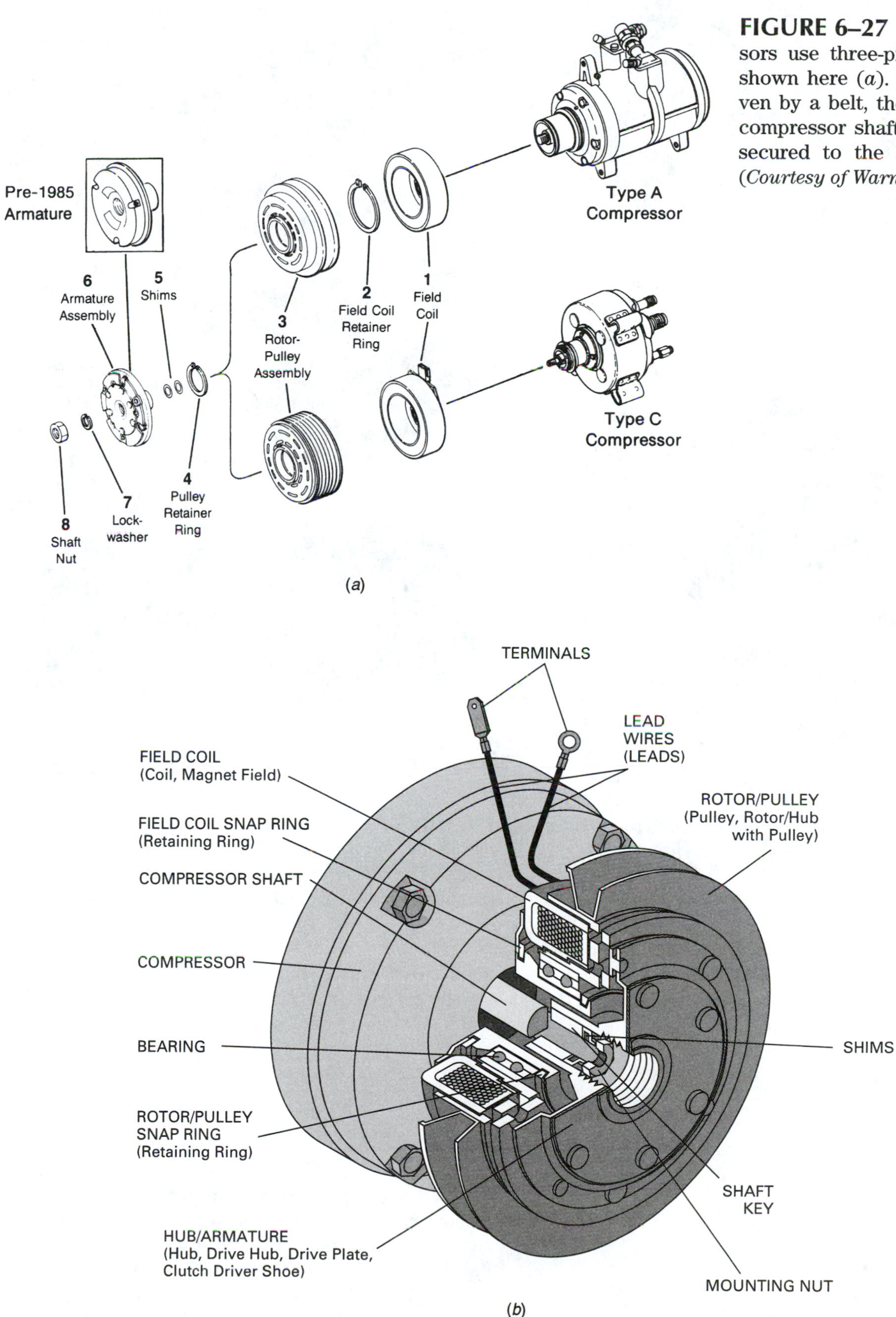

FIGURE 6–27 Most compressors use three-piece clutches as shown here (*a*). The rotor is driven by a belt, the hub drives the compressor shaft, and the coil is secured to the compressor (*b*). (*Courtesy of Warner Electric*)

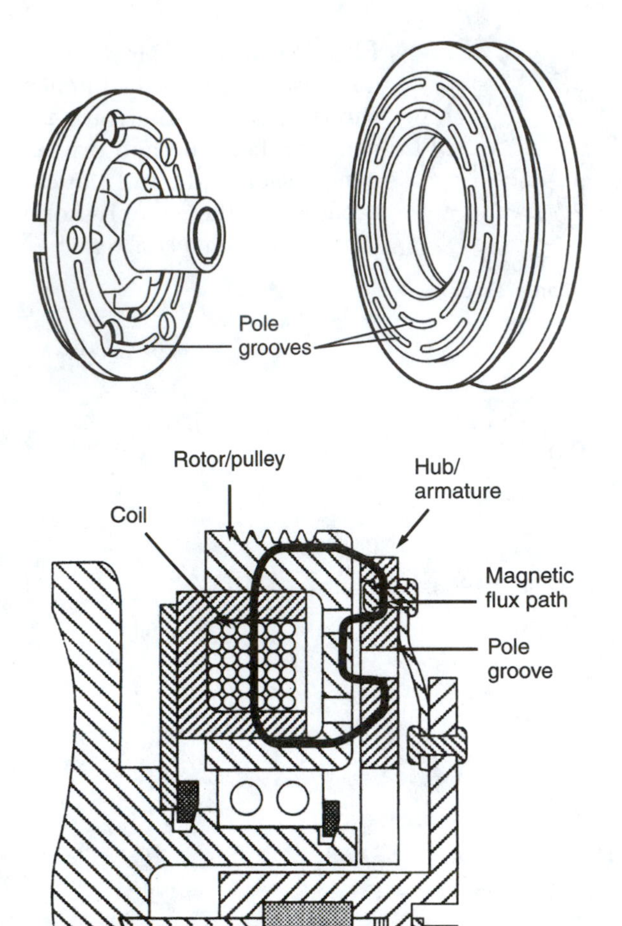

FIGURE 6–28 The magnetic flux path is from the coil and through the metal of the rotor and clutch hub. When it meets a pole groove, it travels from the hub to the rotor or vice versa, which increases the clutch holding power.

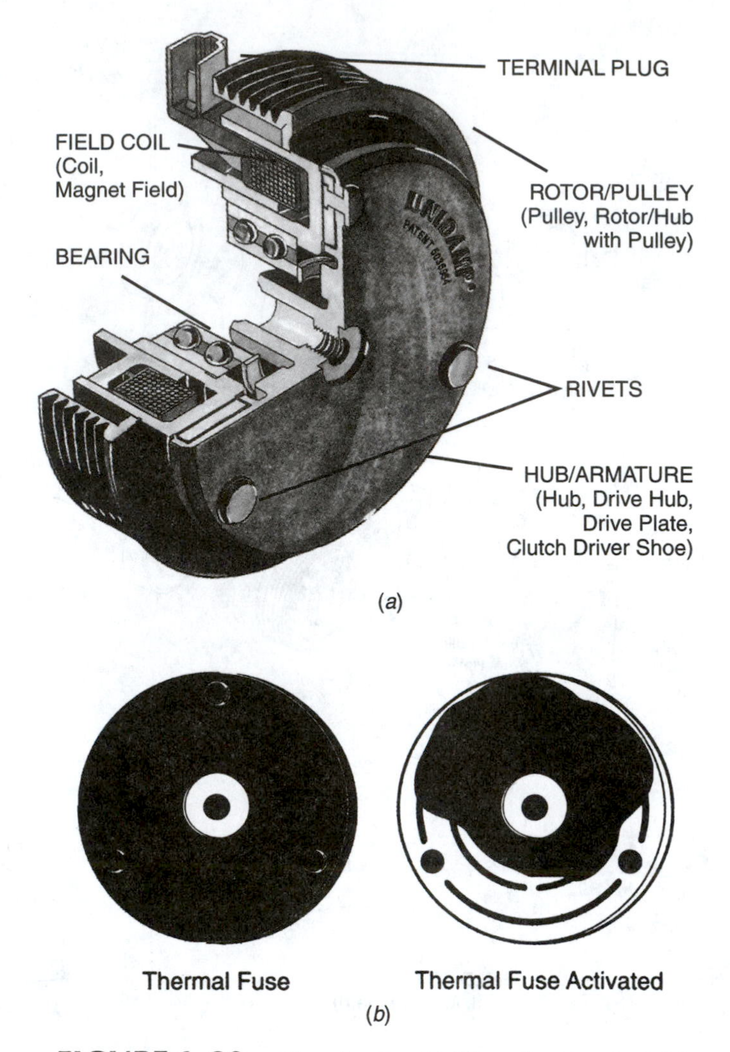

FIGURE 6–29 The plastic shield on the front of this clutch hub (*a*) is a thermal fuse; if it gets too hot, it will melt and cause the clutch to fail before compressor damage occurs (*b*). (*Courtesy of Warner Electric*)

melts the **thermal fuse** in the drive plate (Figure 6-29). In clutches that do not use the thermal fuse, this added heat often causes bearing failure.

The variable displacement compressor used on the new C-class Mercedes Benz does not use a clutch. The compressor is electronically controlled to go to minimum displacement of 2% output when A/C is not used; this displacement requires very little power and is enough to circulate oil through the moving parts. The pulley drive plate includes a rubber shear portion that can break to protect the drive belt in case the compressor should fail and lock up.

6.2.4 Compressor Shaft Seal

Another important part of every compressor is the seal that keeps refrigerant from escaping through the opening where the drive shaft enters. Many compressors use a **rotating seal cartridge** attached to the drive shaft and a **stationary seal seat** attached to the front of the compressor housing. Many compressors have one or two flats on the shaft so that the cartridge is positively driven (Figure 6-30).

A gasket or rubber O-ring is used so that the seal seat makes a gas-tight seal at the housing, and the seat has an extremely smooth sealing face. The carbon-material sealing member is spring loaded so that its smooth face makes tight contact with the seal seat. The cartridge also uses a rubber O-ring or molded rubber unit to seal the carbon to the driveshaft. Another important part of the seal is the compressor oil, which lubricates the surfaces and forms the final seal between the sealing surfaces (Figure 6-31).

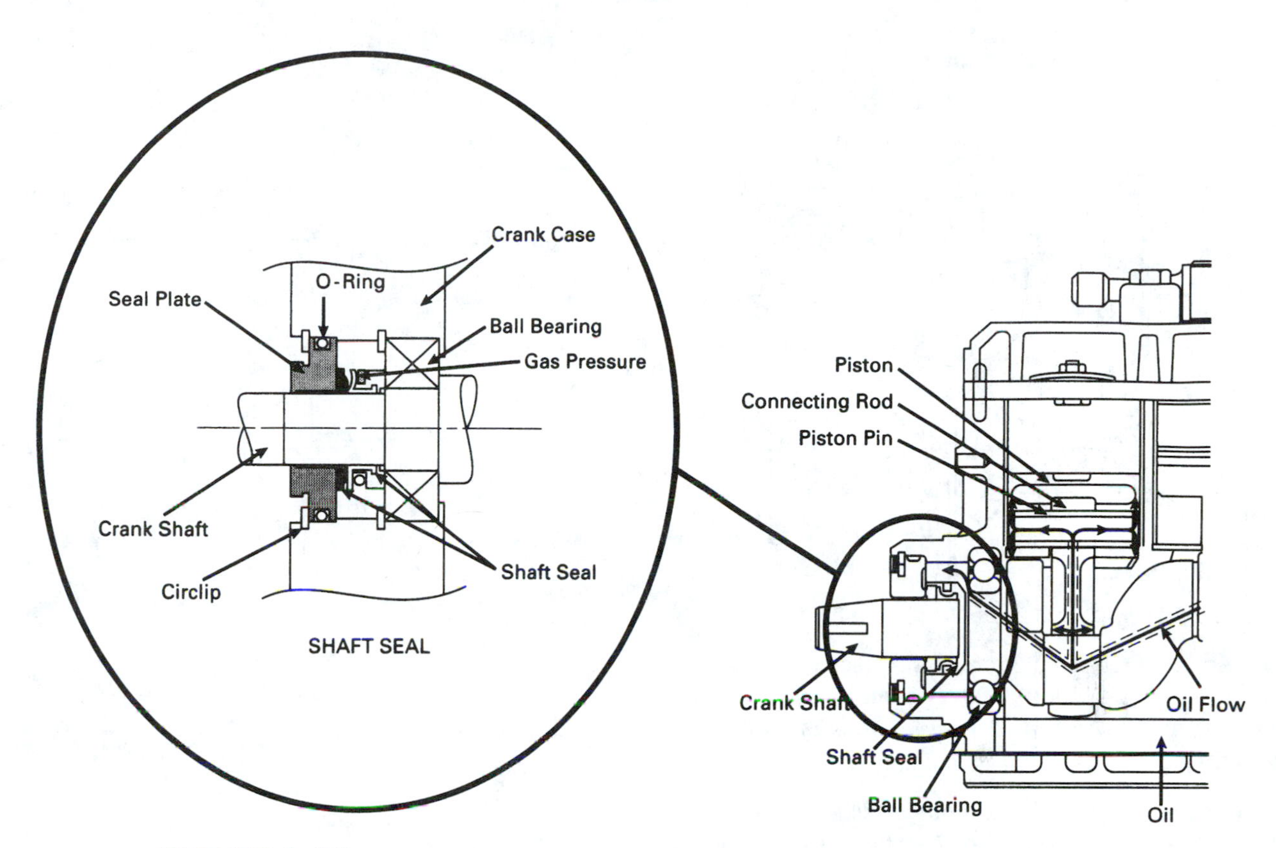

FIGURE 6–30 The shaft seal must keep refrigerant from escaping out the front of the compressor. Most compressors have an oil flow routed to them to reduce wear and improve the sealing action. (*Courtesy of Toyota Motor Sales*)

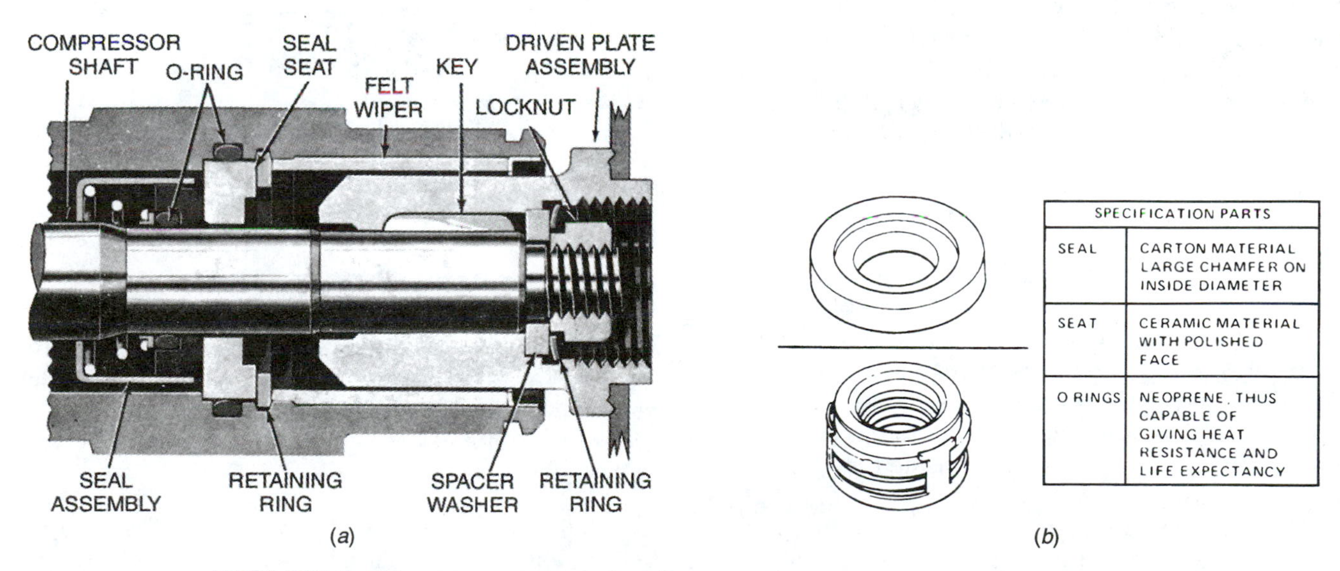

SPECIFICATION PARTS	
SEAL	CARTON MATERIAL LARGE CHAMFER ON INSIDE DIAMETER
SEAT	CERAMIC MATERIAL WITH POLISHED FACE
O RINGS	NEOPRENE, THUS CAPABLE OF GIVING HEAT RESISTANCE AND LIFE EXPECTANCY

FIGURE 6–31 Many shaft seals (*a*) use a stationary plate that is sealed and secured to the compressor; the shaft seal has a carbon ring against the seal plate and a rubber ring to seal to the shaft (*b*). (*Reprinted with permission of General Motors Corporation*)

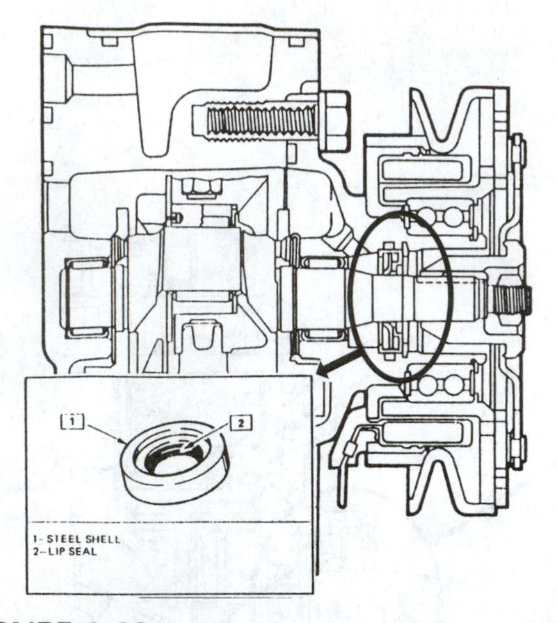

FIGURE 6–32 This compressor uses a lip seal (inset). The lip of the seal fits tightly against a smooth portion of the shaft. (*Reprinted with permission of General Motors Corporation*)

A **ceramic** material has replaced **cast iron** for the seal seat in some compressors. Ceramic is not affected by water or acids, which can cause rust, corrosion, or etching of the iron seats. Ceramic seats are easy to identify because they are white instead of gray.

Some newer compressor designs use a **lip seal**. The lip of the seal is made from Teflon and rides against a smooth portion of the driveshaft. Some shaft seals use double sealing lips. The outer shell of the seal fits into a recess in the compressor housing and is sealed using a rubber O-ring. Gas pressure in the compressor ensures a tight fit between the seal lip and the shaft (Figure 6-32).

6.2.5 Compressor Lubrication

Refrigerant oil serves several purposes, the most important being to **lubricate** the moving parts of the compressor to reduce friction and prevent wear. Refrigerant oil also helps **seal** the compressor shaft seal, the insides of the hoses, and various connections between the parts to reduce refrigerant leakage. In addition, it lubricates the TXV and coats the metal parts inside the system to reduce corrosion (Figure 6-33).

The oil used in a system must be completely compatible with all the materials in the system. It also must be **miscible** or **soluble** with the refrigerant so that it will be circulated throughout the system (Figure 6-34). There are essentially three types of oil: **mineral oil** is used in R-12 systems, PAG (**polyalkylene glycol**) oils are used in most OEM R-134a systems, POE **(Polyol Ester)(ester)** oils are used in a few OEM R-134a systems, and either ester or PAG oils are used in R-12 systems retrofitted to R-134a. PAG and POE are synthetic, human-made oils. Most compressor remanufacturers consider PAG oil a better lubricant than ester oil.

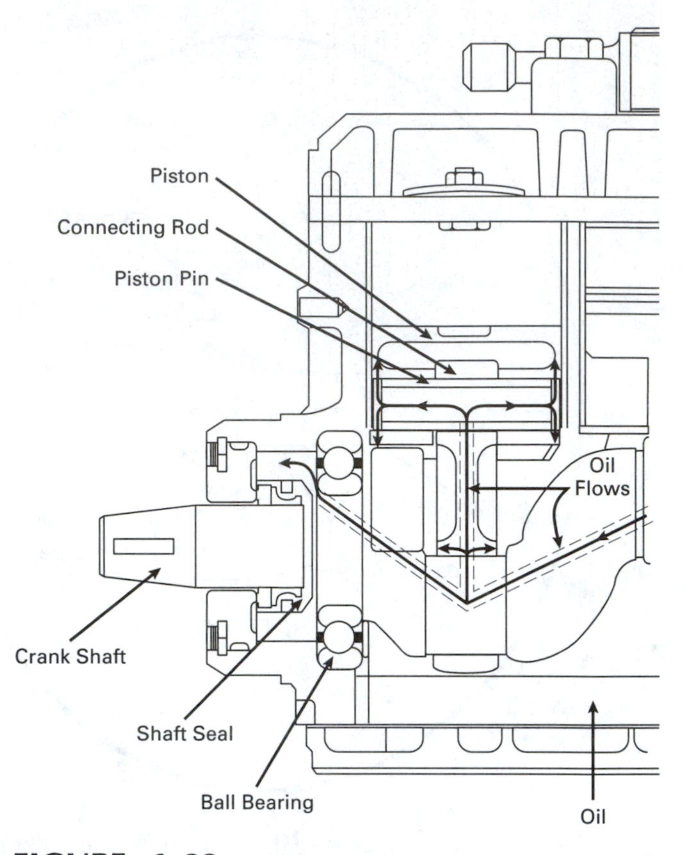

FIGURE 6–33 Older compressors pumped oil through passages in the crankshaft to lubricate moving parts. (*Courtesy of Toyota Motor Sales*)

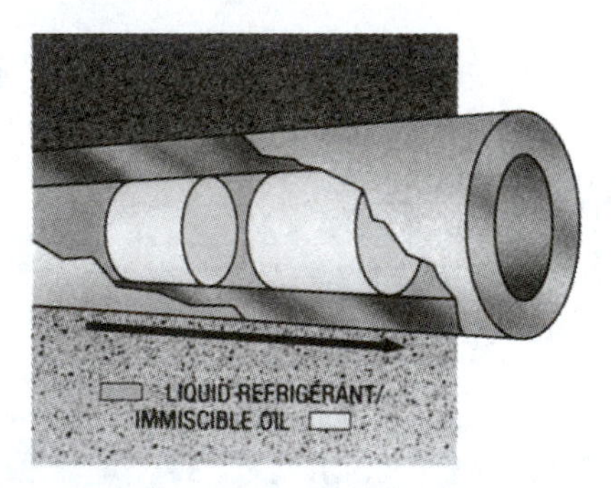

FIGURE 6–34 If the oil is immiscible with the refrigerant, as shown here, it can separate out of the refrigerant. Poor lubrication will result. (*Courtesy of Castrol North America*)

✓SERVICE TIP

It is important to keep oil containers closed so that the oil does not absorb water from the atmosphere. A mineral oil can absorb about 0.005% water by weight. A PAG or ester oil is very **hygroscopic** and can absorb about 2 to 6% water (Figure 6-35). Some synthetic oils undergo hydrolysis if exposed to too much water and revert back to their original components: acid and alcohol.

As mentioned earlier, mineral oils are not compatible with R-134a, and PAG oil is not compatible with R-12.

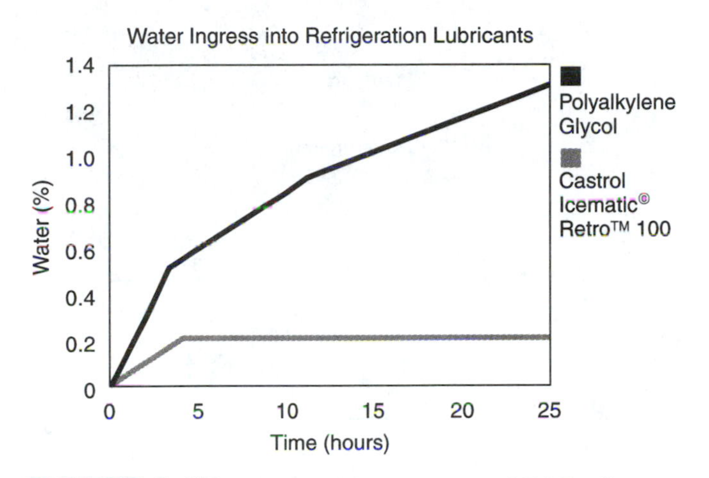

FIGURE 6–35 Some refrigerants and PAG oils are very hygroscopic and absorb moisture more rapidly than other oils, such as the polyol ester Icematic. *(Courtesy of Castrol North America)*

Since R-12 and mineral oil are completely compatible, another important retrofitting step is a thorough system evacuation to ensure that all of the R-12 has a chance to boil out of the oil; this is called **outgassing**. Ester oil is more chlorine tolerant than PAG, which many people believe makes it more suitable for retrofitting R-12 systems.

Manufacturers designate specific blends of oil for R-12 and R-134a systems. There are about 30 varieties of refrigerant oil whose use depends on the requirements of a particular compressor or system. Vane compressors often require a higher-viscosity oil than piston compressors to maintain proper lubrication and a seal at the tips of the vanes.

Most early compressor designs had a sump where oil was stored and used an oil pump to circulate oil to critical areas. Checking the oil level was a normal service operation on these compressors. A dipstick was usually used when systems had service valves, or the oil-level plug could be opened slightly on A-6 compressors to check the level. Newer compressors use internal pressures to circulate the oil. These compressors do not have a sump, but they normally contain a certain amount of oil (about 6 to 13 oz). The oil level on newer compressors can only be checked by removing the compressor and draining all of the oil, then measuring how much oil came out. This is normally done when a compressor is replaced or when major service is performed on the system.

Refrigerant oils are commonly available in several different viscosities: 46, 100, and 150 (Figure 6-36). These are International Standards Organization (ISO) viscosities, measured at 40°C. Viscosity is a measurement of

Refrigerant Oil Types		*Refrigerant Type*	*Viscosities*
Mineral	Mineral Oil	R-12	300, 500/525 (SUS) (about 68 and 100 cSt)
PAG, Polyalkylene glycol	Synthetic	R-134a	46, 100, 150 @ 40° C (ISO)
POE, Polyolester	Synthetic	R-134a	100, 150 @ 100° C (ISO)
PolyAlpha + (ROC 68)	Synthetic	R-12/R-134a	68.2 @ 40° C (cSt)

Note: Most vehicle and compressor manufacturers recommend the use of PAG oil of the proper viscosity.
Note: Oil viscosity is measured using different methods; SUS: Saybolt Universal Seconds, ISO: International Standards Organization, cSt: Centistokes.

FIGURE 6–36 The major types of refrigerant oil, the refrigerant it is used with, and the viscosities in which it is commonly available.

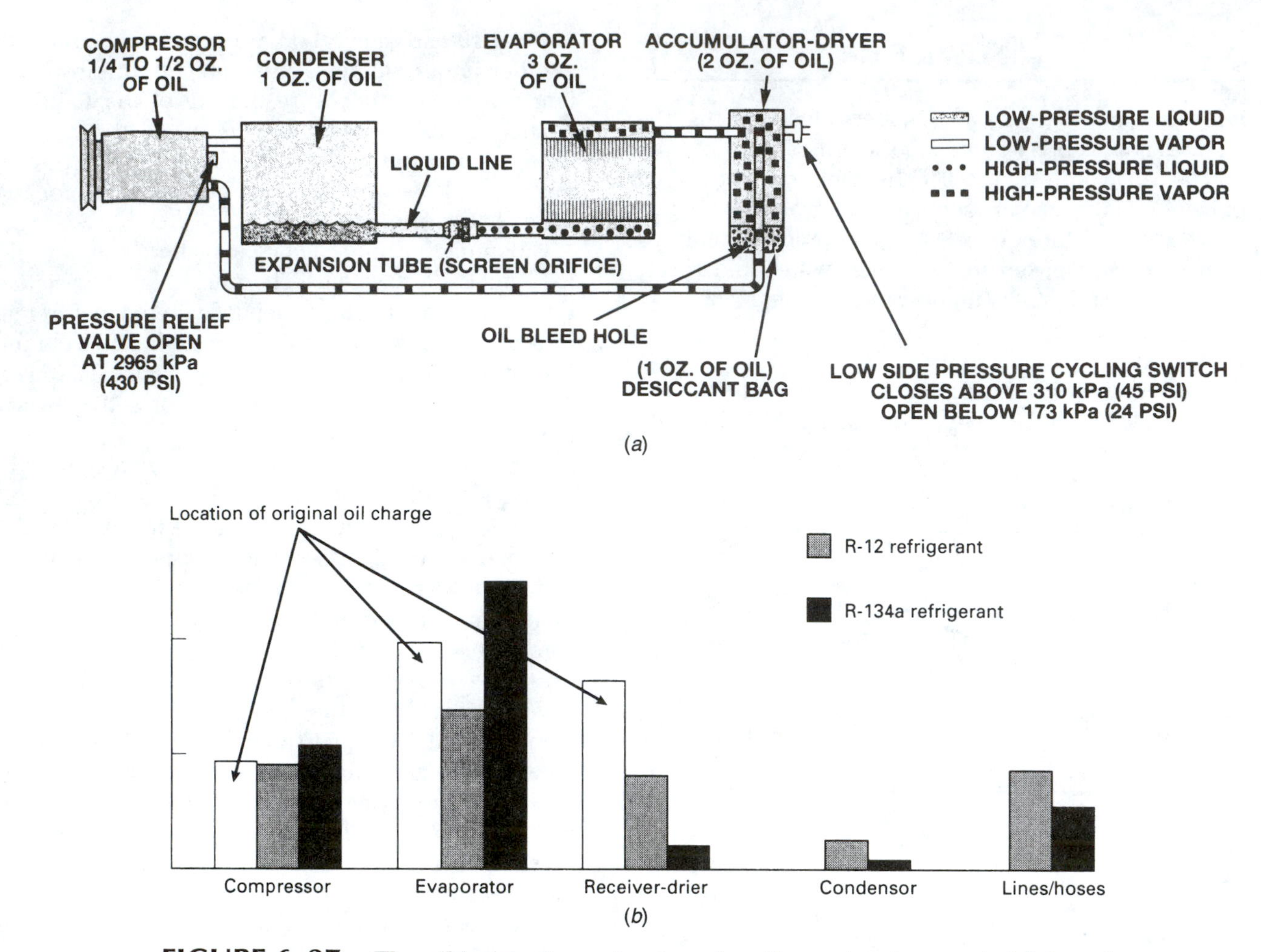

FIGURE 6–37 The oil in a system migrates when the system is operated (*a*), and the migration is slightly different in R-12 and R-134a systems (*b*). (a *is reprinted with permission of General Motors Corporation*)

how thick the oil is and how easily it flows. The oil must be thick enough so that moving parts float on an oil film; wear occurs when moving parts rub against stationary parts. Oil must also be fluid enough to flow into the tiny spaces between parts. Compressor manufacturers normally specify the oil type and viscosity for each type of compressor.

When a system operates, mineral oil is absorbed by R-12 refrigerant and migrates through the system. It does not stay in any particular component, but (as shown in Figure 6-37) a certain amount of oil can be expected in each component, with a difference between the R-12 and R-134a systems. Oil also migrates while a system is shut off because of temperature changes of its various parts. Consequently, it is good practice, if conditions permit, to operate a system until normal conditions occur before checking oil levels.

✓PROBLEM SOLVING

A friend is concerned that something is wrong with her A/C system. It cools well, but she has found some oily dirt on the hoses. When you check it, you find two different spring-lock fittings with oily dirt on them. What is this telling you? Is it a good or bad sign?

6.2.6 Compressor Switches and Relief Valves

Many compressors contain one or more of the following: a pressure relief valve, a low- or high-pressure switch, and/or a low- or high-pressure sensor. The

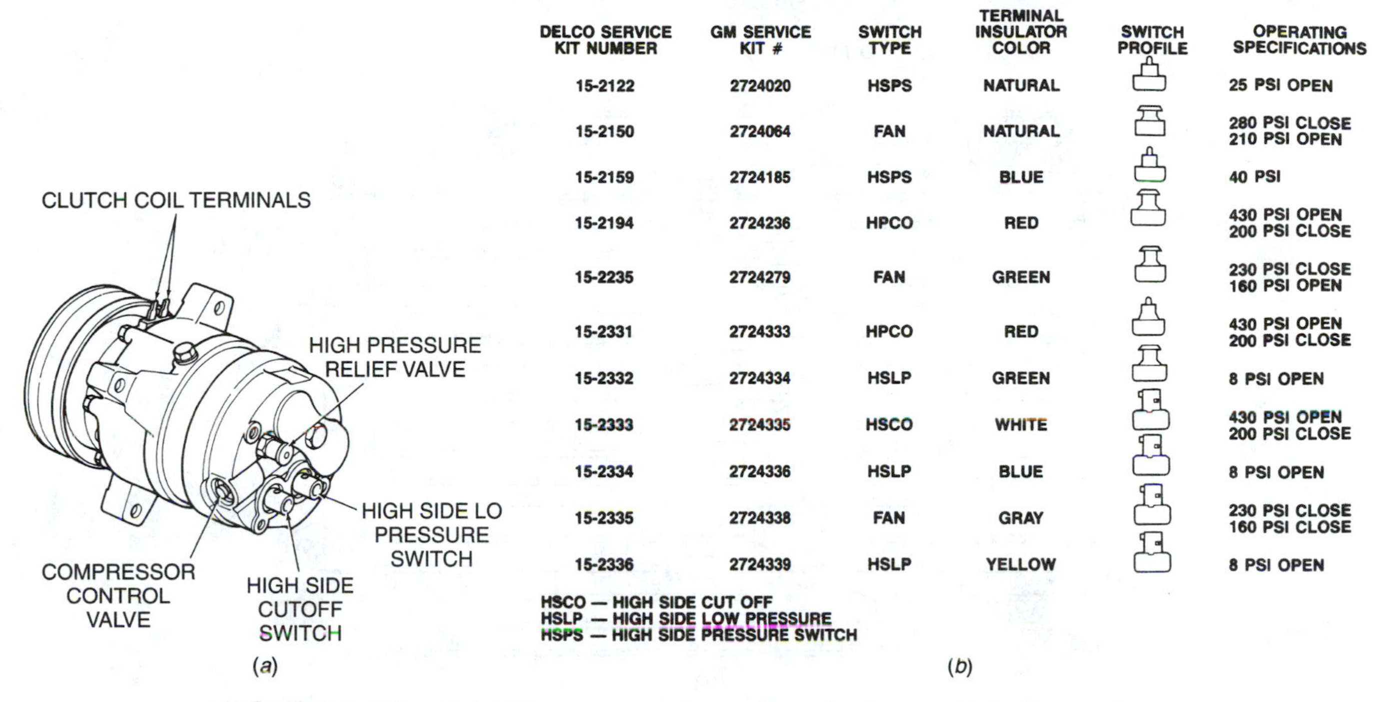

COMPRESSOR SWITCH DETAIL

DELCO SERVICE KIT NUMBER	GM SERVICE KIT #	SWITCH TYPE	TERMINAL INSULATOR COLOR	SWITCH PROFILE	OPERATING SPECIFICATIONS
15-2122	2724020	HSPS	NATURAL		25 PSI OPEN
15-2150	2724064	FAN	NATURAL		280 PSI CLOSE 210 PSI OPEN
15-2159	2724185	HSPS	BLUE		40 PSI
15-2194	2724236	HPCO	RED		430 PSI OPEN 200 PSI CLOSE
15-2235	2724279	FAN	GREEN		230 PSI CLOSE 160 PSI OPEN
15-2331	2724333	HPCO	RED		430 PSI OPEN 200 PSI CLOSE
15-2332	2724334	HSLP	GREEN		8 PSI OPEN
15-2333	2724335	HSCO	WHITE		430 PSI OPEN 200 PSI CLOSE
15-2334	2724336	HSLP	BLUE		8 PSI OPEN
15-2335	2724338	FAN	GRAY		230 PSI CLOSE 160 PSI CLOSE
15-2336	2724339	HSLP	YELLOW		8 PSI OPEN

HSCO — HIGH SIDE CUT OFF
HSLP — HIGH SIDE LOW PRESSURE
HSPS — HIGH SIDE PRESSURE SWITCH

(b)

FIGURE 6–38 This V-5 compressor has a control valve, a relief valve, and two switches mounted in it (*a*). Many of the switches used in GM compressors are shown in (*b*). (*Reprinted with permission of General Motors Corporation*)

pressure relief valve is installed at the discharge port or into the discharge cavity. This valve opens to release excess refrigerant if high side pressure gets too high.

✓SERVICE TIP

A relief valve discharging pressure produces a loud, popping noise, and the oil cloud that escapes with the refrigerant often looks like smoke. Most relief valves reclose after excess pressure has been released.

Newer systems are designed to shut the system off if pressures get too high to avoid venting refrigerant into the atmosphere (Figure 6-38).

Switches can be connected to ports, leading to either the suction or discharge cavities in the compressor. These switches are usually used in circuits either to protect the compressor or system from damage or as sensors for the engine control module. They are described in more detail in Section 6.8.

6.3 CONDENSERS

The condenser has to get rid of the heat that is moved from the passenger compartment; an automotive condenser is a simple device. Many condensers are merely a tube bent back and forth into a serpentine shape, with fins attached. Many modern condensers use a parallel flow with a manifold at each side (Figure 6-39). After the tubes are pressed through the fins, return bends or manifolds are used to connect the tubes to give the desired flow pattern. When flattened tubes are used, corrugated fins are usually attached between tube pairs. Parallel flow using flattened, extruded tubes is much more efficient than round tube and fin designs (Figure 6-40). This design allows a smaller, lighter condenser to be used or the increased efficiency to be used to reduce high side pressure and compressor load.

The condenser of most vehicles is mounted in front of the radiator; ram air is forced through it by the vehicle's forward motion. This airflow is enhanced by the engine's cooling fan and shroud between the fan and radiator. Most rear-wheel-drive (RWD) vehicles

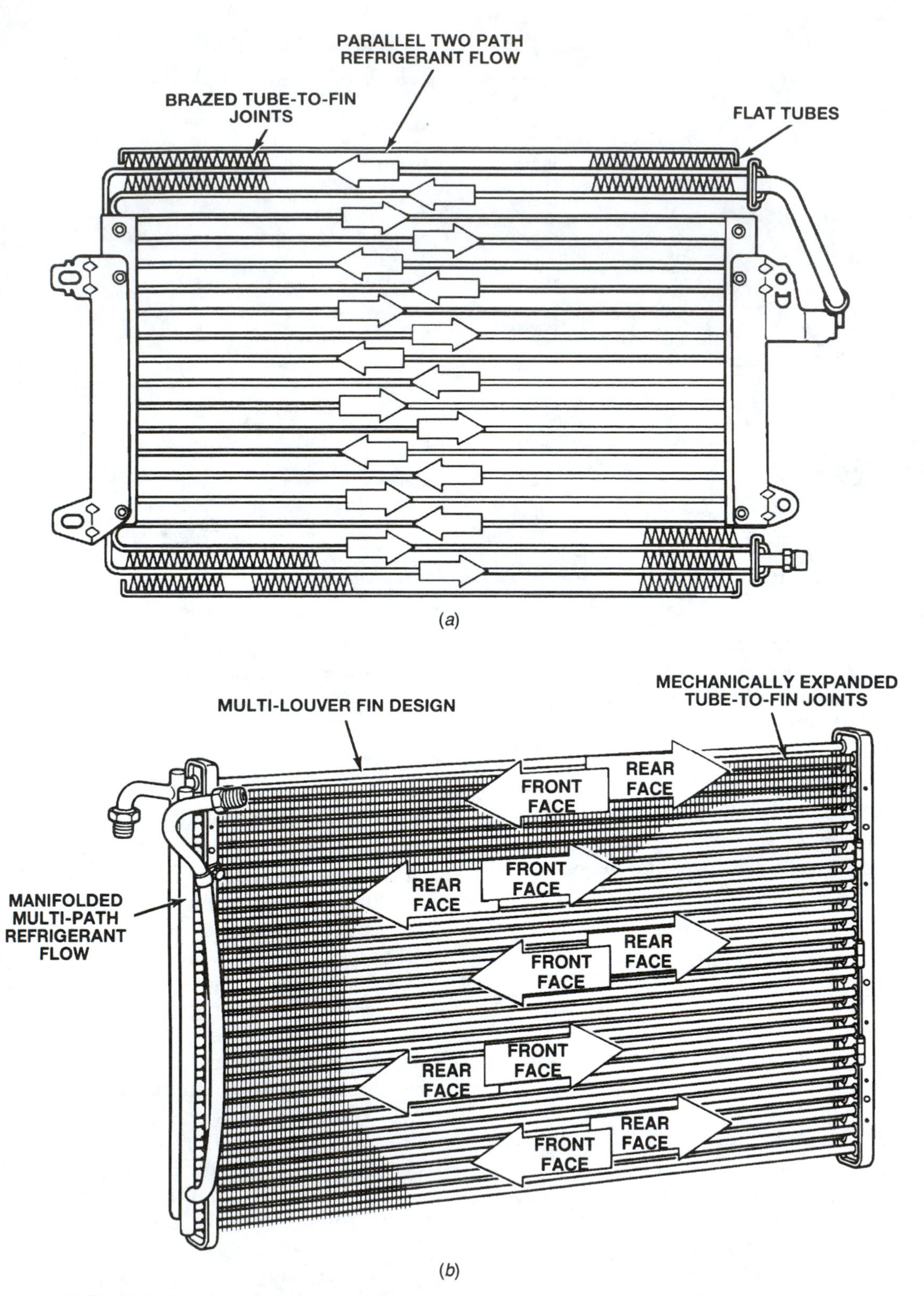

FIGURE 6–39 The serpentine condenser shown in (*a*) uses two flat tubes between the inlet and the outlet. The parallel flow condenser (*b*) uses manifolds at the sides to route the flow directions. Note that in each case, the inlet is a larger fitting feeding into the top, and the outlet is a smaller fitting feeding out of the bottom. *(Reprinted with permission of General Motors Corporation)*

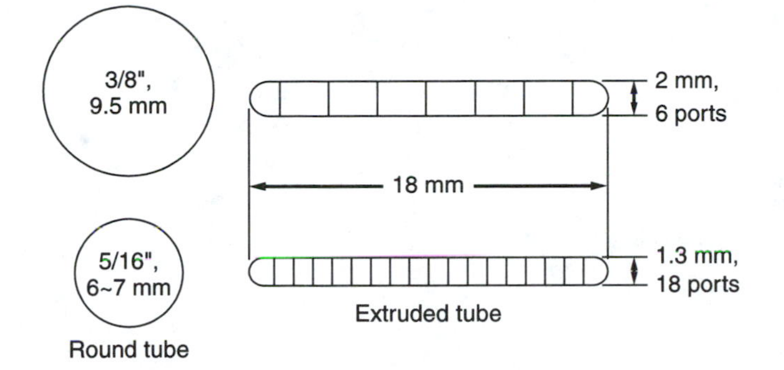

FIGURE 6–40 The common condenser tube sizes. Modern extruded tube condensers have up to 18 small ports or passages through the tube. Actual size is less than one-half of the drawing size.

use a fan driven by a fan clutch mounted on the water pump. Front-wheel-drive (FWD) cars normally use an electric motor to drive the fan. This motor is controlled by two or more switches: the engine coolant temperature switch, which turns the fan on when the coolant reaches a certain temperature, and the A/C control switch or high side pressure switch, which turns the fan on when the A/C is turned on or when the high side pressure reaches a certain point. Fan operation is described in more detail in Chapters 12 and 15 (Figure 6-41).

Many modern vehicles use foam seals around the condenser to block air flow past the condenser (Figure 6-42). This ensures that all of the air flow entering through the front opening is forced to go through the condenser.

6.4 EXPANSION DEVICES

All early A/C systems used a TXV to control refrigerant flow into the evaporator. Many modern systems use an OT to meter this same flow.

6.4.1 Thermal Expansion Valves

A TXV is designed to allow a maximum flow of refrigerant, but this flow must have a few degrees of **superheat** at the evaporator tailpipe. The superheat ensures that all of the liquid has boiled. A TXV **modulates** the flow into the evaporator to match the heat load.

A TXV senses both temperature and pressure (Figure 6-43). Temperature is sensed by using a refrigerant charge

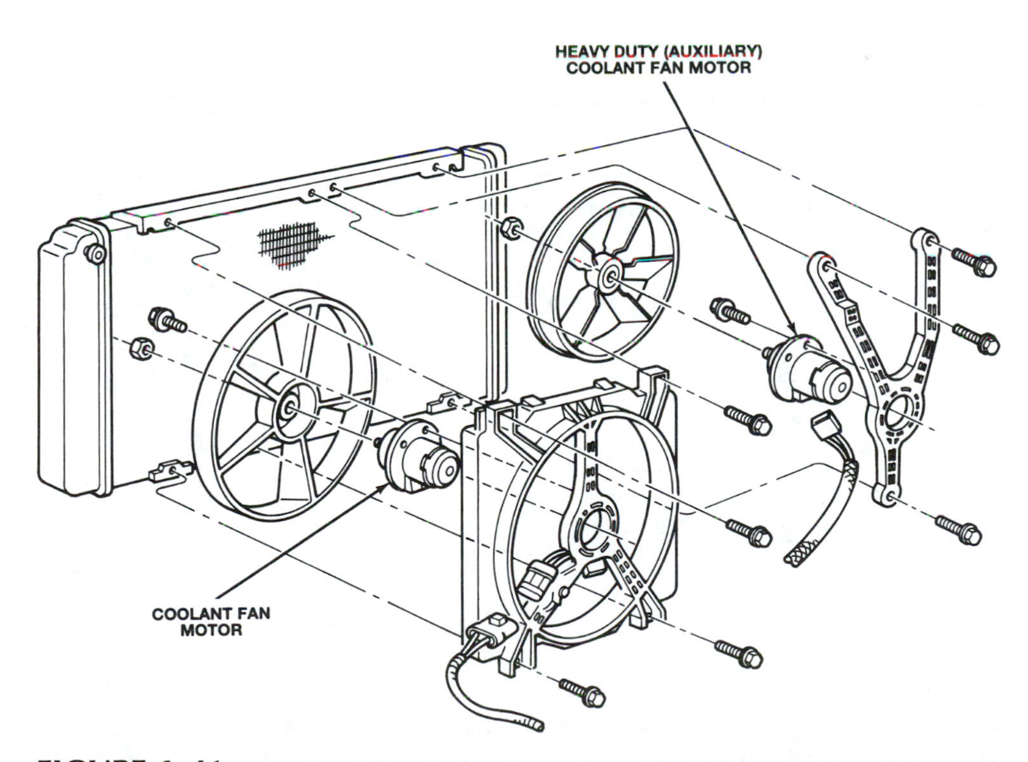

FIGURE 6–41 A pair of electric fans is used to pull air through the radiator and condenser on this FWD car. *(Reprinted with permission of General Motors Corporation)*

FIGURE 6–42 The foam seals at the front sides of this condenser force all of the air to flow through the condenser and prevent any air flow around or past it.

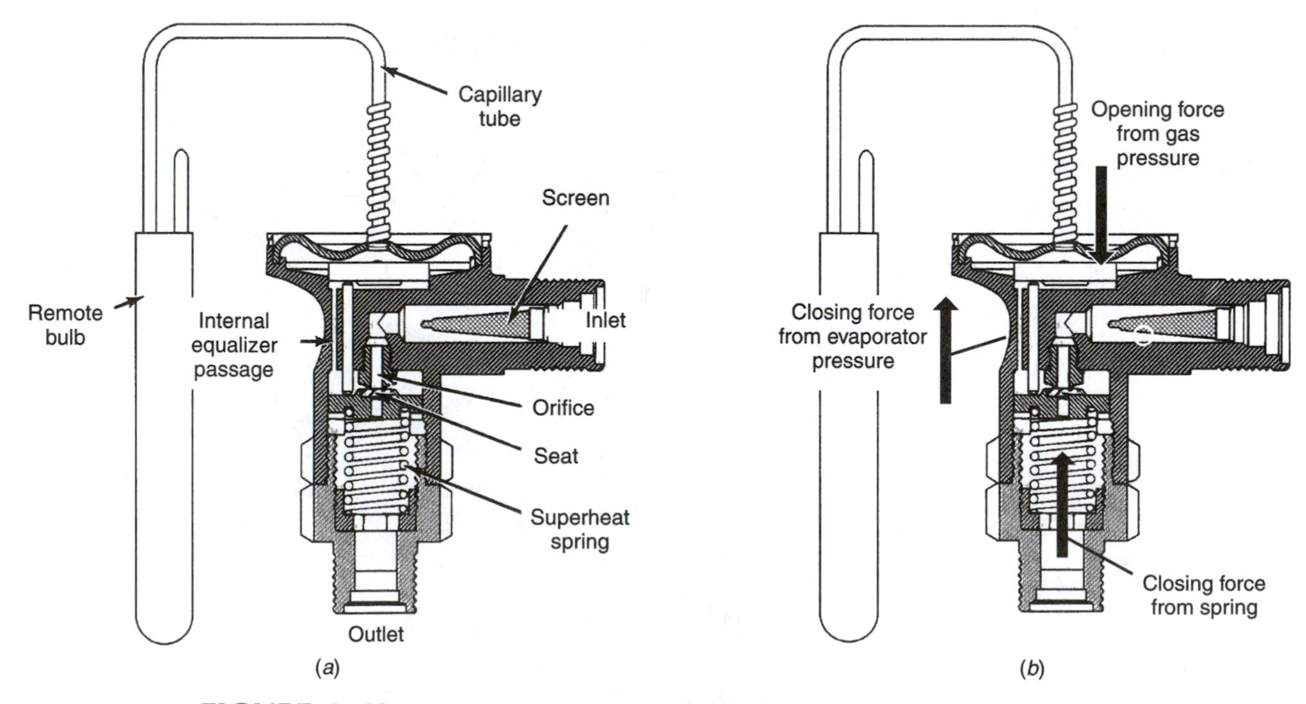

FIGURE 6–43 An internal equalized TXV. Note the internal passage to bring evaporator inlet pressure to the bottom of the diaphragm and the screen, which stops debris that might plug the valve (*a*). The valve is opened by gas pressure on top of the diaphragm and closed by pressure from the evaporator and the superheat spring (*b*). (*Reprinted with permission of General Motors Corporation*)

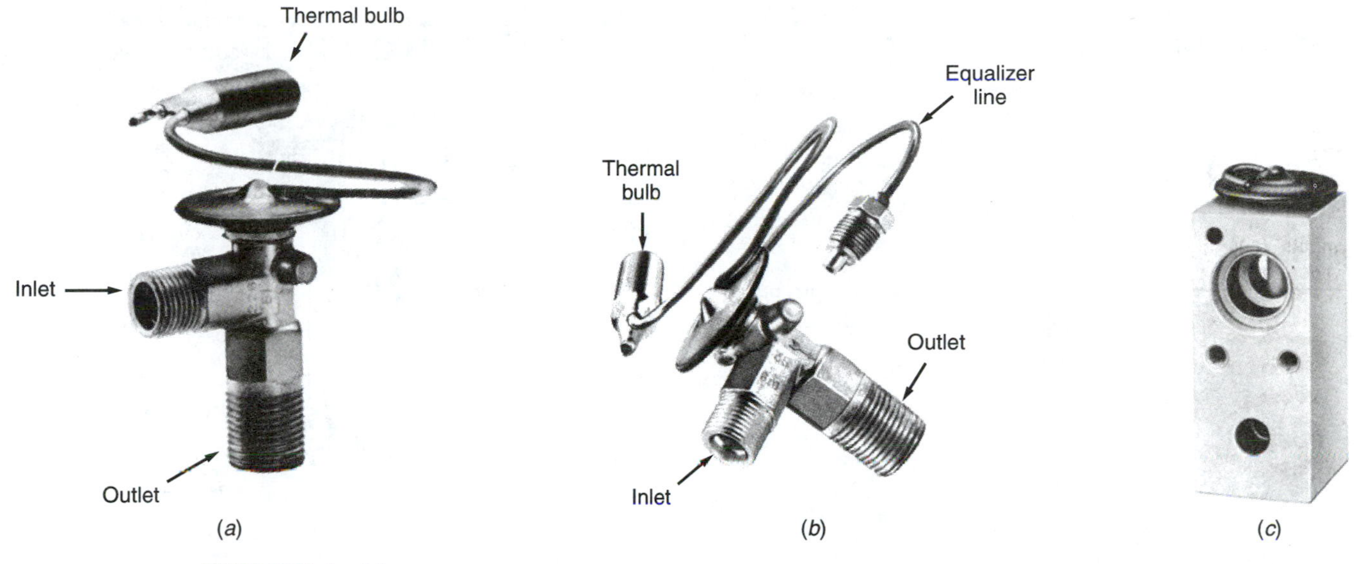

FIGURE 6–44 An internal equalized TXV has two large connectors for the liquid line and evaporator (*a*); an external equalized valve has a smaller line to connect to the evaporator outlet (*b*). A block-type valve with evaporator inlet at the bottom and outlet at the top is shown in (*c*). (*Reprinted with permission of General Motors Corporation*)

sealed in a **sensing bulb** and a **capillary tube**, a very thin metal tube connected to a metal diaphragm. The metal diaphragm is arranged so that it can push against a group of three pins that, in turn, push on the valve plate to open it. Some valves use a single pin that becomes the valve stem with a ball valve at the end. The valve plate or ball also has pressure on it from the superheat spring that pushes to close the valve.

There is also a passage that allows evaporator pressure to act on the bottom of the diaphragm in opposition to thermal bulb pressure. In an *internally balanced* TXV, this passage is open to evaporator inlet pressure. In an *externally balanced* TXV, this passage connects to a length of small tubing that connects to the evaporator tailpipe. Externally balanced valves are used on some larger evaporators and give better response (Figure 6-44).

At room temperature, a TXV is open because the gas pressure in the capillary tube exerts a rather high pressure, greater than the spring. As the system cools, the temperature of the sensing bulb and capillary tube drops, and the pressure on the diaphragm drops with it. When the pressure at the diaphragm drops below the spring pressure, the valve closes. Also, if an excess amount of refrigerant enters the evaporator, high refrigerant pressure can act through the balance passage to close the TXV.

TXVs come in many forms. Most of the early valves threaded onto the evaporator inlet, and the liquid line threaded onto the valve. The thermal bulb, or end of the capillary tube, was clamped onto the evaporator tailpipe or inserted into a well in it. It was important that the thermal bulb be clamped tightly and be well insulated so that it transmitted an accurate temperature signal. If used, the external equalizer line threaded onto a fitting on the tailpipe.

Many newer valves are of the block or H type. These valves are connected with both the liquid and suction lines and the evaporator inlet and outlet. Some block valves use threaded fittings, and some are bolted between manifolds and are sealed using O-rings. Block valves use all of the features of other TXVs in a much more compact unit (Figure 6-45).

A TXV can also be a capsule inserted into a larger assembly. GM (between 1973 and 1975), Audi, and Volvo used a **valves in receiver (VIR)**, a receiver–drier with the TXV and the suction throttling valves built into it. The TXV capsule was mounted so that it controlled the refrigerant flow from the receiver portion to the evaporator port (Figure 6-46).

Most TXVs have a small, very fine screen at their inlet. This screen traps debris that can plug the valve, and it can be removed for cleaning or to install a replacement.

6.4.2 Orifice Tubes

When first used in GM pickups, trucks, and some cars, the first orifice tubes (OTs) were porous brass units that looked somewhat like fuel filters. They are also called **fixed orifice tubes** (FOTs). Newer OTs are longer, more slender plastic units used on GM vehicles since the mid 1970s and on most Ford cars since 1980.

Manufacturers color-code OTs to identify the car make and model for which a tube is used (Figure 6-47).

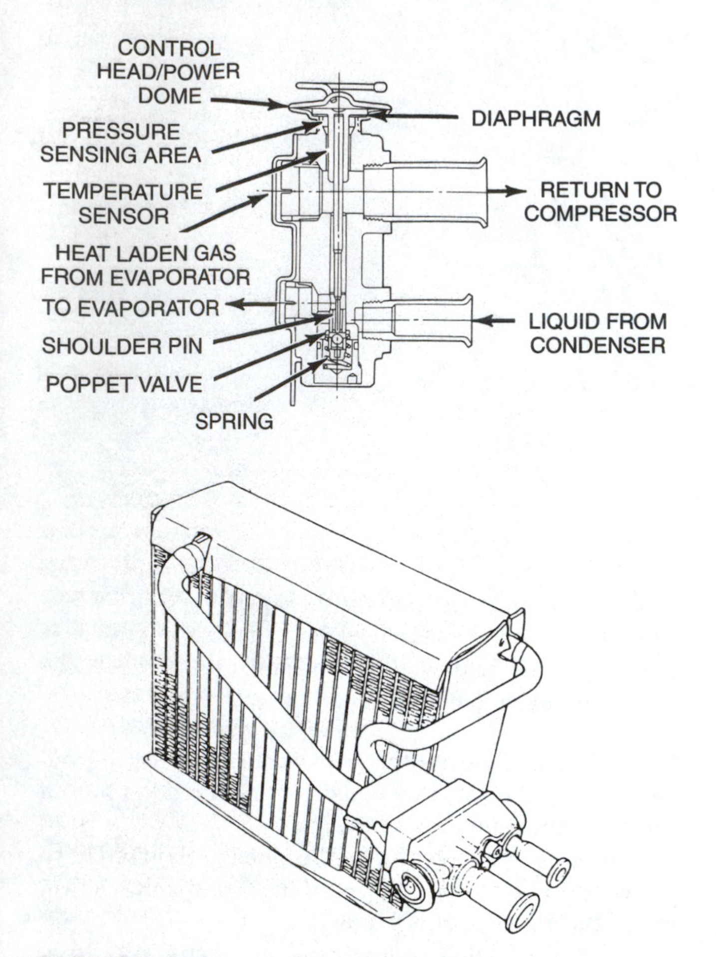

FIGURE 6–45 A block TXV has the control head next to where the cooled gas is leaving the evaporator, eliminating the need for a thermal bulb and capillary tube. Note that this valve uses spring-lock fittings. (*Courtesy of DaimlerChrysler Corporation*)

There are at least eight different sizes of OTs that have a similar appearance but a size range from 0.047 to 0.072 inch (1.19 to 1.8 mm).

An OT is a thin brass tube that is a couple of inches long and has a plastic filter screen around it. This tube is sized to flow the proper amount of refrigerant into the evaporator for maximum cooling loads. As mentioned earlier, the OT floods the evaporator during light cooling loads, so a low side accumulator is always used with an OT. One expert states that under typical operating conditions, the evaporator might contain about 10 to 20% liquid, 70 to 80% saturated vapor, and 10 to 20% superheated vapor. The flow through an OT is also affected by pressure, and excessive high side pressure can cause evaporator pressure and temperature to become too high.

The OT in most early systems was placed into the evaporator inlet tube. An O-ring was used around it to stop refrigerant from flowing past the outside of the OT. Several small indentations, or dimples, were put in the

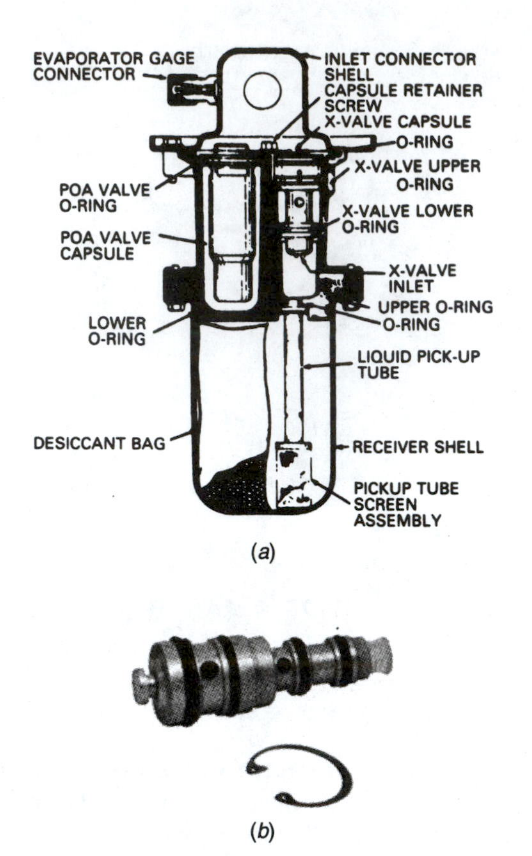

FIGURE 6–46 The VIR assembly (*a*) contains a serviceable cartridge-style TXV (*b*). (*Reprinted with permission of General Motors Corporation*)

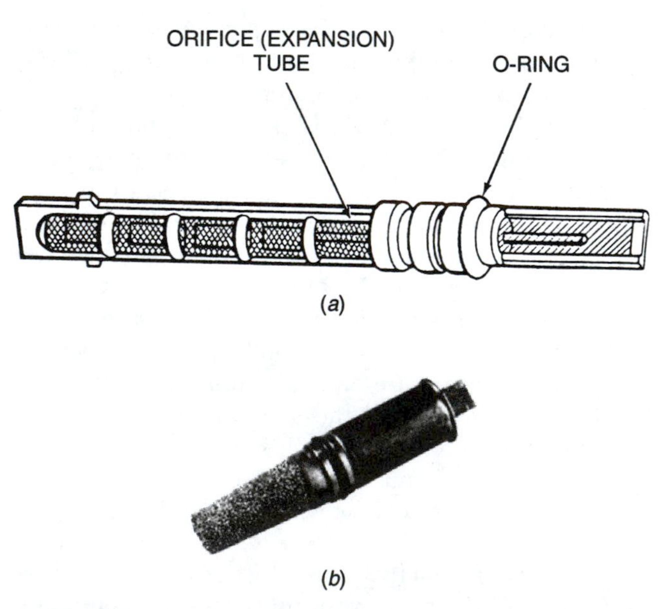

FIGURE 6–47 Most OTs are small tubes with a filter screen over each end (*a*); some older GM vehicles use a sintered bronze OT (*b*). (*Reprinted with permission of General Motors Corporation*)

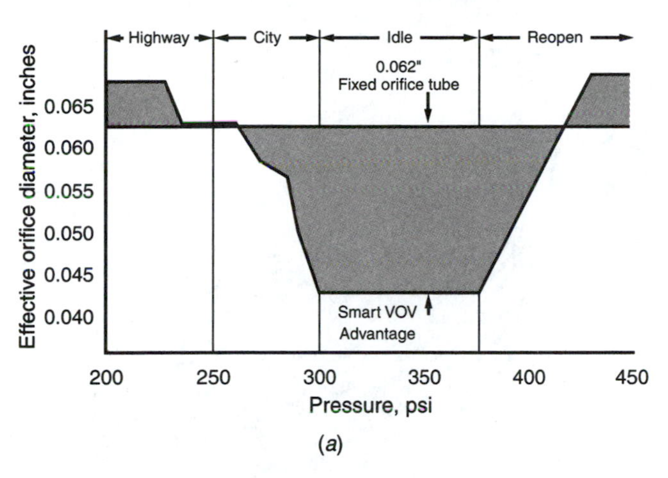

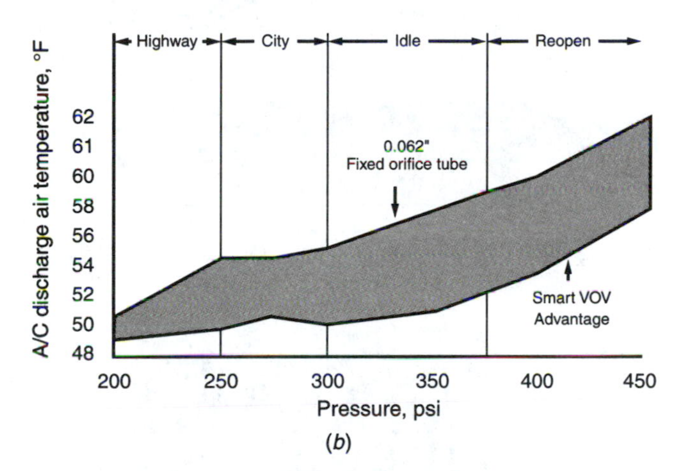

FIGURE 6–48 A flow rate of a VOV as compared to a fixed orifice tube. Note how the VOV is either larger or smaller under certain conditions (*a*) and produces cooler discharge air (*b*). (*Courtesy of Nartron Corporation*)

tubing wall to keep the OT from moving too far into the evaporator. Many newer vehicles place the OT in the liquid line farther away from the evaporator. This is done because of complaints of hissing noises that occur after the vehicle is shut off and high side pressure bleeds down through the OT.

✓SERVICE TIP

In an operating system, the OT position can be located by finding the point where the liquid line changes temperature from hot to cold. If the system is not operating, the dimples in the line show this location.

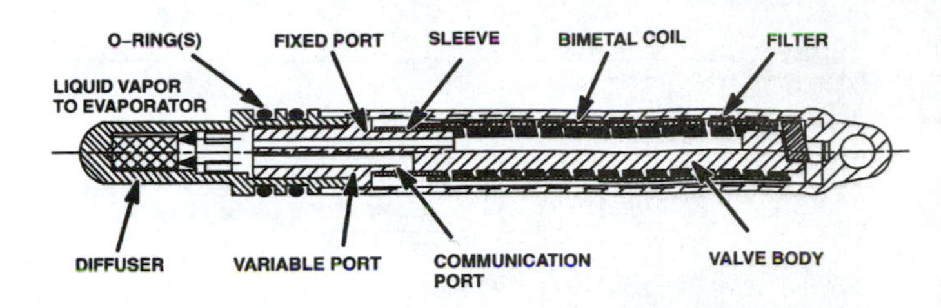

FIGURE 6–49 This VOV uses a bimetal coil to sense the temperature of the refrigerant. A higher temperature will cause the coil to expand and partially close the variable port to increase the restriction. (*Courtesy of DaimlerChysler*)

6.4.3 Variable Orifice Valves (VOVs)

Most OTs have a fixed-size orifice that is sized for the proper refrigerant flow to produce maximum A/C performance during 50- to 60-mph operation. At lower engine speeds, the orifice size is too large. Remember that the orifice size should produce the high side pressure for proper condenser action and the pressure drop into the low side to produce proper evaporator pressure and temperature. Vehicles that spend considerable time idling commonly have poor A/C and experience short compressor life (Figure 6-48).

One vehicle manufacturer uses a variable orifice valve (VOV) that senses temperature to change the size of the orifice (Figure 6-49). A bimetal coil spring senses the temperature of the liquid refrigerant, and when the temperature increases, the spring moves the variable port to a closed position. This increases the restriction at the VOV and reduces the flow to the evaporator.

The VOV is the same external size and configuration as a fixed OT, and it contains an internal valve (Figure 6-50). The valve is calibrated to reduce the orifice size under conditions of high head pressure and low refrigerant flow rate. Installation of the VOV is simply a matter of removing the OT and installing the VOV plus an additional ounce of oil and an additional ounce or two of refrigerant. This device is becoming popular with law enforcement agencies. A VOV is also said to increase the performance of dual systems.

✓PROBLEM SOLVING

A friend is concerned that something is wrong with her A/C system. It cools well, but when the engine and the system are turned off, she hears a hissing noise. You also hear this noise when you check it. What could be causing this? Is this a sign of a problem?

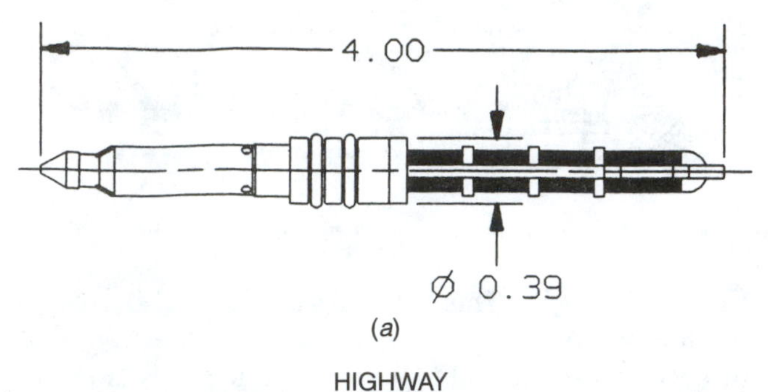

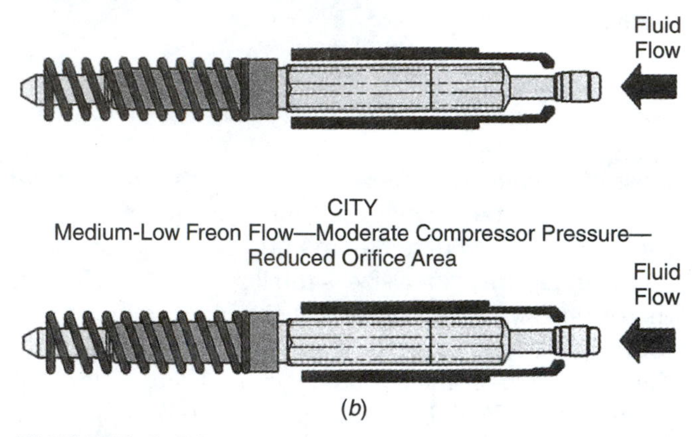

FIGURE 6–50 In a VOV (*a*), the valve is inside the tubular portion at the left; it has the ability to reduce the flow rate as head pressure increases (*b*). (*Courtesy of Nartron Corporation*)

6.5 EVAPORATORS

An evaporator, like a condenser, is a simple device. Some evaporators are merely groups of tubes with fins attached. Most evaporators are a series of plates sandwiched together to form both the refrigerant and air passages. A plate-type evaporator is the most efficient of these two styles (Figure 6-51).

Evaporators have at least two line connections. The smaller line connects to the TXV or OT, and the larger one connects to the suction line and to the compressor. Some evaporators have a third, very small line. This is an oil bleed line and is used in a system with a suction throttling valve (STV). This line allows oil to be pulled from the bottom of the evaporator to ensure compressor lubrication when the STV is throttling (Figure 6-52).

6.6 RECEIVER–DRIERS AND ACCUMULATORS

Receiver–driers and accumulators serve two functions: (1) They store a supply of liquid refrigerant and (2) they contain desiccant. The purpose of refrigerant storage is

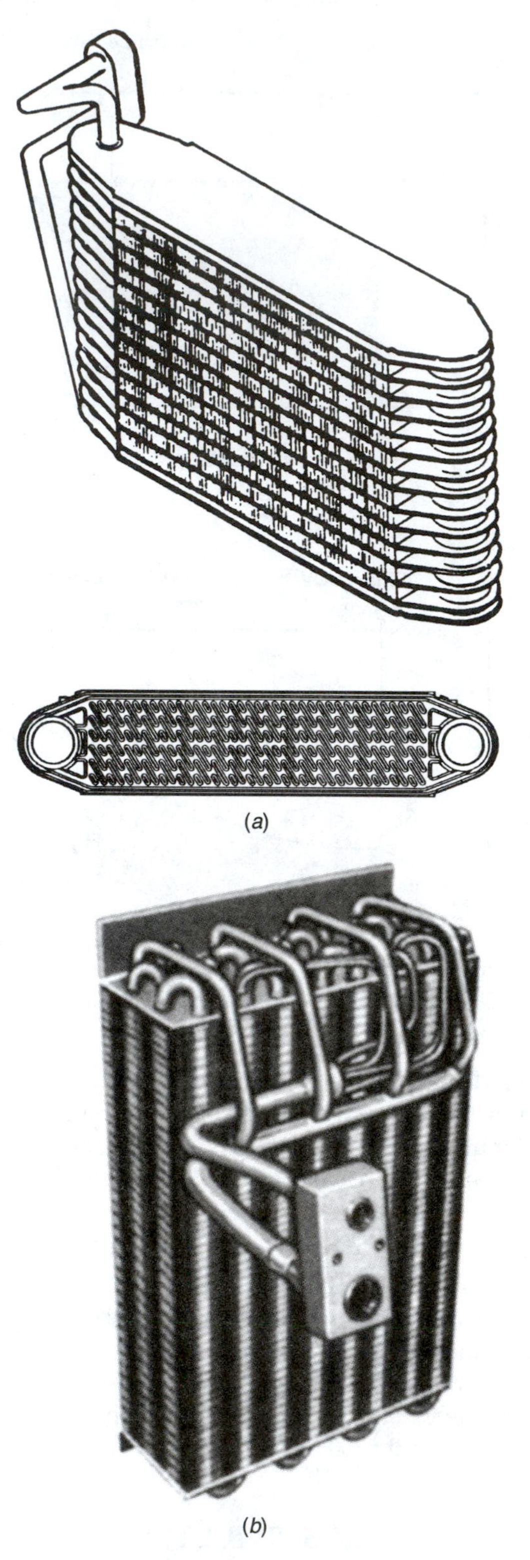

FIGURE 6–51 A plate-type evaporator is made from a group of plates that form the passages for the gas flow (*a*). A fin-and-tube evaporator routes the refrigerant flow through one or more tubes (*b*). (a *reprinted with permission of General Motors Corporation;* b *courtesy of Four Seasons*)

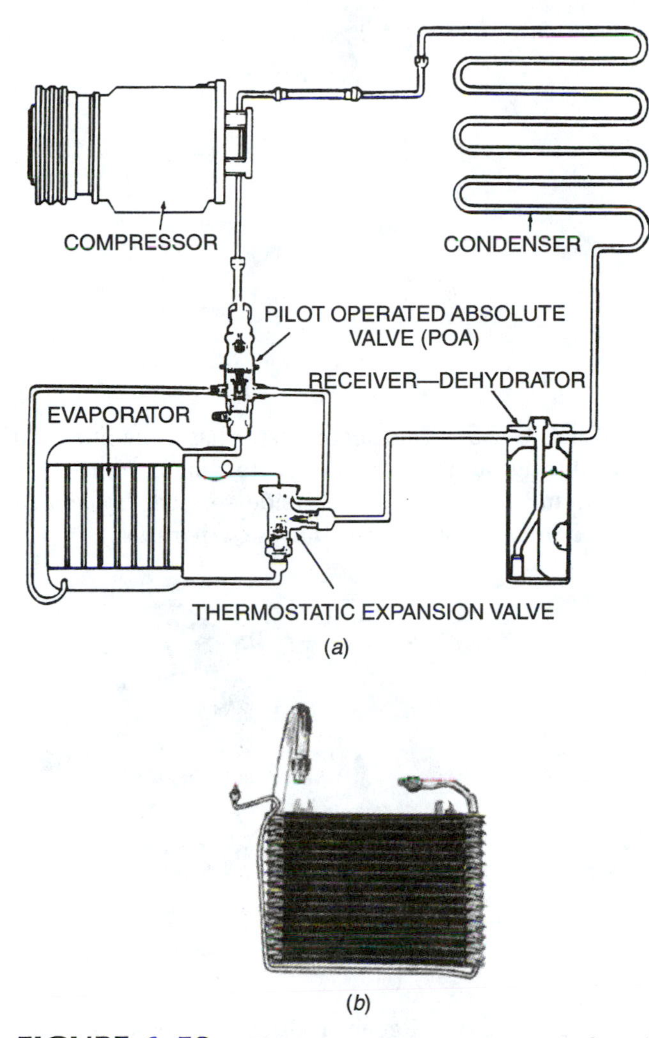

FIGURE 6–52 The compressor can starve for oil while the STV/POA is throttling refrigerant flow (*a*). The evaporator used in the system includes an oil bleed line to prevent oil starvation. The oil bleed line is a small-diameter tube at the bottom of the evaporator core (*b*). (*Reprinted with permission of General Motors Corporation*)

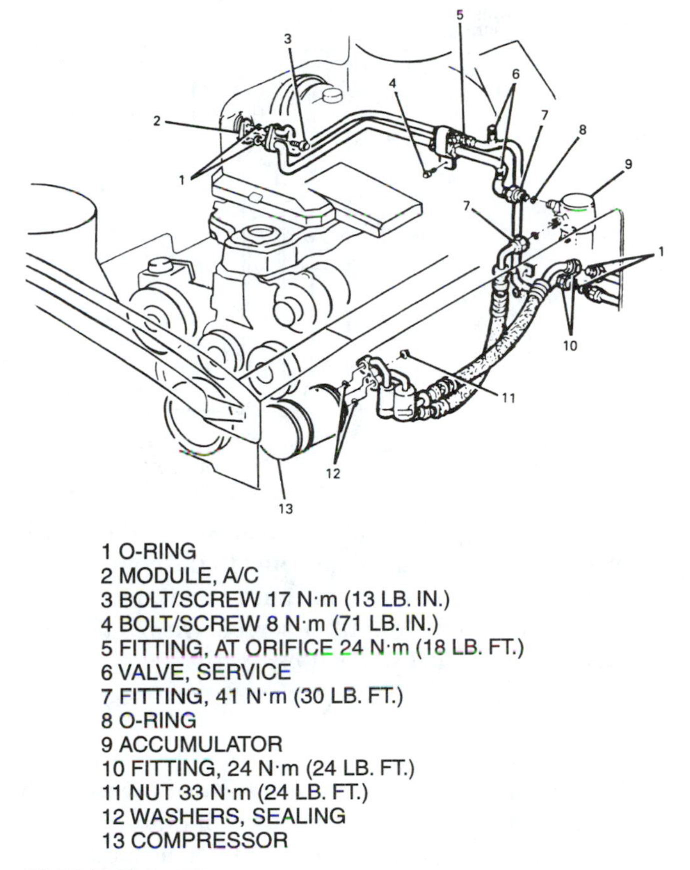

FIGURE 6–53 This accumulator (9) is connected to the evaporator through a metal tube and to the compressor through a rubber suction hose. (*Reprinted with permission of General Motors Corporation*)

to compensate for volume change due to temperature change or refrigerant loss. Desiccant is needed to remove moisture or water, which can cause rusting or corrosion.

As previously mentioned, the receiver–drier or accumulator used in an R-134a system should use XH-7 or XH-9 desiccant. New units contain either of these desiccants, so they are compatible with either R-12 or R-134a. These desiccants are a little more expensive and require a little more volume than XH-5, which was used in R-12 units. R-134a driers must have greater water-absorption capacity because R-134a is more water soluble than R-12.

6.6.1 Accumulators

The accumulator inlet on many vehicles is connected directly to the evaporator tailpipe. On other vehicles it is mounted separately, and a metal tube or rubber hose is used to connect it to the evaporator. The outlet of the accumulator is connected to the compressor inlet through the flexible suction line (Figure 6-53).

The inlet simply dumps the incoming refrigerant vapor and liquid into the container. The outlet begins near the top, runs downward to the oil bleed hole at the bottom, and then exits out of either the top or the bottom. As mentioned earlier, a fine mesh filter screen is used at the oil bleed hole to prevent plugging and compressor oil starvation. Most accumulators have an internal baffle to prevent incoming liquid refrigerant from passing directly into the outlet tube. The accumulator of a properly operating system normally holds liquid refrigerant at the bottom, and it supplies the compressor with refrigerant vapor from the top (Figure 6-54). Remember

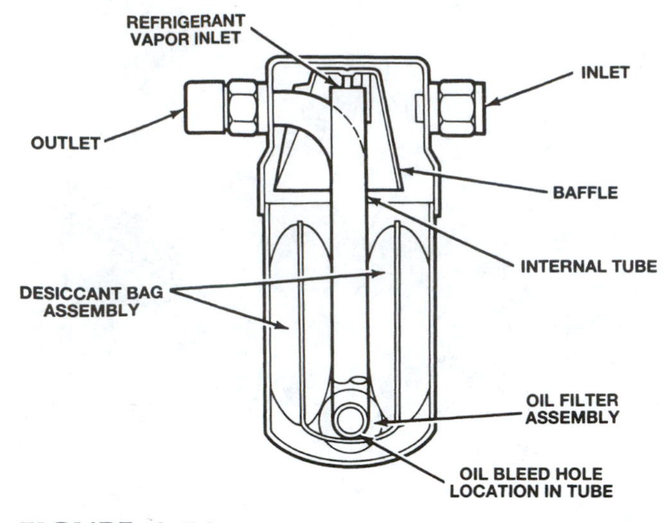

FIGURE 6–54 This cutaway accumulator shows the vapor inlet and outlet connected to the compressor, a baffle to keep liquid out of the inlet, the desiccant bag, and an oil bleed hole with filter screen. *(Reprinted with permission of General Motors Corporation)*

that this liquid refrigerant is at evaporator pressure, so it is at the same temperature as the evaporator.

In addition to the inlet and the outlet, accumulators for an R-12 system are often equipped with one or two smaller ports. These traditionally have been 1/4-inch male flare fittings (the same size as a standard R-12 service fitting). These small fittings are for the low side service fitting and low side pressure switches (Figure 6-55). Replacement accumulators should have the same service and/or switch ports as the original accumulator.

✓SERVICE TIP

The accumulator of some R-134a systems has an insulating jacket to help reduce the heat absorption and lower the air temperature at the discharge ducts.

✓SERVICE TIP

The accumulator is normally replaced if a system has been opened to atmosphere for a period of time because the desiccant is probably saturated with moisture. It is also standard practice to replace the accumulator whenever major service work is done on a system, especially if the compressor is replaced.

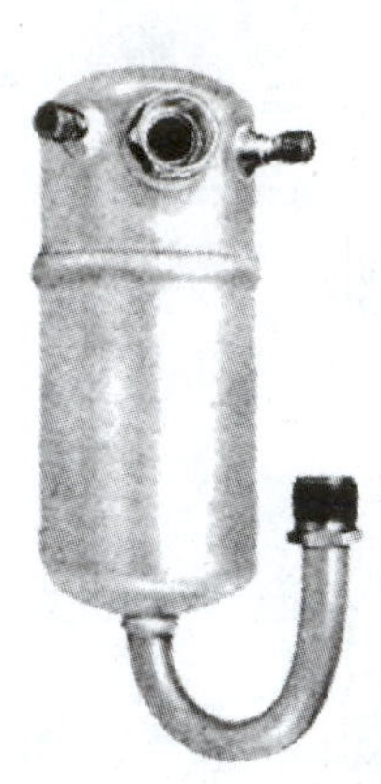

FIGURE 6–55 This accumulator has large inlet and outlet fittings and two smaller ports, one for a pressure switch and the other for low side service. *(Reprinted with permission of General Motors Corporation)*

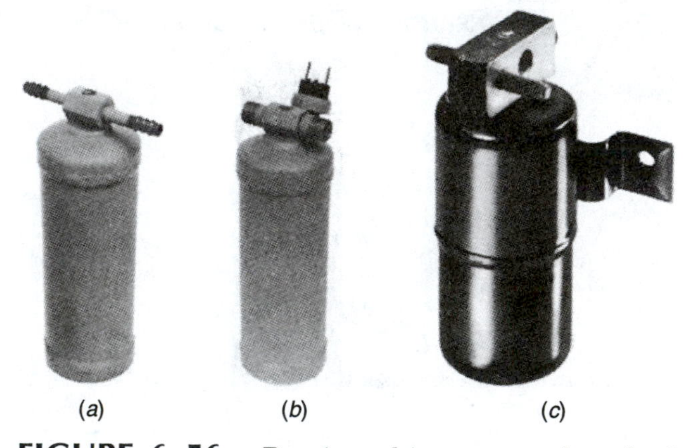

FIGURE 6–56 Receiver–driers use push-on barb fittings (*a*), threaded line fittings (*b*), or harness- or block-type fittings (*c*). Note the various sizes and that some include mounting straps. *(Reprinted with permission of General Motors Corporation)*

6.6.2 Receiver–Drier

The receiver–drier is normally found in the high side liquid line, somewhere between the condenser and the TXV. Factory-installed systems use threaded fittings or block-type fittings with O-rings at the line connections. Receiver–driers for some dealer-installed and aftermarket installations use push-on, barb-type fittings. On these, the hose is slid onto the fitting and secured with a clamp. The shapes and sizes of receiver–driers vary greatly (Figure 6-56).

The receiver–drier inlet dumps incoming refrigerant into the container. The outlet, often called a **pickup tube**, begins near the bottom and usually exits at the top. A fine-mesh filter screen is used at the inner opening to stop debris from passing out of the receiver–drier

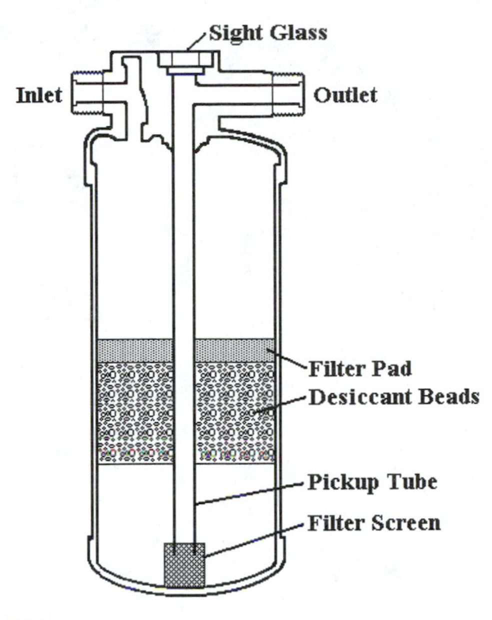

FIGURE 6–57 This cutaway receiver-drier shows the filter pads and desiccant; many units include a filter at the opening of the pickup tube. Note the sight glass at the top of the pickup tube.

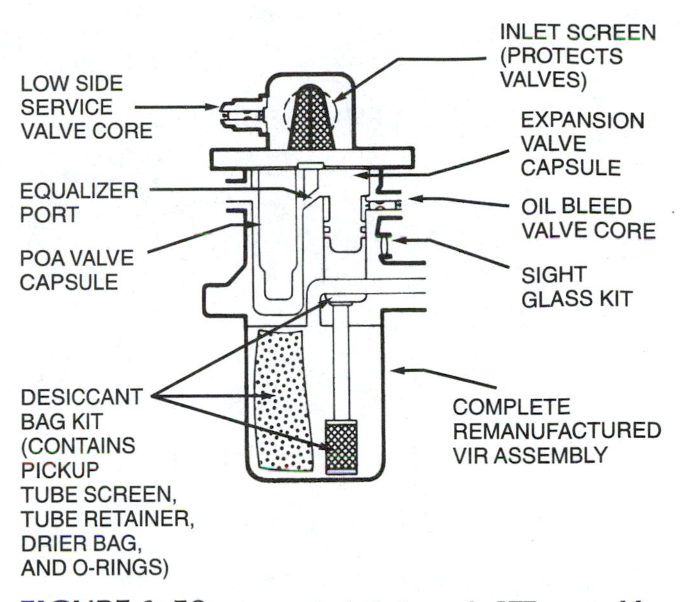

FIGURE 6–58 An exploded view of a VIR assembly; note the two valves and the receiver–drier portions. (*Courtesy of Four Seasons*)

and on to the TXV. The receiver–drier is usually about half full of liquid refrigerant (Figure 6-57).

A receiver–drier can contain a sight glass, pressure release plug or valve, or switch. The **sight glass** allows observation of the refrigerant flow as it leaves the receiver–drier. The flow of R-12 should be invisible; remember that R-12 is clear, and the oil is dissolved in the liquid refrigerant. Bubbles or foam in the sight glass can indicate abnormal operation. R-134a would appear cloudy in a sight glass, so sight glasses are not commonly used in R-134a systems. The pressure relief device is described in Section 5.3.4. The switch is used to sense high side pressure. It can be used to prevent compressor operation if the pressure is either too low or too high.

As with an accumulator, a receiver–drier is normally replaced if the system is left open or if major service work is done, especially if the compressor is replaced.

The VIR assembly is a serviceable unit: Almost every one of its individual components can be replaced. A kit that contains a desiccant bag, pickup tube filter, and O-rings is available (Figure 6-58).

✓SERVICE TIP

When servicing a VIR, removal of the assembly is recommended to get better access to the screws, which tend to seize in the aluminum housing; also, index marks should be placed on the housing parts to ensure proper assembly.

6.6.3 Filters and Mufflers

In-line filters are available from aftermarket sources for installation in the liquid line. These are designed to serve two purposes: (1) to filter the refrigerant to stop debris from plugging the OT or TXV and (2) to add additional desiccant to a system for additional moisture removal.

✓SERVICE TIP

Service problems have occurred because scale and other debris from the condenser and compressor have plugged the OT. Normally if this happens the OT is cleaned or replaced. To prevent recurrence of this problem, a technician has a choice: thoroughly clean out the system or install a filter. If the system is still working well, it can be much simpler and cheaper to add a filter–drier than to flush the system (Figure 6-59).

Mufflers are installed in the discharge or suction line of some systems. These mufflers are usually a simple baffled cylinder and are used to dampen the pumping noise of the compressor (Figure 6-60).

6.7 HOSES AND LINES

As mentioned earlier, because the suction and discharge lines in a system must be flexible, they are usually made of reinforced rubber hose. The liquid line can

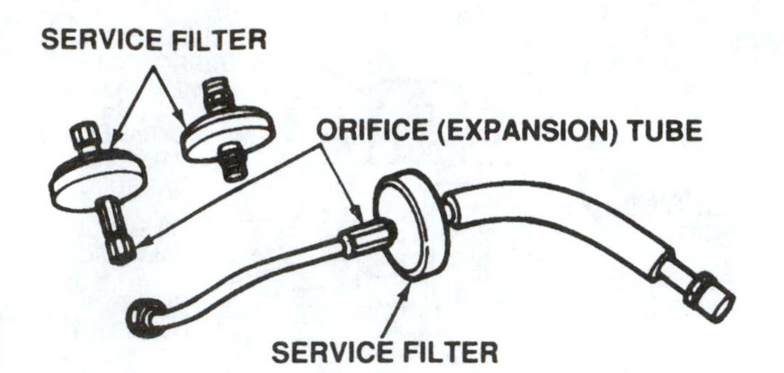

FIGURE 6–59 Aftermarket in-line filters can be added to a system to trap debris. Some include a replacement OT.

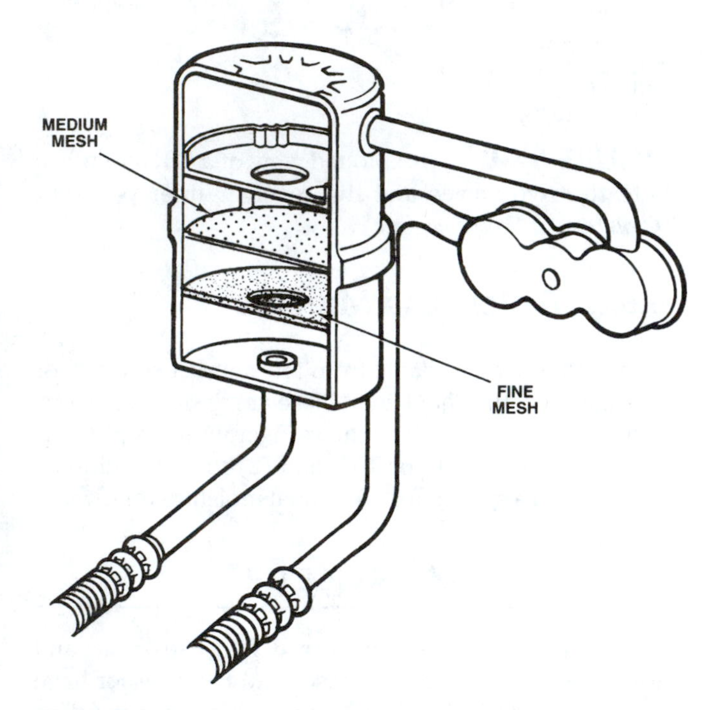

FIGURE 6–60 Some systems include a muffler in the discharge or suction line. (*Reprinted with permission of General Motors Corporation*)

be made from metal because the parts it connects are rigidly mounted, but many systems use hose for this function also.

Various end fittings of different sizes are used for the connections. These fittings include female and male flare fittings, female and male O-ring fittings, block or manifold fittings with O-rings, spring-lock fittings, and simple push-on hose-to-barb connections (Figure 6-61). Early systems traditionally used flare fittings, but a leaky flare fitting is often expensive and difficult to repair. O-ring fittings are much easier to repair, simply by replacing the O-ring. Fitting types and sizes are shown in Appendix H.

Block or manifold fittings are held together by a bolt or nut and are sealed by either a crimped metal

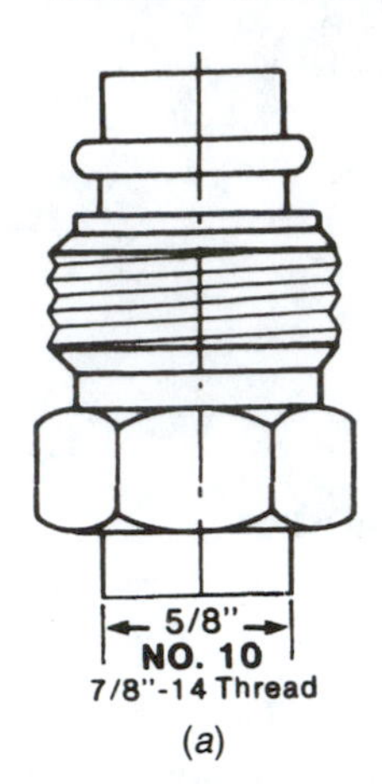

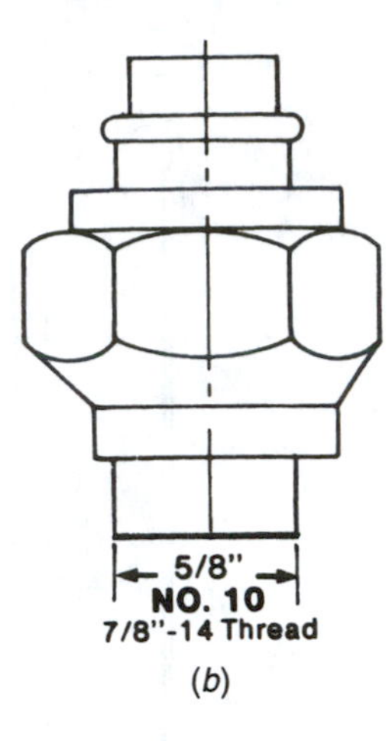

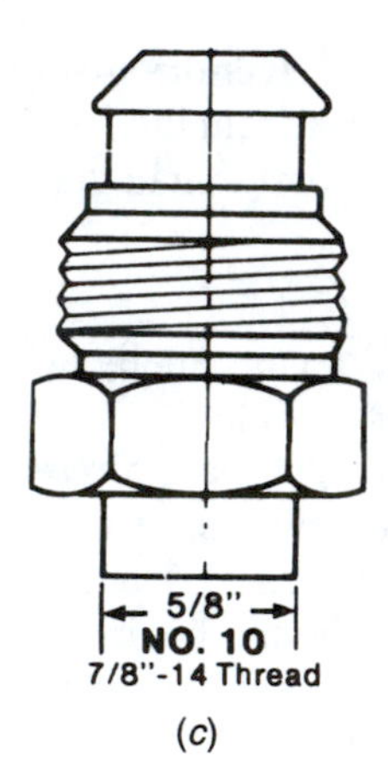

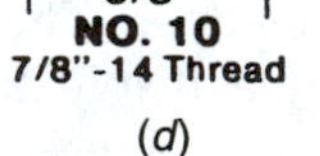

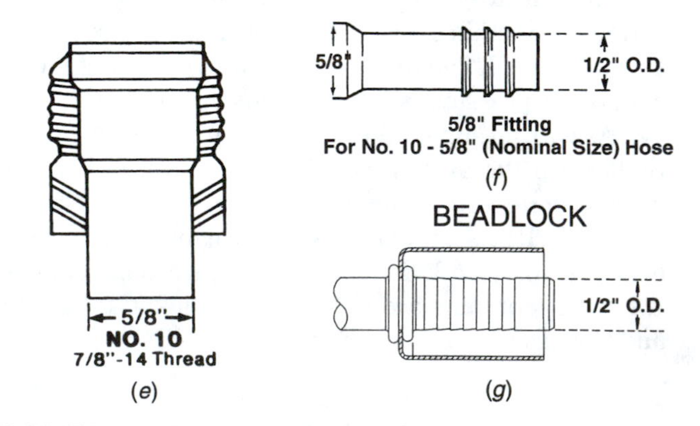

FIGURE 6–61 A/C fitting types include male O-ring (*a*), female O-ring (*b*), male flare (*c*), female flare (*d*), male insert O-ring (*e*), push-on barb (*f*), and beadlock (*g*). (*Courtesy of Four Seasons*)

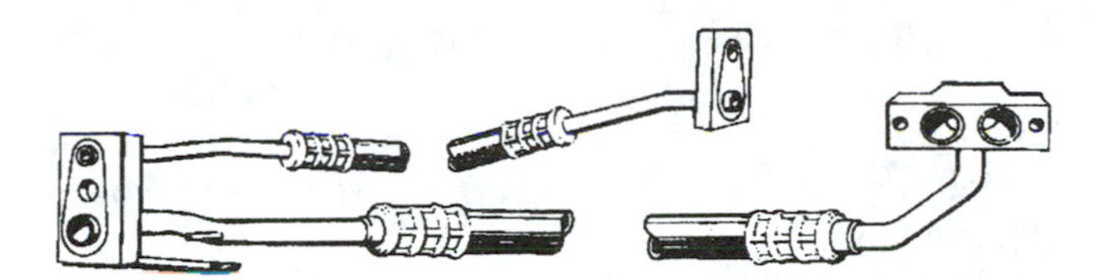

FIGURE 6–62 This suction and liquid hose has block fittings for the connections to the receiver–drier, TXV, and compressor; they are sealed by gaskets or O-rings. (*Courtesy of Four Seasons*)

gasket or O-rings (Figure 6-62). A newer style of fitting, called a P-nut fitting, uses a single O-ring.

✓SERVICE TIP

When making block connections, use a torque wrench. Overtightened connections damage the O-ring or distort the block and cause leaks.

O-ring fittings squeeze a rubber O-ring between the two parts being connected to make the seal. In most cases, the O-ring is slid over the end of the metal tube and is located by a raised metal ring or bead. Some manufacturers use a **captive**, also called a **captured**, O-ring, which locates the O-ring more positively in a groove. Captive O-rings use a larger-diameter cross section than standard O-rings (Figure 6-63). Some designs use two O-rings for a better seal.

O-rings made from Buna or nitrile are commonly used in R-12 systems, but these materials are not compatible with R-134a. High-grade neoprene O-rings (HSN or HNBR), often tinted blue or green, are used in R-134a systems; these also can be used in R-12 systems. There is no color standard for O-rings; some manufacturers use red, yellow, and tan colors. Most manufacturers use the following color designations:

- Black indicates nitrile or neoprene
- Blue indicates neoprene or nitrile
- Green indicates HNBR

Because of this confusion, when you order O-rings, you should purchase them from a reputable source, specify the desired material, and then specify the color.

A spring-lock fitting is a type of quick-disconnect fitting. The female portion of this fitting has a flarelike ridge at the end; this ridge is gripped by a garter spring when connected. One or two O-rings form the seal between the two fitting parts. These O-rings must be

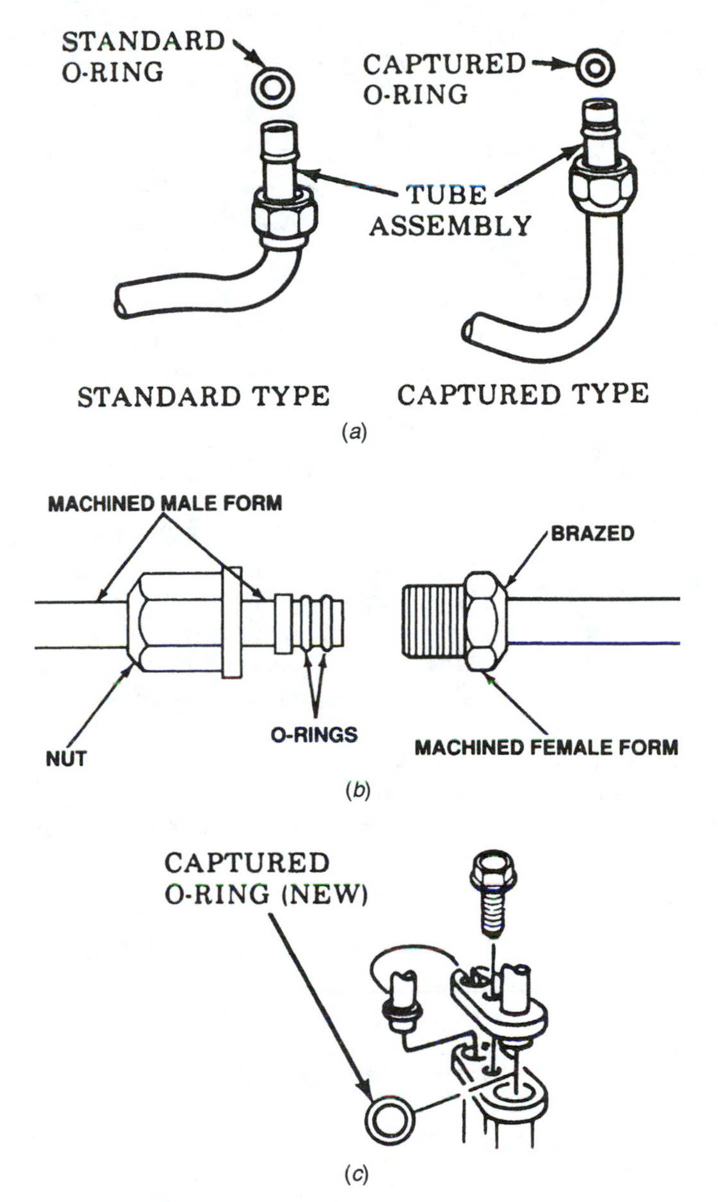

FIGURE 6–63 Standard O-rings merely slide onto the line; captive O-rings are positioned in a groove (*a*). Some fittings use dual O-rings (*b*). O-rings can also be used to seal block fittings (*c*). (*Reprinted with permission of General Motors Corporation*)

resilient enough to compensate for slight movement between the two parts. Spring-lock fittings are connected by pushing one line over the other until the garter spring moves into position. A special tool is inserted into the fitting to expand the garter spring and release the fitting. Refrigerant leakage was a fairly common problem with early spring-lock fittings; newly designed O-rings have improved sealing ability (Figure 6-64).

Metal tubing is sized by its outside diameter (OD). Pipe and hose are sized by the ID; these sizes are often

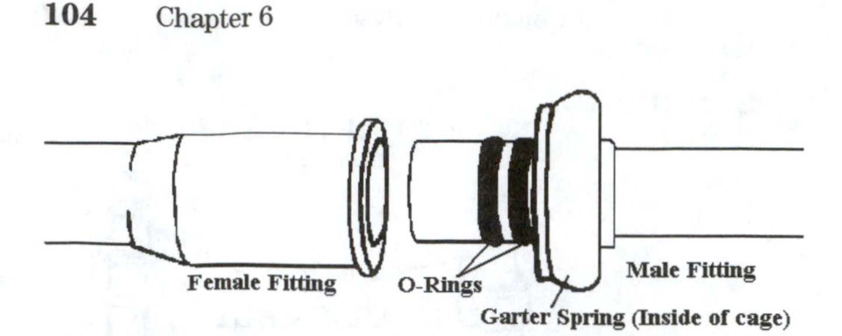

FIGURE 6–64 A spring lock fitting is a type of quick-disconnect fitting that is sealed by two O-rings and held together by a garter spring. A special tool is required to expand the garter spring to release the fitting.

nominal (approximate) sizes. A number sizing is often used for refrigerant hose and fittings, with the most popular sizes being #4, #6, #8, #10, and #12: #4 equals 1/4 inch (4.4 mm), #6 equals 3/8 inch (8 mm), #8 equals 1/2 inch (11.1 mm), #10 equals 5/8 inch (12.4 mm), and #12 equals 3/4 inch (15.5 mm) (Figure 6-65). A new reduced-diameter, lightweight hose is currently being used for some OEM and replacement usage. Reduced-diameter hose has the same ID as standard hose, but the OD is noticeably smaller (Figure 6-66).

A hose must have an impervious inner liner so it can seal, and it must be completely compatible with the refrigerant and oil. Some inner liners are effective with both R-12 and R-134a; a nitrile inner liner, however, works well with R-12 and has a high leak rate with R-134a. The inner layer is usually surrounded by rayon, another reinforcing layer, and the outer, protective layer of rubber. Some older hoses are made from nylon with a single protective layer. Nylon hoses have a smaller OD and are much more rigid. Modern hoses use a thin nylon layer, which is quite effective in containing R-134a; these are called **barrier** hoses (Figure 6-67).

Most OEM hose uses a captive ferrule to swage the hose onto the end fitting. Grooves or raised rings (barbs) at the fitting and the clamping action of the ferrule hold the hose securely in place. Repair of hoses and the fittings used for this repair are described in Chapter 14 (Figure 6-68).

6.8 ELECTRICAL SWITCHES AND EVAPORATOR TEMPERATURE CONTROLS

A variety of electrical switches are used in A/C systems to prevent evaporator icing, protect the compressor, and control fan motors. In addition to electrical controls, some systems use a suction throttling valve (STV) to control evaporator pressure and prevent icing. An STV can be located anywhere between the evaporator tailpipe and the compressor inlet. Control switches can be located anywhere in the system; the most common locations are the compressor discharge or suction cavities, the receiver–drier, and the accumulator.

Some modern systems use a variable displacement compressor to prevent icing. As described in Sections

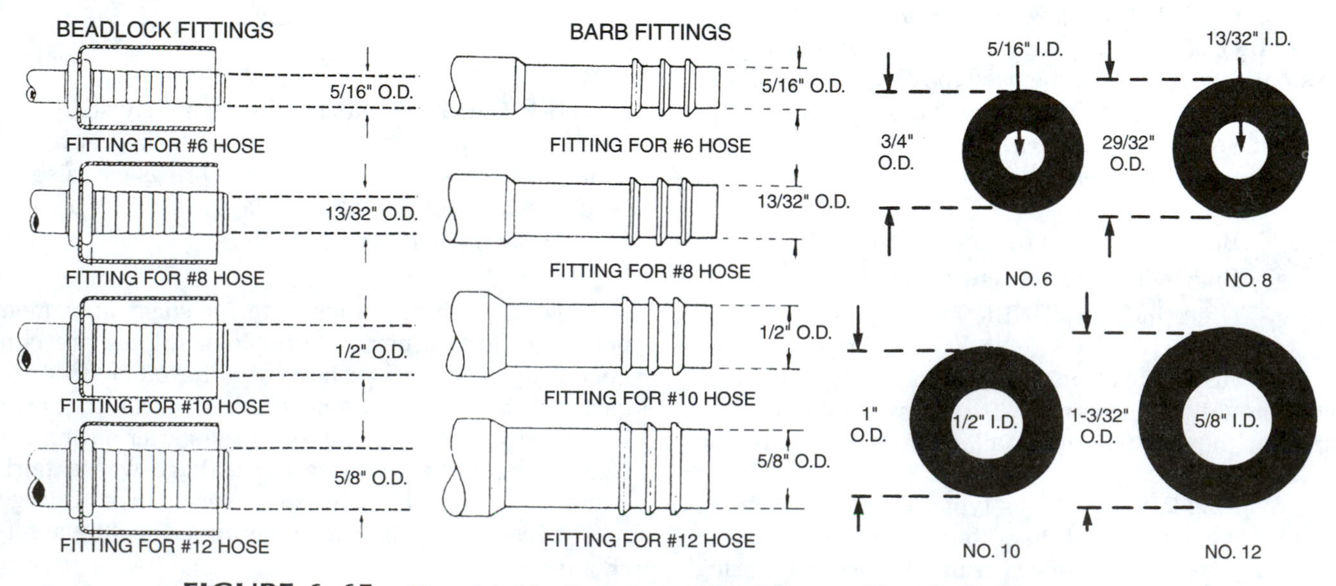

FIGURE 6–65 The outside diameter of the fitting and inside diameter of the hose determine the size. (*Courtesy of Four Seasons*)

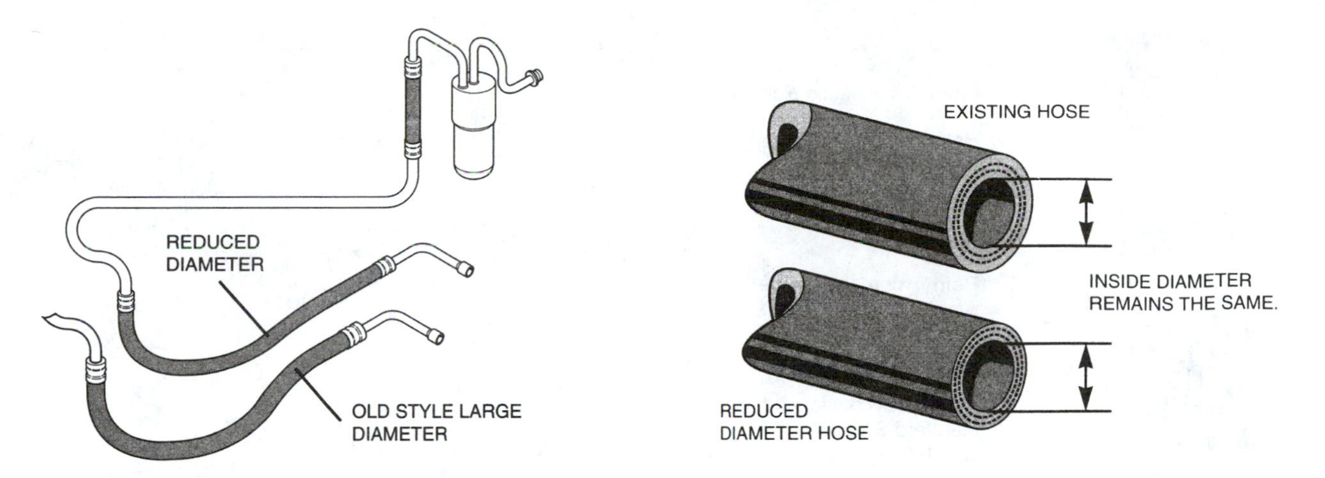

FIGURE 6–66 A reduced-diameter hose has the same inside diameter with a smaller outside diameter and lighter weight, making it more suitable for congested underhood spaces of fuel-efficient vehicles. (*Courtesy of Four Seasons*)

6.2.1.5 and 6.2.2.1, a valve that senses evaporator pressure is used in the compressor, and compressor displacement is reduced in response to that pressure (Figure 6-69).

6.8.1 Switches and Relays

At one time, the A/C electrical circuit was rather simple. As shown in Figure 6-70, a typical circuit connects the evaporator blower motor and compressor clutch at one master switch. The power to the clutch passes through a temperature switch that opens to cycle the clutch when the evaporator gets too cold. The power to the blower motor passes through a speed control switch so that the blower speed can be changed.

Today various switches, sensors, and relays are used. A sensor is usually an input to an electronic control module (ECM), and many sensors provide a variable signal so the ECM will know the actual temperature or pressure at certain points. A relay is essentially a magnetic switch that is controlled by another switch; a relay is used to control a greater amount of current than a switch can handle. Relays also allow computer control modules to control electrical circuits (Figure 6-71). Control switches and relays that you may encounter are as follows:

- Ambient sensor or switch: Senses outside temperature and is designed to prevent compressor operation when ambient temperature is below a certain point, about 35 to 40°F
- Compressor high-pressure sensor or switch: Mounted in compressor discharge cavity; senses high side pressure and is used to cut out the compressor clutch if pressure is too high or too low or provides a signal to another device that pressure is too high
- Compressor low-pressure sensor or switch: Mounted in compressor suction cavity; senses low side pressure and is often used to cut out the compressor clutch if pressure is too low
- Compressor rpm sensor: Provides input to the ECM that the compressor is running; ECM will cut out compressor if rpm is low (possible belt slippage or impending compressor lockup)
- Compressor superheat sensor or switch: Function is similar to compressor low-pressure switch
- Compressor cutoff switch: Mounted at power steering gear or transmission; senses pressure in that system and is used to stop the compressor under certain external conditions
- Coolant temperature sensor, also called the engine coolant temperature (ECT): Mounted near the engine thermostat; senses engine temperature for power-train control module and is also used to turn on cooling fan
- Evaporator pressure sensor: Provides input to the ECM as to the operating pressure in the evaporator
- Evaporator temperature sensor: Mounted at the evaporator; senses temperature and is used to cycle compressor clutch to prevent icing
- High-pressure cutout switch: Mounted at receiver–drier or liquid line; senses high side pressure and is used to cut out the compressor clutch if high side pressure is too high
- High-temperature cutoff sensor or switch: Mounted at the condenser outlet; senses condenser temperature and is used to cut out

FIGURE 6–67 This barrier-type hose is made from five layers of different materials. (*Courtesy of Dayco Products, Inc.*)

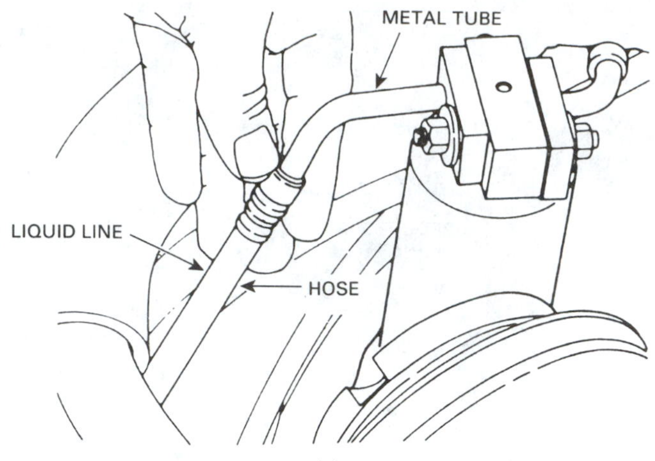

(*a*)

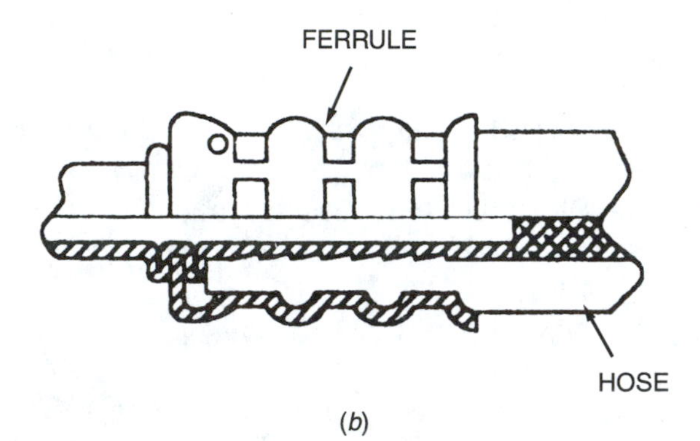

(*b*)

FIGURE 6–68 Most OEM hoses use a captive ferrule (*a*) that greatly increases the holding power; the ferrule is connected to the metal tubing (*b*). (a *is courtesy of DaimlerChrysler Corporation;* b *is courtesy of ACDelco*)

the compressor clutch if the temperature is too high

- Low-pressure cutout sensor or switch: Mounted in receiver–drier or liquid line; senses low pressure and is used to cut out the compressor clutch with low refrigerant charge
- Master switch: Mounted at control head; turns system on and off
- Pressure cycling switch: Mounted at accumulator; senses low pressure and cycles the compressor clutch to prevent evaporator icing
- Radiator fan switch: Mounted at engine thermostat cover or upper radiator hose; senses engine coolant temperature and operates radiator fan motor to prevent overheating
- Thermostatic cycling switch: Mounted at evaporator; senses air temperature and cycles the compressor clutch to prevent evaporator icing
- Trinary pressure switch: Mounted at receiver–drier; senses high side pressures and cuts out compressor clutch if pressure is too high or too low; can also be used to control radiator shutters or fan motor

- Blower relay: Can be used to turn blower on or off; most common usage is to provide high blower speed
- Clutch cutoff relay: Can be used to interrupt compressor clutch
- Condenser fan relay: Used to turn condenser fan motor on or off
- Radiator fan relay: Used to turn radiator fan motor on or off

Evaporator temperature in a cycling clutch system is usually controlled by either a thermostatic (thermal) or pressure switch. Both bellows and bimetal thermal switches are used. The bellows switch uses a capillary tube inserted into the evaporator fins. As with a TXV, the gas pressure in the capillary tube is exerted on a bellows. In a thermal switch, the bellows acts on a set of contact points. A

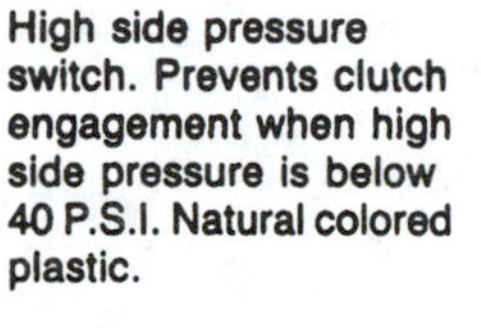

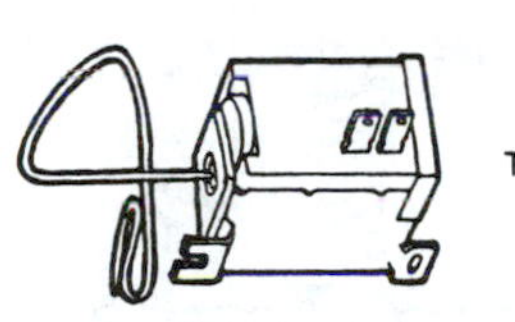

FIGURE 6–69 Some of the switches used in A/C electrical circuits. (*Courtesy of Everco Industries*)

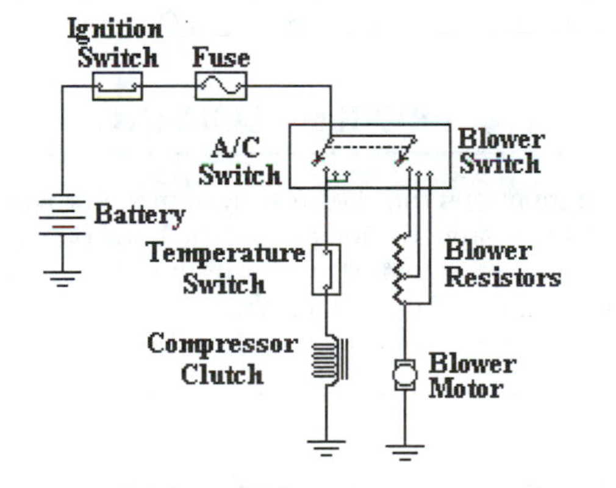

FIGURE 6–70 Many early A/C systems use a simple electrical circuit as shown.

warm evaporator produces a higher bellows pressure, which keeps the points closed; a cold evaporator reduces the pressure, which causes the points to open. A bimetal switch has a contact arm that is laminated from two metals with greatly different thermal expansion rates. When the switch is warm, the metals expand to close the contact points; as the switch cools, the metals in the arm contract to open the contacts. Both of these switch styles are calibrated so they are closed at temperatures above 32°F and open at temperatures below 32°F. When the switch opens, the compressor clutch cycles out. This causes the evaporator to begin warming, and after a few degrees of temperature increase, the switch closes to cycle the compressor in again.

A pressure cycling switch is mounted to sense low side pressure. Ice begins to form when evaporator pressure drops below 30 psi in an R-12 system or slightly less in an R-134a system. When evaporator pressure is above 30 psi, the switch contacts close so the compressor will operate. When the pressure drops below 30 psi, the switch opens and cycles the compressor out. After the compressor stops, evaporator pressure increases; after an increase of about 10 to 20 psi, the pressure switch contacts close to cycle the compressor in again. This switch also prevents the compressor from operating if there is low pressure from a refrigerant loss (Figure 6-72).

Many systems include a low-pressure switch to prevent compressor operation if there is a loss of

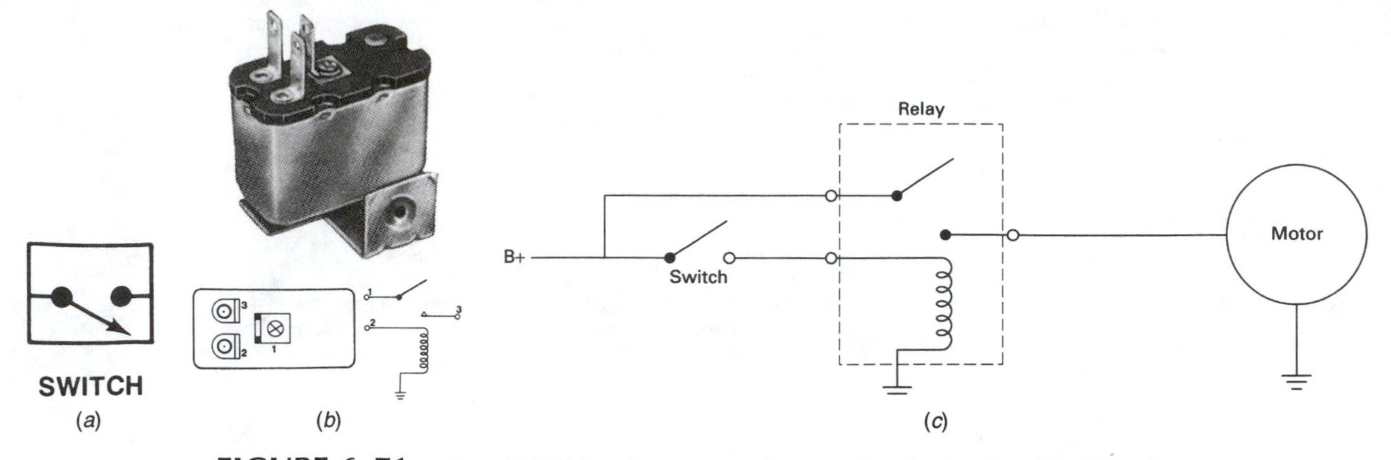

FIGURE 6–71 A switch (*a*) can open or close a circuit. A relay (*b*) does the same thing but can control more current. Most relays are controlled through switches (*c*). (a *courtesy of Everco Industries;* b *courtesy of Four Seasons*)

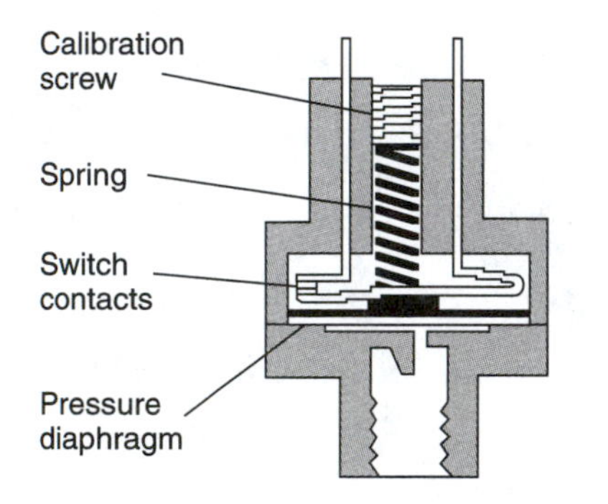

FIGURE 6–72 A pressure switch. The contacts are closed by gas pressure on the diaphragm; they are opened by the spring.

refrigerant. Remember that damage will occur if the compressor is operated without refrigerant or the oil that refrigerant circulates through it.

✓SERVICE TIP

If the system is low on refrigerant, it will **short cycle**, or cycle out and back in fairly rapidly. This is the result of the compressor pulling the refrigerant out of the low side quickly to open either the cycling switch or the low-pressure switch. With the compressor off, the flow into the evaporator raises the pressure enough to reclose the switch and restart the compressor.

Modern systems use solid-state sensors that do not use switch contacts, a source of failure. The most common are **thermisters** and **transducers**. A thermistor is commonly used to sense temperatures. It is basically an electrical resistor that changes resistance in direct inverse relationship to its temperature. Most automotive therminsters are of the negative temperature coefficient, NTC, type. The resistance will increase as the temperature drops and vice versa. A transducer senses pressure; it changes a variable pressure signal into a variable electrical signal. A transducer can let the ECM know the actual pressure in the low and/or high side of the system. Sensors provide an electrical signal to a control module that, in turn, controls compressor clutch or condenser fan operation. A/C system, blower motor, and radiator fan motor switches, sensors, and relays are described in more detail in Chapters 8 and 12.

✓PROBLEM SOLVING

A friend mentions that her A/C system is not working at all. With the system turned on and the hood open, you see that the compressor short cycles; it runs for a few seconds and then turns off for a while. What could be causing this? What will you need to do to confirm this opinion?

6.8.2 Suction Throttling Valves

An STV, also called a **control valve, evaporator pressure regulator valve,** or a **pressure regulator**, is used in the low side of some systems. STVs were common on domestic systems manufactured in the 1960s and 1970s. STVs are still used in some imported vehicles. Placed somewhere between the evaporator outlet and the compressor inlet, an STV is sensitive to evaporator pressure and, in one case, temperature. An STV is wide open when evaporator pressure is above 30 psi. When the evaporator cools and the pressure begins to drop below 30 psi, the valve closes to keep the pressure

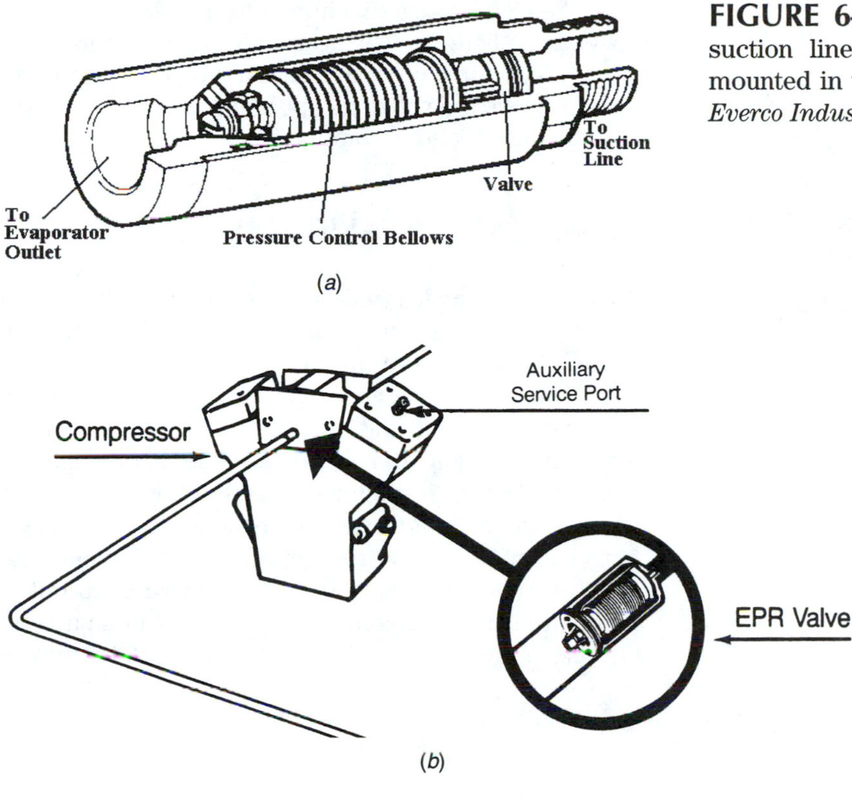

FIGURE 6–73 Some STVs are connected into the suction line (*a*), some to the TXV, and some are mounted in the compressor inlet (*b*). (b *is courtesy of Everco Industries*)

from dropping further. Most valve types modulate and close just enough to maintain correct evaporator pressure. Some valves close almost completely and cycle between open and closed to maintain the correct pressure. A system with an STV does not cycle the clutch except when the system is turned on or off (Figure 6-73).

These valves have gone by different names, depending on manufacturer and year. Major models are as follows:

- Robotrol: Early 1960s; aftermarket
- Selectrol: Early 1960s; aftermarket
- EPR: Chrysler Corporation, 1961 to 1979; mounted in compressor inlet
- EPRV: Toyota and Lexus; mounted in suction line of ATC systems
- ETR: Chrysler Corporation; mounted in compressor inlet
- POA: GM, 1965 to 1975; mounted at the evaporator outlet
- POA (capsule form): GM, 1973 to 1978 (Audi and Volvo); mounted in the VIR
- STV (early types): GM, 1962 to 1966; mounted at evaporator outlet; allows driver control of temperature through mechanical cable or vacuum control
- STV (modern types): Ford Motor Company, 1972 to 1979; mounted at evaporator outlet or suction line
- STV (capsule): Ford Motor Company, 1976 to 1981; combined with TXV in a unit called the combination valve

EPR, POA, and other STVs contain an internal bellows that expands or contracts in response to evaporator pressure. When pressure begins to drop, the expansion of the bellows closes the passageway to trap the pressure upstream in the evaporator (Figure 6-74). Downstream pressure in the suction line will be quite low, and frost and ice will often form on the suction line of a properly operating system.

✓PROBLEM SOLVING

You have been driving your friend's 1976 Oldsmobile at freeway speeds for about an hour, and the day is hot so you have had the A/C on. You stop for gasoline. When you open the hood to check the oil, you notice that the A/C suction hose is covered with frost, from the compressor inlet to a valve at the rear of the engine compartment. Is this a good sign or something to be concerned about?

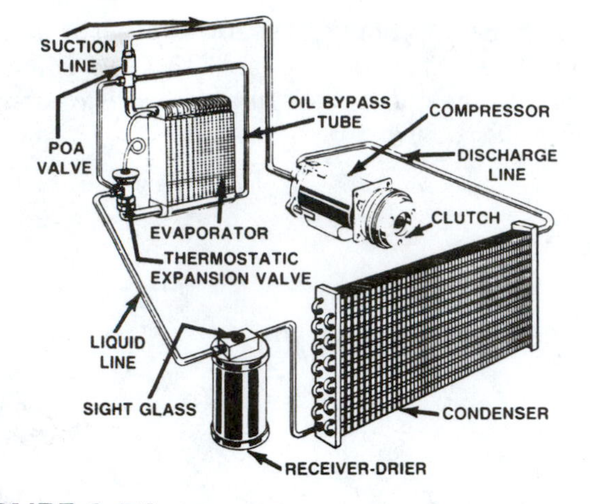

FIGURE 6–74 The POA valve in this system closes when evaporator pressure begins to drop too low. Evaporator temperature is kept just above icing. (*Courtesy of Everco Industries*)

Some STVs use a diaphragm for pressure to act on instead of a bellows. The evaporator temperature regulator (ETR) valve uses an internal electric solenoid to open or close the valve; its operation is controlled by a thermal switch at the evaporator.

6.9 REAR A/C SYSTEMS

Some larger vehicles (vans and small buses) have dual heat and A/C assemblies, with the rear unit mounted in a rear side panel or in the roof. The rear A/C unit consists of an evaporator and TXV that operates in parallel flow with the front unit (Figure 6-75). Tee fittings are placed in the liquid and suction lines so that refrigerant can flow through both units, with the flow through the rear unit dependent on the cooling load. Many dual systems use an OT to control the refrigerant flow into the front evaporator and a TXV with the rear evaporator. The rear evaporator is normally mounted in an assembly that includes a blower, heater core, and doors to

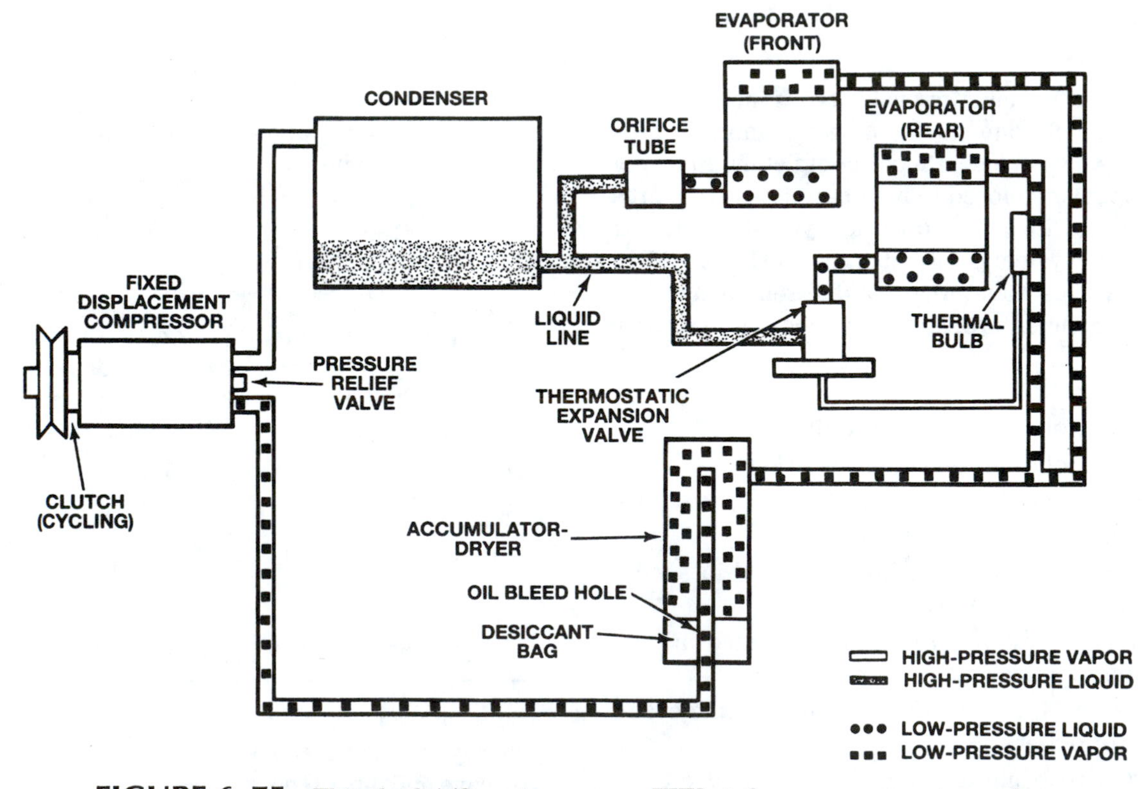

FIGURE 6–75 This dual A/C system uses a TXV at the rear evaporator and an OT at the front evaporator. Note how the liquid and suction lines split to both units. (*Reprinted with permission of General Motors Corporation*)

FIGURE 6–76 Aftermarket A/C systems are available that fit the evaporator under the dash of cars and pickups (*a*), under the seats (*b*), in the side panels (*c*), or in the roofs of vans (*d*). (*Courtesy of Acme Radiator & Air Conditioning*)

control the air temperature and where the air returns to the passenger compartment.

The rear TXV shuts off the flow through the rear evaporator when the rear system is shut off, but a potential problem is created. A TXV does not ensure a complete shutoff, and refrigerant can leak through the valve. Because the blower is shut off, the evaporator will chill, and liquid refrigerant can puddle in this portion of the suction line. The puddle can flow down the suction line and slug the compressor causing a knock and possible damage. Some systems use a solenoid valve to stop any flow through the rear system.

6.10 AFTERMARKET A/C UNITS

Aftermarket A/C units are installed in vehicles after they are sold and leave the dealership. These units are designed for cars, trucks, vans, recreational vehicles (RVs) and motor homes, ambulances, and tractors and farm equipment. Some units are packages with the entire system, including all of the A/C parts. Many units for vans, RVs, and ambulances include just the evaporator, blower, and air ducts to supplement the A/C unit already in the vehicle. Some of these include a supplementary heater (Figure 6-76).

The evaporator in some aftermarket units is in an assembly that includes a blower, air registers, and controls. This unit is mounted under the dash and is often called a **hang-on unit**. The evaporator unit for van conversions mounts in the rear, with air discharge registers running forward or overhead. On some units, the evaporator case is incorporated into the factory heater system and uses the original blower and controls.

The evaporator case for trucks and farm equipment is often a **roof pack**, located in the vehicle's roof. The evaporator and its blower are positioned to circulate air through the roof, and the condenser and its fans are located outside (Figure 6-77). The compressor is mounted on the engine in a normal manner and connected to the roof unit by long suction and discharge lines.

Complete units include a compressor with a mounting kit that includes mounts and drive pulleys, hoses, receiver–driers, and all other necessary parts. Supplementary units use tee fittings to connect into the existing liquid and suction lines.

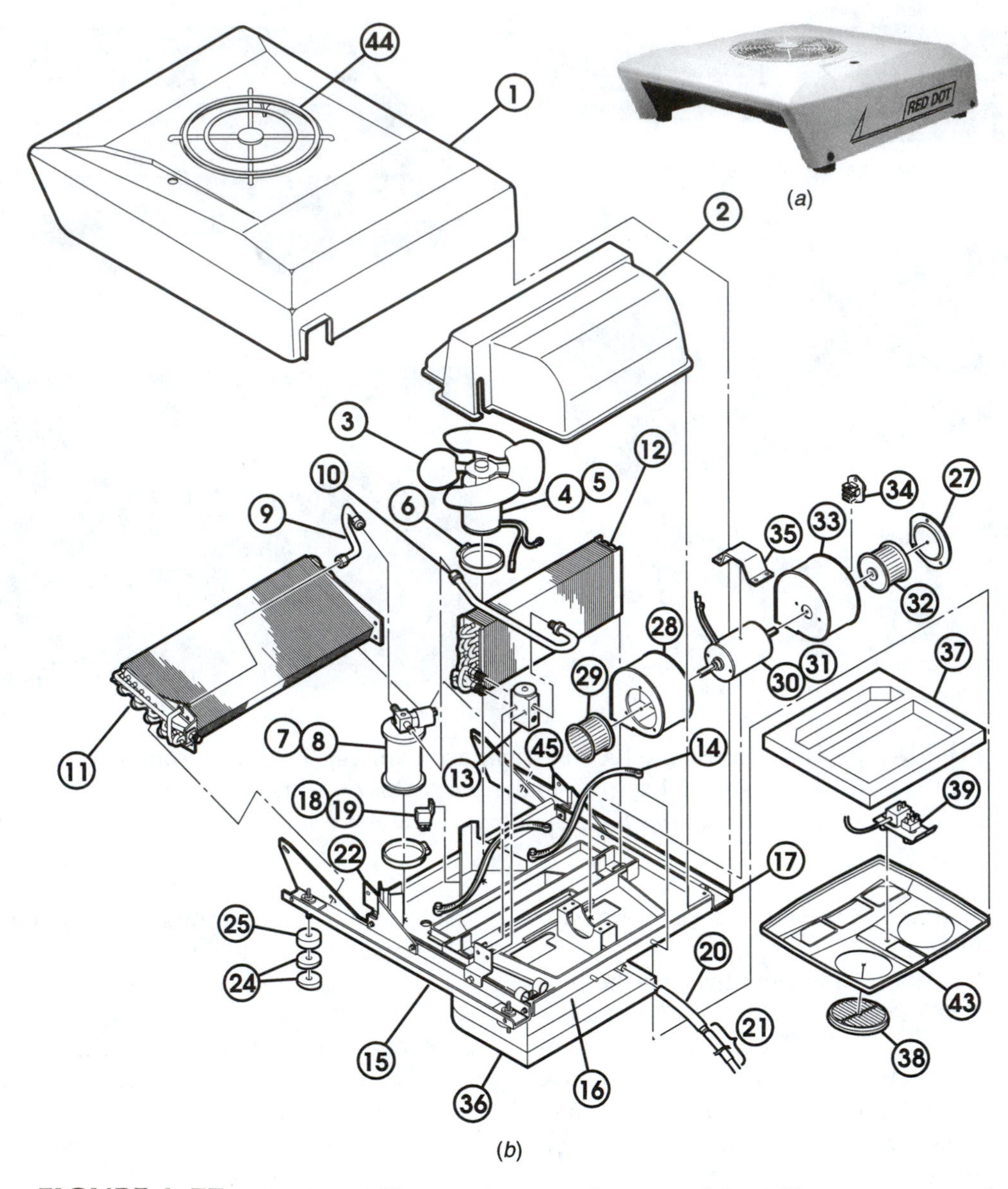

FIGURE 6–77 Rooftop A/C units that contain most of the A/C system are used in trucks, tractors, and various other vehicles. Assembled (*a*) and exploded (*b*) are shown. (*Courtesy of Red Dot Corporation*)

CHAPTER QUIZ

These questions help you study this chapter. Enter the proper word(s) in the blanks to complete each statement.

1. A swash plate compressor uses __________ __________ pistons, and a wobble plate compressor uses __________ __________ pistons.
2. __________ __________ moves the refrigerant through the suction reed as the piston moves __________, and piston pressure moves the refrigerant through the __________ during the upward stroke.
3. The wobble plate will move to a low-angle position when the __________ gets __________.
4. For pumping members, a scroll compressor uses a(n) __________ __________ and a(n) __________ __________ that travels in a small orbit.
5. Vane and scroll compressors are more __________, which makes them more suitable for modern vehicles with tight engine compartments.
6. The three major parts of a compressor clutch are the __________ __________, __________, and __________.

7. A simple way for the manufacturer to increase the strength of a clutch is to make slots called __________ __________ in the amature disc.
8. Oil is transported by the refrigerant because it is __________ with the refrigerant.
9. __________ oil is used with R-12, and __________ or __________ is used with R-134a.
10. Refrigerant oils are __________ so the oil container should be kept __________ except when removing oil.
11. Most RWD vehicle condensers use __________ __________ flow to move air through the condenser.
12. A TXV is opened by __________ on the __________ coming from a warm __________ __________.
13. An OT is mounted at the __________ __________ or in the __________ __________ close to the condenser outlet.
14. A(n) __________-__________ evaporator is made from a series of stamped aluminum plates.
15. A TXV system will have a(n) __________-__________ mounted in the liquid line, and an OT system will use a(n) __________ mounted in the suction line.
16. A(n) __________ __________ fitting has the most potential for leaks, and a(n) __________ __________ is the most difficult type to repair if leaking.
17. The hose type with an impervious inner layer is called a(n) __________ __________.
18. The minimum temperature of an evaporator is controlled by __________ the __________ off and on, by __________ the __________ compressor, or by __________ the flow back to the compressor with a __________ __________ __________.

REVIEW QUESTIONS

These questions allow you to check what you have learned. Select the answer that correctly completes each statement.

1. In a reciprocating piston compressor, the pistons are driven by a
   - a. crankshaft.
   - b. swash plate.
   - c. wobble plate.
   - d. All of these
2. Most York compressors use cylinders made from
   A. cast iron. B. aluminum.
   Which is correct?
   - a. A only
   - b. B only
   - c. Both A and B
   - d. Neither A nor B
3. The reciprocating piston compressor used on many older General Motors vehicles is designated by which of the following?
   - a. R-6
   - b. D-6
   - c. DA-6
   - d. A-6
4. Two students are discussing variable displacement piston compressors. Student A says that the wobble plate is moved to the high-angle position for maximum output when the cooling load is high. Student B says that the wobble plate angle is controlled by the pressure in the crankcase. Who is correct?
   - a. Student A
   - b. Student B
   - c. Both A and B
   - d. Neither A nor B
5. The General Motors variable displacement compressor has _____ cylinders.
   - a. 4
   - b. 5
   - c. 6
   - d. 7
6. Which of the following is an advantage offered by a scroll-type compressor?
   - a. Smooth engagement
   - b. Low engagement torque
   - c. Ability to use a small clutch
   - d. All of these
7. A _____ is used to seal the opening for the compressor driveshaft.
   A. carbon seal cartridge B. lip seal
   Which is correct?
   - a. A only
   - b. B only
   - c. Both A and B
   - d. Neither A nor B
8. Two students are discussing refrigerant oils. Student A says that the mineral oil used in R-12 systems is completely compatible with R-134a. Student B says a synthetic oil should be used with R-134a. Who is correct?
   - a. Student A
   - b. Student B
   - c. Both A and B
   - d. Neither A nor B
9. Two students are discussing TXVs. Student A says that the TXV senses evaporator temperature to control the valve opening. Student B says that the TXV senses evaporator pressure to control the valve opening. Who is correct?
   - a. Student A
   - b. Student B
   - c. Both A and B
   - d. Neither A nor B
10. Two students are discussing A/C systems. Student A says that the TXV systems always use a receiver–drier mounted in the suction line. Student B says that the OT systems always use an accumulator in the suction line. Who is correct?
    - a. Student A
    - b. Student B
    - c. Both A and B
    - d. Neither A nor B
11. The fitting found on an accumulator is for which of the following?
    - a. High side service
    - b. Leak detection
    - c. Flushing
    - d. Low side service
12. A filter can be installed in the liquid line for which of the following?
    - a. When the system is contaminated
    - b. When extra moisture removal is needed
    - c. Instead of flushing the evaporator
    - d. All of these

13. The pressure cycling switch does which of the following?
    - **a.** Closes the TXV
    - **b.** Senses condenser temperature
    - **c.** Senses high side pressure
    - **d.** Senses low side pressure

14. Two students are discussing line fittings. Student A says that many early A/C systems used flare fittings. Student B says that modern systems commonly use O-rings to seal the fittings, and these are much easier to repair than flare fittings. Who is correct?
    - **a.** Student A
    - **b.** Student B
    - **c.** Both A and B
    - **d.** Neither A nor B

15. The hose that connects the compressor to the evaporator is commonly called the
    - **a.** discharge line.
    - **b.** liquid line.
    - **c.** suction line.
    - **d.** cold gas line.

CHAPTER

Heating Systems

LEARNING OBJECTIVES

After completing this chapter, you should:

- Understand how the common automotive heating system works.
- Know what parts make up the heating system.
- Understand how heater temperature is controlled.

TERMS TO LEARN

Bowden cable
cellular
control valve
heater core
hoses
quick-connect coupling
thermal cycling
three-way thermostat
vacuum motor

7.1 INTRODUCTION

As previously mentioned, the heating system resembles a small version of the engine's cooling system. Some people consider the heater the most efficient part of the vehicle because it uses waste heat to warm the interior (Figure 7–1). The heating system is made up of the **heater core, hoses,** and, in some systems, a **control valve**.

It should be noted that Federal Motor Vehicle Safety Standard (FMVSS) 103 requires that every vehicle sold in the U.S. must be able to defrost or defog certain areas of the windshield in a specified amount of time. It also requires that this system must remain operable. The defrost/defog system in most vehicles uses the heater and diverts the air flow to the base of the windshield.

7.2 OPERATION

The **inlet hose** to the heater core connects to an outlet near the engine thermostat, or an area of the engine with the hottest coolant. The **outlet hose** from the heater core runs to a connection near the inlet of the engine's water pump, the area with the lowest coolant pressure. When the engine runs, coolant flows through the engine's water jackets and heater core. The heated coolant warms the heater core and the air passing through it. A **three-way-design thermostat** is used in a few engines to shut off coolant flow to the heater core when the engine gets very hot (see Chapter 15) (Figure 7–2).

Some vehicles have a valve in the heater inlet hose that allows coolant flow to be shut off and keeps the core from heating up. Some vehicles use a heater valve to control the heat of the core and therefore the temperature of the air entering the passenger compartment. Most newer cars do not use a control valve, so the core is always the same temperature as the coolant. With these systems, the temperature of the air to the passenger

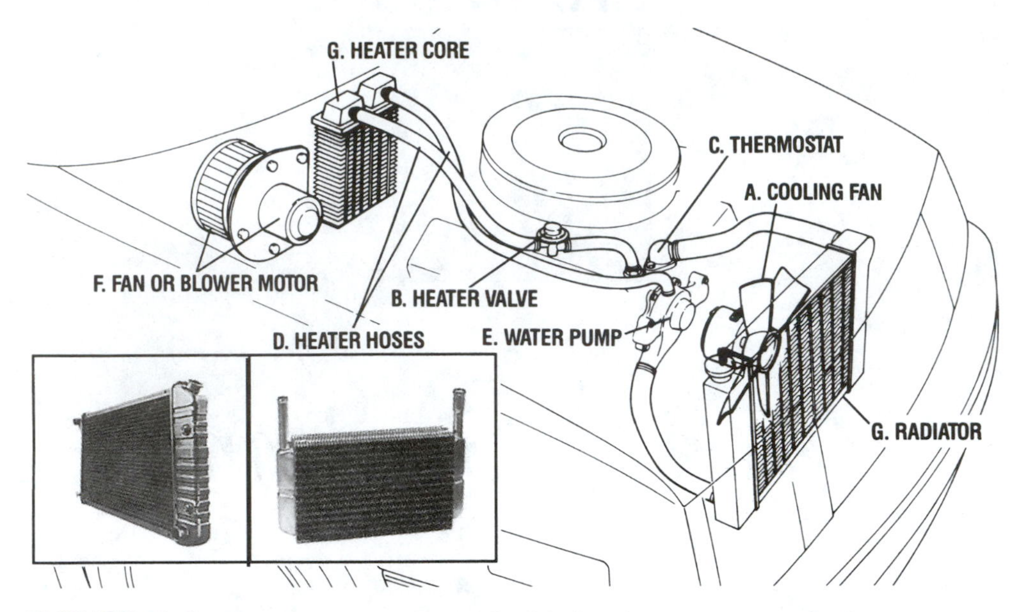

FIGURE 7–1 The main parts of a vehicle's heating system are the heater core, blower, heater hoses, and, in some cases, heater valve. (*Courtesy of Everco Industries*)

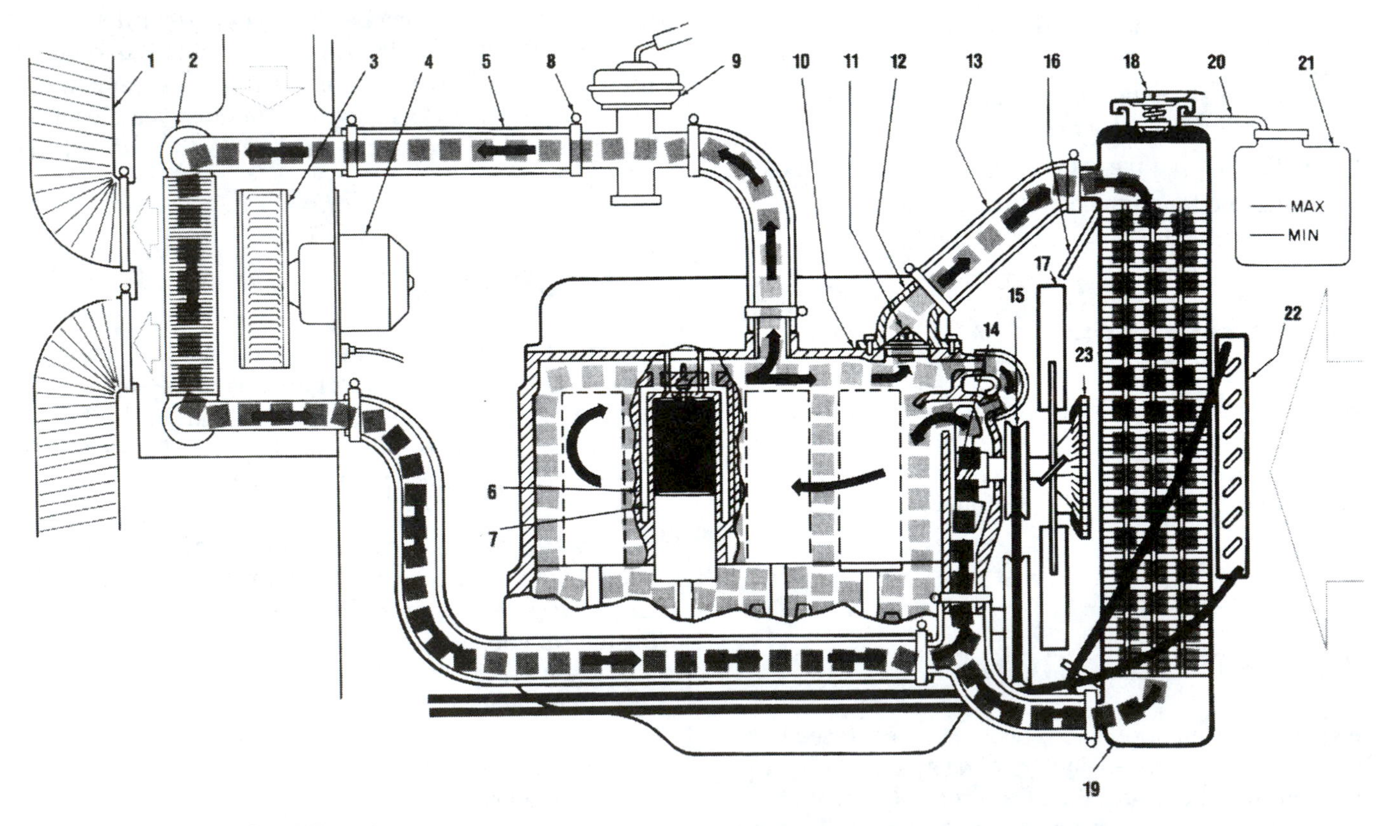

FIGURE 7–2 Some heater systems use a valve (9) that can shut off the flow of hot coolant to the heater core (2). (*Courtesy of Stant Manufacturing*)

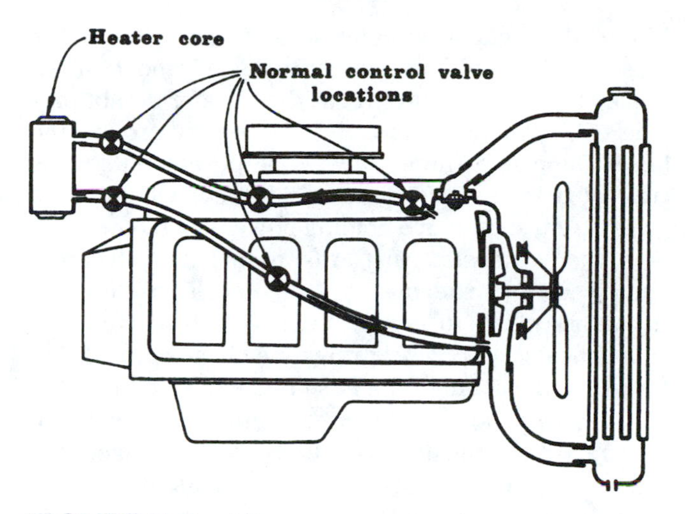

FIGURE 7–3 A heater control valve can be located at any of the locations shown.

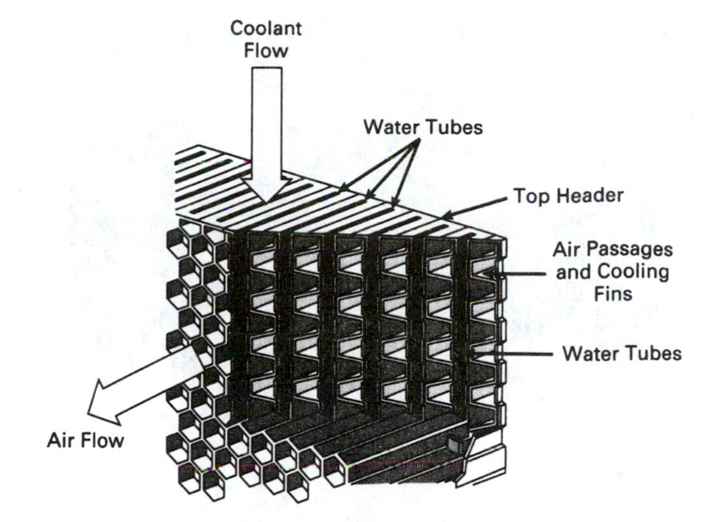

FIGURE 7–4 Heat is transferred from the hot coolant flowing through the water tubes to warm the air flowing through the fins of the core.

compartment is controlled by an air temperature-blend door in the heater plenum. The operation of this door is described in Chapter 8 (Figure 7–3).

7.2.1 Electric and Hybrid Vehicle Heating Systems

Electric vehicles do not have an engine, and hybrid vehicles use an engine that does not run long enough to produce enough heat in the coolant for a heater. Some ultra-low emission vehicles do not produce enough waste heat to supply a heater.

Electric vehicles that use a heat pump for A/C include a valve that switches the flow between the condenser and evaporator. This switches heat transfer from the condenser to the evaporator (see Figure 5–30). This is similar to what happens when a heat pump used in a home is asked to heat the residence. Some electric and hybrid vehicles supplement the heater with an electrical resistance heater. This greatly increased electrical draw from the batteries reduces the operating range significantly.

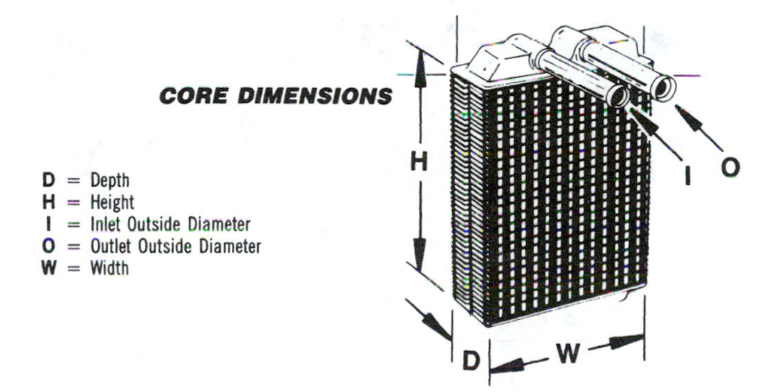

FIGURE 7–5 The critical dimensions needed when replacing a heater core. (*Courtesy of Four Seasons*)

7.3 HEATER CORE

The heater core is a heat exchanger much like the condenser, evaporator, and radiator. Heat transfers from the coolant, to the fins, and to the air passing through the core. As with other heat exchangers, there is a large area of fin-to-air contact to allow sufficient heat transfer and airflow (Figure 7–4).

Most heater cores use a cellular form of construction that is somewhat like a plate-type evaporator. The tubes are made by joining two corrugated brass or aluminum plates, and the corrugated fins are attached between pairs of tubes. In most cores, the tanks at the ends of the core serve as manifolds to direct the flow back and forth through the core.

A core is usually designed to fit a particular car model. The critical dimensions needed to ensure correct replacement are the depth, height, and width of the core and the diameters and shapes of the inlet and outlet tubes (Figures 7–5 and 7–6).

7.4 HOSES

Most heater hoses are made of reinforced rubber, which allows the flexibility needed to connect to a movable engine (Figure 7–7). Common hose sizes are 1/2, 5/8,

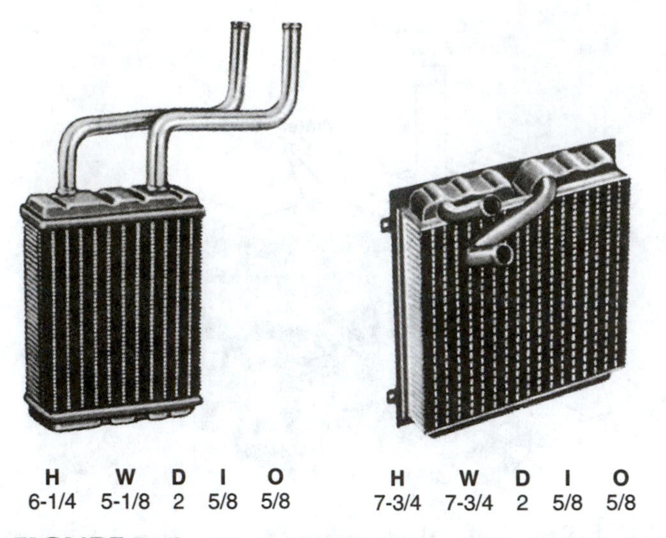

FIGURE 7–6 Two examples of the many shapes and sizes of heater cores. *(Courtesy of Four Seasons)*

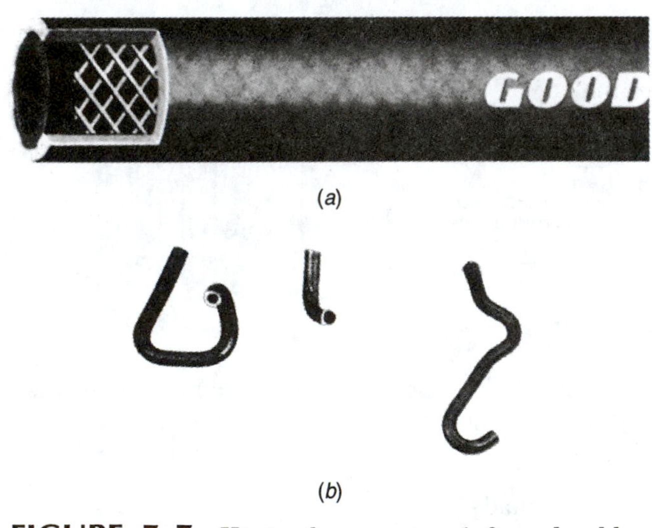

FIGURE 7–7 Heater hose uses reinforced rubber construction (*a*) and is available in straight or molded (*b*) shapes. *(Courtesy of Goodyear Tire & Rubber Company)*

and 3/4 inch (12.7, 15.8, and 19 mm). Some systems connect the hose to a metal tube that can be connected to the engine using either a threaded connector or a short hose section. This allows easier routing through congested or very hot areas around the engine that might cause hose failures.

Many systems use a smaller hose diameter for the heater inlet than for the outlet. This allows an easier exit of the coolant and reduces the pressure inside the core, which in turn reduces the possibility of leaks.

A hose is usually clamped to the connector using one of several styles of clamps. The screw-type or Whittek-type clamp is considered by many to be the most reliable and the easiest to install and remove. Some clamp styles tend to cut the hose or do not provide even clamping pressure all the way around the hose. The clamp should always be positioned so it is right next to the raised area of the connection. This gives the most effective holding power and reduces one area where corrosion can form. Another clamping problem is caused by **thermal cycling** that causes the hose, fitting, and clamp to expand and contract each time the engine heats and cools. Spring-type clamps provide automatic retensioning. The hose tends to seize onto the connector so the technician needs to be careful when disconnecting a heater hose to prevent damaging the rather fragile heater core or its connectors.

Some newer vehicles use a **quick-connect coupling** at the hose-to-heater-core connections. Care and a suitable tool are required to disconnect these couplings to prevent damage to the very expensive hose assemblies (Figure 7–8).

7.4.1 Restrictors

Some systems include a restrictor to slow the coolant velocity as it passes through the heater core. The restrictor can be part of the manifold fitting or the inlet heater hose assembly. The major purpose of the restrictor is slow the flow rate in order to reduce internal heater core erosion.

7.5 CONTROL VALVES

Most heater control valves are on–off valves and are used to make the core either hot or cold. Some valves are designed to modulate and adjust the flow so the core temperature can be controlled to all points between hot and cold. Some valves allow a return flow, so the coolant still circulates, bypassing the core when the valve is shut off.

The control for most valves is through a **vacuum diaphragm** or **motor**. These systems use a vacuum signal from the HVAC control head to close the valve and shut off the heater. This feature, using a normally open valve, allows heater and defroster operation if the valve or vacuum control system fails. Some valves are operated mechanically through a **Bowden cable,** a steel wire that slides through a housing. Bowden cables are commonly used for hood release mechanisms and in the past were used for carburetor choke control. A few older cars use manual control valves, which require someone to turn the valve stem to open or close the valve (Figure 7–9).

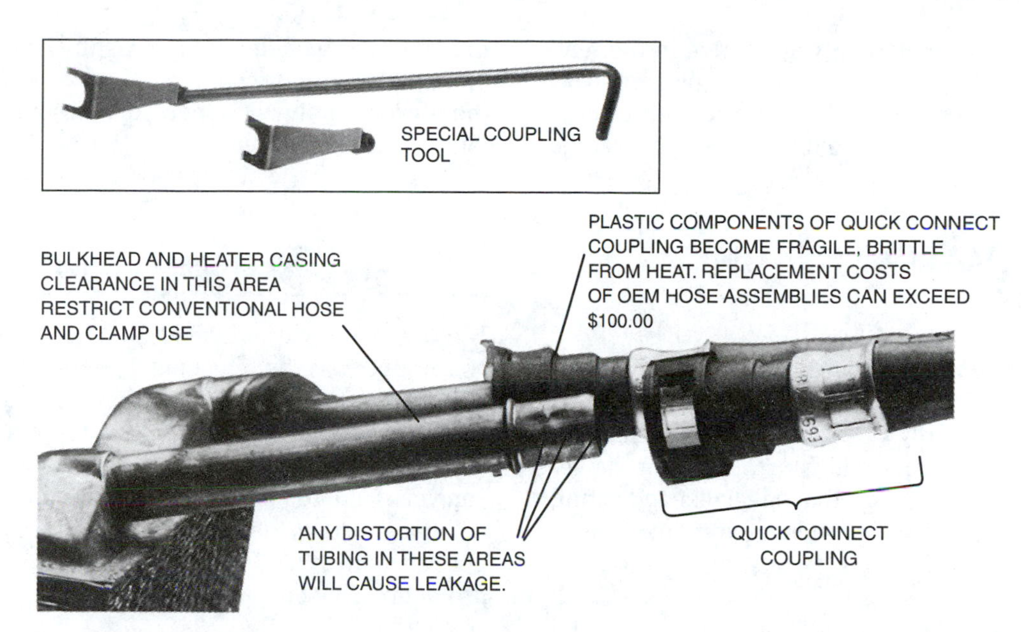

FIGURE 7–8 Some vehicles use quick-connect couplings for the heater hose connections. The hose is merely slid firmly onto the connector to make the connection. It should be disconnected carefully using a suitable tool. (*Courtesy of Four Seasons*)

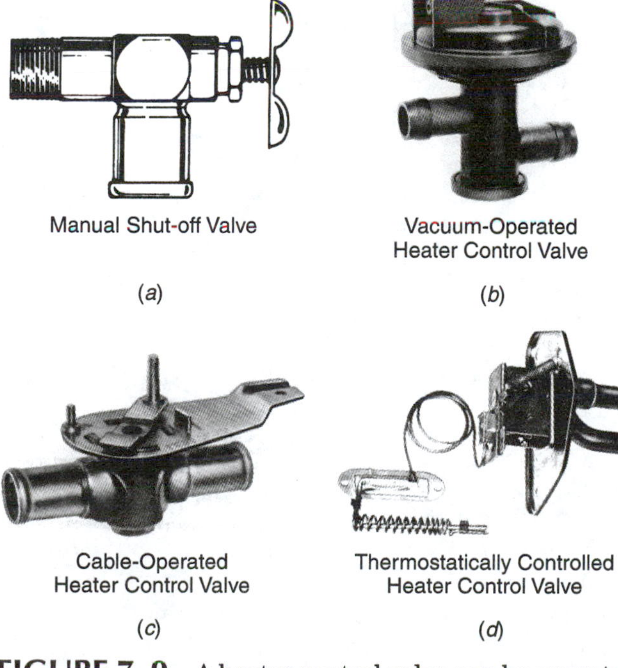

FIGURE 7–9 A heater control valve can be operated manually (*a*), by vacuum control (*b*), through a mechanical cable (*c*), or by a thermostatic element (*d*). (a *is courtesy of Four Seasons;* b, c, *and* d *are courtesy of Stant Manufacturing*)

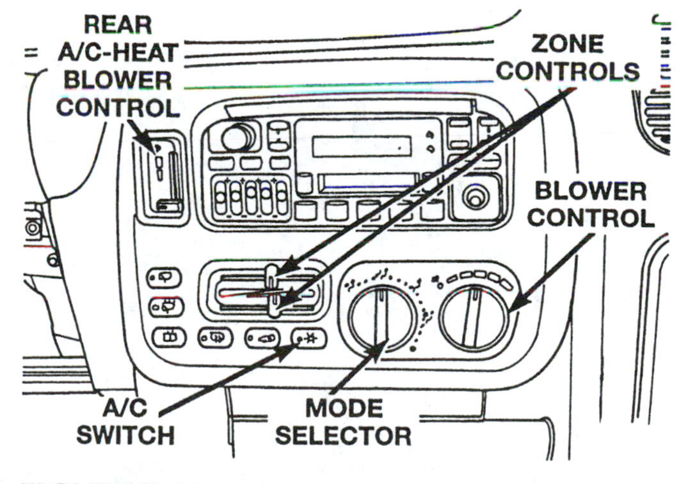

FIGURE 7–10 The control panels for a vehicle with front and rear climate control systems. (*Courtesy of DaimlerChrysler Corporation*)

7.6 DUAL HEATING SYSTEMS

Many vehicles with rear A/C systems include a heater in the rear unit. These rear units include a heater core and temperature-blend door (Figure 7–10). The heater hoses include a tee fitting in each hose so heated coolant can flow through either or both heater cores.

Some manufacturers include a water valve in the rear heater core so that hot coolant can be kept out of the core during A/C operation to improve A/C efficiency. The heater operation on these units is the same as that for a front unit.

7.7 AFTERMARKET HEATING SYSTEMS

Before heaters became standard equipment, aftermarket heating systems were installed in many cars. Today, these units are primarily designed for RVs, vans, and motor homes. They are normally installed after the vehicle has been built, by shops that specialize in heating and A/C service or RV van conversion and repair (Figure 7–11).

The units are basically a heater core in a case with a blower. After locating and mounting the case, the heater hoses are connected to the engine connectors or to tee fittings installed in the existing heater hoses. The blower switch is mounted at a convenient location and the power supply is connected to a source of *B*+ (battery positive). In some units, the heater is combined with an A/C unit, as described in Section 6.9.

✓PROBLEM SOLVING

You are riding with a friend on a cold winter day, and you notice that the air coming from the heater is only slightly warm. You mention this to him, and he tells you that it takes a long time for the heater output to become hot. What do you think might be wrong? What checks can you make to confirm your suspicions?

Imagine that the heater output stays cold and does not warm up at all. What do you think could cause this problem? What checks would you make to confirm your suspicions?

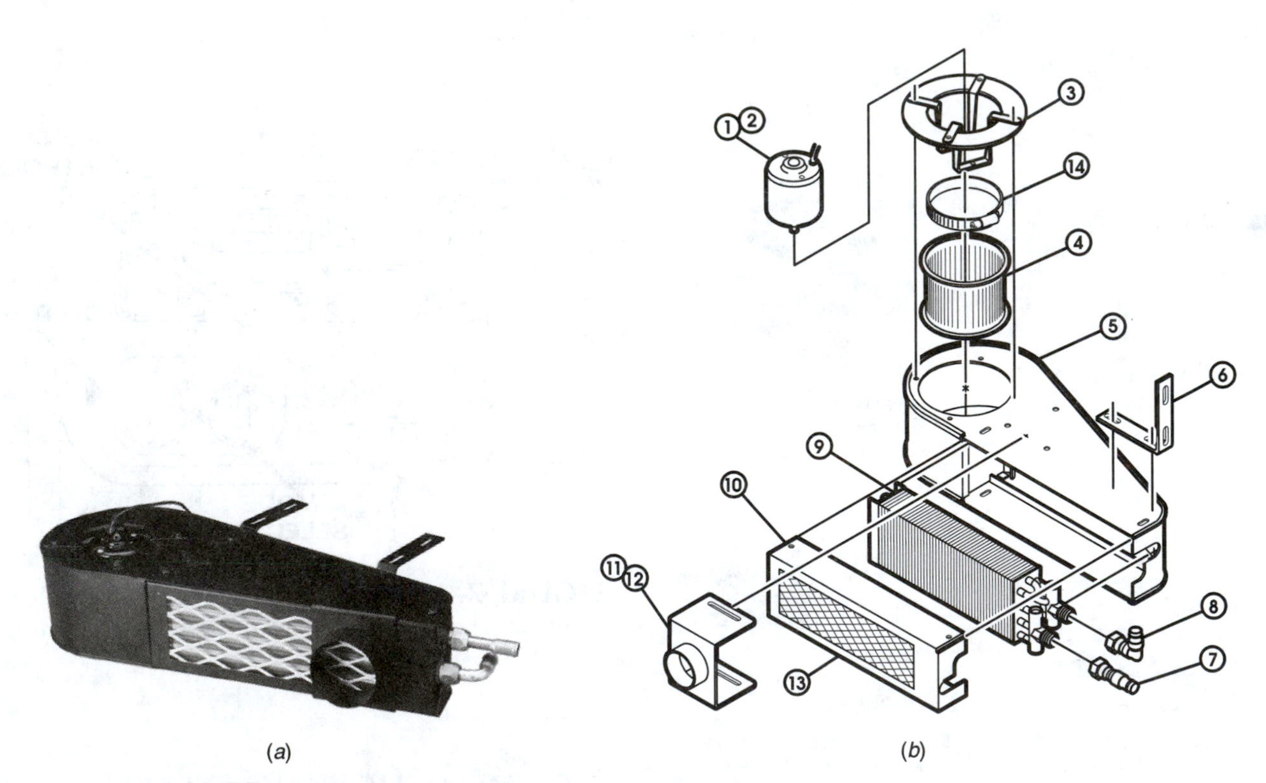

FIGURE 7–11 This aftermarket heater assembly *a* contains a heater core (9) and blower (1–4). An exploded view is at *b*. (*Courtesy of Red Dot*)

CHAPTER QUIZ

These questions help you study this chapter. Enter the proper word(s) in the blanks to complete each statement.

1. The heater core inlet hose connects to the engine close to the __________, and the outlet hose is connected close to the water pump __________.
2. The heat from a vehicle's heater comes from the engine __________.
3. A(n) __________ __________ heater core is made from a set of stamped metal plates.
4. Critical dimensions of a heater core include the __________, __________, and __________ of the core plus the __________ and __________ of the tubes.
5. A hose clamp of good design will not __________ the hose, clamp completely __________ the hose, and compensate for __________ and __________.
6. Some modern heater hoses use __________ - __________ couplings at the heater core.
7. Temperature of the air leaving the heater ducts is controlled by a(n) __________ that regulates the coolant flow through the core or by a(n) __________ - __________ door.
8. Some vans will use __________ __________ systems to help warm up the rear compartment.

REVIEW QUESTIONS

These questions allow you to check what you have learned. Select the answer that correctly completes each statement.

1. Two students are discussing the coolant flow through a heater core. Student A says that heated coolant enters the heater inlet hose close to the engine thermostat. Student B says the coolant from the heater core returns close to the water pump outlet. Who is correct?
   - **a.** A only
   - **b.** B only
   - **c.** Both A and B
   - **d.** Neither A nor B
2. Two students are discussing the heater core. Student A says that a heater core is a heat exchanger much like a radiator. Student B says the coolant from the engine brings the heat to the heater core. Who is correct?
   - **a.** A only
   - **b.** B only
   - **c.** Both A and B
   - **d.** Neither A nor B
3. Two students are discussing heater hoses. Student A says that most hoses have a diameter of about one inch. Student B says the heater core inlet hose is usually larger than the outlet. Who is correct?
   - **a.** A only
   - **b.** B only
   - **c.** Both A and B
   - **d.** Neither A nor B
4. Heater hoses are attached to the core using
   - **A.** clamps.
   - **B.** quick-connect couplings.

   Which is correct?
   - **a.** A only
   - **b.** B only
   - **c.** Both A and B
   - **d.** Neither A nor B
5. Two students are discussing the heater output temperature. Student A says that a valve is used to reduce coolant flow through the core when less heat is desired. Student B says that most modern systems use a blend air door to control the coolant flow. Who is correct?
   - **a.** A only
   - **b.** B only
   - **c.** Both A and B
   - **d.** Neither A nor B
6. The control for most heater control valves is either a Bowden cable or mechanical.
   - **a.** True
   - **b.** False
7. Two students are discussing heater temperature control. Student A says that a control valve will be wide open when full heat is desired. Student B says the coolant control valve should go wide open if the vacuum line is disconnected. Who is correct?
   - **a.** A only
   - **b.** B only
   - **c.** Both A and B
   - **d.** Neither A nor B
8. An aftermarket heater system consists of
   - **a.** a core.
   - **b.** a blower.
   - **c.** control switches.
   - **d.** All of these

CHAPTER

Air Management System

LEARNING OBJECTIVES

After completing this chapter, you should:

- Understand the function of the air control doors in the A/C and heating duct system.
- Understand how the temperature of the air entering the car is controlled.
- Be familiar with the methods used to control the blower speed.
- Be familiar with manual, semi-automatic temperature control (SATC), and automatic temperature control (ATC) systems.
- Be familiar with the sensors and controls used with ATC and SATC systems and their function in controlling air temperature and flow.

TERMS TO LEARN

air distribution
air doors
air inlet
air management
ambient sensor
aspirator
automatic temperature control (ATC)
bleeds
blend door
blower motor
cabin air filter
ducts
dual-zone
flap doors
function door
HVAC air filter
in-car sensor
manual system
mode door
plenum
pulse width modulation (PWM)
ram air pressure
recirculation
resistor
restrictor
semi-automatic temperature control (SATC)
sensor
thermister
vacuum actuator
vacuum bleed

8.1 INTRODUCTION

A system that contains the HVAC plenum, ducts, and air doors, called the **air management system** or **air distribution system,** controls the airflow to the passenger compartment. Air flows into the case that contains the evaporator and heater core from two possible inlets. From the case, the air can pass on to one or more of three possible outlets. Proper temperature control to enhance passenger comfort during heating should maintain an air temperature in the footwell about 7 to 14°F (4 to 8°C) above that around the upper body. This is accomplished by directing the heated airflow to the floor. During A/C operation, the upper body should be cooler, so the airflow is directed to the instrument panel registers. Airflow is controlled by three or more doors, which are called **flap doors** or **valves** by some manufacturers (Figure 8–1).

A multispeed blower is included in this system to force air through the ductwork when the vehicle is moving at low speeds or to increase the airflow at any speed. At freeway speeds, most systems have an airflow from **ram air pressure.** This is the pressure generated at the

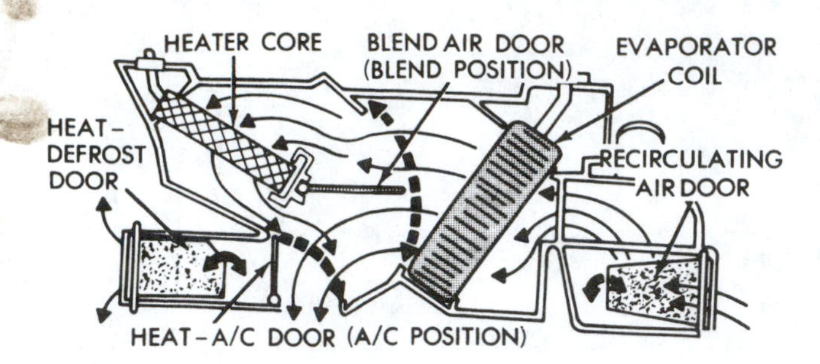

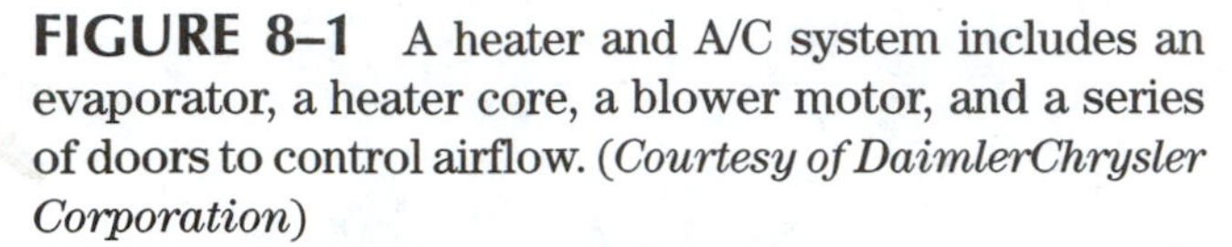

FIGURE 8–1 A heater and A/C system includes an evaporator, a heater core, a blower motor, and a series of doors to control airflow. (*Courtesy of DaimlerChrysler Corporation*)

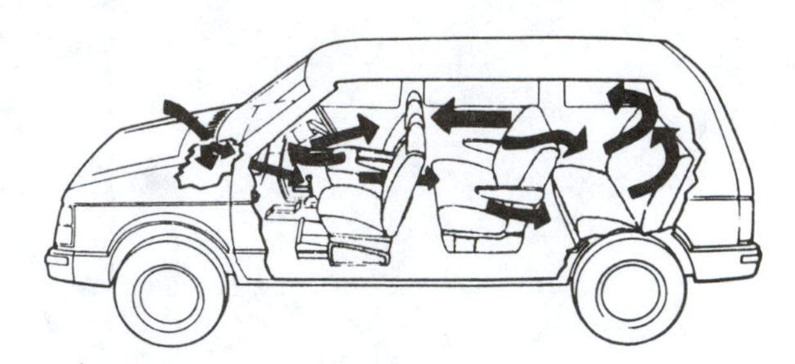

FIGURE 8–2 Fresh air enters through the grill below the front of the windshield; it becomes a high-pressure area when the car is moving forward. (*Courtesy of DaimlerChrysler Corporation*)

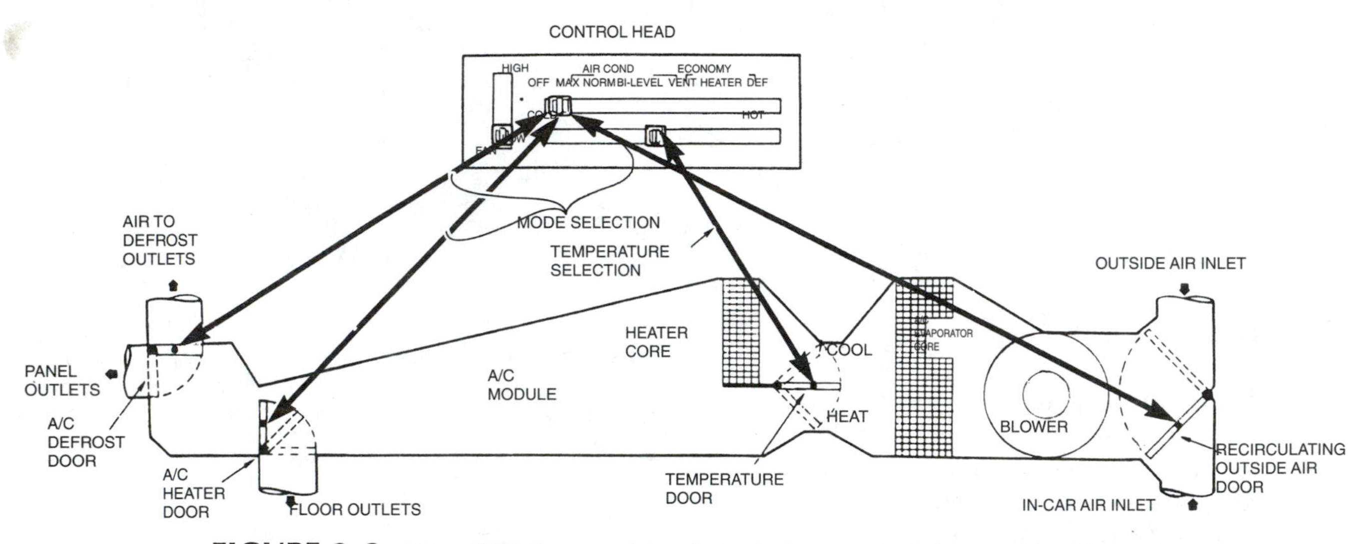

FIGURE 8–3 Most HVAC control heads include a control for turning things on and setting the mode of operation, a control for adjusting the temperature, and a control for the fan speed. (*Reprinted with permission of General Motors Corporation*)

base of the windshield by the speed of the vehicle. This airflow is improved in some vehicles by exhaust registers placed in low-pressure areas toward the rear of the vehicle (Figure 8–2).

The HVAC control head or panel is mounted in the instrument cluster. The control system provides the switches and levers needed to control the different aspects of the heating and A/C system, which include

- System on and off
- Fresh or recirculated air
- A/C, defrost, or heating function
- Temperature desired
- Blower speed

The control head is connected to various parts through electrical connections, vacuum connections, mechanical Bowden cables, or a combination of these (Figure 8–3).

In vehicles using a **manual system,** the driver moves the temperature lever or dial to change the temperature setting, selects the air inlet and discharge locations, selects the blower speed, and turns the A/C compressor on or off. With a **semi-automatic temperature control system (SATC),** the driver selects the desired temperature using buttons or a dial; in some systems, the automatic controls adjust the temperature door and blower speed while the driver selects the air inlet and discharge locations. Different SATC systems allow different types of control for the driver to override certain automatic functions. With an **automatic temperature control system (ATC),** the driver can turn the automatic controls on or off and select the desired temperature. The ATC will adjust the blower speed, temperature door, air inlet door, and mode door to achieve the proper temperature.

A vehicle with an ATC system uses many of the same parts as a manual system. Automatic features include two or more temperature sensors, a microcomputer, and either a group of switches and relays or vacuum valves to control the vacuum or electric motors that operate the air control doors.

In this text, vehicles equipped with both heaters and A/C are described. A vehicle with a heater only is essentially the same in all respects except for not having an evaporator and the rest of the A/C components and the upper-level air discharge registers in the instrument panel.

8.2 CASES AND DUCTS

The evaporator and heater plenum is molded from reinforced plastic or stamped sheet metal and contains the evaporator, heater core, and most of the air control doors. Sometimes the evaporator is in a case separate from the heater core, and these cases are connected by a duct. The plenum is connected to the air inlets and outlets using formed plastic or sheet-metal ducts. These often use portions of the body bulkhead and instrument panel sheet metal for mounting points, as well as parts of the ducts. Some cars use round, wire-reinforced, flexible ducts. These parts are required to contain and direct airflow, be quiet, keep outside water and debris from entering, and isolate engine fumes and noises. Their design is complicated by the limited space allotted to them and competition with other items for this space (Figure 8–4).

The duct system can be divided into three major sections: **air inlet, plenum,** where the cold and hot air are mixed, and **air distribution** (Figure 8–5).

8.2.1 Air Inlet and Control Door

Air can enter the duct system from either the **plenum chamber** in front of the car's windshield (fresh air) or from the **recirc** (short for "recirculation") or **return register.** The return register is often positioned below the right end of the instrument panel. (The right and left sides of the car are always described as seen by the driver.) The fresh air plenum often includes a screen to keep leaves and other large debris from entering with the air. Incoming fresh air is usually routed to the right side of the car, then downward; a sharp bend back toward the blower serves to throw water and other heavier objects out of the airstream. A drain is usually located at this point (Figure 8–6).

The air inlet control door is also called the **fresh air, recirculation,** or **outside** air door. This door is positioned so it can allow airflow from one source while it shuts off the other. It can be positioned to allow fresh air to enter while shutting off the recirculation opening; to allow air to return or recirculate from inside the vehicle while shutting off fresh air; or, in some vehicles, to allow a mix of fresh air and return air. In many modern vehicles, the door is set to the fresh air position in all function lever positions except off, max heat, and max A/C; max A/C and max heat positions the door to recirculate in-vehicle air (Figure 8–7).

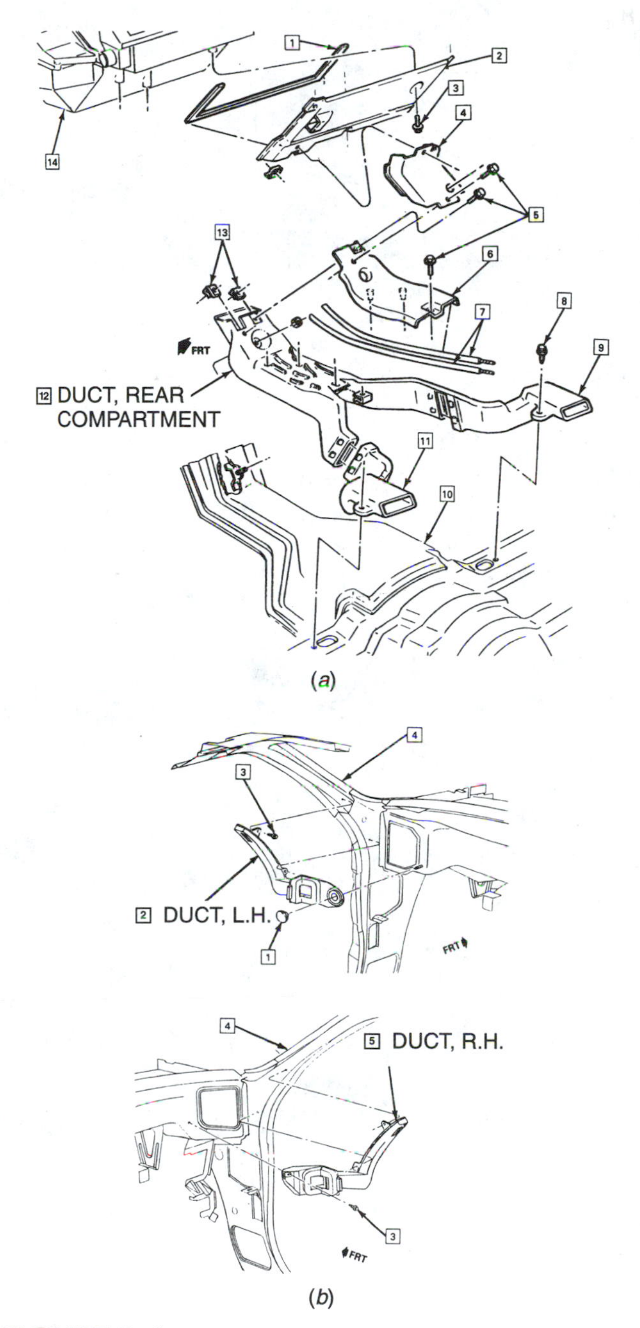

FIGURE 8–4 Some systems include ducts to move air to the rear seat area (*a*) and side windows (*b*). (*Reprinted with permission of General Motors Corporation*)

8.2.1.1 Air Filtration Many modern systems include an **HVAC air filter** in the air distribution system.

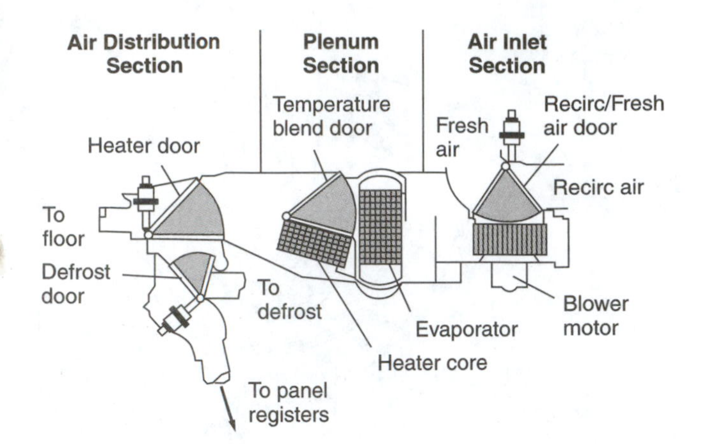

FIGURE 8–5 The three major portions of the A/C and heat system are air inlet, plenum, and air distribution. The shaded portions show the paths of the four control doors.

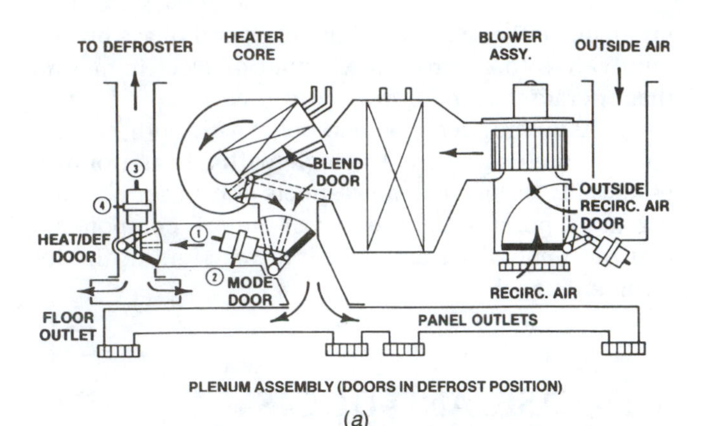

(a)

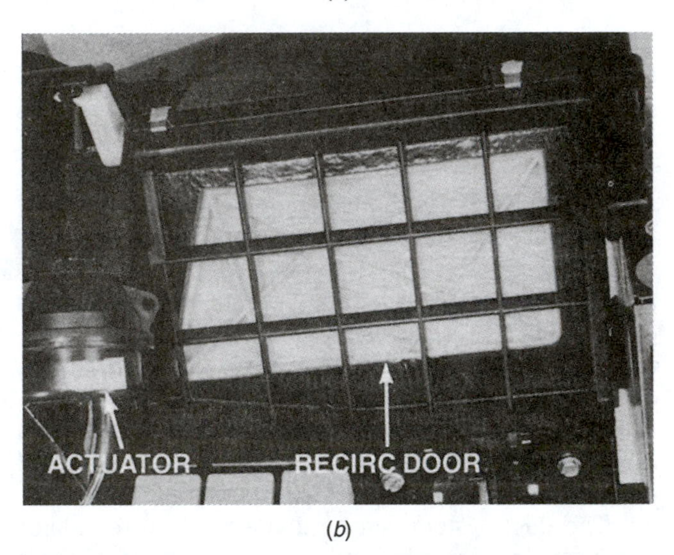

(b)

FIGURE 8–7 This outside-recirculation (air inlet) door is moved from one position to the other by a vacuum actuator (motor); here, it is in the fresh air position (*a*). The actual unit is shown in (*b*). (*Courtesy of DaimlerChrysler Corporation*)

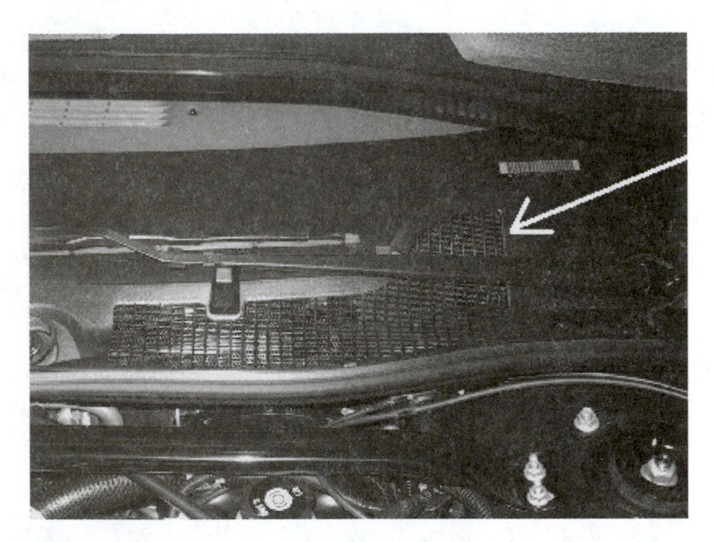

FIGURE 8–6 The fresh air inlet for most vehicles is at the base of the windshield. On this car, the fresh air inlet is under the hood (arrow).

This filter, also called a **cabin, interior ventilation, micron, particulate,** or **pollen filter,** removes small dust or pollen particles from the incoming airstream. (Figure 8–8). These filters require periodic replacement; if they are not serviced properly, they will cause an airflow reduction when they become plugged.

There are two types of filter media: **particle filters** and **adsorption filters.** Particle filters remove solid particles like dust, soot, spores, and pollen using a special paper or nonwoven fleece material; these particles are about 3 micron or larger. The filter media can have an electrostatic charge to make it more efficient. Adsorption filters remove noxious gases and odors using an activated charcoal media. These two filter types can be combined into a combination or two-stage filter.

8.2.2 Plenum and Control Door

Most systems position the evaporator so all air must pass through it. This allows removal of moisture and dust particles by the evaporator's cold temperature. (Many systems operate the A/C when defrost is selected to dry the air.) The heater core is placed downstream so that air can be routed either through or around it; one or two doors are used to control this airflow. This door is called the **temperature-blend** door; some of its other names are **air mix door, temperature door, blend door, diverter door,** and **bypass door** (Figure 8–9).

The temperature-blend door is normally connected to the temperature lever at the control head using a mechanical Bowden cable; in some modern vehicles, the temperature-blend door position is adjusted through an electric servomotor. When the temperature lever is set to the coldest setting, the temperature-

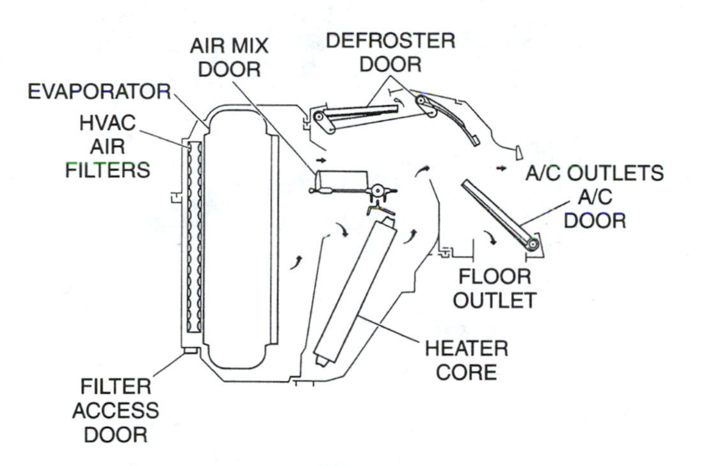

FIGURE 8–8 Some systems include a filter to remove small particles from the fresh air; this filter requires replacement every 12 months or 12,000 miles. *(Reprinted with permission of General Motors Corporation)*

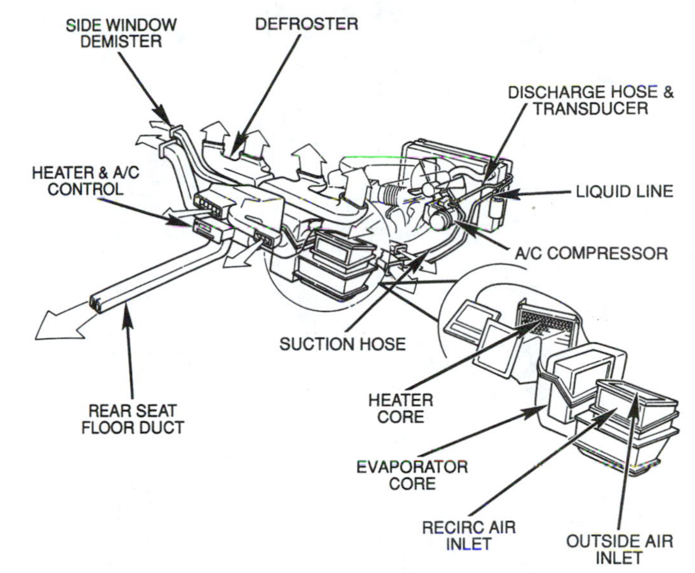

FIGURE 8–9 A complete climate control system. *(Courtesy of DaimlerChrysler Corporation)*

blend door routes all air so it bypasses the heater core; air entering the passenger compartment is at its coldest, coming straight from the evaporator. When the temperature lever is set to the hottest setting, the temperature-blend door routes all air through the heater core, and heated air goes to the passenger compartment. Setting the control lever to somewhere between cold and hot will mix or temper cold and hot air, allowing the driver to adjust the temperature to whatever is desired (Figure 8–10).

In the past in some vehicles, the evaporator was placed or stacked right next to the heater core. These systems controlled the air temperature by regulating the amount of reheat at the heater core. A few modern vehicles are using a stacked heater and evaporator and reheating the heater core to adjust outlet temperature. The dual-zone version uses a two-part heater core to allow temperature control of each zone.

8.2.3 Air Distribution, Control Doors, and Outlets

Air from the plenum can flow into one or two of three outlet paths: (1) the A/C registers in the face of the instrument panel, (2) the defroster registers at the base of the windshield, and (3) the heater outlets at the floor under the instrument panel. Some vehicles include ducts to transfer air to the rear seat area, and some vehicles include ducts to demist the vehicle's side windows with warm air.

Airflow to these ducts is controlled by one or more mode doors controlled by the function lever or buttons. Mode doors are also called **function, floor–defrost,** and **panel–defrost** doors. Setting the function lever to A/C or max A/C sets the doors to deliver air to the in-dash registers, setting the function lever to heat sets the doors to deliver air to the floor level, and setting the function lever to defrost sets the doors to deliver air to the defroster registers. Many control heads also provide for in-between settings, which combine these operations (Figure 8–11). In many systems, a small amount of air is directed to the defroster ducts when in the heat mode, and while in defrost mode a small amount of air goes to the floor level.

8.2.4 Dual-Zone Air Distribution

Dual-zone air distribution allows the driver and passenger to select different temperature settings. The temperature choices can be as much as 30°F different. Dual-zone systems split the duct and airflow past the heater core and use two air mix valves or doors; each air mix valve is controlled by a separate actuator (Figure 8–12).

8.2.5 Horizontal-Split Air Flow

One vehicle manufacturer has introduced a two-layer, split airflow system. This system allows fresh air to be sent through the case and to the defrosters while recirculated air is also brought through for heating. Fresh, outside air is drier during severely cold weather so there will be much less window fogging. For heating purposes, recirculated air conserves heat that is inside the vehicle. The split airflow comes from a blower that has two fan chambers; one of them has a slightly larger fan wheel. Each fan inlet has a door that allows selection of fresh or recirculated air. The duct work from the blower to the registers also has a divider.

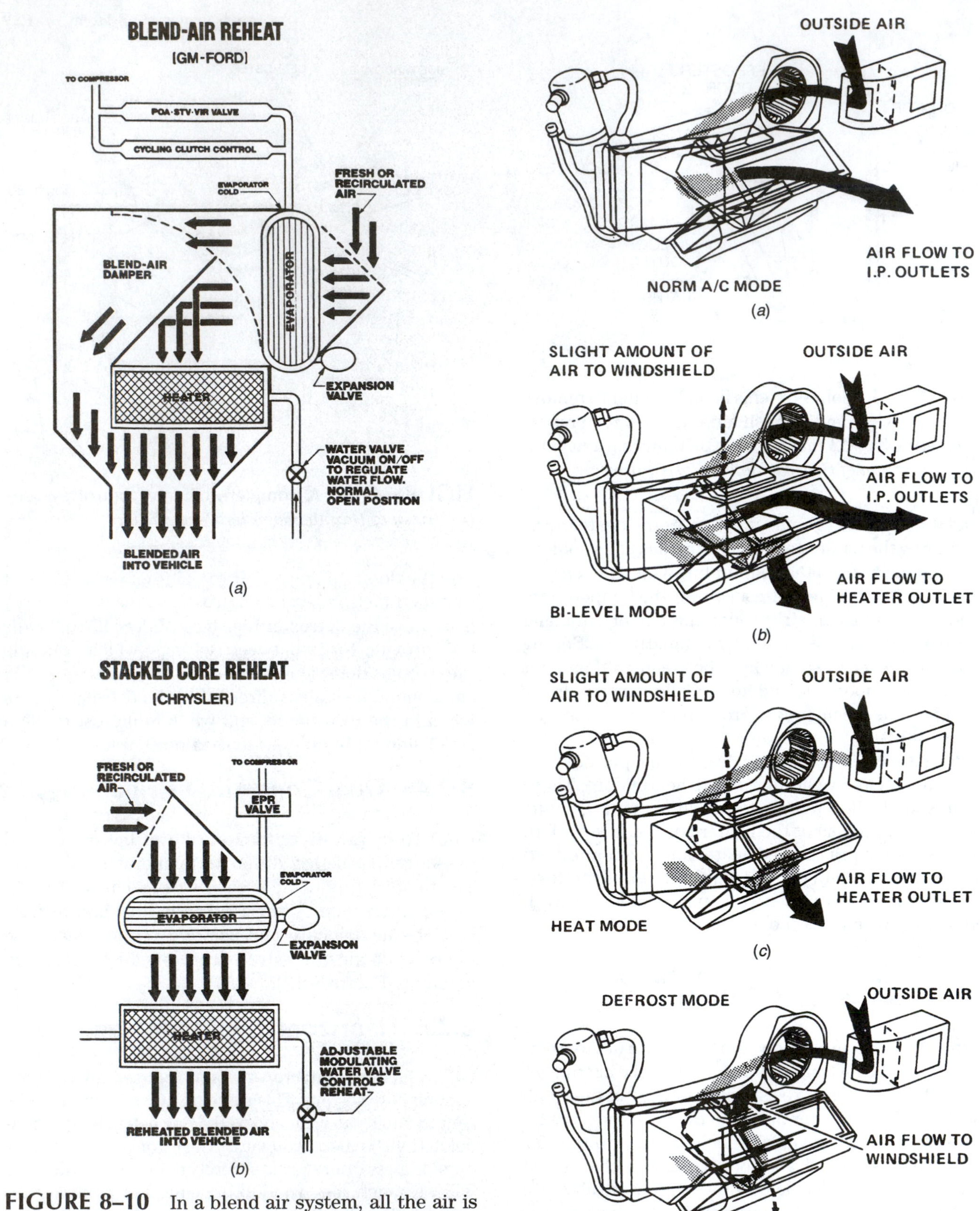

FIGURE 8–10 In a blend air system, all the air is cooled. Then some of it is reheated and blended with the cool air to get the right temperature (*a*). In a reheat system, all of the air is cooled and then reheated to the correct temperature (*b*). (*Courtesy of Everco Industries*)

FIGURE 8–11 The mode door controls the airflow to the instrument panel and floor outlets for normal A/C (*a*), bilevel (*b*), heat (*c*), and defrost (*d*). (*Reprinted with permission of General Motors Corporation*)

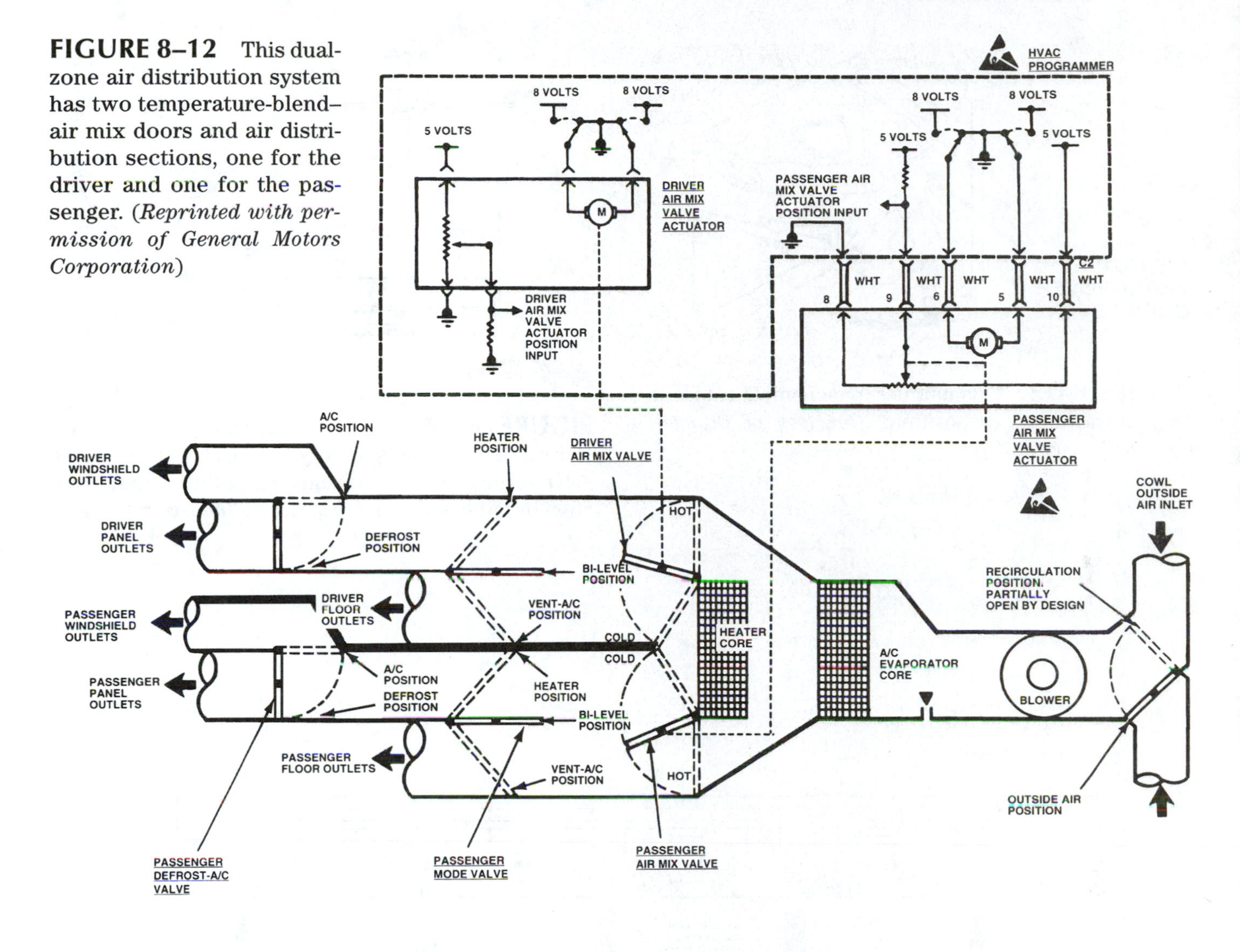

FIGURE 8–12 This dual-zone air distribution system has two temperature-blend–air mix doors and air distribution sections, one for the driver and one for the passenger. *(Reprinted with permission of General Motors Corporation)*

8.3 CONTROL HEAD

Many types of control heads are used on today's vehicles. These control heads can vary greatly between manufacturers and even between vehicle models from a single manufacturer. All control heads do essentially the same thing, but they use different methods to do it.

Mechanical systems are the least expensive. Most early control heads used purely mechanical operation for the doors, and one or more Bowden cables connected the function lever to the air inlet and mode doors. The temperature lever was also connected to the temperature blend door by a Bowden cable. These mechanical levers were rather simple and usually trouble free, but they had some disadvantages. They tended to bind and could require a good deal of effort to operate. A cable length adjustment was usually required to ensure complete door operation; modern systems can use automatic cable adjustments (Figure 8–13).

Many vehicles use **vacuum actuators,** sometimes called **vacuum motors,** to operate the air inlet and mode doors. These are controlled by a vacuum valve that is operated by the control head. Vacuum controls operate more easily than Bowden cables, and vacuum hoses are much easier to route through congested areas than cables. A Bowden cable is used for the temperature-blend door in most of these systems because of its ability to move the door to the exact position desired. When vacuum actuators operate, they alter the air–fuel mixture in the engine. Because vacuum controls affect engine operation and therefore emissions, modern vehicles are being designed to use electric control systems (Figure 8–14).

Some modern vehicles use **electrical function switches** at the HVAC control head. These are often called **electromechanical controls.** These switches operate a group of solenoid valves that control the vacuum flow to the vacuum motors at the doors. The vacuum actuators are the same as those just described, but the vacuum switches have been changed (Figure 8–15).

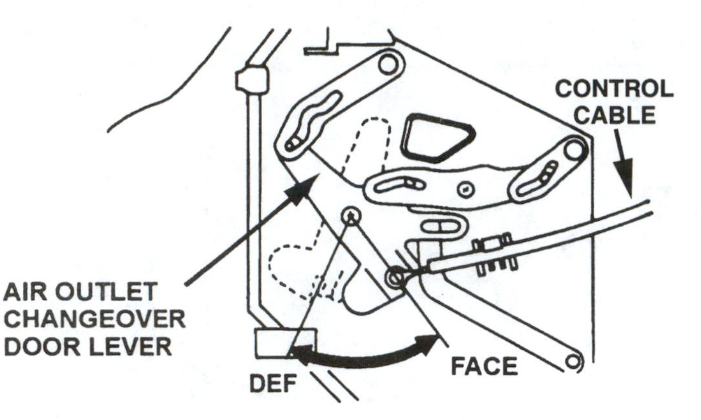

FIGURE 8–13 This unit uses mechanical cables to change the air door position. (*Courtesy of Daimler-Chrysler Corporation*)

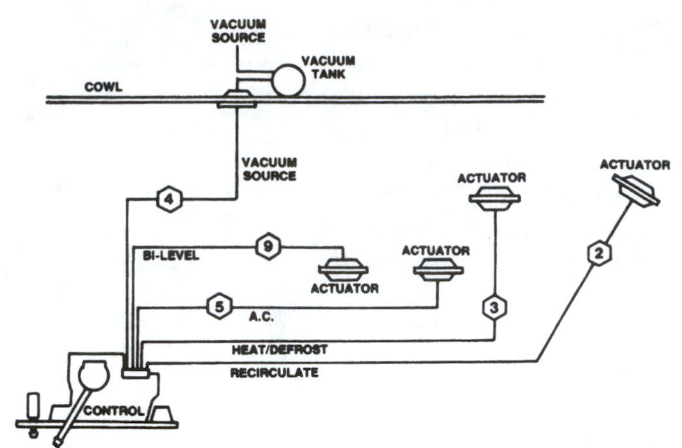

FIGURE 8–14 A typical vacuum control circuit starts at the vacuum source (engine), is controlled by a valve at the control head, and ends at the vacuum actuators or motors. (*Courtesy of Everco Industries*)

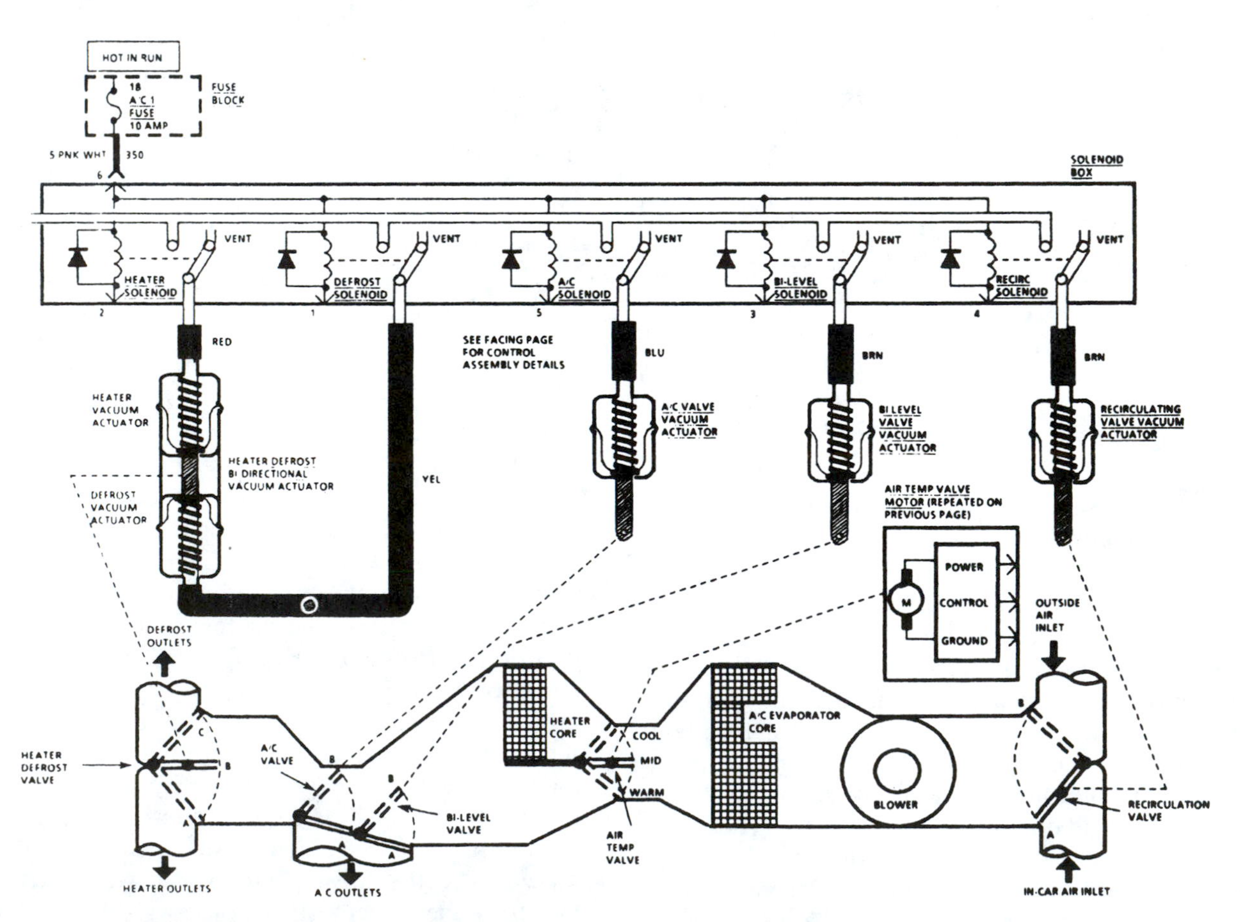

FIGURE 8–15 This system controls the vacuum actuators with a series of solenoid valves, one valve and control solenoid for each actuator operation. (*Reprinted with permission of General Motors Corporation*)

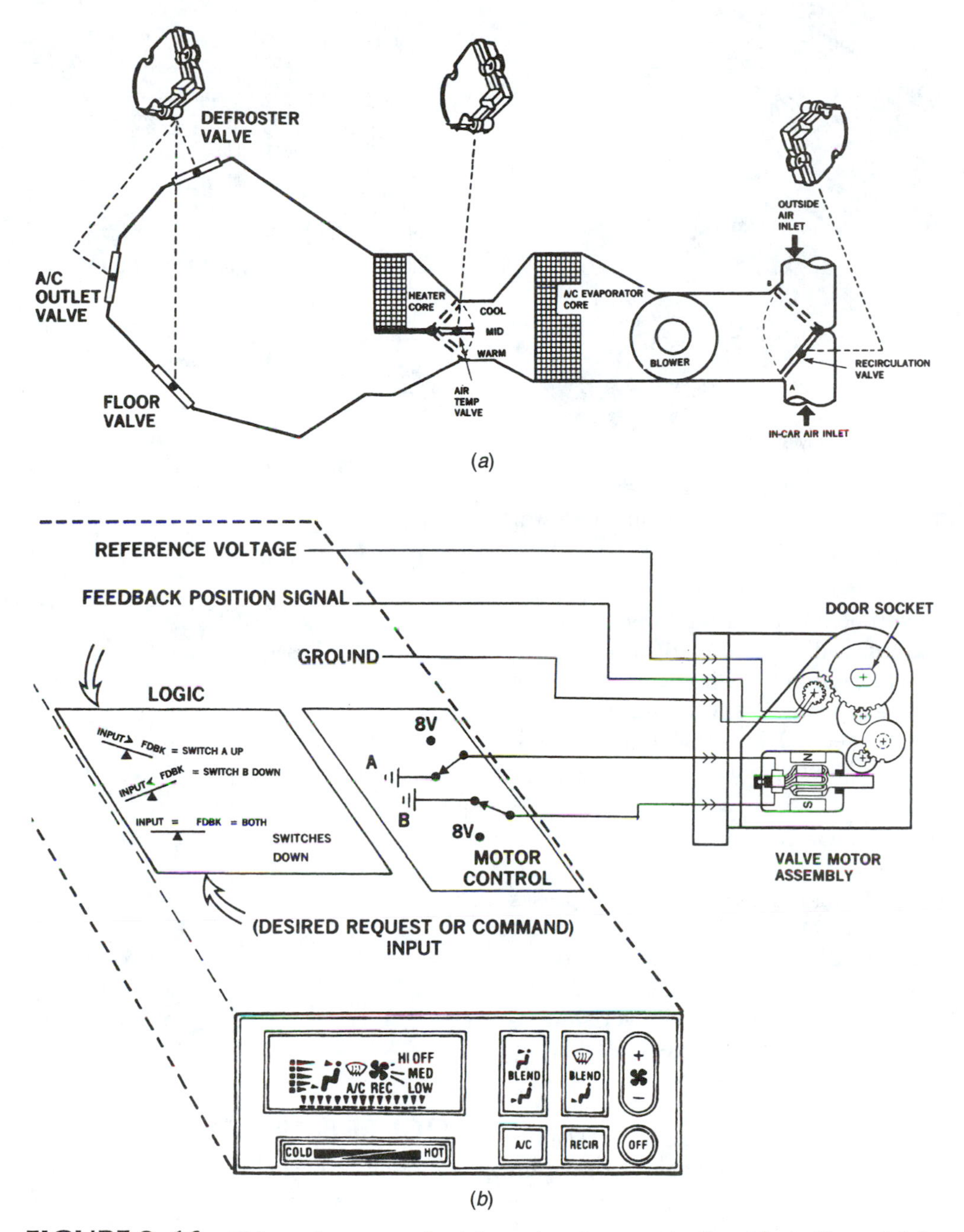

FIGURE 8–16 This system uses electric motors to operate the valves (doors) (*a*). The motors include circuits so that they can go in either direction and so that the controls know where the motor and door are positioned (*b*). (*Reprinted with permission of General Motors Corporation*)

Many modern cars use electric motors to operate the air distribution and temperature-blend doors. The temperature lever at the HVAC control head for these systems is an electrical **potentiometer.** The potentiometer is a type of variable resistor; the electric motor responds to the electric signal by turning a portion of a revolution. This system is similar to a TV antenna direction motor (Figure 8–16). Proper control motor position is determined using a feedback circuit or control module that counts commutator bar movement. A feedback circuit is shown in Figure 8–16.

Blower speed control in many of these systems is through a multiposition electrical switch and a group of resistors or electronic controls. The position of the switch determines the amount of resistance in the blower circuit and therefore the speed of the motor

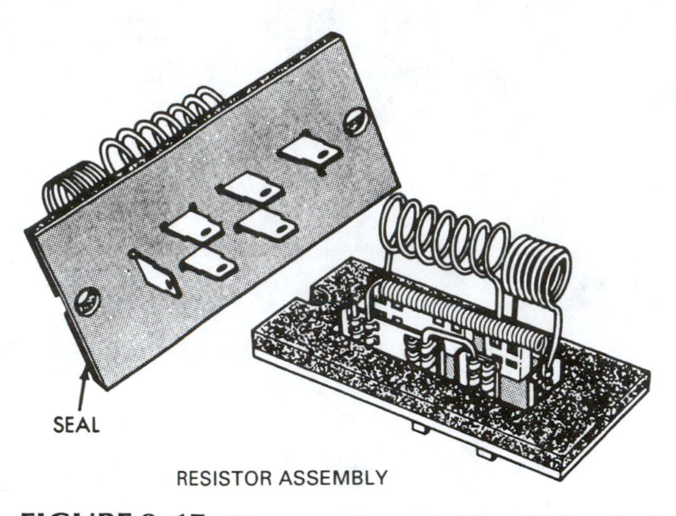

FIGURE 8–17 Most systems route blower electrical feed through one or more resistors to obtain the slower speeds. (*Courtesy of Chrysler Corporation*)

(Figure 8–17). Electronically controlled systems use a **pulse width-modulated (PWM) blower motor,** which is described in Section 8.5. Blower speed is controlled by electronically switching the motor off and on, up to 40,000 times per second. Increasing the length of the on time produces higher speeds.

8.3.1 Vacuum Control Circuit

The vacuum control circuit begins at the engine, passes through small hoses, is controlled by one or more valves, and ends at one of several vacuum motors (actuators). The vacuum source is the intake manifold of the engine. Remember that a vacuum is really very low atmospheric pressure, and a vacuum is generated in the intake manifold of a gasoline engine operating with the throttle partially closed. Most systems use a one-way check valve and vacuum reservoir in the hose from the engine to maintain a supply of vacuum for operation during wide-open throttle conditions (Figure 8–18).

Early systems used rubber hose with a 3/16- or 1/4-inch (4.7- or 6.3-mm) ID. This material often became hard and brittle and then broke. Newer systems use 1/8-inch (3.1-mm) plastic hose, which is much more durable. To make a connection, rubber hose is slipped over the tight-fitting connector; a pliable end is attached to plastic hoses to make the connection. Many vehicles combine the hoses at a multiport connector at the control valve and bulkhead connection. Both types of hoses are coded with a color stripe so that, with the aid of a vacuum diagram, they can be easily identified.

Control valves normally have two major positions: The closed position vents the vacuum motor hose to atmosphere and allows the spring to move the motor to the off position; the open position connects the source vacuum to the motor. Many systems control all of the vacuum motors with a single valve, which has a position for each control function. Other systems use a valve block with a separate valve for each circuit. Some newer systems use an electric solenoid to function as the valve.

A vacuum motor consists of a flexible diaphragm in a metal canister. One side of the canister is sealed and connected to the vacuum source; this side usually includes a spring capable of pushing the diaphragm toward the other chamber. The other chamber has the diaphragm stem passing through it; this side is usually vented to atmosphere (Figure 8–19).

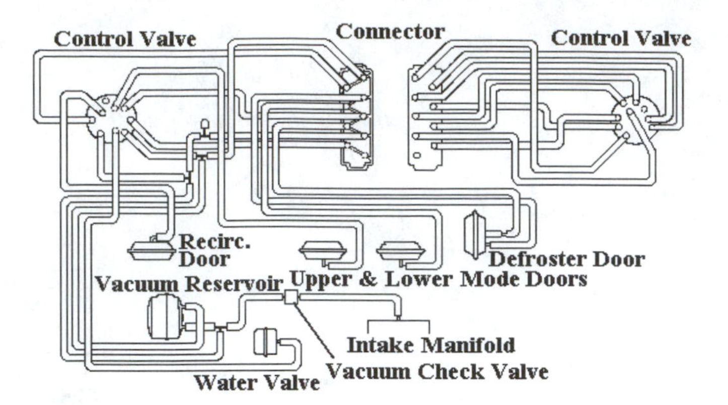

FIGURE 8–18 This vacuum control circuit includes a check valve and reservoir to maintain constant vacuum during acceleration. Most vacuum harnesses are color-coded to help in locating a particular hose during repair procedures.

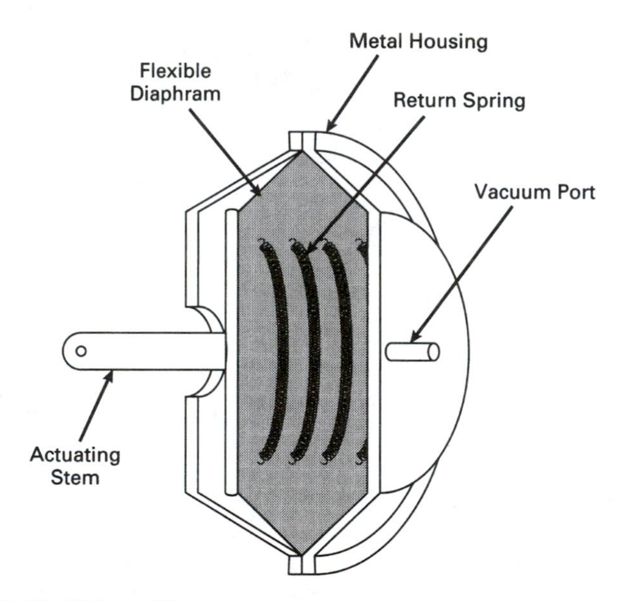

FIGURE 8–19 A vacuum actuator. This unit uses an internal spring to return the diaphragm when there is no vacuum signal.

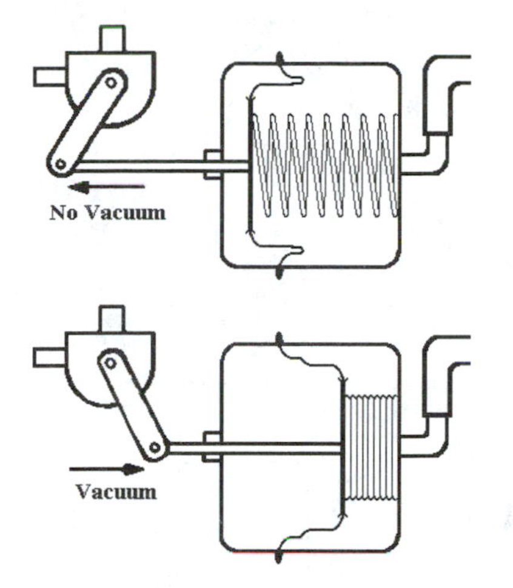

FIGURE 8–20 With no vacuum signal, the spring extends the actuator shaft to place the door in a certain position (top). A vacuum signal pulls the shaft inward and moves the door to the other position (bottom).

When the control valve connects the sealed chamber to the source vacuum, air pressure is pulled out of that chamber. Atmospheric pressure on the vented side pushes on the diaphragm to move the motor to the on position. When the sealed chamber is vented to atmosphere by the valve, the spring moves the motor to the off position. Air has to enter this chamber for the diaphragm to move (Figure 8–20).

Some vacuum motors use two sealed chambers. These dual chamber motors have three operating positions.

A vacuum circuit can also include **bleeds** and **restrictors.** Bleeds are calibrated leaks into the circuit and are used to release a motor at a very slow rate. Restrictors are very small orifices that slow down the rate that a motor can apply.

8.3.2 Electrical Control Circuits and Blower Motors

Automotive electrical circuits begin at the battery or alternator. This connection is often called B+, for battery positive. The electricity passes through **wires,** and the flow is controlled by **switches** or **relays.** In the heat and A/C system, the actuators that do the work are either **solenoid valve(s),** a **blower motor, servomotor(s),** or a **compressor clutch** (described in Section 6.2.3).

In most systems, the heat and A/C circuit begins at B+, connects to a **fuse** or **circuit breaker,** passes to the **master on–off switch** at the HVAC control head, and then branches off to the various parallel circuits. The master switch in many systems is operated by the function lever (Figure 8–21).

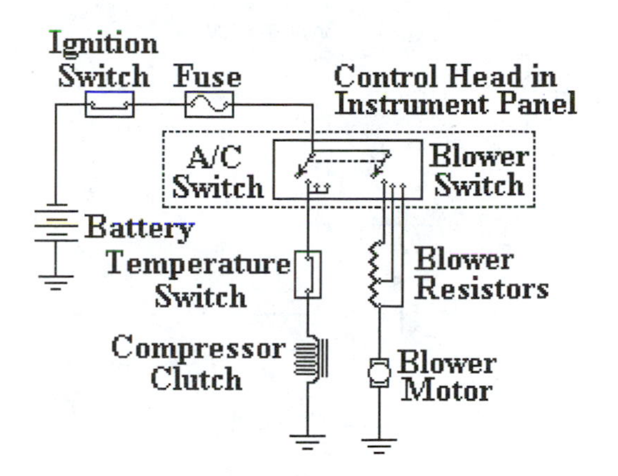

FIGURE 8–21 Most early HVAC systems had simple electrical circuits and placed the blower and A/C switch in the control head as they do today.

One of these circuits is through the **multiposition switch** to the blower. The low and intermediate switch positions route the electrical path through one or more **resistors.** The resistors cause a drop in the voltage at the blower and reduce blower speed.

In a typical system, the lowest speed might have a resistance of 1.5 to 2 ohms and a blower voltage of about 4 volts. The medium-low speed might have a resistance of 1 ohm and a blower voltage of about 6 volts. The medium-high speed might have a resistance of 0.3 ohm and a blower voltage of 8 or 9 volts. High speed will have no resistance and a blower voltage of 12 to 14 volts (Figure 8–22). A typical blower motor at full speed will draw about 20 amps of current. The high blower speed in some systems operates a relay, so full battery voltage goes from a separate B+ connection to the blower.

Another variation is that while most systems control the B+ voltage to the motor, which is mounted with a ground connection, some systems are **ground-side switched.** The blower motor is connected directly to B+ through the fuse or circuit breaker, with the speed switch and resistors in the ground circuit. Some systems use electronic blower speed control, described in Section 8.5.3.

Most blower motors drive a squirrel-cage–type blower wheel, which is much quieter and more efficient than a fan. There are many types of blower motors, some of them fitting only one make or model of vehicle. The mounting method and any flanges, length and diameter of the motor, length and diameter of the shaft, and direction of rotation are all important when trying to locate a replacement motor. Blower motors are not serviceable and are replaced if they become faulty (Figure 8–23).

Electrical theory, trouble diagnosing, and service are described more completely in Chapter 12.

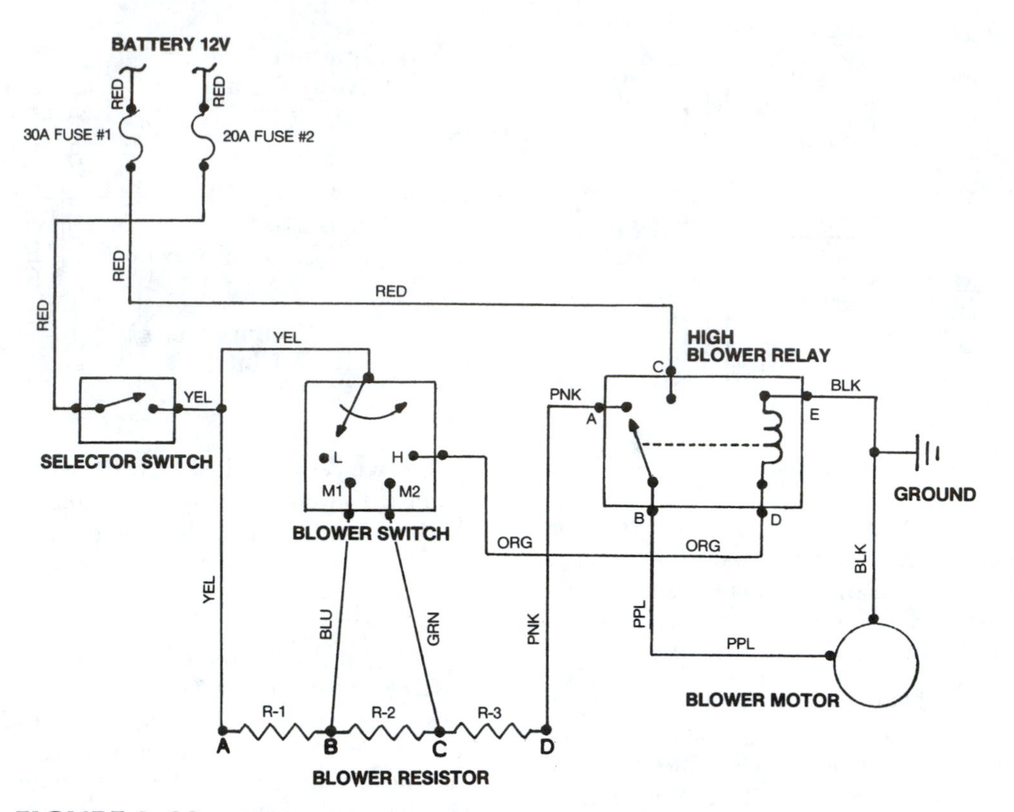

FIGURE 8–22 This blower circuit routes the current through resistors R-1, R-2, and R-3 for low speed; through R-2 and R-3 for M1 speed; and through R-3 for M2 speed. High speed actuates the relay to bypass the resistors for high speed. *(Courtesy of Everco Industries)*

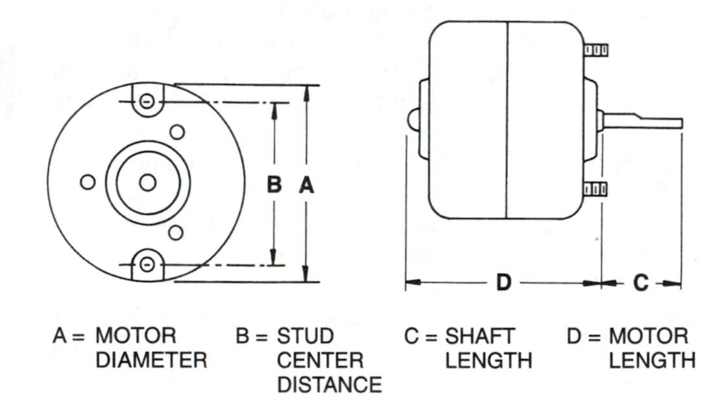

FIGURE 8–23 The critical dimensions needed when replacing a blower motor are shown here. *(Courtesy of Four Seasons)*

8.4 REAR AIR DISTRIBUTION SYSTEMS

Some larger vehicles (vans and small buses) have dual heat and A/C units with an air distribution system that is completely separate from the front heat–A/C unit (Figure 8–24). The unit is usually mounted in a rear side panel or in the roof. It normally contains its own evaporator, heater core, blower, and doors to control temperature and airflow. With the exceptions that there is no fresh air intake so it can only recirculate and that there is no defog or defrost, most rear units perform just like front units. Depending on the vehicle's usage, the controls can be mounted in the instrument panel or in the rear.

✓PROBLEM SOLVING

One of your friends has asked you about a problem she is having with her car. The A/C system works because the air gets cold, but it only comes out of the defroster registers and at the floor. Another problem with this same car is that the blower does not have a high speed; it stops when the switch is moved to the high position. What do you think might be wrong? What would you do to confirm your suspicions?

Another friend has an odd problem when he tows a trailer with his pickup. His A/C system usually works quite well, but when he has to drive up a long, uphill grade, the cold air discharge moves from the panel registers to the floor level. What might be wrong? What would you do to find the cause of this problem?

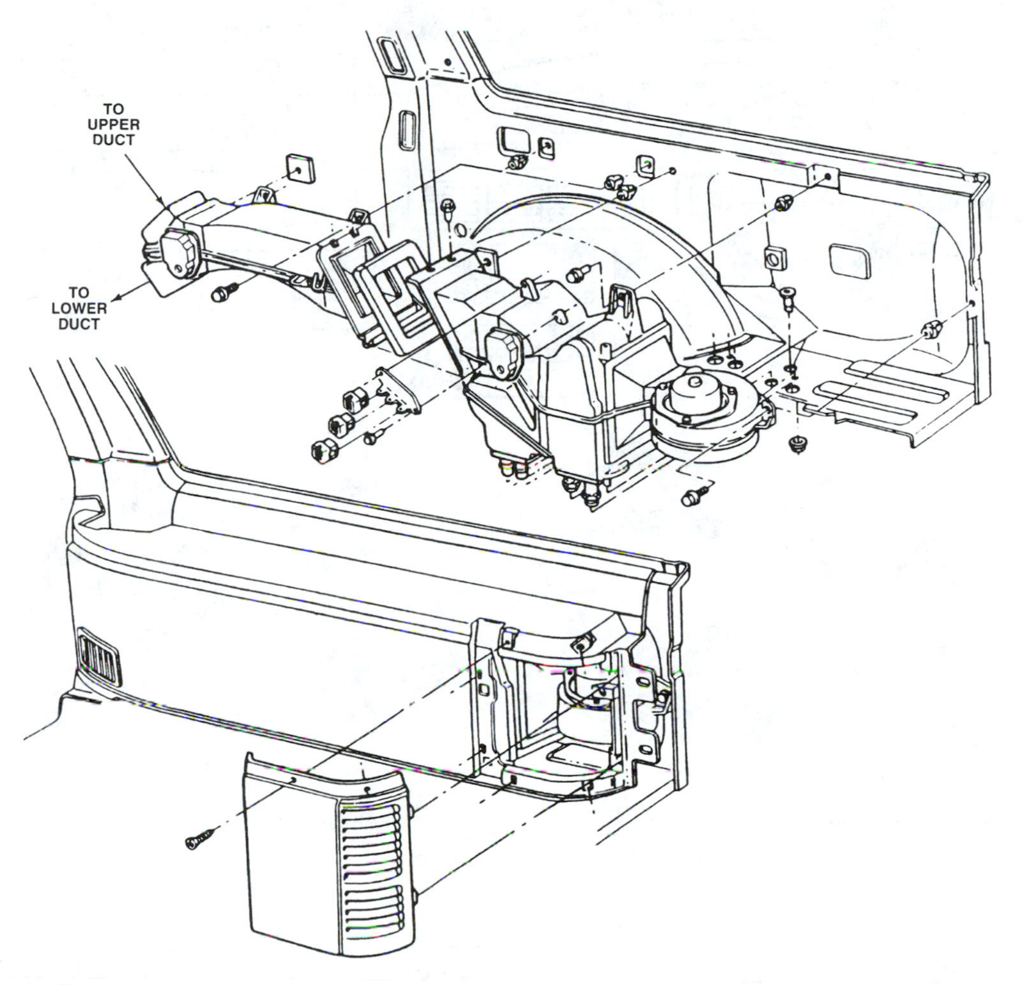

FIGURE 8–24 The rear air distribution system for a van can be mounted in the roof or rear side panel as shown here. (*Reprinted with permission of General Motors Corporation*)

8.5 AUTOMATIC TEMPERATURE CONTROL AND SEMIAUTOMATIC TEMPERATURE CONTROL

ATC is essentially a system of sensors and controls that allows the driver to set a desired temperature at the HVAC control head and let the system take care of maintaining that temperature and selecting the proper air discharge location and blower speed. A **semiautomatic temperature control (SATC)** system is similar, but the driver selects the air discharge location. These systems are basically manual heat and A/C systems with automatic controls. These controls can move the temperature-blend door, adjust the blower speed, and change the function door and inlet settings (Figure 8–25).

There are a variety of ATC systems, but they have some things in common. All use a group of **sensors,** a **control device,** and a set of **actuators** (Figure 8-26). The sensors determine the temperature inside and outside the car and the temperature the driver wants. The control device, often an electronic control module (ECM) or electronic control assembly (ECA), compares these temperatures and resets the actuators as necessary to reach the desired temperature.

8.5.1 ATC Sensors

Most systems use electrical temperature sensors called **thermistors.** A thermistor is a resistor that changes resistance value inversely to temperature changes;

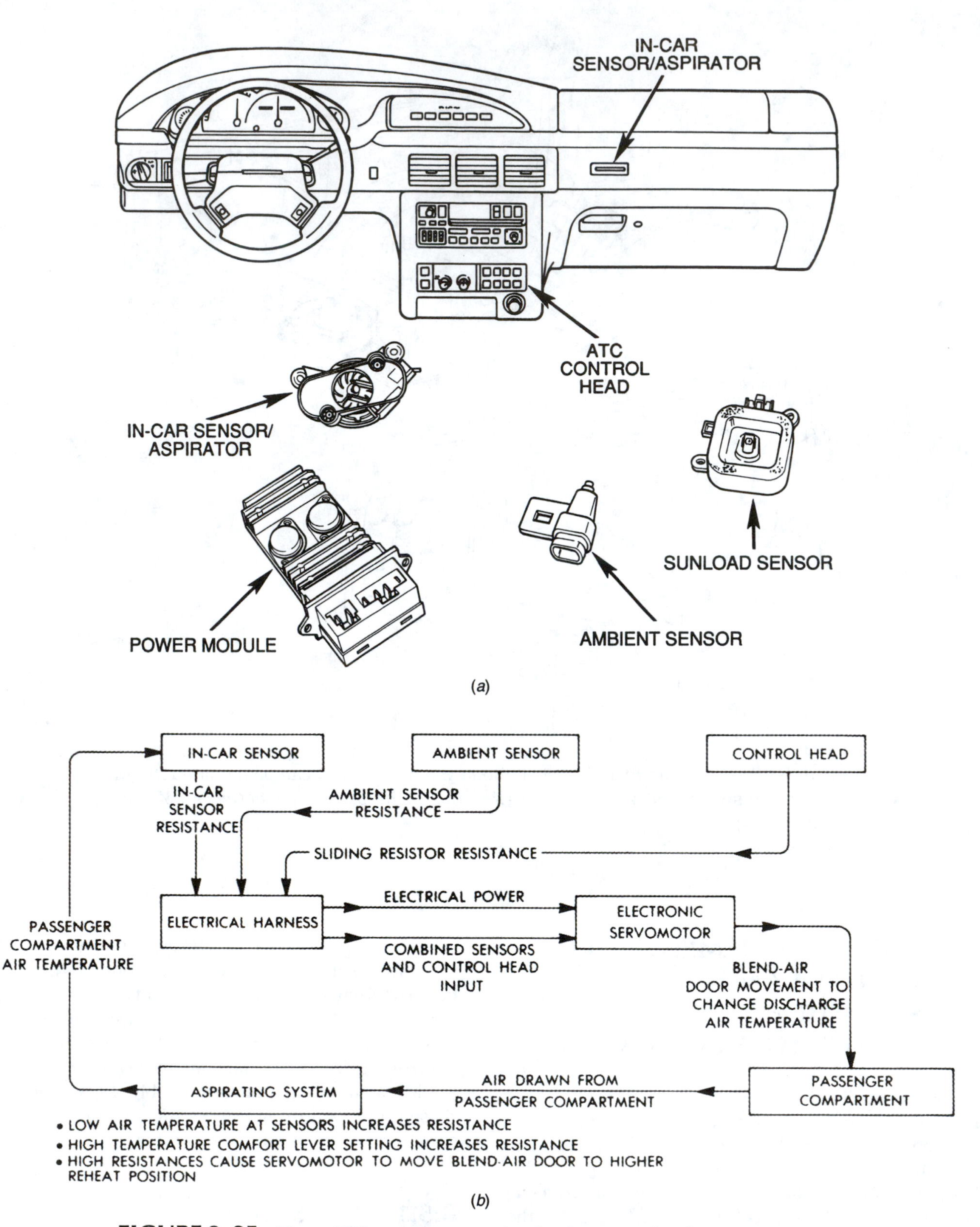

FIGURE 8–25 Many ATC systems are a standard system plus these components (*a*). They are combined as shown in (*b*). (*Courtesy of DaimlerChrysler Corporation*)

Component Locations

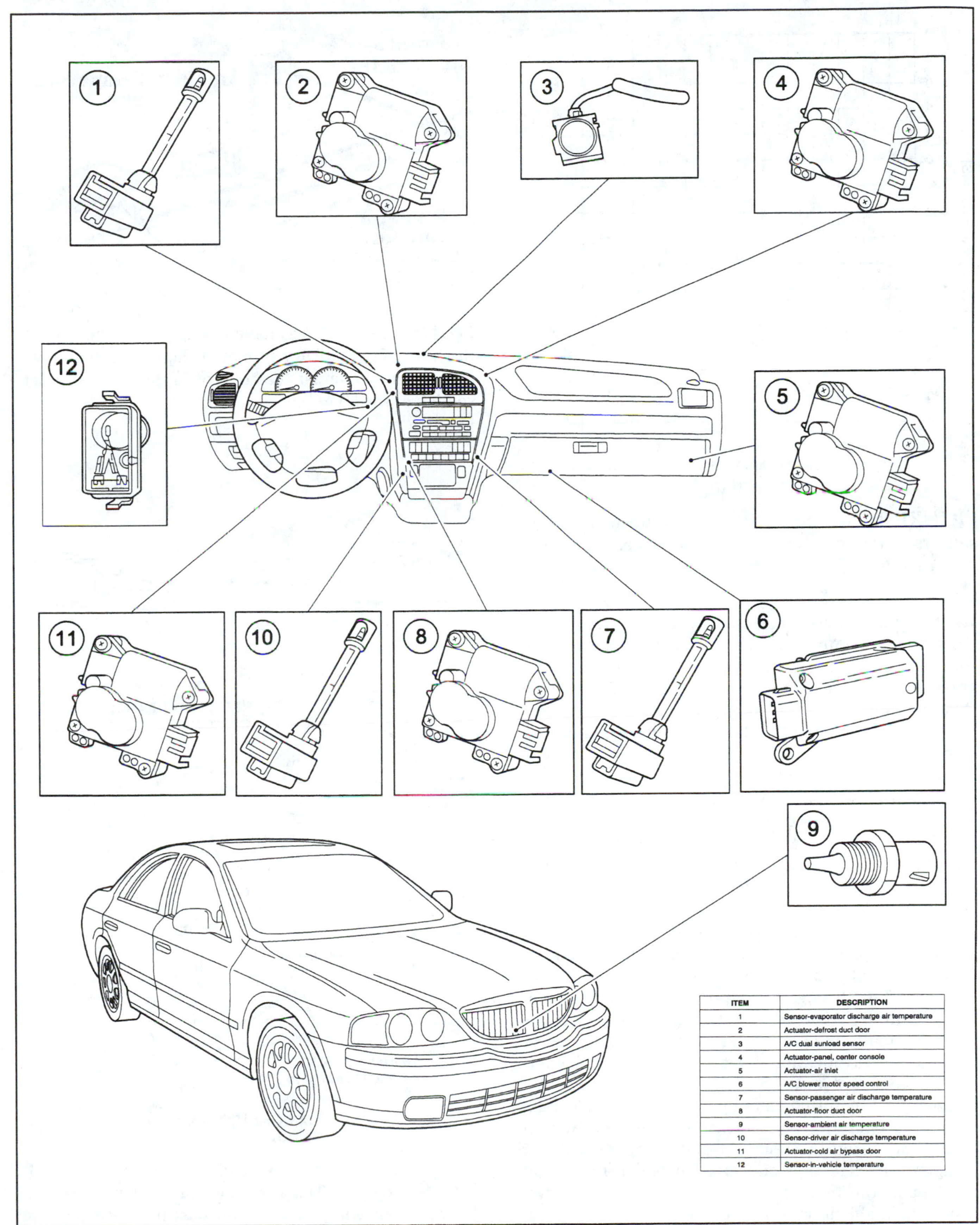

ITEM	DESCRIPTION
1	Sensor-evaporator discharge air temperature
2	Actuator-defrost duct door
3	A/C dual sunload sensor
4	Actuator-panel, center console
5	Actuator-air inlet
6	A/C blower motor speed control
7	Sensor-passenger air discharge temperature
8	Actuator-floor duct door
9	Sensor-ambient air temperature
10	Sensor-driver air discharge temperature
11	Actuator-cold air bypass door
12	Sensor-in-vehicle temperature

FIGURE 8–26 This ATC system uses twelve major components, in addition to the control head, at various locations. *(Courtesy of Visteon)*

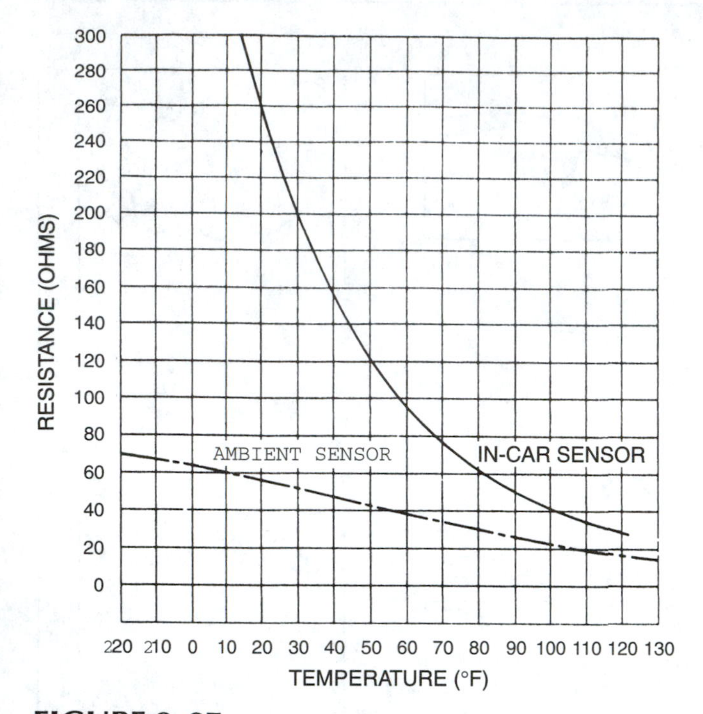

FIGURE 8–27 The resistance of ambient and in-car sensors varies directly with temperature. *(Reprinted with permission of General Motors Corporation)*

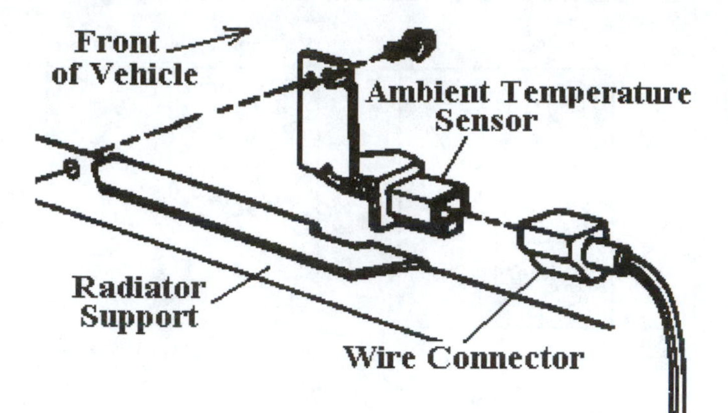

FIGURE 8–28 This ambient sensor is mounted onto the front of the radiator support; others are mounted at other locations under the hood or in the fresh air inlet section of the HVAC case.

as the temperature increases, resistance decreases. As shown in Figure 8–27, these values change at a precise rate, and the two sensors have different temperature-to-resistance curves. One new system uses infrared temperature sensing. This dual-zone system uses infrared sensing, similar to a modern, hand-held, non-contact thermometer, to monitor the surface temperature of and around the driver and passenger.

The **ambient sensor** measures outside temperature and is often mounted in the fresh-air portion of the duct. On some cars, it is mounted at the radiator shroud or in the area behind the front grill (Figure 8–28). On many older vehicles, the **in-car sensor** is mounted behind the instrument panel, and a set of holes or a small grill allows air to pass by it. Most newer cars place the in-car sensor in an **aspirator** connected to the blower case. Blower operation produces airflow through the aspirator and past the sensor (Figure 8–29).Some vehicles use an in-car sensor with a small, integrated fan. These units can be mounted in any location like the vehicle's head liner.

In older systems, the driver selected a particular temperature by turning a temperature dial or sliding a lever. This operation adjusted a potentiometer. Modern electronic systems use push buttons or touch pads for temperature selection and give an electronic digital readout of the temperature setting. In the early systems, the ambient sensor, in-car sensor, and temperature potentiometer were connected in series between B+ and the electrical amplifier (Figure 8–30). In the

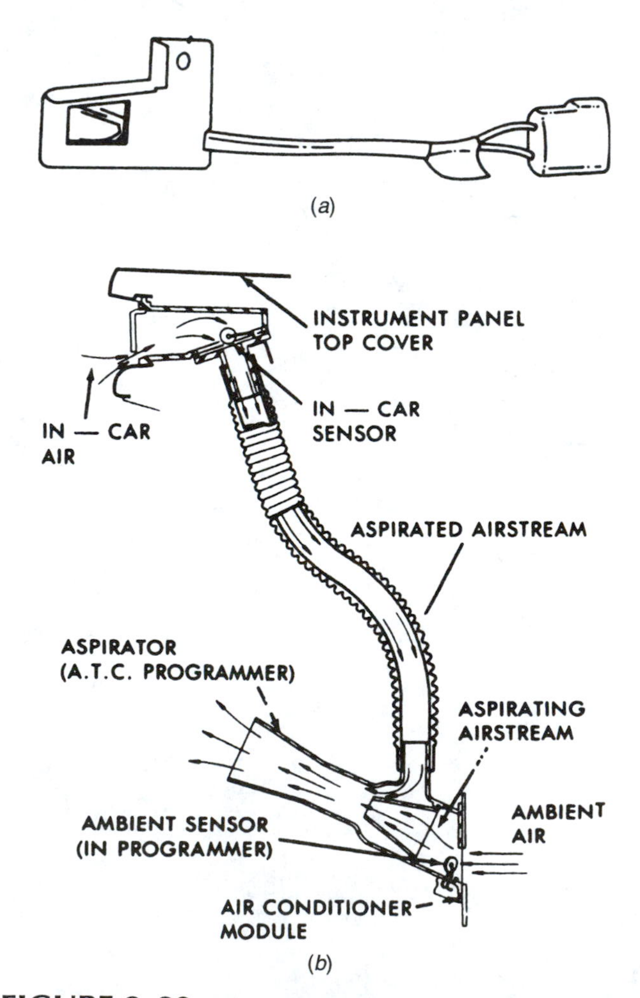

FIGURE 8–29 This in-car sensor (*a*) is mounted so the aspirator will pull in-car air past it (*b*). *(Reprinted with permission of General Motors Corporation)*

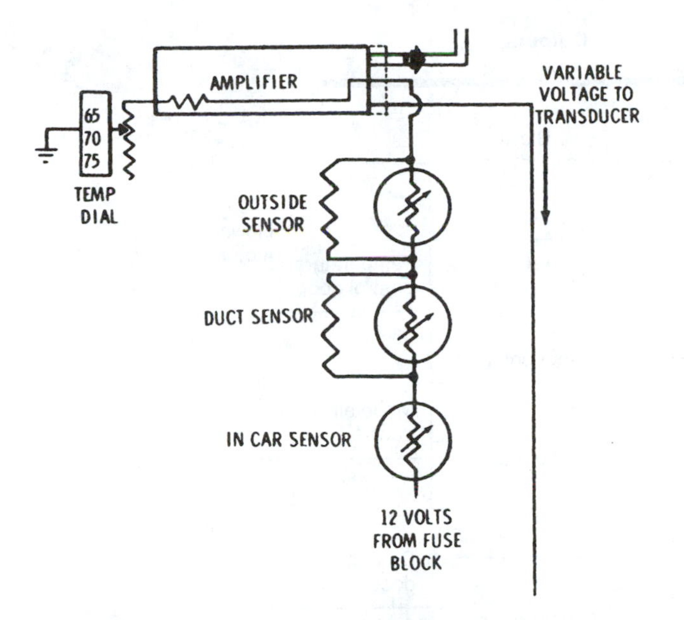

FIGURE 8–30 This sensor string routes current through the in-car sensor, duct sensor, outside (ambient) sensor, amplifier, and temperature dial and to the ground. An increase in resistance causes the amplifier to reduce the vacuum signal to the programmer. *(Reprinted with permission of General Motors Corporation)*

newer systems, these have become parallel input circuits to the ECM (Figure 8–31).

Many modern systems also use an engine temperature sensor. This sensor keeps the system from turning on the heater before the coolant is warmed up and is often called **cold engine lockout.** Some systems also include a **sunload sensor** mounted on top of the instrument panel that measures radiant heat load, which might cause an increase in in-vehicle temperature (Figure 8–32).

Some modern vehicles use an A/C compressor speed sensor so the ECM will know if the compressor is running, and by comparing the compressor and engine speed signals, the ECM can determine if the compressor clutch is slipping excessively. This system is often called a **belt lock** or **belt protection system.** It prevents the possibility of a locked-up compressor from destroying the engine drivebelt, which, in turn, can cause engine overheating or loss of power steering. If the ECM detects an excessive speed differential for more than a few seconds, it will turn the compressor off.

Some systems use a **mechanical vacuum sensor** that uses a bimetallic strip to measure temperature. This strip bends in response to temperature change; this bending action is used to control a vacuum valve. In one system, the bimetal sensor is positioned in a unit with both ambient and in-vehicle air passing by it. The driver control is attached so it acts on a pivot bar of the bimetal sensor. This unit provides a variable vacuum signal that is sensitive to ambient temperature, in-car temperature, and the driver's temperature setting. The signal is zero vacuum when heat is called for, engine vacuum when full A/C is called for, or somewhere between.

Most systems are designed so they return to a full heat position when turned off. This way, if they fail, they will fail to that position. A person in the southern United States would probably be uncomfortable if the A/C and heat system failed, but in the northern states a system failure could be life threatening in the winter.

8.5.2 Control Device

The **ECM,** which is a microcomputer, acts on signals from the sensors to produce the desired action at the actuators. The ECM has an output to each item that is controlled, which can include the compressor clutch, blower speed, temperature-blend door, inlet air door, and mode doors. Some of these—the temperature-blend door, for example—have a feedback circuit that tells the ECM what position the door is in. This allows the ECM to move a door to any position within its operating range. Some actuators are two position: on–off or open–closed.

One modern system has the ability to count the actuator motor commutator segments so it has the ability to determine how far the motor revolves. A small current flow reduction occurs as the space between the commutator bars passes under the brushes of a DC motor. This system needs no feedback circuit, but a calibration procedure must be performed if a motor is replaced.

Modern systems have an output from the ECM to each of the controlled actuators or outputs. If vacuum motors are used for the recirculation and mode doors, the vacuum control solenoids and vacuum valve are often combined into a single assembly.

Older systems use an electronic amplifier that generates a signal to drive an electric servomotor or vacuum servomotor to the proper position (Figure 8–33). Systems that use vacuum servomotors also need a transducer to change the electrical signal from the amplifier into a vacuum signal for the servomotor. Both styles of servomotors provide a linear motion. They are positioned at one location for full heat, another position for full A/C, or anywhere in between. The servomotor is connected to electric switches, vacuum valves, and the temperature blend door.

When the sensors indicate cold in-vehicle conditions, below the driver's settings, the servomotor moves to its full heat position. Here the A/C compressor is turned off; the temperature blend door is at the full heat position; the mode doors direct the heated air to floor level; the blower motor is at high speed; and,

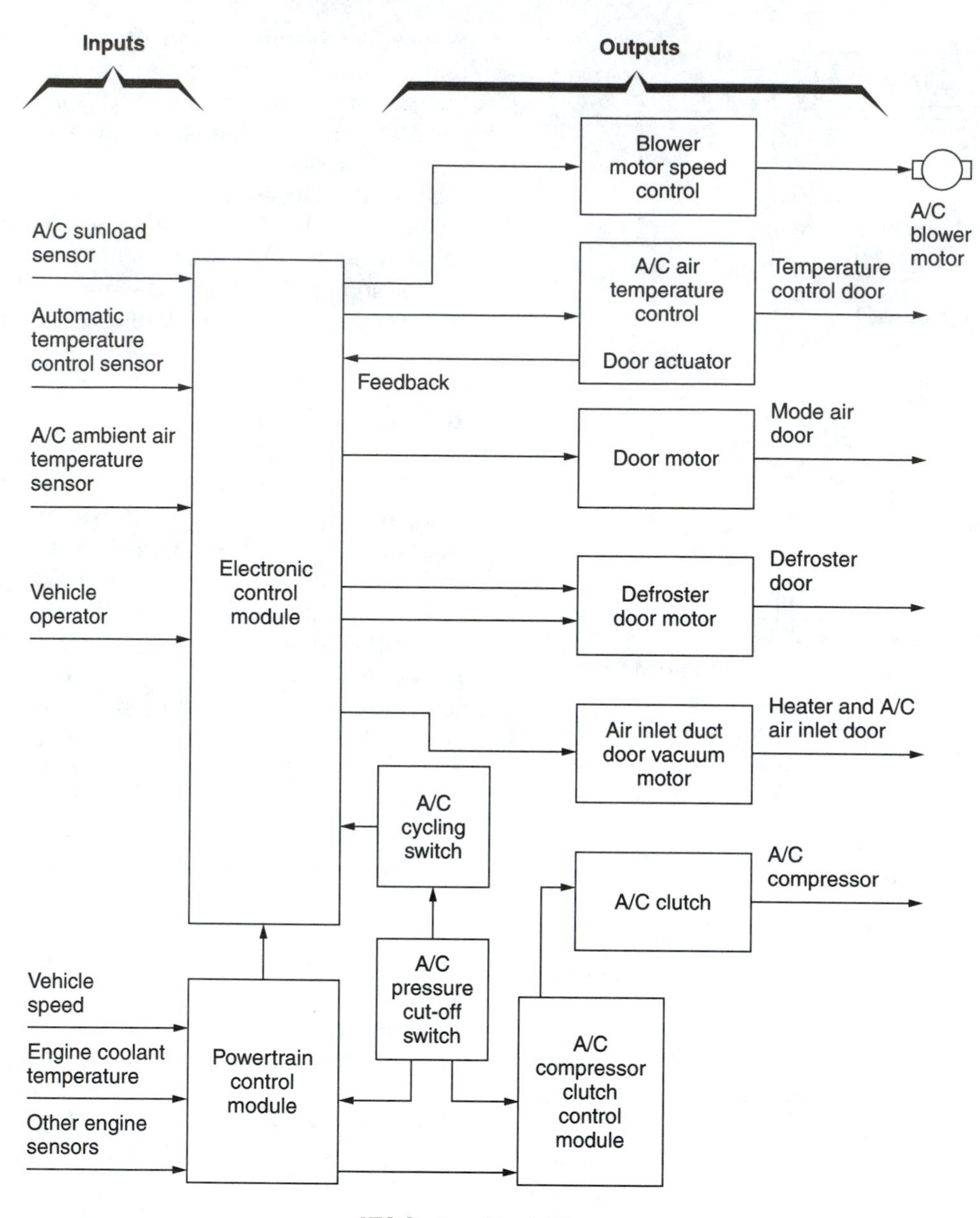

ATC System Block Diagram

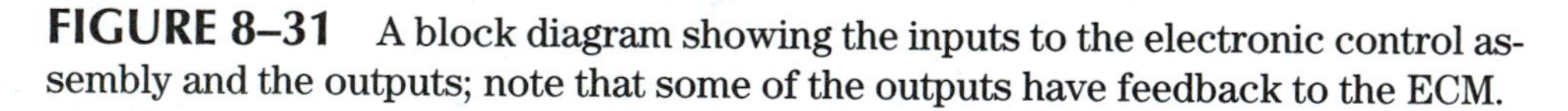
FIGURE 8–31 A block diagram showing the inputs to the electronic control assembly and the outputs; note that some of the outputs have feedback to the ECM.

in some systems, the air inlet door is set to recirculate. When the sensors indicate an in-vehicle temperature that is too high, the servomotor moves to its full A/C position; the compressor is turned on; the temperature blend door is set to full cold; the mode doors are set to the instrument panel registers; the blower is set to high speed; and, in some systems, the air inlet door is set to recirculate. When the sensors indicate moderate temperatures, the servomotor moves to a mid position. Depending on the temperature, the compressor may operate, the temperature blend door will be in the middle, the mode doors may be mixing between floor and instrument panel, the air inlet door is at fresh air, and the blower operates at low speed (Figure 8–34).

A system with a mechanical vacuum sensor connects the sensor directly to the vacuum servomotor. This servomotor is essentially the same as that in a system using electrical sensors.

8.5.3 Actuators

Some ECM-controlled ATC systems use vacuum motor door actuators and control them through solenoid valves; some use electric door actuators controlled di-

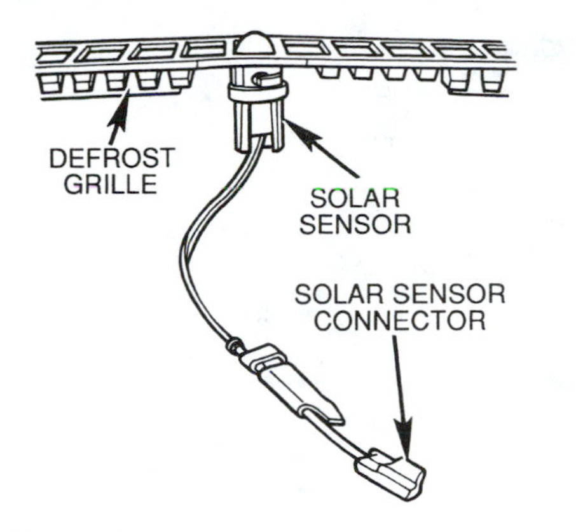

FIGURE 8–32 The solar sensor provides a signal to the control module when the sun shines on it. *(Courtesy of DaimlerChrysler Corporation)*

FIGURE 8–33 This electronic servomotor converts the signal from the sensors and control head and moves to adjust the temperature-blend door, mode doors, and blower speed. *(Courtesy of DaimlerChrysler Corporation)*

	OPERATION		
Air Inlet Door	Recirc		Fresh Air
Heater Water Valve	Closed		Open
Mode Door	A/C Mode	Bi–Level	Heater Mode
Temperature Door	Max Cold	Blend	Max Heat
Blower Speed	High – I3 – I2 – Slow – I2 – I3 – High		
Temperature Control Setting	Coolest		Warmest
	With high temperatures, system will start here.		With cold temperatures, system will start at this end.

FIGURE 8–34 When the system is cold, the servomotor produces the operations on the right; when the system is hot, servomotor movement changes to produce the operations on the left.

rectly from the ECM. The ECM sends an electric signal to a particular solenoid or door actuator, which causes either the solenoid to operate the vacuum valve (which in turn causes the vacuum motor to operate the door) or the electric door actuator to move the door. Electric door actuators can be either continuous-position or two-position units (open or closed). Continuous position actuators can stop anywhere in their range and need a feedback circuit so the ECM will know their position. The temperature blend door is operated by an electric servomotor (continuous-position actuator) that can move the door to any position called for to produce an air mix of the desired temperature (Figure 8–35).

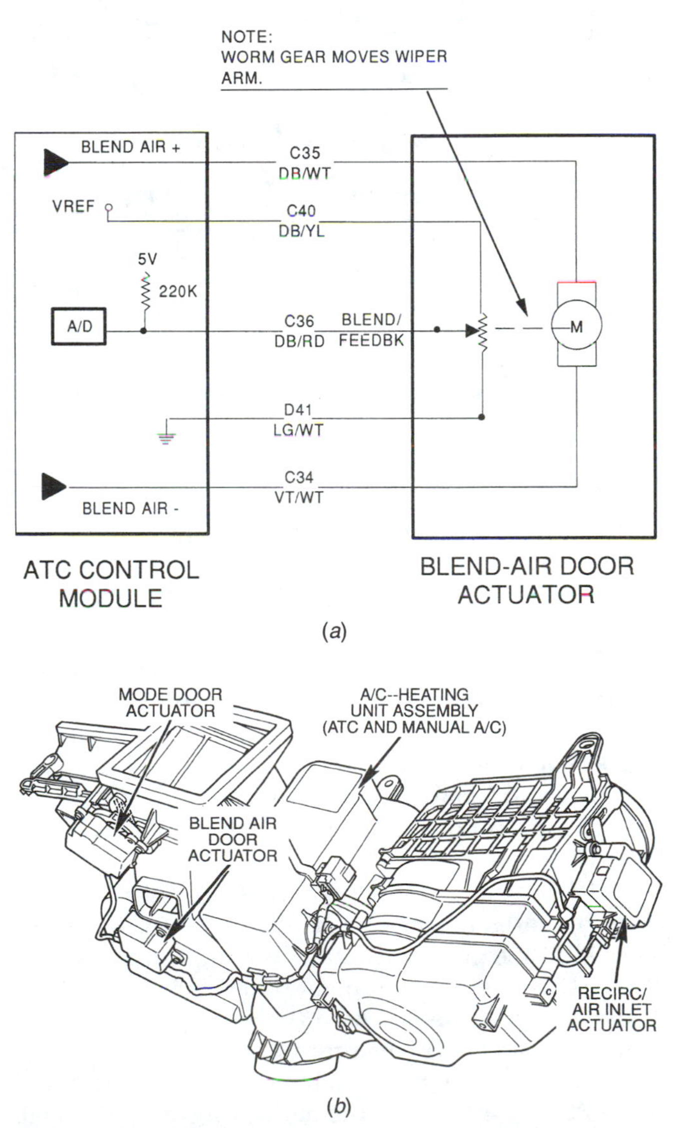

FIGURE 8–35 These mode door, blend air door, and recirculation–air inlet actuators are electric motors (*a*). The circuit for the blend air door is shown in (*b*). *(Courtesy of DaimlerChrysler Corporation)*

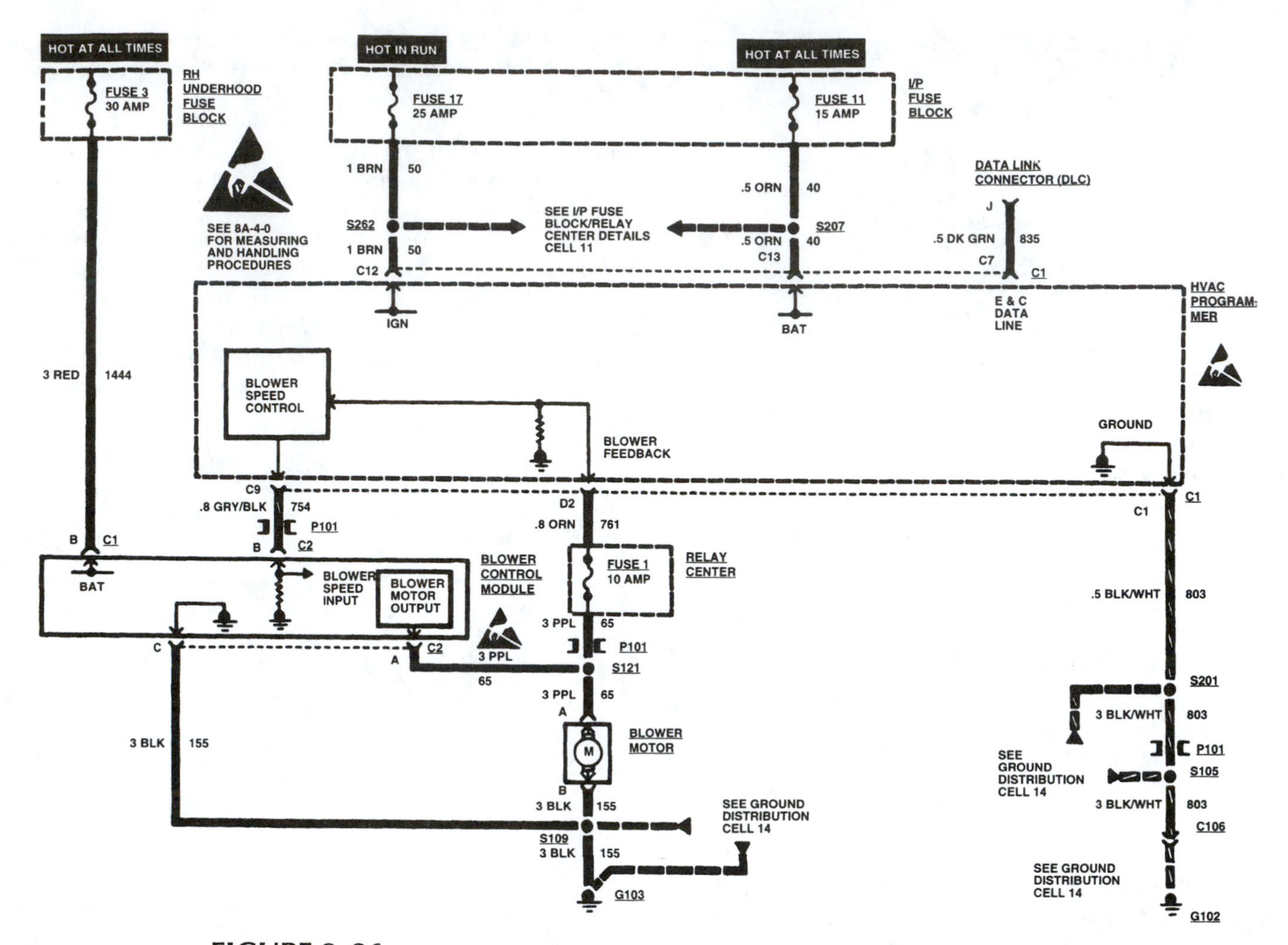

FIGURE 8–36 The blower control module with input from the HVAC programmer controls the speed of this blower motor. *(Reprinted with permission of General Motors Corporation)*

Many ATC systems operate the air inlet and mode doors using vacuum motors, the same as a manual system. The temperature-blend door is connected to the servomotor through a connecting link. The servomotor also operates a slide valve to control the vacuum to the door motors and a switch for blower speed control.

The blower motor in some ECM-controlled systems operates through a set of different-value resistors with a high-speed relay, like most manual systems. Other systems operate through a PWM or duty-cycle process in which the controller turns the voltage on and off many times a second. If the on cycle and the off cycle are each one-half the time, there is a 50% duty cycle, and the motor runs at half speed. If the off portion is greater than the on portion (say 25% on and 75% off), there is a 25% duty cycle, and the motor runs slowly. As the on portion gets longer and the off portion shorter (say 75% on and 25% off for a 75% duty cycle), the motor runs at three-fourths speed (Figure 8–36). A system that uses resistors usually has four definite stepped speeds. The duty-cycle system has an infinite number of speeds with a gradual change between them (Figure 8–37).

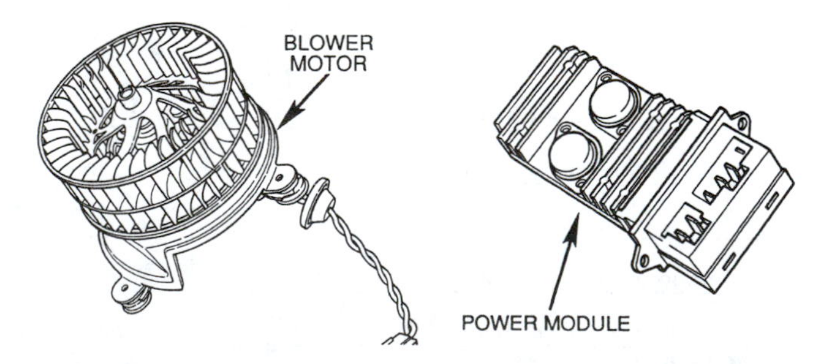

FIGURE 8–37 The power module is used to switch the current feed to the blower and produce infinite blower speeds. *(Courtesy of DaimlerChrysler Corporation)*

8.6 REAR WINDOW DEFROSTER

The rear window defroster is an electrical resistance heater designed to clear fog and frost from the rear window. The wires of the electric grid are visible in the rear glass. When current from the car's electrical system flows through the grid, it heats up and warms the glass (Figure 8–38).

The system consists of the connection to B+ through a fusible link or circuit breaker, a control switch and timer at the instrument panel, an indicator light, and the grid. When the switch is operated, the timer turns the system on and then shuts it off after 10 minutes. Most systems can be reactivated if necessary (Figure 8–39).

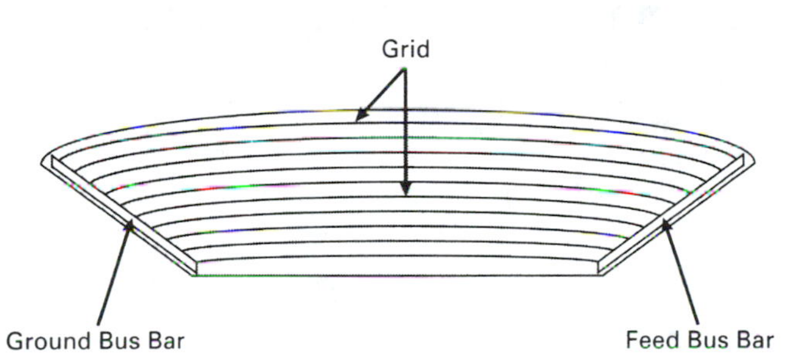

FIGURE 8–38 A rear window defroster routes electricity from the feed bus bar at one side through the grid to the ground bus bar to warm and defrost or de-ice the window.

✓PROBLEM SOLVING

An older friend has asked you about a problem he is having with his A/C system. It is a late-model vehicle with ATC. On a hot day, the air comes out of the panel registers with good air movement, but it is about the same as the outside temperature. What do you think might be wrong with this car? What checks can you make to confirm your suspicions?

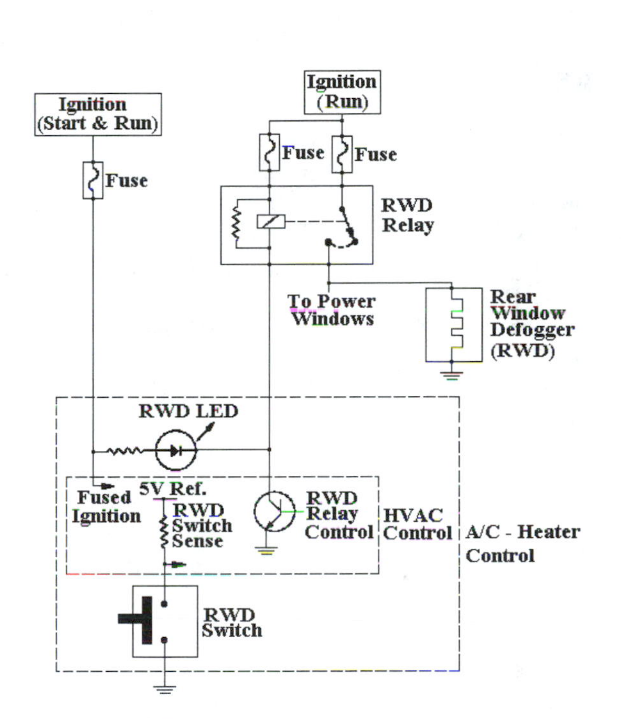

FIGURE 8–39 This circuit for a rear window defogger (RWD) includes an LED light in the instrument panel and three fuses.

CHAPTER QUIZ

These questions help you study this chapter. Enter the proper word(s) in the blanks to complete each statement.

1. Airflow will enter an HVAC system from either the __________-__________ __________ at the base of the windshield or through the __________ __________.
2. The HVAC system case can be divided into three major parts which are the __________ __________ section, __________ where cold and hot air are mixed, and __________ __________ section.
3. The air inlet door is normally set to recirculation when the mode controls are set to __________ __________ or __________ __________.
4. When the temperature control is at full heat, the __________ __________ will direct all of the airflow through the __________ __________.
5. Many modern systems use a(n) __________ __________ to remove dust and pollen from the airstream.
6. A(n) __________ __________ system requires two sets of temperature doors and mode doors.
7. Mode doors can be operated by a(n) __________ __________, __________ __________, or __________ __________.
8. Low blower operation routes the current flow through a __________, and the current for high blower operation often comes directly from a(n) __________.

9. Electric door control motors usually include a(n) __________ __________ so the controls will know what the motor position is.
10. The two major sensors used with ATC systems are the __________-__________ and __________ __________ sensors; some systems also use a(n) __________ __________ sensor.
11. Many ATC systems use a(n) __________ to bring an airflow across the __________-__________ sensor.
12. Some systems use __________ __________ __________ blower motor speed control to provide an infinite number of blower speeds.

REVIEW QUESTIONS

These questions allow you to check what you have learned. Select the answer that correctly completes each statement.

1. Which of the following methods is used to connect the A/C control head to the airflow control doors?
   - a. Electrical
   - b. Vacuum
   - c. Bowden cable
   - d. All of these
2. Air can enter the duct system through which of the following?
   - a. Fresh air plenum
   - b. Recirculate return
   - c. Defrost duct
   - d. Both a and b
3. Two students are discussing A/C door operation. Student A says that when the control head is set to heat, the mode door should direct air discharge from the instrument panel ducts. Student B says the air discharge should be at floor level. Who is correct?
   - a. A only
   - b. B only
   - c. Both A and B
   - d. Neither A nor B
4. Most vacuum control valves have __________ operating position(s).
   - a. one
   - b. two
   - c. three
   - d. four
5. The __________ blower speed has the least electrical resistance.
   - a. low
   - b. high
   - c. medium-low
   - d. medium-high
6. Two students are discussing A/C blower operation. Student A says that modern systems use a squirrel-cage type blower. Student B says that a plain fan-type unit is used with systems having cable type controls. Who is correct?
   - a. A only
   - b. B only
   - c. Both A and B
   - d. Neither A nor B
7. An ATC system has the ability to automatically
   - a. move the temperature-blend door.
   - b. adjust the blower speed.
   - c. change the function door setting.
   - d. All of these
8. Two students are discussing ATC sensors. Student A says that a NTC thermister has more resistance as it warms up. Student B says an increase in temperature will cause a NTC thermister to have lower resistance. Who is correct?
   - a. A only
   - b. B only
   - c. Both A and B
   - d. Neither A nor B
9. Many modern systems use an engine temperature sensor that
   - a. allows heater operation as soon as the engine is started.
   - b. turns on the A/C compressor after the engine has warmed up.
   - c. blocks A/C operation when the engine is cold.
   - d. prevents heater operation until the coolant is hot.
10. An ATC system uses a(n) __________ sensor to determine the temperature of the incoming air.
    - a. in-vehicle
    - b. ambient
    - c. anticipation
    - d. sunload

HVAC System Inspection and Trouble Diagnosis Procedures

LEARNING OBJECTIVES

After completing this chapter, you should be able to:

- Inspect an HVAC system to determine whether it is operating correctly, and if not to make further tests to locate the subsystem at fault and the cause of the problem.
- Check system temperatures and determine if they are correct.
- Follow a logical procedure to locate the cause of a problem in an HVAC system.
- Complete the Automotive Service Excellence (ASE) tasks related to HVAC inspection and test procedures.

TERMS TO LEARN

after blow module
comeback
diagnostic trouble code (DTC)
functional test
ground out
technical service bulletin (TSB)
trouble tree
visual inspection

9.1 INTRODUCTION

Except for preventative maintenance operations, vehicle service work should begin with a systematic procedure to determine exactly what is wrong. If trouble diagnosis is thorough and accurate, all problems can be located and repaired at the same time.

✓SERVICE TIP

Refrigerant must be recovered from an A/C system before repairs can be made; after the repairs are complete, the system must be recharged. An incomplete or inaccurate diagnosis usually leads to an incomplete repair, which may mean going through the recovery and recharge steps twice, wasting time and possibly some refrigerant.

A still bigger concern for technicians is the repair that fails after the vehicle has left the shop and the vehicle comes back. In the vehicle service field, **comebacks** are usually repaired at the expense of the shop and the technician. It is much better to fix the problem the first time a vehicle comes into the shop. A comeback can cause an immediate loss of time and money, but the biggest loss is often the shop's reputation.

Diagnosis procedures vary greatly, depending on the nature of the problem and the experience of the technician. If the vehicle shows a familiar set of symptoms, the experienced technician can often shortcut the procedure and go directly to the steps that confirm the cause for that particular problem. After technicians determine the probable cause, they often make other checks to verify or confirm their conclusion.

Chapter 10 describes the diagnosis procedure for all of the problems normally encountered in the A/C subsystem of the HVAC system. Diagnoses of the heater, air management, and control subsystems are described in Chapter 11. Electrical and electronic circuits and their diagnostic procedures are described in Chapter 12. Heater core and hose problems and cooling system problems and repair are described in Chapter 16.

An important step in the early part of the diagnostic procedure is to ensure that there is a problem and, if there is, to determine which HVAC subsystem is at fault. It is possible to get a complaint such as **insufficient cooling** from an A/C system on a very hot and humid day, and under these conditions, the cooling load often exceeds the capacity of the system. The system can be operating properly at full design capacity. In this case, the only thing a technician can do is to try and educate the customer about what is happening.

✓SERVICE TIP

A competent technician normally makes several diagnosis checks of a system. Diagnosis usually begins with a thorough under-hood inspection using sight, sound, touch, and even smell. Under-hood checks usually begin with an inspection while the engine is off, and parts of this check are repeated with the engine running. The inspection procedure also includes a check of the HVAC system controls, blower, and air management doors both while the engine is off and while it is running. After these checks, the technician should know whether the system is operating properly or which subsystem has a problem.

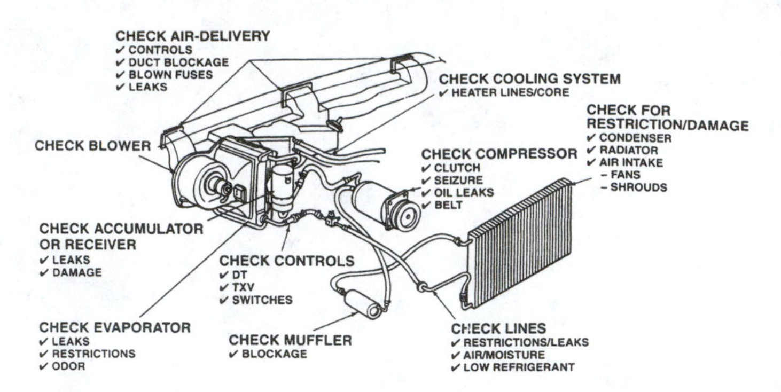

FIGURE 9–1 A visual inspection of an HVAC system includes these checks. *(Reprinted with permission of General Motors Corporation)*

9.2 HVAC SYSTEM INSPECTION

HVAC system inspection, often called a visual inspection or a functional test, is the quickest way to locate obvious problems (Figure 9–1). It also gives the technician a good idea of the overall condition of the system. During this step, the technician determines whether the A/C system is an R-12, retrofitted from R-12, or R-134a system; a TXV or OT system; and a cycling clutch, STV, or variable compressor system. Although this is called a visual inspection, the competent technician listens for things like clutch or idler bearing problems; feels for system temperatures and under hoses for damage, excessive vibration, and evidence of leaks; shakes parts such as compressor mounts while feeling for looseness; and is aware of any unusual smells such as from an antifreeze leak or a moldy evaporator.

A technician often follows a service checklist like that shown in Figure 9–2 while doing a visual inspection. This list keeps important checks from being overlooked, and it also gives the customer an idea of what was done. Many motorists seldom open the hood, and of those who do, only a few understand what they are looking at.

9.2.1 A/C System Inspection

With the engine off, the under-hood checks are as follows:

1. Check the condition of the belts. With a V belt, roll the belt so you can see the sides. Cracked, frayed, highly glazed, or otherwise damaged belts should be replaced. On belts that show wear or damage, check the condition and alignment of the pulleys. Faulty belts should be replaced (Figure 9–3).
2. Check the tension of the belt by pushing against the center at a longer span. Any belts that seem loose should be readjusted. A belt tension gauge is the most accurate way of checking belt tightness. With serpentine belt systems, you should be able to see and feel the automatic tensioner operate as you push on the belt.
3. Inspect the refrigerant hoses and lines for oily residue and damage. Oil residue with caked-on

AIR CONDITIONING SERVICE CHECK LIST

Customer ______________________________ Date ______________

Address ______________________________

City ______________________ State ____________ Zip ____________

Year ____________ Make ______________________ Model ______________

License No. ______________ Engine ____________ V.I.N. ______________

Description of Problem ______________________________

GAUGE READINGS	INITIAL	FINAL
High		
Low		
Aux.		

TEMPERATURE READINGS	INITIAL	FINAL
Ambient		
Duct		

COMPONENT CHECKS

COMPONENT	OK	REPAIR
1. Drive Belts: Proper Tension	☐	☐
Condition	☐	☐
2. Charging System: Volts	☐	☐
Amps	☐	☐
Volt Drop	☐	☐
3. Fan Clutch: Bearing	☐	☐
Leaks Front/Rear	☐	☐
4. Idler Pulley: Bearing	☐	☐
Groove/Surface	☐	☐
5. Compressor Clutch: Air Gap	☐	☐
Bearing	☐	☐
Field Coil	☐	☐
Electrical Connections	☐	☐
Surge Diode	☐	☐
6. Compressor: Leaks-Shaft Seal/O Rings	☐	☐
Mounting Hardware	☐	☐
Operation	☐	☐
7. Discharge Line: Muffler	☐	☐
Hose	☐	☐
Fitttings	☐	☐
8. Condenser: Fittings/Leaks	☐	☐
Fins Clean	☐	☐
Mounting Hardware	☐	☐
Temperature Change	☐	☐
9. Receiver/Drier: Mounting Hardware	☐	☐
Fittings	☐	☐
Even Temperature	☐	☐
Sight Glass	☐	☐
High Pressure Valve	☐	☐
10. Liquid Line: Routing/Mounting Hardware	☐	☐
Even Temperature	☐	☐
Fittings	☐	☐
11. Thermostatic Expansion Valve:		
Leaks	☐	☐
Inlet/Outlet Temperature	☐	☐
Sensing Bulb Clean/Tight	☐	☐
Insulation Tape	☐	☐
11A. Orifice Tube:		
Fittings/Leaks	☐	☐
Inlet/Outlet Temperature	☐	☐
12. Evaporator: Leaks	☐	☐
Mounting Hardware	☐	☐
Air Flow	☐	☐
Discharge Air Odors	☐	☐
Condensation Drain	☐	☐
13. Suction Throttle Valve:		
Leaks	☐	☐
Inlet/Outlet Temperature	☐	☐
Operating Pressure	☐	☐
14. Thermostatic Clutch Cycling Switch:		
Sensing Bulb—Clean/Tight	☐	☐
On/Off Cycling Times	☐	☐
Insulation Tape	☐	☐
Electrical Connection	☐	☐

COMPONENT	OK	REPAIR
15. Pressure Cycling Switch: Fitting/O Ring Leaks	☐	☐
On/Off Cycling Time	☐	☐
Electrical Connection	☐	☐
16. Accumulator: Mounting Hardware	☐	☐
Inlet/Outlet Temperature	☐	☐
Fittings/Leaks	☐	☐
17. Suction Line: Mounting and Routing	☐	☐
Fittings/Leaks	☐	☐
Hose Condition	☐	☐
Temperature	☐	☐
18. Radiator: Leaks	☐	☐
Air Fins Clean	☐	☐
Hose and Clamps	☐	☐
Pressure Cap	☐	☐
Over Flow Tank and Hose	☐	☐
19. Cooling System:		
Antifreeze 2 years or 30,000 miles	☐	☐
Freeze Protection (−20° minimum)	☐	☐
Heater Hoses, By-Pass Hose and Clamps	☐	☐
Heater Control Valve	☐	☐
Thermostat	☐	☐
20. Electric Cooling Fan: Bearings	☐	☐
Fan Blade	☐	☐
Electric Connection	☐	☐
Amp Draw	☐	☐
Coolant Sensor	☐	☐
Relays	☐	☐
21. Blower Motor Circuit:		
Blower Motor Amp Draw	☐	☐
Electrical Connections	☐	☐
Harness Connections	☐	☐
Blower Switch	☐	☐
Blower Resistor	☐	☐
22. Vacuum Control Circuit:		
Vacuum Canister	☐	☐
Vacuum Hoses	☐	☐
Vacuum Motors	☐	☐
Vacuum Control Switches	☐	☐
Vacuum Check Valve	☐	☐
23. Dash Control Switches:		
Proper Mode Changes	☐	☐
Temperature Control Cable	☐	☐
Duct Doors	☐	☐
24. System Control and Protection:		
Low Pressure Protection Switches	☐	☐
High Pressure Protection Switches	☐	☐
Wide Open Throttle Switch or TPS	☐	☐
Power Steering Cutoff Switch	☐	☐
Coolant Temperature Switch	☐	☐
Coolant Temperature Relay	☐	☐
Power Brake Delay Relay	☐	☐
Throttle Kick Solenoid	☐	☐
Isolation Relay	☐	☐
A/C Relay	☐	☐
Time Delay Relay	☐	☐
Constant Run Relay	☐	☐

Remarks ______________________________

Repair Estimate ______________________________

Part No. A9578 Printed in U.S.A.

FIGURE 9–2 Many technicians use a checklist like the one shown here while inspecting a system. It provides a good record of the inspection to show the customer. *(Courtesy of Everco Industries)*

FIGURE 9–3 A drive belt should be replaced if it has any of the problems shown. (*Courtesy of Dayco Products, Inc.*)

dirt indicates a probable leak (Figure 9–4). Each of the A/C test ports should be capped.

4. While checking the hoses and lines, determine whether you are dealing with a TXV or OT system and whether an STV or a variable displacement compressor is used. On GM cars, a variable compressor can be identified by its shape. Chrysler Corporation cars use differently colored TXVs to identify variable compressors.
5. Check the compressor mounting bolts to make sure they are tight.
6. Check to make sure there is an air gap at the compressor clutch. If there is good access, turn the clutch plate while you feel for smooth compressor operation (Figure 9–5).
7. Check the electrical wires to the clutch, blower motor, and any A/C switches for good, tight connections and possible damage.
8. Check the vacuum hoses between the intake manifold and bulkhead, looking for kinked, cracked, or loose hoses.
9. Check the condition of the radiator cooling fan, fan clutch, and fan shroud.
10. Check the condition of the radiator and heater hoses, looking for swollen, cracked, kinked, or leaky hoses.

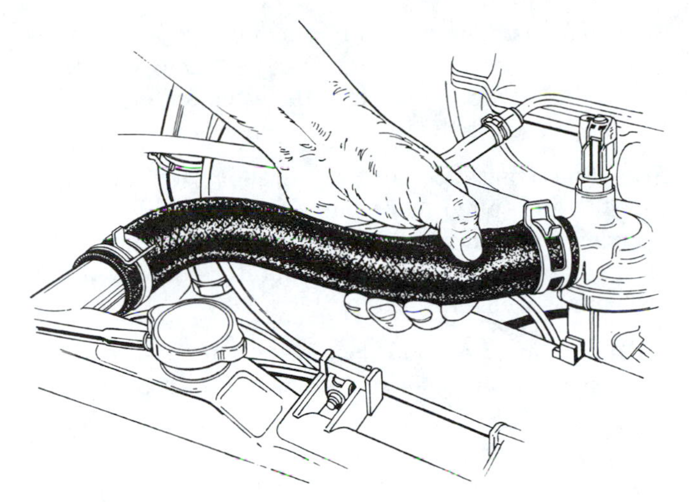

FIGURE 9–4 Hose inspection can be difficult in tight quarters. Check for cracks and cuts and squeeze the hose to check for hardening or excessive softness. Squeeze the hose close to the ends to check for softening. (*Courtesy of The Gates Rubber Company*)

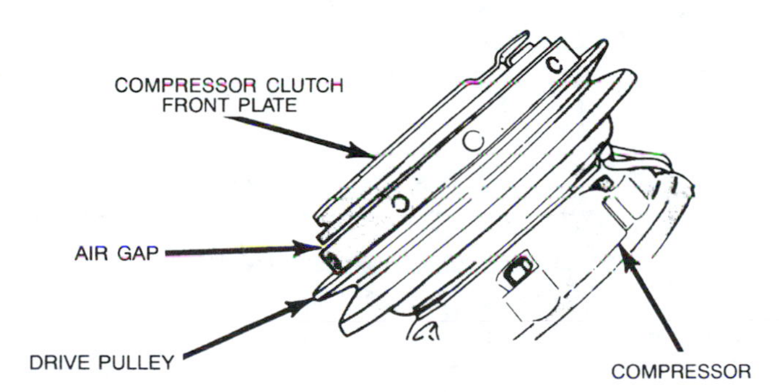

FIGURE 9–5 With the engine off, compressor clutch inspection includes a check for air gap between the clutch plate and pulley. The hub and compressor shaft should rotate smoothly, without runout of the clutch plate. (*Courtesy of DaimlerChrysler Corporation*)

11. If the engine and radiator are cool, remove the cap and check the appearance of the coolant, which should be clean and have a bright color.
12. Check the faces of the condenser and radiator core; they should be clean, with no restriction to airflow. There should be no apparent leaks.

✓SERVICE TIP

A missing A/C service cap can allow a refrigerant leak of up to 1 lb (2.2 kg) of refrigerant per year.

✓REAL WORLD FIX

The 1992 Dodge Caravan (140,000 miles) had a failure of the serpentine belt that unraveled and caused a failure of the A/C drive belt. The belts have been replaced, but now the A/C clutch will not engage. The fuses are okay, but there is no power to the clutch.

Fix: Following advice, the technician checked the pressure switch that is mounted on an A/C line below the radiator and found the flapping belt had disconnected the wire connections. Replacement of these wires onto their connectors fixed this problem.

With the engine off, the in-car checks are as follows:

1. Operate the blower switch through its various speeds while you listen to the fan and motor for unusual noises. Note that some systems do not have blower operation unless the heat or A/C controls are on or unless the engine is running (Figure 9–6).
2. Move the temperature lever of mechanically operated doors to both ends of its travel. It should move smoothly and stop before making contact at the ends. A late stop indicates that adjustment is needed.

✓SERVICE TIP

On modern systems, poor air flow can be caused by a dirty cabin filter that needs changing.

✓SERVICE TIP

On many cars, a "thunk" noise occurs as the temperature door contacts the stop. If you are in an area with cold winters, make sure it "thunks" at the full heat stop; if you are in an area with hot summers, make sure it "thunks" at the full cool stop.

With the engine running, the in-car checks are as follows:

1. Note the alternator charge indicator; it should show a normal charge level or voltage after the engine runs a short while.
2. Set the blower speed to high, move the function control to all of its positions, and note the air

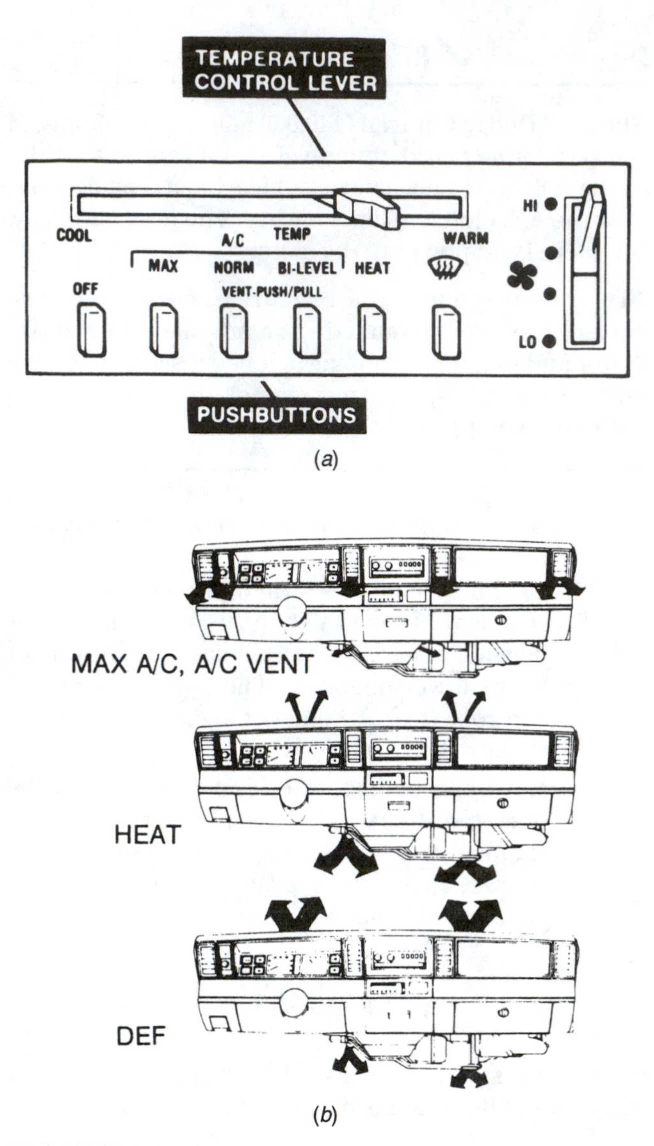

FIGURE 9–6 Inspection includes a check for proper blower operation through each of the speeds (*a*) and air discharge from each location for the different functions (*b*). (*Courtesy of DaimlerChrysler Corporation*)

discharge. The airflow should change to the various outlets and should be about the same from each outlet.

3. Turn on the A/C to determine whether the compressor clutch engages and the compressor operates.

✓REAL WORLD FIX

The 1988 Buick (73,000 miles) had very little airflow from the registers. The programmer was checked, and the solenoids appeared to be functioning. A new programmer was installed, but this did not help. Changing the operating mode produces the same amount of airflow from each outlet.

Fix: Following advice, the evaporator core was checked and found to be plugged almost solid. It had been leaking, and the refrigerant oil had collected a lot of dirt. Replacement of the evaporator core fixed this problem as well as an A/C problem.

Author's Note: A good review of HVAC basics would have saved the unneeded replacement of the programmer.

✓SERVICE TIP

Many modern HVAC systems include a cabin air filter, and many vehicle owners are not aware that these require periodic replacement. A plugged filter will cause a severe airflow reduction and also could be the source of a bad odor.

✓SERVICE TIP

Reduced airflow can also be caused by a plugged evaporator core. This can be checked by measuring the blower motor current draw. Any restriction to the airflow will reduce the power requirement for the blower motor and the amount of current required. A more complete explanation can be found in Section 12.5.

✓REAL WORLD FIX

The 1996 Ford Explorer (50,000 miles) had a faulty blend door actuator motor. A new motor was installed, but it did not operate the door. The power supply and ground checked good, and the motor can be made to operate. After determining that the new motor was the wrong part for this vehicle, even though it looked the same, the correct motor was installed, but it still would not operate the door.

Fix: Closer inspection revealed that the blend door operating shaft had broken off. Installation of a new, updated heater box/blend door assembly fixed this problem.

✓REAL WORLD FIX

The 1986 Saab (170,000 miles) HVAC system has no temperature control; it stays at a constant, hot temperature. The stepper motor operates the arm attached to the blend door in what appears to be a normal manner.

Fix: Following advice from other technicians, the system was disassembled. The sealing foam from the blend door had disintegrated, and a large piece had fallen down and jammed the door. This caused the door operating arm to break. Replacement of the plenum section fixed this problem.

✓REAL WORLD FIX

The 1993 Pontiac Bonneville (120,000 miles) has a very poor airflow from the instrument-panel registers. This is a manual system so there is no scan tool information. The technician is concerned about removing the instrument panel to locate the problem.

Fix: Following advice, the glove compartment seals were checked and found to be faulty. Another faulty seal was also found after the radio was removed. Repair of these two air leaks fixed this problem.

Author's Note: Some vehicles route the instrument panel air through the glove compartment, and some radios are easy to remove, providing a way to check hidden areas.

✓SERVICE TIP

With the hood and driver's door or window open, on most vehicles you can hear an audible click as the compressor clutch engages when you turn the A/C system on. Set the system to A/C, cold temperature, and high blower speed. With electronically controlled systems, the engine must be running for the A/C clutch to operate.

With the engine running, the under-hood checks are as follows:

1. Make sure the compressor clutch is engaged and the compressor is running. Listen for any signs of improper compressor operation. If there are any, feel the compressor for harsh, rough motions.
2. Disconnect the clutch to make sure it releases smoothly. With the clutch released, listen for proper clutch bearing operation. If there are unusual noises, carefully feel the compressor next to the clutch for harsh, rough operation. Reconnect the clutch and note the engagement (Figure 9–7).

FIGURE 9–7 With the engine running, compressor clutch inspection includes a rapid engagement as the clutch wire is connected, a clean disengagement with no drag as the wire is disconnected, smooth and quiet pulley operation, and very little runout of the pulley.

3. Feel the temperature of the A/C lines and hoses. Be cautious on the high side because the lines should be warm to hot, with the temperature increasing. All the lines on the low side should be cool to cold, with the temperature getting colder (Figure 9–8).
4. Feel the temperature of the heater hoses. If the engine is at operating temperature, both hoses should feel hot.
5. Check the drive belts. They should be running straight and smooth, with a slight rippling allowed in the belt section returning to the crankshaft pulley. Excess belt whip or slap indicates that adjustment is needed.
6. Check the fan operation (if running). The fan should be turning smoothly, with good airflow. Depending on the temperature and type of fan clutch, a mechanical fan in an RWD car should deliver moderate to high airflow. An electric drive fan is probably operating to produce airflow.
7. Check the evaporator drain. By this time, there should be a small puddle of water under the evaporator area and drops of water coming from the drain. With some vehicles, the drain is routed through a frame member; check the vehicle service information if you cannot locate the drain.

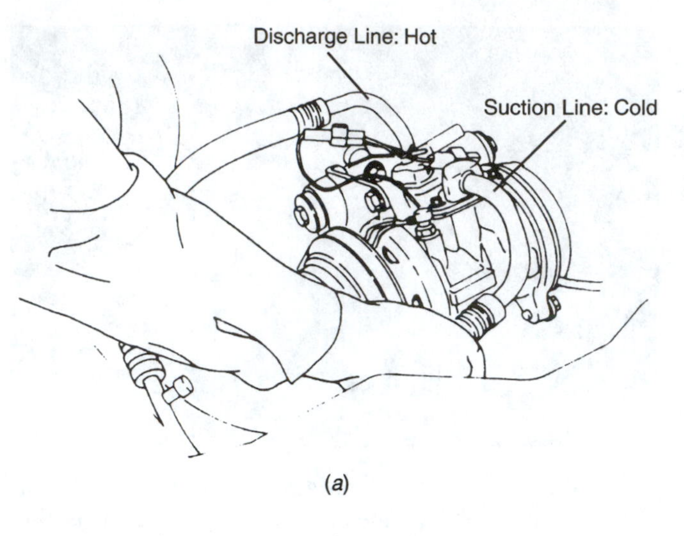

(a)

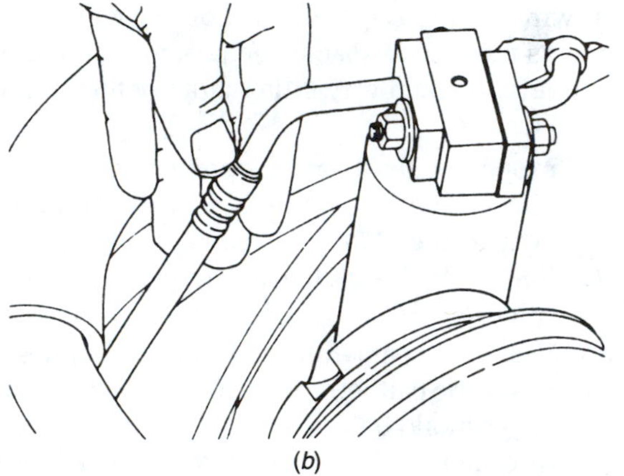

(b)

FIGURE 9–8 When a system is operating properly, the suction line to the compressor should be cool, and the discharge line should be hot to very hot (*a*). The liquid lines should also be hot (*b*). (*Courtesy of DaimlerChrysler Corporation*)

✓SERVICE TIP

An infrared, noncontact thermometer is a fast and easy way to check these temperatures (see Figure 10–12).

✓SERVICE TIP

To determine if there is adequate airflow through a condenser, many technicians place a sheet of paper or a dollar bill in front of the condenser; with the engine running at idle speed, it should stick there.

With the engine still running, check next inside the vehicle for overall operation. By now the A/C system should be delivering cool to cold air from the instrument panel registers. Moving the temperature control should cause a warming in the airflow. Changing the function control to heat should move the air discharge to floor level.

✓SERVICE TIP

Automatic operation of ATC can be checked by warming the in-vehicle sensor with a heat gun (it should shift to a colder operation), cooling the sensor with a quick-freeze aerosol (it should shift to a warmer operation), or bringing a lightbulb over the sunload sensor (it should shift to a colder operation).

✓SERVICE TIP

An in-car sensor must have an airflow past it in order to measure the temperature. This airflow is caused by the aspirator or an integrated fan. You can test the airflow by placing a piece of tissue paper next to the sensor; the paper should be held in place by the air pressure.

9.3 PROBLEM DIAGNOSIS

A visual inspection often locates the exact cause of a problem. It nearly always reveals the nature of the problem, showing in what area of the system the problem lies. If there is no blower motor operation along with no compressor clutch operation, for example, an experienced technician knows that in many vehicles these are branches of a single electrical circuit. Therefore, the most probable cause is a blown fuse; the second most probable cause is either a faulty switch at the control head or a bad connection to that switch. With modern vehicles, there can be a problem in the electrical control circuit or control modules.

The most important aspect of diagnosis is knowledge of how a system works. If we understand a system, we know what to expect from each part of that system, and we also know what to expect when that part does not work. Classroom instruction and information sources such as service manuals and this text can help you in this regard.

Also important in diagnosing problems is knowing where failure is most likely to occur. In an A/C system, refrigerant leaks are more common than compressor failures, and leaks are much more likely to occur at line connections than in metal tubing. Experience and troubleshooting charts can help you master diagnosis (Figure 9–9).

Complete diagnosis is concerned with locating the exact cause of a problem relatively quickly. For example, let's say that the cause of no blower motor and compressor operation was a blown fuse. A fuse might blow from old age or a high-voltage spike from the system, but it is more probable that excess current flow caused by a shorted or grounded circuit is the cause. If the only repair is to replace the fuse, the system will probably come back with another blown fuse.

Another example is a system with poor A/C operation that is caused by a low charge level. The low charge level was probably caused by a leak.The leak should be located and repaired to fix the cause of the problem rather than simply topping off or recharging the system.

✓REAL WORLD FIX

The 1988 Dodge Caravan (130,000 miles) had a complaint of no heat from the heater. The engine thermostat was working, and the engine coolant was getting hot. The heater hoses are hot, but the technician is concerned that there is no vacuum signal to the heater control valve.

Fix: A check of a vacuum diagram shows that a vacuum signal is only at the valve during A/C to close the valve. He was advised to check the temperature blend door, and it was found to be stuck. A small hole was drilled into the plenum case at the lower bearing for the door, and penetrating oil was sprayed into the hole. This freed the door and fixed the problem.

Diagnosis charts, also called **trouble trees**, help guide the technician through the procedure to locate the cause of a problem in the shortest time possible. In most cases, they use a logical, systematic series of tests and checks to isolate the area causing the problem. Charts are available from vehicle manufacturers and suppliers of A/C parts. Most vehicle manufacturers develop charts for the features of a particular system. For example, as you will find out, the normal low side pressure of a cycling clutch system is different from that of a system that uses an STV or variable displacement compressor, and the diagnostic procedure for each of these systems is slightly different. The diagnostic chart helps eliminate those portions of a system that are working properly so we can concentrate our tests and further checks on those portions where the problem might be.

With a common problem, such as **no or insufficient cooling,** an experienced technician knows the problem can be caused by a number of faults in the A/C system, faulty door or blower operation in the air management system, or faulty compressor switching by the controls. Most vehicle manufacturers recommend that problem-solving or troubleshooting procedures be done in an organized manner (Figure 9–10).

Step 1. Verify the Complaint Verifying the complaint ensures that the technician understands the nature of the problem. In many cases, this can be done in the service bay; in others, a road test is required to duplicate the conditions under which the fault occurs. This step requires that the technician be familiar with the system and how it should perform. It is sometimes necessary for the technician to learn the exact nature or operating conditions; this is especially important with intermittent problems that happen intermittenly only once in a while.

Step 2. Perform Preliminary Checks Some problems require preliminary checking to determine which avenue to take. For example, one of the early checks for a **no cooling** complaint is to see whether the compressor operates. If it does, the problem is probably caused by a fault in the A/C system; if it does not operate, the problem can be caused by a fault in the air distribution system or the controls.

Step 3. Check Service Bulletins With our modern and complex systems, more things can go wrong, and some of these are design or component problems. When these occur often enough on a particular make or model, the manufacturer develops a **technical service bulletin (TSB)** that describes the problem symptoms and the recommended repair procedure. Technicians in independent shops use service manuals, in printed and computer versions, including TSBs. Many new-car dealerships have on-line, dealership-to-factory computer connections for up-to-date service bulletins and diagnostic information.

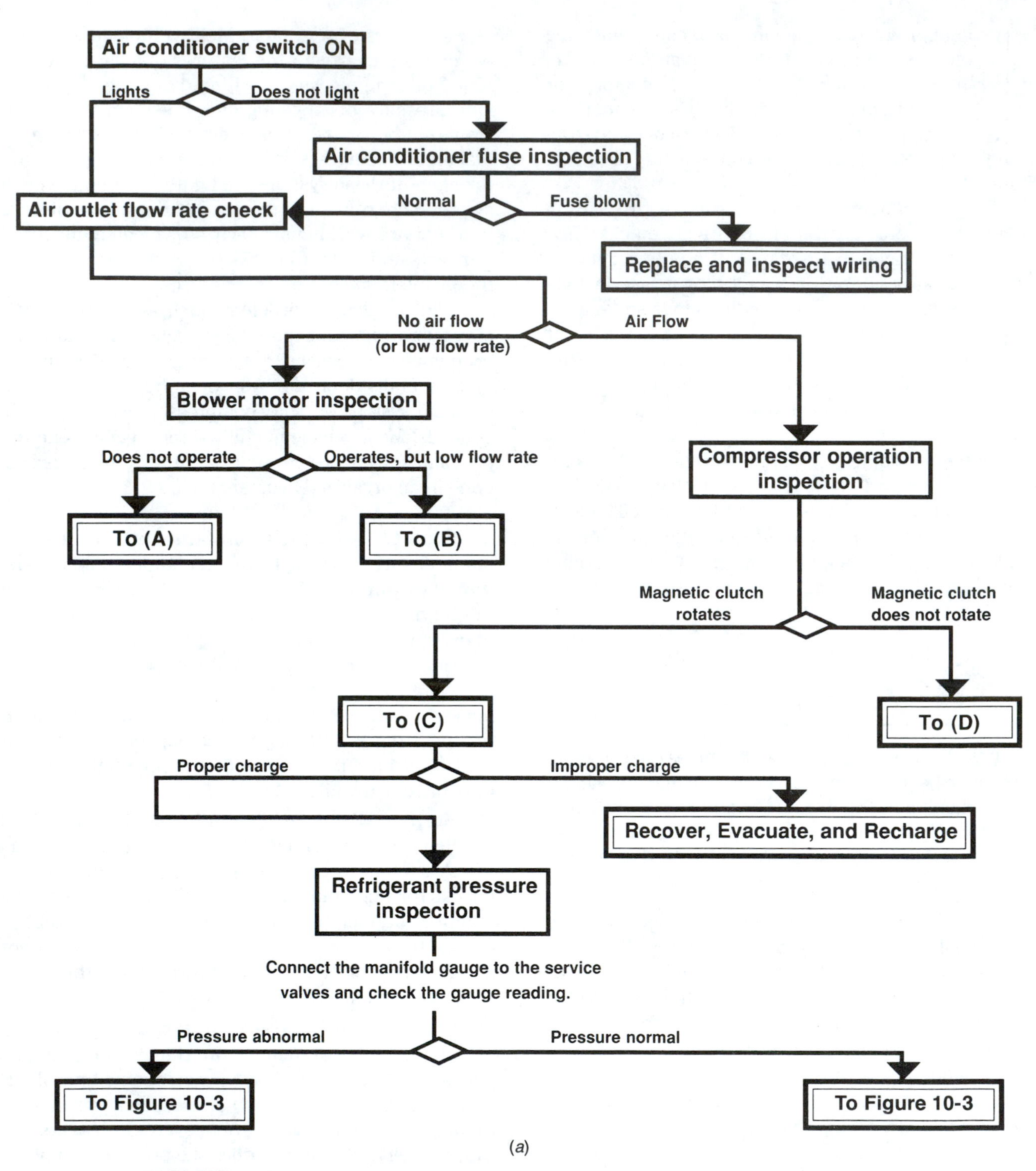

(*a*)

FIGURE 9–9 A trouble tree can be followed to determine the cause of the problem (*a*). Note that some checks lead to further checks that are listed on the troubleshooting charts (*b*). (*Courtesy of Zexel USA Corporation*)

TROUBLESHOOTING CHART

(A) Blower motor does not operate

Possible cause	Inspection	Remedy
1. Blown fuse.	Inspect the fuse/wiring.	Replace fuse/repair wiring.
2. Broken wiring or bad connection.	Check the fan motor ground and connectors.	Repair the wiring or connector.
3. Fan motor malfunction.	Check the lead wires from the motor with a circuit tester.	Replace.
4. Broken resistor wiring.	Check resistor using a circuit tester.	Replace.
5. Fan motor switch malfuction.	Operate the fan switches in sequence and check whether the fan operates.	Replace.

(B)Blower motor operates normally, but air flow is insufficient

Possible cause	Inspection	Remedy
1. Evaporator inlet obstruction.	Check the inlet.	Remove the obstruction and clean.
2. Air leak.	Check the cooling unit case joints.	Repair or adjust.
3. Defective thermo switch, (frozen evap.).	Check the switch using a circuit tester.	Replace.

(C) Insufficient cooling although air flow and compressor operation are normal

Possible cause	Inspection	Remedy
1. Insufficient refrigerant.	There will be little temperature difference between the low and high-pressure sides.	Repair any leaks and recharge the refrigerant to the correct level.
2. Excessive refrigerant.	Verify by gauge reading.	Utilize your refrigerant recovery equipment to capture excess refrigerant. Charge to the correct refrigerant level.

(D)The compressor does not operate at all, or operates improperly

Possible cause	Inspection	Remedy
1. Loose drive belt.	The belt oscillates considerably.	Adjust the tension.
2. Internal compressor malfunction.	The drive belt slips.	Replace compressor.
	• ***Magnetic clutch related***	
3. Low battery voltage.	Clutch slips.	Recharge the battery.
4. Faulty coil.	Clutch slips.	Replace the magnetic clutch.
5. Oil on the clutch surface.	The magnetic clutch face is dirty, causing it to slip.	Replace, or clean the clutch surface.
6. Excessive clearance between the clutch plate and clutch disk. The clutch plate clings when pushed.	Check clutch gap according to specifications.	Adjust the clearance, or replace the clutch.
7. Open coil.	Clutch does not engage and there is no reading when a circuit tester is connected between the coil terminals.	Replace.
8. Broken wiring or poor ground.	Clutch will not engage at all. Inspect the ground and connections.	Repair.
9. Wiring harness components.	Test the conductance of the pressure switch, thermoswitch, relay, etc.	Check operation, referring to the wiring diagram, and replace defective parts.

(*b*)

FIGURE 9–9 (Continued)

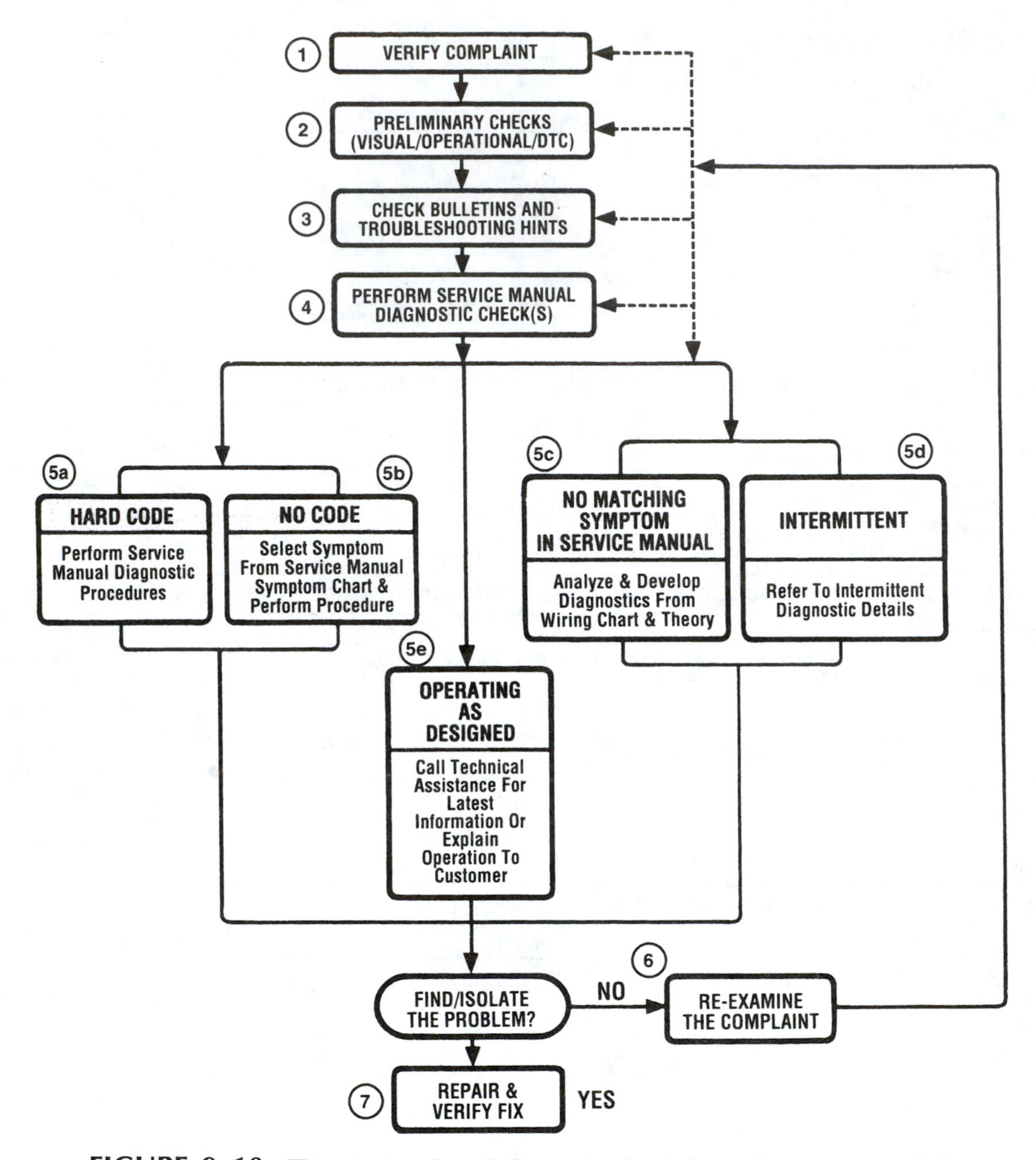

FIGURE 9–10 This strategy-based diagnosis chart shows the recommended procedure to locate a system problem so that it can be fixed right the first time it is repaired. (*Reprinted with permission of General Motors Corporation*)

Step 4. Perform Diagnostic Checks Many information sources, both printed and electronic, include a checking procedure to isolate the causes of various problems. This information can be in the form of charts showing the probable and possible faults or in the form of flowcharts (Figure 9–11). Flowcharts are often called **trouble trees** because of the way they branch off to check other possible causes.

This step includes checking for **electronic diagnostic trouble codes (DTCs).** DTCs for older vehicles are specific to the vehicle make and model, making service information necessary to determine the procedure as well as the meaning of the codes. DTCs are described more completely in Chapter 12.

Step 5. Interpret Test Results While making checks, you often have to make decisions. Is the

X — Most Probable Cause(s)
• — Possible Causes

COMPLAINT–CONDITION / CAUSE	Sec. HVAC Controls	Damage Temp. Cable	Damage Mode Cable	HVAC Module Temp. Door	HVAC Module Mode Door	HVAC Module Linkage	Cable(s) Disconnected	Damage Air Ducts	Case Drain Plugged	Mode Cable Adjust	Temp. Cable Adjust	Damage Air Outlets	Debris in Evap. Housing	
Mode Lever Binds/Sticks			X		•	X								
Temp. Lever Binds/Sticks		X		•		X					•			
No Mode Change			•			•	X			•				
No Temp. Change		•				•	X							
Unequal Air Distribution								X		X		•		
Odor from Air Outlets									X				X	

(a)

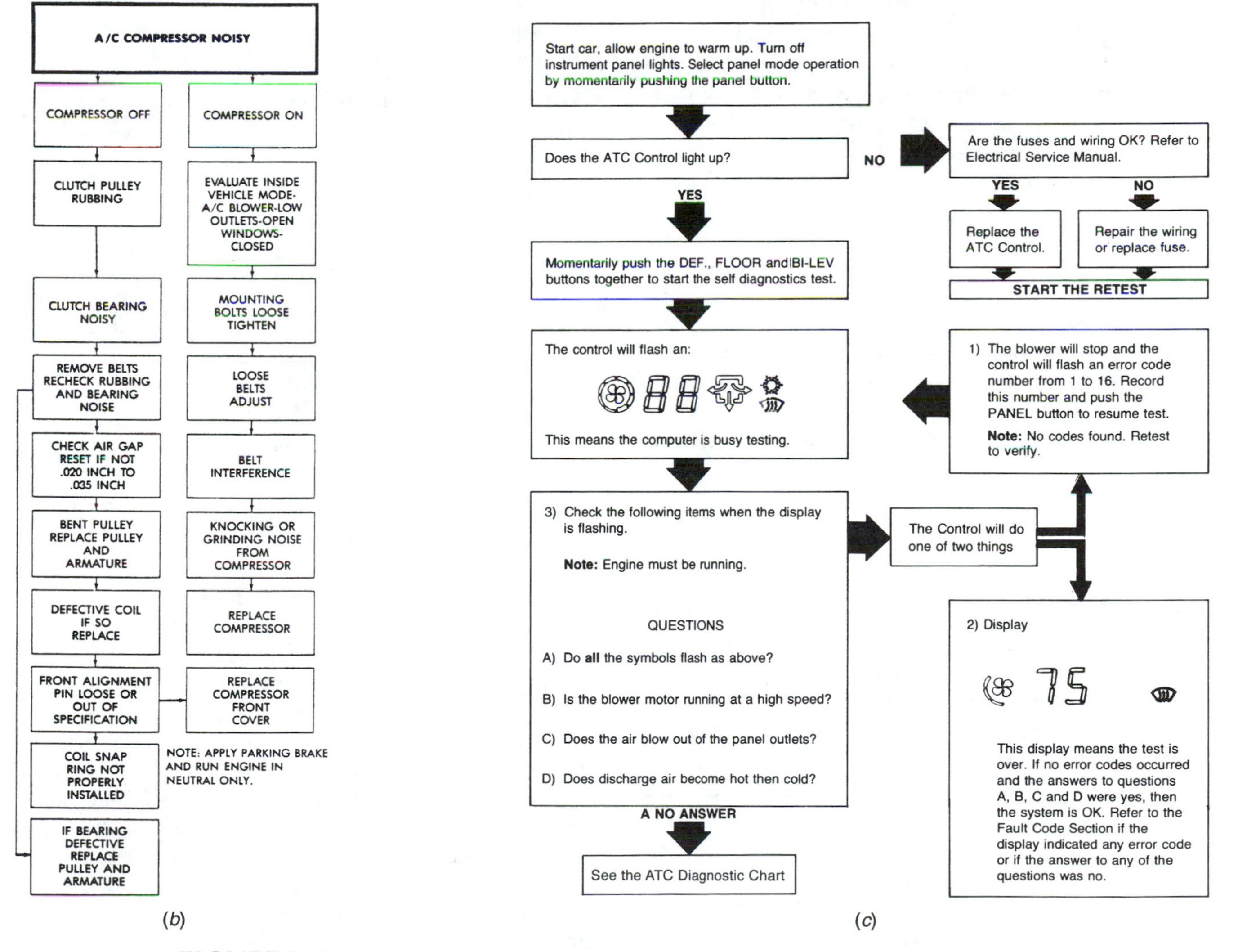

FIGURE 9–11 Three problem diagnosis charts. Chart *a* shows the most probable and possible causes of a group of problems. Chart *b* shows the procedure for locating a noisy compressor problem. Chart *c* is a flowchart for locating an ATC problem. (a *is courtesy of Saturn Corporation;* b *and* c *are courtesy of DaimlerChrysler Corporation*)

operation right or wrong or good or bad? Does your answer lead you to the need for additional checks? You might have to return to step 3 or 4 for more information. At the end of this step, you should have determined that

- There is nothing wrong, and the system is operating normally.
- The system is almost operating properly, but some part of it needs adjustment or reprogramming.
- A faulty part needs to be repaired or replaced.

Step 6. Make the Repair Repair operations for the HVAC subsystems are described in Chapters 12, 13, 14, and 16. They are also covered in vehicle service manuals.

9.4 HVAC SYSTEM PROBLEMS

As a technician diagnoses a system problem, it is usually very easy to determine the faulty subsystem. For example, a **no cooling from A/C** complaint is often diagnosed with a check to ensure compressor operation and then a check of system pressures. If there is no pressure, the system probably has a leak that needs to be fixed; this would be an A/C system problem. If the compressor does not operate, then the technician checks for a voltage signal at the clutch. If there is a signal, the clutch must be faulty; this is also an A/C system problem. If there is no voltage, there must be an open switch, relay, or fuse interrupting the signal, and this could be a problem in the control circuit.

Some problems cross the line between HVAC subsystems or do not fit neatly with a subsystem.

✓SERVICE TIP

A complaint of uneven air discharge temperature from the instrument panel registers (cold on one side and warm on the other) can be caused by a low charge level. This will cause some parts of the evaporator to be cold while others are warm. It is possible for the air from the cold side to flow a single register.

✓REAL WORLD FIX

The 1995 Chevrolet Lumina (60,000 miles) came in with a complaint of no A/C. The system appeared to be low on refrigerant so it was leak tested, the refrigerant recovered and recycled, the system evacuated and recharged. The high and low side pressures were now okay, but the air coming from the ducts was not cold. An under-dash check shows vacuum at the proper location, and the blend door seems to be working.

Fix: A closer under-dash inspection revealed an ink pen blocking complete movement of the blend door lever; removal of the ink pen fixed this problem.

✓REAL WORLD FIX

The 1999 GMC Yukon (24,000 miles) has a wet floor on the passenger side. The A/C is not draining to the outside, and the technician cannot find the drain.

Fix: Following advice, the technician checked the proper location and found that the drain hose was missing, and the drain was plugged. Cleaning the drain and installing a new drain hose fixed this problem.

Author's Note: Some vehicles have drains that are very difficult to locate; one vehicle drains the HVAC case to the inside of a body frame member.

9.4.1 System Odors

Some systems develop a musty, moldy smell, which is not really a fault of the system. Some sources classify these odors into two types:

- "Dirty socks/gym locker" odor, which has an organic cause
- "Refrigerator, cement, or dusty room" odor, which is caused by chemicals

The organic odor problem is most common in areas with high relative humidity, and it is caused by mildew-type fungus growth on the evaporator and in the evaporator plenum. Modern evaporators have more fins that are closer together, and they tend to trap more moisture and bacterial growth. The cool surface of the evaporator collects moisture as it dehumidifies the air, and most of this moisture runs out of the bottom of the case. After a vehicle is shut off, the moist surface of the evaporator warms up, and this warm, wet area becomes an ideal environment for fungus and bacteria growth. A coating is applied to many evaporators to speed up water runoff; this coating helps dry the evaporator and reduce bacterial growth. Airborne bacteria also collect on this surface, and if the surface stays moist, these bacteria will live and grow, creating the unpleasant smell.

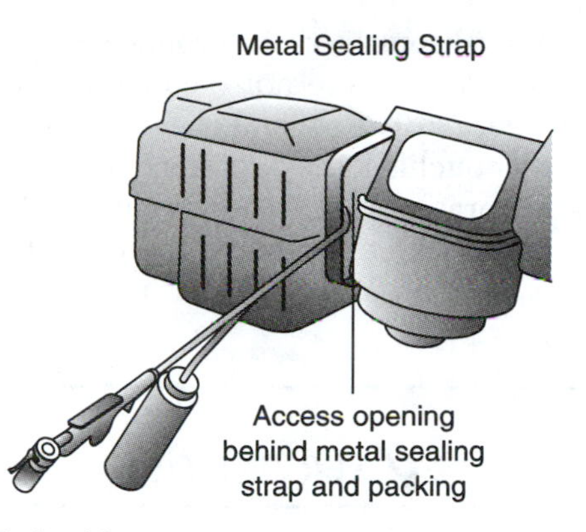

FIGURE 9–12 A foul smell from the A/C can be cured by spraying a cleaning solution or fungicide onto the evaporator to either clean it thoroughly or kill the bacteria. (*Courtesy of Airsept, Inc.*)

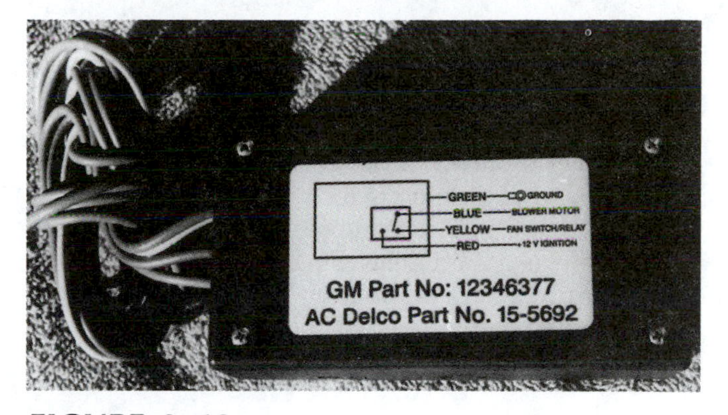

FIGURE 9–13 An after blow module can be connected into the blower motor circuit. It will operate the blower motor after the system is shut off to dry the evaporator and prevent bacterial growth. (*Courtesy of Airsept, Inc.*)

When we turn on the air conditioning or even the ventilation, we blow that smell into the car.

Several companies market chemicals, essentially fungicides, to kill the bacterial growth, or detergents to clean the evaporator core. Some of these chemicals and a procedure to use them have been approved by vehicle manufacturers. These chemicals are sprayed into the ductwork or onto the evaporator fins.

✓SERVICE TIP

In some systems, spraying into the inlet of ductwork simply wets the blower fan and does very little good. It is much better to spray the material onto the evaporator fins. With some systems, fairly good access can be obtained by removing the blower motor resistor; other systems require drilling a properly placed hole, which must be covered when you are finished. Be sure to follow the manufacturer's procedure and wear the required safety gear; most require a face shield or goggles, and some require a respirator (Figure 9–12).

✓SERVICE TIP

Bacterial growth can be prevented by regularly drying off the evaporator. Running the system on vent or heat with the A/C off is a good way, but most people do not want to do this on a hot day. Sometimes just turning the A/C off shortly before the car is parked will let the high side pressure bleed off, so the evaporator will warm up slightly and dry off faster.

✓SERVICE TIP

Bacterial growth odor can be prevented by operating the A/C in fresh air mode rather than recirculation. The moisture from the outside air collects on the evaporator core and can wash the growth off the core.

Some manufacturers install an **after blow module,** also called an **electronic evaporator dryer module.** In some vehicles, the A/C control module is programmed to operate the blower for a drying cycle after the car is shut off. This device turns on the blower (with the ignition switched off for 30 to 50 minutes) and lets it run long enough to dry off the evaporator after the vehicle has been shut off. One system waits 10 minutes, runs the blower on high for 10 seconds, shuts down for another 10 minutes, and then repeats the 10-second operation and 10-minute pause for 10 cycles. The object is to blow the moist air out of the evaporator without discharging the battery. After blow modules are available for vehicles with both ground or B+ side blower switch systems (Figure 9–13).

9.4.2 Noise Problems

The A/C system is the potential source for several noise problems, and the compressor and clutch are the main culprits. When diagnosing noise problems, it helps to remember that noise pitch is related to the frequency of the item that is producing the noise. A moaning or growling noise is a relatively low-frequency noise caused by something moving slowly; a whine or squeal is a high pitch and frequency noise produced by something moving rapidly.

Another problem can be **groundout,** in which a vibrating metal A/C line contacts another surface (flat surfaces produce the most noise). Groundout problems also occur when the exhaust pipes make metal-to-metal contact with the vehicle body.

✓SERVICE TIP

Groundout problems are cured by either separating the A/C lines from the other surface or clamping them together.

✓SERVICE TIP

When there is a suspected noise problem with a compressor, the first check is to note the noise change with the clutch engaged and disengaged. If the noise stops with the clutch disengaged, the problem must be caused by the compressor; if the noise continues, it could be the clutch bearing. The second check is to (1) feel for a vibration by touching the area next to the clutch (a noisy part vibrates at a frequency that matches the noise frequency) or (2) remove the drive belt (if the noise stops, one of the items driven by that belt was making the noise).

✓SERVICE TIP

A high pitched noise from the area of the compressor clutch that occurs during engagement can indicate a weak clutch. Check for proper voltage at the clutch, an excessive clutch air gap, or excessive high side pressure. This noise can also be caused by a loose or misaligned drive belt. With serpentine belts, check to make sure the automatic tensioner is operating correctly.

Figure 9–14 can help you locate the cause of other noise problems.

NOISE	SOURCE	CURE
Moan, growl	Compressor	Tighten mounting brackets Adjust drive belt Check for internal damage Check for excessive high side pressure
	A/C lines	Cure groundouts Change hose length or material
	Blower motor	Replace blower motor
Chatter (at idle)	Compressor	Add ½ lb. R-134a & 3 oz PAG oil to check effect Check for internal damage
Whine, squeal	Compressor P/S pump Alternator Belt idler	Check for faulty bearings, interference with rotating parts Check drive belt alignment Check drive belt for edge wear
Hiss, gurgle	Orifice tube TXV Heater core	Normal if occurs at A/C shut down Isolate or insulate liquid line Bleed air from heater core
Chirp, thump	Compressor clutch	Click is normal at clutch engagement Chirp indicates slippage
Knock	Compressor	Tighten mounting brackets
Rattle of constant frequency	Blower fan	Remove interference with blower wheel

FIGURE 9–14 The causes of most HVAC noises are shown here.

CHAPTER QUIZ

These questions help you study this chapter. Enter the proper word(s) in the blanks to complete each statement.

1. The goal of automotive technicians is to fix vehicle problems __________ the __________ __________.
2. HVAC system inspection normally begins __________ the hood with the __________ __________.
3. During the under-hood inspection, the technician should determine what __________ of __________ the vehicle has.
4. __________ __________ on a refrigerant hose or fitting can indicate a leak.
5. With the system off, there should be a(n) __________ __________ at the compressor clutch plate.
6. The in-vehicle HVAC system checks include operation of the __________ and __________ __________ for air-flow control.
7. Compressor engagement is usually indicated by an audible __________.
8. An early step in determining how to make a repair is to check for __________ __________ __________.
9. __________ __________ will often lead the technician to the cause of many HVAC problems.
10. Many stinky HVAC system problems are caused by __________ __________ on the evaporator.

REVIEW QUESTIONS

These questions allow you to check what you have learned. Select the answer that correctly completes each statement.

1. Technician A says that you should begin HVAC problem diagnosis with a thorough inspection. Technician B says that you can determine from the inspection what type of system you are dealing with. Who is correct?
   - **a.** A only
   - **b.** B only
   - **c.** Both A and B
   - **d.** Neither A nor B
2. Technician A says that sometimes nothing is wrong with a system when there is a complaint of insufficient cooling. Technician B says that you should begin diagnosing this complaint by checking the A/C system pressures. Who is correct?
   - **a.** A only
   - **b.** B only
   - **c.** Both A and B
   - **d.** Neither A nor B
3. Technician A says that the compressor clutch should cycle out at a low side pressure of 25 psi or lower in all systems. Technician B says that too much air gap can cause clutch slippage during engagement. Who is correct?
   - **a.** A only
   - **b.** B only
   - **c.** Both A and B
   - **d.** Neither A nor B
4. Technician A says that oily residue on a hose can indicate a refrigerant leak. Technician B says that this residue is normal for most connections. Who is correct?
   - **a.** A only
   - **b.** B only
   - **c.** Both A and B
   - **d.** Neither A nor B
5. Technician A says that a quick check of A/C system operation is to feel the temperature of the suction and discharge lines. Technician B says that the suction line should be cold and the discharge line hot after a minute of system operation. Who is correct?
   - **a.** A only
   - **b.** B only
   - **c.** Both A and B
   - **d.** Neither A nor B
6. Technician A says that replacement of the fuse is the only repair necessary for a blown fuse. Technician B says that a blown fuse is caused by a system fault such as high resistance. Who is correct?
   - **a.** A only
   - **b.** B only
   - **c.** Both A and B
   - **d.** Neither A nor B
7. Technician A says that the first step in diagnosing a problem is to verify what the complaint is. Technician B says that a TSB can tell you the cure for some problems. Who is correct?
   - **a.** A only
   - **b.** B only
   - **c.** Both A and B
   - **d.** Neither A nor B
8. Technician A says that a misadjusted blend door can cause poor A/C operation. Technician B says that you can check this by clamping off a heater hose. Who is correct?
   - **a.** A only
   - **b.** B only
   - **c.** Both A and B
   - **d.** Neither A nor B
9. Technician A says that foul-smelling air from the A/C ducts is caused by bacteria growing on the evaporator fins. Technician B says that this problem can be cured by installing a control that operates the blower motor after the vehicle is shut off. Who is correct?
   - **a.** A only
   - **b.** B only
   - **c.** Both A and B
   - **d.** Neither A nor B
10. Technician A says that a refrigerant line groundout problem makes a whining noise at moderate speeds. Technician B says that the only way to cure a knocking noise from the compressor is to install a new one. Who is correct?
    - **a.** A only
    - **b.** B only
    - **c.** Both A and B
    - **d.** Neither A nor B

A/C System Inspection and Diagnosis

LEARNING OBJECTIVES

After completing this chapter, you should be able to:

- Inspect an A/C system to determine whether it is operating correctly, and if not make further tests to locate the cause of the problem.
- Connect a manifold gauge set to a system and check system pressures.
- Determine whether system pressures are normal, and if abnormal determine the cause of the fault.
- Check a TXV and determine whether it is operating properly.
- Locate the source of a refrigerant leak.
- Complete the ASE tasks related to A/C system diagnosis.

TERMS TO LEARN

back seat
charging station
delta T
front seat
hand valves
manifold gauge set
mid seat
rub through
service unit
vacuum pump

10.1 INTRODUCTION

If the HVAC inspection performed (as outlined in Section 9.2) has determined a fault in the A/C system, further checks can be made to determine the exact cause of the problem. This further evaluation usually consists of a check of system pressures and temperatures while also checking for unusual noises, vibrations, and smells (Figure 10–1).

✓SERVICE TIP

Remember that the purpose of the A/C system is to make the evaporator cold, slightly above 32°F (0°C). We determine its temperature by feeling the suction line where it reenters the engine compartment; it should be cold to touch (Figure 10–2). The suction line warms up as it goes to the compressor, but it still should be cool to touch. The discharge line leaving the compressor should be hot to very hot; use caution when you feel this line. Remember that the condenser must give up heat to the ambient air, so it must be hotter than ambient temperature.

System pressures are closely linked to temperatures. The hot part of the system is the high side, and the pressure is about 150 to 350 psi, rather high. High side pressure is directly linked to ambient temperature and the

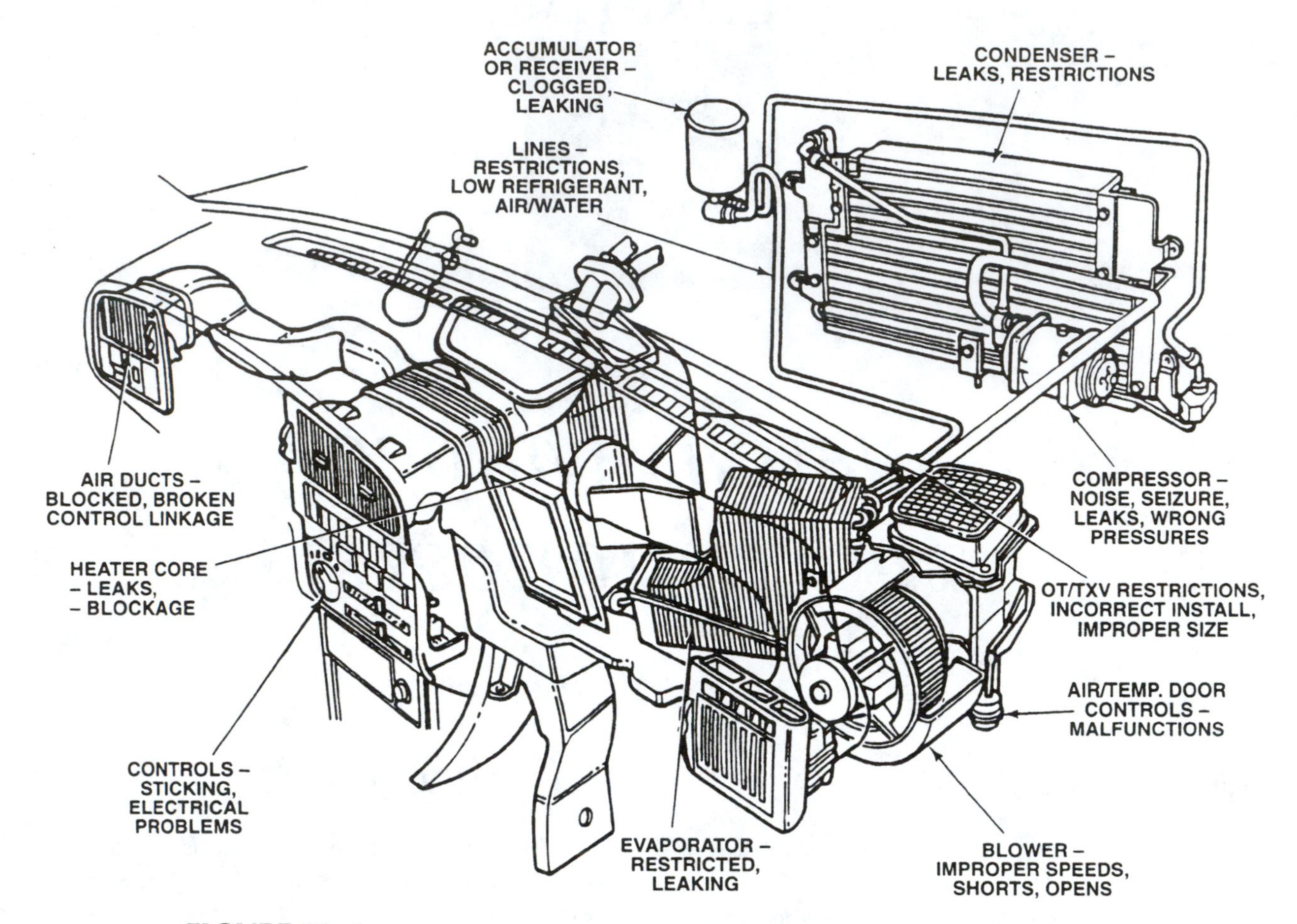

FIGURE 10–1 Possible A/C system problems. (*Reprinted with permission of General Motors Corporation*)

airflow through the condenser. The cold part of the system is the low side, where the pressure is lower, about 30 psi.

Leak checks to find the location of refrigerant loss are another important test that the A/C technician performs. Refrigerant loss allows refrigerant to enter our environment, which it may harm, and it also hinders the performance of the A/C system.

✓SERVICE TIP

When using an infrared thermometer, remember that the sensing beam spreads out and increases in width as it travels outward. If you are too far from a small object, you will also be measuring the temperature of objects behind it. Infrared thermometer readings are also effected by the emissivity of the item being checked. Shinny, light colored items should be marked with a dark marker or covered by a single layer of dark colored tape.

10.1.1 A/C Service Regulations

At one time, anyone could service or repair A/C systems for remuneration or profit. Since the passage of the Clean Air Act in 1992, American A/C technicians must be certified and certain pieces of service equipment must meet required specifications. The Federal EPA has the power to enforce the technician certification and shop equipment requirements.

As specified by the Clean Air Act since January 1992, automotive A/C technicians must complete an approved training program and use approved equipment. Approved training programs are offered by the *International Mobile Air Conditioning Association (IMACA), the Mobile Air Conditioning Society (MACS)*, the *National Institute for Automotive Service Excellence (ASE)*, as well as private and public sources. Technicians who successfully complete a training program receive a certificate bearing their certificate number.

Some states in the United States have additional requirements. For example, the California Bureau of Automotive Repair (BAR) requires that all shops performing automotive A/C service or repair that involves evacuation and full or partial charging must have the following equipment:

- Current service information
- Refrigerant identification equipment that meets or exceeds SAE Standard J1771

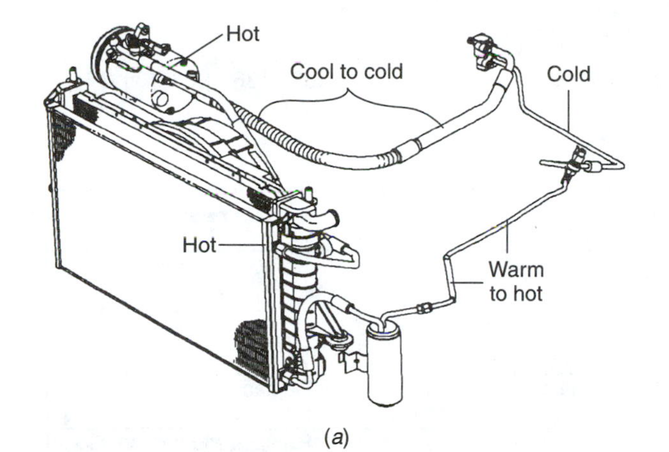

(a)

(b)

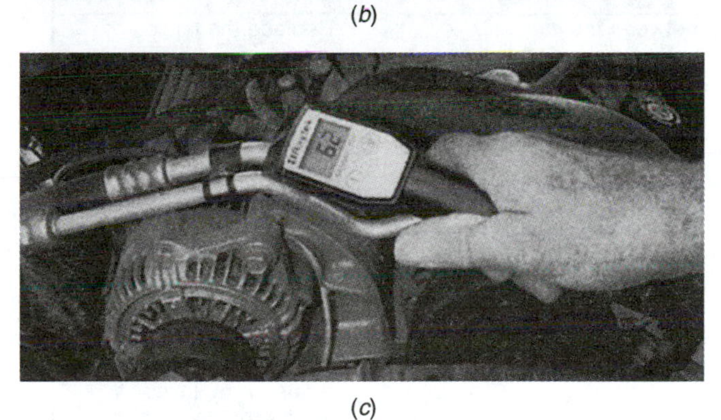

(c)

FIGURE 10–2 Many technicians begin system checks by feeling the temperature of the lines with the system operating (*a*); an infrared thermometer makes this check faster and more accurate (*b* and *c*). Some of the lines should be hot, and some should be cold. (a *is courtesy of Saturn Corporation;* b *is courtesy of Raytek Corporation*)

- Refrigerant leak detection equipment that meets or exceeds SAE Standard J1627
- Refrigerant recovery equipment that meets or exceeds SAE Standards J1732, J1770, J1990, and J2209
- Low- and high-pressure A/C system gauges (The low-pressure gauge must be capable of measuring from 0 to 30″ Hg and from 0 to 250 psi. The high-pressure gauge must be capable of measuring from 0 to 500 psi.)
- A/C system vacuum pump that is capable of reducing system pressure to a minimum of 29.5″ Hg
- Thermometer that is capable of measuring air temperatures from 20 to 100°F

If the shop claims to inspect, service, diagnose, or otherwise lead the customer to believe that it repairs A/C systems, the repair procedure must include checking at least 13 items for such problems as damage, leaks, missing parts, etc. The checked items must include the hoses, tubing, and connections; compressor and clutch; compressor drive plate rotation; service ports and caps; condenser; expansion device; accumulator, receiver–drier, and/or inline filter; drive belt(s) and tensioner; fan clutch; cooling fan; all accessible electrical connections; and the refrigerant to ensure that it is not contaminated. The system must be checked for leakage while it has an internal pressure of 50 psi or greater; the compressor clutch, blower, and air control doors must operate properly; and the low and high side pressures along with the center duct temperature must be recorded on the final invoice.

10.2 A/C PRESSURE CHECKS

The pressures in the high and low sides of a system are normally a factor of the refrigerant type and the temperature of the evaporator and the condenser. The temperature of a saturated vapor is tied to the pressure and vice versa (Figure 10–3).

Pressure tests are made using either a portable manifold gauge set or a service unit. Self-contained service units go by various names like **charging station, refrigerant handling systems, refrigerant management center**, or **recovery/recyling/recharging units.** They usually include a pressure gauge set, vacuum pump, and refrigerant supply for recharging; some include refrigerant recovery and recycling capability along with computer controls. Newer versions include refrigerant identifiers and the ability to perform refrigerant calculations like superheat and to analyze the pressure and temperatures to provide diagnostic information for problem solving (Figure 10–4).

The introduction of R-134a and the incompatibility between R-12 and R-134a and their respective oils have made different service equipment for each of these systems necessary (Figure 10–5). R-12 and its oil can contaminate an R-134a system and vice versa. The gauge sets and other equipment are quite similar, but different sets must be used to prevent mixing. To prevent cross-contamination, the equipment must be dedicated to either R-12 or R-134a, but not both. The service fittings for

GAUGE PRESSURE RELATED TROUBLESHOOTING

Normal compressor suction and discharge pressures at an atmospheric temperature of 30 - 38°C {86 - 96°F} and engine speed of approx. 1,500 r/min are:

High-pressure side pressure: 1.5 - 1.7 MPa {15 - 17kgf/cm², 213 - 242 psi}

Low-pressure side pressure: 0.13 - 0.2 MPa {1.3 - 2.0kgf/cm², 18 - 28 psi}

Possible cause	Inspection	Remedy
Low-pressure side too high.	The low-pressure side pressure normally becomes too high when the high-pressure side pressure is too high. As this is explained below, the following inspection is only used when the low-pressure side pressure is too high.	
1. Defective thermoswitch.	The magnetic clutch switch turns off before the outlet air temperature is sufficiently low.	Replace the thermoswitch.
2. Defective compressor gasket.	The high-and low-pressure side gauge pressures equalize when the magnetic clutch is turned off.	Replace the compressor.
3. Poor expansion valve temperature sensor contact.	Frost has adhered to the suction hose/pipe.	Install the temperature sensor against the low-pressure pipe.
4. The expansion valve opens too far.	Same as above.	Replace.
5. Clogged compressor suction filter.	The compressor fitting is cool but the low-pressure hose is not.	Remove and clean the filter.
Low-pressure side too low.		
1. Insufficient refrigerant.	There will be little temperature difference between the low- and high-pressure sides.	
2. Clogged liquid tank.	Considerable temperature difference between the inlet and outlet sides, or the tank is frosted.	Replace the liquid tank.
3. Clogged expansion valve.	The expansion valve's inlet side is frosted.	Replace the expansion valve.
4. Expansion valve temperature sensor gas leak (damaged capillary tube, etc.).	The expansion valve's outlet side is chilled, and low pressure gauge indicates low pressure.	Clean or replace the Expansion valve.
5. Clogged or blocked piping.	When the piping is clogged or blocked,the low-pressure gauge reading will decrease,or a negative reading may be shown. A frost spot may be present at the point of the restriction.	Clean or replace piping.
6. Defective thermoswitch.	The evaporator is frozen.	Adjust or replace the thermoswitch.
High-pressure side too high.		
1. Poor condenser cooling.	Dirty or clogged condenser fins. Cooling fans do not operate correctly.	Clean, and/or repair the fan.
2. Excessive refrigerant.	Verify by gauge reading.	
3. Air in the system.	Pressure is high on both High and Low sides.	Evacuate and recharge with refrigerant.
High-pressure side too low.		
1. Insufficient refrigerant.	Refer to "1. Insufficient refrigerant," above.	

FIGURE 10–3 This troubleshooting chart relates possible system problems to abnormal gauge pressures. (*Courtesy of Zexel USA Corporation*)

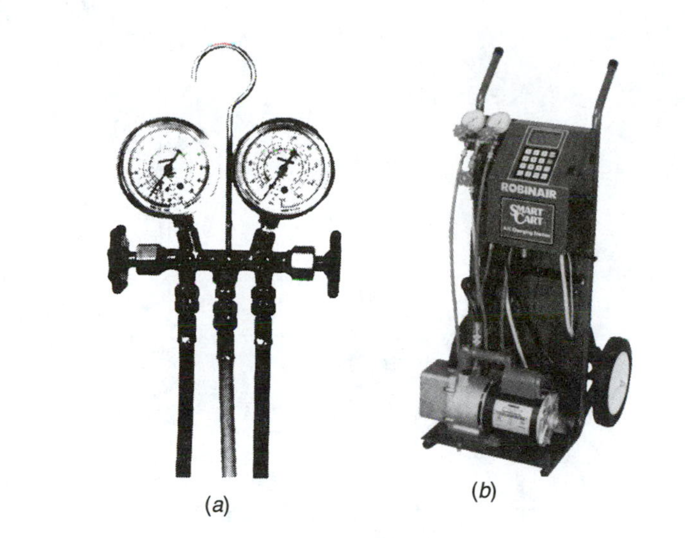

(*a*) (*b*)

FIGURE 10–4 System pressure checks can be made using a simple manifold gauge set (*a*) or a charging station (*b*). (*Courtesy of Robinair, SPX Corporation*)

R-134a						*R-12*					
Temperature °C (°F)		*Pressure kPa (Psi)*	*Temperature °C (°F)*		*Pressure kPa (Psi)*	*Temperature °C (°F)*		*Pressure kPa (Psi)*	*Temperature °C (°F)*		*Pressure kPa (Psi)*
–9 (16)		106 (15)	38 (100)		857 (124)	–9 (16)		127 (18)	38 (100)		808 (117)
–8 (18)		115 (17)	39 (102)		887 (129)	–8 (18)		136 (20)	39 (102)		893 (121)
–7 (20)		124 (18)	40 (104)		917 (133)	–7 (20)		145 (21)	40 (104)		859 (125)
–6 (22)	Evaporator Range	134 (19)	41 (106)		948 (137)	–6 (22)	Evaporator Range	155 (22)	41 (106)		893 (129)
–4 (24)		144 (21)	42 (108)		980 (142)	–4 (24)		165 (24)	42 (108)		917 (133)
–3 (26)		155 (22)	43 (110)	Condenser Range	1,012 (147)	–3 (26)		175 (25)	43 (110)	Condenser Range	940 (136)
–2 (28)		166 (24)	44 (112)		1,045 (152)	–2 (28)		185 (27)	44 (112)		969 (140)
–1 (30)		177 (26)	46 (114)		1,079 (157)	–1 (30)		196 (28)	46 (114)		997 (145)
0 (32)		188 (27)	47 (116)		1,114 (162)	0 (32)		207 (30)	47 (116)		1,027 (149)
1 (34)		200 (29)	48 (118)		1,149 (167)	1 (34)		219 (32)	48 (118)		1,057 (153)
2 (36)		212 (31)	49 (120)		1,185 (172)	2 (36)		230 (33)	49 (120)		1,087 (158)
3 (38)		225 (33)	50 (122)		1,222 (177)	3 (38)		249 (36)	50 (122)		1,118 (162)
4 (40)		238 (35)	51 (124)		1,260 (183)	4 (40)		255 (37)	51 (124)		1,150 (167)
7 (45)		272 (40)	52 (126)		1,298 (188)	7 (45)		287 (42)	52 (126)		1,182 (171)
10 (50)		310 (45)	53 (128)		1,337 (194)	10 (50)		322 (47)	53 (128)		1,215 (176)
13 (55)		350 (51)	54 (130)		1,377 (200)	13 (55)		359 (52)	54 (130)		1,248 (181)
16 (60)		392 (57)	57 (135)		1,481 (215)	16 (60)		398 (58)	57 (135)		1,334 (194)
18 (65)		438 (64)	60 (140)		1,590 (231)	18 (65)		440 (64)	60 (140)		1,425 (207)
21 (70)		487 (71)	63 (145)		1,704 (247)	21 (70)		484 (70)	63 (145)		1,519 (220)
24 (75)		540 (78)	66 (150)		1,823 (264)	24 (75)		531 (77)	66 (150)		1,618 (235)
27 (80)		609 (88)	68 (155)		1,948 (283)	27 (80)		580 (84)	68 (155)		1,721 (250)
30 (85)		655 (95)	71 (160)		2,079 (301)	30 (85)		633 (92)	71 (160)		1,828 (265)
32 (90)		718 (104)	74 (165)		2,215 (321)	32 (90)		688 (100)	74 (165)		1,940 (281)
35 (95)		786 (114)	77 (170)		2,358 (342)	35 (95)		746 (108)	77 (170)		2,057 (298)

Note: Evaporator pressure represents gas temperatures inside the coil and not at the coil surfaces. Add to temperature for coil and air-off temperatures (4 to 6°C or 8 to 10°F). Condenser temperatures are not ambient temperatures. Add to ambient (19 to 22°C or 35 to 40°F) for proper heat transfer; then refer to chart.

Example:

32°C
+ 22°
54°C condenser temperature = 1,377 kPa (R-134) or 1,248 kPa (R-12), based on 30-mph airflow.

Conditions vary for different system configurations. Refer to the manufacturer's specifications.
Reprinted with permission of General Moror Corporation.

FIGURE 10–5 System temperatures and pressures for R-134a and R-12. Note that the low side (evaporator range) pressures are close to each other, with a wider difference on the high side (condenser range), especially at higher temperatures. (*Reprinted with permission of General Motors Corporation*)

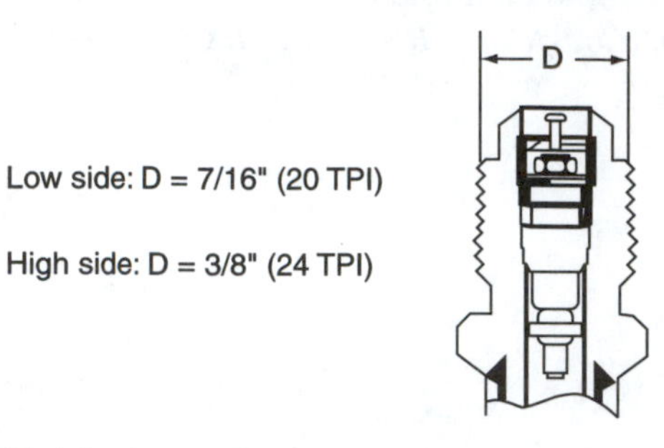

FIGURE 10–6 R-12 service ports use male flare fittings with two different sizes sealed by a Shrader valve. The low side fitting is larger.

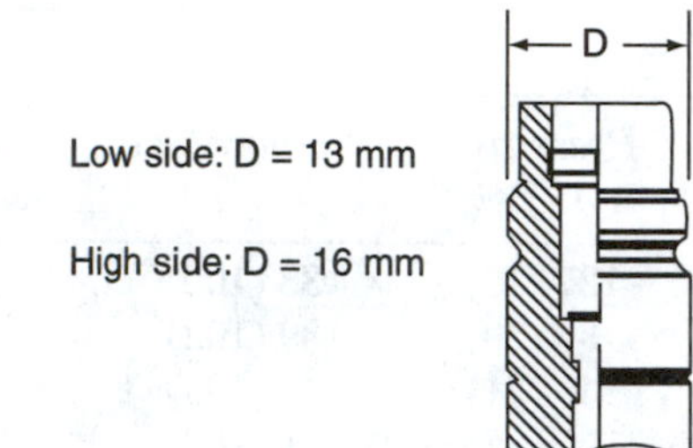

FIGURE 10–7 R-134a service ports use quick-disconnect–type fittings, and the high side fitting is larger.

R-134a systems have been changed, so R-12 equipment cannot be connected to them (Figures 10–6 and 10–7). Some service units are dual-gas units with R-12 and R-134a units side by side (Figure 10–8).

When R-12 and R-134a mix, they act like a single compound; the mix becomes physically and chemically different from both R-12 and R-134a. The mixture of R-12 and R-134a exhibits higher pressures (almost 10%), and there will be much greater corrosion and deterioration in the system (Figure 10–9). Mixing a blend or flammable refrigerant into R-12 is equally problematic. A/C service technicians must always be aware that someone may have charged R-22, R-134a, or a blend into an R-12 system.

Refrigerant that is cross-contaminated cannot be recycled; it must be disposed of and not reused. The price of refrigerant has increased to the point where it is too valuable to waste R-12 by contaminating it, and cross-contaminated refrigerant creates a disposal problem.

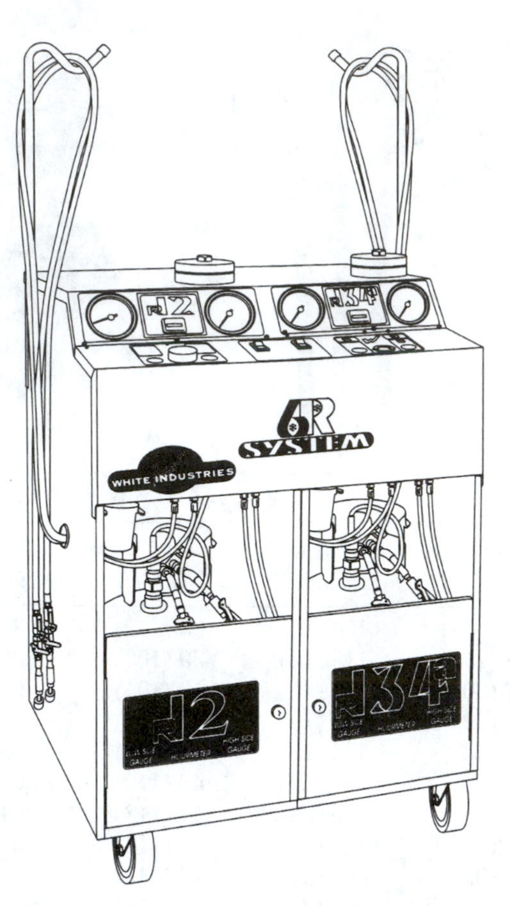

FIGURE 10–8 A dual A/C service unit that can be used on both R-12 and R-134a systems. Note that is essentially two side by side units. *(Courtesy of White Industries)*

FIGURE 10–9 Different types of refrigerants should not be mixed. The resulting contamination can cause excessive pressures and damage the system. *(Courtesy of Zexel USA Corporation)*

10.2.1 Gauge Set

A manifold gauge set is a unit with two pressure gauges mounted on a manifold that has two hand valves and usually three hoses. Two of these hoses connect to the low and high side service ports of a system to allow the pressures to be read on the low and high side gauges. The low side gauge and service hose are color-coded blue, and the high side gauge and service hose are color-coded red. The third, center hose is coded yellow, white, or green. Yellow and white are the preferred colors for R-12, and solid yellow with a black stripe is preferred for R-134a. One of the hand valves controls the flow between the low side hose and the center hose, and the other hand valve

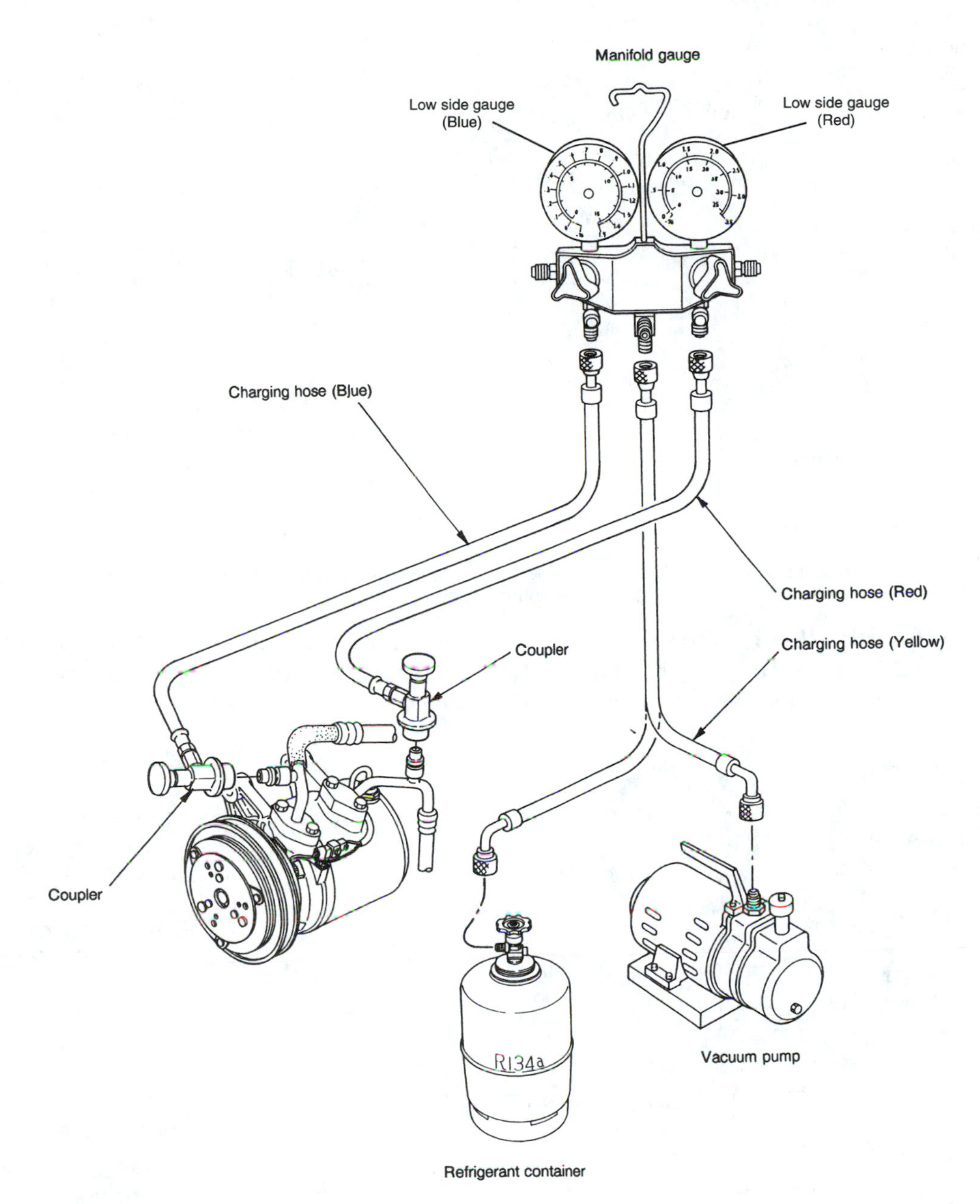

FIGURE 10–10 The blue, low side hose is connected to the low side of the system; the red, high side hose is connected to the high side of the system; and the center hose(s) is connected to a refrigerant container, vacuum pump, or recovery unit (not shown). *(Courtesy of Zexel USA Corporation)*

controls the flow between the center hose and the high side hose (Figure 10–10).

Gauges are printed to give pressure readings in pounds per square inch, kilopascals, BAR, or a combination of these, with pounds-per-square-inch readings being the most popular. The pressure readings are normally printed in black. Most gauges are also printed with red scales that show the corresponding refrigerant temperature (Figure 10–11).

The low side gauge is a **compound gauge**; it can read pressures into a vacuum as well as pressures above atmospheric. The readings start at 0 (atmospheric pressure); they go downward to 30″ Hg (a vacuum) and upward to about 150 psi (pressure).The high side gauge is a pressure gauge that reads from 0 to about 500 psi (Figure 10–12). Most high side gauges have a restricting orifice in the passage leading to the gauge. This orifice dampens pressure pulsations from the compressor, which can cause very rapid needle movements. The manifold is drilled so a gauge will always read the pressure in its respective hose. Many gauges include a **calibration adjustment**. This screw

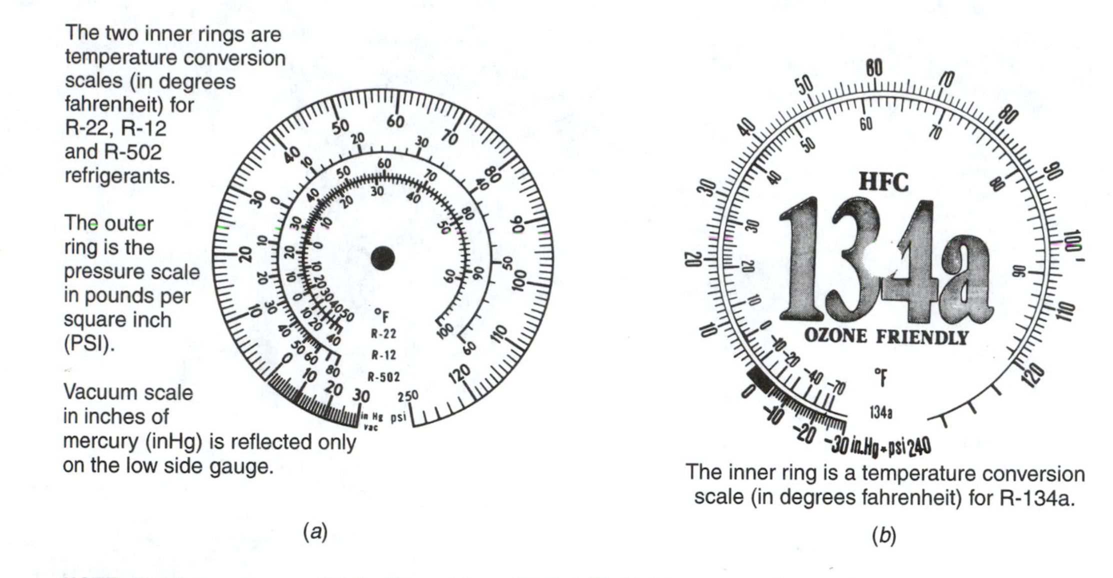

NOTE: Metric gauges are slightly different from (*a*) and (*b*). Metric pressure is stated in either BAR, KG/Cm2, or KPA scales, and temperature is stated in centigrade.

FIGURE 10–11 The faces of a low side gauge for R-12, R-22, and R-502 (*a*) and for R-134a (*b*). The outer ring shows the pressure; the inner rings are temperature–pressure relationships. (*Courtesy of TIF Instruments*)

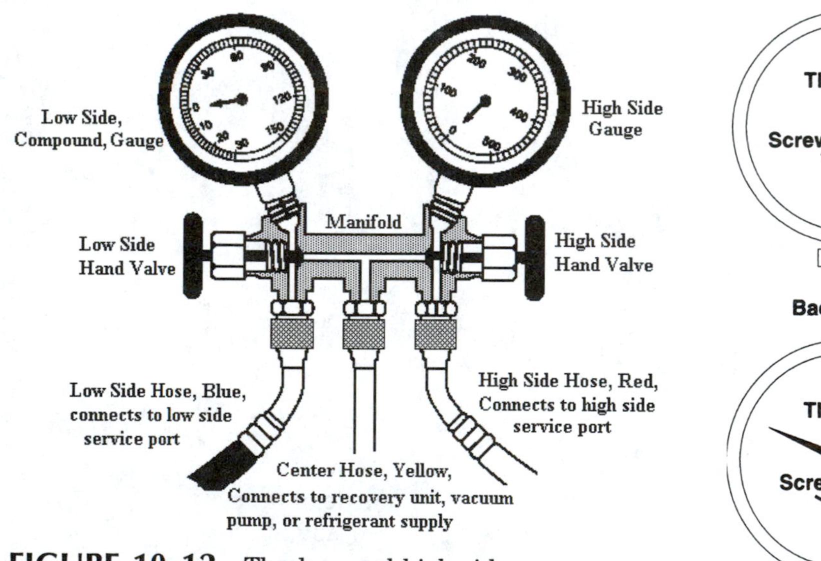

FIGURE 10–12 The low and high side gauges are mounted onto the manifold so they always read the pressure in their respective hoses. The manifold hand valves control the flow to and from the center hose.

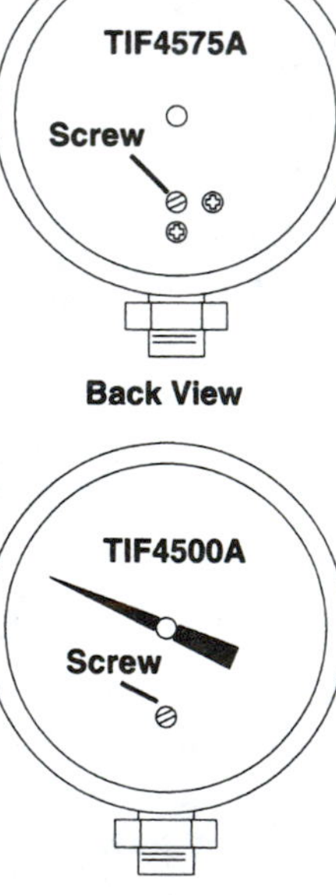

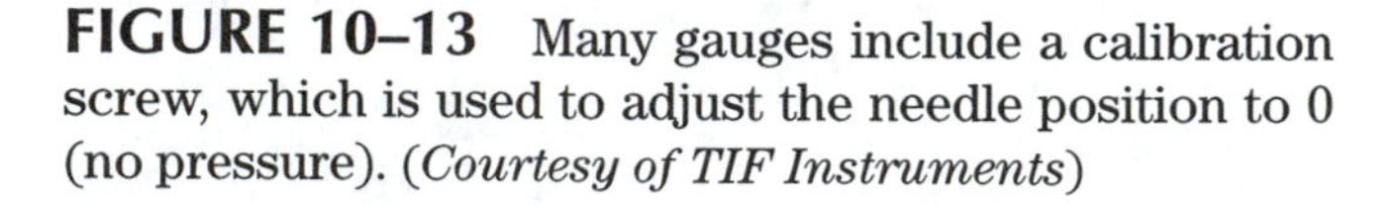

FIGURE 10–13 Many gauges include a calibration screw, which is used to adjust the needle position to 0 (no pressure). (*Courtesy of TIF Instruments*)

is turned to adjust the gauge needle to 0 (when there is atmospheric pressure at the gauge) (Figure 10–13).

The center hose is the **service hose**. Depending on the service operation, it can be connected to a recovery unit, vacuum pump, or refrigerant supply. As mentioned, the two hand valves allow the technician to control the operation and the flow to or from the low side, high side, or both hoses (Figure 10–14). Some gauge sets have two center hoses so the manifold can be connected for two operations at the same time—a vacuum pump and refrigerant supply, for example. Some gauge sets include a sight glass at the center hose port so that the technician can observe the refrigerant flow through the manifold.

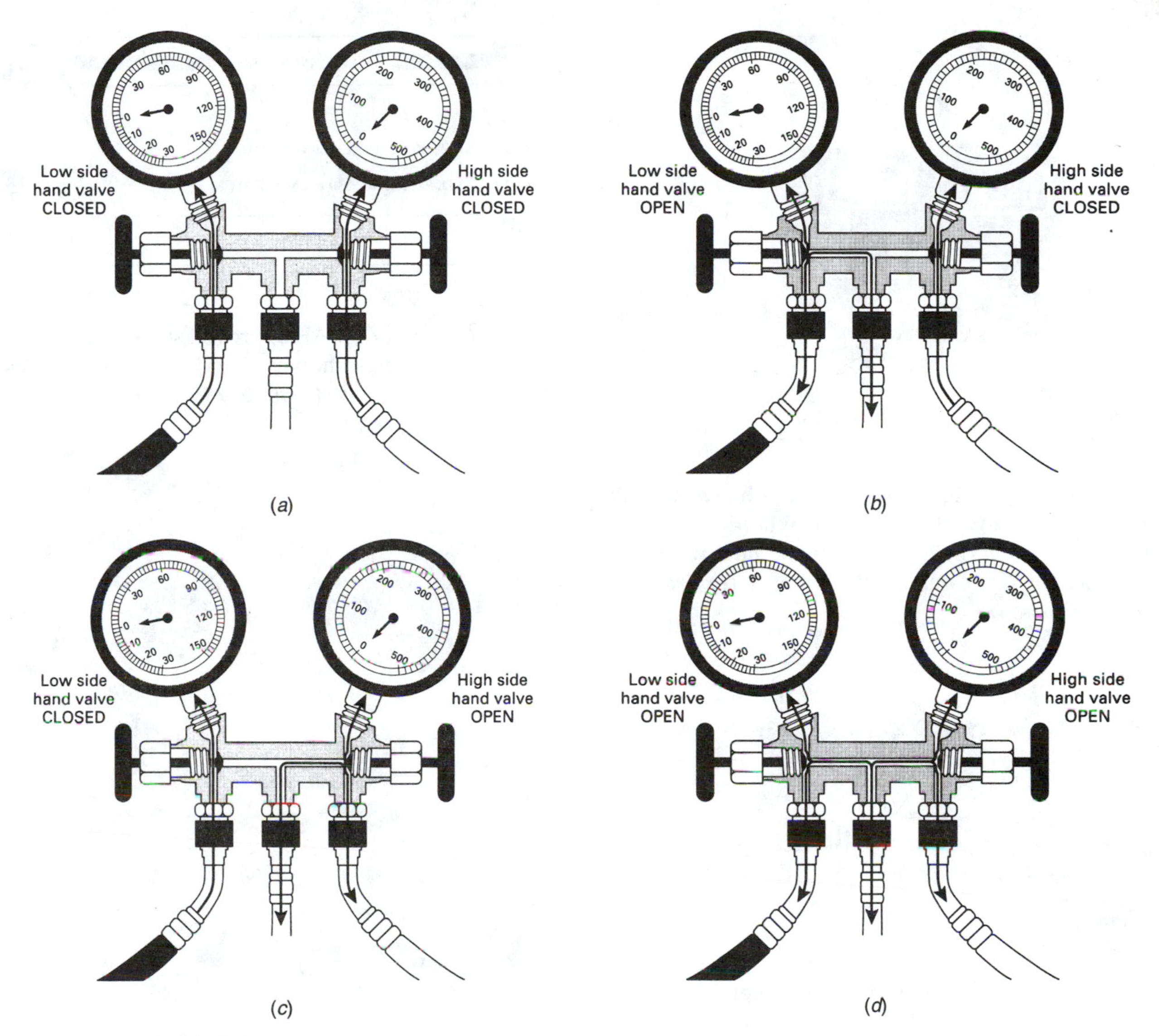

FIGURE 10–14 The low and high side manifold valves are used to control flow through the manifold. (*a*) is used when pressure checking a system; (*b*),(*c*), and (*d*) are used for various service operations.

Many manifolds include a hook so the manifold and gauges can be hung up, and many also have hose holders for the service ends of the hoses. Hose holders keep the hose ends clean and neat (Figure 10–15).

Hoses are available in different lengths: 36, 48, 60, and 72 inches (90, 120, 150, and 180 cm) are common. With gauge sets, many technicians prefer the shorter length for the low and high side hoses and either a 48- or 60-inch length for the center hose. A manifold set is often hung from the car's hood or placed on a fender cover; long low and high side hoses become cumbersome and can get into the moving parts of the engine. The longer center hose is needed to reach to the floor, where the vacuum pump is usually placed. Most charging stations need the longer hoses to reach the service fittings.

✓SERVICE TIP

Longer hoses can hold a substantial amount of refrigerant when charging a system. As much as an ounce of refrigerant can be lost because of this. A longer time period is required for the refrigerant to pass through the hose.

The manifold end of the hose is normally straight and close coupled. The system end of the hose usually uses a longer, bent-metal section to allow easier connection to the system's service fittings. The system end also includes a **valve core depressor.** The knurled nut on some R-12 hoses includes grooves in the internal threads to control the path of escaping refrigerant,

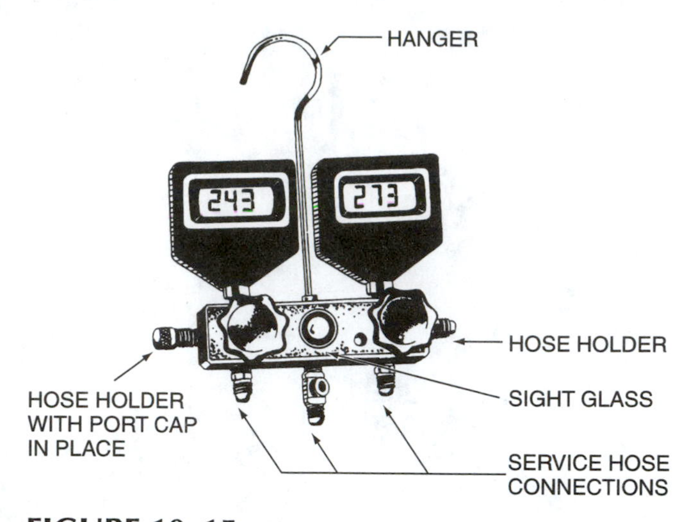

FIGURE 10–15 This gauge set has a hanging hook for suspending the gauge set and hose holders to keep the working ends of the hoses clean and neat when out of use. Note the electronic, digital gauges. (*Courtesy of TIF Instruments*)

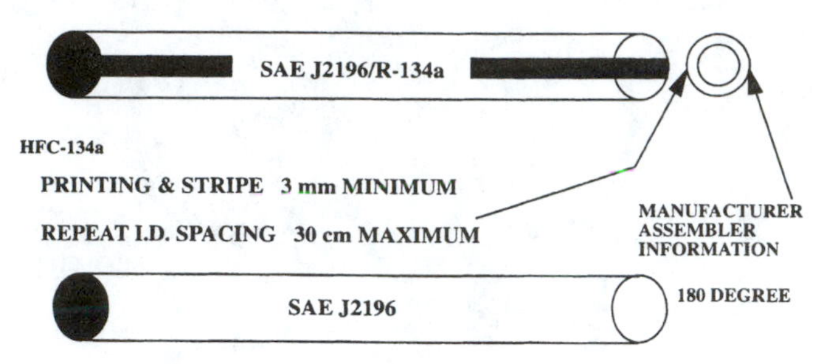

FIGURE 10–17 SAE-approved service hoses should have the markings shown. (*Reprinted with permission from SAE Document M-106, © 1992, Society of Automotive Engineers, Inc.*)

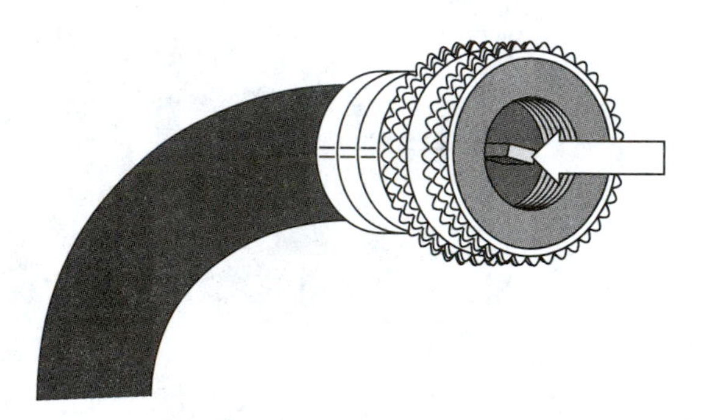

FIGURE 10–16 The working, system end of a hose for R-12 systems should have a gasket and Schrader valve depressor (arrow) in it; the manifold end should have only a gasket. The working end has a slight bend; the manifold end is straight.

FIGURE 10–18 R-12 system service hose adapters include quick seal (stops backflow out of the hose when disconnected) (*a*), 90° (shown) and 45° (*b*), flexible (*c*), quick disconnect (*d*), and straight (*e*). Adapter *e* is designed to fit GM Positive Seal valves. The female side is made in different sizes to fit the different high side ports. (a,b,c, *and* d *courtesy of Robinair, SPX Corporation;* e *courtesy of Kent-Moore*)

directing any sprayback away from the technician (Figure 10–16). SAE standard J 2196 requires a **shutoff valve** within 12 inches (30 cm) of the end of the hose. This valve is closed to trap refrigerant in the hose when it is disconnected. Some hose brands use a seal feature that automatically closes as the hose is disconnected to prevent refrigerant loss and sprayback. A service hose for R-134a must have a black stripe along its length and be marked with "SAE J2196/R-134a" (Figure 10–17).

In the past, all R-12 hose connections used 1/4-inch flare fittings, with normal 7/16-20 threads. Male fittings are used at the manifold and the system service fittings, and female fittings are used at each end of the hose. These fittings are sealed by a rubber ring in the hose end as the fitting is tightened. Knurled nuts are used at the hose ends and tightened by hand. Some high side R-12 service fittings are a quick-coupler style or a 3/16- or 1/8-inch flare fitting. A hose adapter is required to make these high side connections (Figure 10–18).

Systems that use R-134a require a metric service fitting with a 0.500 × 16-2G Acme thread. This completely different fitting is used to prevent technicians from connecting equipment used for R-12 service. The system end of the hose uses a quick coupler to make the hose

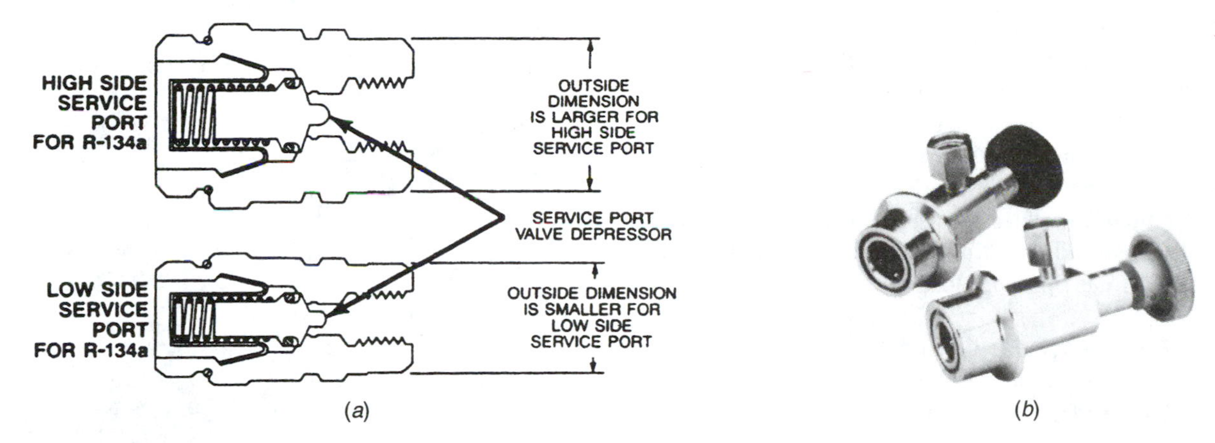

FIGURE 10–19 A cutaway view of R-134a service ports (*a*). These service hose couplers use a quick-connect attachment to the fitting and a knob that is turned inward to depress the valve (*b*). (a *is courtesy of DaimlerChrysler Corporation;* b *is courtesy of Robinair, SPX Corporation*)

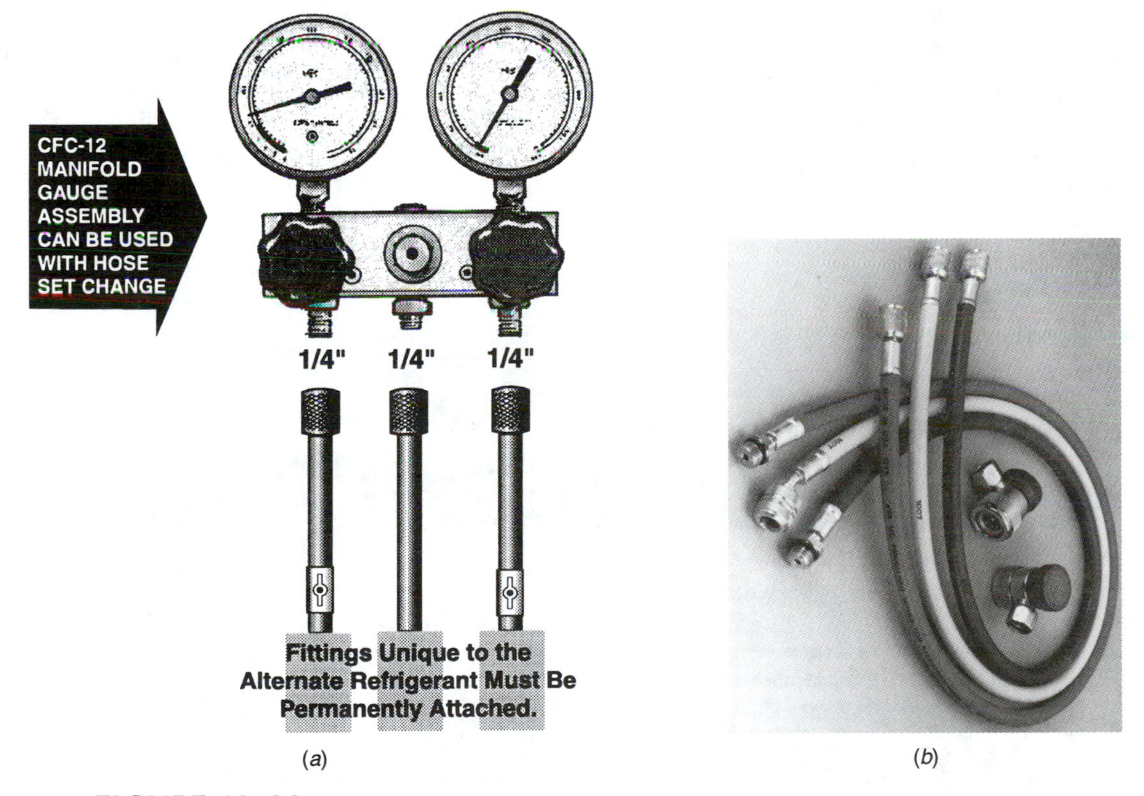

FIGURE 10–20 A gauge set may be used with different refrigerants by changing the hoses (*a*). Each hose set must have unique fittings, permanently attached, for that refrigerant. Hose set *b* connects R-134a service fittings to an R-12 manifold. (a *is courtesy of the International Mobile Air Conditioning Association [IMACA];* b *is courtesy of Mastercool*)

connection and can include a hand knob to depress and open the internal valve (Figure 10–19).

As mentioned earlier, a service fitting that uses unique threads to prevent improper connections is required for each of the different blends. These fitting sizes are shown in Appendix D. EPA regulations do not allow using an adapter so an R-12 or R-134a hose can be hooked onto these ports. However, the regulations do allow a standard gauge set to be converted by changing the hoses. Each hose set must have permanently attached fittings for each refrigerant being checked (Figure 10–20).

Some systems that use an STV have a second low side service port to allow the use of a **third gauge**. A third gauge is a single, compound gauge, similar to the low side gauge, and is often attached directly to a service hose. The third gauge is usually attached to the third service port, which is downstream from the STV, closer to the compressor. Comparison of the low side pressure with that of the third gauge allows the technician to determine whether the STV is open or closed (Figure 10–21).

10.2.1.1 Electronic Gauge Sets Some modern gauge sets use electronic gauges and pressure transducers in place of hoses (Figure 10–22). One set has inputs for low side pressure, high side pressure, temperature, and vacuum sensing in microns. This allows the gauge to display both low and high side pressures, temperature from selected points, and vacuum in microns; it will also calculate and display refrigerant superheating and subcooling.

10.2.2 Service Ports and Valves

An R-12 system has a service port, which is simply a male flare fitting equipped with a Schrader-type valve

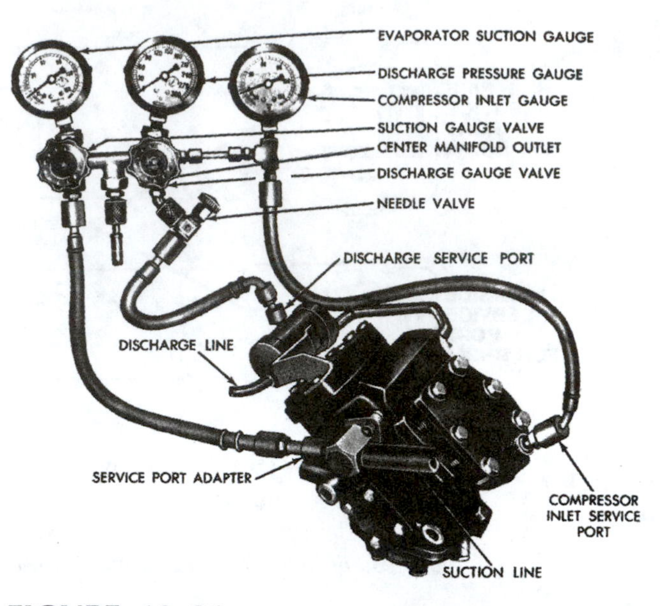

FIGURE 10–21 A third gauge (compressor inlet gauge) is being used to check the operation of the EPR valve located inside the suction line port. Many technicians do not use the solid bar holding the third gauge to the manifold and keep it separate. *(Courtesy of DaimlerChrysler Corporation)*

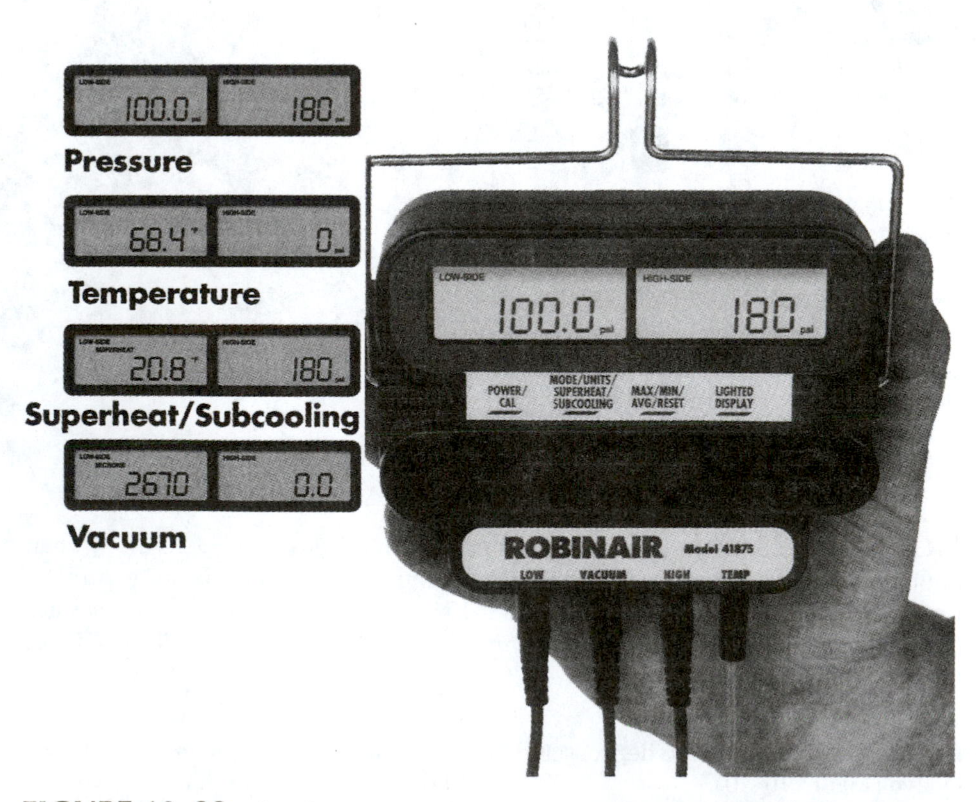

FIGURE 10–22 An electronic gauge set can measure low and high side pressures, vacuum (in microns), and temperature. It can also calculate superheating and subcooling. *(Courtesy of Robinair, SPX Corporation)*

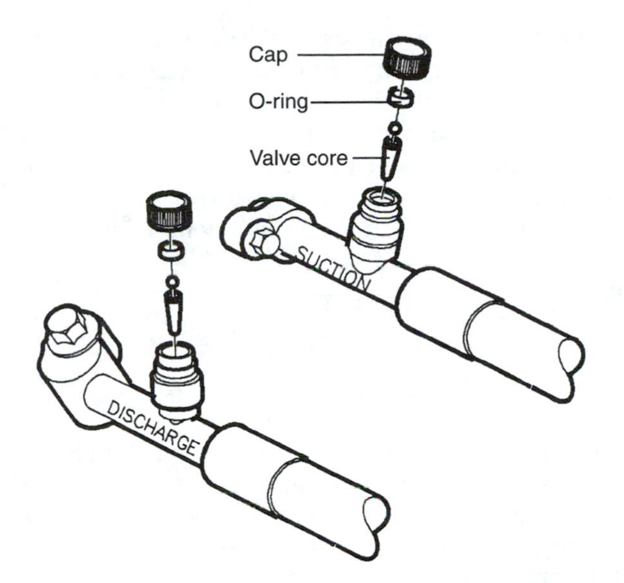

FIGURE 10–23 Each service port should have a cap to keep out dirt with a sealing O-ring to help prevent refrigerant loss. The valve core is the primary seal, and the cap and O-ring are the secondary seal. *(Courtesy of Saturn Corporation)*

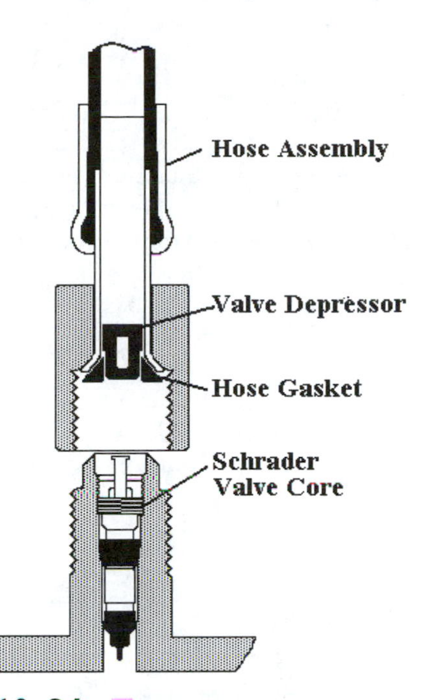

FIGURE 10–24 The valve depressor opens the Schrader valve as the service hose is connected to the fitting.

core. This valve core is like the ones used in tire valves; the actual valve core is one designed for refrigerant. The suction service port uses a 1/4-inch flare fitting with 7/16-20 threads; the discharge port is usually a 3/16-inch flare fitting with 3/8-24 threads. The service port should be covered with a protective cap that also serves as a secondary seal (Figure 10–23).

✓SERVICE TIP

If the protective cap is missing, dirt will enter the port and valve core, being attracted and trapped by the oil film. When the system is serviced, this dirt will enter either the gauge set or the system. Dirt particles can damage the valves and the gauges in the gauge set, damage the compressor, or plug small orifices if they enter the system. An unprotected port should be cleaned before connecting a hose to it.

The port is normally closed by the spring in the valve core. This is sometimes called a **back-seat** position, a term held over from stem-type service valves. When an R-12 service hose is connected to the port, the depressor in the end of the hose pushes inward on the valve stem to open the valve (Figure 10–24).

Service ports for R-134a systems use a unique configuration; the low side service port is smaller than the high side port (13 and 16 mm [0.51 and 0.63 inch] OD). R-134a service hoses use color-coded quick couplers. These couplers automatically close the hose end when they are disconnected to prevent refrigerant loss. The valve in the service port is opened and closed as the coupler is connected and disconnected (Figure 10–25).

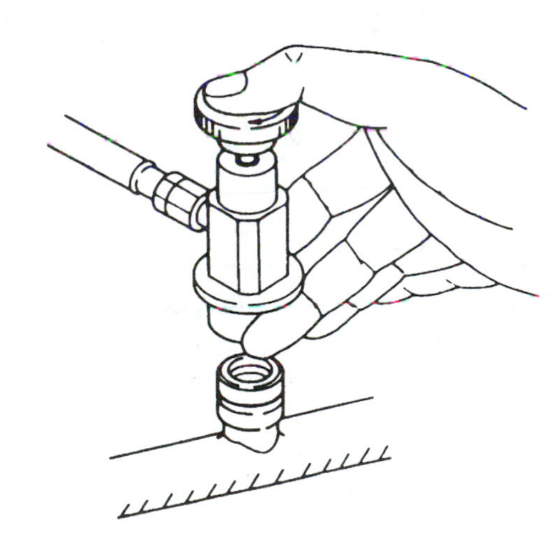

FIGURE 10–25 The locking ring is lifted as the R-134a service coupler is connected, and the coupler is pushed firmly inward until locked. Next, on some couplers, the knob is turned to depress the valve. *(Courtesy of Zexel USA Corporation)*

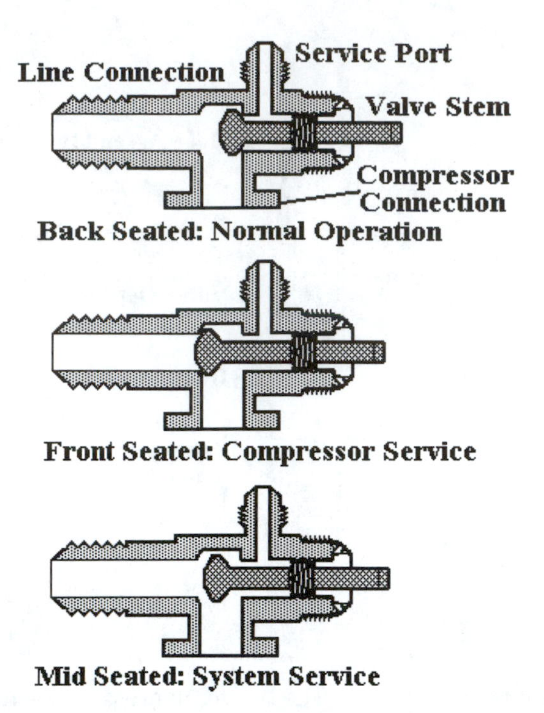

FIGURE 10–26 A service valve is normally kept in its back seated position (top). Turning the valve stem completely inward, using a special wrench, will front seat the valve to shut off the hose connection (center). Turning the valve a turn or two inward will mid seat the valve (bottom); this opens the service port so system pressure can be read on a gauge set.

✓SERVICE TIP

Some modern systems have only one service port, and some of these use an electronic transducer to sense the internal pressure. With these systems, the pressure can be read using a scan tool. If necessary, a service port can be installed using a saddle clamp (see Figure 13–47).

Some early A/C systems (R-12) use service valves at the service ports. A service valve has a stem that can be rotated using a 1/4-inch square socket or box wrench. This valve is set to one of three positions: back seat, **mid seat**, or **front seat**. During normal operation the valve is kept in the back-seat position: This closes the service port while the compressor port is left open. During service operations the valve is mid seated: This opens both the compressor port and the service port. During some special operations the valve is front seated: This closes the compressor port while the service port is left open. With both service valves front seated, the compressor can be disconnected from the service valves and removed from the system with the refrigerant trapped in the rest of the system by the service valves. While working with service valves, open-end wrenches, adjustable wrenches, and pliers should never be used on the valve stem. These tools will ruin the 1/4-inch square stem and prevent the use of the proper wrench. A service valve is fitted with protective caps for both the service port and the valve stem (Figure 10–26).

10.2.3 Connecting and Disconnecting Gauges to a System

A gauge set is connected to a system so that diagnostic pressure checks can be made, to allow recovery of the refrigerant, and so a system can be evacuated and recharged. The first step in this procedure is to locate and identify the service ports.

The low side service port is located between the evaporator outlet and the compressor inlet. In some vehicles, the low side service port is located just downstream from the OT. Some systems have two low side ports, one on each side of the STV. The low side hose is connected to the upstream service port (closest to the evaporator), and the third gauge is connected to the downstream service port if needed. Some OT systems do not have a separate low side service port. These systems require the removal of the low side pressure switch from the accumulator and the installation of a tee fitting. A Schrader valve is used at the accumulator port, so there should be only a very small amount of refrigerant loss.

The high side service port is located between the compressor outlet and the expansion device, TXV or OT (Figure 10–27). Inspection of these ports tells the technician whether it is an R-12 or R-134 a system and, on R-12 systems, whether an adapter is needed to connect the high side hose. All ports should be covered with a cap.

✓SERVICE TIP

As the R-12 style service hose is being connected to a Schrader valve port, refrigerant often escapes because the valve depressor usually makes contact with the valve stem before the seal ring contacts the seat. This leak can be reduced by using the following procedure: Thread the knurled nut on the port one or two turns, grip the hose and push it quickly and firmly against the port, and tighten the nut to hold the hose in place. The knurled nut should be tightened only finger tight. The seal and seat can be damaged if pliers are used. If the seal leaks while finger tight, the sealing ring in the hose end should be replaced (Figure 10–28).

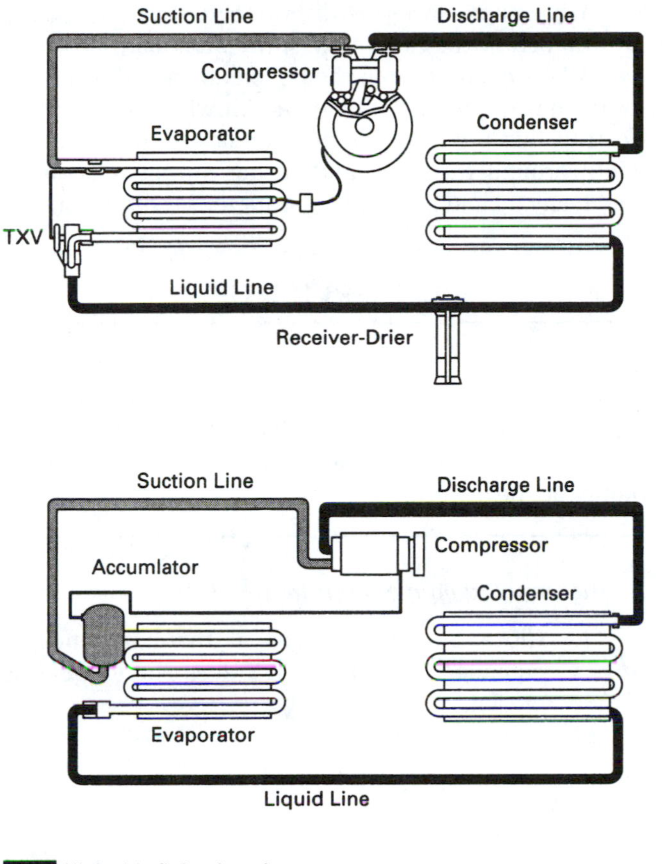

FIGURE 10–27 The low side service port is normally located between the evaporator outlet and the compressor inlet. The high side service port is between the compressor outlet and the evaporator inlet.

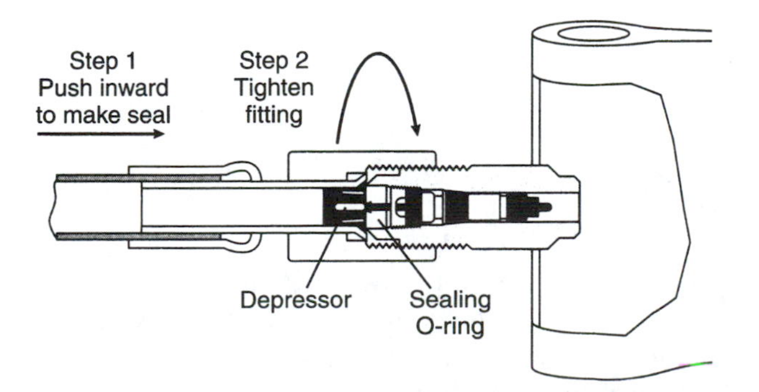

FIGURE 10–28 To reduce leakage when connecting a hose to an R-12 service port, push the hose firmly against the fitting and then finger tighten the fitting nut.

It was once common practice to purge the air from the hoses by making refrigerant flow through them for a few seconds. This vented refrigerant to the atmosphere, which is no longer an approved practice. With zero-loss hoses, refrigerant remains in the hoses while they are disconnected: There is no longer any need to purge the air from them. If using older-style hoses (which are no longer approved), a vacuum pump should be used to evacuate the air from the hoses as they are connected.

SERVICE CAUTION

Eyes and skin should always be protected when you are performing operations during which refrigerant might escape.

CAUTION

With the increased possibilities of contaminated refrigerant, some experts recommend using a refrigerant identifier before connecting a gauge set to a system to determine the purity of the refrigerant in that system. Refrigerant identification is described in Section 13.3.2.

To connect a gauge set to a system, you should:

1. Make sure all of the valves on the manifold and hoses are closed. Also check to ensure that the sealing ring and depressor are in the service ends of the hoses.
2. Remove the low side protective cap and connect the low side hose to the low side service port. In many cases, one or both of the caps can be stored on the manifold's hose holder (Figure 10–29).
3. Remove the high side protective cap. Connect the high side service hose to the port or adapter and store the protective cap. On R-12 systems determine whether an adapter is needed. If needed, many adapters have a valve core so they can be connected to the system; then the hose is connected to the adapter. If the adapter does not have a valve, it must be connected to the hose before connecting to the service port.
4. Open the valve in each hose (leave the manifold valves closed) and read the system pressure on the gauges (Figure 10–30).
5. On systems equipped with service valves, mid seat the valve by turning the stem inward one or two turns using a service valve wrench. This valve must be back seated before disconnecting the hose.

FIGURE 10–29 An R-12 service port cap can be stored on the manifold's hose holder.

When some hoses are disconnected, refrigerant and oil trapped in the hose may spray back and blow outward. The hand valve in the hose end should be closed or the hose equipped with an antiblowback check valve to trap this pressure. Some technicians wrap a shop cloth around the hose and fitting during removal to catch any oil and liquid refrigerant that is blown out.

✓ SERVICE TIP

Before connecting the hoses to R-134a service ports, check for a sharp edge or burr at the edge of the port: This burr can cut the seal in your hose connector. A sharp edge or burr can be removed with a file, sharp knife, or special tool.

To disconnect a gauge set from a system, you should:

1. Close the hand valves at the hose ends, if equipped; the manifold valves should also be closed.

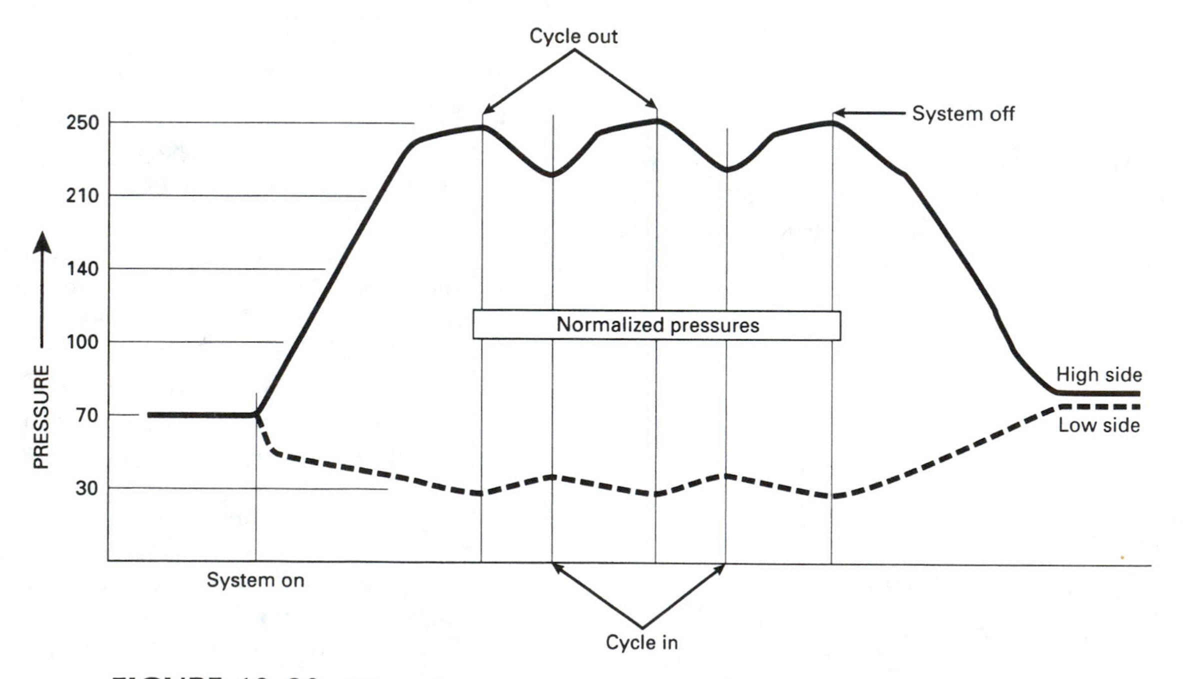

FIGURE 10–30 When the gauges are connected, they should both show the same pressure, which is dependent on the temperature. When the system is turned on the pressures will change, depending on evaporator temperature (low side) and ambient temperature (high side). Pressures will equalize again after the system is shut off.

2. Disconnect the low side service hose and replace the protective cap. Most protective caps include a sealing ring or O-ring. These should be threaded onto the port until the O-ring makes contact and then tightened finger tight. Replace the hose end on the hose holder.
3. Disconnect the high side hose, replace the protective cap, and replace the hose end on the hose holder.

✓SERVICE TIP

Some technicians prefer to disconnect the low side hose while the system is still operating and the pressure is at its lowest. The high side hose is disconnected after the system has been shut off for a while and the pressure drops.

✓PROBLEM SOLVING

The technician has connected a gauge set to a system, and both gauges show a pressure of about 10 psi. Is this the correct pressure for a system that is not running? If not, what might be wrong?

10.3 PERFORMANCE TEST

A performance test is used to determine whether the A/C system is operating properly and to indicate what is wrong if the system is not operating normally. The inspection checklist and report recommended by IMACA is shown in Figure 10–31.

Performance tests begin with a reading of the pressure of the system with the engine off. The low and high sides of the system should be equal, and the pressure should be about the same as the **pressure–temperature (PT)** relationship for that refrigerant. For example, the pressure in an R-12 system at 70°F should be about 70 psi. An R-134a system should have a slightly higher pressure of about 71 psi. A lower pressure indicates a possible leak; a higher pressure indicates possible refrigerant contamination. A refrigerant identifier should be used to confirm the contaminants.

The pressures should change as the system is operated. When the system is started, the low side pressure drops and the high side pressure increases. After a rather quick pressure change as the compressor starts, the pressures change more slowly as the temperatures of the evaporator and condenser change. The low side pressure reflects the evaporator temperature; the high side pressure reflects the temperature of the liquid refrigerant leaving the condenser. The pressures normalize or stabilize when system temperatures reach normal operating points (Figure 10–32).

Most technicians try to simulate real-world operating conditions during a performance test. This is hard to do with the vehicle empty and sitting in a shop. The engine speed is adjusted to cruising speeds, about 1,500 to 2,000 rpm. The windows and doors are usually closed to reduce the heat load, but if a high heat load is desired, they are opened. Some manufacturers recommend closing the hood as far as possible.

A performance test should be run long enough to thoroughly chill the evaporator so the low-temperature deicing control is activated. At this point the clutch cycles, the STV throttles, or the compressor reduces displacement. This time can be shortened by reducing the heat load, closing the doors and windows, and adjusting the blower speed to slow. Some manufacturers recommend measuring the clutch cycle rate and timing. Charts are then used to determine whether the cycle rate is wrong, and if so what problem is indicated.

Cold air from the discharge ducts usually indicates good performance. Some manufacturers specify duct discharge temperatures from their various systems.

✓SERVICE TIP

Many technicians use this rule of thumb: Any system should be able to drop the air temperature at least 20°F as the air passes through the evaporator. In other words, if 75°F air enters the ductwork, 55°F or cooler air should leave the discharge registers.

To make a performance test, you should:

1. Connect the gauge set to the system, hang the gauge set from the hood or some convenient location or place it on a fender cover, and make sure the hoses are routed away from the moving engine parts and exhaust manifold.
2. Place a thermometer at the center of the radiator grill to record ambient temperature and one

International Mobile Air Conditioning Association

Code of Professional Practice Inspection Checklist

This Inspection Checklist is a summary of the steps detailed in the IMACA Code of Professional Practice manual. For detailed information, consult the manual.

VISUAL INSPECTION - Engine Compartment

1) Hoses, tubing and connections (Suction, Discharge & Liquid lines)
A. Examine exterior for deterioration, blistering, bubbling refrigerant, oil stains and battery acid damage or burns. Oil stains could indicate leakage.
B. Check for incorrect routing, rubbing, missing hardware or loose hoses, bent or collapsed tubing.

2) Compressor
A. Examine exterior for damaged or missing bolts / hardware, broken housing, oil stains.
B. Internal - rotate compressor 2 complete turns by hand to determine if seized or locked up.

3) Clutch
A. Examine for broken springs, burnt face, damaged grooves, oil stains from seal leak.

4) Service Ports
A. Check size and thread to determine type of refrigerant - confirm with engine compartment label.
B. Examine ports for missing caps, damaged threads and leaking Schrader valves.

5) Condenser
A. Check for loose or damaged connections, loose or missing hardware or air dams.
B. Examine coil for bent or damaged fins, restriction due to debris or dirt, oil stains.

6) Expansion Device (if possible)
A. Examine for physical damage or oil stains.

7) Evaporator Pressure Regulator (early models)
A. Check POA, EPR, STV (if equipped) for physical damage or oil stains.

8) Cabin Air (Evaporator) Filter (if equipped)
A. Check for physical damage, oil stains and for proper installation.

9) Accumulator or Receiver/Drier
A. Check for physical damage, loose or missing hardware, loose connections or oil stains
B. Examine sight glass (Receiver/Drier only) for stains.

10) Drive Belts
A. Check for missing or damaged pulleys and tensioners; routing, tension and alignment.
B. Examine condition of belts for cracking, checking and excessive wear.

11) O-Rings, Gaskets, Spring Locks (if equipped)
A. Examine all connections not previously inspected for loose or missing parts and oil stains.

12) Inline Filter
A. Check for physical damage or oil stains.

13) Fan Clutch & Blade
A. Examine for fluid leakage or excessive bearing wear.
B. Check for damaged or bent blades on fan.

14) Electrical Components
A. Examine connectors for loose, burnt, broken parts or corrosion.
B. Examine wiring harness for burns, cracks or rubbing on insulation.

VISUAL INSPECTION - Passenger Compartment

1) Air Flow
A. Check all louvers for directional movement and air flow.

2) Control Head
A. Check all blower speeds.
B. Check controls for proper movement and function.
C. Check operation for heater, defrost and A/C.

3) Interior Condition
A. Check evaporator / heater case for water leakage.
B. Check carpet for water damage.

LEAK CHECK

1) Preparation (Engine Off)
A. A refrigerant identification check is recommended to verify the refrigerant in use or to detect flammables, unknown or contaminated refrigerant.
B. Connect manifold gauge - refrigerant pressure must read 50 psig or more. (Add refrigerant to increase the pressure if necessary).
C. Clean all connections using a clean dry rag.

2) Leak Check (Engine Off)
A. Start at the compressor discharge port and follow the flow of refrigerant through the system. Move the detector sensor completely around each connection.
B. Refrigerant is heavier than air; leak-check the underside of the hoses, clean and leak-check the condesate drain tube(s).
C. Leak check the compressor shaft seal.
D. Leak check evaporator through dash vents.

REPAIR SYSTEM PER CUSTOMER APPROVAL

FINAL PERFORMANCE EVALUATION

1) Functional Inspection (Engine On)
A. Check the compressor clutch for proper operation.
B. Check evaporator blower at all speeds
C. Check operation of function control doors for fresh air/recirculate (A/C - Max A/C), Dash louvers, Floor outlets, Defroster outlets
D. Check operation of heater flow control (if equipped).
E. Check fan clutch (if equipped). Once engine has reached normal operating temperature, turn off engine and "soak" fan clutch for 2 minutes. Restart engine, fan clutch should be engaged. With engine OFF, spin fan, should rotate maximum of 2 turns.
F. Check electrically driven condenser/radiator fan(s) (if equipped).

2) System Checkout (1,200 rpm's, condenser air flow = 35 mph)
A. Measure and record (on Inspection Report) temperature 2" in front of center of radiator. (Ambient temperature.)
B. Set A/C controls to OEM specs. Allow system to stabilize 5-10 minutes.
C. Record the high- and low-side pressure.
D. Record the center louver temperature and interior temperature.
E. Check operation of temperature controls.

3) Post Service Inspection
A. A final refrigerant check is optional to verify purity of refrigerant and absence of air.
B. Install service port caps and perform final leak check.
C. Perform final visual inspection. Check for tools and loose components.
D. Record results on Inspection Report.

***CAUTION:* Safety Glasses *must* be worn during any A/C diagnosis or repair procedure. Refrigerant may cause blindness if it comes in contact with eyes.**

Customer Name:		License No.:
Automobile Year/Make:	Model:	Engine Size:
Inspection Performed By:		Date:

International Mobile Air Conditioning Association

CODE OF PROFESSIONAL PRACTICE **Inspection Report**

Procedure	Recommendations	Estimated Cost of Repairs	
		Parts	Labor
VISUAL INSPECTION - Engine Compartment			
1) Hoses, tubing and connections (Suction, Discharge & Liquid Lines)			
2) Compressor			
3) Compressor Clutch			
4) Service Ports			
5) Condenser			
6) Expansion Valve/Orifice Tube			
7) Evaporator Pressure Regulator (POA, STV or VIR)			
8) Cabin Air (Evaporator) Filter (if equipped)			
9) Accumulator/Drier			
10) Drive Belts, Pulleys and Tensioners			
11) O-rings, Gaskets, Seals and Spring Locks			
12) Inline Filter			
13) Electric Fan, Fan Clutch & Fan Blade			
14) Electrical Components			
VISUAL INSPECTION - Passenger Compartment			
1) Air ducts, louvers, sensors, control knobs and cables			
2) Control Head			
3) Interior Condition			
LEAK CHECK - Engine Compartment (NOTE: Engine must be off during this procedure)			
1) Refrigerant Check			
2) Results of Leak Check			
	Subtotal Of Estimated Repair Costs		

INITIAL PERFORMANCE EVALUATION			
Type of Refrgerant:	Purity ❑ Yes ____ % ❑ No	High-Side Press.:	Low-Side Press.:
Louver Temperature:	Interior Temperature:		Ambient Temperature:
Amount of Refrigerant Added to system:	Amount of Refrigerant Recovered from system:		
FINAL PERFORMANCE EVALUATION			
Type of Refrigerant:	Purity ❑ Yes ____ % ❑ No	High-Side Press.:	Low-Side Press.:
Louver Temperature:	Interior Temperature:		Ambient Temperature:
Amount of Refrigerant Added to system:			

Total Estimated Cost of Repairs Based On This Inspection:

The above inspection was done in accordance with the IMACA Code Of Professional Practice procedures manual. If repairs are recommended you will be provided an estimate and only the repairs authorized by you (the customer) will be made. If further repairs are necessary you will be informed of and approve the additional parts and labor costs before the repairs are performed. *Thank You for your business!*

Manager:______________________

Date:______________

FIGURE 10–31 An A/C system checklist and inspection report. (*Courtesy of the International Mobile Air Conditioning Association [IMACA]*)

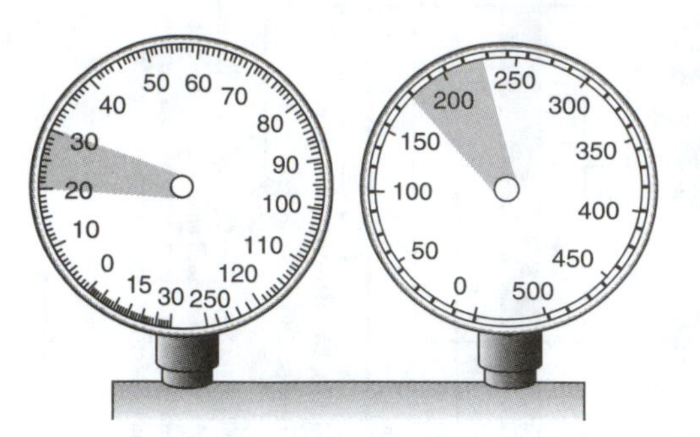

FIGURE 10–32 During normal operation, most systems will have a low side pressure around 20 to 30 psi and a high side pressure that is dependent on ambient temperature. Service information should be checked to determine the pressure for a specific system.

in the center A/C register to record system temperature (Figure 10–33).

3. Start the engine, turn the A/C on to full cold, open the registers, and adjust the blower speed to medium or medium-high. Adjust the engine speed to 1,500 to 2,000 rpm. On most vehicles, a small screwdriver can be inserted at some point in the throttle linkage to hold the throttle open.
4. Run the system for 5 to 10 minutes, or until the pressure readings stabilize.
5. Record the low and high side pressure readings and the ambient and system temperature readings. Check the evaporator outlet tube temperature; it should be cold. If the clutch is cycling, note the length of the on and off times

(*a*)

(*b*)

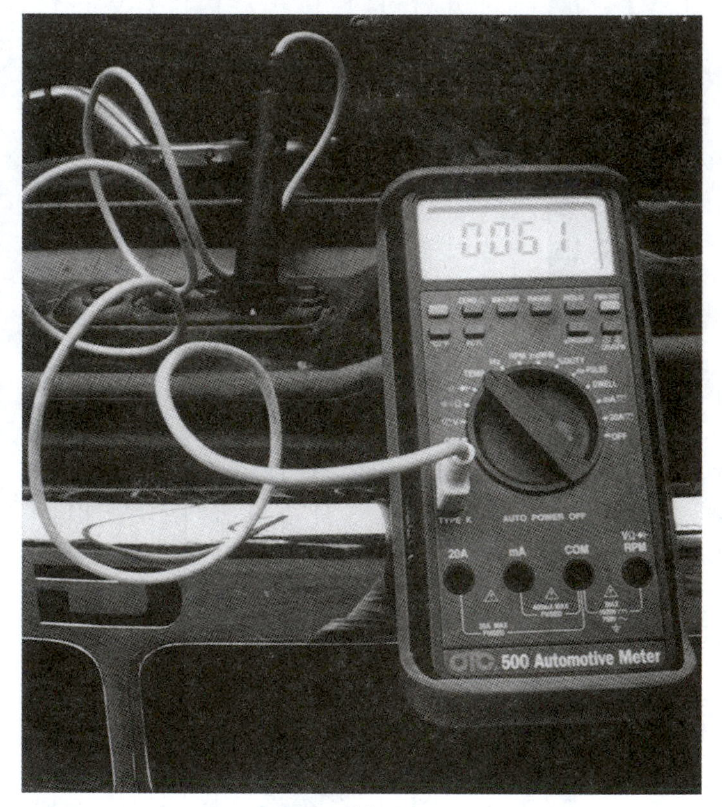

(*c*)

FIGURE 10–33 Proper high side pressure can only be determined from the temperature of ambient air entering the condenser. A dial pocket thermometer (*a*), an infrared, noncontact thermometer (*b*), or a digital thermometer (*c*) can be used. (b *is courtesy of Raytek Corporation*)

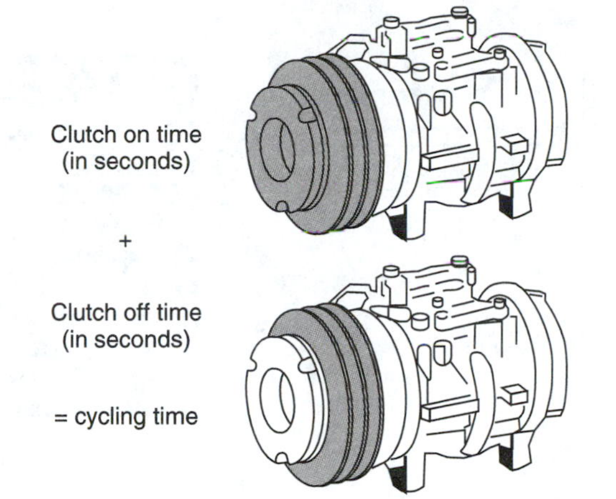

FIGURE 10–34 Clutch cycle time is determined by adding the time that the clutch runs to the time that it is stopped.

(Figure 10–34). If the system uses an STV, check for a temperature drop between the STV inlet and outlet. In some systems, a frost line forms from the middle of the STV to some point on the suction hose when it is modulating pressure. If low side pressure or STV operation seems abnormal and the system has a third port, install a third gauge and record the pressure.

6. If the system has a sight glass, note the condition of the refrigerant flowing past the sight glass. Cycle the clutch as you watch the sight glass. Shortly after cycle in, a flash of bubbles should flow past, and then the refrigerant should run clear. About 30 to 60 seconds after cycle out, another bubble flash should appear (Figure 10–35). With systems that have been retrofitted to R-134a, the sight glass normally has a cloudy or milky appearance and shows bubbles in a fully charged system. On OT systems, feel the temperatures of the OT outlet and the bottom of the accumulator. They should be equal.
7. Reduce the engine speed to normal idle speed for 5 to 10 seconds before shutting the engine off.
8. Compare the readings of this test with those for a normal system.

✓SERVICE TIP

On very humid days, the clutch may not cycle, even with the blower on low. To make sure the system will cycle, you might need to disconnect the blower or spray water on the condenser to force the evaporator temperature low enough to cycle the clutch.

FIGURE 10–35 A sight glass can be used to determine the condition of the refrigerant passing through the liquid line. *(Courtesy of DaimlerChrysler Corporation)*

10.3.1 Measuring Relative Humidity

Some manufacturers compare system pressures with relative humidity because the system must work harder when relative humidity is high. Relative humidity is measured by taking two temperature measurements, a dry bulb and a wet bulb temperature, and referencing these to a chart. A wet bulb is the same thermometer or electronic temperature probe with a damp sock on it. If the humidity is low, water evaporates from the wet bulb, and the cooling effect causes a lower temperature reading (Figure 10–36).

To measure relative humidity, you should:

1. Measure the ambient temperature using a dry thermometer or temperature probe and record this temperature.
2. Measure the ambient temperature using a wet bulb. Clean water is used to wet the bulb. Fan air across the wet bulb or move the bulb through air. Take several wet bulb readings until the temperature stabilizes at its lowest reading and then record this temperature.
3. Locate the dry and wet bulb readings on a psychrometric chart: The relative humidity is where these two lines intersect (Figure 10–37).

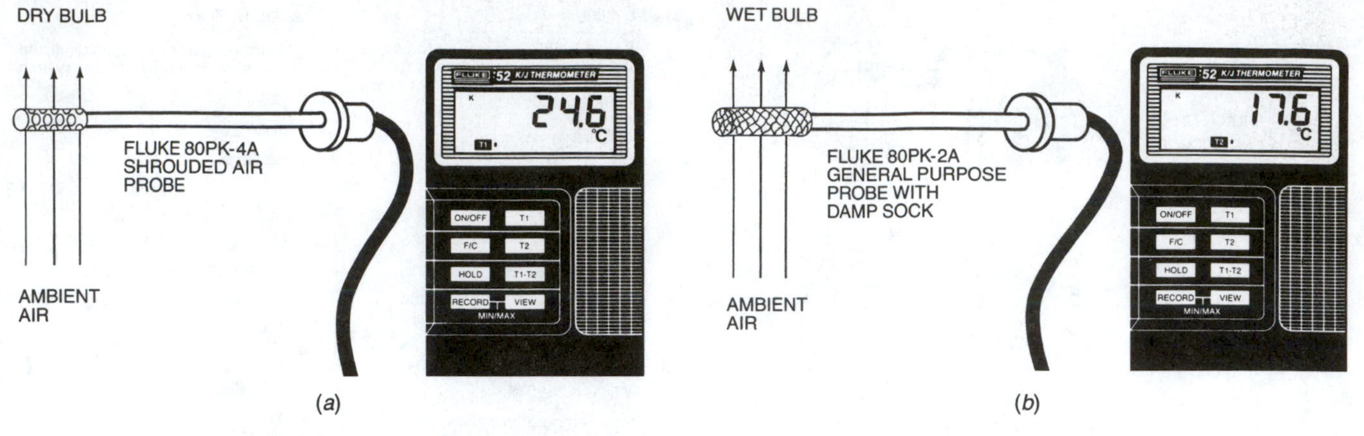

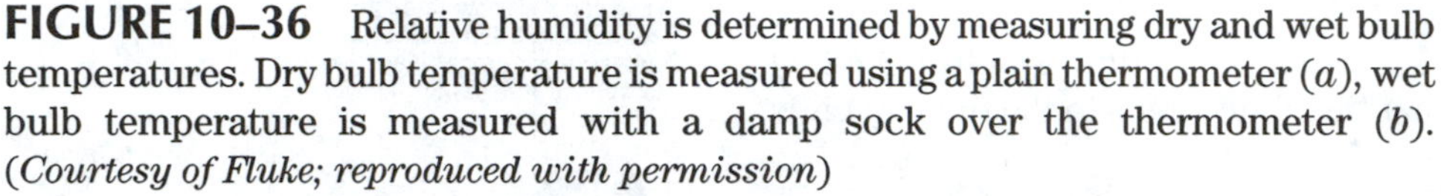
FIGURE 10–36 Relative humidity is determined by measuring dry and wet bulb temperatures. Dry bulb temperature is measured using a plain thermometer (*a*), wet bulb temperature is measured with a damp sock over the thermometer (*b*). (*Courtesy of Fluke; reproduced with permission*)

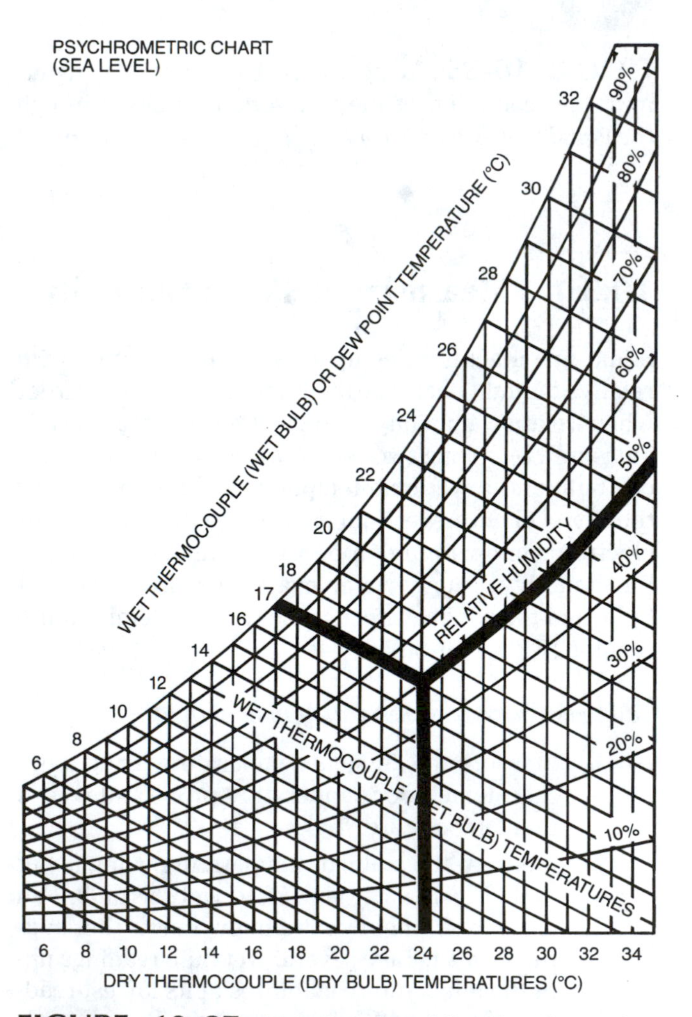

FIGURE 10–37 Relative humidity is the point where dry and wet bulb temperatures intersect on a psychrometric chart. In this example, a 17°C wet bulb and a 24°C dry bulb indicate 50% RH. (*Courtesy of Fluke; reproduced with permission*)

✓SERVICE TIP

With some thin electronic temperature probes, a shoelace can be cut and used as a sock over the probe. Wet the sock and you have a wet bulb.

10.3.2 Interpreting Sight Glasses

Most experienced technicians realize that a sight glass can give false readings. Depending on the design of the pickup location in the receiver–drier and the temperature, a sight glass can show bubbles in a system that is fully charged. It is also normal for an STV system to pass bubbles in the sight glass while the STV is operating to control evaporator pressure. The sight glass of an R-134a system normally has a milky appearance.

An electronic sight glass can be connected to the liquid line of a system without a sight glass (Figure 10–38). This unit sends an ultrasonic signal through the refrigerant and indicates whether bubbles pass through.

Bubbles can be expected to appear in the sight glass of a fully charged system

- Slightly after the clutch cycles in
- Slightly after the clutch cycles out
- At temperatures below 70°F
- When an STV is throttling

Therefore, a technician uses system pressures and temperatures to indicate correct charge level; the condition of the sight glass or output of an electronic sight glass should be considered just supplementary, helpful information.

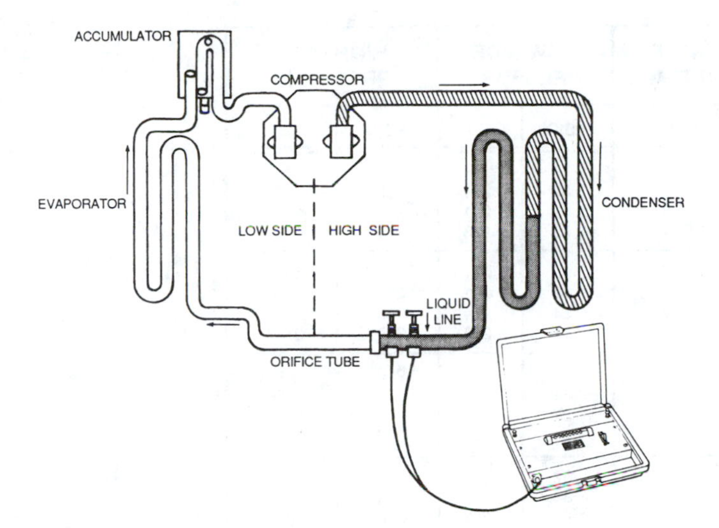

FIGURE 10–38 An electronic sight glass connected to a system. Bubble indicator lights light up and a beep sounds to indicate liquid or gas passing through the liquid line. (*Courtesy of TIF Instruments*)

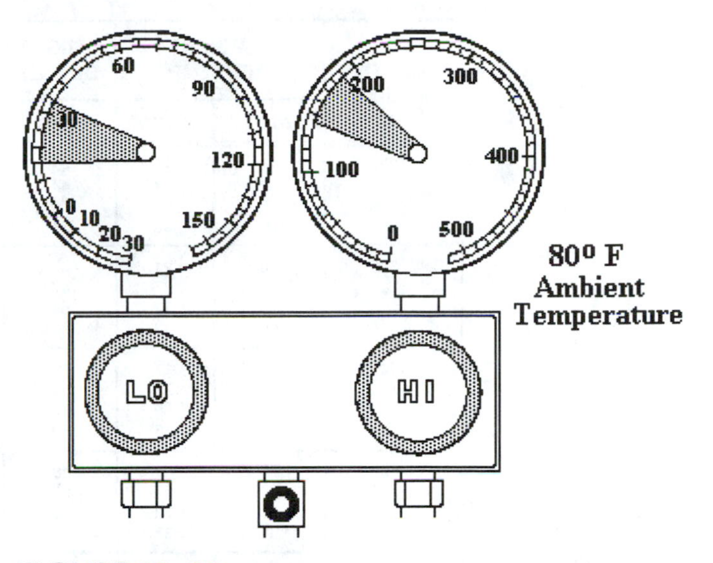

FIGURE 10–39 The low and high side pressures for a TXV system at 80°F. Note that they will vary about 10 to 15 psi on the low side and about 50 psi on the high side. High side pressures are also dependent on ambient temperature.

10.3.3 Interpreting Performance Test Readings

For most drivers, the most important indicator of A/C performance is the discharge air temperature at the center A/C register. For a technician, the most important indicators are the low and high side pressures and temperatures. Many vehicles with borderline A/C performance can produce cool air with the vehicle parked in the shade, the windows closed, and the passenger compartment empty. The temperature range for the center register air discharge temperature is specified by most manufacturers; it is about 35 to 55°F, depending on ambient temperature and humidity.

✓SERVICE TIP

If no specifications are available, some technicians note the drop in air temperature as it passes through the evaporator; this drop should be at least 20°F. A technician can check this temperature by placing the back of his or her hand in front of the register; the air should have a cold, sharp feel to it.

Low side pressure, also called **suction pressure,** should tell evaporator temperature. The actual pressure varies slightly between R-12 and R-134a systems and the style of evaporator icing control. With cycling clutch systems, the low side pressure should drop to the point at which the clutch cycles out; then the pressure increases to the point at which the clutch cycles in again. This pressure range should be about 10 to 20 psi and be between 15 and 35 psi for R-12 systems. With R-134a systems, normal low side pressure is about 18 to 40 psi. On systems that use an STV, the low side pressure should drop to the point at which the STV starts controlling it; from this point, the pressure should vary only a few psi. Normal low side pressure for most STV systems is about 30 to 35 psi. For an older Chrysler EPR system, evaporator pressure should be about 22 psi (152 kPa); with a Toyota EPR system, the evaporator pressure should be 27 psi (186 kPa). With a variable displacement compressor, the low side pressure should also drop to the control point, and then it should remain fairly constant. The exact pressure varies somewhat between system designs, but it should be within a range of 32 to 55 psi. The clutch does not cycle in either an STV or variable displacement compressor system (Figure 10–39).

Remember that the orifice in an OT system is merely a restriction between the low and high sides. Increased high side pressure can cause the pressure in the evaporator to increase, which in turn causes higher evaporator temperatures. Also, the accumulator is an extension of the evaporator and has the same pressure and temperature as the evaporator.

High side pressure, also called **discharge** or **head pressure,** should tell us condenser temperature; this pressure varies directly with ambient temperature and condenser airflow. After noting the ambient temperature on the thermometer at the vehicle's radiator grill, a technician often consults a chart to determine the pressure range for normal high side pressure (Figure 10–40).

RELATIVE HUMIDITY %	AMBIENT AIR TEMP.		RIGHT CENTER OUTLET AIR TEMP.		LOW SIDE PRESSURE		HIGH SIDE PRESSURE	
	F DEG.	C DEG.	F DEG.	C DEG.	PSIG	kPa	PSIG	kPa
20	70	21	43	6	32	221	175	1207
	80	27	44	7	32	221	225	1551
	90	32	50	10	32	221	275	1896
	100	38	51	11	33	228	275	1896
30	70	21	45	7	32	221	190	1310
	80	27	47	8	32	221	235	1620
	90	32	54	12	34	234	290	2000
	100	38	57	14	38	262	310	2137
40	70	21	46	8	32	221	210	1448
	80	27	50	10	32	221	256	1758
	90	32	57	14	37	255	305	2103
	100	38	63	17	44	303	345	2379
50	70	21	48	9	32	221	225	1551
	80	27	53	12	34	234	270	1882
	90	32	60	16	41	283	326	2241
	100	38	89	21	49	338	380	2620
60	70	21	60	10	32	221	240	1855
	80	27	66	13	37	255	290	2000
	90	32	63	17	44	303	340	2344
	100	38	75	24	55	379	395	2724
70	70	21	52	11	32	221	255	1758
	80	27	59	15	40	276	305	2103
	90	32	67	19	48	331	355	2448
	100							
80	70	21	53	12	36	248	270	1862
	80	27	62	17	43	286	320	2206
	90	32	70	21	52	356	370	2551
	100							
90	70	21	55	13	40	278	286	1965
	80	27	65	18	47	324	335	2310
	100							

FIGURE 10–40 This chart shows how outlet temperatures and system pressures are affected by ambient temperature and relative humidity. *(Reprinted with permission of General Motors Corporation)*

✓SERVICE TIP

Some technicians use a rule of thumb that high side pressure for an R-12 system should equal ambient temperature (°F) plus 100 psi. The range is ± 20 psi from this point. Using this method, the high side pressure on a 70°F day should be 170 ± 20 psi, or between 150 and 190 psi. With R-134a systems, high side pressure is slightly higher.

System pressure in modern vehicles is affected more by design variables than it was in earlier cars. Increasing engine speed usually increases pressure, and with many RWD cars, discharge pressure drops when a fan clutch engages to pull more air through the condenser. Discharge pressures are often higher in FWD cars because of their fan types and sizes. Condenser design and size also affect this pressure. With a modern system, the vehicle manufacturer's pressure chart for that particular make and model should be used for reference.

A technician places both temperature and pressure readings into one of three categories: **normal, low,** or **high**. Normal readings indicate that the system is operating within design limits, and it is okay. Low or high readings indicate a fault.

A cycling clutch system meets the following conditions if it is operating correctly:

Low side pressure: Normal: 15 to 35 psi with the clutch cycling; it will be on for about 45 to 90 seconds and off for about 15 to 30 seconds.

High side pressure: Normal: within the range of pressure for that ambient temperature.

Center register discharge temperature: Normal: between 35 and 55°F (in-vehicle temperature minus 20°F); cold and sharp feel.

Evaporator outlet: Cold, often with dew or sweat forming.

Sight glass: Clear, no foam, very few bubbles; the sight glass in an R-134a system has a milky appearance.

OT and accumulator comparison: Bottom of accumulator is as cold as OT, with dew or sweat forming on both.

NORMAL SYSTEM PERFORMANCE, CYCLING CLUTCH

Ambient Temperature	*Suction Pressure*	*Discharge Pressure*	*Center Outlet Temperature*
80°F	15–35 psi	145–190 psi	35–55 psi

A good STV system has most of the same conditions except the following:

Low side pressure: Normal for that type of pressure control (usually 28 to 32 psi); no clutch cycling; dew, sweat, or possibly frost forming on the valve and extending on the suction hose toward the compressor; a third gauge shows a pressure drop from low side pressure (Figure 10–41).

NORMAL SYSTEM PERFORMANCE, STV SYSTEM

Ambient Temperature	*Suction Pressure*	*Discharge Pressure*	*Center Outlet Temperature*
80°F	28–32 psi	145–190 psi	35–55 psi

A system that uses a variable displacement compressor should have the same conditions, except the following:

Low side pressure: Normal for that type of compressor; low side pressure drops as the evaporator cools but levels off and does not drop below a certain point; no clutch cycling.

✓SERVICE TIP

Remember that if low side pressures tries to drop too far, system low pressures can be affected by the system controls trying to operate normally. This can cause a clutch to cycle out or a variable displacement compressor to go to minimum output as the pressure drops.

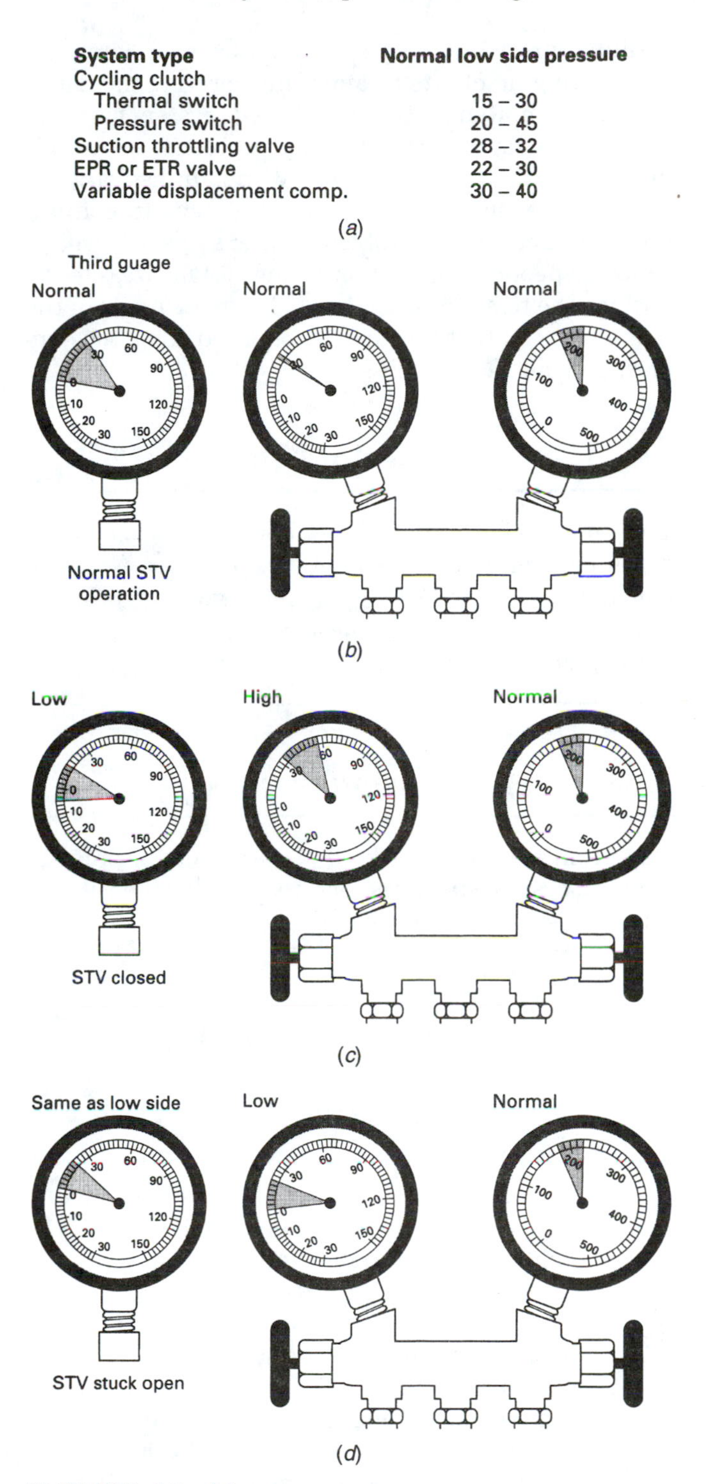

FIGURE 10–41 The low side pressures of various systems (*a*). If a third gauge is used on a properly operating STV system, it will show a pressure drop across the STV (*b*). An STV that is stuck closed (*c*), and an STV that is stuck open (*d*).

If conditions are not normal, faults usually occur in certain combinations that indicate the nature of the problem. An experienced technician often recognizes a problem from a few of these clues. Most vehicle manufacturers and larger aftermarket suppliers publish diagnostic charts to aid in interpreting the cause of a problem. Remember that the A/C system can only make the evaporator cold, and problems in the air management system can cause warm air discharge.

Many technicians perform extra steps to confirm the exact nature of a problem. These steps are usually visual inspection of a part or feeling certain areas to determine the temperature. The goal is to locate all of the faulty parts so that they can be repaired using the simplest and quickest process.

✓SERVICE TIP

If the compressor clutch either does not engage or engages and then quickly releases, there could be excessive clutch clearance, an A/C system problems, or a problem in the electrical controls. These problems are described in Chapter 11 and 14.

✓SERVICE TIP

One prominent aftermarket A/C parts supplier recommends a system stress test before releasing the vehicle. This test is performed with:

- The engine idling at normal operating temperature.
- System set to max A/C, max cold
- Blower on high
- Vehicle doors open

The checks are to:

1. Measure the condenser inlet and outlet line temperature (within 4 inches or 100 mm of the condenser). The outlet line temperature should be cooler with a difference of 20 to 60°F (11 to 33°C). Less than 20°F difference indicates inadequate condenser performance.
2. Measure the ambient air temperature in front of the condenser and the discharge air temperature at the center registers in the instrument panel. The discharge temperature should be at least 30°F (17°C) colder if the system is working properly.
3. Measure the evaporator inlet (after the OT) and outlet (before the accumulator) line temperatures. The outlet line temperatures should be within 5°F (3°C). An excessive temperature difference indicates a possible charge level, OT, blend door, or condenser problem.

10.3.4 Abnormal Conditions

All of the following combinations of symptoms can cause poor A/C performance. Poor performance is most often the result of an evaporator that is not cold enough, but it can also be caused by an evaporator that is too cold and is frosty (Figure 10–42).

ABNORMAL SYSTEM PERFORMANCE,
LOW CHARGE LEVEL OR OT/TXV BLOCKAGE

Ambient Temperature	*Suction Pressure*	*Discharge Pressure*	*Center Outlet*
70 to 100°F	Low	Low	Cool to warm

Possible cause: Low refrigerant charge level.

Confirmation: Warm evaporator tailpipe; very little temperature drop across TXV or OT; bubbles or foam in sight glass; bottom of accumulator is warmer than OT; slightly low refrigerant on OT systems causes discharge air temperature to go cool to warm to cool with engine at higher speeds; air discharge temperature can be cool at one register and warm at another; with a faulty thermostatic switch, the clutch does not cycle; with pressure switch, the clutch short cycles; the refrigerant recovery process is quite short.

Cure: Locate leak, repair leak, and recharge.

Possible cause: Blocked TXV or OT.

Confirmation: Warm evaporator tailpipe; noticeable temperature drop across TXV or OT; usually shows frost; clear sight glass; low side pressure drops very fast when the clutch cycles in; low side may go into a vacuum; normal length of refrigerant recovery time.

Cure: Remove and clean or replace TXV screen; check for TXV stuck closed; remove and replace (R&R) OT.

✓SERVICE TIP

A compressor is cooled by the incoming refrigerant so a properly operating system will have a compressor with a case temperature of about 130 to 150°F (54 to 66°C) in a much hotter engine compartment. Compressor case temperature can be easily checked using an infrared thermometer, and an excessive operating temperature indicates a low charge level or internal friction from a low oil charge. Some compressors include a temperature switch to shut the compressor off if an overheat condition should occur.

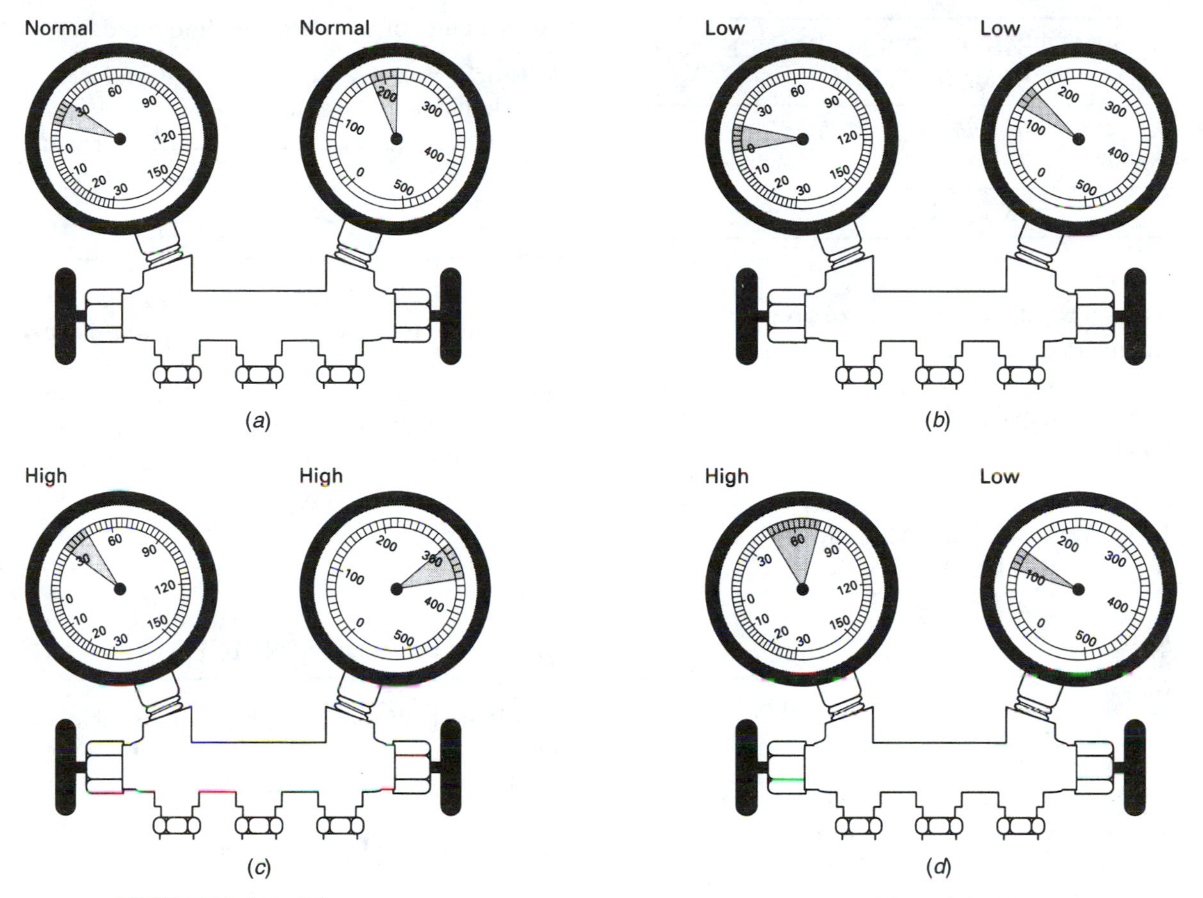

FIGURE 10–42 Typical gauge readings for a normal system (*a*) and faulty systems (*b*, *c*, and *d*).

✓REAL WORLD FIX

The 1996 Ford van (115,000 miles) had warm air from the A/C ducts. The pressures were almost normal at 60 psi on the low side and 250 psi on the high side. The refrigerant was recovered and the system evacuated and recharged, but this did not help.

Fix: A temperature check using an infrared thermometer showed a small temperature drop across the OT, so it was removed for inspection. Some shavings partially plugged it; since such parts are inexpensive, it was replaced using a new one. Replacement of the OT fixed this system; after recharging, there was a temperature drop across the OT of 35°F.

✓SERVICE TIP

If a blocked TXV or OT problem is intermittent and only occurs after the system has been run for awhile, a possible cause is ice forming at the TXV or OT. This can be confirmed by shutting the system off for a few minutes. If the system returns to normal for awhile, the ice has probably melted. In this case, the refrigerant should be recovered, the receiver–drier or accumulator should be replaced, and the system should be recharged with new or recycled refrigerant.

✓REAL WORLD FIX

The A/C in the 1997 Pontiac Grand Am (73,000 miles) is not working. The car came in with no refrigerant, and the system was recharged. The system pressures are normal—25 psi low side and 125 psi high side—but there is no cool air from the registers, and both high and low side lines are the same temperature.

Fix: Following advice that this system uses a variable displacement compressor that will try to maintain a 27 psi low side pressure, the technician checked the orifice tube and found it plugged. Replacement of the OT along with the accumulator fixed this system.

ABNORMAL SYSTEM PERFORMANCE, THERMAL SWITCH

Suction Pressure	*Discharge Pressure*	*Center Outlet*
Low	Normal	Cool to warm

Possible cause: Faulty thermostatic or pressure switch.

Confirmation: Frost on evaporator tailpipe with evaporator icing and continuous compressor operation indicates a shorted or closed switch.

Cure: R&R switch.

ABNORMAL SYSTEM PERFORMANCE, LINE RESTRICTION

Suction Pressure	*Discharge Pressure*	*Center Outlet*
Low	Low to normal	Cool to warm

Possible cause: Restriction in liquid line.

Confirmation: Noticeable temperature drop in liquid line or receiver–drier; can have frost at this location.

Cure: R&R restricted item.

✓SERVICE TIP

Many systems place the high side service port in the liquid line, downstream from the condenser. A plugged condenser in these systems will show up as low high side pressure while the compressor is showing signs of excessive pressure, noise and very high temperatures.

✓REAL WORLD FIX

The 1993 Acura (107,000 miles) had a low side pressure of 10″ Hg. The expansion valve was replaced, but the pressure stayed low. If the TXV temperature bulb is warmed up, the pressure increases. The suction line is very cold.

Fix: After talking to the customer, the technician was told about an earlier accident. With this knowledge, a more thorough inspection was performed that revealed a crimped suction line under the air cleaner. Replacement of this damaged line fixed this problem.

Author's Note: A more thorough under-hood inspection should have revealed this fault.

✓SERVICE TIP

A low side restriction *usually* causes low high side pressure while a high side restriction *usually* causes excessive high side pressure. Remember that an A/C system is a closed loop, and a compressor can only pump the refrigerant that returns to it. System pressures will depend on the location of the restriction, amount of restriction, and the location of the service ports relative to the restriction.

✓SERVICE TIP

If a restriction is suspected, on some vehicles (depending on how accessible the orifice tube is), it is a good practice to recover the refrigerant and inspect the OT. The nature of debris caught by the OT can indicate faults like a failed compressor.

✓REAL WORLD FIX

The 1997 Taurus (80,000 miles) that came into the shop has no A/C. The O-rings in the liquid line and evaporator connections had been recently replaced to repair leaks. Both the high and low side pressures are 25 psi with the compressor shutting off after a few seconds. The compressor and orifice tube were replaced and the system was flushed, but this did not help. With a jumper wire across the low pressure switch, the pressure will drop to 10″ of vacuum on both sides. The various components were tested by blowing air through them, and they could not blow through the condenser. The condenser was replaced, but this did not help.

Fix: Closer inspection revealed that the discharge hose was kinked so the flow was pinched off. Loosening the hose connections allowed better hose alignment and fixed this problem.

Author's Note: When interpreting gauge readings, remember when high side service ports are downstream from the condenser, they do not give true compressor discharge pressures. Service operations must be done properly even though cramped engine compartments make proper inspection and service operations difficult.

ABNORMAL SYSTEM PERFORMANCE, OVERCHARGE		
Suction Pressure	*Discharge Pressure*	*Center Outlet*
High	High	Cool to warm

Possible cause: System overcharge.

Confirmation: Pressures become normal after some refrigerant has been removed.

Cure: Recover refrigerant, evacuate, and recharge with correct amount.

ABNORMAL SYSTEM PERFORMANCE, CONTAMINATED REFRIGERANT OR CONDENSER BLOCKAGE		
Suction Pressure	*Discharge Pressure*	*Center Outlet*
High	High	Cool to warm

Possible cause: Contaminated system: air or foreign refrigerant in system.

Confirmation: Use an identifier to determine the type and purity of the refrigerant. With air contamination, the problem is corrected by cure: recover, recycle, recharge, and check for leaks.

Cure: With contaminated refrigerant, recover refrigerant and send off for disposal. With air contamination, recover and recycle refrigerant, evacuate, and recharge.

Possible cause: Poor airflow through condenser.

Confirmation: Inspect front and back of condenser and radiator for debris; cool the condenser with water and watch for a pressure drop; check for proper fan and fan clutch operation and airflow.

Cure: Clean condenser and restricted surfaces; repair fan as needed. Make sure the condenser seals are properly installed.

Possible cause: Plugged condenser (internal).

Confirmation: Measure temperature at condenser inlet and outlet; 40 to 50°F or greater differential indicates blockage.

Cure: Flush or R&R condenser.

✓REAL WORLD FIX

The high pressure relief valve of the 1996 Mazda (37,000 miles) pops offs. The low side pressure is 25–40 psi, but the high side pressure steadily increases until the valve pops off. The compressor, condenser, TXV, receiver–drier, and evaporator have all been replaced, and air has been blown through the lines to show no restriction. None of this has helped. Both cooling fans operate.

Fix: Following advice, the cooling fan airflow was checked, and it was found that one of the fans was running backward. Reversing the wires in the connector to this fan corrected the rotation and fixed this problem.

Author's Note: If the technician had remembered the basic principles of A/C, this fault would have been found much quicker.

✓REAL WORLD FIX

The 1992 Honda Civic (173,000 miles) came in with an empty A/C system. The receiver–drier was replaced, the TXV and system were flushed, and the system was recharged with the proper amount of R-12. The system worked great, and the car was released to the customer. Several days later, the customer used the system, and after about 4 minutes, the engine began bogging down and the pressure relief valve popped off. The compressor was replaced, the system was flushed, and then recharged. After 30 minutes of operation, the high pressure problem returned.

Fix: A condenser temperature check showed only a 20°F difference from the top to the bottom, indicating an internal restriction. A new condenser was installed, and the temperature difference was now 50°F.

Author's Note: The reason for replacing the compressor is unknown.

✓SERVICE TIP

Check for a plugged condenser by measuring the temperature drop from the inlet to the outlet using an infrared thermometer. A properly operating condenser will have a temperature drop of 11 to 33°C. (20 to 60°F) between the inlet and the outlet. Check a serpentine condenser from top to bottom and a parallel flow condenser from one side to the other. The temperature should drop evenly and gradually.

✓SERVICE TIP

If diagnosing a problem of excessive high side pressure on a system using multi-speed electric fans, be sure that the fans will operate at their high speeds. Insufficient fan speed will cause excess high side pressure.

ABNORMAL SYSTEM PERFORMANCE, FAULTY COMPRESSOR, OT, OR TXV

Suction Pressure	*Discharge Pressure*	*Center Outlet*
High	Low	Cool to warm

Possible cause: Faulty compressor.

Confirmation: Feel compressor suction and discharge hoses for temperature change. Rotate the compressor drive plate and feel for the amount of internal resistance and damage.

Cure: R&R compressor and receiver–drier or accumulator.

✓SERVICE TIP

When a modern compressor fails, it will often produce black, sludgelike material that is commonly called **black death.** An inspection of the OT will usually show this material. System repair will require compressor and accumulator replacement, plus a thorough cleaning of the high side between the compressor and OT.

✓REAL WORLD FIX

The 1990 Lexus (165,000 miles) A/C does not work. Both the low and high side pressures are 70 psi with the compressor clutch engaged. The technician questioned whether this problem could be caused by a faulty TXV.

Fix: Further checking showed that the compressor shaft had broken. Replacement of the compressor and receiver–drier fixed this problem.

✓REAL WORLD FIX

A 1992 Saturn has a low side pressure of 60 psi and a high side pressure of 70 psi. It looks like the vane compressor has failed.

Fix: With the engine running at 2,000 rpm, the technician disconnected and reconnected the clutch about six times, and the internal vanes became free. The compressor started normal operation with good pressures and cooling.

ABNORMAL SYSTEM PERFORMANCE, FAULTY TXV OR OT

Suction Pressure	*Discharge Pressure*	*Center Outlet*
High	Low to normal	Warm

Possible cause: TXV stuck open, missing OT or faulty OT seal rings.

Confirmation: With TXV the suction line will be colder than expected for this pressure; TXV does not respond to TXV test; no temperature drop across OT.

Cure: R&R TXV or OT seal rings.

✓REAL WORLD FIX

The customer had replaced the condenser in the 1988 Plymouth Horizon (174,000 miles) with a used one, and the shop evacuated and recharged the system. The low side pressure was 25″ Hg, with a high side pressure of 100 psi.

Fix: Checking revealed a faulty H-valve (TXV). Replacement of the H-valve and receiver–drier fixed this problem.

Author's Note: It is not known why the condenser needed replacement or if the H-valve was plugged up with debris.

✓REAL WORLD FIX

The 1997 Mercury Mountaineer SUV (54,000 miles) came into the shop for a cooling problem caused by a faulty fan clutch. The A/C system was also checked, and the low side was 70 psi with a high side pressure of 180–200 psi. A faulty compressor was suspected, but the customer said that the A/C worked properly at highway speeds.

Fix: Further checking revealed that another shop had replaced the evaporator core. Someone had forgotten to install the OT into the new evaporator. Installation of the OT fixed this problem.

Author's Note: It often helps to get the history of any previous vehicle repairs.

At this point, the technician usually makes further checks or begins the repair. Further checks can be rather simple or much more involved. For example, if the diagnosis indicates poor compressor operation, you can shut off the engine and rotate the compressor clutch by hand. A very low turning resistance confirms a faulty compressor. The diagnosis of a TXV being stuck open is harder to confirm because it must be removed from the system. An on-vehicle check for TXV operation can be made on most TXV systems. Several methods can be used to locate refrigerant leaks.

10.3.5 Verifying Refrigerant Charge Level

Three different checks, two of them quite accurate, can be made to verify whether a system has the correct amount of refrigerant. The simplest is to feel the temperature of the accumulator and OT using your hand. The other checks include measuring the difference in temperature, called the **Delta T**, across the evaporator and measuring the amount of subcooling of the liquid leaving the condenser. If using system pressure, either static or dynamic, as a guide to charge level, remember that temperature has a definite role in the pressure of pure refrigerants and that contaminants—air or foreign refrigerants—greatly affect the pressure.

The Delta T method is preferred for OT–accumulator systems. The accumulator should contain some liquid refrigerant at the same pressure and temperature as the evaporator. Also, in a properly charged system, liquid leaves the OT, boils, and absorbs heat as it passes through the evaporator, so the evaporator outlet should be slightly cooler than the OT. This method requires two electronic thermometers or a digital voltmeter–ohmmeter with thermocouples to provide the needed accuracy. An accurate infrared thermometer can be used.

The subcooling method is preferred for TXV systems. The refrigerant leaving the condenser should be subcooled, or about 10 to 25°F cooler than its condensing

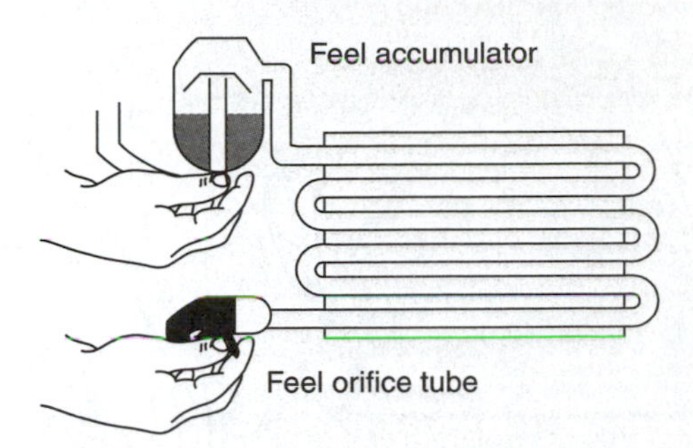

FIGURE 10–43 In a fully to slightly undercharged OT system, the bottom of the accumulator is as cold as the line just downstream from the OT. A warmer accumulator indicates an undercharge.

point. This method requires an electronic thermometer and high side gauge pressure to determine subcooling.

The recommended service procedure for improper charge level is to recover the refrigerant, service (an undercharge indicates a leak), evacuate, and recharge.

To make a hands-on check, you should:

1. Start the engine and run it at normal idle speed.
2. Open the hood and doors.
3. Set the selector to normal A/C, full cold, and high blower.
4. Feel the temperature of the evaporator inlet tube and accumulator surface (Figure 10–43). (a) If both are the same temperature, the charge level is okay; (b) if inlet is cooler, the system is undercharged; (c) if inlet is warmer, the system is overcharged.

To measure Delta T, you should:

1. Connect a gauge set to the system, as described in Section 10.2.3.
2. Connect an electronic thermometer pickup to the evaporator inlet and outlet lines (Figure 10–44).
3. Place a jumper wire across the connector to the clutch cycling switch so the compressor will run continuously.
4. Start the engine, turn the system on to high blower and normal A/C, and adjust the engine speed to 1,000 rpm.
5. Measure the temperatures, record them, and reduce the engine speed to idle.
6. Compare the temperatures to the chart in Figure 10–45. Note that this chart is for a specific vehicle make and model, and the amount of refrigerant that needs to be added is not necessarily the same for all vehicles.

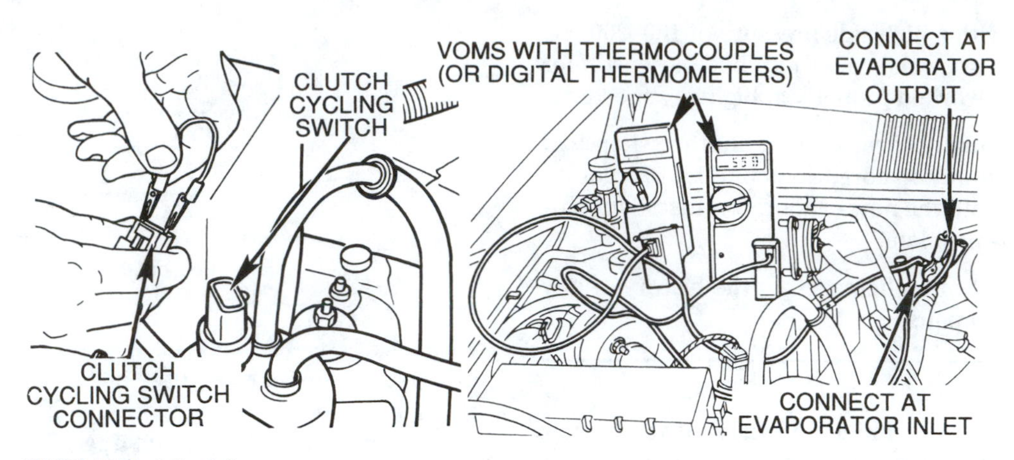

FIGURE 10–44 When measuring the Delta T across the evaporator, one electronic thermometer is attached to the inlet and another to the outlet. The temperature difference is the Delta T. The jumper wire at the clutch cycling switch keeps the compressor operating. (*Courtesy of DaimlerChrysler Corporation*)

Evaporator Outlet and Inlet Temperature Differential

- If Outlet is WARMER than Inlet, temperature differential is plus (+).
- If Outlet is COLDER than Inlet, temperature differential is minus (-).

See the example in the Refrigerant Charge Check (Alternative Method).

Added Amount of R134a to Properly Charge A/C System	Ambient Temperature				
	21°C (70°F)	27°C (80°F)	32°C (90°F)	38°C (100°F)	43°C (110°F)
	Differential Temperature				
0.90 lbs. (14 oz.)	+22°C (+40°F)	+23°C (+42°F)	+24°C (+43°F)	+25°C (+45°F)	+26°C (+47°F)
0.75 lbs. (12 oz.)	+12°C (+22°F)	+12°C (+23°F)	+13°C (+24°F)	+15°C (+26°F)	+16°C (+28°F)
0.60 lbs. (10 oz.)	+4°C (+8°F)	+5°C (+9°F)	+6°C (+10°F)	+7°C (+12°F)	+8°C (+13°F)
0.50 lbs. (8 oz.)	0°C (0°F)	+0°C (+1°F)	+1°C (+2°F)	+2°C (+3°F)	+3°C (+4°F)
0.40 lbs. (6 oz.)	-1°C (-2°F)	-1°C (-1°F)	+0°C (-0°F)	0°C (0°F)	0°C (0°F)
Recommended Charge	-2 to -6°C (-3 to -10°F)				

FIGURE 10–45 This chart indicates the amount of refrigerant that should be added to this particular system, depending on the Delta T. With any system, a Delta T of –3 to –10°F indicates a full charge of refrigerant. (*Courtesy of DaimlerChrysler Corporation*)

To measure subcooling, you should:

1. Connect a gauge set to the system as described in Section 10.2.3.
2. Connect an electronic thermometer pickup to the liquid line near the condenser (Figure 10–46).
3. Start the engine, turn the system on to high blower and normal A/C, and adjust the engine speed to 1,000 rpm.
4. Block off the airflow into the condenser using cardboard or shop cloths so that the high side pressure increases to 260 psi. Do not allow the pressure to increase much above this level.
5. Measure the high side pressure and the liquid line temperature and record them. Remove the blocking material from the condenser and reduce the engine speed to idle.

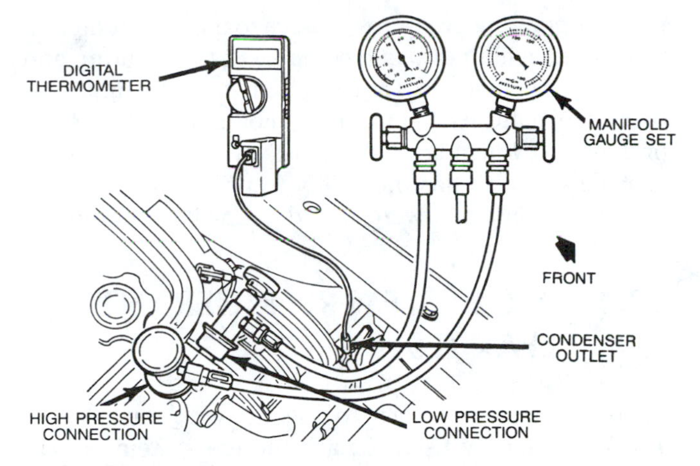

FIGURE 10–46 Subcooling is determined from the high side pressure and an accurate measurement of the condenser outlet. *(Courtesy of DaimlerChrysler Corporation)*

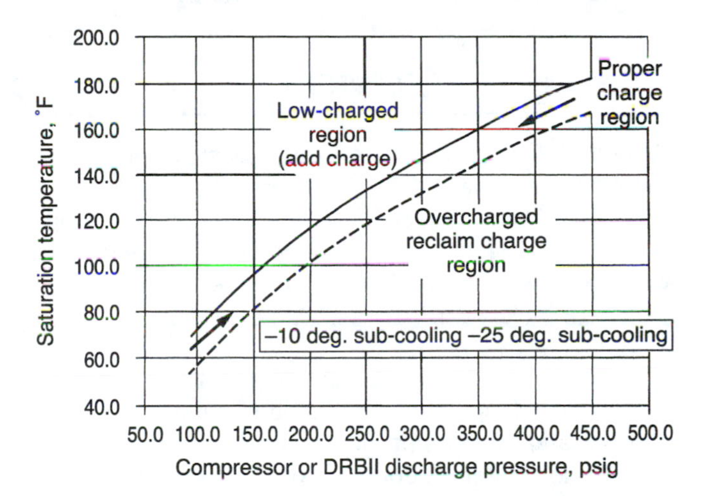

FIGURE 10–47 The point of intersection between the temperature and pressure indicates the relative charge level on this chart. Above the solid line is an undercharge; below the dashed line is an overcharge. *(Courtesy of Chrysler Corporation)*

6. Compare the pressure and temperature with the chart in Figure 10–47 to determine charge level.

10.3.6 On-Vehicle TXV Tests

If tests indicate that a TXV is not working correctly, a technician can make further checks while the system is still together. The first checks are to make sure the thermal bulb is attached securely to the evaporator tailpipe and well wrapped with insulation. Poor contact or insulation can cause the TXV to get a signal that is too warm. This causes the valve to open too far, which produces

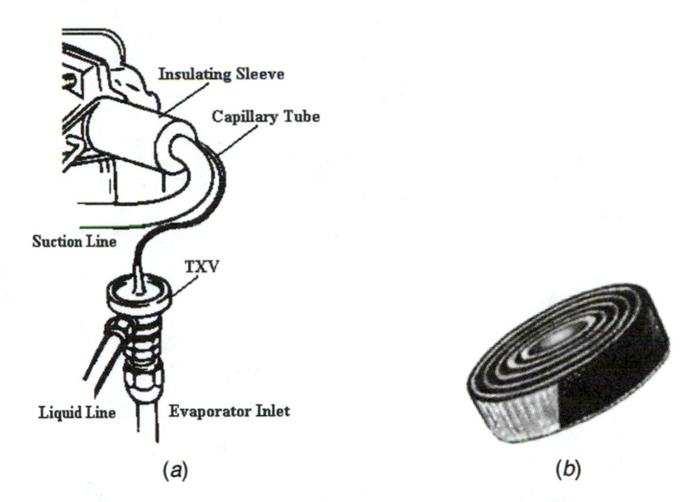

FIGURE 10–48 The thermal bulb of the TXV should be clamped tightly to the evaporator tailpipe and insulated with a foam sleeve (*a*) or wrapped with insulating tape, a thick, pliable tape (*b*). (*b is courtesy of Four Seasons*)

high evaporator pressure. This type of problem cannot occur with H, block, or capsule TXVs (Figure 10–48) . TXVs with capillary tubes can be checked for proper response while the system is operating.

To check TXV operation:

1. Disconnect the TXV capillary tube or thermal bulb from the evaporator tailpipe.
2. With the engine running at test speed and the A/C system operating, insert the thermal bulb into a container of ice water. This should cause the TXV to close down, and a noticeable drop in low side pressure should occur (Figure 10–49).
3. Remove the thermal bulb from the ice water and warm it in your hand or warm water. This should cause the TXV to open up, and a noticeable increase in low side pressure should occur.

With block-type TXVs, liquid CO_2 can be poured onto the diaphragm portion to cool it. This should cause the valve to close.

✓PROBLEM SOLVING

While checking for the cause of a system with a low side pressure that is too high, you make the checks for a faulty TXV, but cooling or heating the thermal bulb does not change the pressure readings. Does this system have a faulty TXV? What else could cause this problem?

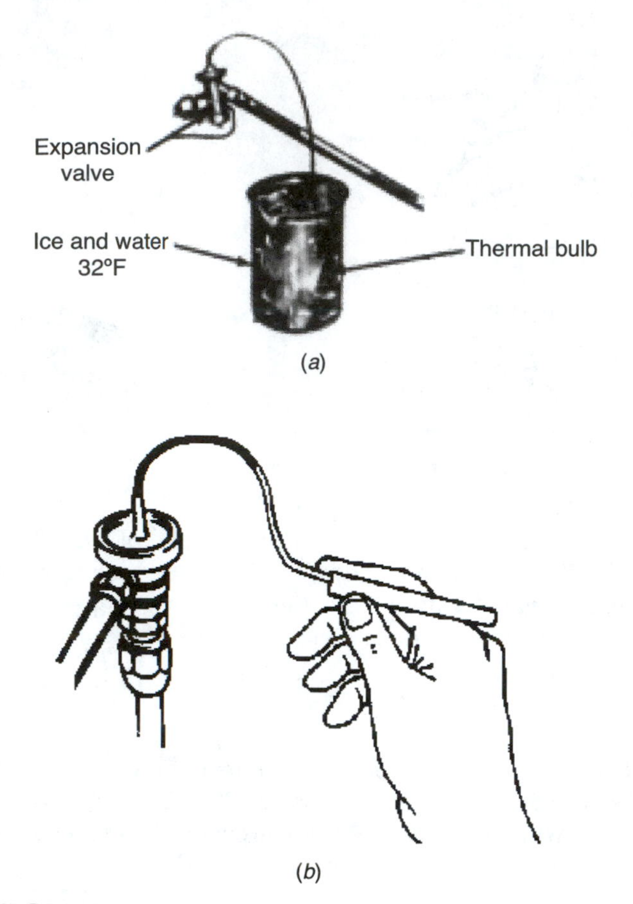

FIGURE 10–49 Chilling the thermal bulb with ice water (*a*) or CO_2 should cause the valve to close and the low side pressure to drop. Warming the bulb with your hand (*b*) should cause the valve to open and the low side pressure to increase. (a *is courtesy of Daimler-Chrysler Corporation*)

10.3.7 Electronic Diagnostic Tool

A/C diagnostic tools have been developed by several companies to aid technicians in testing A/C sysems (Figure 10–50). One tool is based on a small handheld PDA (personal digital assistant) computer, and it compares the high and low side pressures to ambient temperature and relative humidity. The technician also inputs the temperature of various components, such as the condenser, evaporator, and hoses, using a flexible temperature probe.

After being connected to the A/C service ports, the diagnostic tool uses the software in the PDA to guide the technician through the test procedure. It prompts the technician to input needed component temperatures at the inlet and outlet of key components using the completely flexible temperature probe. It will determine good or bad A/C performance. It does this by recognizing proper operating pressures relative to ambient temperature, relative humidity, and component temperature and the clutch cycling rate. The tool also recognizes improper operating patterns for the most common A/C system problems. It will indicate the nature of about 24 different problems. It also has the ability to print out a detailed copy of the data that were logged during its operation.

10.4 A/C SYSTEM REFRIGERANT LEAK TESTS

There are several ways to locate refrigerant leaks. Those commonly used for automotive systems range from very simple soapy water to flame-type detectors and electronic units. Dye or trace solutions are used when trying to locate very difficult or slow R-12 leaks. A fluorescent tracer solution has become a popular method of leak location in R-134a systems.

✓SERVICE TIP

Many R-12 leaks are easy to locate because refrigerant oil tends to escape with the refrigerant. This oil leaves a telltale spill that collects dust. The PAG and POE oils used with R-134a tend to dissipate or be washed away and don't always leave this telltale trace.

An experienced technician begins leak detection with a visual inspection for this oil residue at line connections and other points and for oil throw-off from the compressor clutch. A compressor shaft seal leak often carries oil to the clutch plate, and centrifugal force throws the oil outward and onto the hood or other parts next to the clutch. The most probable leak locations are any line connection (especially spring lock connections), the compressor seal, any other compressor sealing surfaces, the service ports, any place where a refrigerant hose rubs against something, both ends of a hose where it joins a metal fitting, and any kink or dent in a metal refrigerant line. Also check for leaks at the evaporator and condenser.

✓SERVICE TIP

Some technicians use children's bubble soap for leak testing.

The simplest and least expensive leak detector is **soapy water**. Commercial liquid solutions are avail-

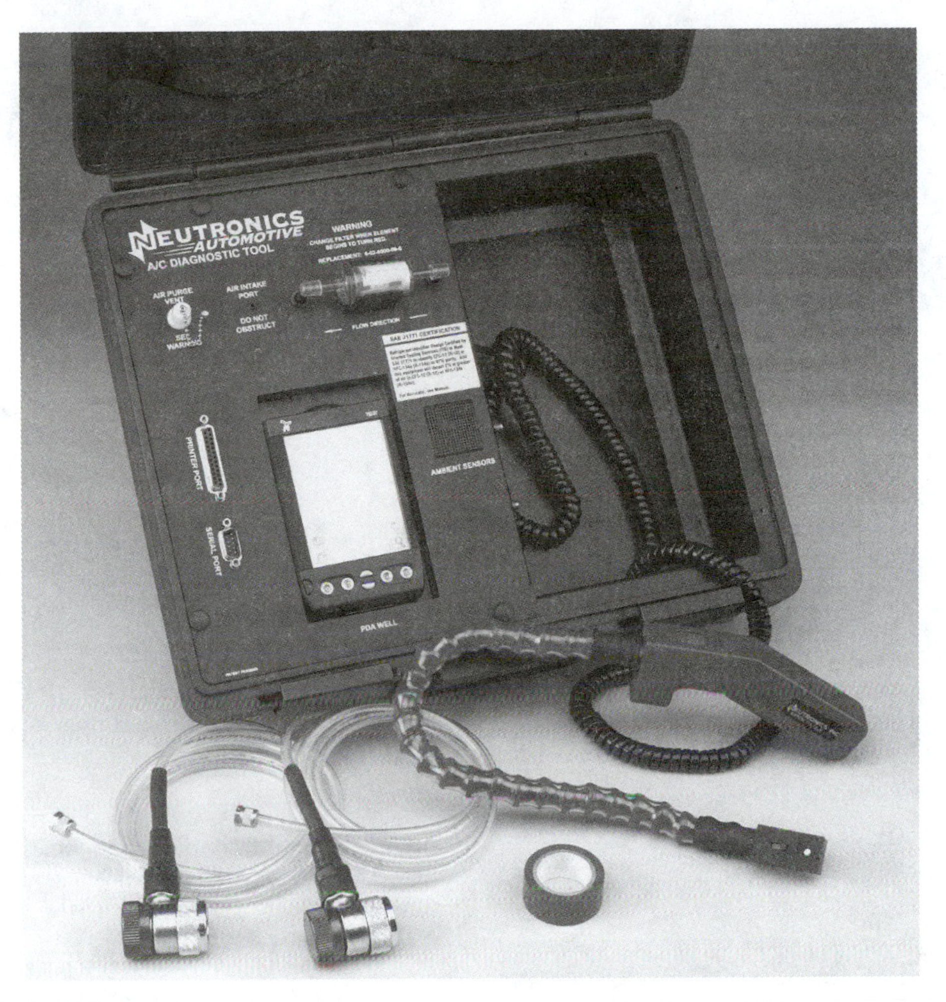

FIGURE 10–50 This A/C diagnostic tool uses low and high side pressure, ambient temperature, relative humidity, and A/C component temperature readings from various locations to input into the PDA. The PDA guides the technician through the test procedure, analyzes the data, and then determines if the system is operating properly or, if not, what is wrong. *(Courtesy of Neutronics, Inc.)*

ter and one-half liquid dish soap. The liquid solution is applied to a suspected area with a dauber, small brush, or spray bottle; leaks make bubbles or foam. Liquid solutions have the advantage of showing the exact location to pinpoint a leak at a service port or line connection, but this method only finds leaks of about 40 oz per year and it is too slow to use it on large areas (Figure 10–51).

✓SERVICE TIP

To determine if a compressor shaft seal is leaking, slip a 3 × 5 card into the clutch gap and, with the engine off, cycle the clutch on and off several times. If there is oil on the card when you remove it, the shaft seal is leaking.

Flame-type leak detectors *are not* recommended by most experts. They use a small propane bottle (usually 1 lb) and a burner to heat a copper reaction plate or wires. Gas flow to the burner passes through a venturi, where the search hose is attached. This creates a vacuum that pulls air through the search hose. If a CFC comes in contact with red-hot copper, it changes the flame color.

CAUTION

Many people do not like the flame-type detector because it produces a poisonous gas as well as it being a potential fire hazard. Flame-type leak detectors should not be used on R-134a systems. R-134a will not cause flame color change. These leak detectors are fairly inexpensive and quite accurate.

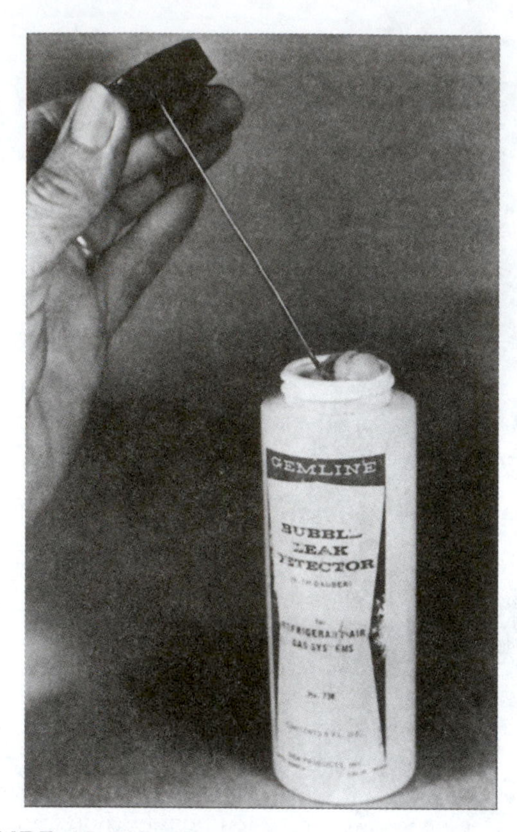

FIGURE 10–51 Bubble leak detector. This is like a liquid soap mixture.

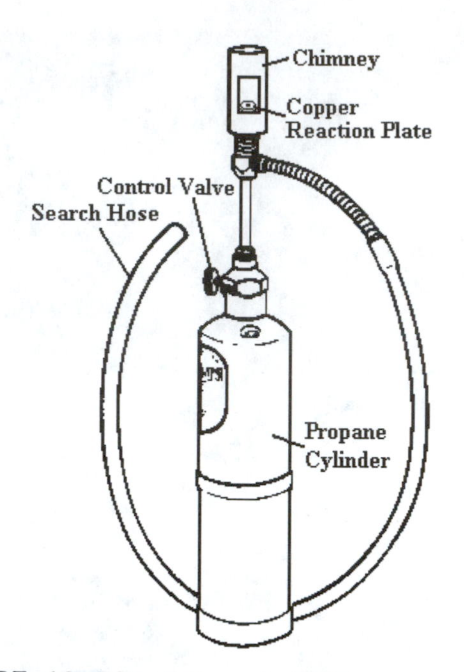

FIGURE 10–52 A flame leak detector (no longer recommended) heats the copper reaction plate to red-hot. The flame color will change if a CFC touches this plate.

CAUTION

If using a flame-type leak detector, make absolutely certain that there is not a flammable refrigerant, hold the unit at arm's length away from your face so you will not breathe the fumes, and be aware of the location of the flame at all times (Figure 10–52).

There are many **electronic leak detectors;** these are the most popular style of leak detector. Most are rather expensive, very accurate, and completely safe to use. Many electronic leak detectors can be used for both R-12 and R-134a systems.

✓SERVICE TIP

When buying a leak detector, it should be certified as meeting SAE standard J 1667.

Most electronic leak detectors emit a clicking noise that starts slowly and gets faster as it detects a leak. Some units use a flashing light that flashes more quickly as the leak is found. Some units use a combination of noise and light to indicate the location and rate of leakage. These leak detectors use a search probe where air and any leakage are drawn into the unit; this probe must be kept clean (Figure 10–53).

Another tester type that is based on electronics is the **ultrasonic leak detector** (Figure 10–54). This tester "listens" for the sound of gas escaping. This sound is at a frequency higher than the human ear is capable of hearing. The tester converts that sound to an audible frequency. The unit is pointed at the suspected leak location, and the technician listens for the noise on a headset or watches the indicator scale. It can pick up sound from a distance, and the sound level will grow louder as the detector is moved closer to the leak.

Dyes and tracer leak detectors are usually charged into the system with the refrigerant. Dye was once sold in small cans of R-12 but is not used any more; the dye leaves a visible stain when it leaks out. The red dyes can ruin interior carpeting and upholstery if they come into contact with the vehicle's interior. Fluorescent trace solutions are commonly used now; these solutions are available in small containers and special capsules. They are propelled into a system by either a refrigerant charge, system pressure, or a hand-powered pump. The fluorescent tracers or dyes glow under ultraviolet light so that they show up more easily. These solutions can pinpoint the location of small leaks relatively quickly. Some vehicle manufacturers dislike trace solutions because they may contaminate a system (Figure 10–55). Other manufacturers add fluorescent tracer dye with the original refrigerant charge.

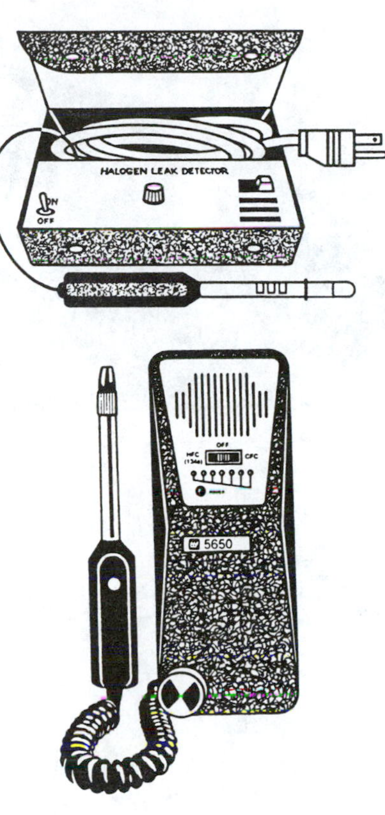

FIGURE 10–53 Some electronic leak detectors can be used on both R-12 and R-134a systems. *(Courtesy of TIF Instruments)*

When you need to locate a refrigerant leak, you should remember the following guidelines:

- Make sure there is enough refrigerant in the system to produce normal pressures (at least 50 psi). (With an empty system, this will be about 7 to 10% of the total charge.)
- Conduct leakage tests in an area free of wind and drafts.
- If the area is contaminated with refrigerant, use a fan to blow the excess away.
- Operate the system long enough to circulate the refrigerant and produce normal pressures (Figure 10–56).
- Shut the engine off while searching for leaks.
- Clean oily spots with a dry shop cloth; solvents may leave a residue that could confuse the leak tester.
- Hold the leak detector probe under the point being checked, because refrigerant is heavier than air (Figure 10–57).
- Move the probe about 1 inch (2.5 cm) per second while checking along a line or hose at a distance of 1/4 inch or less from the item being checked.

FIGURE 10–54 An ultrasonic leak detector converts the sound of a gas leaking through a small opening into a sound that we can hear using a set of headphones. It also displays the leak rate on a scale. *(Courtesy of Robinair, SPX Corporation)*

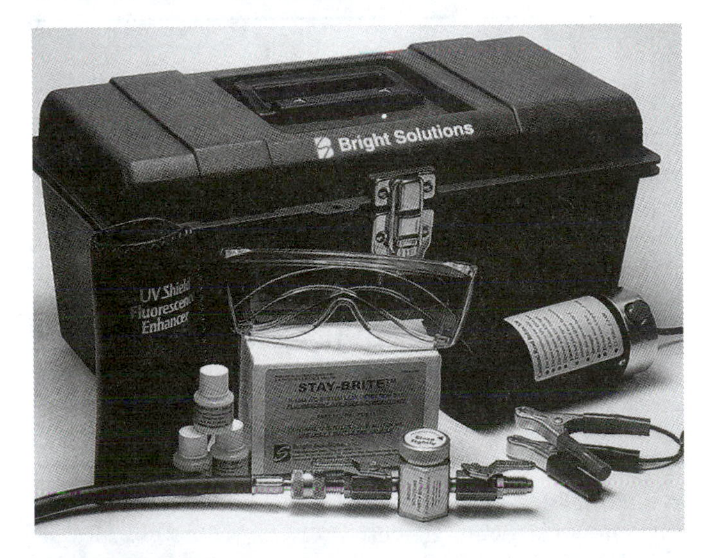

FIGURE 10–55 A tracer solution leak detection kit. Tracer solutions are charged into the system; a color stain visible under an ultraviolet light indicates the leak area. *(Courtesy of Bright Solutions)*

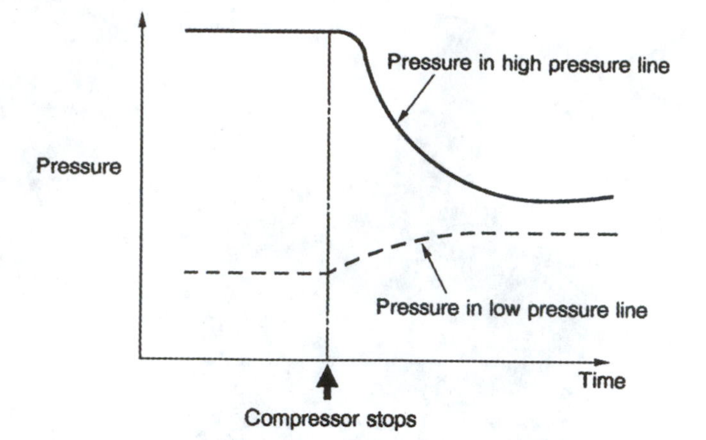

FIGURE 10–56 A leak in the high side of a system shows up best right after the system is shut off, before the pressure drops off. (*Courtesy of Zexel USA Corporation*)

- Make sure the probe of an electronic leak detector does not touch the item being checked.
- When checking a specific location, hold the probe stationary for about 5 seconds.
- If a leak is indicated, move the probe away, blow out the area with shop air, and then return to verify the problem.

✓SERVICE TIP

Check the high side for leaks right after shutting the engine and system off while the pressure is the highest.

✓SERVICE TIP

A faulty Schrader valve can be removed by unscrewing it just like a tire valve; in many cases, a tire valve service tool can be used. A special service tool is used in very tight locations. An A/C Schrader valve should be installed; tire valves will not work properly.

✓SERVICE TIP

A leaking high-pressure relief valve can be replaced using a new valve and gasket.

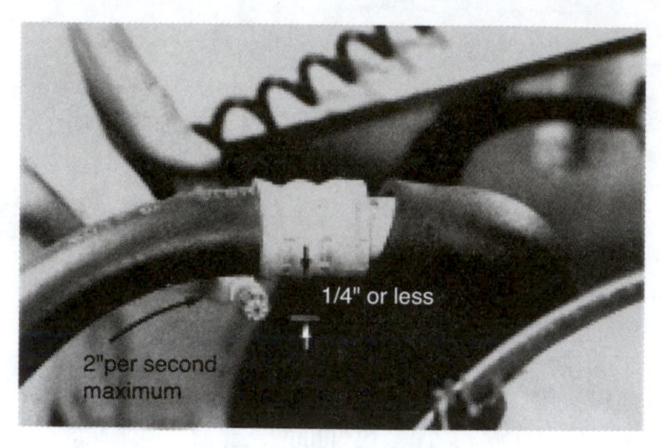

(*a*)

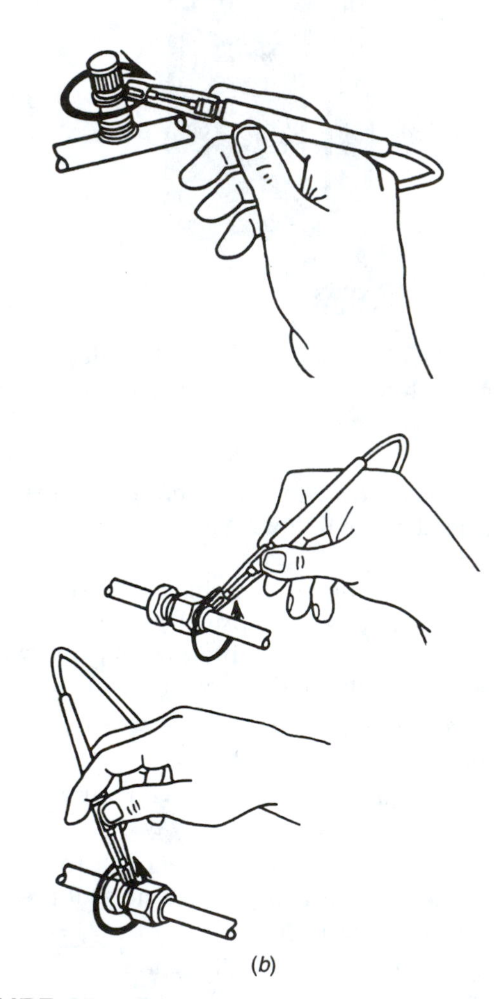

(*b*)

FIGURE 10–57 The leak detector probe should be kept under the line or component, within 1/4 inch, and when checking a hose, moved along at a rate of 2 inches per second (*a*). Move the probe around each line fitting and service port (*b*). (b *is reprinted with permission of General Motors Corporation*)

✓SERVICE TIP

At least one vehicle manufacturer uses a painted steel accumulator with a foam insulating wrap. The foam retains condensed water next to the accumulator, which promotes rust, and after a few years, it can rust through, producing a refrigerant leak. Although accumulator leaks are fairly rare, a painted accumulator is probably steel, and these should be checked for rust and leaks.

✓SERVICE TIP

A leak in an R-134a system is generally harder to locate and requires more patience and care than a leak in an R-12 system. Many technicians search using an electronic detector and following the recommended SAE J-1628 procedure, and if they cannot locate the leak, they use fluorescent dye, which tends to locate the smaller leaks.

✓SERVICE TIP

A leak that occurs at the service port fitting may be temporarily sealed if you connect a gauge set and hoses to it. Disconnect the hoses and check the exposed port and valve.

Other possible leak locations that are often overlooked are the pressure switches; unplug the electrical connectors so you can check the entire switch body.

✓SERVICE TIP

During a hot soak period on a hot day, evaporator pressure can reach 250 to 400 psi, much higher than the pressure normally found during leak checks. Fluorescent dye is the most effective leak-checking process for very small evaporator leaks, but inspection is very difficult.

✓SERVICE TIP

If you think a compressor shaft seal is leaking but the leak does not show up using your detector, rotate the compressor drive plate by hand to another position.

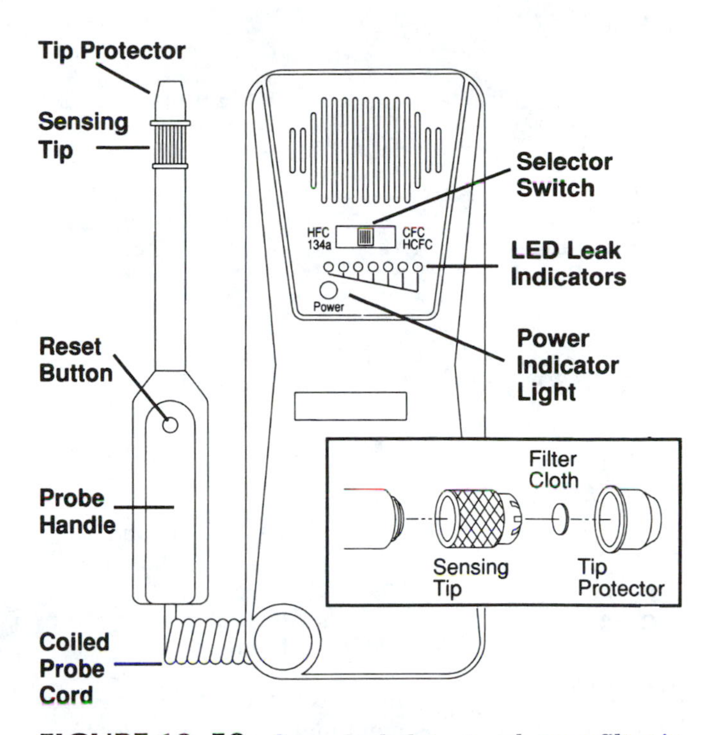

FIGURE 10–58 Some leak detectors have a filter in the tip; the tip and the filter must be kept clean and free of dirt and grease. (*Courtesy of TIF Instruments*)

Sometimes corrosion or a small pit in the shaft sealing surface will produce a leak with the shaft in certain positions but not others.

To locate a refrigerant leak using an electronic leak detector or soap solution, you should:

1. Wipe away any oil and dirt from line connections, hoses, and so forth with a shop cloth.
2. Check to make sure the leak detector probe tip and filter are clean (Figure 10–58).
3. Turn on the electronic leak detector and adjust or calibrate it according to the unit's instructions (Figure 10–59).
4. Begin at a convenient location. Some technicians prefer to follow the normal path through a system and move the probe under each of these points of possible leakage, staying within 1/4 inch of the part being checked.
   a. *Compressor:* Line connections, surfaces where parts are joined, any switches and valves, the clutch and pulley (Figure 10–60)
   b. *Condenser:* Line connections, all welded joints, any visible damage

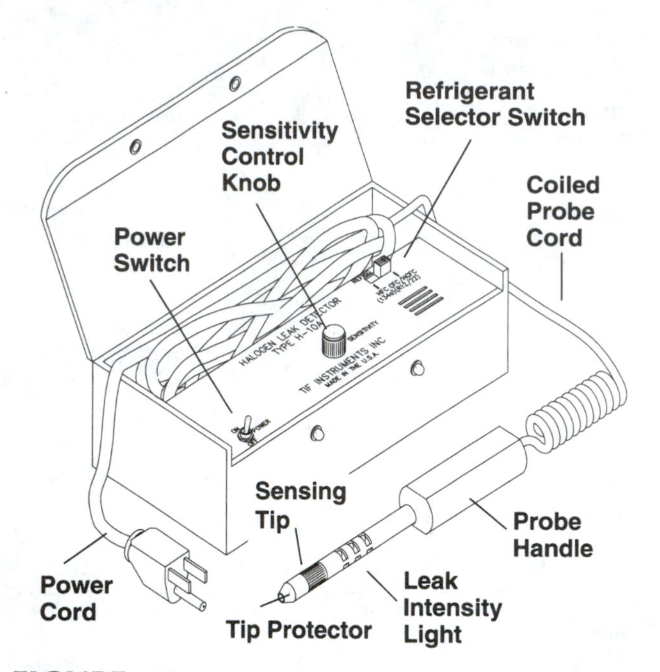

FIGURE 10–59 Some electronic leak detectors have a sensitivity adjustment that is usually adjusted to give a slow noise or light response. (*Courtesy of TIF Instruments*)

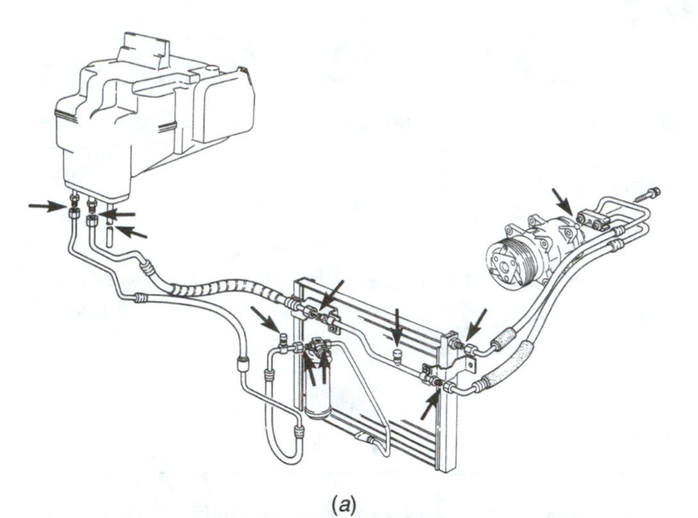

(*a*)

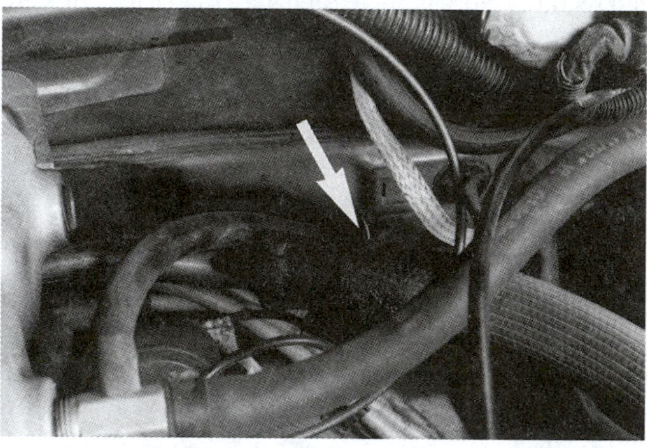

(*b*)

FIGURE 10–60 The most probable leak areas are indicated by arrows, but leaks can occur anywhere along a hose and at the compressor (*a*). The oily, dirty coating (arrow) on this hose indicates a probable leak (*b*). (a *is courtesy of Zexel USA Corporation*)

c. *Receiver–drier:* Line connections; sight glass; any switches, plugs, or relief valves
d. *TXV or OT:* Line connections
e. *Evaporator:* Line connections, at the evaporator drain tube and instrument panel outlet with the blower at the lowest speed setting.
f. *Accumulator:* Line connections, switches, or service port
g. *Hoses:* Line connections, where the hose end meets the metal connector, any area that shows damage or rubbing contact
h. *Service ports:* Valve core, remove cap or disconnect gauge hose

5. If the electronic leak detector indicates a probable leak at a point such as a line connection or joint in the compressor shell, daub a wet film of soap solution over the point. Observe for bubbles, which confirm the point of leakage (Figure 10–61).

✓SERVICE TIP

If an evaporator leak is suspected, run the system long enough for the engine and HVAC system to warm up to normal operating temperatures. With the engine and system off, run the blower motor for about 10 minutes to warm up the evaporator and increase the temperature and low side pressure. Shut off the blower motor, wait for 10 to 15 minutes, and check for a refrigerant leak at the evaporator drain.

✓SERVICE TIP

If you need to check for a leak in the area between the condenser and radiator, place a length of plastic hose over the tester probe. This extension will let you extend the reach through small areas.

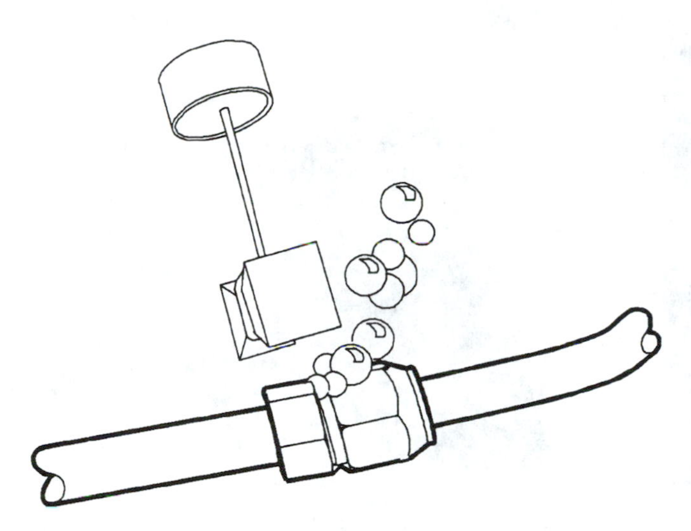

FIGURE 10–61 Refrigerant leaks cause bubbles to form when liquid leak detector is spread over the area. This pinpoints the leak location. (*Courtesy of Saturn Corporation*)

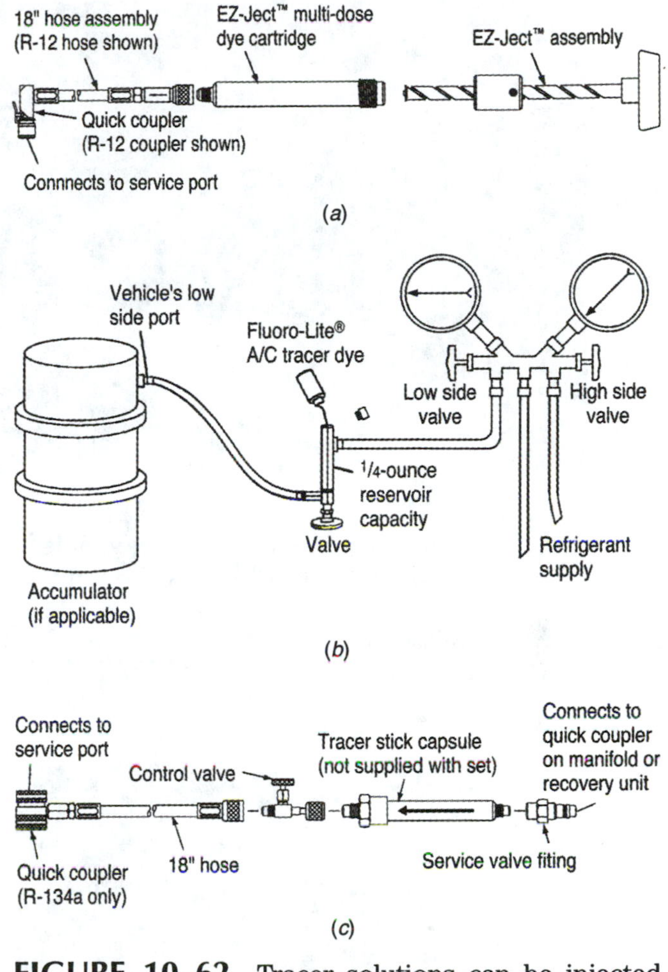

FIGURE 10–62 Tracer solutions can be injected into a system that is shut off using a hand-powered injector (*a*) or added into an operating system by using a dye injector (*b*) or Tracer-Stick capsule (*c*). (*Courtesy of Tracer Products, a division of Spectronics Corporation*)

✓SERVICE TIP

When checking hoses, pay special attention at places where the hose contacts other objects that might cause a **rub-through**.

✓SERVICE TIP

When checking for leaks at the evaporator drain, place a clear plastic hose over the probe of the tester to help ensure that you do not contaminate the probe with water.

To locate a leak using fluorescent tracers, you should:

1. Add the proper amount of fluorescent additive into the dye injector tool and, *if using a hand powered injector system*, connect the injector to a service port, operate the injector to inject the proper amount of dye, and disconnect the injector tool (Figure 10–62). *If using a system with dye capsules or containers:*
   a. Connect the tool in series in the low side service hose *or* connect the disposable Glo-Stick capsule into the low side service hose (Figure 10–63).
   b. If the system is low on refrigerant, connect a refrigerant source to the center hose, open the low side handwheel, and allow refrigerant pressure to force the dye into the system. If the system has normal pressure, start the engine, turn on the A/C, open both low and high side service valves, and allow high side pressure to force the additive into the system.
2. Run the system for a few minutes to circulate the fluorescent dye; then shut off the system and engine.
3. Shine a high-intensity ultraviolet light over the system as you carefully search for a glowing fluorescent trail or puddle, which indicates a leak. Wearing yellow goggles enhances the visibility of the leak, making it easier to find. Follow the trail to the exact location of the leak.

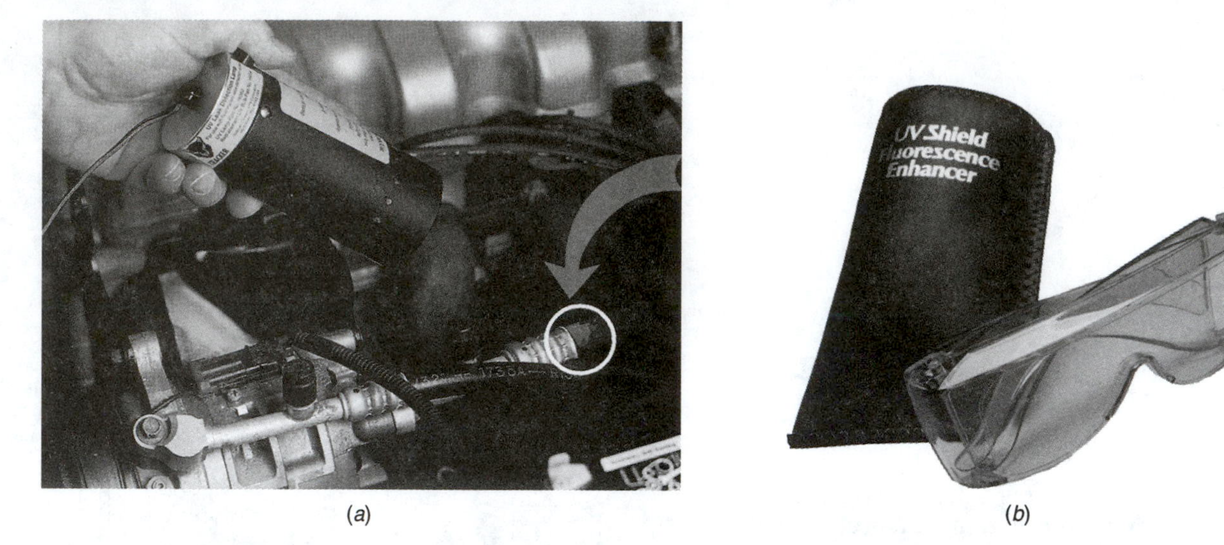

(a) (b)

FIGURE 10–63 A refrigerant leak shows up as a yellow-green glow under ultraviolet light (*a*). A leak shows up easier if yellow fluorescent enhancer glasses are worn (*b*). (a *is courtesy of Robinair, SPX Corporation;* b *is courtesy of Bright Solutions*)

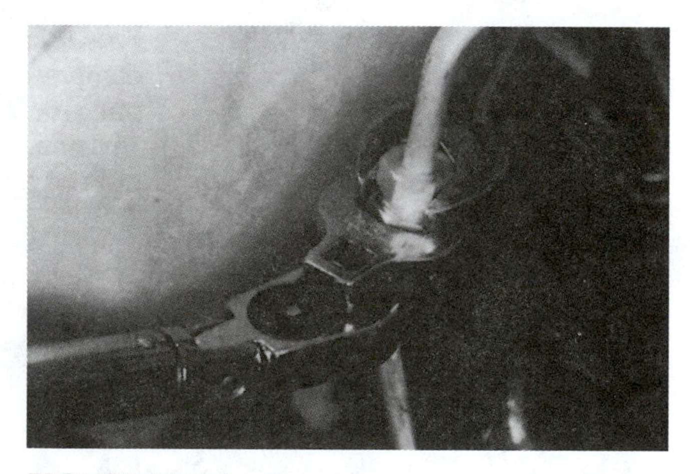

FIGURE 10–64 Line connections must be tightened to the correct torque. A crow's-foot adapter is being used on the torque wrench as a backup wrench keeps the other half of the fitting stationary. (*Courtesy of Nissan Motor Corporation in USA*)

✓SERVICE TIP

In cases where the leak is very slow, the fluorescent dye is injected into the system and the vehicle given back to the customer to give the leak time to show up. After a few days, the vehicle is returned for leak checking.

The leakage location determines the needed repair. Occasionally a leaky connection can be corrected simply by tightening the fitting. But the fitting should not be overtightened—that can damage the seal seat, making the repair more serious than necessary. If the fitting still leaks when tightened to the correct torque, replace the O-ring (Figure 10–64).

✓REAL WORLD FIX

The 1993 Dodge Colt (128,000 miles) has no A/C. The system was found to have no refrigerant. The system would hold a vacuum, so the system was recharged after evacuating it and dye was added. System pressures were normal, so the car was given back to the customer. The car came back the next day with an empty system. A check using a blacklight showed no sign of the dye. The system was recharged, but a check with an electronic tester showed no leak. The system was charged again, and the next day, all of the refrigerant had leaked out. The evaporator core was removed and inspected, but no sign of a leak was found.

Fix: While checking for a leak again, a soft spot on the discharge hose was found. When the bubble was cut open, dye was found inside the bubble. Cutting the hose open revealed that the hose inner liner had failed. Replacement of this hose fixed the leak problem.

✓SERVICE TIP

A fluorescent dye must be circulated to the leak, and it is transported by refrigerant oil. Water will wash it away along with the dye. This helps cleanup after the repair has been made, but it also can remove the ability of finding a leak if the vehicle is operated where water can splash on portions of the A/C system.

10.4.1 Vacuum Pump Leak Tests

During the evacuation and recharging stage of system service, it is a common practice to determine whether the system holds a vacuum as a final check for leaks. This process takes only a few minutes: Simply shut off the service valves after the system is down to a full vacuum (almost 30" Hg), wait 5 minutes, and recheck the vacuum. If the vacuum stays at 30" Hg, the system does not have a leak.

✓SERVICE TIP

Checking the vacuum of a system may indicate a leak that does not exist. The O-rings and compressor shaft seal are designed to keep pressure in, not out. They can allow a slight leakage into the system and still seal perfectly under pressure. If the system's refrigerant has just been recovered, a pressure rise can be caused by refrigerant outgassing or boiling out of the oil. When experienced technicians check the vacuum of a system, they interpret a slight pressure rise as an indicator of either a small leak or normal conditions.

Adapters can be fitted to the compressor or evaporator, so these parts can be individually checked for leaks. These adapters can be commercial fittings or can be made in the shop by connecting the proper line fitting to a 1/4-inch male flare adapter. The check is usually a quality control check for a compressor that has been repaired or a final check before removing an evaporator suspected to be leaking. Evaporator replacement is a difficult, time-consuming job that should not be done unless absolutely necessary (Figure 10–65).

To test a single A/C component for leaks:

1. Recover any remaining refrigerant from the system and disconnect the lines to that component. (Refrigerant recovery is described in Section 13.3.1.)
2. Attach an adapter to each line connection; connect the low and high side service hoses to these adapters).

(*a*)

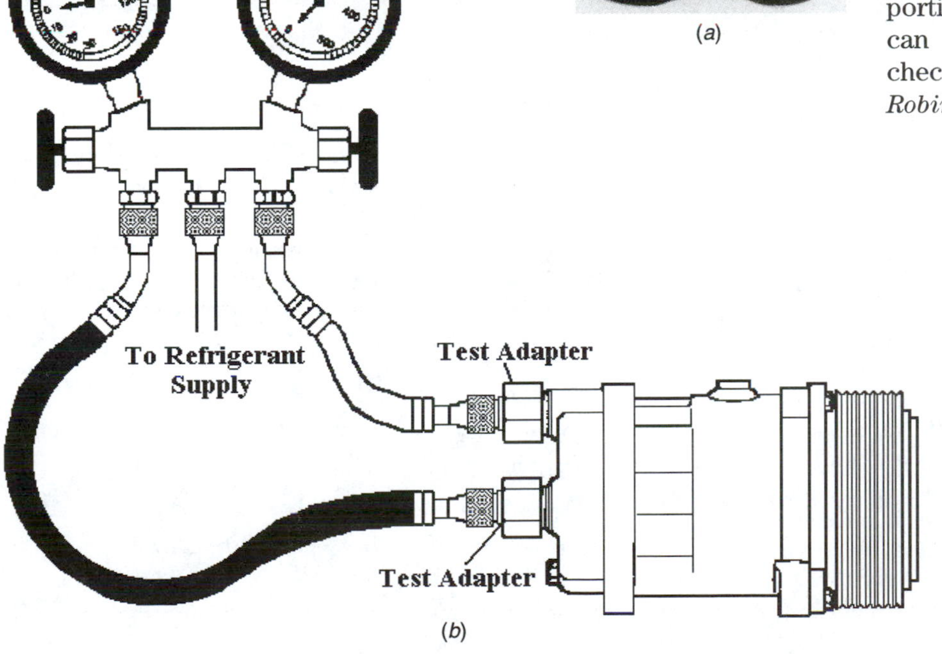

(*b*)

FIGURE 10–65 Special adapters (*a*) are available to allow connection of a gauge set to a compressor (*b*) or portions of a system. The compressor can then be pressure or vacuum checked for leaks. (a *is courtesy of Robinair, SPX Corporation*)

3. Connect the center service hose to a vacuum pump, open both manifold valves, and operate the vacuum pump until there is an almost 30" Hg vacuum for 2 minutes.

4. Shut off all hand valves, wait 5 minutes, and recheck the vacuum. If it has not lost vacuum, the component is good. If it has lost vacuum or if the vacuum will not go this deep, check the line fittings and pump. If they are okay, a leak is indicated.

CHAPTER QUIZ

These questions help you study this chapter. Enter the proper word(s) in the blanks to complete each statement.

1. An A/C gauge set contains a(n) __________ __________ and a(n) __________ __________ pressure gauge, at least two hand valves, and __________ __________.
2. An A/C service unit will include __________ __________, a(n) __________ __________, and a(n) __________ __________.
3. An A/C service unit can also include equipment to __________ and __________ refrigerant.
4. A low side, R-12 service port is the same as a 1/4-inch __________ __________.
5. A high side, R-134a service port is __________ than the low side port, and both ports use a(n) __________-__________ style coupler.
6. The low side gauge and hose are colored __________, and the high side gauge and hose are colored __________.
7. When the hoses are connected to an R-12 system on a 70°F day, both gauges should read about __________ __________.
8. In an R-134a system with normal operation, the low side pressure should be about __________ psi, and the suction line should be __________.
9. In an R-134a system with normal operation, the discharge line should be __________, and the high side pressure will __________ as ambient temperature increases.
10. Bubbles are expected to appear in the sight glass during __________ __________.
11. High relative humidity can cause __________ high side __________.
12. Low pressure on the low and high sides is commonly caused by __________ __________.
13. A faulty compressor can cause __________ low side pressure and __________ high side pressure.
14. The refrigerant charge level can be verified by checking the __________ and __________ of the refrigerant at the condenser outlet.
15. The most popular refrigerant leak detectors used today are __________ __________ and __________ __________.
16. A system should have a pressure of at least __________ __________ when checking for a leak.
17. The leak detector probe should be moved along about __________ __________ from the location being checked at a maximum rate of __________ inch(es) per second.
18. A fluorescent trace will show up better if you wear __________ __________.

REVIEW QUESTIONS

These questions allow you to check what you have learned. Select the answer that correctly completes each statement.

1. Technician A says that you should begin A/C problem diagnosis with a thorough HVAC system inspection. Technician B says that from the inspection it can be determined what type of system the vehicle has. Who is correct?
   - **a.** A only
   - **b.** B only
   - **c.** Both A and B
   - **d.** Neither A nor B
2. Technician A says that oily residue on a fitting or hose can indicate a refrigerant leak. Technician B says that this residue will not affect a leak detector. Who is correct?
   - **a.** A only
   - **b.** B only
   - **c.** Both A and B
   - **d.** Neither A nor B
3. When a system is operating correctly, the lines and hoses on the high side should be cool to the touch.
   - **a.** True
   - **b.** False
4. Technician A says that pressure checks are one method of diagnosing A/C problems. Technician B says that trouble trees are used only for mechanical problems. Who is correct?
   - **a.** A only
   - **b.** B only
   - **c.** Both A and B
   - **d.** Neither A nor B
5. While discussing a manifold gauge set, technician A says that the high side service hose is blue. Technician B says that the low side pressure gauge goes up to 500 psi. Who is correct?
   - **a.** A only
   - **b.** B only
   - **c.** Both A and B
   - **d.** Neither A nor B

6. Technician A says that the manifold gauge set for R-12 can be used on all systems. Technician B says that R-134a systems use service ports with metric flare threads. Who is correct?
   a. A only
   b. B only
   c. Both A and B
   d. Neither A nor B
7. On a system that uses service valves, the valve is kept in the __________ seated position for normal operation.
   a. front
   b. back
   c. mid
   d. None of these
8. An R-12 system is shut off, and the ambient temperature is 70°F. The evaporator pressure should be about
   a. 50 psi.
   b. 70 psi.
   c. 83 psi.
   d. 100 psi.
9. In most R-12 systems that use an STV, the normal low side pressure should be about
   a. 15 to 20 psi.
   b. 15 to 30 psi.
   c. 28 to 30 psi.
   d. over 35 psi.
10. High side pressure is affected by
   a. ambient temperature.
   b. fan clutch operation.
   c. refrigerant charge level.
   d. All of these
11. A faulty A/C system has a high low side pressure and a high high side pressure. Technician A says that the system is probably starved. Technician B says that the system has a bad thermostatic switch. Who is correct?
   a. A only
   b. B only
   c. Both A and B
   d. Neither A nor B
12. __________ can be used to locate refrigerant leaks.
   a. Soap and water
   b. Flame-type testers
   c. Electronic testers
   d. Any of these
13. The engine should be running while the technician checks for leaks.
   a. True
   b. False
14. When troubleshooting a vacuum control circuit, a technician uses a vacuum
   a. pump.
   b. gauge.
   c. diagram.
   d. All of these

Heating and Air Management Systems Inspection and Diagnosis

LEARNING OBJECTIVES

After completing this chapter, you should be able to:

- Inspect a heater system to determine whether there is proper coolant flow through the core.
- Inspect a heater system to determine whether there are any coolant leaks.
- Check the operation of mechanical and vacuum controls and determine whether there are any faults.
- Complete the ASE tasks related to heater system diagnosis.
- Complete the ASE tasks related to air management system diagnosis.

TERMS TO LEARN

air bleed
cable adjustment
corrosion
electrolysis
erosion
technical service bulletin (TSB)
temperature cycling
quick-connect coupling

11.1 INTRODUCTION

Problems in the heater system or air management system can show up with improper operation of the components or during the HVAC system inspection (see Section 9.2). Further tests can be done to determine the exact causes of these problems.

11.2 HEATER INSPECTION

Most heater complaints are of **insufficient heat, wet carpet**, **excessive window fogging**, or **air discharge from the wrong location**. Occasionally a few other problems can occur; most of these problems are listed in Figure 11–1. The most commonly encountered problems are not enough heat output and leaks. Not enough heat output is usually caused by either poor coolant circulation through the heater core or coolant that is not hot enough.

With copper or brass heater cores, leaks are often the result of **cycling;** this is the expansion and contraction of the core that takes place between a cold start with zero coolant pressure and operating temperatures and pressures. The most common causes of failure with

PROBLEM	CAUSE	CURE
Insufficient heat	Plugged heater core Closed water valve Kinked heater hose Faulty blend door position Faulty heater core seals Low coolant level Open engine thermostat	Replace heater core Repair or adjust valve control Repair or replace valve Realign or replace hose Repair or adjust door control R&R core seals Correct coolant level & repair leak Replace thermostat
Excess heat	Faulty blend door position Open water valve	Repair or adjust door control Repair or adjust valve control Repair or replace valve
Wet carpet/window fogging	Leaky core Hose leaks	Replace heater core Repair hose connections
Noisy operation	Air trapped in core	Bleed air out

FIGURE 11–1 Most heater system problems can be placed in one or more of these categories.

aluminum heater cores are **corrosion** and **erosion**. Corrosion is caused by chemical action and electrolysis. The chemicals eat away the metals; **electrolysis** speeds up this chemical change if the coolant becomes acidic. Erosion is wear from the coolant flowing past the metal until the metal is too thin to function properly, and erosion increases if the coolant becomes contaminated with abrasive materials. Having clean coolant with an effective inhibitor package is extremely important in systems with aluminum components; coolant maintenance is covered more completely in Chapters 15 and 16.

✓REAL WORLD FIX

The heater core in the 1994 Pontiac Bonneville (109,000 miles) was leaking, and the core had been replaced four different times at another shop. The cooling system was checked for combustion leaks and none were found. A new, top-quality core was installed, but the vehicle returned in two and a half months with the same problem, a leaking core.

Fix: Following advice from a technical service bulletin (TSB), the technician thoroughly flushed the cooling system to remove abrasive silicates from the old antifreeze. A new heater core was installed, and the system was filled with new antifreeze and water coolant. This repair fixed the problem.

In all modern engines, coolant temperature is controlled by the thermostat, and the thermostat is designed to bring the engine temperature quickly up to the operating point for proper emission control. An engine that is too cold does not operate efficiently, and besides losing power and wasting fuel, the exhaust contains more pollutants than if it is at the proper temperature. Some modern vehicles with **on-board diagnostics (OBD)** turn on the check engine lamp, also called the **malfunction indicator light (MIL)**, if the engine stays cool too long after start-up; this also sets a **diagnostic trouble code (DTC).** Thermostat tests are described in Section 16.3.4.

11.2.1 Coolant Circulation Checks

Most technicians begin checking the insufficient heat complaint by feeling the heater hoses to check for coolant circulation. With the engine at operating temperature, they should both be too hot to hold comfortably. If the hoses seem too cool, feel the upper radiator hose; it should be hot (as hot as the thermostat opening temperature), about 190 to 200°F (88 to 94°C). If the upper hose is too hot to hold, the thermostat is probably working properly. If the heater hoses are not hot, something is blocking circulation through the heater system.

✓SERVICE TIP

An infrared thermometer makes coolant circulation checks easier, safer, and more accurate. Simply point the unit at the heater hoses and read their temperature (Figure 11–2).

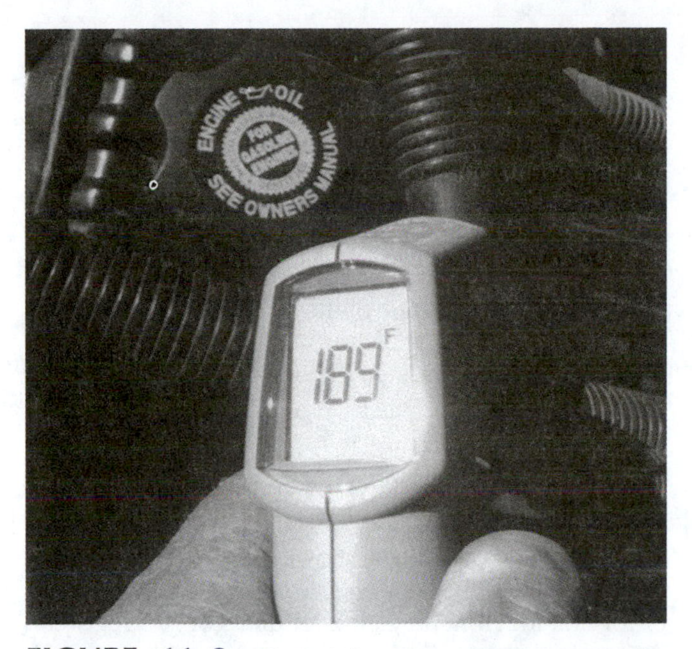

FIGURE 11–2 Heater hose temperature can be checked quickly with an infrared thermometer. This return hose is 189°F, which indicates good flow through the heater core.

✓SERVICE TIP

Some vehicles have a coolant restrictor at one of the heater core connectors. This restrictor orifice can become plugged.

✓SERVICE TIP

Make sure that the system is full of coolant. A low coolant level might not supply a full flow to the heater core.

If the problem is poor circulation, check to see whether there is a water valve in the system; if there is, make sure it is open. Most water valves are vacuum operated and are normally open. A vacuum signal closes the valve. Vacuum control system checks are described in Section 11.4.3.

If you do not find a restricted water valve, you need to disconnect the heater hoses to check for blockage. Use caution and allow the system to cool to comfortable temperatures before removing the radiator cap or disconnecting a hose. The reasons for these cautions are described in Chapter 16.

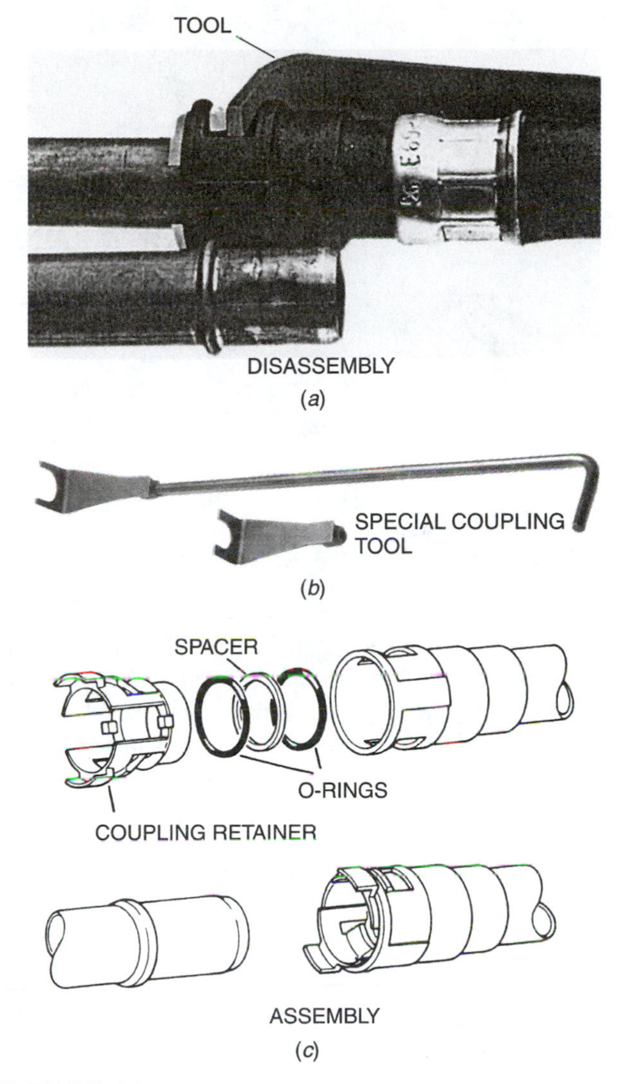

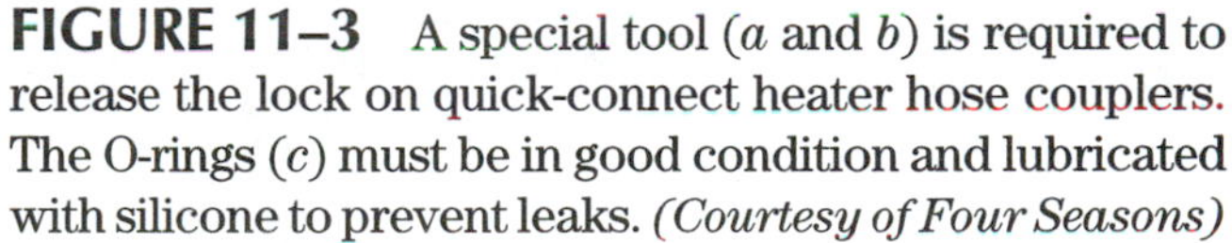

FIGURE 11–3 A special tool (a and b) is required to release the lock on quick-connect heater hose couplers. The O-rings (c) must be in good condition and lubricated with silicone to prevent leaks. *(Courtesy of Four Seasons)*

✓SERVICE TIP

Use care when removing a hose from a heater core connection; do not force hoses on or off a connector. The hose will probably stick to the connector, and the twisting and pulling to try to remove it can easily damage the connector. On systems that use quick-connect couplings, use the proper tool to release the coupling and then pull the coupling straight off the heater core connector (Figure 11–3). Damage to the coupling or connector can necessitate expensive repairs. Proper hose removal methods are described in Section 16.2.5.

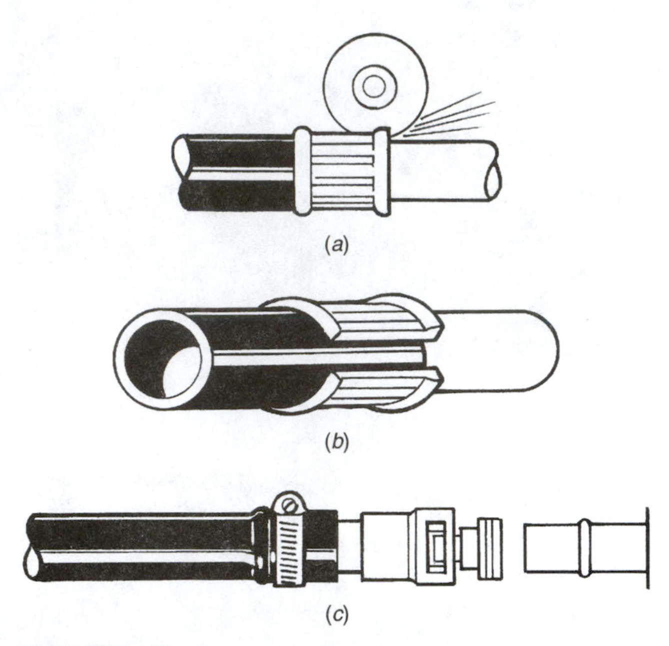

FIGURE 11–4 An alternate repair method for damaged heater hoses using quick-connect couplers is to cut off the ferrule (*a*), remove the damaged hose from the quick-connect nipple (*b*), and clamp a new hose onto the nipple (*c*). *(Courtesy of The Gates Rubber Company)*

✓SERVICE TIP

A less expensive option to repair a leaky heater hose using quick connectors is to remove the damaged hose from the connector by cutting the ferrule. If using an abrasive wheel, wear eye protection. Be careful not to cut too deeply and damage the nipple (Figure 11–4). Carefully remove the ferrule and old hose and clamp a new hose to the connector.

To check heater core coolant circulation, you should:

1. Place a drain pan to catch the escaping coolant and disconnect the engine ends of the heater hoses.
2. Blow into one of the hoses; air along with coolant should come from the other hose.
3. *If you cannot blow through the system*, disconnect the other end of one of the hoses and blow through it. The hose is faulty if you cannot blow through it.
4. Repeat step 3 on the other hose.
5. With the hoses removed, blow into one of the heater core connections; *if you cannot blow through it*, the core is plugged and should be replaced.

✓SERVICE TIP

Heater core flushing is described in Section 16.4, but flushing is not effective if the core is completely plugged. Some sources suggest filling a plugged core with ammonia, but this is not recommended because ammonia can eat away some heater cores very rapidly. The best repair is to replace the core, and many heater cores are mounted inside of the evaporator–heater case (Figure 11–5). Following the manufacturer's replacement procedure is strongly recommended.

11.2.2 Heater Core Leak Check

A leaky heater core usually shows up as coolant dripping from the A/C–heater case and a wet carpet or an oily mist or fog coming from the defroster ducts or A/C registers. This leak can be confirmed by pressure loss during a system pressure test.

✓SERVICE TIP

It is possible for the leak to be located at the heater hose connection that allows coolant to run along the connector and into the case. For this reason, and also because heater core replacement is usually difficult and expensive, it is recommended that an additional check be made to confirm that the core is bad. This test can be made with the core still in place or on the bench. If bench testing, the core can be placed under water and the air bubbles from a leak will show the exact location of the leak.

To test a heater core for leaks, you should:

1. Partially drain the coolant and disconnect the hoses from the heater core.
2. Plug one of the connections and connect a vacuum pump or a hand pressure tester to the other connector (Figure 11–6).
3. Depending on the tester, pump a 28″ Hg vacuum or a 30 psi pressure into the core and observe the gauge. If the core holds either the vacuum or pressure for at least 20 seconds (some sources recommend 3 minutes), it is good. If the vacuum or pressure does not hold, the core should be replaced.

11.2.3 Bleeding Air from Core

As a cooling system is filled, air can get trapped in some places inside the system. Some systems are equipped

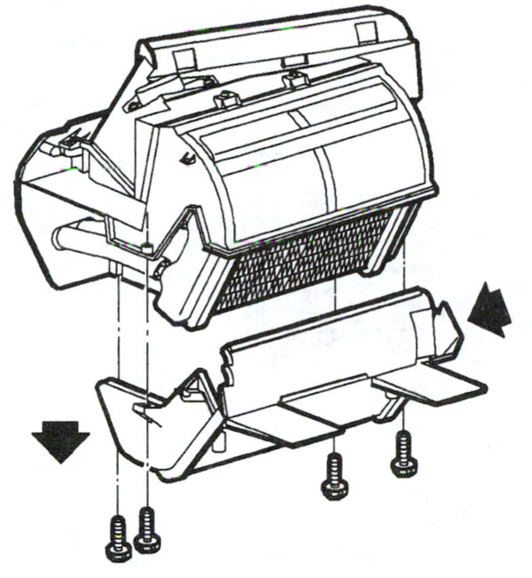

(a)

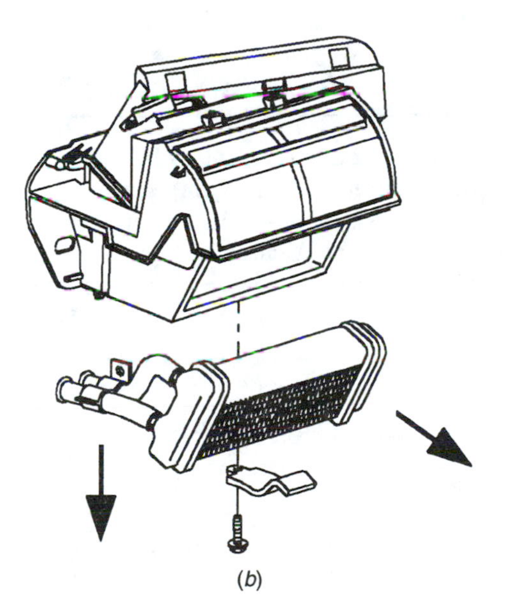

(b)

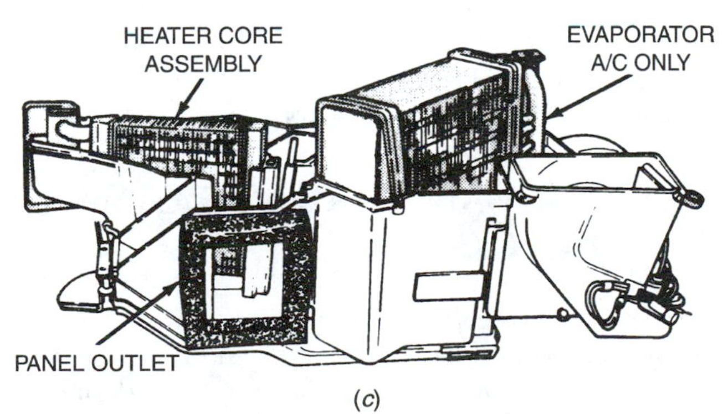

(c)

FIGURE 11–5 Some heater cores can be removed and replaced with the case still mounted (*a* and *b*). Many require that the assembly be removed; then the core is removed and replaced from the assembly (*c*). *(*a *and* b *are courtesy of Saturn Corporation;* c *is courtesy of DaimlerChrysler Corporation)*

FIGURE 11–6 One heater core connector is plugged as the vacuum tester pumps the core into a vacuum to test for leaks. A pressure tester can also be used in the same way. *(Courtesy of Stant Manufacturing)*

with air bleed valves that are opened during the filling process to let the air escape. The manufacturer's coolant fill process should be consulted.

✓SERVICE TIP

After filling a cooling system, it is a good practice to run the engine until the thermostat opens and add additional coolant as it becomes necessary. Thermostat opening is indicated by the upper radiator hose becoming warm. During the waiting period, turn on the heater to full heat or defrost, which should open any water valves, and feel for a warm air discharge, which indicates that the core is full of coolant. If the air does not warm up, bleed the air from the core by loosening the uppermost heater hose at the core. A "burp" of air indicates the release of an air lock. When coolant starts coming from the loosened hose, replace the hose and retighten the connection (Figure 11–7).

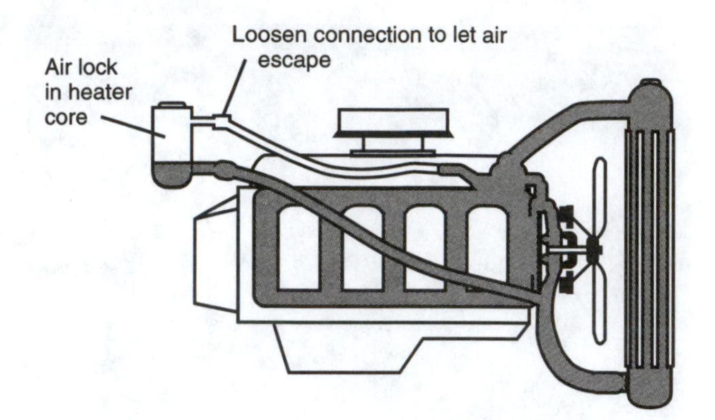

FIGURE 11–7 In many systems the heater core or its hoses are above the engine's water jackets. Loosening the upper heater hose allows the air lock to leave and coolant to enter the core.

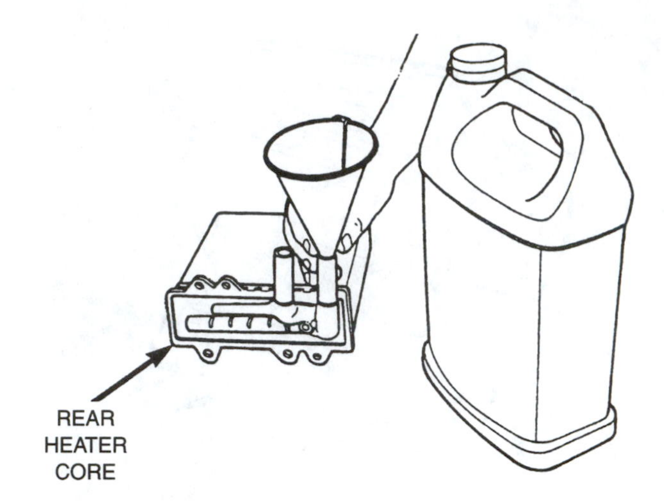

FIGURE 11–8 This manufacturer recommends filling this rear-mounted heater core with coolant before installing it in the vehicle. *(Courtesy of DaimlerChrysler Corporation)*

✓SERVICE TIP

Some technicians prefer to cut the heater hose and install a flushing tee when changing coolant. The tee fitting allows the attachment of a garden hose to the system for flushing, and the cap on the tee allows complete filling of the system. Other technicians dislike the tee because it alters the system and can cause problems.

When replacing a rear-mounted heater core, the length of the hoses and the placement of the core make it difficult to bleed the air out of the core and lines.Some heater cores are mounted in a location that is higher than the radiator cap. If the connections are at the top of the core, which is the normal location, it is a good practice to fill the core with coolant before installing it (Figure 11–8).

✓REAL WORLD FIX

The 1989 Dodge Caravan has no heat from the front heater with good heat output from the rear unit. If the hose to the rear unit is clamped off, there is good heat output from the front unit.

Fix: It was discovered that a restrictor should be located in the hose to the rear unit. Installation of a restrictor fixed this problem.

✓SERVICE TIP

Some cooling system service machines evacuate all the air out of a cooling system using a vacuum pump. A check is then made to see if the system is leak-free and will hold the vacuum. A valve is then turned so coolant is pulled in to fill the cooling system. The entire procedure takes about 5 or 6 minutes. (See Section 16.2.4.1.)

11.3 HVAC AIR FILTER REPLACEMENT

In many vehicles filter replacement is a simple matter of opening the access, and removing and replacing the filter element. With some vehicles, parts such as a portion of the instrument panel or the center console must be removed (Figure 11–9). The exact procedure is described in the service information.

11.4 AIR MANAGEMENT SYSTEM CHECKS

Most air management system problems are related to the blower motor operation and the air doors not moving to the proper position. The basic styles of door control are as follows:

- Manual, using levers and mechanical cables
- Vacuum switches, hoses, and motors
- Electronic actuators and ATC or SATC using electric solenoid–operated vacuum motors or electronic actuators

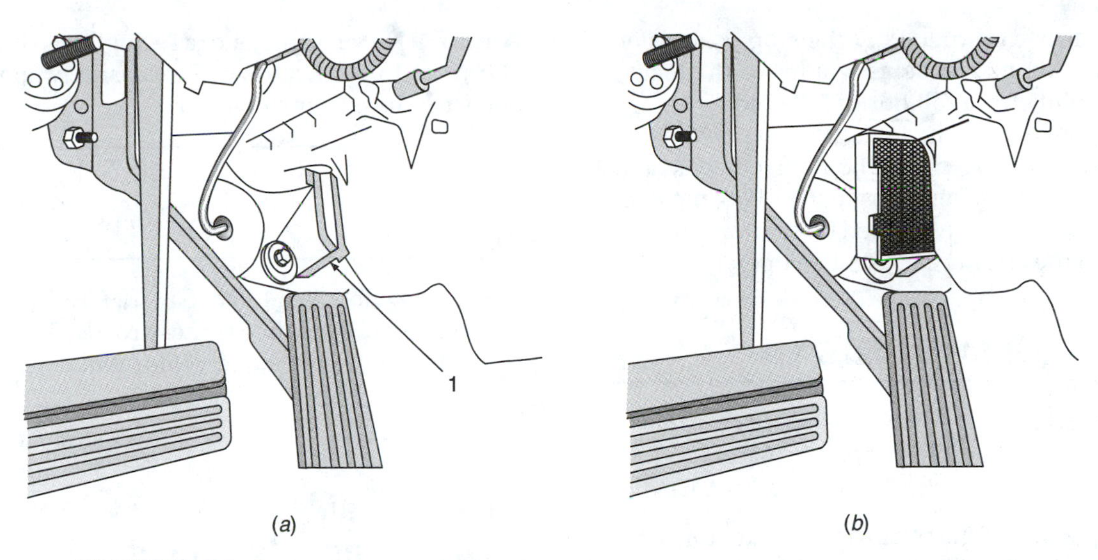

FIGURE 11–9 This HVAC cabin filter is removed by first removing the filter access door (a) and then pulling the three filters out (b). For replacement, the procedure is then reversed making that the filters are properly aligned. *(Reprinted with permission of General Motors Corporation)*

Some systems use a combination of cable, vacuum, and electronic actuators.

The biggest obstacle that a technician normally encounters when troubleshooting the air management system is access to the components. The under-dash area is very cramped and congested, often filled with wires, braces, and ductwork.

✓REAL WORLD FIX

The 1988 Buick has poor air flow coming from the ducts. The blower sounds normal at all speeds, but the air flow is not correct for the different speeds. The vacuum servos all seem to be working properly, and the temperature cable and its door are moving.

Fix: Inspection showed that the evaporator core was covered with oily debris. Replacement of the leaking evaporator core fixed this problem.

✓SERVICE TIP

Caution should be exercised when pulling, pushing, or probing wires to prevent the unexpected. At least one case of a technician accidentally triggering the air bags has been reported. On many cars, the control head can be partially disconnected and moved out far enough from the instrument panel to get to the cables or vacuum selector valve and hose connections behind it (Figure 11–10).

FIGURE 11–10 After removing the fascia panel and two retaining screws, this control head was pulled outward to gain access to the mechanical, electrical, and vacuum connections.

Electric or electronic circuit checks for blower motor, switches, solenoids, and electronic actuators are described in Chapter 12.

11.4.1 Blend Door Operation

If the temperature-blend air door does not close off completely to the full cool position, some heated air will mix in the plenum and the air discharge will be warmer than it should be. With older, mechanical cable-operated doors, it is fairly easy to look at the stop position or to listen for the "thunk" noise as you move the temperature lever rapidly from full hot to full cold.

With many modern, electric actuator-operated doors, we can no longer feel the door movement, and many of them operate almost silently. Also, most modern under-dash areas are so filled up that we cannot see or reach into the door area to feel its operation.

✓REAL WORLD FIX

The 1994 Taurus A/C system was recharged, and the compressor was operating normally with good pressures, but it would not switch from heat to A/C.

Fix: On inspection of the blend door and motor, signs of an earlier repair for the same problem were found. Someone had cut a hole into the housing and manually moved the broken blend door to full heat. At the customer's request, the technician moved the door to full cold.

Author's Note: It is expected that this vehicle will have a temperature control problem when the weather turns cool.

✓SERVICE TIP

A fairly accurate test for whether the blend door is operating properly is to clamp off the heater hose. If the air register discharge temperature drops more than a few degrees (some technicians use 5°F or 3°C as the cutoff point), you know the heated air is mixing in, probably from faulty door operation.

✓REAL WORLD FIX

The 1996 Ford Explorer (58,000 miles) stayed on full heat all the time. The rest of the system seems to operate okay. Following advice and working through the glove compartment, the technician removed the blend door motor and determined that the operating shaft of the blend door was broken.

Fix: This blend door was replaced from under the hood by recovering the refrigerant, removing the evaporator, and removing the blend door from a new plenum assembly. The operating shaft had to be altered slightly in order to install it into the old plenum, but it performed satisfactorily after installation.

Author's Note: Experienced technicians often develop shortcuts to certain jobs. These usually speed the job and save the customer money.

✓SERVICE TIP

It is also a good idea to check OEM TSBs; at least one vehicle model's manufacturing process left plastic material that blocked complete door movement.

11.4.2 Cable Control System Checks and Adjustments

Cable systems are the simplest and most trouble free of all air management systems. The most common problems are (1) the need for adjustment so door travel matches the movement of the control lever and (2) sticky, binding cables that do not move easily. Some cable systems include a slotted bellcrank that operates two doors at the same time, and some of these can be freed up by lubricating them (Figure 11–11). Replacement of a faulty cable varies between vehicle makes and models. The manufacturer's recommended procedure is usually the quickest and easiest method.

Cable adjustment is checked by moving the temperature control lever to the hot and cold ends and either feeling or listening for the temperature-blend door to stop before the lever. The control lever should not reach its stop at either end (Figure 11–12). The adjustment on many early systems was at the attachment point for either the cable or the housing. Some newer systems use a self-adjusting clip that slips along the cable to the correct location. This clip must be preset to the proper position as the cable is connected.

✓SERVICE TIP

Many technicians give priority of the door adjustment relative to their geographic location and climate. In areas that have cold winters, make sure the door "thunks" when moved to full heat; customers in these areas are more sensitive to lack of heat output. In areas that have hot summers, make sure the door "thunks" when moved to full cool; customers in these areas are more sensitive to lack of A/C.

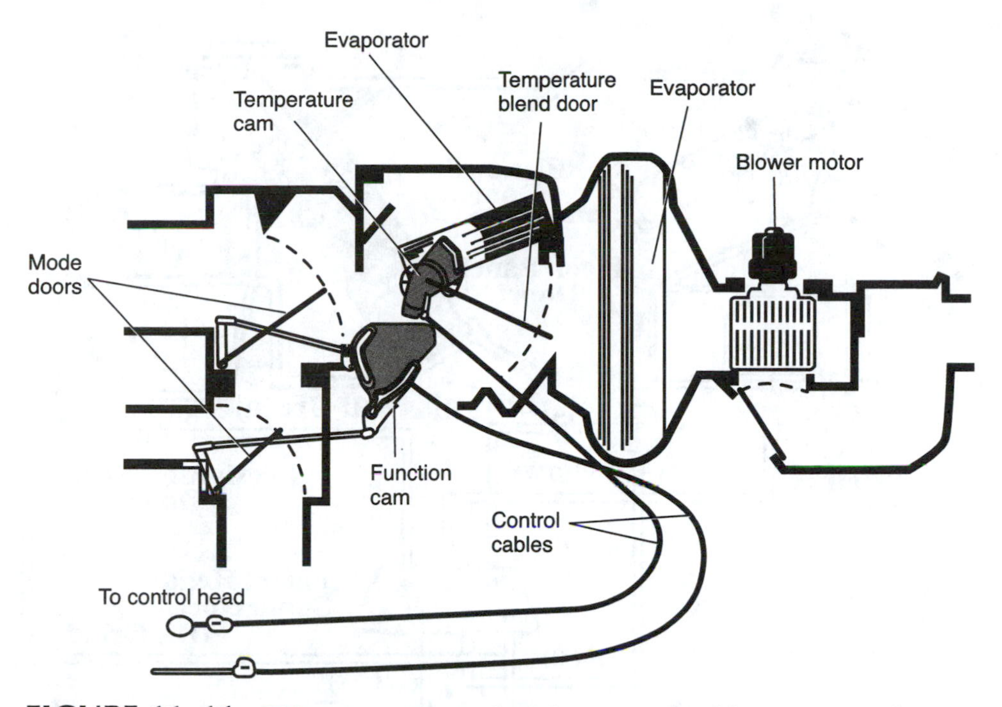

FIGURE 11–11 This system uses a function control cable to operate the two mode doors and a temperature cable to operate the temperature-blend door.

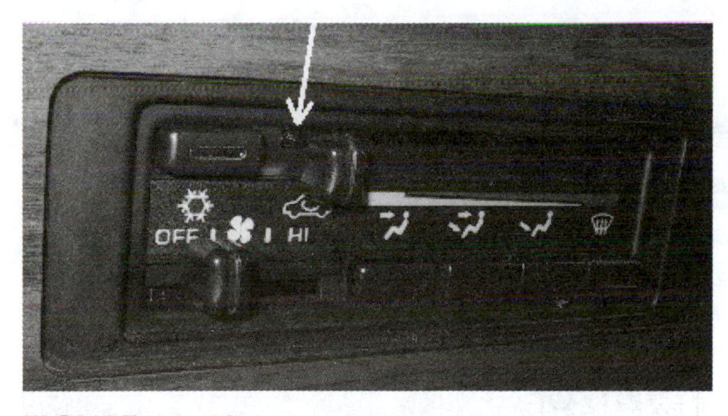

FIGURE 11–12 A small gap (arrow) should be between the temperature lever (full cold position) and the end of its slot if the cable is adjusted properly.

11.4.3 Vacuum Control System Checks

If the problem complaint and the visual inspection show fresh air door or mode door problems, a fault in the vacuum control system is indicated. As mentioned earlier, most older systems use vacuum actuators with a vacuum selector valve at the HVAC control head to control the operation of these doors. Some vehicles use vacuum actuators controlled by electric solenoids.

Most vacuum control systems are quite reliable. The most common problems are pinched or leaking vacuum lines, faulty vacuum motors, faulty selector switches, and vacuum supply problems. The vacuum circuits for most systems start at the engine's intake manifold and end at the various vacuum motors (Figure 11–13). Most manufacturers provide diagrams to show vacuum switching for the various positions of the control valve (Figure 11–14).

Troubleshooting a vacuum system usually begins by ensuring that the manifold vacuum is present at the selector valve. Hose color coding, connections, and any air bleeds or restrictors are shown in vacuum diagrams. The system shown in Figure 11–13 uses a black hose to connect the selector valve or vacuum switch to the engine vacuum source, a small reservoir, check valve, and the manifold connection. This hose should have normal engine vacuum, about 18 to 20″ Hg, whenever the engine is running and for a few minutes after the engine is shut off. The basic test tools are a vacuum gauge and a hand vacuum pump, and these are connected into the system to measure the amount of vacuum (Figure 11–15). Occasionally a leak can be located by holding one end of a 3/8- or 1/2-inch hose

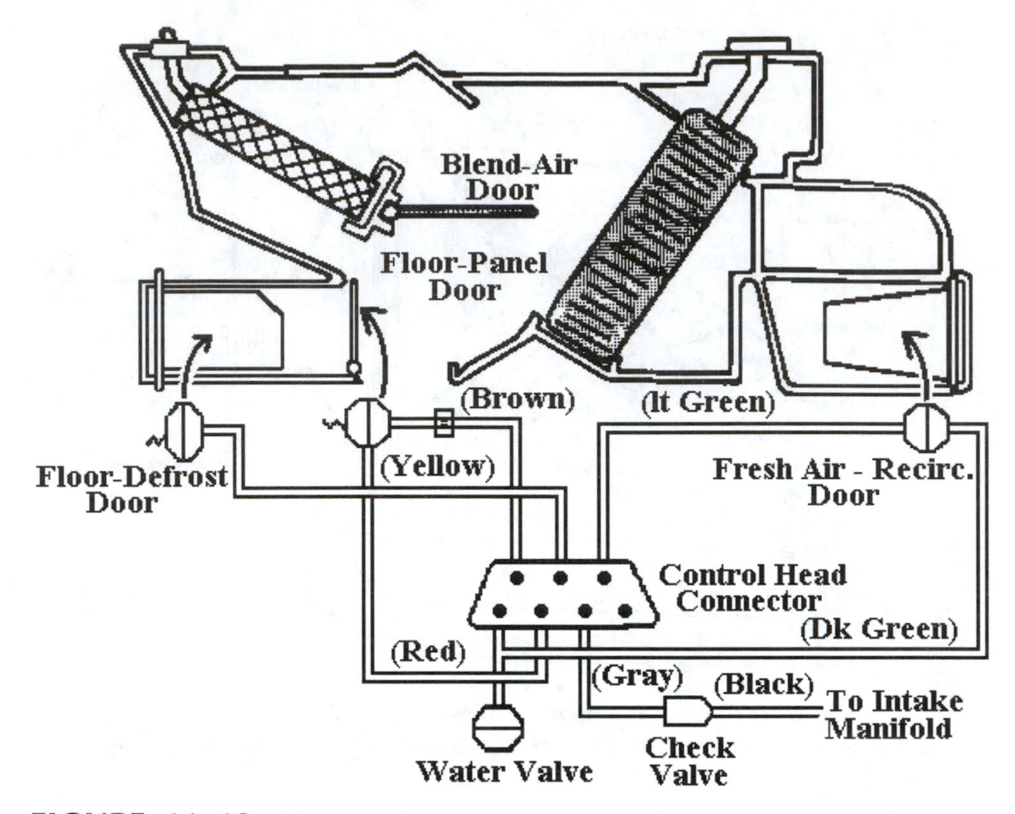

FIGURE 11–13 Most of the components of this vacuum control circuit are under the instrument panel.

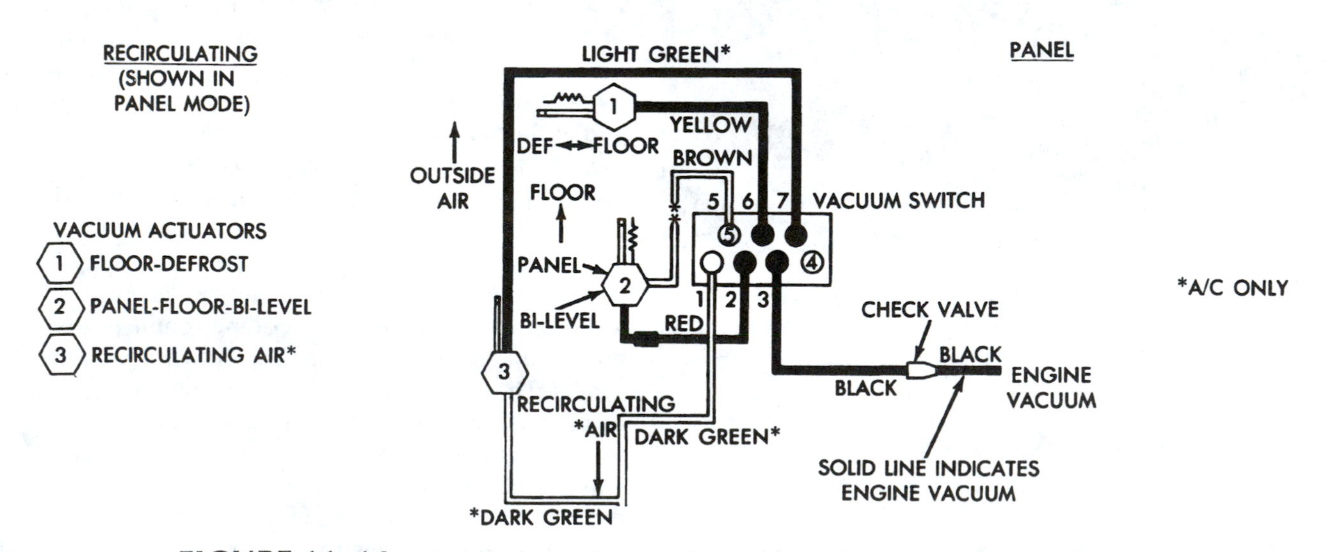

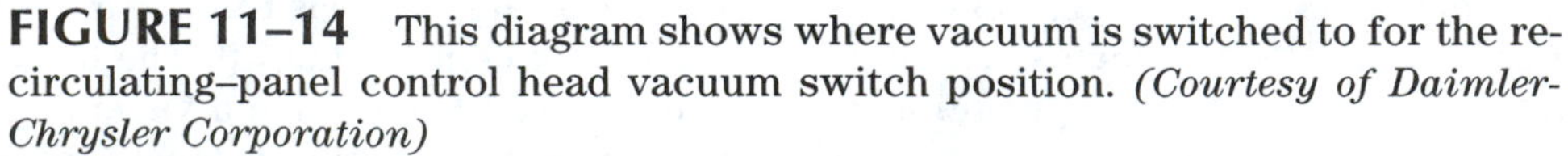

FIGURE 11–14 This diagram shows where vacuum is switched to for the recirculating–panel control head vacuum switch position. *(Courtesy of DaimlerChrysler Corporation)*

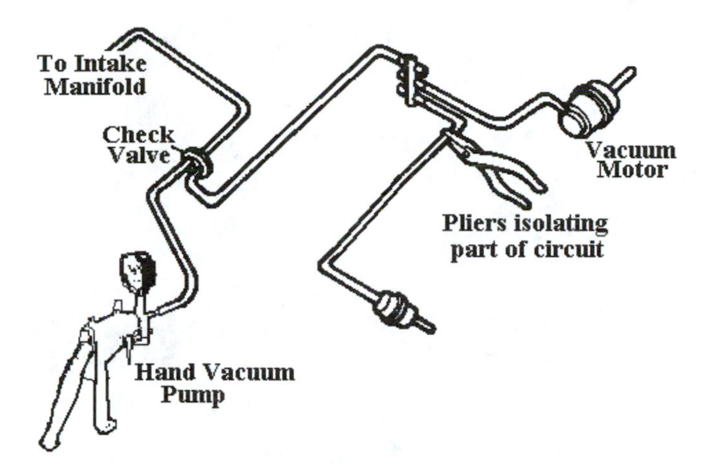

FIGURE 11–15 Vacuum leaks can be located using a hand vacuum pump and gauge while isolating portions of the circuit using plugs or pliers.

FIGURE 11–16 Vacuum leaks can be located by using a hand vacuum pump with a gauge while isolating portions of the circuit.

next to your ear as you pass the other end along the expected leak area (Figure 11–16). A hissing noise indicates the leak.

✓REAL WORLD FIX

The 1992 Pontiac Bonneville (109,000 miles) HVAC system stays on floor/defrost no matter where the controls are set. The technician is not sure how the doors operate, but he does not hear a vacuum leak.

Fix: Following advice from other technicians, the under-hood vacuum lines were inspected, and the plastic vacuum storage canister was found to be cracked. Replacement of this item fixed this problem.

✓SERVICE TIP

A tight-fitting piece of rubber hose can be used to splice plastic lines.

✓SERVICE TIP

In the past, with rubber hoses, it was possible to kink or pinch a line closed to shut off vacuum flow. This practice should be avoided with modern, hard plastic lines because the kink can cause a break and later failure. Hard plastic hoses should be isolated by disconnecting them and plugging the opening. Rubber vacuum hoses can be squeezed closed using pliers.

To troubleshoot a vacuum control system, you should:

1. Gain access to the vacuum selector valve, disconnect the vacuum hose harness from the selector valve, and identify the various hoses.
2. Connect a vacuum gauge to the hose coming from the vacuum source and start the engine. There should be at least 15 inches of vacuum. If the vacuum is less than 15 inches, repeat this check at the source vacuum locations, the intake manifold, check valve, and tank. The problem is located between the areas of adequate vacuum and inadequate vacuum. If the vacuum is 15 inches or greater at the control head, shut off the engine and note the reading. It should hold for a short period (Figure 11–17). Vacuum drop-off indicates a leak at the hose connections, tank, or check valve.
3. Connect the hand vacuum pump to the vacuum hose, either for each of the vacuum actuators or just for the door actuator that is not working right, depending on the nature of the problem. Operate the pump to generate 15 to 20 inches of vacuum and, if possible, observe and listen for operation of the door actuator (Figure 11–18).

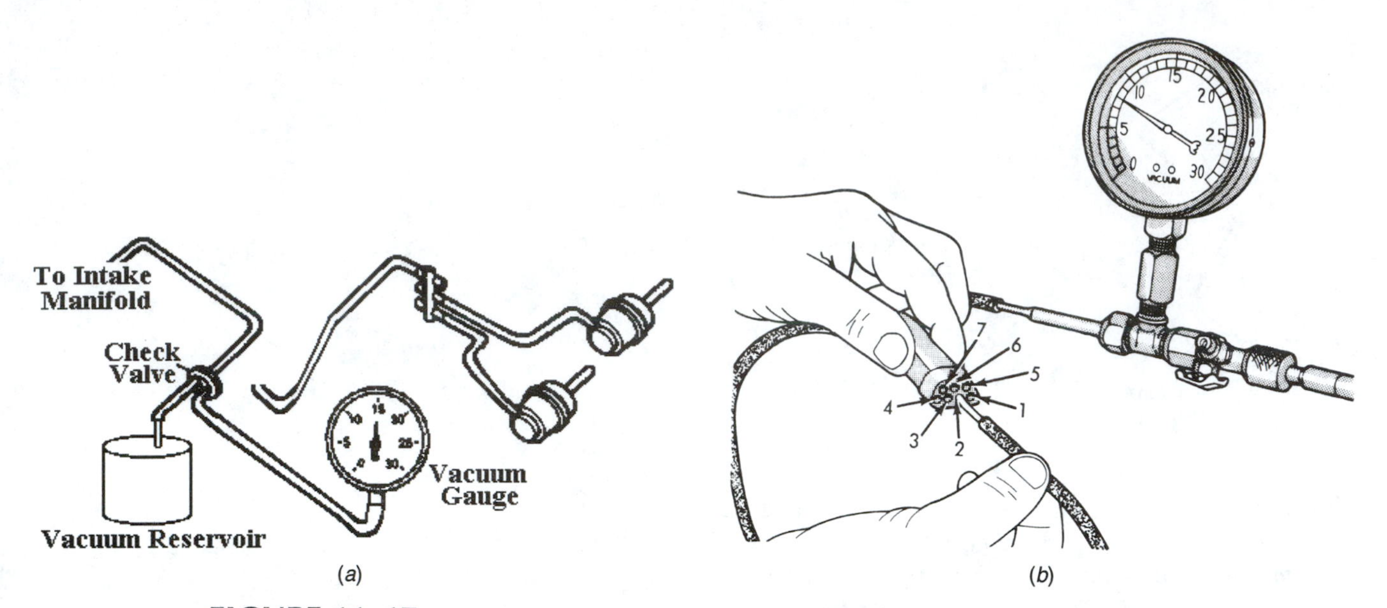

FIGURE 11–17 With the engine running, there should be at least 15 inches of supply vacuum at the control head (*a*). Vacuum can be measured by probing the connector with a vacuum gauge (*b*). A vacuum actuator circuit should also hold the vacuum. *(b is courtesy of DaimlerChrysler Corporation)*

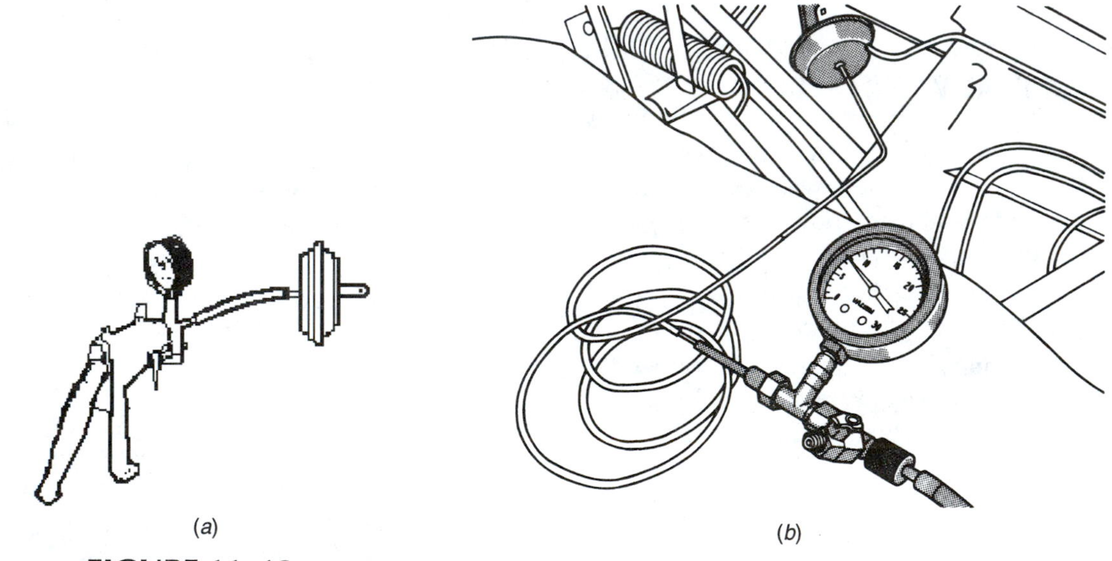

FIGURE 11–18 An actuator can be tested to see whether it operates and holds a vacuum using either a hand or motorized vacuum pump. *(Courtesy of Daimler-Chrysler Corporation)*

4. Reconnect the vacuum harness to the selector valve with the hand vacuum pump connected in place of the source vacuum hose. Move the selector valve to each of its control positions and operate the pump to determine whether a vacuum can be generated. A faulty selector valve is indicated if a circuit operates during the check in step 3 but not in this step.

✓SERVICE TIP

Vacuum leaks can also be detected using an ultrasonic leak detector or smoke machine. The ultrasonic tester (see Figure 10-54), can "hear" a vacuum leak and indicate its location. The smoke machine pumps machine-made smoke into the vacuum circuit so you can see the smoke escaping through the leak.

✓SERVICE TIP

If a vacuum is generated but it bleeds off rapidly, a small leak is indicated. Check the diagram to determine whether a vacuum bleed is contained in that part of the circuit; if so, this might be a normal condition.

If a vacuum cannot be generated, a leak in the hose or actuator is indicated.

If a vacuum is generated but the actuator does not operate, a plugged hose or faulty motor is indicated. In either case, move the vacuum pump directly to the actuator connection and retest.

✓REAL WORLD FIX

While driving the 1993 Dodge Caravan at cruise speeds with the A/C on, the system will start discharging all of the air out of the defroster ducts. It also does this under a hard acceleration. The vacuum storage tank and all of the vacuum lines under the hood have been checked and are okay. The vacuum check valve and control head have been replaced, but this did not help.

Fix: Following advice, the mode door actuator was tested and found to have an internal leak. Replacement of this vacuum motor fixed this problem.

CHAPTER QUIZ

These questions help you study this chapter. Enter the proper word(s) in the blanks to complete each statement.

1. Wet carpet and excessive window fogging are signs of a heater __________ __________.
2. Early heater core failure can be caused by __________ or __________ resulting from __________ coolant.
3. Poor heater output can be caused by __________ engine __________ or __________ through the heater core.
4. When a vehicle is operating, __________ heater hoses should be __________.
5. The hose connections at the heater core of some modern vehicles use __________-__________ connectors that can require a(n) __________ __________ to disconnect them.
6. A heater core can be checked for leakage by testing it to see if will hold either a(n) __________ or __________.
7. A(n) __________ __________ can prevent a heater core from filling with coolant.
8. An improper blend door adjustment can cause reduced __________ or __________ from the HVAC system.
9. When a control cable is replaced, the __________ should be checked.
10. A(n) __________ __________ is indicated if an actuator will not hold a vacuum.

REVIEW QUESTIONS

These questions allow you to check what you have learned. Select the answer that correctly completes each statement.

1. Technician A says that faulty heater operation can be caused by a faulty engine cooling system. Technician B says that a complaint of insufficient heat can be caused by someone not filling a cooling system properly. Who is correct?
   - **a.** A only
   - **b.** B only
   - **c.** Both A and B
   - **d.** Neither A nor B
2. Technician A says that a heater core leak can be the result of the coolant wearing a hole through the core. Technician B says that a core leak can be caused by chemical action eating a hole through the core. Who is correct?
   - **a.** A only
   - **b.** B only
   - **c.** Both A and B
   - **d.** Neither A nor B
3. Technician A says that the heater core inlet hose is connected to the water pump inlet. Technician B says that the heater inlet hose is often bigger than the outlet hose. Who is correct?
   - **a.** A only
   - **b.** B only
   - **c.** Both A and B
   - **d.** Neither A nor B
4. Insufficient heat output from a heater can be caused by a(n)
   - **a.** plugged core.
   - **b.** faulty thermostat.
   - **c.** air lock.
   - **d.** any of these
5. Technician A says that faulty cable adjustments can cause improper heat and A/C operation. Technician B says that a broken vacuum hose can do the same thing. Who is correct?
   - **a.** A only
   - **b.** B only
   - **c.** Both A and B
   - **d.** Neither A nor B
6. A vacuum motor should
   - **a.** move when vacuum is applied to the port.
   - **b.** hold a vacuum.

   Who is correct?
   - **a.** A only
   - **b.** B only
   - **c.** Both A and B
   - **d.** Neither A nor B

HVAC System Electrical or Electronic Controls, Theory, Inspection, and Diagnosis

LEARNING OBJECTIVES

After completing this chapter, you should:

- Understand the relationship between volts, amperes, and ohms.
- Understand the circuits required for electrical units to operate.
- Be aware of the types of circuit problems that may be encountered.
- Be familiar with the procedure used to locate and repair heat and A/C electrical problems.
- Complete the ASE tasks related to electrical or electronics diagnosis and repair.

TERMS TO LEARN

alternating current (AC)
ampere (amp)
belt-lock strategy
body control module (BCM)
conductor
diagnostic trouble code (DTC)
direct current (DC)
engine control module (ECM)
electro-static discharge (ESD)
ground
malfunction indicator light (MIL)
ohm
parallel circuit
pulse width modulation (PWM)
scan tool
schematic
series circuit
volt (V)
watt
wiring diagram

12.1 INTRODUCTION

At one time, the electrical control circuit for the heat and A/C system was rather simple; a few checks with a test light located most problems. Today's systems have evolved into complex electronic circuits with connections to the **engine control module (ECM)** and/or **body control module (BCM).** The HVAC control head on some vehicles features touch panel operation and a digital readout of in-car and outside temperature and can call up **diagnostic trouble codes (DTCs).**

These features have made it necessary for the heat and A/C technician to have a working knowledge of automotive electricity and basic electronics. Instruction in basic automotive electronics is required; this chapter provides a brief review.

12.2 BASIC ELECTRICITY

Three measurable aspects of electricity concern a technician: **volts, amperes (amps),** and **ohms**.

Voltage is **electrical force,** comparable to the pressure and suction generated by the A/C compressor: Voltage forces electrical flow much like pressure forces refrigerant flow. In a car, the source of this voltage is the battery or alternator.

An ampere is the amount of **current** flowing through a circuit. It can be compared with the amount of refrigerant that flows through an A/C circuit.

An ohm is a unit of electrical **resistance.** Ohms are used to control the amount of current flow, much like an OT restricts refrigerant flow. The symbol Ω **(omega)** is commonly used to indicate ohm. Zero, or a very few, ohms of resistance allow a large current flow; a high amount of resistance reduces or stops current flow (Figure 12–1).

Electrical power is measured in **watts;** this is the product of volts multiplied by amps.

Like an A/C system, electricity requires a **complete circuit** from the energy source, through the appliance or component doing the work, and back to the energy source (Figure 12–2). This circuit is made up of the battery or alternator, protection devices, wires, switches, and electrical components—all electrical **conductors.** A modern vehicle has many different electrical circuits, with about a mile (1.6 km) of wire completing them. Each circuit begins at a positive battery or alternator (B+) con-

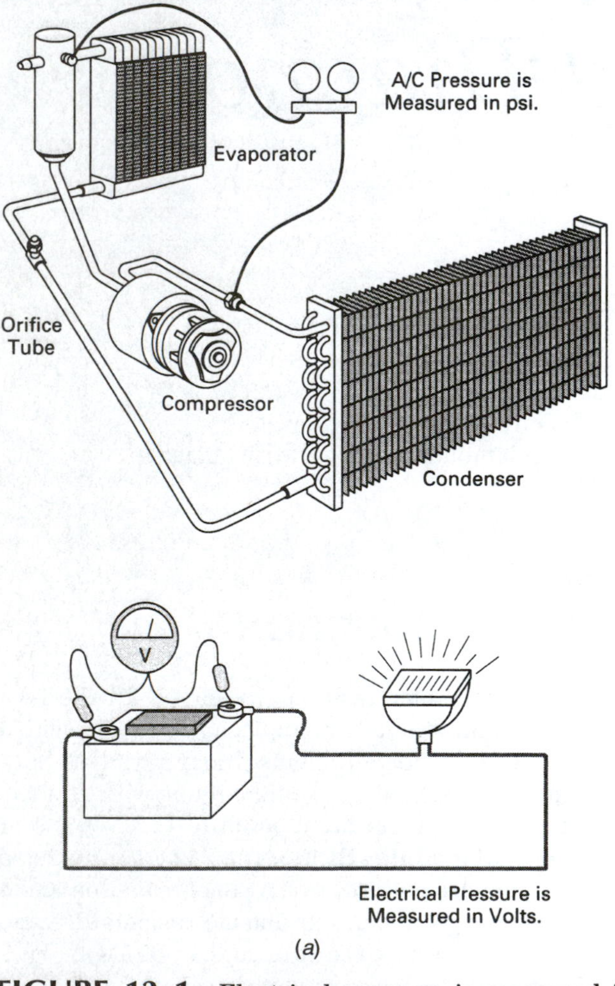

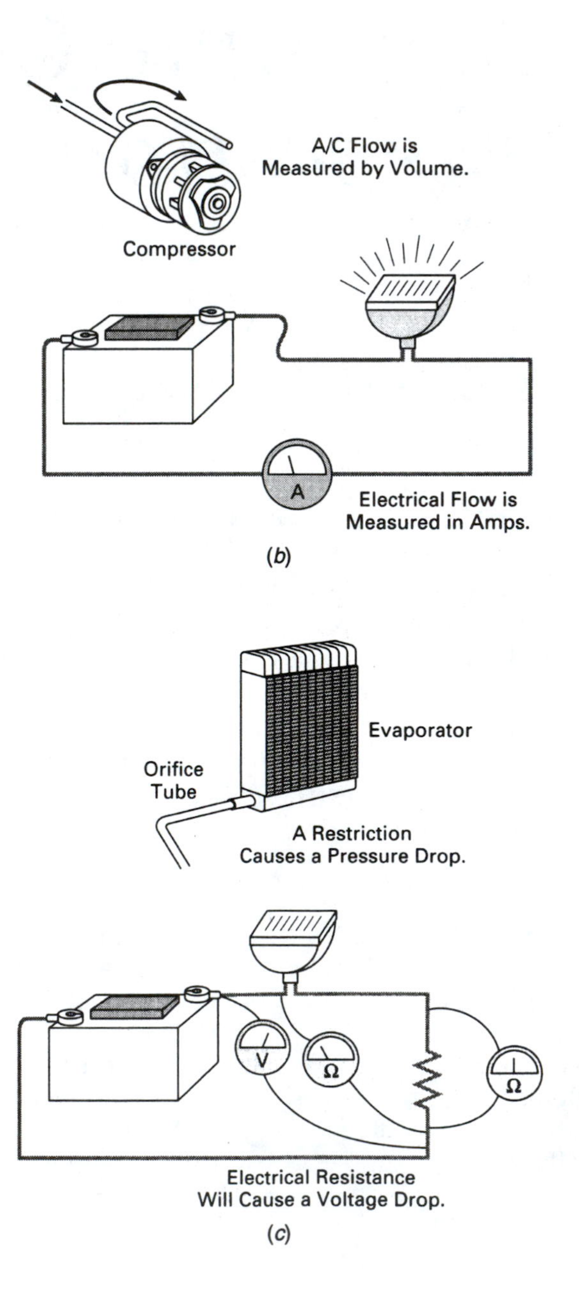

FIGURE 12–1 Electrical pressure is measured in volts; refrigerant pressure is measured in pounds per square inch (*a*). Electrical current flow is measured in amps; refrigerant flow is measured by volume (*b*). Electrical resistance causes a voltage drop; resistance in an A/C system causes a pressure drop (*c*).

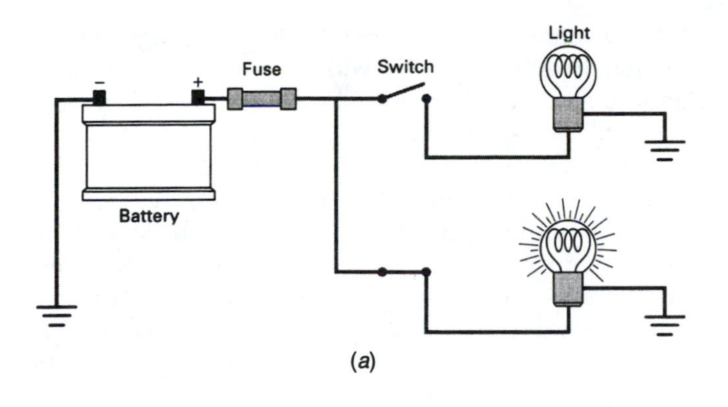

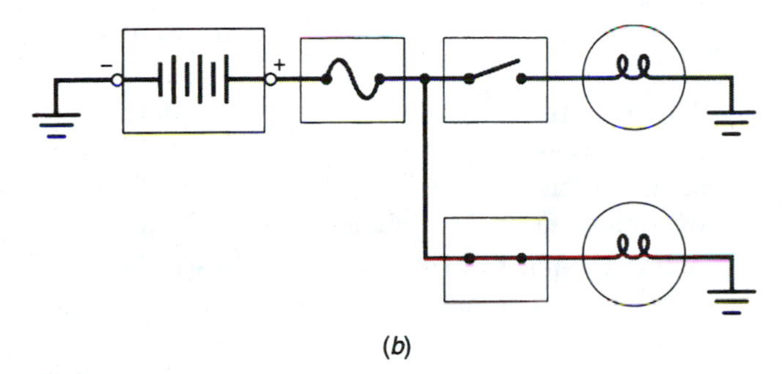

FIGURE 12–2 An electric circuit must be complete for current to flow. Circuits are shown in the form of a diagram (*a*) or schematic (*b*).

nection and ends at a battery or alternator **ground** connection (B−). Many are simple, one-component circuits; some use a **series circuit** with several components connected in a string, and some use a **parallel circuit** with branches that allow current flow through separate paths (Figure 12–3). Vehicle manufacturers provide **wiring diagrams** that are maps of these circuits. These wiring diagrams or **schematics** often use symbols for the components to help simplify the drawings (Figure 12–4).

Most vehicle circuits use **direct current (DC).** DC always travels in one direction; the direction of electron flow is thought to be from negative to positive. Most commercial and household electricity is **alternating current (AC);** AC switches direction many times a second.

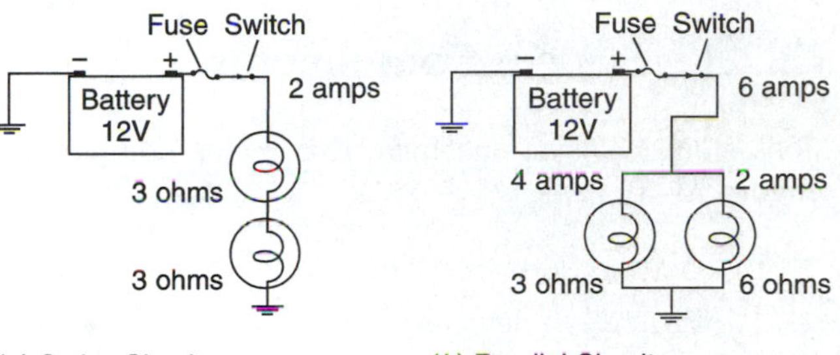

FIGURE 12–3 A series circuit (*a*) and a parallel circuit (*b*).

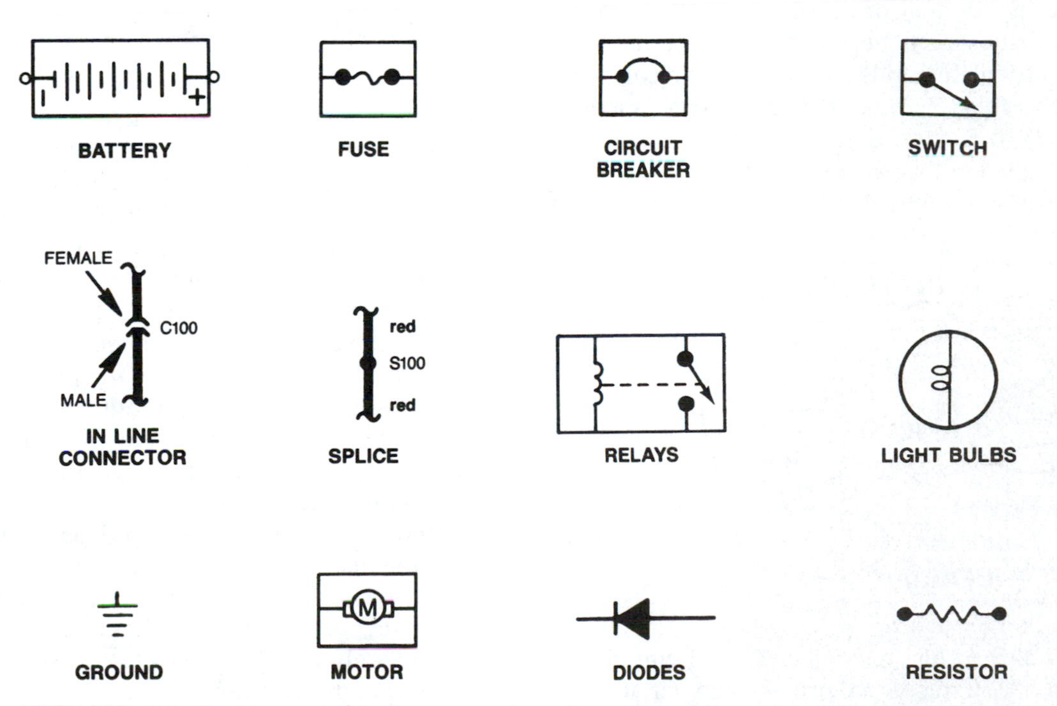

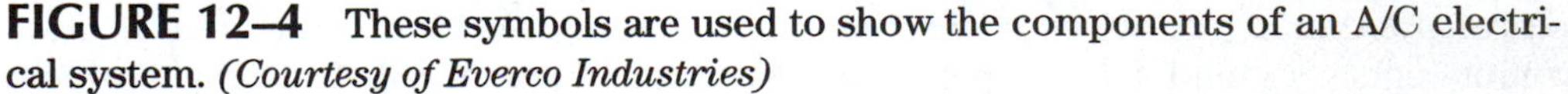

FIGURE 12–4 These symbols are used to show the components of an A/C electrical system. *(Courtesy of Everco Industries)*

Most vehicles use 12-volt circuits, and most household and commercial circuits are either 110 (really 117) or 220 volts.

Many technicians see a strong similarity between electrical circuits and A/C systems or the vacuum control circuits used to control the air ducts. Voltage can be compared with either compressor pressure or the pressure differential in a vacuum motor. These forces produce movements, such as amperage, refrigerant, or vacuum flow (movements created by pressure differences). Electrical circuits and vacuum circuits are both paths of that movement. A switch and a vacuum valve can start or stop flow, an electrical resistor and an OT can restrict or reduce flow, a variable resistor and a TXV can adjust the amount of flow, and a diode and a vacuum check valve allow flow in one direction only.

12.2.1 System Components

Most electrical systems have five major components (Figure 12–5):

1. Power source
2. Protection devices
3. Control devices
4. Connecting wires
5. Output devices

As mentioned, the power source for vehicle systems is either the battery or the alternator. The battery is used primarily when the engine is not running or when the electrical demands exceed the rating of the alternator. When fully charged, battery voltage is a little more than 13 V. If electricity is used without the engine running, the voltage drops. Before vehicles had computers, most circuits would not work properly with a battery voltage of less than 9 V; vehicles using computers require at least 10.5 V. With the engine running, the alternator raises the voltage applied to the circuit to a regulated voltage between 13.6 and 15.6 V.

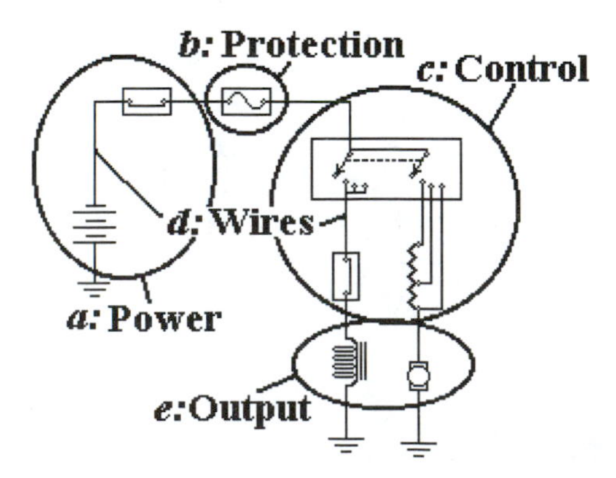

FIGURE 12–5 This A/C electrical circuit has a power source (*a*); fuses to protect the circuit (*b*); switches to control the circuit (*c*); wires connecting each of the components (*d*); and two outputs: the blower motor and the compressor clutch (*e*).

If too much current flows through a circuit, it can cause serious overheating of the wires and increase the likelihood of burnout or fire. Protection devices allow enough current flow through a circuit for normal operation, but if a short or ground circuit occurs, excessive current flow causes the protection device to open the circuit. These protection devices are **fuses, fusible links, PTCs**, or **circuit breakers.** (Figure 12–6)

A **fuse** is a device that melts at a certain current flow. Fuses are designed to be easily replaced if they melt or blow out.

A **fusible link** is a one-time protection device. It is a short piece of wire, about four wire gauge sizes smaller than the wire used in that circuit. A current overload causes the fusible link to burn out.

A **PTC, Positive Temperature Coefficient** is a thermister that acts like a self-resetting circuit breaker. The circuit is opened by the temperature increase as the current reaches the maximum value. The circuit will reclose when the cause of the excess current flow is corrected.

A **circuit breaker** senses current flow. If amperage becomes excessive, a set of contacts cycles open to protect the circuit. Some circuit breakers reclose after cooling down; some must be reclosed mechanically by pushing a button or lever.

The **switch** is a common control unit used to **break (open)** a circuit to stop current flow or to **make (close)** a circuit to allow current flow. Most switches offer no resistance (0 Ω) when they are closed and **infinite resistance** (∞ Ω) when they are open. Some switches are combined with **rheostats** or **variable resistors** so that the amount of resistance can be changed. These switches can be used to control the brightness of instrument panel lights or the speed of a motor. A switch can control a unit directly, like the pressure switch in a compressor clutch circuit, or indirectly through a relay.

A **relay** is essentially an electromagnet and a set of switch contacts. The electromagnet is controlled by another switch and requires only a small current flow. When the magnetic coil is energized by its control circuit, the magnetic pull closes the switch contacts, which, in turn, control a larger current flow. These are called **normally open (NO) relays.** Some relays are **normally closed (NC);** they are opened by the control circuit (Figure 12–7).

The wires used to complete an electrical circuit are normally composed of a copper **conductor** surrounded by a plastic **insulator.** Copper is a good electrical con-

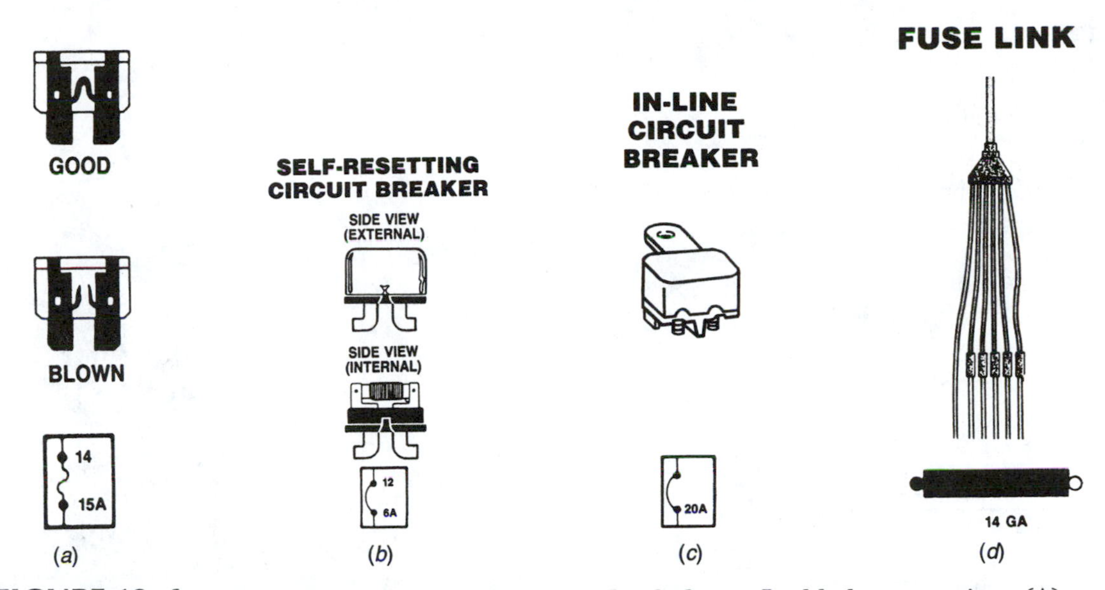

FIGURE 12–6 Circuit protection can be fuses with tubular or flat blade connections (A), a circuit breaker (B and C), or a fusible line (D). *(Courtesy of DaimlerChrysler Corporation.)*

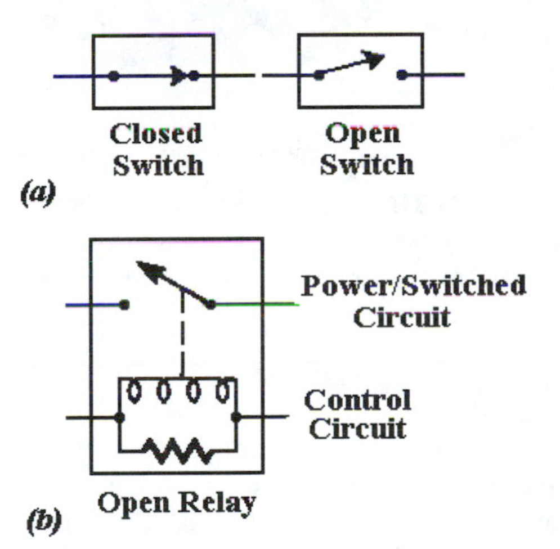

FIGURE 12–7 A simple switch can be normally open or normally closed (*a*). Relays use a magnetic coil to close or open switch contacts (*b*).

ductor and offers a relatively small amount of resistance. The amount of metal in the wire portion determines the **gauge** of the wire and the amount of current it can safely carry. Plastic is a poor conductor, which makes it a very good electrical insulator; this insulation contains the current flow in the wire. Insulation allows two or more wires to be placed next to each other or alongside ground without having the current bypass the desired electrical path (Figure 12–8).

In most vehicles, the metal body, frame, transmission, and engine provide part of the electrical circuit called the **ground circuit** (Figure 12–9). The negative side of the battery and alternator are connected to ground. Each electrical component is also connected to ground, normally the metal case or frame of the component being bolted in place. Some components use a separate ground wire fastened to a ground connection to complete a circuit. At one time, some cars used a positive ground.

WIRE SIZE CONVERSION TABLE			
Metric Size mm²	AWG Size	Metric Size mm²	AWG Size
.22	24	5.0	10
.5	20	8.0	8
.8	18	13.0	6
1.0	16	19.0	4
2.0	14	32.0	2
3.0	12	52.0	0

(*a*)

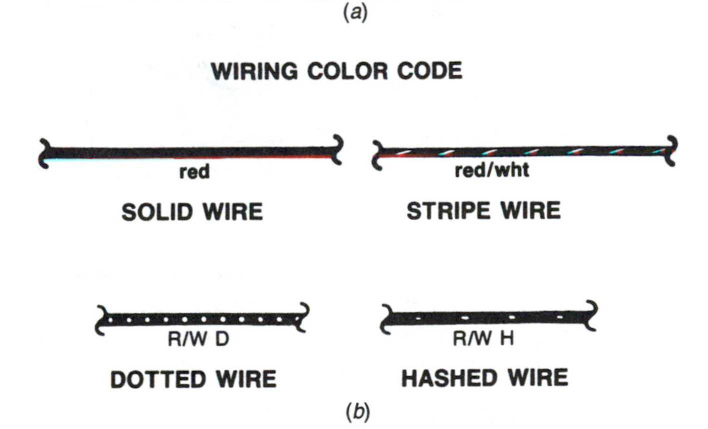

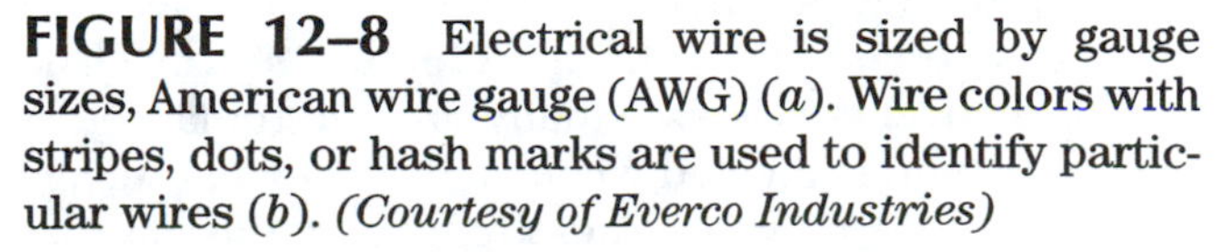

(*b*)

FIGURE 12–8 Electrical wire is sized by gauge sizes, American wire gauge (AWG) (*a*). Wire colors with stripes, dots, or hash marks are used to identify particular wires (*b*). *(Courtesy of Everco Industries)*

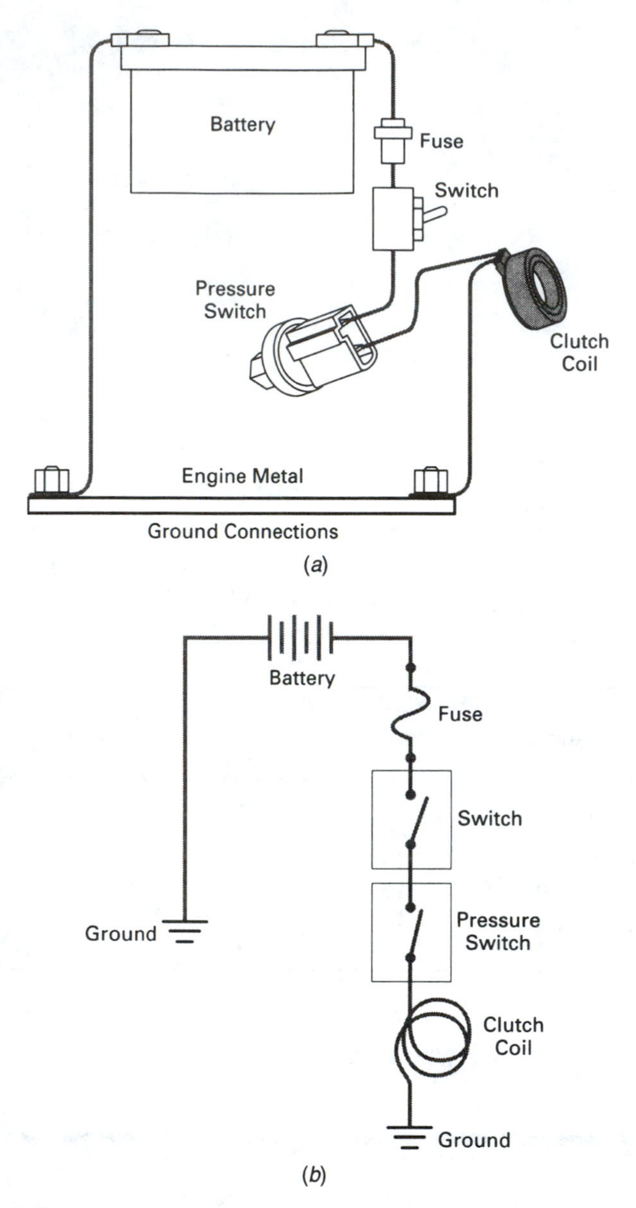

FIGURE 12–9 Insulated wires are used to conduct electricity to the clutch coil. The metal of the engine forms the ground circuit to complete the circuit back to the battery. A circuit is shown in a diagram (*a*) and a schematic (*b*).

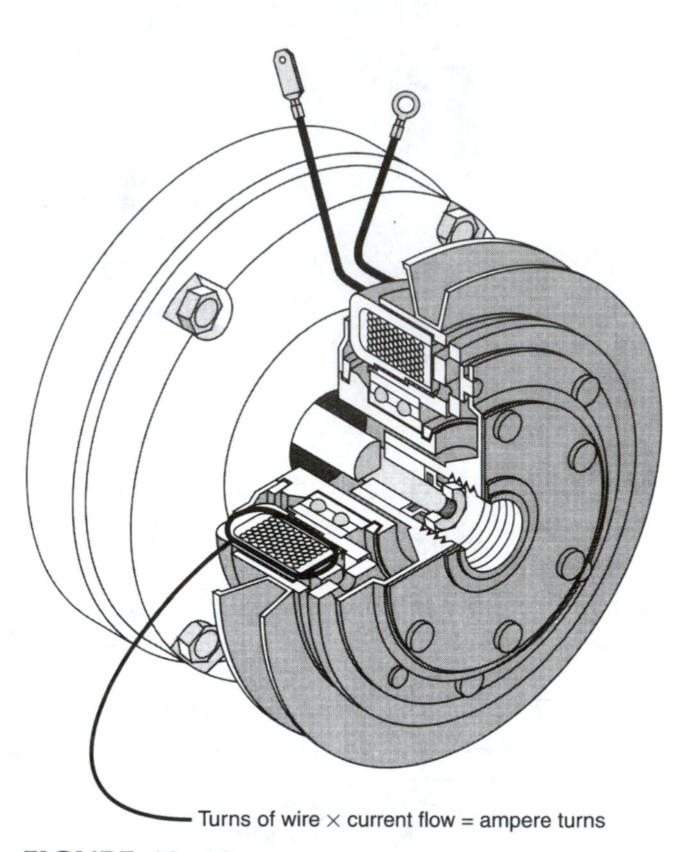

FIGURE 12–10 The strength of a magnetic coil is determined by multiplying the number of wire turns by the current flow. *(Courtesy of Warner Electric)*

Output devices in most HVAC circuits are the **compressor clutch** and the **blower motor.** The compressor clutch is essentially an electromagnet. A magnetic field is created when electric current travels through a coil of wire. The strength of this magnet is determined by the number of turns of wire in the coil and the amount of current, called **ampere-turns.** Increasing either the number of turns or the amperage increases the magnetic pull. Current flow is determined by the resistance in the wire of the coil (Figure 12–10).

When current flows through the wire coil of an electromagnetic device such as a relay or clutch coil, it creates a magnetic field. When the current stops flowing in the coil, the magnetic field collapses, possibly creating problems in computer-controlled circuits. When a magnetic field collapses over a coil of wires, high voltage is induced in the coil. This is the operating principle of an ignition coil or transformer: Whenever the current flow to an A/C clutch or relay is shut off, the magnetic field collapses, inducing voltage in the circuit. This voltage can spike to a very high voltage, high enough to damage solid-state electronic controls. To prevent this damage, a **diode** is included in the circuit. Connected to each end of the coil, a diode allows current to pass in one direction but not the other. During normal operation, the diode does nothing but block unwanted current flow past the coil. When the system is shut off, the diode allows the induced current to bleed off around the coil, eliminating high voltage spikes (Figure 12–11).

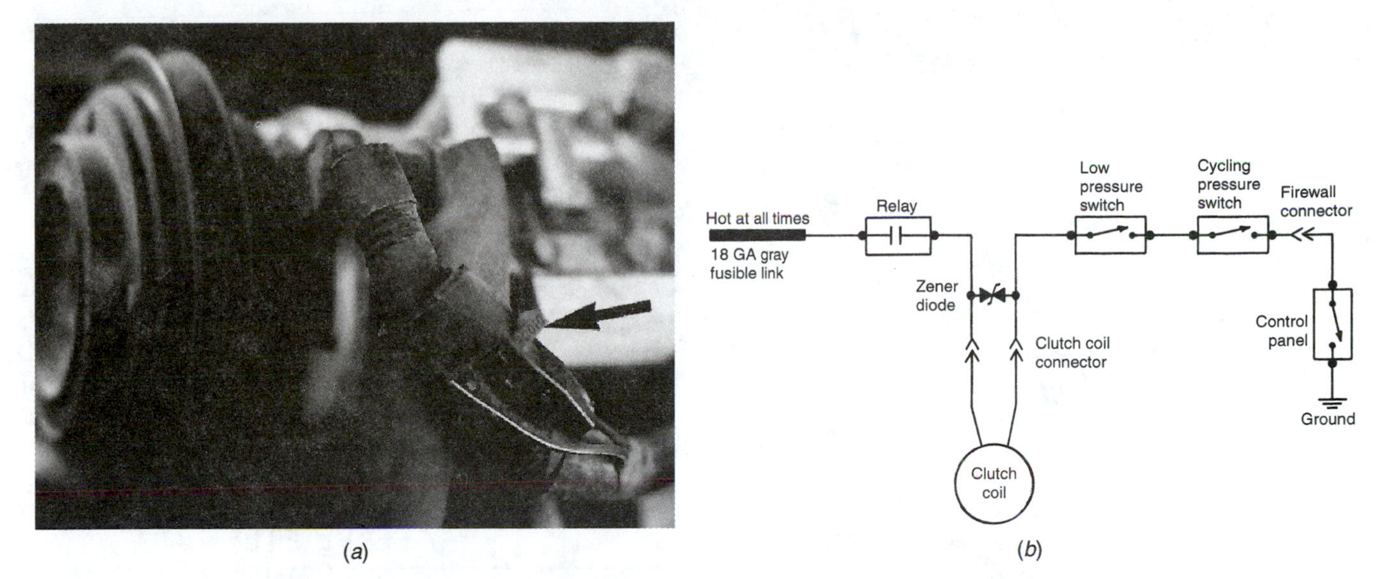

FIGURE 12–11 The zener diode at the clutch coil connector prevents a high voltage spike when the clutch releases. (b *is courtesy of Everco Industries*)

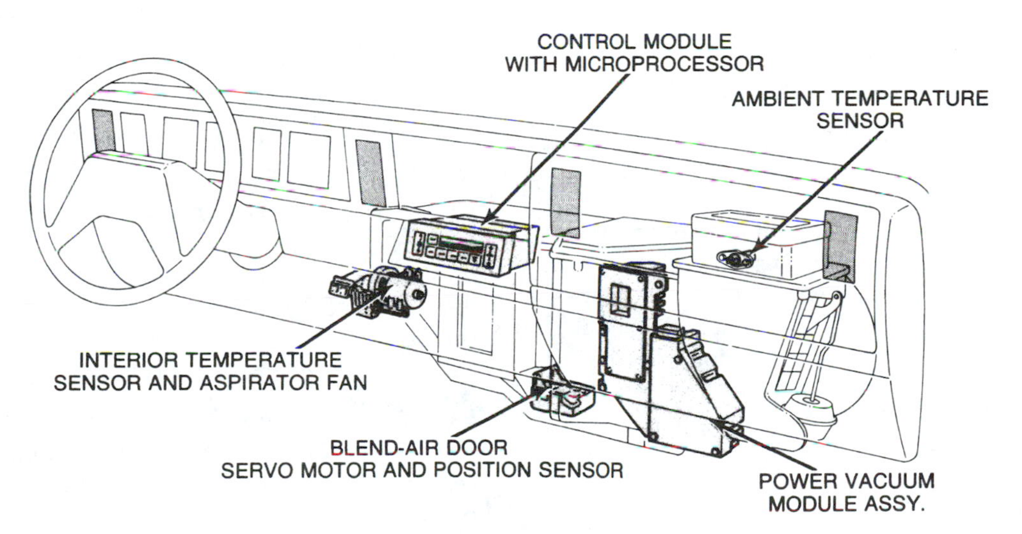

FIGURE 12–12 Most of the control components of an ATC system are solid-state electronic devices—diodes, thermistors, and transistors. *(Courtesy of Daimler-Chrysler Corporation)*

12.3 BASIC ELECTRONICS

Solid-state electronics is the basis of modern computers and automotive control modules. Solid-state electrical devices are units in which nothing moves except electrons. They include diodes, transistors, and integrated microchips. These units are quite reliable because nothing in them wears out. They can, however, be easily damaged by rough handling and vibration, high temperatures, high current flow, and high voltage spikes from the circuit or **electrostatic discharge (ESD).** Electronic devices are usually used for sensing and control circuits.

Automatic temperature control (ATC) lends itself to electronic control circuits (Figure 12–12). These systems can be divided into **sensors,** the **control panel and module,** and **actuators.**

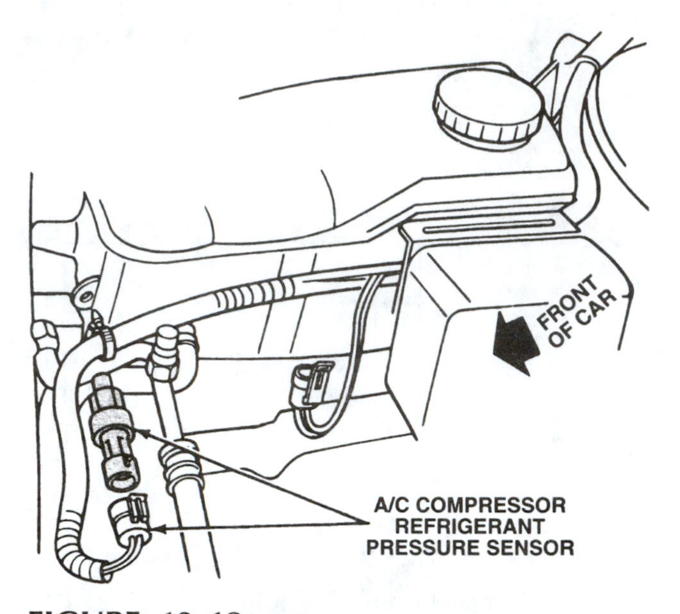

FIGURE 12–13 This pressure sensor is a transducer that changes the high side pressure into an electrical signal for an ECM input. *(Reprinted with permission of General Motors Corporation)*

12.3.1 Sensors

As mentioned in Chapter 8, sensors monitor **ambient temperature, in-car temperature, driver input**, and, maybe, **engine coolant temperature, compressor rpm, compressor temperature and sun load.** In some vehicles, A/C low side and high side switches or pressure transducers are also inputs to the control module. These can be a switch that is made to open or close at certain pressures, a transducer that senses pressure, or a thermistor that senses temperature.

Many vehicles with a single accessory drive belt use a **belt-lock strategy** to prevent the possibility of a seized compressor stopping the drive belt. A compressor rpm sensor generates a signal that lets the ECM know how fast the compressor shaft is turning. This speed is compared to the engine rpm signal, and if there is 30% or greater difference, the ECM will disengage the compressor clutch. Excessive belt slippage or compressor lockup could cause total belt failure that could result in stopping the power steering pump and the engine's water pump and causing an overheated engine.

Temperature sensors are **thermistors,** units that change electrical resistance relative to temperature. One new system uses infrared temperature sensing. A **transducer** produces a variable electrical signal that is relative to pressure, allowing the control module to monitor actual pressure in that part of the system (Figure 12–13). A pressure signal allows the ECM to monitor system operation, determine whether the compressor is operating, and turn on the radiator or condenser fan when necessary.

The control module sends a low-voltage (about 5 V) signal to some sensors; the sensor returns a voltage that is relative to the condition of the system. For example, a particular ambient sensor has a resistance of 7.2 k (7,200) Ω at 32°F and 3.4 k (3,400) Ω at 100°F: Because sensor resistance has more than doubled, the voltage returning from the sensor to the control module will be much lower at 32°F than at 100°F (Figure 12–14).

✓SERVICE TIP

Suspect an RPM sensor problem if the compressor starts but disconnects after a short or very short operation. With a relay or switch circuit fault, the compressor will not start at all.

✓REAL WORLD FIX

Under hard acceleration, the 1992 Mercedes (95,000 miles) A/C compressor shuts off, and it will not restart until the engine is cycled off and then on. This problem does not occur during normal acceleration. The compressor rpm sensor was checked, and it is good. Voltage supply to A/C components is good.

Fix: A closer check of the rpm sensor revealed that it lost connection when the engine lifted due to torque reaction. Repair of this wire connection fixed this problem.

✓REAL WORLD FIX

The 1989 Honda (80,000 miles) compressor was replaced, and the system was recharged. The system operates, but the compressor shuts off after 5 seconds. If the A/C is turned off and then on, the compressor will operate for another 5 seconds.

Fix: Following advice, the technician checked the rpm sensor using an ohmmeter, and this sensor had an open circuit. Replacement of the rpm sensor fixed this problem.

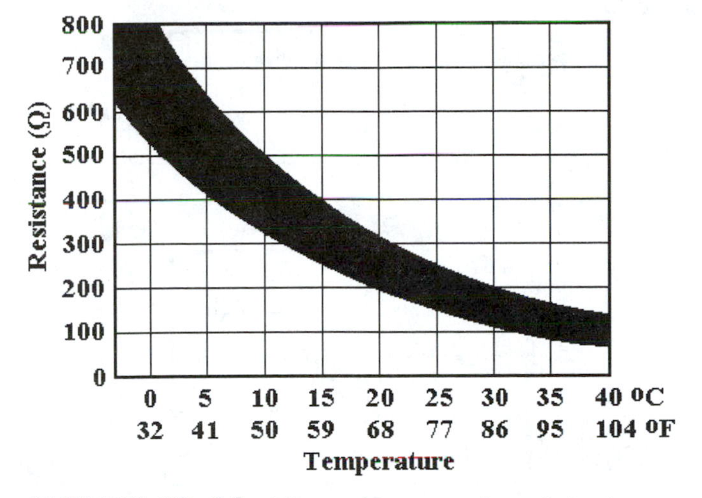

FIGURE 12–14 The resistance value of the in-car sensor or ambient sensor is inversely related to the temperature. As the temperature increases, the resistance decreases.

12.3.2 Control Module

An ECM, BCM, or HVAC control module is programmed to open or close circuits to the actuators based on the values of the various sensors. Although it is unable to handle the electric current for devices such as the compressor clutch or blower motor, the control module can operate relays. These relays in turn control the electric devices. Units that use small current flows, such as a light-emitting diode (LED) or digital display, operate directly from the control module (Figure 12–15).

An ECM is programmed with various strategies to suit the requirements of the particular vehicle. For example, when A/C is requested by the driver of a vehicle with a relatively small engine, the ECM will probably increase the idle rpm. On this same vehicle, the ECM will probably shut off the compressor clutch during wide-open throttle (WOT). On other vehicles, the ECM might shut off the compressor clutch at very high speeds to prevent the compressor from spinning too fast. In some cases, part of the A/C and heat operating strategy is built into the control head; these are called **smart control heads.** In some cases, door operating strategy is built into the door operating motors, and these are called **smart motors.**

With some modern control modules, various calibration values can be reprogrammed by technicians in the field. These calibration values include the high side pressure at which the cooling or condenser fan operates or the coolant temperature at which the cooling fan operates. Improper fan operation can lead to engine overheating, poor performance from the A/C or heating system, or a control module that denies the driver's request for A/C.

Many control modules are programmed to run a test sequence at start-up (when the ignition key is turned on); if improper electrical values are found, the ECM indicates a failure by turning on the **malfunction indicator light (MIL),** often the A/C indicator light at the control head. Some also monitor the system during operation and indicate a failure or stop the compressor if there is a problem. One system, for example, notes the frequency of clutch cycling; remember that short cycling indicates a low refrigerant charge level. If there are too many clutch cycles during a certain time period, the control module will shut off the compressor and set a DTC. A technician can read the DTC, also called an **error code,** by following a certain procedure. Looking up codes in a manual shows where the problem is and what tests are needed to determine its exact cause (Figure 12–16).

Wire connections at the sensors and control module must be clean and tight to prevent any change in value. A small amount of corrosion can produce enough resistance to cause a significant value change in these low-voltage, low-current circuits. Most manufacturers use mechanical-locking, **weather-tight connectors.** Many also use waterproof conductive compound (dielectric silicon) on these connections to help protect them (Figure 12–17).

✓ REAL WORLD FIX

The 1991 Acura (126,000 miles) has a problem of intermittent A/C operation. The voltage to the A/C compressor relay checks good. An A/C leak has been repaired and the system recharged.

Fix: Following advice, the timer fan control was checked and found to be bad. Replacement of this part fixed this problem.

✓ REAL WORLD FIX

The 1995 Ford Explorer (90,000 miles) has a problem of the A/C cutting out after the vehicle runs for a while and sitting still at stoplights. A scan tool check indicates the ECM is sending a wide-open throttle (WOT) signal to the A/C relay.

Fix: On a test drive, the technician noticed that this problem occurs when the temperature gauge is close to hot. Replacement of the engine thermostat fixed this A/C problem.

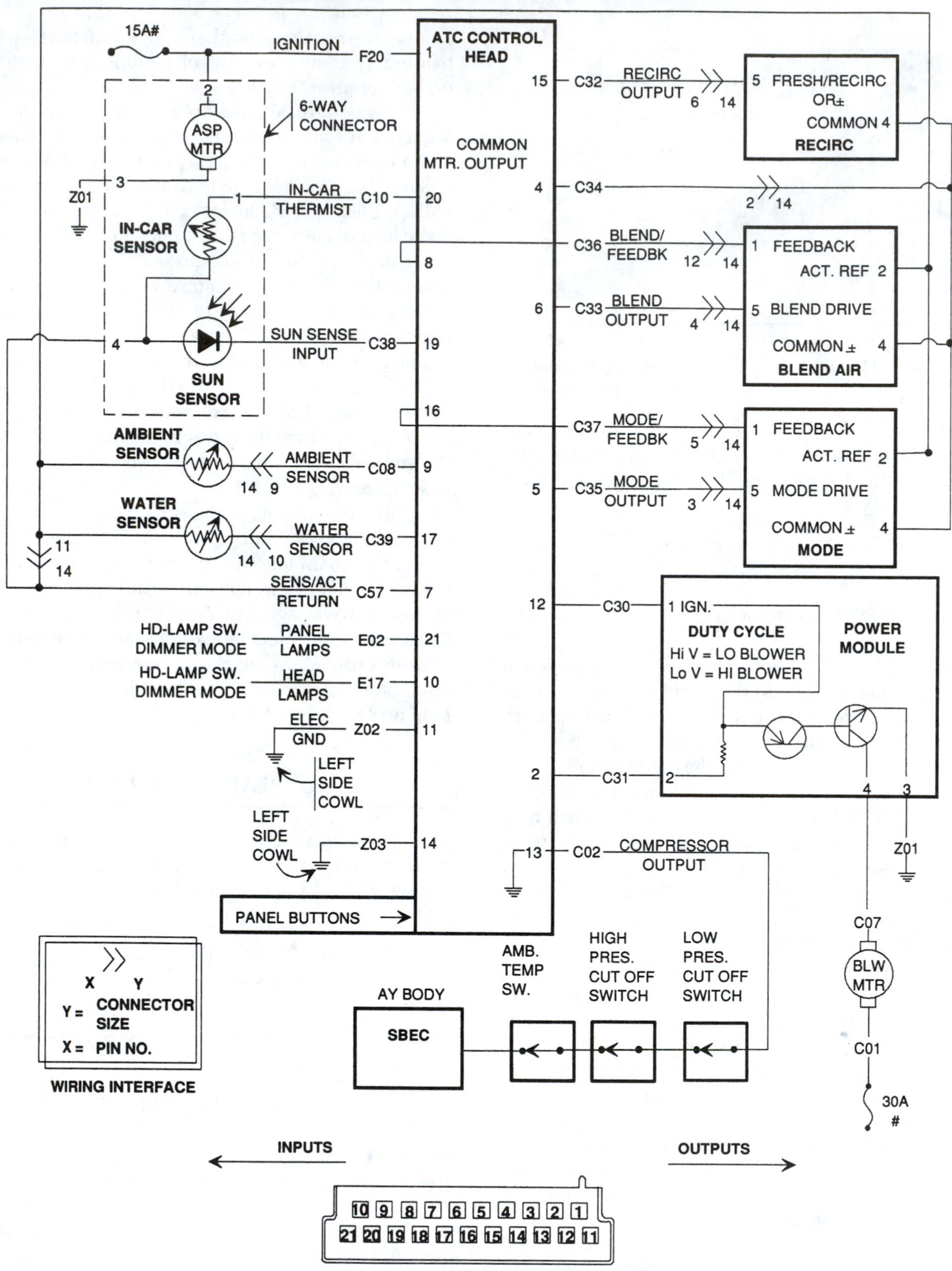

FIGURE 12–15 This electronic control head receives B+ power through terminal 1 and input through the other terminals shown at the left. It controls the components shown at the right. *(Courtesy of DaimlerChrysler Corporation)*

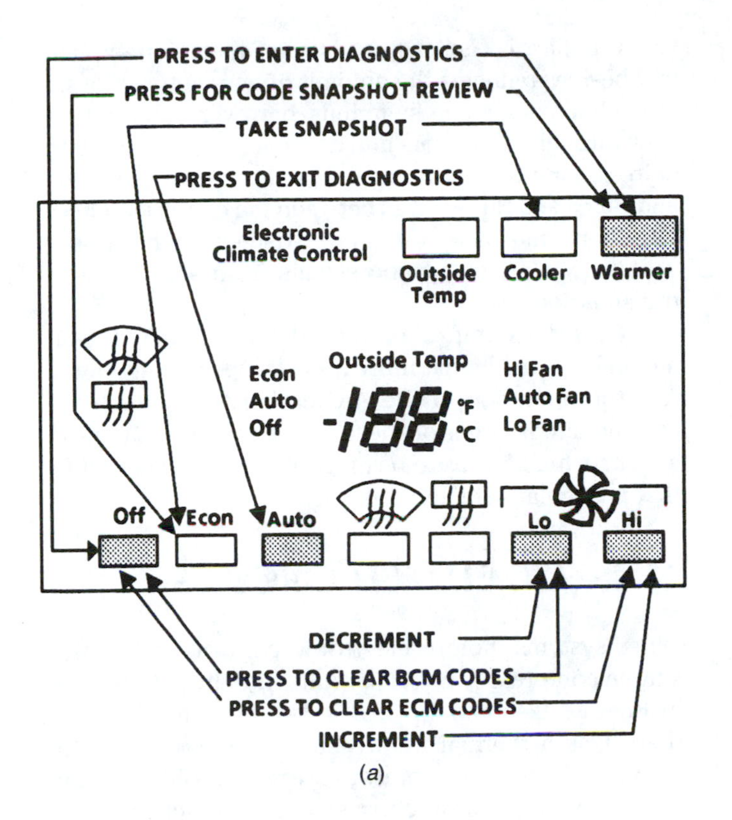

(a)

Code	Description
0	No ECC system faults
10	Outside air temperature sensor circuit problem
12	Evaporator inlet sensor circuit problem
13	In-car temperature sensor circuit problem
19	Blower motor problem
32	ECM-ECC data problem*
40	Air mix valve problem
48	Very low A/C refrigerant pressure condition (comp. off)*
49	High engine temperature (comp. off)*
55	Shorted coolant temperature sensor†

Note: History codes are indicated when a 1 is displayed in front of the code.

*5.7 liter only

†5.0 liter only

(b)

FIGURE 12–16 This control head can be used to enter diagnostic functions and to clear codes (*a*). Diagnostic codes that it can display are shown in (*b*). *(Reprinted with permission of General Motors Corporation)*

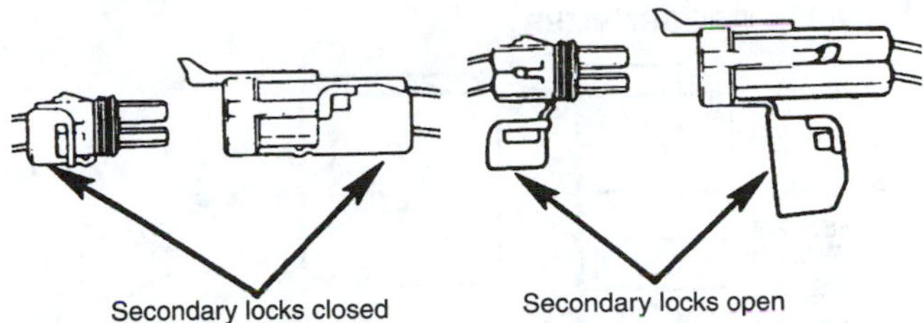

FIGURE 12–17 Modern electrical components use weather-tight connectors with locks to keep the contacts clean and tight. *(Reprinted with permission of General Motors Corporation)*

✓REAL WORLD FIX

The 1991 Camry (130,000 miles) A/C system shuts off the compressor after driving for about 30 minutes. The engine cooling fans run properly and continue to run after the compressor shuts off. The A/C indicator light in the instrument panel stays on. The A/C system pressures check out normal.

Fix: A check of the wiring diagram showed that the A/C circuit included an engine temperature sensor, and this sensor was found in the upper radiator hose. The radiator core was checked, and about half the tubes were plugged up. Replacing the radiator and backflushing the engine fixed this problem.

12.3.3 Actuators

The **actuators** for a system are the relays, motors, and lights operated by the control module. Lights are the display at the A/C control head and trouble indicators. The relays in an ATC system can control the blower speed, compressor clutch, radiator fan motor, condenser fan motor, and motors for the temperature-blend door, fresh air door, and mode doors. Modern systems operate the air inlet, mode, and temperature doors using bidirectional (reversible) electric motors. These motors run one direction to open a door and the opposite direction to close the door. The ECM chooses the direction and operates relays or transistors to provide current flow in the proper direction. With the air doors, operation is usually in a complete motion, from one door stop to the other, but the temperature door will be placed in any position to suit the desired output temperature. Door position motors also include a feedback (sensor) circuit to the control module to inform it of

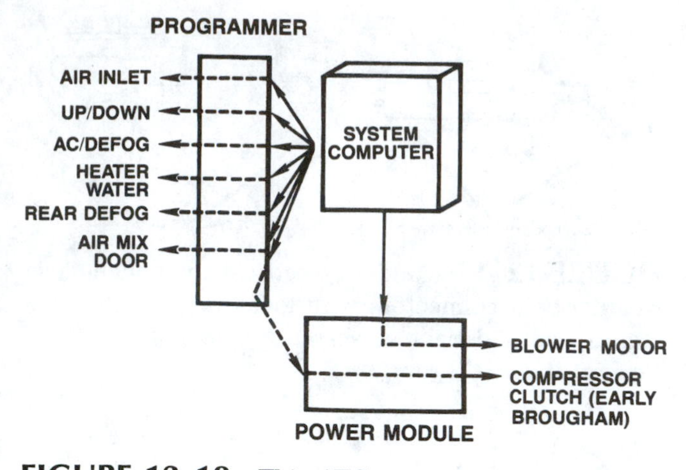

FIGURE 12–18 This ATC system uses a programmer to control the temperature and function doors and a power module to control the blower motor and compressor clutch. *(Reprinted with permission of General Motors Corporation)*

actual door position (Figure 12–18). The feedback circuit starts at a potentiometer (variable resistor) that is rotated by the motor. Motor operation changes the amount of resistance. A voltage signal from the ECM will be reduced depending on the amount of this resistance (see Figure 8–16). A smart motor will include part of this door positioning strategy.

The position of some door operating motors is controlled by a different strategy. DC motors use a segmented commutator to conduct electricity into the armature. When these commutator bars pass the brushes, there is a brief change in the current flow to the motor. The ECM notes these small pulses and can count them. The number of pulses relate to revolutions of the motor and can tell the ECM the position of the motor and door.

Most systems require a calibration process so the ECM can learn where the doors and operating motors are.

12.3.4 Multiplex

Multiplex allows one wire to work for more than one circuit; in most cases, every vehicle circuit uses a separate wire or circuit board part to complete the circuits. **Multiplex** switches a pair of wires so they complete multiple circuits, which greatly reduces the number of wires in a car. Multiplexing is usually done between two control modules, such as the power train control module (PCM) and the BCM. The two modules use some of the same sensors, and the output of one often affects the output of the other (Figure 12–19). For example, when the driver turns on the A/C, this becomes an A/C request signal to the PCM, which might actually turn on the A/C and use this signal as a value for when to begin cooling fan operation or what the engine idle speed should be.

Multiplex is also called a **bus information system,** and besides reducing the amount of wire connections, it also eliminates some switching hardware and current flow through the sensors and improves diagnostics. It requires a chip or microprocessor in each control module and Bus+ and Bus− wires between the control modules. The Bus+ and Bus− wires are twisted together to prevent stray electromagnetic signals from altering signal transmission.

The biggest impact of multiplex for the HVAC technician is probably that more and stranger things can affect the operation of the A/C compressor clutch and cooling fan. If either of these two do not operate properly, a technician has to be more aware of the circuit for that particular vehicle.

12.4 A/C CLUTCH CIRCUITS

Older systems, before electronic control, used rather simple compressor clutch circuits like those illustrated in Figures 12–5 and 12–9. Besides the control switch, there was a thermal or pressure switch to cycle the clutch to prevent evaporator icing and probably a low- or high-pressure protection switch to block the compressor from operating when conditions in the system were not right. These switches are normally connected in a series circuit so that the opening of any of the switches will stop compressor operation.

Many modern vehicles use a relay to supply power to the clutch (Figure 12–20). Power input to this relay is normally from ignition run (B+ voltage or hot when the ignition is in the run position) and passes through a fuse for circuit protection. The ECM or PCM provides the ground for the control circuit through the relay so that the compressor is turned on when the ECM or PCM provides the ground connection. The A/C request signal to turn on A/C comes to the ECM or PCM from the instrument panel control switch, and this request circuit can pass through the same type of switches used on older systems. More modern sensors, thermistors, or pressure transducers are used in some vehicles to replace switches.

✓SERVICE TIP

Operating voltage at a clutch should be almost the same as battery voltage (within 0.2 V). If it is lower, perform a voltage drop check: connect the + voltmeter lead to the + battery post and the − lead to the B+ connector at the clutch. The reading should be 0.2 V or less. Next connect the − voltmeter to the − battery post and the + lead to the ground connection at the compressor. Again, the reading should be 0.2 V or less. Excessive voltage drop indicates a problem in that part of the circuit.

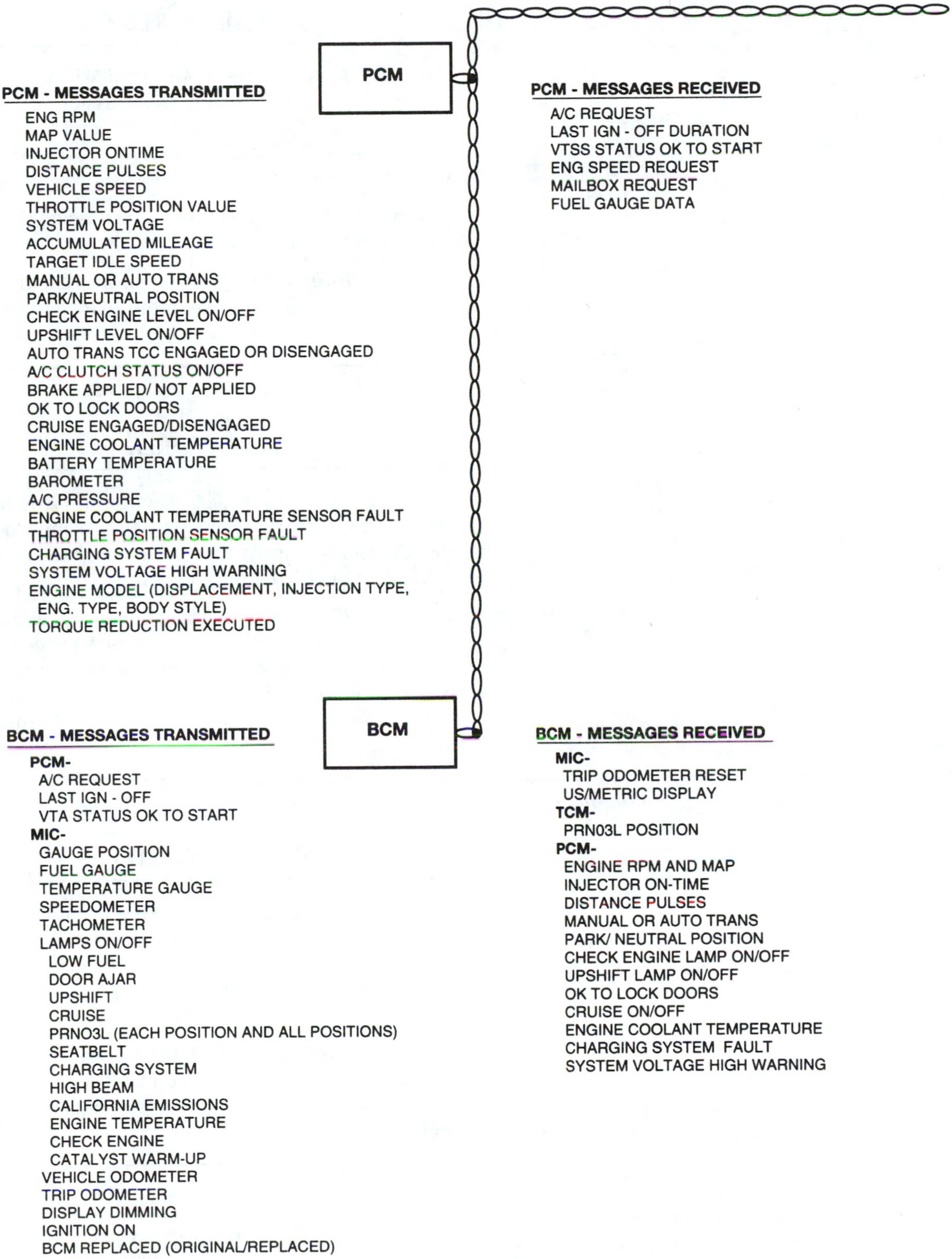

PCM

PCM - MESSAGES TRANSMITTED

ENG RPM
MAP VALUE
INJECTOR ONTIME
DISTANCE PULSES
VEHICLE SPEED
THROTTLE POSITION VALUE
SYSTEM VOLTAGE
ACCUMULATED MILEAGE
TARGET IDLE SPEED
MANUAL OR AUTO TRANS
PARK/NEUTRAL POSITION
CHECK ENGINE LEVEL ON/OFF
UPSHIFT LEVEL ON/OFF
AUTO TRANS TCC ENGAGED OR DISENGAGED
A/C CLUTCH STATUS ON/OFF
BRAKE APPLIED/ NOT APPLIED
OK TO LOCK DOORS
CRUISE ENGAGED/DISENGAGED
ENGINE COOLANT TEMPERATURE
BATTERY TEMPERATURE
BAROMETER
A/C PRESSURE
ENGINE COOLANT TEMPERATURE SENSOR FAULT
THROTTLE POSITION SENSOR FAULT
CHARGING SYSTEM FAULT
SYSTEM VOLTAGE HIGH WARNING
ENGINE MODEL (DISPLACEMENT, INJECTION TYPE,
ENG. TYPE, BODY STYLE)
TORQUE REDUCTION EXECUTED

PCM - MESSAGES RECEIVED

A/C REQUEST
LAST IGN - OFF DURATION
VTSS STATUS OK TO START
ENG SPEED REQUEST
MAILBOX REQUEST
FUEL GAUGE DATA

BCM

BCM - MESSAGES TRANSMITTED

PCM-
A/C REQUEST
LAST IGN - OFF
VTA STATUS OK TO START
MIC-
GAUGE POSITION
FUEL GAUGE
TEMPERATURE GAUGE
SPEEDOMETER
TACHOMETER
LAMPS ON/OFF
LOW FUEL
DOOR AJAR
UPSHIFT
CRUISE
PRNO3L (EACH POSITION AND ALL POSITIONS)
SEATBELT
CHARGING SYSTEM
HIGH BEAM
CALIFORNIA EMISSIONS
ENGINE TEMPERATURE
CHECK ENGINE
CATALYST WARM-UP
VEHICLE ODOMETER
TRIP ODOMETER
DISPLAY DIMMING
IGNITION ON
BCM REPLACED (ORIGINAL/REPLACED)

BCM - MESSAGES RECEIVED

MIC-
TRIP ODOMETER RESET
US/METRIC DISPLAY
TCM-
PRN03L POSITION
PCM-
ENGINE RPM AND MAP
INJECTOR ON-TIME
DISTANCE PULSES
MANUAL OR AUTO TRANS
PARK/ NEUTRAL POSITION
CHECK ENGINE LAMP ON/OFF
UPSHIFT LAMP ON/OFF
OK TO LOCK DOORS
CRUISE ON/OFF
ENGINE COOLANT TEMPERATURE
CHARGING SYSTEM FAULT
SYSTEM VOLTAGE HIGH WARNING

FIGURE 12–19 The multiplex between the BCM and PCM is the path for each to send messages or signals to each other as well as the transaxle control module (TCM), electromechanical instrument cluster (MIC), air bag control module (ACM), and data link connector (DLC). *(Courtesy of DaimlerChrysler Corporation)*

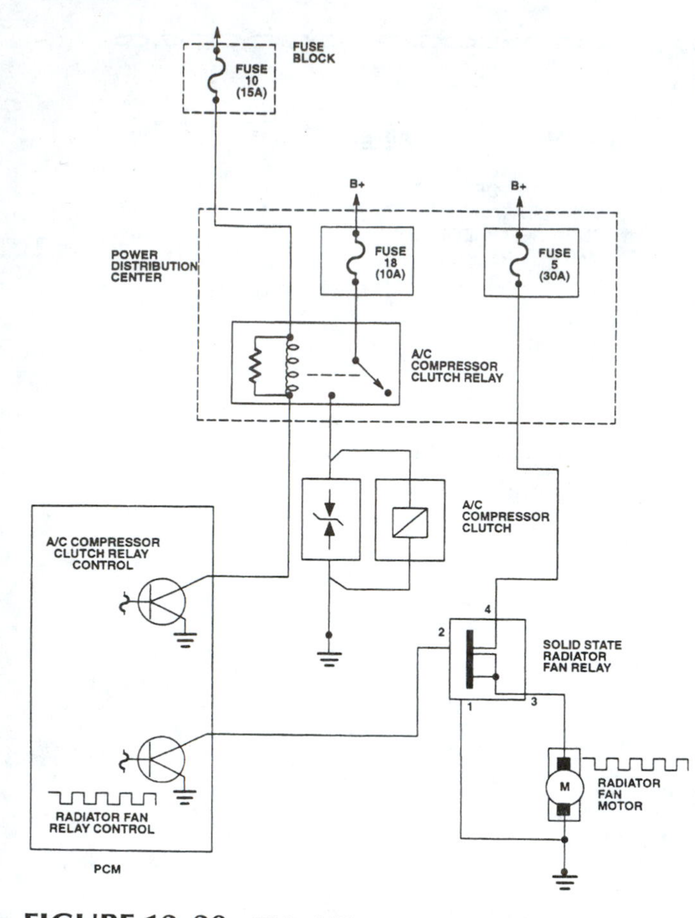

FIGURE 12–20 This A/C compressor clutch relay is controlled by the PCM. Note that the radiator fan motor uses a solid-state relay that is also controlled by the PCM. *(Courtesy of DaimlerChrysler Corporation)*

✓SERVICE TIP

It is also a good practice to repeat voltage tests with the engine running. Actual voltage should increase from about 12 V to about 14 V.

Clutches in modern vehicles still need battery voltage and a current flow just like the older systems. The relay must be energized to turn the clutch on and deenergized to turn it off; this function is taken care of by the fuse providing B+ and the ECM or PCM providing ground. The ECM or PCM receives the request for turning on the clutch, receives operating conditions, and contains the strategies for turning the clutch on or off. The ECM or PCM can also receive input from other control modules such as the BCM or electronic ATC modules.

✓REAL WORLD FIX

The 1984 Mercedes-Benz A/C fails when the driver forces a downshift from the transmission. This action causes the A/C fuse to blow. The A/C and transmission work normally during all other driving operations. An inspection of the wiring and fuse box shows no problem.

Fix: Replacement of a shorted transmission downshift solenoid fixed this problem.

Author's Note: Apparently, the transmission downshift solenoid shares the same fuse as the A/C compressor, and the faulty downshift solenoid drew excessive current.

✓REAL WORLD FIX

The 1991 Mitsubishi (89,000 kilometers) came in for a stereo radio installation, and the owner complained that the A/C comes on every time the ignition is switched on, even when the A/C controls are off.

Fix: The A/C circuit was traced on a wiring diagram, and it was found that the A/C circuit is dependent on the vehicle's interior light. The interior light operation is often not noticed during daylight, and this circuit had a blown fuse. Replacement of the interior light fuse fixed this A/C problem.

✓REAL WORLD FIX

The 1988 Corvette (123,000 miles) has no A/C; the compressor clutch does not engage. The clutch operates if the technician connects a jumper wire for ground to the green wire at the clutch connector. The technician does not have a schematic of this clutch circuit.

Fix: Advice from fellow technicians directed the technician to the blower control module that was the cause of the problem. Replacement of this module fixed this problem.

✓REAL WORLD FIX

The 1994 Volkswagen Jetta (75,000 miles) came in with a request to check the A/C. A low charge level was found, so the refrigerant was recovered from the system and then evacuated and recharged. But, when the system was to be checked, the compressor clutch would not engage; there was no voltage to it. Using a jumper wire engaged the clutch, and the system worked normally. The fuses are good, but the technician did not have a wiring diagram for this system.

Fix: Given a wiring diagram, the technician learned that this system has a high temperature switch mounted in the compressor, and this switch was open. Replacement of the high temperature switch fixed this system.

✓REAL WORLD FIX

The 2001 Chevrolet Malibu (5,000 miles) came from a body shop after a front-end accident repair. The compressor and discharge hose had been replaced by the body shop. The A/C clutch would not engage because of no signal to the A/C clutch relay. A scan tool would not activate the relay so the BCM and control head were replaced, but this did not help. The relay and compressor clutch will operate if the proper pin on the relay or BCM is grounded.

Fix: Following advice, the technician double-checked the refrigerant pressure sensor and found that jumping the sensor would activate the clutch. The sensor was replaced, but this did not help. Further checking revealed that the pressure port in the new hose was not drilled completely through. Carefully drilling the hole completely allowed the sensor to receive a pressure signal and fixed this problem.

12.5 BLOWER MOTOR CIRCUITS

As partially described in Chapter 8, the blower motor that moves air through the air distribution system can have three or four fixed speeds through a set of resistors or an infinite number of speeds through pulse width modulation (PWM). Blower motors are fairly powerful so they can move the air volume needed; during high blower operation they will draw about 20 amps.

Older systems used a multiposition rotary or slide switch with a position for off plus each of the speeds. Each of the output positions is connected to a resistor or the relay for high blower speed. Using a high blower relay reduced the current flow through the blower speed switch. A variation of this circuit was to use ground-side switching. B+ was sent directly to the motor, and the resistors for speed selection were in the circuit between the motor and ground (Figure 12–21).

Another variation is to use a relay to complete the circuit through each of the resistors (Figure 12–22). This reduces the current flow through the switching device and allows an ECM strategy to control the blower speed. These circuits will vary; one circuit uses three

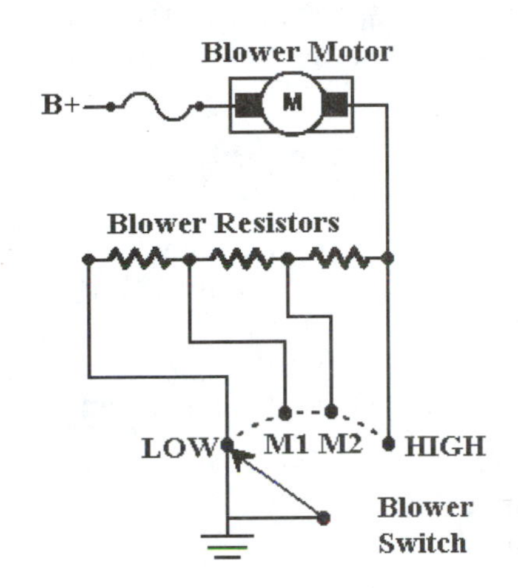

FIGURE 12–21 This blower uses ground-side switching. Note that the blower switch and resistors are between the blower motor and the ground connection.

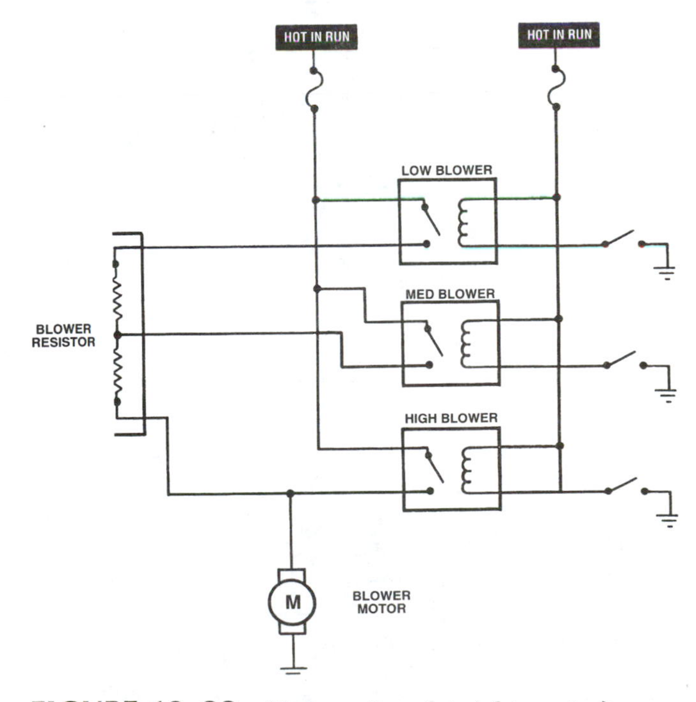

FIGURE 12–22 Three relays (at right center) control the current flow for the three blower speeds. Note that the flow for low and medium speeds pass through a resistor. *(Reprinted with permission of General Motors Corporation)*

relays to control four blower speeds (Figure 12–23). By operating these relays in the proper sequence, off, low, medium, and high speeds are produced. Three relays produce four blower speeds by operating the relays in the proper sequence.

PWM speed control is also controlled electronically; the control head can send a pulsed signal to the blower power module that turns the blower on and off to match the signal pulses (see Figure 8–45).

✓SERVICE TIP

A blower motor works the hardest when it is moving the most air. A current draw of under 18 amps (at 14 V) on high blower indicates an airflow restriction. You can prove this by measuring the current draw with the blower motor assembly off the car and mounted in a vise with no restriction to the inlet and outlet. Use a controlled voltage of 14 V. The current draw will be about 20 amps or greater. Now place a section of cardboard over the inlet and/or the outlet, and note the change in current draw.

12.6 COOLING FAN CIRCUITS

The engine cooling fan(s) of most modern FWD vehicles are also controlled through one or two relays using circuits much like the compressor clutch. The relays are often mounted at the power distribution center with the other relays and fuses for the vehicle (Figure 12–24). Some vehicles use a pair of relays, one for high speed and one for low speed. Some vehicles use two separate

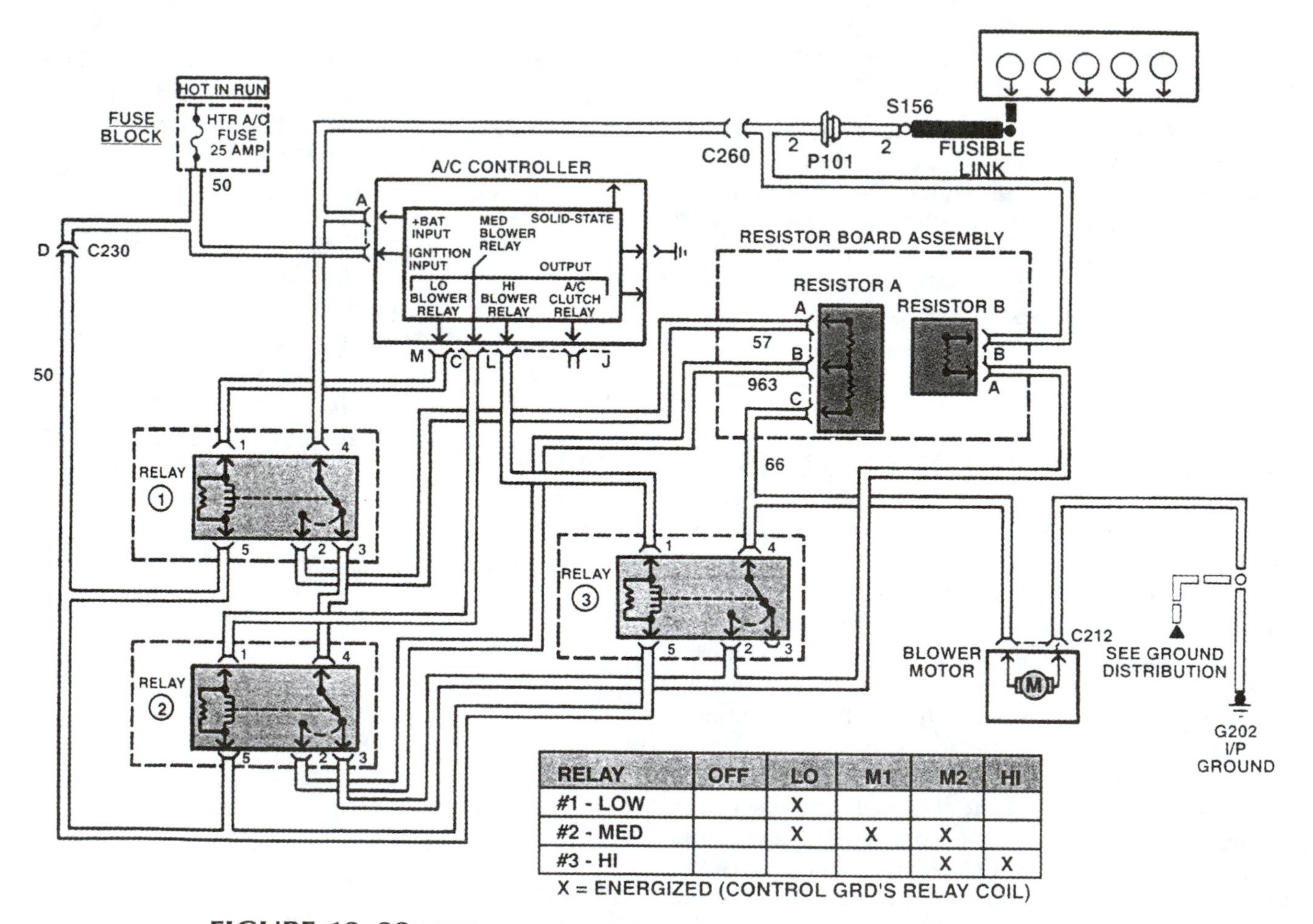

RELAY	OFF	LO	M1	M2	HI
#1 - LOW		X			
#2 - MED		X	X	X	
#3 - HI				X	X

X = ENERGIZED (CONTROL GRD'S RELAY COIL)

FIGURE 12–23 This system uses three relays to provide four blower speeds. Note that the relays are energized by A/C controller in the proper order to provide those speeds. *(Reprinted with permission of General Motors Corporation)*

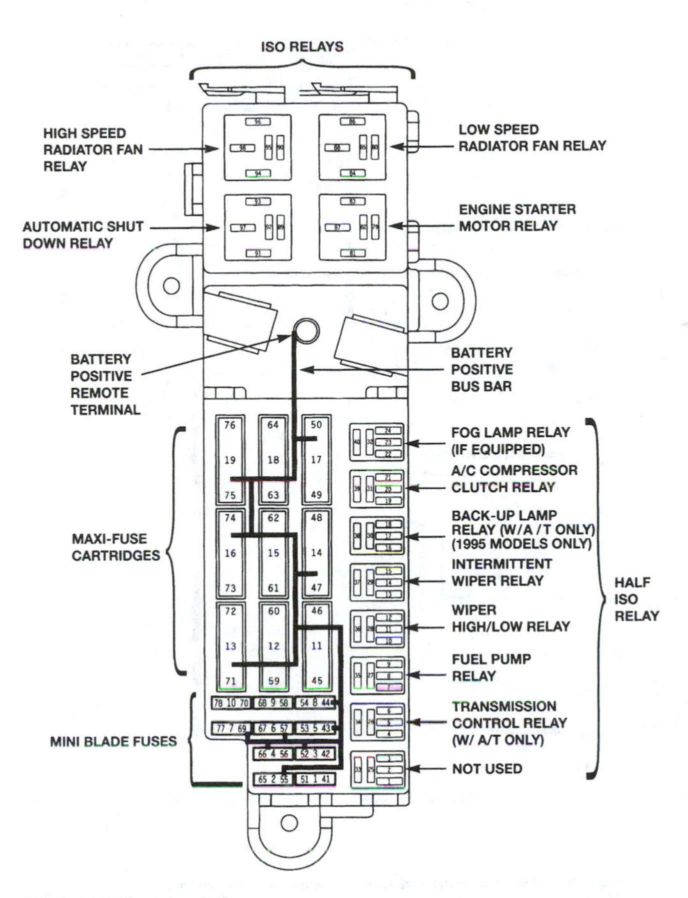

FIGURE 12–24 This power distribution center includes two radiator fan relays and the compressor clutch relay, as well as fuses to protect their operation. *(Courtesy of DaimlerChrysler Corporation)*

fans. In some vehicles one or both of these relays are connected directly to B+, so fan operation can occur with the engine shut off.

Relay operation is controlled by the ECM or PCM with a strategy based on both engine coolant temperature and A/C high side pressure (Figure 12–25). The cooling fan must begin operation when the engine reaches a specified temperature (about 212°F) to prevent overheating. Two-speed systems turn the fan on first at low speeds and then high speeds as engine temperature increases. Some systems start fan operation whenever the A/C is switched on; others switch the fan on when the high side pressure reaches a certain point.

ECM or PCM fan operating strategy on some modern vehicles can be reprogrammed in the field; this allows the technician to change the point at which the fan is switched on. Changing the program strategy of an ECM or PCM requires a compatible scan tester with the proper internal capability, and the procedure is very specific to the vehicle model.

✓REAL WORLD FIX

The 1990 Cadillac (60,000 miles) control head displays a message of "A/C Overheating" while driving around town. On the highway, the engine temperature reached 253°F, and an A/C pressure check shows 400 psi and rising. The pressure drops to about 200 to 275 psi when water is sprayed on the condenser. The refrigerant was recovered, and it was determined that it was overcharged about 1 pound. The system was evacuated and recharged, but with the engine at idle speed, the compressor locked up. A new compressor was installed, but the pressures were still too high. The electrical connector to the cooling fan had been replaced some time ago.

Fix: A check of the cooling fan showed that it was running backward. Reversing the wires at the connector fixed this problem.

✓SERVICE TIP

Airflow direction from a fan can be easily checked by holding a piece of string or paper into the airstream and seeing which way it is blown.

12.7 ELECTRICAL CIRCUIT PROBLEMS

Most electrical problems fall into one of four categories: open, high resistance or weak, shorted, or grounded. These problems can occur in either a continuous or intermittent manner; intermittent problems are usually much harder to locate and are often the result of the vehicle's movement, vibration, and changes in temperature. With the proper equipment and knowledge, most electrical components can be easily checked to determine whether any of these problems are present.

12.7.1 Open Circuits

An unwanted **open circuit** is an incomplete, broken circuit in which no current can flow and is usually caused by a broken or disconnected wire, a blown fuse or fusible link (probably caused by a short or ground), or a broken filament in a lightbulb. Source voltage is present up to the faulty point. An open circuit causes a complete loss in voltage and current. An open or weak circuit can occur at any point between the B+ and ground connections (Figure 12–26).

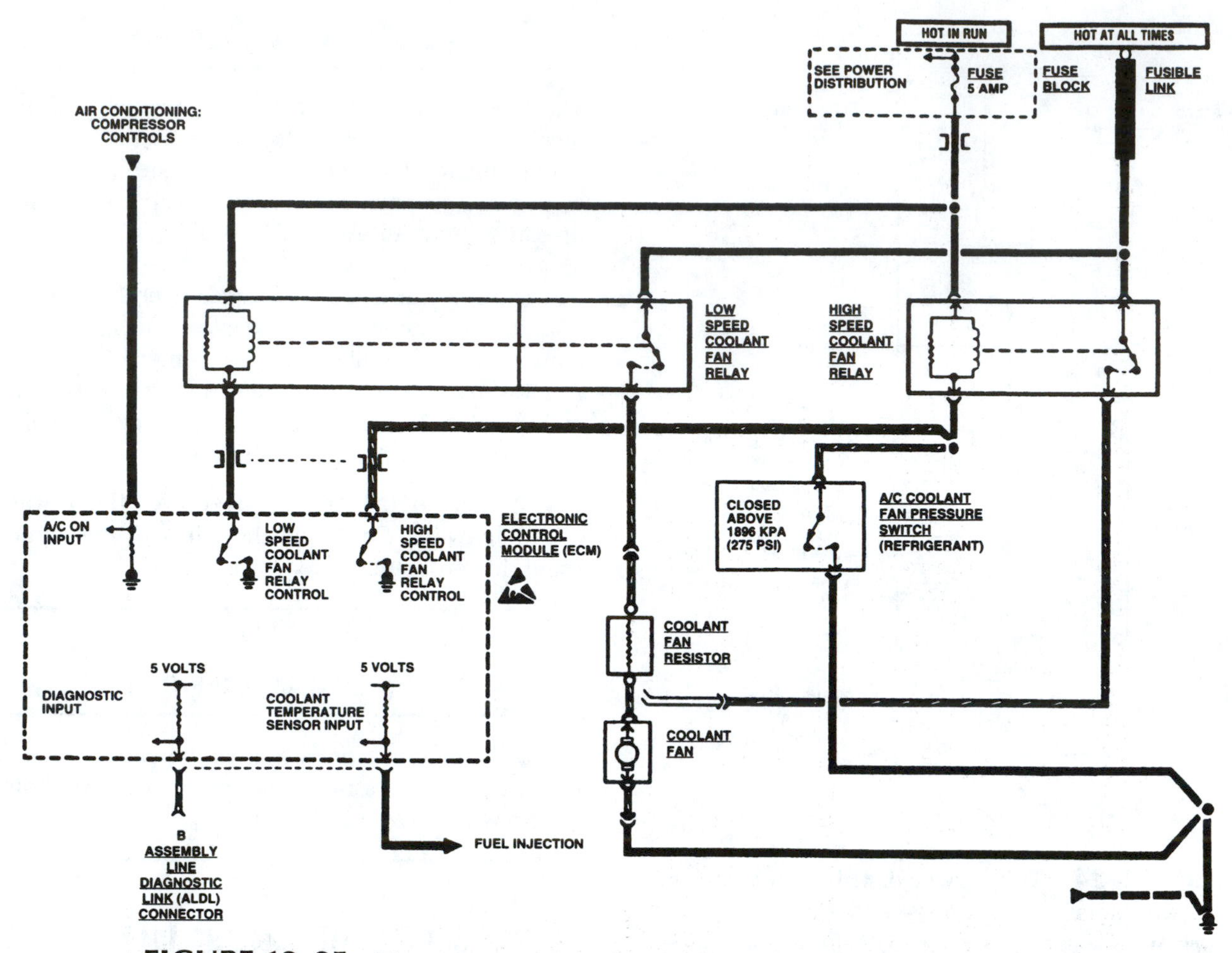

FIGURE 12–25 The current to this two-speed coolant fan is controlled by two relays, which are in turn controlled by an A/C pressure switch and the ECM. *(Reprinted with permission of General Motors Corporation)*

✓REAL WORLD FIX

The 1998 Jeep Cherokee (37,000 miles) has an intermittent problem of the HVAC blower, speedometer, and tachometer all shutting off at the same time. The new car dealer replaced the low pressure cycling switch and PCM and cleaned the electrical connectors, but this did not help.

Fix: Following advice, the technician checked the instrument panel electrical ground stud and found that the nut was missing. Cleaning the terminals and installing a nut on this stud provided the proper ground connection and fixed this problem.

12.7.2 High-Resistance Circuits

A **high-resistance circuit** is similar to an open circuit, except a reduced amount of current (not enough to do the job) flows. This circuit is often caused by a corroded, loose, or dirty connection. A high-resistance circuit causes reduced voltage; this is called a **wasted voltage drop.** Excess voltage drop in turn reduces the current flow.

12.7.3 Shorted Circuits

Although some people use the term **shorted circuit** to describe any electrical problem, a short circuit is most often found in a wire coil. If the wires lose their insulation and the metals of the wires touch, current flows on the shortest path with the least resistance and bypasses some of the coils. The effect of the short is a lower coil resistance because of the shortened path, allowing an increase in current flow. The strength of the magnet is also reduced because of fewer ampere-turns. A short can also occur between the wires of two separate circuits if their insulation is damaged and cause an unwanted current flow between the circuits (Figure 12–27). Some people call a short a **copper-to-copper connection.**

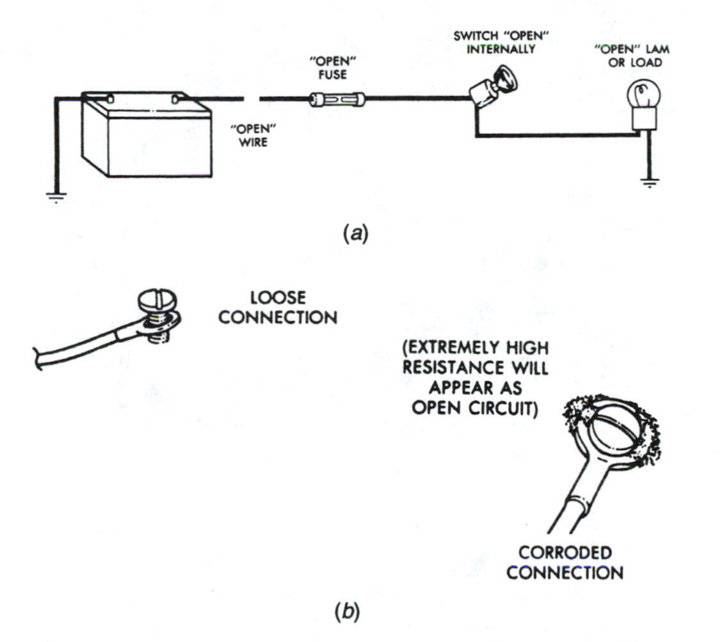

FIGURE 12–26 An open circuit is a break in the circuit that will stop the current flow (*a*). Corroded or loose connections will cause high resistance that will reduce the current flow (*b*). *(Courtesy of DaimlerChrysler Corporation.)*

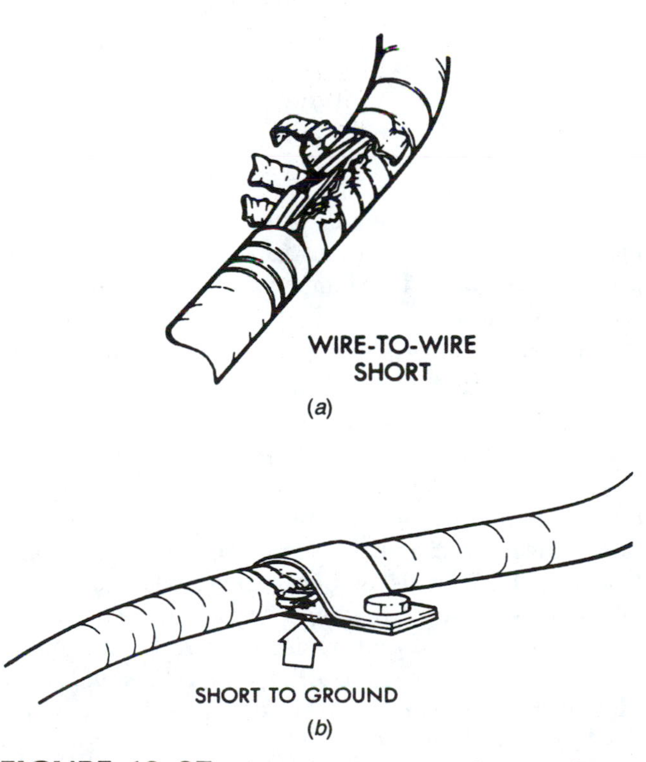

FIGURE 12–27 A short circuit is often a wire-to-wire connection that can reduce magnetic coil strength or allow current to flow to the wrong circuit (*a*). It can also be a short to ground (*b*). (*Courtesy of DaimlerChrysler Corporation*)

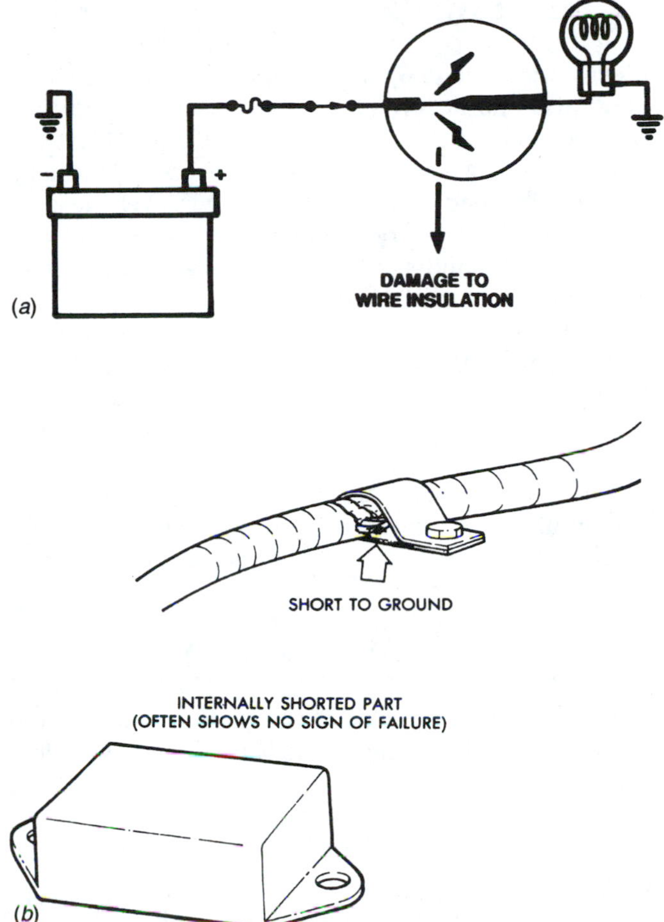

FIGURE 12–28 A ground or short-to-ground circuit occurs when damage to the insulation allows an electrical path to the metal of the vehicle (*a*). It can occur at a wire or inside a component (*b*). (*Courtesy of DaimlerChrysler Corporation*)

12.7.4 Grounded Circuits

A **grounded circuit** is similar to a short except the bare wire touches ground. This is sometimes called a **short-to-ground** or **copper-to-iron** connection. The grounded circuit completes a path directly back to battery ground (B−), so the rest of the circuit is bypassed. Again, current follows the path of least resistance. Depending on where the connection occurs, a grounded circuit can have zero resistance, and the current flow will instantly increase to the limit of the fuse, wire, or battery. This usually causes rapid burnout, and the resulting smoke or burned-out wire is often easy to locate (Figure 12–28).

12.7.5 EMI, RFI

Electronic circuits can be influenced by **electromagnetic induction (EMI),** and **radio frequency interference (RFI).** RFI is a form of EMI. This is the interaction between ferrous metal objects and the magnetic field around a current carrying wire. Vehicle manufacturers twist electrical wires together, use shielded coaxial cable, and install metal shielding to protect some circuits from EMI interference.

EMI can cause strange electronic circuit malfunctions and an ECM to set a DTC when there is no real fault. If this is encountered, first check the vehicle for improperly installed aftermarket electronic accessories. These include a cellular telephone, radar detector, remote starter system, radio, or any other non-OEM electrical accessory. Discarded shields or misrouted wires from an improperly performed repair can also cause EMI problems.

✓REAL WORLD FIX

The customer had replaced the blower motor in the 1995 Cadillac (72,000 miles) with a new OEM part, and this motor failed. The blower available voltage and ground were checked and found to be good. A new blower motor was installed, but it soon failed also.

Fix: Following advice, the engine spark plug wires were replaced and repositioned as far away from the blower motor and wiring as possible. Another new blower motor was installed, and this one has not failed.

✓SERVICE TIP

It is recommended that the vehicle's air bags be deactivated before making any checks under the instrument panel. Accidental shorting of the air-bag circuit wires can cause the air bags to deploy. This can cause possible injury to the technician and bystanders and will require expensive repairs.

12.8 MEASURING ELECTRICAL VALUES

At one time, technicians used a **test light, jumper wire,** and **analog voltmeter–ohmmeter** or **multimeter** (a combination ammeter, ohmmeter, and voltmeter) for troubleshooting automotive electrical problems. Today, the weather-tight connectors on modern

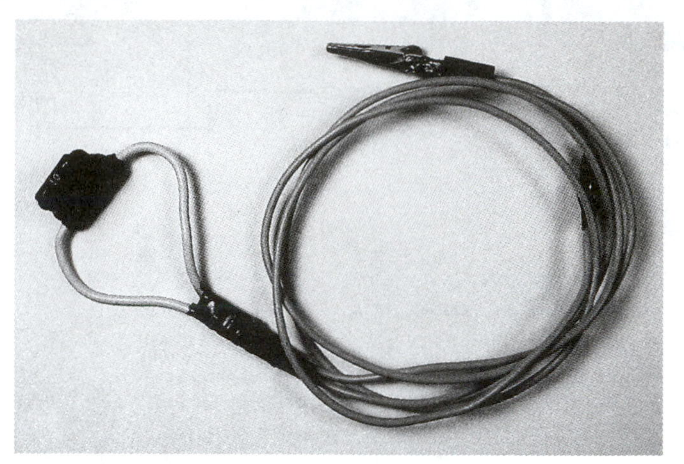

(*a*)

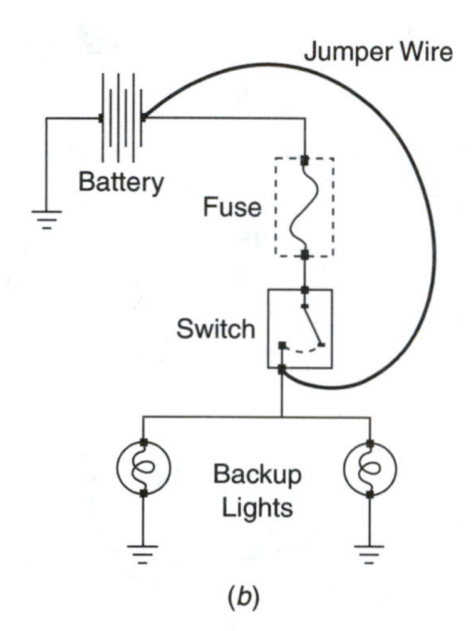

(*b*)

FIGURE 12–29 Jumper wires can often be used to bypass portions of a circuit to determine where the problem is located. This jumper wire has a fuse to limit the current flow (*a*). If the lights light with the jumper wire installed, the problem is the fuse, switch, or wires to the battery (*b*).

cars make the use of a jumper wire very difficult (Figure 12–29). The common test light and analog meter should not be used on solid-state circuits because they cause the circuit to draw too much amperage from the ECM. This amperage overload can cause faulty sensor meter readings or damage. The meter must have an internal resistance of 10 megohms (10,000,000 Ω) or greater. **LED test lights** and most **digital multimeters (DMM)** can be safely used on any automotive circuit.

Most test lights are simple, cheap, durable, and easy to use. They have one wire, which is normally connected to ground, and when the probe contacts B+, the bulb lights up. Some test lights are self-powered by an internal battery. Self-powered lights can be used to check **continuity** (a continuous, complete circuit) through a component such as a clutch coil. Most test lights indicate either a pass or fail; because they cannot make measurements, they do not tell us how faulty the component is (Figure 12–30).

Meters make measurements and come in two types: **analog** (a needle sweeps across a scale) and **digital** (actual value is displayed in a series of digits). As mentioned earlier, analog meters should not be used on electronic circuits because they draw too much power and can damage a circuit. The modern DMM is more accurate, easier to read, and can be used on all modern computer circuits. On circuits with changing values, however, the sweep of an analog needle is much faster and much easier to follow than the rapidly changing display of digits. For that reason, some digital multimeters include a **bar graph** for a quicker response to value changes.

Multimeters have a selector switch that must be set to the desired function. Some meters also have a range switch set to the area where the readings are expected to fall. The range selector should always be set above the highest reading expected and then reset downward until readings occur (Figure 12–31).

Computer-controlled circuits also require care because of possible damage through electrostatic discharge (ESD). ESD is the shock you feel or the spark you see sometimes when you touch metal after walking across carpet or sliding across a seat covering. The electrostatic charge of electricity felt is several hundred volts; such an ESD can easily damage solid-state electronic devices. The symbol shown in Figure 12–32 is placed on some components and wiring diagrams to indicate components that can be damaged by ESD. Unless you are connected electrically to the vehicle's ground, do not touch that component or its electrical connections. Do not make electrical measurements or checks on these components unless instructed to do so, and always follow directions exactly. If directed to make voltage checks, connect the negative probe to ground first.

Modern weather-tight connectors present a difficult problem when probing for electrical values. You can disconnect a connector and probe the ends, but sometimes this is an inaccurate reading without normal circuit load. Thin probes are available that allow back-probing a connection by sliding the probe into a connector and through the weather-tight connectors.

✓SERVICE TIP

Corrosion can enter a wire that has the insulation pierced. Fingernail polish or battery sealant spray can be safely used to seal wire insulation so that water does not leak into the connector or component and cause future corrosion and resistance.

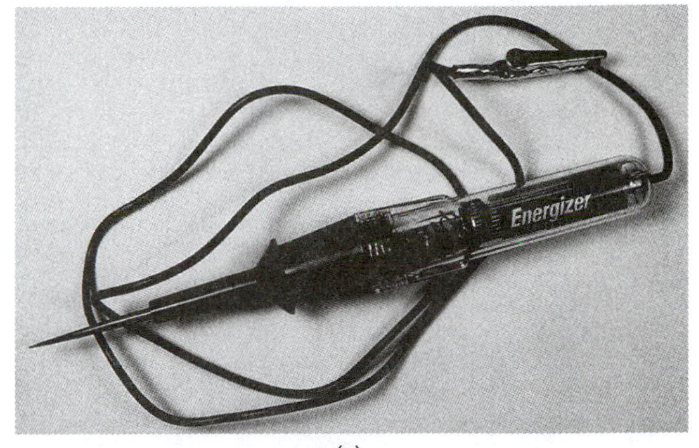

(*a*)

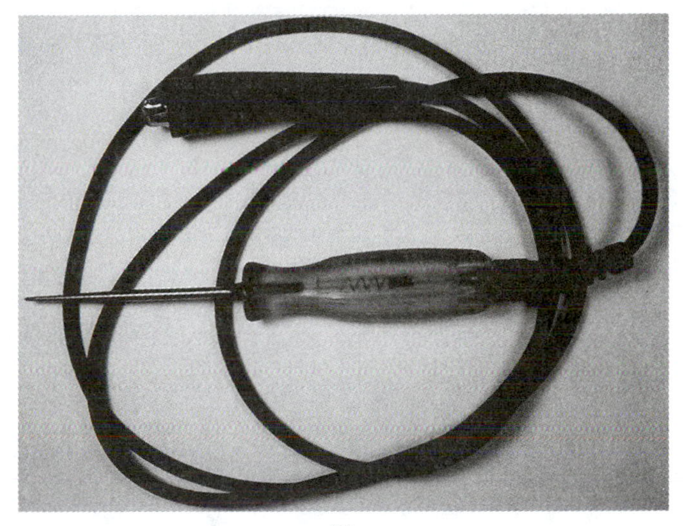

(*b*)

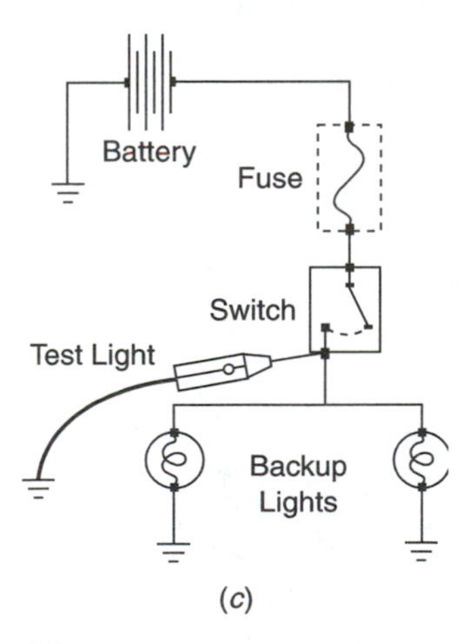

(*c*)

FIGURE 12–30 A self-powered test light includes a battery and can be used to check for continuity (*a*). An ordinary nonpowered test light (*b*) can be used to determine if a point in a circuit has voltage (*c*).

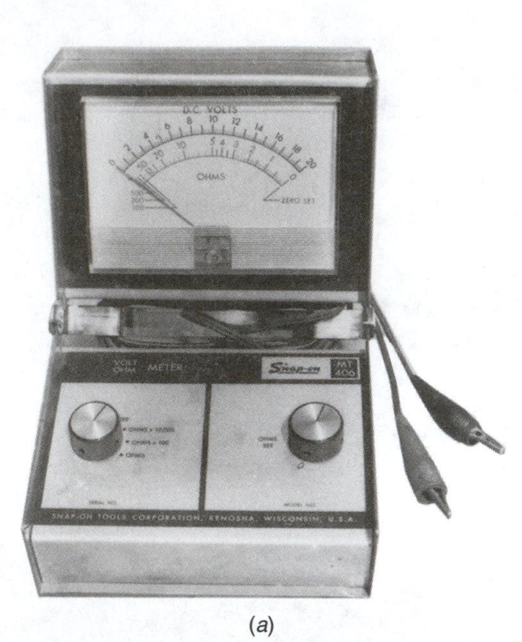

(a)

(b)

FIGURE 12–31 An analog (*a*) and a digital (*b*) multimeter. Do not use analog meters when testing electronic circuits. (b *is courtesy of OTC Division, SPX Corporation*)

FIGURE 12–32 Extreme care should be used when testing components that display this ESD symbol.

✓REAL WORLD FIX

The 1996 Dodge Caravan (134,000 miles) with a dual zone system has temperature control only on the left side. The control unit was replaced with a unit known to be good, but this did not help. The door actuator was replaced, but this did not help. The temperature door can be easily moved by hand so it is not binding. A wire insulation rub-through from contacting a sharp piece of metal was found and repaired, but this did not help. The system was recalibrated and returned to the owner, but after several weeks it returned with intermittent good operation along with fluctuating temperature, switching from cool to heat.

Fix: A closer inspection of the wiring revealed two more places where the insulation and wiring had rubbed through. Replacement of the harness and padding of the sharp edges fixed this problem.

Author's Note: Rub-throughs occur at places where the wiring harness contacts sharp edges of the instrument panel.

12.8.1 Measuring Voltage

A voltmeter is used to measure voltage and voltage drop by connecting the negative lead to ground and probing various points along the circuit with the positive lead. The meter displays the actual voltage for that point in the circuit; depending on the circuit and its components, this reading should be B+ (around 12 V) or some lesser value. A 0 reading indicates an open circuit between the probe and B+. A reading less than B+ indicates a voltage drop; this can be good or bad. If you have full battery voltage to the component and a full battery voltage drop across the component, the component is either using all the electric power or has an open circuit.

Voltage can also be measured by connecting the negative lead to the ground side of a component and the positive lead to the B+ side. The reading is the amount

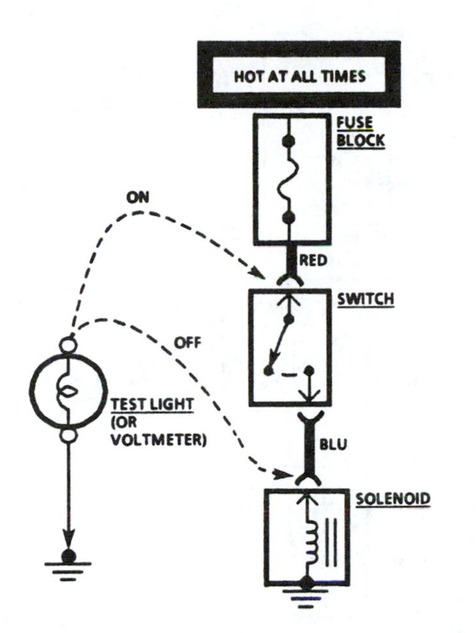

FIGURE 12–33 Voltage is measured by connecting one voltmeter lead (normally the negative) to ground and probing the wire connections with the other lead. *(Reprinted with permission of General Motors Corporation)*

of voltage dropped across that component. It should be noted that voltage drop only occurs while the circuit is on and under load. The **total voltage drop** in any circuit is always **equal to the sum of each voltage drop**, and this is **equal to the source voltage.** Usually, all of the voltage drop occurs across the actuator or output device. A 12 V drop across an element may be okay, *or* it may indicate that the element is open (Figure 12–33).

In a simple, old-fashioned A/C compressor clutch circuit, there should be B+ voltage right up to the insulated clutch connection and a 12 V drop across the clutch. A small drop across a connection or switch is allowed, but this is usually limited to about 0.2 or 0.3 V per connection. Any greater drop indicates a high-resistance problem that will cause a circuit to function below normal and should be corrected (Figure 12–34).

A heating and A/C blower fan circuit is a different example in which we expect a significant drop across the blower speed resistors. Voltage readings at the blower motor connection should be about 12 V on high, 9 V on medium-high, 6 V on medium-low, and 4 V on low. This tells us that 0 V was dropped before the motor on

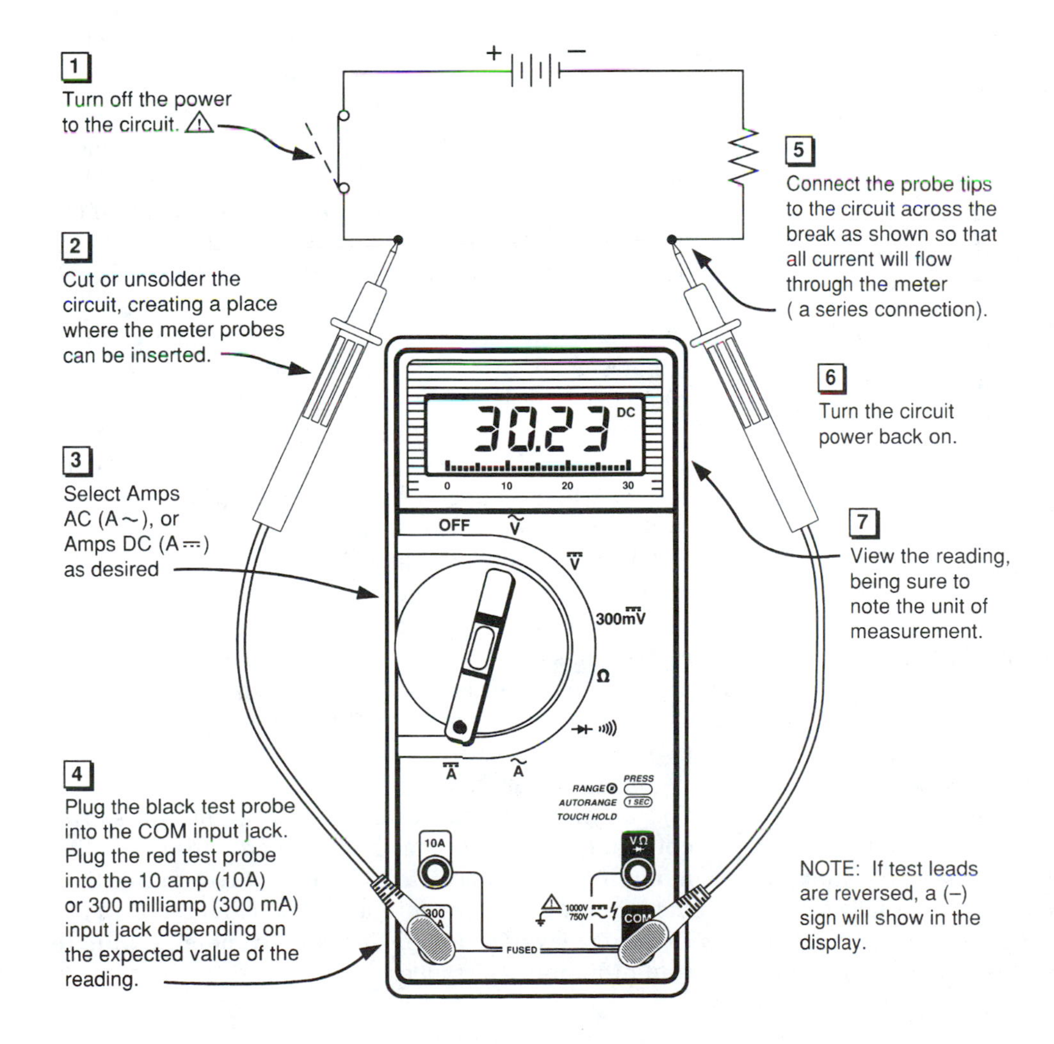

FIGURE 12–34 This meter is connected to measure the voltage drop across the resistor. *(Courtesy of Fluke; reproduced with permission)*

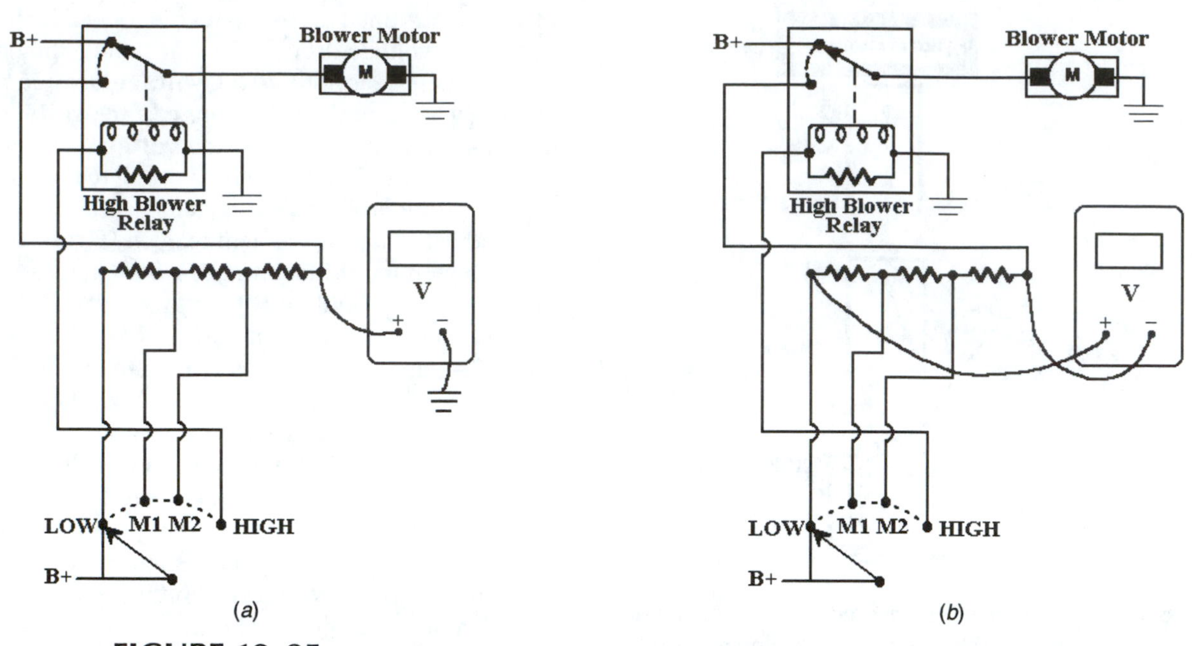

FIGURE 12–35 Voltage drops across resistors can be measured indirectly (*a*) or directly (*b*). Source voltage at the input side minus the indirect measurement will equal the direct measurement.

high, 3 V (12 − 9 = 3) were dropped on medium-high, 6 V were dropped on medium-low, and 8 V were dropped on low (Figure 12–35).

12.8.1.1 Voltage Source HVAC system voltage usually begins at the vehicle's electrical power distribution center (Figure 12–24). Some vehicles will use a relay that is energized by the ignition switch or through the ECM; other vehicles will use a fuse for the HVAC circuit. A separate fuse or relay is often used for the A/C clutch, HVAC blower motor, and engine cooling fan.

A wise technician begins voltage checks at the fuse or relay to ensure that there is proper system voltage at the beginning, at least 12 V, and this should be higher (about 14 V) when the engine is running.

12.8.2 Measuring Resistance

An ohmmeter is used to measure the resistance of electrical components. The two ohmmeter leads are connected to the two ends of a wire or connections of a component, and the meter displays the resistance value of that component. Ohmmeters are self-powered by an internal battery. *They must never be connected to a circuit that has power* because the usually higher voltage from the circuit will damage the ohmmeter. Some meters have a protective fuse that will burn out to save the meter. Always turn the power off or disconnect the power lead when checking a circuit with an ohmmeter to prevent damage to the meter. Disconnecting the circuit will also keep the meter from reading other parts of the circuit. When checking components with parallel circuits, remember that the meter does not know which path you wish to measure. It will include all possible paths. To be safe, disconnect one end of the component you are measuring. Many DMMs are designed not to affect solid-state components when measuring resistance.

Many DMMs are self-ranging and give a reading of any resistance value within their ability (Figure 12–36). Older analog meters require setting the meter to the range (1 k or 1,000 Ω, 10 k or 10,000 Ω, 1 m or 1,000,000 Ω) and recalibrating the meter for each range. An analog ohmmeter is calibrated by connecting the leads and rotating the calibration knob until the meter reads 0.

An ohmmeter is useful for checking an item such as a clutch coil when it is off the vehicle; connect the two ohmmeter leads to each end of the coil. If the resistance value matches the specifications (about 3 to 5 Ω), the unit does not have a short circuit and shows continuity. Moving one of the leads to the unit's mounting point checks for a grounded coil; at this time the reading should be infinite (∞ Ω) (Figure 12–37). A DMM displays *OL* for open leads to indicate infinite (out-of-limits) resistance.

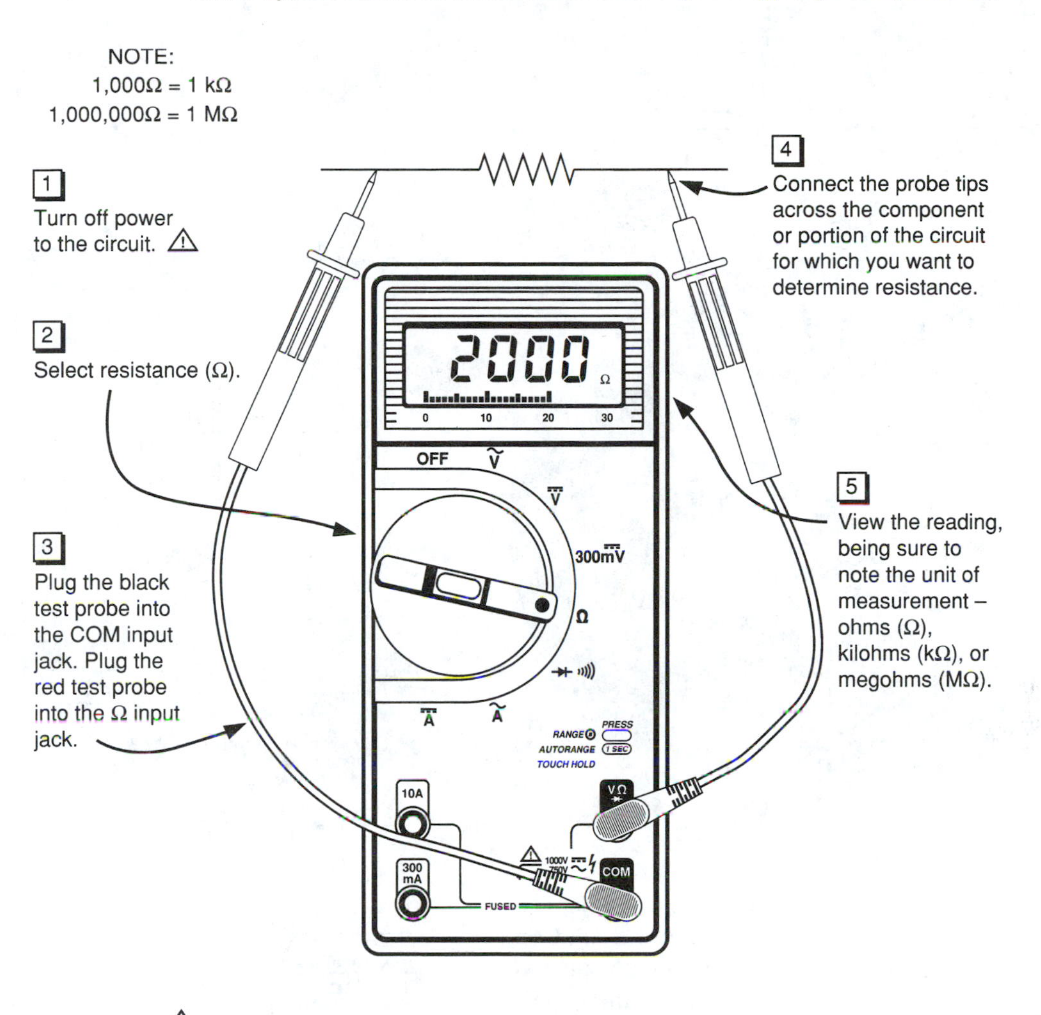

FIGURE 12–36 A digital multimeter being used to measure resistance. Be sure to turn off or disconnect the electrical power to the circuit when using ohmmeter functions. *(Courtesy of Fluke; reproduced with permission)*

✓SERVICE TIP

Dirty, slightly corroded switch contacts can be cleaned by rubbing them with a Pink Pearl eraser.

✓SERVICE TIP

It is possible for a clutch coil to have a weak internal connection and test out showing continuity and correct resistance with an ohmmeter. This weak connection causes reduced current flow, so many technicians prefer to check circuit resistance using an ammeter. If a circuit has the correct current flow with the correct input voltage, it has the proper resistance and therefore is good.

12.8.3 Measuring Amperage

An ammeter is used to measure amperage, and it is often connected by breaking the circuit and connecting the ammeter in series with the circuit. Some ammeters use a transformer-type, inductive pickup, which is simply placed over or around the wire. Most technicians prefer the inductive pickup because it is much easier to connect, and it does not change the circuit in any way. If the current readings measured in the circuit are less than specifications, a weak circuit with excessive resistance is indicated. If the current readings are higher than specifications, a shorted or grounded circuit is indicated (Figure 12–38).

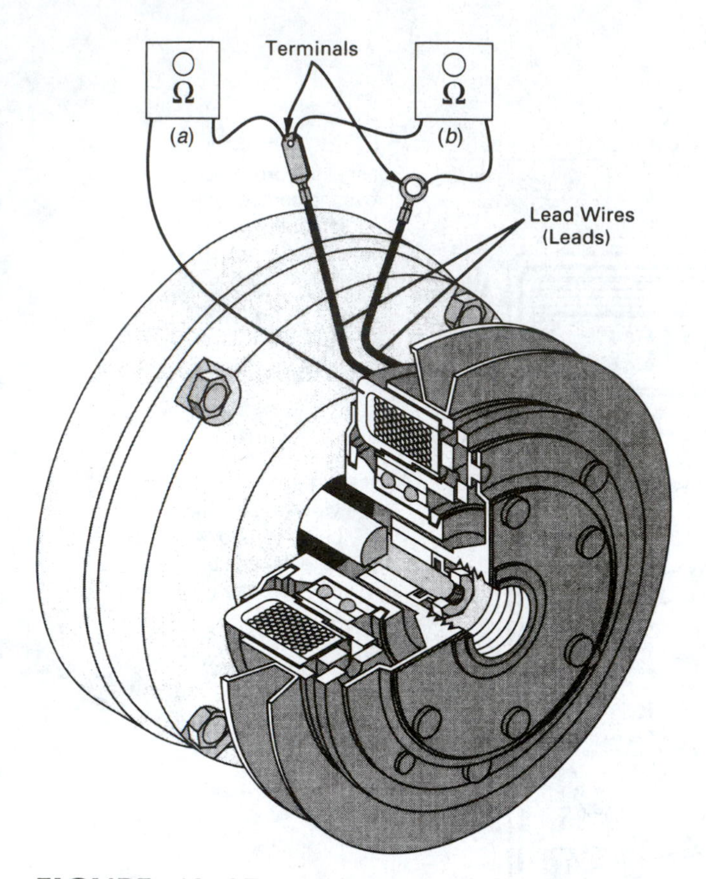

FIGURE 12–37 Moving one of the ohmmeter leads to the coil housing checks for a ground (*a*). Connecting the two leads of the ohmmeter to the two leads of a clutch coil measures resistance and checks for shorts (*b*). The resistance reading should match the manufacturer's specification; the ground reading should be 0. (*Courtesy of Warner Electric*)

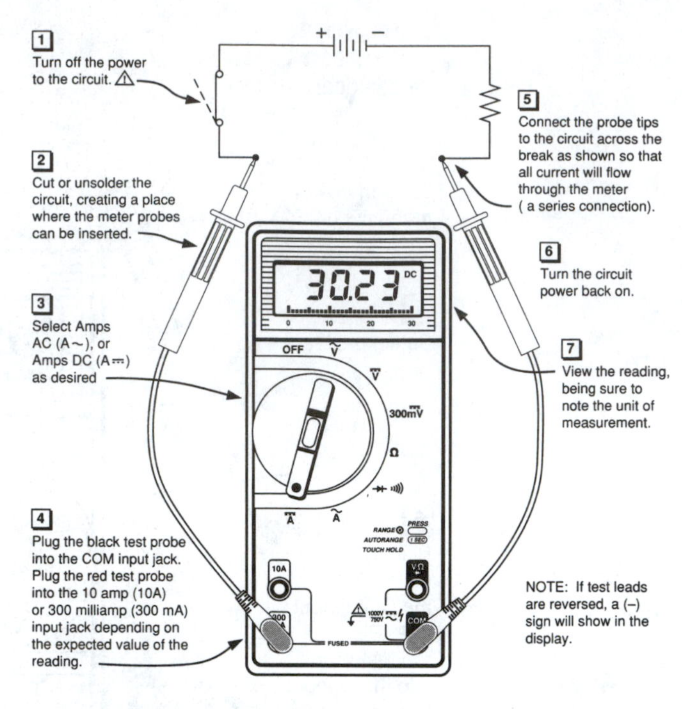

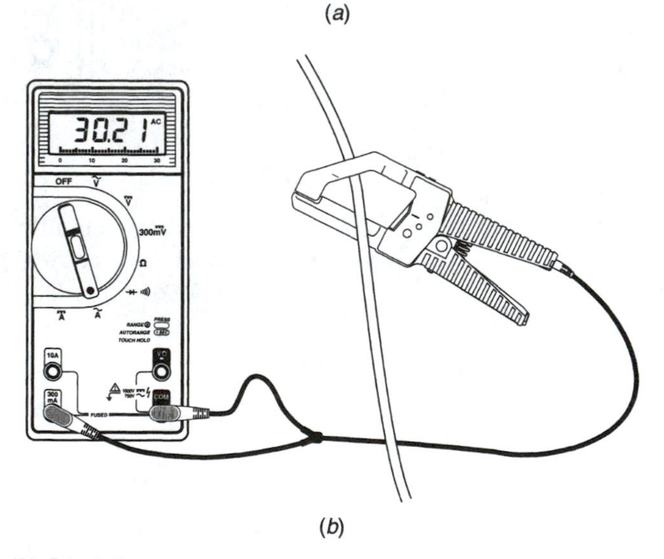

FIGURE 12–38 A digital multimeter being used to measure current flow (*a*). A transformer probe can be used to check current flow without breaking the circuit (*b*). (*Courtesy of Fluke; reproduced with permission*)

✓REAL WORLD FIX

The Toyota Supra (56,000 miles) has no A/C, and a check shows a blown 10 amp fuse. The fuse was replaced. The technician tried to get a current reading, but the fuse blew before the reading could be read.

Fix: The technician measured the coil resistance using an ohmmeter, and the reading was 1.2 ohms, too low. Replacement of the clutch coil fixed this problem.

✓SERVICE TIP

Some technicians diagnose a circuit by measuring the current flow through it. As components are switched on or off, the current flow should change to match the added or decreased load. For example, we know that a compressor clutch circuit is complete if the current flow increases by about 4 amps when the A/C is turned on.

✓SERVICE TIP

If using a inductive pickup and the current flow is very low, making current flow reading difficult, and there is sufficient slack in the wire, make a loop in the wire and close the induction pickup over the doubled wire. This will double the current reading.

✓SERVICE TIP

Many current flow checks are made with the engine off using B+ V. When troubleshooting problems caused by excessive current flow, make the critical current checks with the engine running using alternator voltage.

✓REAL WORLD FIX

The A/C fuse (10 amps) blows when the engine of the 1988 Camry is accelerated; it does not blow while the engine is idling. The wires were inspected, and they appear good.

Fix: The clutch coil was replaced, and this fixed the problem.

Author's Note: Apparently the coil had a short, and when the engine speed was increased, the higher voltage from the alternator increased the current flow.

✓SERVICE TIP

Some technicians make extended time period current flow checks using a fused jumper wire to supply current to the suspected component. This jumper is used to replace the current feed wire to that component using a fuse that matches the maximum current rating for it. The vehicle can then be driven for a period of time, and if the fuse burns out, you know that it used too much current.

✓REAL WORLD FIX

The compressor clutch on the 1995 Ford Windstar (112,000 miles) does not engage, and a voltage check shows no power to the clutch. Voltage checks at the integrated control module show good B+ power and grounds at the proper pins. Use of a scan tool shows normal system voltages, except there is a fluctuating WAC (wide open throttle cutoff) voltage accompanied with a fluctuating idle speed, and if all the outputs are turned on using the scan tool, the voltage drops to 7 V.

Fix: Checking the current flow through the low pressure switch shows 0.35 amps, but using a jumper around the switch increases the current flow to 3 amps, and this causes the clutch to engage. Replacement of the low pressure switch fixed this problem.

Author's Note: It is unknown why a voltage drop at the low pressure switch showed only a 0.18 V drop.

✓SERVICE TIP

A shorted clutch coil diode will cause the circuit to draw too much current, so the clutch circuit fuse will blow as soon as it is turned on. Diodes can be checked using an ohmmeter, but the easiest way to check this problem is to cut one of the leads to the diode. If the current flow is now okay, the diode is faulty and should be replaced. If the problem still exists, it is caused by different fault.

✓SERVICE TIP

Current draw can increase as a component reaches operating temperature. In one case, the current draw through a clutch coil was in the normal range, about 3 amps, but when warmed up using a heat gun, it increased to about 10 amps.

12.8.4 Interpreting Measurements

The technician must be familiar with the circuit and its components so he or she knows what to expect while measuring electrical values. Wiring diagrams can be used to follow the current flow through a circuit, much like a road map shows us how to get from one point to another. At one time these circuits were quite simple and easy to follow; with many modern vehicles, the technician has to study the diagram and any information available on that particular circuit to identify what he or she is working with and what path the current follows.

✓SERVICE TIP

Some students have difficulty reading wiring diagrams. The following tips should make this easier:

- Begin by locating the component (compressor clutch, blower, etc.) you are concerned with.
- Find the path to ground for the component, and note any switches, wire connectors, wire splices in the ground path.
- Find the path to B+ from the component.
- Note any fuses, relays, switches, wire connectors, and splices in the B+ connection.
- Note how the relays are controlled, whether the relay is NO or NC, or if it is a double-contact relay, and note the path of the relay control circuit.
- If the circuit is complex, make a copy of it or cover it with a clear sheet of plastic; now, you can use a marker pen to clearly identify its paths.

Exact specifications are often not available for many automotive circuits. The resistance value or current draw for a clutch or relay coil is not always given. When testing with a meter, the technician often has to guess what a usable range should be. In most circuits, B+ voltage should be found up to the major output device, except for an allowance for slight voltage drop at the connectors.

✓SERVICE TIP

A possible cause for no blower operation is ice locking the blower wheel. It is possible for the HVAC case of a vehicle that is parked in the rain with a plugged evaporator drain to partially fill with water. If the temperature drops, the water can freeze. This problem can cause early and total failure of the blower.

✓REAL WORLD FIX

The 1993 Saab had a faulty blower replaced. Four months later it came back with intermittent blower operation and a buzzing noise from the blower during right turns. It was determined that water in the evaporator case was getting into the blower wheel, but checking showed that the evaporator drain was open.

Fix: On closer inspection, another drain, in a hard-to-see location, was found. This hose was twisted, causing it to block draining.

✓SERVICE TIP

A rule of thumb for most clutch coils is about 3 to 5 Ω of resistance and a 3- to 5-amp current flow.

12.9 ELECTRICAL SYSTEM REPAIR

Normally a faulty electrical component such as a switch, relay, blower motor, or clutch is repaired by removing and replacing it. The R&R operation is usually a relatively simple process of disconnecting the wires or connectors, removing the component, installing the new component, and reconnecting the wires and connectors. When replacing a blower motor, make sure the replacement is correct. Some variables and critical dimensions are shaft diameter and length, motor diameter and length, location and spacing of mounting bolt holes, and direction of rotation (Figure 12–39). Clutch coil replacement requires partial disassembly of the compressor (described in Chapter 14).

Occasionally a technician must replace a faulty connector or wire by splicing the wire. A few connectors have the wires molded into them, so replacing the connector requires splicing the new connector to each wire. In most cases, however, individual wires can be removed from the connector. These wires have an end terminal that has a locking tang that expands to hold the terminal in the connector. Various special terminal disconnecting tools are available. The tool is pushed against the locking tang and depresses the tang so the terminal can be pulled out (Figures 12–40 and 12–41).

✓SERVICE TIP

Be aware that it is possible to cause the vehicle's air bags to deploy if the wrong electrical wires are connected. This deployment can possibly cause injury to the technician or bystanders; it will cause an expensive repair. It is recommended that the air bags be deactivated using the manufacturer's recommended procedure before doing any testing or work under the instrument panel.

A wire can be spliced by following a fairly common procedure:

1. If replacing a wire, make sure the new wire is the same size or larger than the original. Strip off an amount of insulation slightly longer than the splice clip, or about 3/8 to 1/2 inch (10 to 13 mm) long (Figure 12–42).

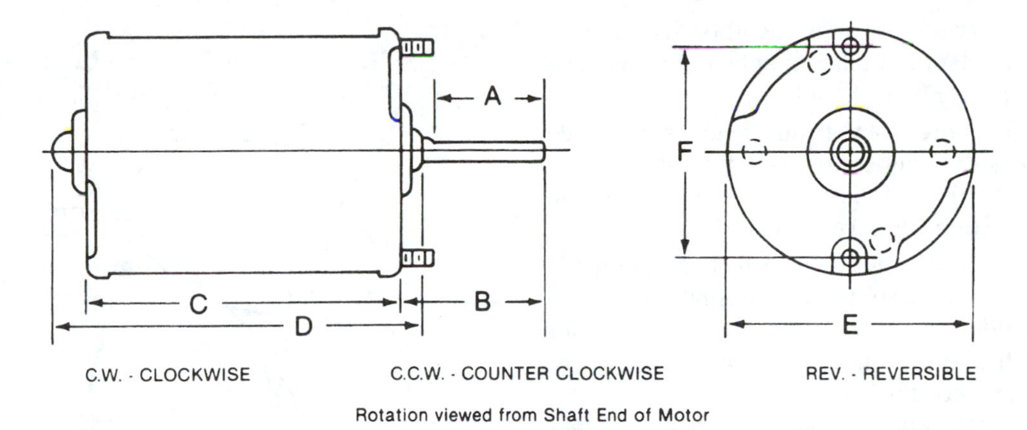

FIGURE 12–39 These dimensions should be checked when obtaining a replacement blower motor to make sure you have the correct replacement. *(Reprinted with permission of General Motors Corporation)*

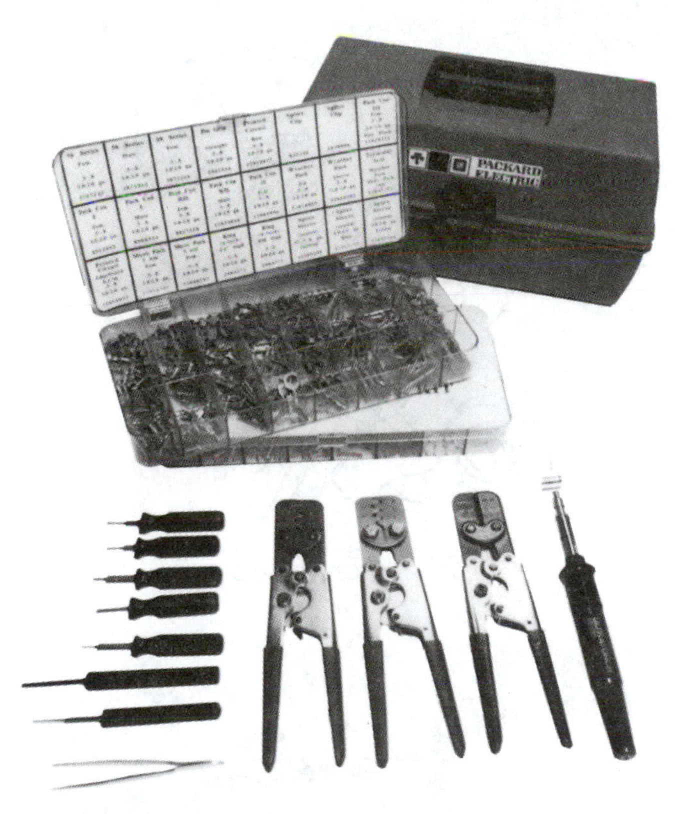

FIGURE 12–40 A terminal repair set. It includes replacement connectors and special pliers to install them, as well as a group of tools to remove terminals from weather-tight connectors. *(Courtesy of Kent-Moore)*

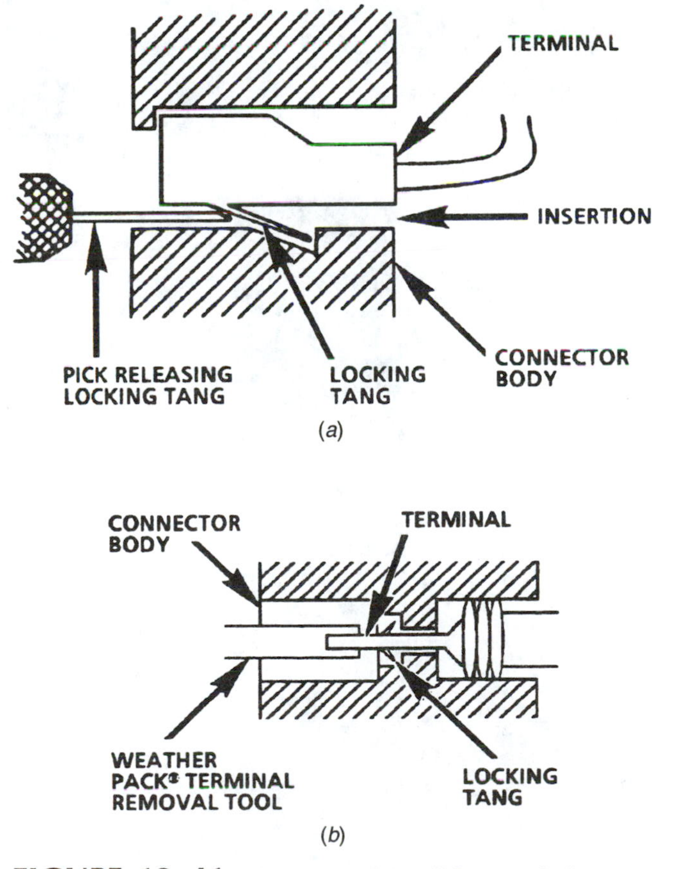

FIGURE 12–41 A terminal is slid out of the connector body (*a*) after compressing the locking tang with a special probe (*b*). *(Reprinted with permission of General Motors Corporation)*

2. Push the two wire ends together so the bare wires overlap and place a splice clip over the connection (Figure 12–43).
3. Use a crimping tool to firmly squeeze the splice clip onto the connection. *Or,* if a splice clip is not being used, twist the connection so the wires are tight (Figure 12–44).
4. Use a soldering gun or iron to heat the wires enough to melt solder, and apply 60/40 rosin core solder to the hot wires until the solder flows through the joint. Do not use acid core solder (Figure 12–45).
5. Insulate the splice by wrapping it with either plastic electrical tape or a shrink tube. A shrink tube is a plastic tube slid over the splice and heated with a match so it shrinks tightly in place.

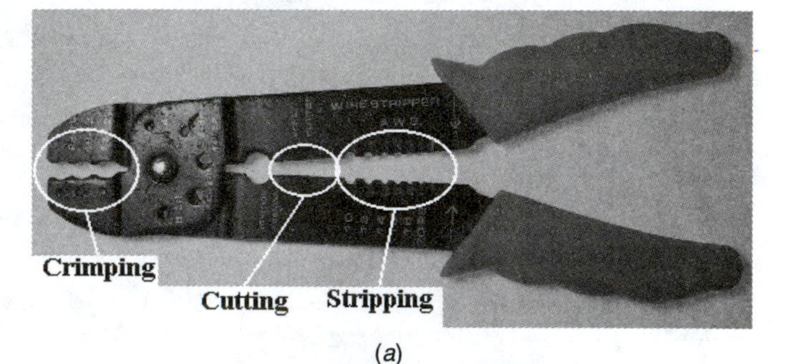

(*a*)

(*b*)

FIGURE 12–42 A wire stripping and crimping tool (*a*). The stripping area is used to cut the insulation and pull it off the wire (*b*).

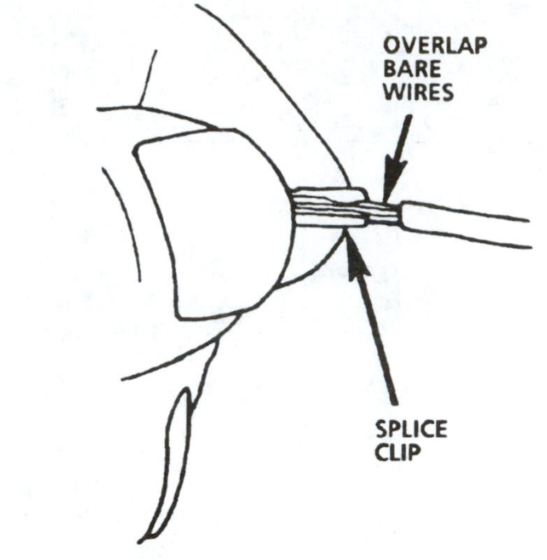

FIGURE 12–43 When splicing a wire or installing a terminal, the stripped section of wire is slid into the connector. (*Reprinted with permission of General Motors Corporation*)

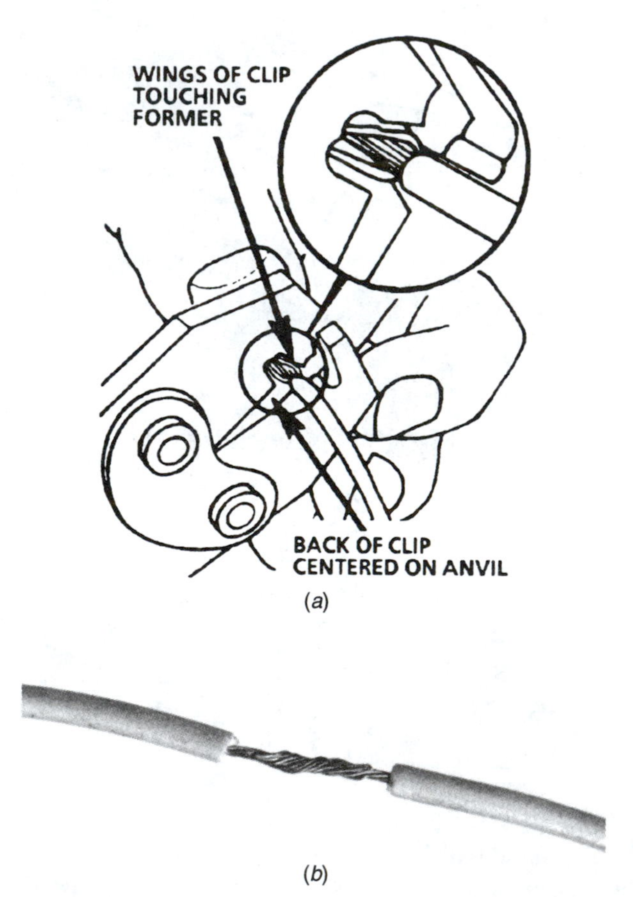

(*a*)

(*b*)

FIGURE 12–44 A splice clip or terminal is locked in place by squeezing it (*a*). If a clip is not used, the bare wires are twisted together before soldering (*b*). (a *is reprinted with permission of General Motors Corporation*)

✓SERVICE TIP

The wire stripper can be used as a gauge: The smallest opening that cleanly strips the insulation without nicking or cutting the wire strands is the wire gauge.

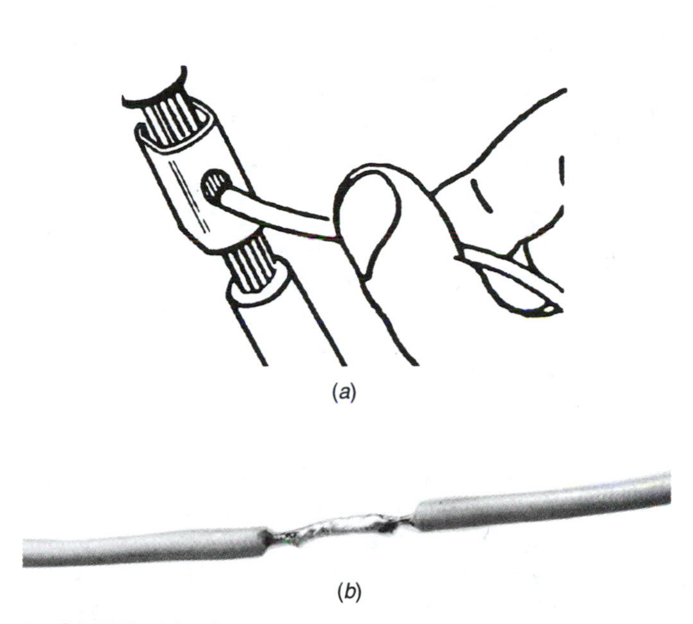

FIGURE 12–45 It is good practice to solder the wires to the terminal after crimping (*a*). The wires must be soldered if a connector is not used (*b*). (a *is reprinted with permission of General Motors Corporation*)

✓SERVICE TIP

If using a shrink tube, slide a piece of tube, about 1/2 inch longer than the splice, over the wire before connecting it in step 2. Slide the tube away from the connection while soldering it. After soldering the connection, slide the tube to the proper location and heat it to shrink the tube to lock it in place.

If using tape, make sure the tape is wrapped tightly and smooth and looks neat (Figure 12–46).

✓REAL WORLD FIX

The 1993 Honda Prelude came in with a complaint of the A/C shutting off and blowing hot air. This problem is intermittent, occurring as many as three times a day. The A/C relay was cycled many times with no malfunction. The A/C pressures are good, and the compressor clutch gap is 0.028 inch, within specifications.

Fix: Following advice from other technicians, the A/C switch control circuit board was inspected for bad solder joints; none were found, but tapping and twisting the board caused a momentary loss of compressor operation. All of the solder joints that looked suspect were resoldered, and this fixed the problem.

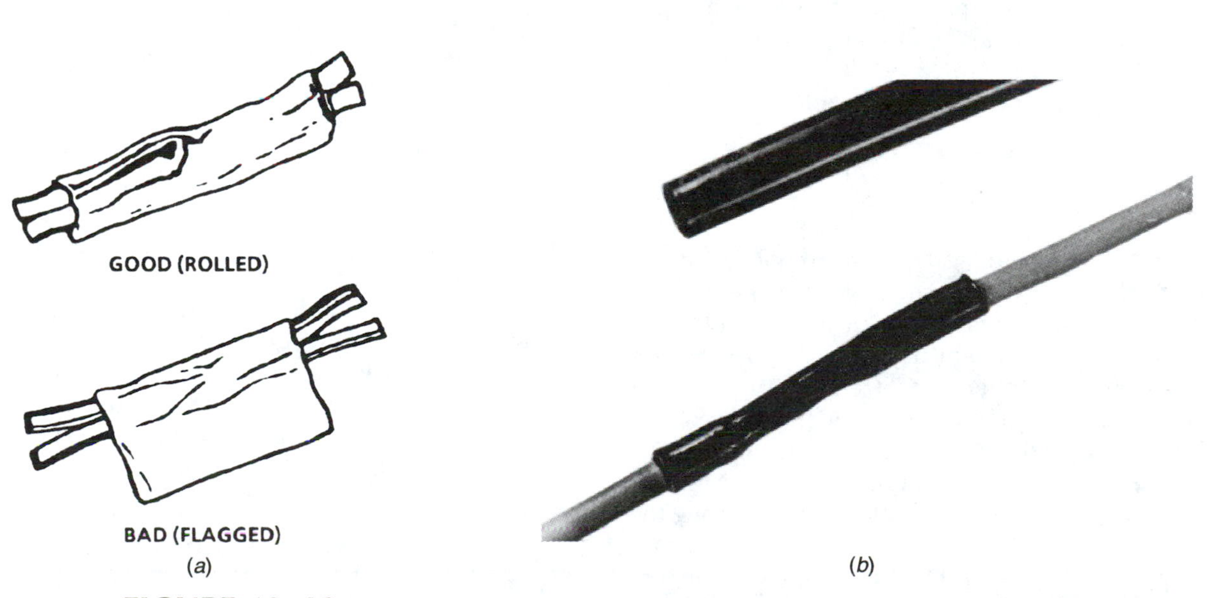

FIGURE 12–46 A wire connection should be insulated by wrapping it tightly with tape (*a*) or using shrink tubing (*b*). (a *is reprinted with permission of General Motors Corporation*)

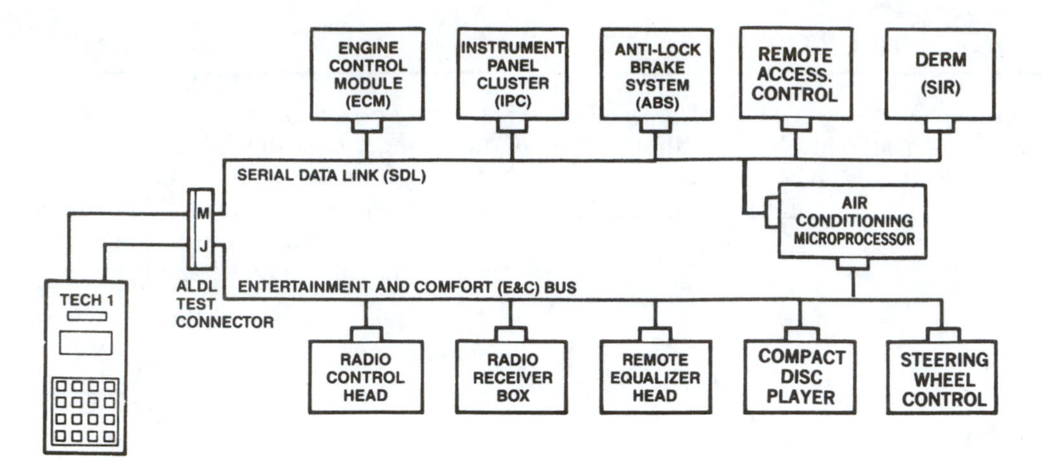

FIGURE 12–47 A scan tool (left) can be connected to the assembly line data link (ALDL) connector to obtain test information from these different sources. *(Reprinted with permission of General Motors Corporation)*

12.9.1 Electrical Component Replacement

Most faulty electrical components are simply removed by disconnecting the electrical connections and removing the mounting fasteners. They are replaced by reversing the procedure. The procedure to remove and replace a compressor clutch coil is described in Section 14.2.2.

A wise technician will always inspect the electrical connector and all of its terminals to make sure that they will make good contact and make sure that the connector latches lock securely.

12.9.2 Electronic System Self-Diagnosis

Most modern electronic systems use a control module that can perform a trouble diagnosis on itself, its sensors, and its output circuits each time it starts up. If it locates an electrical problem, it will set and display a DTC. This can be either **soft** (temporary) or **hard** (semipermanent). A soft code is erased from the control module's memory when the ignition key is turned off; a hard code is erased by performing a special operation, pressing certain control head buttons, or removing the fuse for the control module. This is called **clearing codes.** Some systems can record and display the history of past failures. Self-diagnosis is very specific to the model of vehicle. Follow the exact directions given by the vehicle's manufacturer or, if using a portable scanner, the directions given by the manufacturer of that scanner (Figure 12–47).

In most systems, self-diagnosis is entered (started) by pushing a particular combination of buttons on the control head. The codes resulting from self-diagnosis are displayed (1) on the control head display, (2) on a handheld scanner unit, or (3) on a voltmeter or test light. A scanner is a tool used to enter self-diagnosis, display DTCs, and perform diagnostic checks (Figure 12–48). In most systems, the DTC is read as a one- or two-digit number or a combination of letters and numbers; some systems display the code by a pattern of pulses from a voltmeter or flashes of light . This is often called a blink code.

The DTC number indicates the nature of the problem. This number is keyed to a series of tests that must be performed to locate the exact fault. These tests usually involve measuring the voltage or resistance of particular portions of the circuit and are found in any good service manual.

After the fault is located and repaired, the codes must be cleared. With soft codes, this is easily done by turning off the key. Hard codes are cleared by pushing a particular combination of buttons on the control head, performing certain operations with the scan tool, or removing the control module fuses.

✓REAL WORLD FIX

The A/C compressor on the Suzuki Sidekick (56,000 miles) does not operate. There is voltage at the A/C fuse

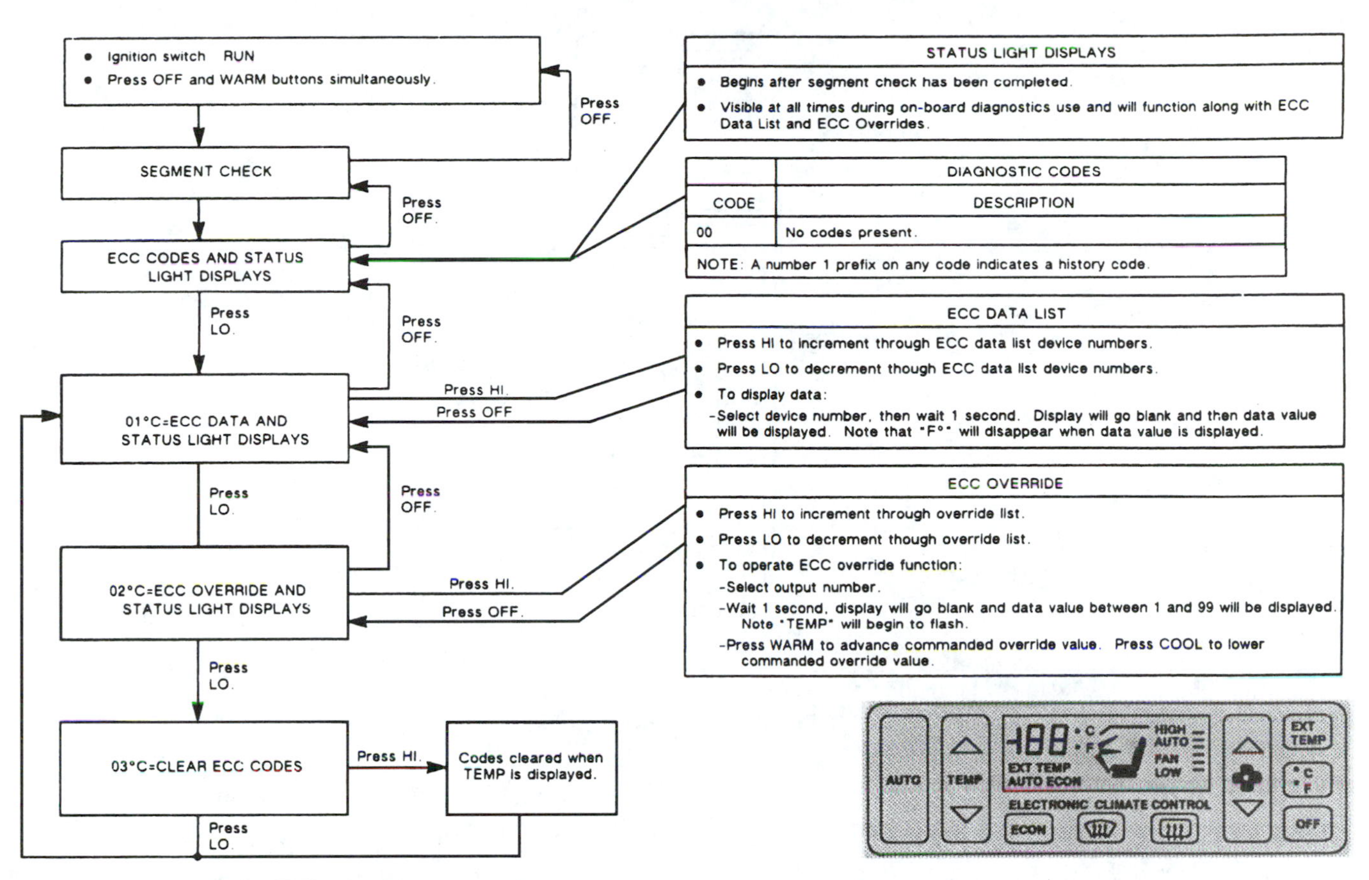

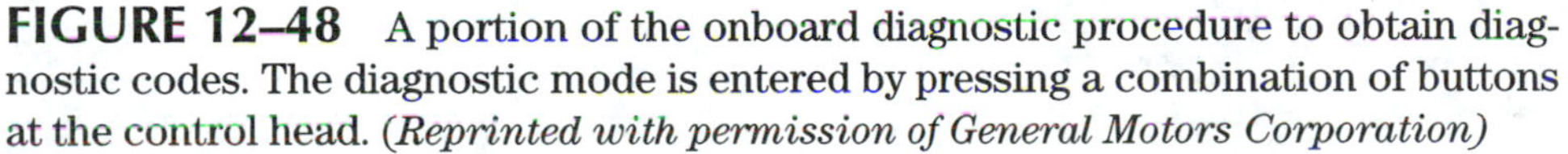

FIGURE 12–48 A portion of the onboard diagnostic procedure to obtain diagnostic codes. The diagnostic mode is entered by pressing a combination of buttons at the control head. *(Reprinted with permission of General Motors Corporation)*

but not at the compressor. The PCM increases the idle speed when the A/C is turned so it is receiving the A/C request signal.

Fix: A check of the wiring using a wiring diagram revealed that someone had removed a stereo system, disconnected the B+ lead, and tucked the lead under other wires. Reconnecting the wires fixed this problem.

Author's Note: It is a good practice to keep an eye out for previous repairs and modifications.

✓REAL WORLD FIX

The 1995 Lincoln Town Car (115,000 miles) has no A/C; the compressor clutch does not engage. This is an ATC system, and all the fuses are good. There is power to the control module but no power out. A jumper wire between the proper connections will produce clutch operation.

Fix: Following advice, the technician checked the WOT relay and found it to be good. He then back-probed the control module output using a test light as he tapped on the side of the module. Power came on, and this indicated a faulty module. Replacement of the control module fixed this problem.

✓REAL WORLD FIX

The automatic transmission of the 1995 Oldsmobile 98 (85,000 miles) only has third gear when shifted to drive. The technician has found that a fuse blows, and this fuse is also for the A/C clutch. This transmission technician has a concern that this problem could be caused by the clutch clamping diode.

Fix: Following advice, the technician found the clamping diode and found it has a short. Replacement of the A/C clamping diode fixed this transmission problem.

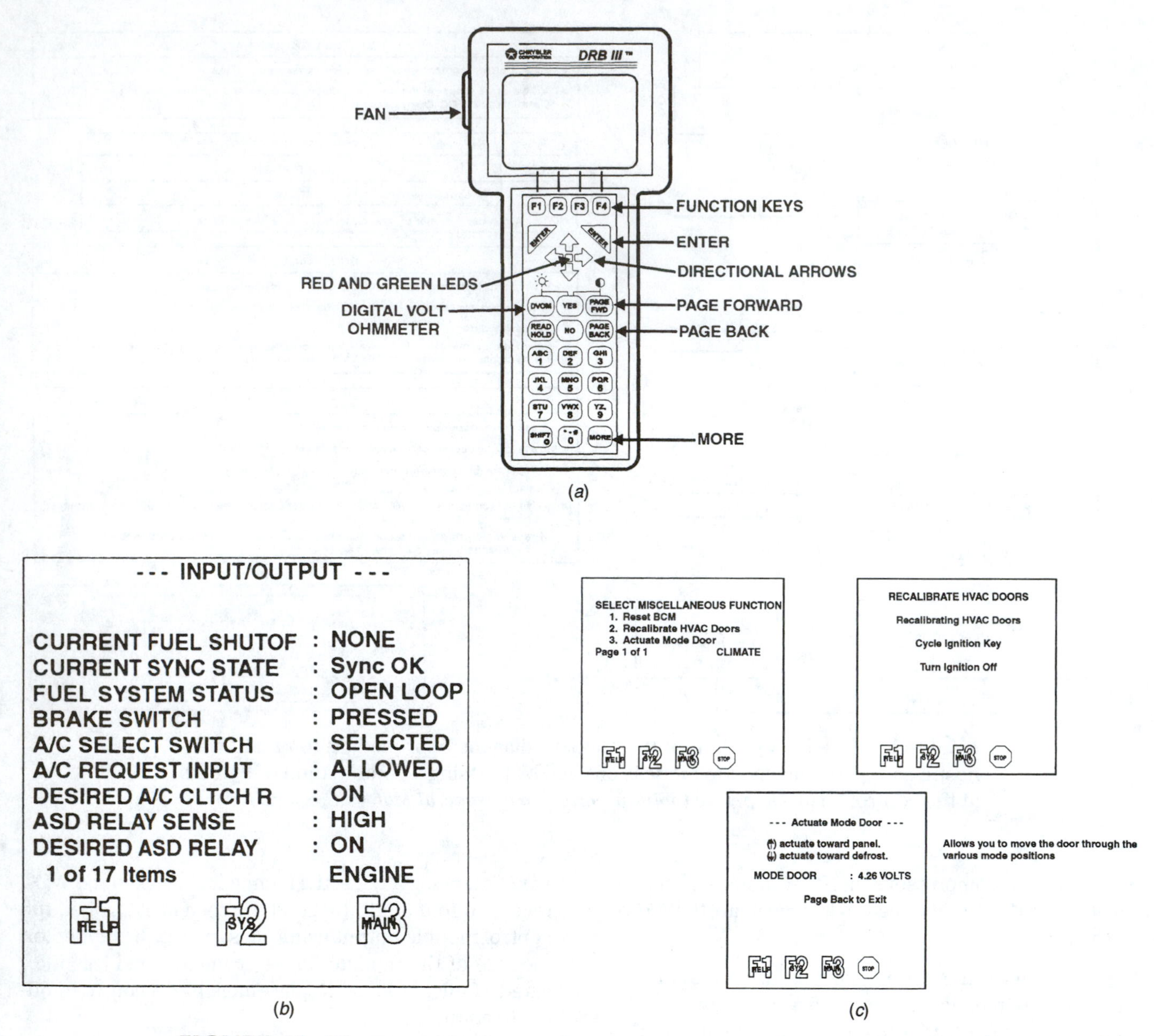

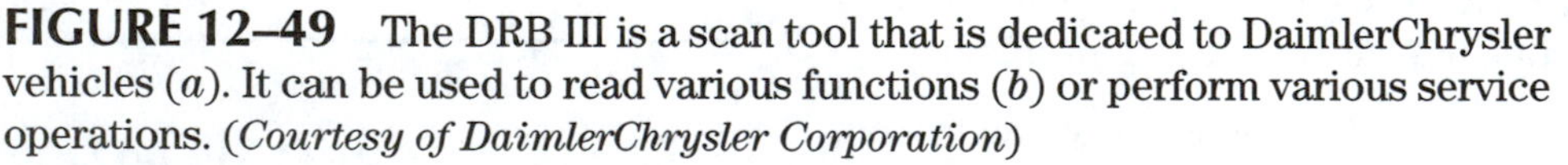

FIGURE 12–49 The DRB III is a scan tool that is dedicated to DaimlerChrysler vehicles (*a*). It can be used to read various functions (*b*) or perform various service operations. (*Courtesy of DaimlerChrysler Corporation*)

✓SERVICE TIP

A scan tool lets you erase codes without affecting any other electrical circuits. Disconnecting the battery erases all of the electronic memory of most vehicles, including clock settings and station presets for the radio, which may irritate the vehicle owner. Removing the control module fuses erases the memory of only that control module but can also cause problems. Some A/C systems are controlled through the ECM, and some of these have adaptive learning for the engine and transmission. On them, the ECM adjusts certain engine functions to particular geographic locations or driving styles; this takes up to 100 miles of driving. If you pull the fuse on these cars, it can take about 100 miles of driving before engine functions are switched from a normal setting to the owner's location and driving style. During this time the vehicle can run poorly.

After the faults have been repaired and the codes erased, rerun the self-diagnosis to make sure that all of the faults have been corrected.

✓SERVICE TIP

At least one system counts the time that the clutch cycles on and off; it determines if it cycles too fast or is off too long relative to the on time. This system is programmed to call this "low refrigerant" and will shut off the clutch. After the leak is repaired and the system recharged, some systems will not reenergize the clutch until certain fault clearing operations are performed with the control head.

✓REAL WORLD FIX

The 1993 Jeep Cherokee (123,000 miles) has a problem of constant heat at floor level. A scan tool check shows codes 21 and 23, "Mode and blend doors not moving." The doors operate if voltage is supplied to them, and the grounds at both motors are good. The controller was replaced, but this did not fix the problem.

Fix: Following advice, the technician checked the multi-pin connector below the glove box, and it was found to be partially plugged in. Making a complete connection at this connector fixed this problem.

12.9.2.1 Scan Tools. Scan tools are available from vehicle manufacturers and aftermarket sources (Figure 12–49). These units are connected to the vehicle's diagnostic connector and are bidirectional test units that can save considerable time in locating electrical faults. Scan tools allow the technician to:

- Read DTCs and, in some cases, serial data.
- Read signals from the sensors to the control module.
- Read signals from the control module to the actuators.
- Command the control module to operate the actuators.
- Record a snapshot of a problem occurrence showing both the input and output signals.
- Clear the DTCs.

Scan tools can display only electrical or electronic signals and can only indicate if a problem occurs in a particular circuit. Further checks are required to locate the exact cause of a problem.

✓REAL WORLD FIX

The 1994 Cadillac (96,000 miles) came in with no refrigerant. The leak was found and repaired. The system was recharged, but the compressor clutch would not engage. The battery was disconnected to clear any codes, but this did not help. A scan tool check shows the compressor relay is not getting a ground signal from the ECM.

Fix: A check revealed that the evaporator inlet sensor had been broken because someone had installed the OT backward. Replacement of the sensor and proper installation of the OT fixed this problem.

✓REAL WORLD FIX

The 1992 GMC pickup (115,000 miles) came in with the A/C light blinking on the control panel and indicating a low charge level. The faulty compressor front seal was replaced along with the accumulator, OT, and high-pressure cutoff switch. The system was recharged, and it worked well, except the A/C light was still blinking. Disconnecting the battery cable stopped the blinking light, but after a few days it began blinking again.

Fix: After checking TSB #92-1B-90, the technician disconnected the green wire from the control head, and this cured the blinking light.

✓REAL WORLD FIX

The 1992 Buick (101,000 miles) A/C system does not work; the compressor clutch does not engage. A jumper wire across the A/C relay causes the clutch to engage, and the system works normally. All three of the system switches check out as good, and the system has a full refrigerant charge.

Fix: A scan tool was connected to the system, and the DTCs were cleared. This fixed the system.

Author's Note: The DTCs were probably set by an earlier fault that was repaired, but the trouble codes were not cleared.

✓ REAL WORLD FIX

The 1999 Buick (48,000 miles) was hit hard in the rear end, and this accident deployed both air bags. The instrument panel had been partially disassembled to replace the air bags. The car was returned to the body shop that did the repairs, with a complaint of low heat output from the left registers (dual zone system). After attempting repairs, the body shop sent the car to another repair shop. The repair shop checked and repaired all the DTCs, and these appeared to have been caused by the body shop's repair attempts. The two blend door motors were removed and bench checked, and they seem to be operating correctly. Supply and reference voltages at both motors are correct. Watching the door operation using a scan tool shows normal operation of the right blend door motor, but the left motor goes to mid-position and stays there. The technician attempted to reprogram the system but could not.

Fix: Following advice, the technician replaced the programmer, and this fixed the problem. The calibration procedure was to turn on the ignition for three minutes, and the programmer went through the procedure, operating the doors and fan motor.

12.9.2.2 Actuator Motor Calibration Some systems require a calibration procedure to ensure that the electric motor will operate the door completely, and so the ECM knows what position the motor is in. This procedure should be performed each time a motor or control head is replaced. The calibration procedure is described in the service information.

CHAPTER QUIZ

These questions help you study this chapter. Enter the proper word(s) in the blanks to complete each statement.

1. Electrical force is called ________; a(n) ________ is the amount of electrical current flow; and ________ are the units of electrical resistance.
2. Most electrical circuits include a(n) ________ ________, ________ ________, ________ ________, ________, and ________ ________.
3. Electric wires transmit electricity through the ________ while the current flow is contained by the ________.
4. The strength of a compressor clutch coil is determined by the number of ________ ________ and the amount of ________ ________.
5. The diode at the clutch coil protects the electronic ________ ________ from voltage ________.
6. An in-vehicle sensor is a type of ________ that increases resistance when the temperature ________.
7. A(n) ________ converts a(n) ________ signal into an electrical signal.
8. If the ECM finds a problem in an electronic control system, it will turn on the ________ ________ ________ and set a(n) ________ ________ ________.
9. Most actuator motors contain a(n) ________ ________ so the ECM will know the motor position.
10. A smart control head contains ________ for motor ________.
11. ________ allows a pair of wires to be used for more than one ________.
12. A strategy that prevents potential compressor problems from damaging the accessory drive belt is called ________ ________ strategy, and it works when the compressor ________ ________ signal shows a slower speed than the engine.
13. A(n) ________ ________ is when electric current leaves the wire it is supposed to be in to go to another wire, and a(n) ________ ________ is when electric current leaves the wire it is supposed to be in and goes to ground.
14. A blown fuse is often an indication of ________ ________ flow.
15. An ECM can be damaged by ________ if you touch one of the terminals with your finger.
16. A clutch coil should have at least ________ ________ at the B+ terminal when the engine is running.
17. A ________ ________ allows you to read DTCs, the signals from the sensors to the ECM, and the signals from the ECM to the actuators.
18. The best way to clear a DTC without erasing memory of other electronic circuits is to use a(n) ________ ________.

REVIEW QUESTIONS

These questions allow you to check what you have learned. Select the answer that correctly completes each statement.

1. Which of the following is the unit for electrical pressure?
   a. Amp
   b. Ohm
   c. Volt
   d. Watt
2. Most automotive electrical systems use AC current.
   a. True
   b. False
3. A fusible link is usually four wire sizes ________ than the wire used in that particular circuit.
   a. larger
   b. smaller
   c. Either a or b
   d. Neither a nor b
4. Which of the following is considered an output device in an ATC system?
   a. Compressor clutch
   b. Blower motor
   c. Both a and b
   d. Neither a nor b
5. Which of the following is monitored by a sensor in an ATC system?
   a. Ambient temperature
   b. In-vehicle temperature
   c. Sunload
   d. All of these
6. Electrical mode doors in an ATC system usually have a feedback sensor string connected to them.
   a. True
   b. False
7. Which of the following test instruments can be used on any automotive electrical circuit?
   a. Test light
   b. Ohmmeter
   c. Analog multimeter
   d. DMM
8. Modern electronic circuits can be damaged by ________.
   a. electrostatic discharge
   b. excess heat
   c. rough handling
   d. Any of these
9. Which of the following is true concerning ohmmeters?
   a. Ohmmeters are self-powered.
   b. Ohmmeters can be used on live circuits.
   c. Ohmmeters cannot be used to check continuity.
   d. None of these
10. A rule of thumb for clutch coil resistance is ________.
    a. 10 ohms
    b. 3 to 5 ohms
    c. about 5 amps
    d. 5 to 10 ohms
11. A vehicle with an ATC system has heater and defroster operation, but the A/C is not working properly. To diagnose this problem, the technician should
    a. check for trouble codes.
    b. clear the trouble codes.

    Which is correct?
    a. A only
    b. B only
    c. Both A and B
    d. Neither A nor B
12. An example of a short circuit is a
    a. clutch coil with a resistance of 0.5 Ω.
    b. burned-out lightbulb.
    c. burned wire with bad insulation that touches ground.
    d. Any of these
13. The blower motor runs on high speed only. This indicates a faulty
    a. high blower relay.
    b. resistor bank.
    c. switch.
    d. None of these
14. The vehicle overheats because the cooling fan does not operate. This could be caused by a
    a. faulty fan relay.
    b. bad blower motor ground connection.
    c. faulty blower motor.
    d. Any of these

CHAPTER

13

Refrigerant Service Operations

LEARNING OBJECTIVES

After completing this chapter, you should:

- Be familiar with the preventive maintenance operations and the adjustments necessary to keep a heating and A/C system operating properly.
- Be able to perform basic A/C service operations, given the operating manual for the equipment.
- Be familiar with how to retrofit R-134a into an R-12 system.
- Be able to complete the ASE tasks related to refrigerant handling.

TERMS TO LEARN

black death
cfm
charging cylinder
evacuate
flush
gas fingerprint
identifier
in-line filter
micron
out-gasses
partial charge
pressure-temperature (PT)
recover
recycle
retrofit
tank certification
topping off
vented

13.1 INTRODUCTION

The service and repair of heating and A/C systems consist of preventive maintenance operations; the adjustment, repair, overhaul, or replacement of system components; and standard A/C service operations. These standard operations include **identifying the refrigerant** in a system to determine what it is and if it is contaminated, **recovery and recycling of good refrigerant, recovery and disposal of contaminated refrigerant, evacuation** of a system, **recharging** a system with refrigerant, **checking the oil level** in a compressor or system, and **retrofitting an R-12 system** with an alternate refrigerant, preferably R-134a.

Service steps are often performed along with repair operations; for example, replacement of a faulty compressor begins with recovery of the refrigerant and ends with recharging the system. Normally, repair operations are necessary because of improper system operation or failure. The diagnostic checks described in Chapter 10 show the cause of the failure and what repair operations are needed. (These repair operations are described in Chapter 14.) Before proceeding with any service or repair procedures, be sure to protect your eyes and skin.

13.2 PREVENTIVE MAINTENANCE AND ADJUSTMENT OPERATIONS

An often neglected area, preventive maintenance and adjustment operations are comparable to the engine-off, under-hood checks described in Section 9.2. The purpose of preventive maintenance is to locate potential problems before they cause a failure.

The major wearing item in a heating and A/C system is the compressor drive belt; with many late-model vehicles, this belt also drives the alternator and water pump. Most vehicles will not operate for long if this belt fails, so it is usually replaced at the first signs of fatigue. Because the first indication of possible failure is usually a broken belt, many service technicians recommend replacing the drive belts every 4 or 5 years. This might be sooner than really necessary, but the cost of belt failure is usually much greater in both time and money (Figure 13–1) (See also Figure 9–2). Belt inspection and replacement are described in more detail in Chapter 16.

Another preventive maintenance operation is to check for debris and bent fins at the condenser. Debris is usually cleaned off using an air nozzle or water spray. A stiff fiber brush, such as a denture brush, can also be used; brushing is in a direction parallel to the fins (Figure 13–2). Bent-over fins can be straightened using a small probe, fin comb, or fin straightener.

FIGURE 13–1 As you inspect a drive belt, roll the belt so you can check the sides and inner face. Any signs of possible failure mean the belt should be replaced.

13.3 A/C SERVICE OPERATIONS

CAUTION

Remember that there is a possibility of injury from refrigerant contact. Goggles or a face shield should be worn to keep liquid refrigerant from getting in the eyes, and gloves should be worn to protect the hands. Avoid breathing refrigerant vapors.

Before any repair operations can be done on an A/C system (except for electrical and some compressor clutch operations), the refrigerant must be recovered from the system. Refrigerants must not be released or vented into the atmosphere; a good technician performs the service and repair operations in such a way as to prevent refrigerant venting during or after the repair (Figure 13–3). At one time, refrigerant was simply **vented** or bled out, discharged, and released, the major concern being the amount of oil that was lost. Now venting is prohibited by the Clean Air Act. New, pure (or virgin) R-12 was charged back into the system. Today's A/C service usually begins with the recovery of whatever refrigerant is left in the system. This refrigerant will be recycled into that or another A/C system; refrigerants have become too valuable to waste (Figures 13–4 and 13–5).

Refrigerant identification has become extremely important. R-12 is out of production and the existing stocks are gone or nearly gone. Bootleg R-12 is available, but some of this is of a counterfeit and unknown composition. With R-134a available in small cans at

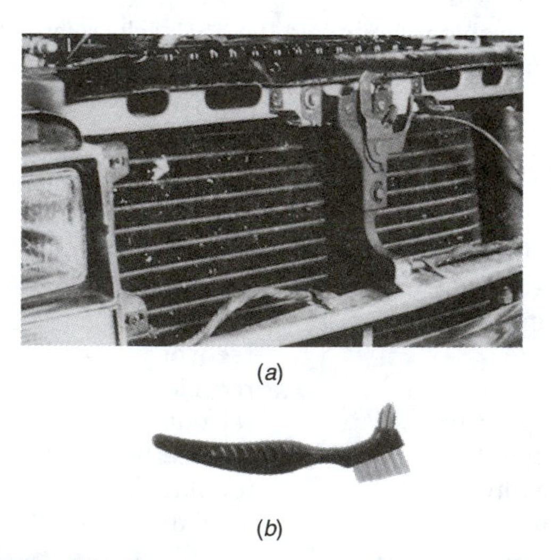

(*a*)

(*b*)

FIGURE 13–2 The condenser should be checked for debris at the front and rear of the core (*a*). Debris can be removed using a denture brush (*b*). (a *is courtesy of Everco Industries*)

many stores and quite a few blends available, a wise technician does not recover refrigerant unless he or she is sure of what is being recovered from the system. Imagine recovering a refrigerant contaminated by R-22 or a hydrocarbon into your recovery bottle that is half full of R-12. Even worse, imagine that you recycled this refrigerant, and you used it to charge another system. A recycling machine is designed to remove water, oil, solids, and air, not another refrigerant. The result of this action is that you have contaminated your recovery and recycling machine, contaminated and ruined the R-12 in the bottle, and contaminated the other system(s). Many compare contaminated refrigerant to a virus that can sometimes spread at a very rapid rate.

Another practice that has changed is system flushing. It used to be common practice to flush portions of an A/C system using a liquid to remove debris and foreign particles, usually the result of a failed compressor. The most common flushing agents were R-11, R-113, and petroleum solvents. Because the two refrigerants contain CFCs, they should not be used or released into the atmosphere. Petroleum solvents present a potential fire hazard, and it is extremely difficult to remove all of the solvent from the system when flushing is complete; the detrimental effects solvents have on the system are

FIGURE 13–3 Refrigerant must not be released or vented into the atmosphere.

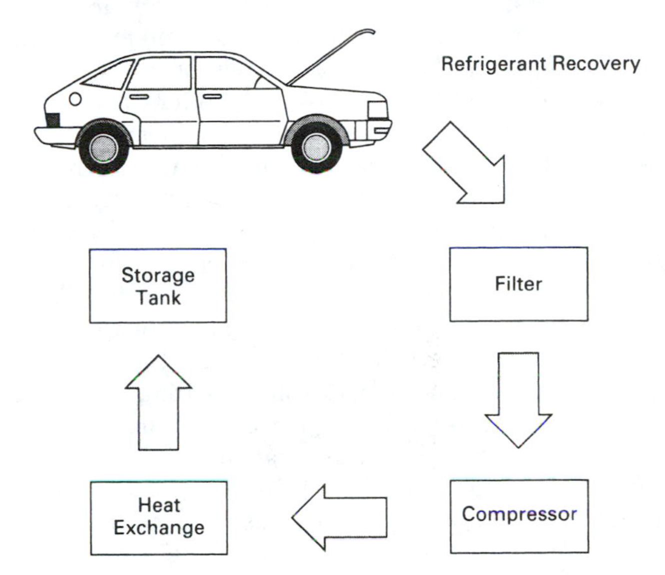

FIGURE 13–4 A recovery unit removes refrigerant vapor from the vehicle. It then filters the refrigerant before compressing it so it condenses and can be stored as a liquid in the storage tank.

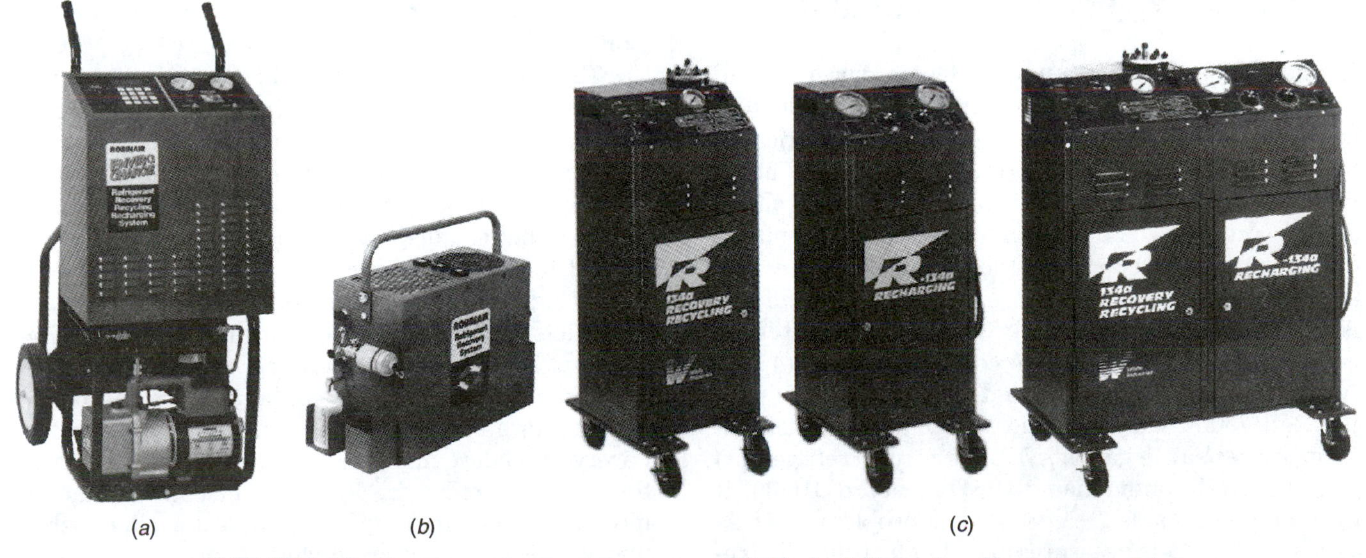

FIGURE 13–5 Recovery machines are available in full-service form (in combination with recycling, vacuum pump, and recharging functions) (*a*), recovery only units (*b*), and recovery and recycling combinations (*c*). This particular unit (*c*) can be "married" into a single unit or "divorced" into two separate units. (a *and* b *are courtesy of Robinair, SPX Corporation;* c *is courtesy of White Industries*)

now known. In early R-134a retrofit tests, more failures occurred in systems that were flushed with R-11 or R-113 than in systems that were not flushed. These failures were attributed to contamination from flushing agents that remained in the system. Several effective flushing chemicals are now available from commercial sources. An alternative to flushing is to install an in-line filter to trap any debris in a system. Filter installation is recommended by several vehicle manufacturers. The filter is installed ahead of the TXV, OT, or compressor.

After the repair has been made, the system must be evacuated to remove all of the air and any water droplets or vapor that might be in it. Both air and water are contaminants. Air is considered a noncondensable gas, and it causes excessively high discharge pressures and temperatures. Besides the added load on the compressor, clutch, condenser, and hoses, air and water can break down the oil in the system. Water can cause rusting and corrosion, the formation of acids that will damage the interior portions of a system, and probable icing at the expansion device.

Recharging a system involves putting the right amount and type of refrigerant into a system. Recharging uses new or recycled refrigerant or a mixture of the two. (At one time, small cans [400 g or 1 lb] were commonly used, but now most shops use larger containers. Since November 1992, small containers [less than 20 lb] of R-12 have been available for purchase only by certified technicians.) Recharging equipment must accurately measure the quantity of refrigerant that is charged into a system.

13.3.1 Refrigerant Contamination

At one time, contamination of refrigerant was rare, and the majority of the problems were caused by air from improper service procedures. Today, with the variety of blends, hydrocarbons, and other refrigerants, along with recycled refrigerants, contamination is a much greater problem. The A/C industry defines contaminated refrigerant as anything less than 97% pure; 98% and better is considered pure, and between 97 and 98% is questionable. Purity standards (SAE J1991 and J2099 and ARI 700-88) permit some contamination with new (virgin) and recycled refrigerant.

Refrigerant with 2% or greater of another material (**noncondensable gases** *[NCG]* or foreign refrigerant) is considered contaminated (SAE standard J1661). If the contamination is greater than 5%, problems such as excessive high side pressures and clutch cycling rate result. Other unseen problems of oil breakdown, seal deterioration, or compressor wear can also occur.

NCG contamination can be reduced by recycling the refrigerant, purging the air out, or diluting the contaminated refrigerant with pure refrigerant. Contamination from a foreign refrigerant requires that the refrigerant be sent off for reclaiming or disposal (Figure 13–5).

✓REAL WORLD FIX

The A/C system in the 1984 Chevrolet motorhome (58,000 miles) was recharged using 3 pounds of R-12. The TXV had been replaced because of suspected restriction. The system still does not cool very well, and the technician has asked for help.

Fix: Following advice, the refrigerant in the A/C service unit was checked and found to be contaminated with air. After purging the air from the refrigerant supply, the A/C system was recharged, and this fixed the problem.

13.3.2 Refrigerant Identification

Two methods are used to identify the refrigerant in a system. One is to look at the service fittings and refrigerant label, which tells the technician what refrigerant *should* be in the system. The other is to use a refrigerant *identifier* to determine what is actually in the system whether it is R-12 or R-134a, or whether it is contaminated with a foreign refrigerant or air. The second method is much better when we consider the problems that can be created if we recover a blend or hydrocarbon refrigerant. Recovering a hydrocarbon refrigerant using an electric-driven machine can cause an explosion inside the machine.

There are two types of refrigerant identifiers. One type, referred to as a **Go-no-go identifier** (accuracy to about 90% or better), indicates the purity of the refrigerant with lights that indicate pass or fail. The other type, referred to as a **diagnostic identifier** (accuracy to 98% or better), displays the percentage by weight of R-12, R-134a, R-22, hydrocarbons, and air (displayed as NCG) (Figure 13–6). Knowing the nature of the contaminants lets you know what approach to take. If the contaminant is air, you can safely recover and recycle the refrigerant for reuse. If the contaminant is another refrigerant or a hydrocarbon, you must use a special recovery procedure and send off the recovered mixture for disposal or recycling. Recycling machines cannot remove a foreign refrigerant, only air, water, oil, or particulates. Newer identifiers include a printer port, so a hard-copy readout can be printed for the customer's use (Figure 13–7).

At times, slightly contaminated refrigerant can be saved. The standard for recovered R-12 purity is 98%.

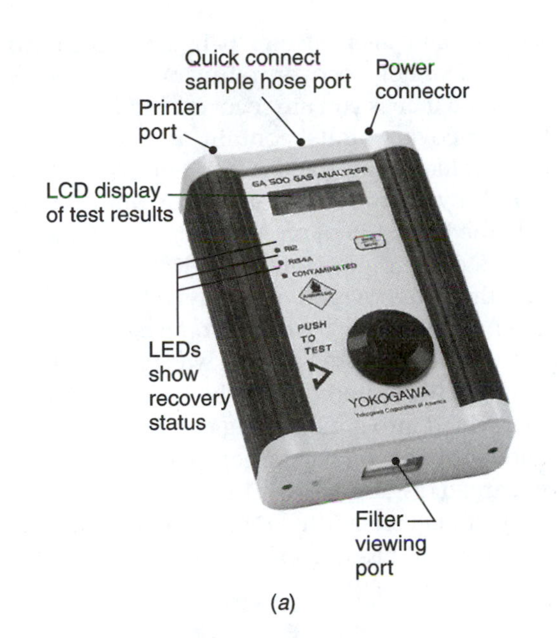

(a)

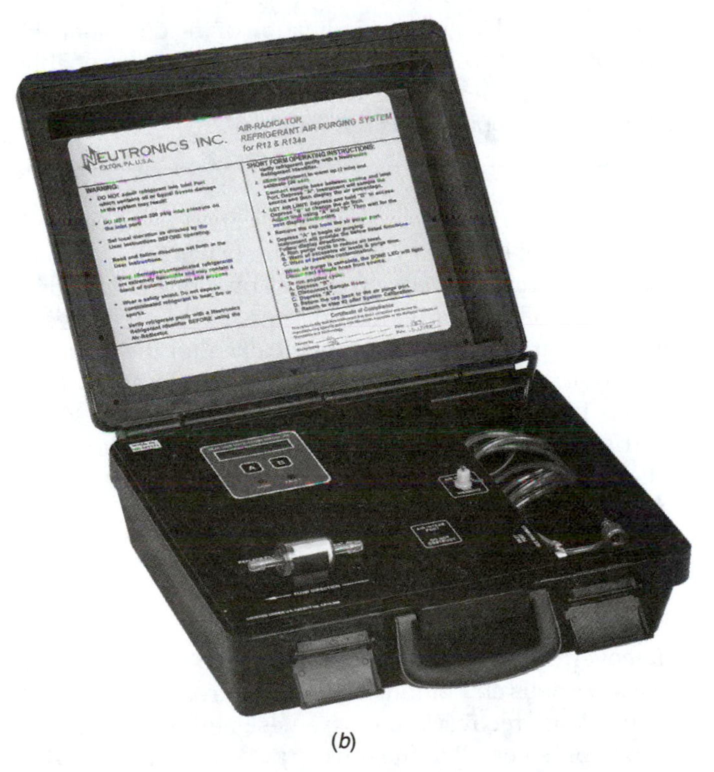

(b)

FIGURE 13–6 A refrigerant identifier or gas analyzer uses a sample of refrigerant from a system and determines what it is (*a*). Identifiers can have a display showing the results, LEDs that light up to show the refrigerant status, an alarm for hydrocarbon, a viewing port to check the filter for liquid or contaminants, and a port so a printer can be connected (*b*). (a *is reprinted by permission of Yokogawa Corporation of America;* b *is courtesy of Neutronics Inc.*)

Refrigerant Analysis

Refrigerant R-134a	=	3.8%
Refrigerant R-12	=	96.2%
Refrigerant R-22	=	0.0%
Hydrocarbons	=	0.0%
Conclusion:	>>>FAIL<<<	
Date:		
Technician:		
Car Model:		
VIN:		

FIGURE 13–7 This printout from an identifier shows that the R-12 in this system is cross-contaminated with 3.8% R-134a. This refrigerant is unusable.

✓SERVICE TIP

The readout in Figure 13–7 shows R-12 contaminated with 3.8% R-134a. If this amount of contaminated refrigerant is diluted with two to three times as much pure R-12, the contamination would be less than 2%.

Some shops are using the diagnostic-style identifier as a quality-control check of new refrigerant they purchase. New R-12 that has been in storage is available through reputable suppliers. Badly contaminated new (virgin) refrigerant is being found, some of which is bootleg and being imported illegally. This refrigerant should not be charged into a system. You will pay a high price to purchase it and also to dispose of it.

To identify the refrigerant in a system, you should:

1. With the engine and the system shut off, connect the identifier to the low side service port using the correct hose assembly for the system's refrigerant type.
2. Check the filter for the incoming gas; it will show a color change if contaminated.
3. Allow a gas sample to enter the unit. Some units include a warning device to make sure liquid refrigerant does not enter it. Many units include a warning device to indicate a flammable refrigerant.
4. Read the display to determine the nature of the refrigerant. Some units allow printing of the results at this time. If the refrigerant is good or R-12 or R-134a contaminated with air, it can be safely recovered and recycled.

5. When the analysis is complete, some units display instructions to disconnect the sampling hose and then bleed out the gas that was sampled. If the hose is not disconnected when prompted, the identifier can bleed excess refrigerant from the system.

✓SERVICE TIP

An identifier cannot identify a blend directly, but by looking at the combination on the display, you can sometimes tell what the refrigerant is. An identifier can only identify the refrigerants contained in the blend. This is referred to as a **fingerprint;** for example, one source says that FR-12, FRIGC, will appear as 2% HC, 26% R-12, 3% R-22, and 69% R-134a. It is recommended that you use your identifier on each of the blends that you encounter and record the readings (fingerprints) for later reference.

✓PROBLEM SOLVING:

A vehicle came in with a complaint of a noisy compressor along with poor cooling. When you put gauges on and tested the system, you found high low side pressure and very high pressure on the high side. Your identifier indicates 12% R-134a, 80% R-12, and 8% NCG. Do you know what caused the customer's complaint? What should you do now?

13.3.3 Refrigerant Recovery

Some recovery units have two service hoses and low and high side pressure gauges and are connected directly to both low and high side service ports. Other units are connected to the center hose of a manifold gauge set. Remember that a recovery unit must be dedicated to either R-12 or R-134a; it cannot be used for both. Blends or contaminated mixtures should be recovered using a different, separate machine. The hoses should be equipped with shutoff valves within 12 inches of the ends and have end fittings to match the refrigerant used.

At this time, there is concern about recovering refrigerant from a system that has a refrigerant blend or the wrong refrigerant added to it. A wise technician is careful to check the service history of the vehicle. Many refuse to service a vehicle if there is a chance that it contains contaminated refrigerant. Identifiers should be used. If the correct machine is available, R-12 or R-134a can be recovered. Do not recover refrigerant that is cross-contaminated unless you are recovering it for disposal.

Most recovery units contain an oil separator, a compressor-like pump, and a condenser-like heat exchanger. They draw refrigerant vapor out of a system and, like the high side of an A/C system, convert it into liquid for storage. Some units weigh the amount of refrigerant that is recovered, which tells the technician if all the refrigerant from a fully charged system has been recovered or whether the system was fully charged. Oil removed and separated during the recovery process is usually drained into a measuring cup and noted so that this amount of new oil can be replaced in the system as it is recharged (Figure 13–8). This feature is helpful during a retrofit, because it tells the technician how much of the mineral oil has been removed.

✓SERVICE TIP

Oil recovery also gives an indication of the oil volume in the system; if no oil is recovered, the system is probably low on oil. If an excessive amount of oil is recovered, the system probably has too much oil.

Some refrigerant in a system is absorbed in the oil and does not leave the oil immediately when a system is emptied. This trapped refrigerant **out-gasses** or boils out of the oil later, after the pressure has been removed. Recovery units shut off automatically after the main refrigerant charge has been removed and the system drops into a slight vacuum. To completely remove the refrigerant, run the normal recovery procedure and then recheck the pressure after a 5-minute wait. If the pressure has increased, restart the recovery process.

✓SERVICE TIP

Recovering refrigerant from a system that has a leak can cause serious air contamination in the recovered refrigerant. Most recovery units are designed to shut themselves off when all of the refrigerant is removed and the system develops a slight vacuum. If there is a loose hose connection or a leak in the system, it will not reach a vacuum, and the unit becomes an air pump that pulls in air through the leak and moves it into the recovery tank. The recovery process should be monitored so it can be stopped manually if the system pressure does not drop properly. It is very difficult to remove excess air from refrigerant that is overly saturated with air.

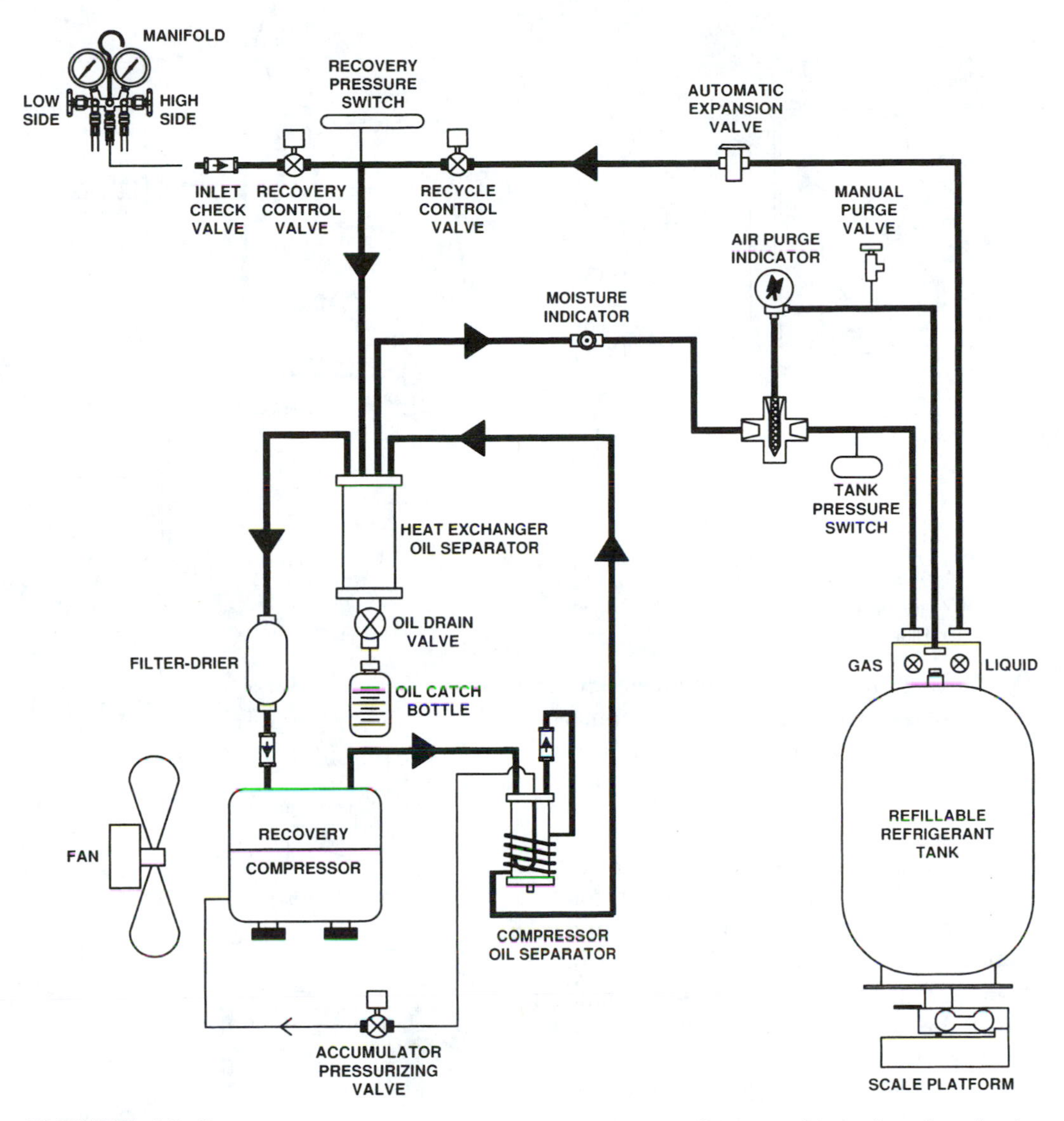

FIGURE 13–8 The flow diagram for a recovery–recycling machine showing the internal components and the path that refrigerant takes during recovery and recycling. *(Courtesy of Robinair, SPX Corporation)*

SAFETY NOTE

Some vehicles with electronic fuel injection have a fuel pressure test port that uses a 1/4-inch flare fitting, the same as an R-12 service port. Make sure that the refrigerant recovery unit is connected into the A/C system and not the fuel system.

The storage container for recovered refrigerant must be approved by the Department of Transportation (DOT) and carry the letters *DOT* and the certification numbers. Also note the date of tank manufacture; as described in Section 3.3.2, recovery tanks must be inspected and certified to be in good condition every 5 years. The container in Figure 13–9 has two hand valves, a blue valve for liquid and a red valve for vapor.

To recover the refrigerant from a system, you should:

1. Identify the refrigerant in the system.
2. Make sure the hoses have the proper shutoff valves and are compatible with the refrigerant in the system. Check to make sure all valves are shut off.

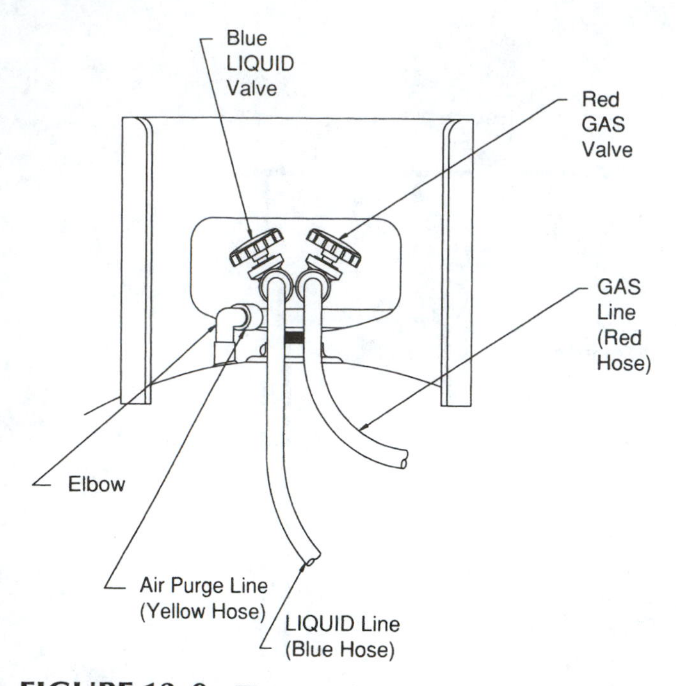

FIGURE 13–9 The container used for recovered refrigerant must be approved by the DOT; the connections to the recovery unit are shown here. (*Courtesy of Robinair, SPX Corporation*)

3. Connect the recovery unit to the system or to the center hose of the manifold gauge set, following the directions of the manufacturer and the precautions in Section 10.2.3 (Figure 13–10).
4. Open the required valves and turn the machine on to start the recovery process, following the directions of the machine's manufacturer (Figure 13–11). Note the receiver–drier or accumulator; frost formation indicates that refrigerant is boiling out of the oil contained in the receiver–drier or accumulator.
5. Continue the recovery until the machine shuts off or the pressure reading has dropped into a vacuum. If the system does not drop into a vacuum, it probably has an air leak; manually stop the recovery process.
6. Verify complete recovery by shutting off all valves and watching the system pressure. If pressure rises above 0 psi within 5 minutes, repeat steps 4 and 5 to recover the remaining refrigerant.
7. With all of the valves shut off, remove and disconnect the recovery unit. If the unit is also equipped to evacuate and recharge the system, disregard this step.
8. Drain, measure, and record the amount of oil removed from the system with the refrigerant; dispose of it properly (Figure 13–12). This amount of new oil should be added during the recharging process.

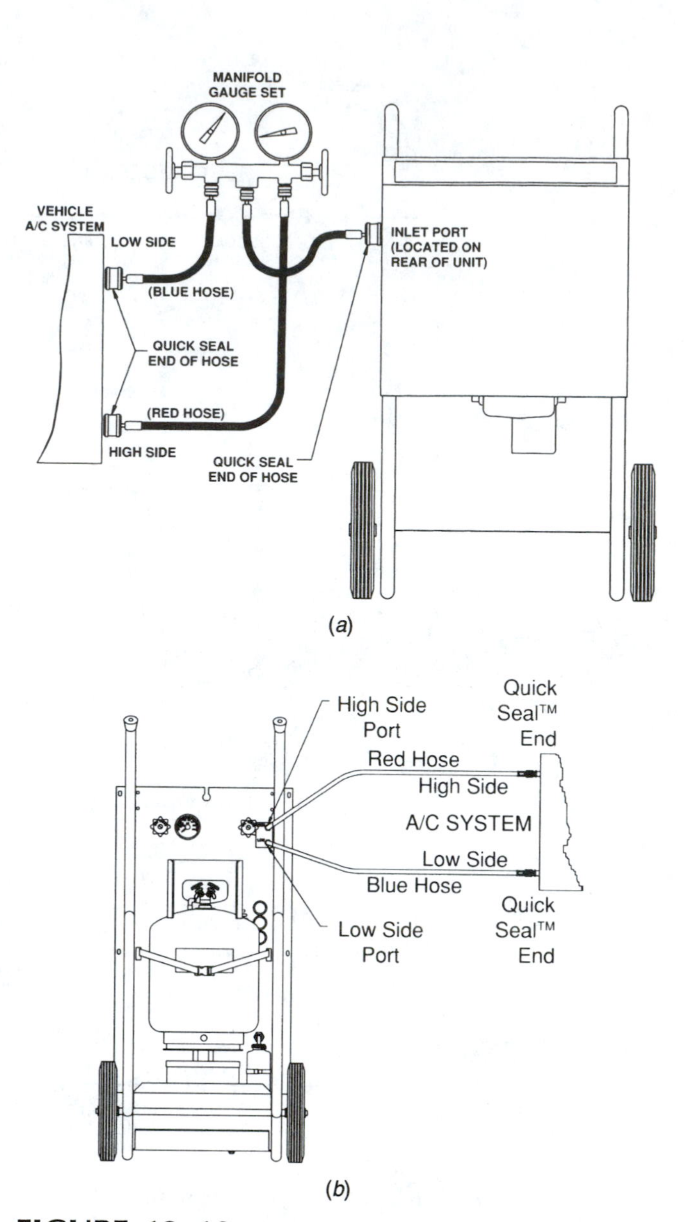

FIGURE 13–10 Some recovery units use a single hose connection for the center service hose of a manifold gauge set (*a*). A full-service unit has two service hose connections to the low and high sides of the A/C system (*b*). (*Courtesy of Robinair, SPX Corporation*)

✓SERVICE TIP

Recovery can be speeded up and made more complete by heating the accumulator or receiver–drier with a heat gun or hair dryer.

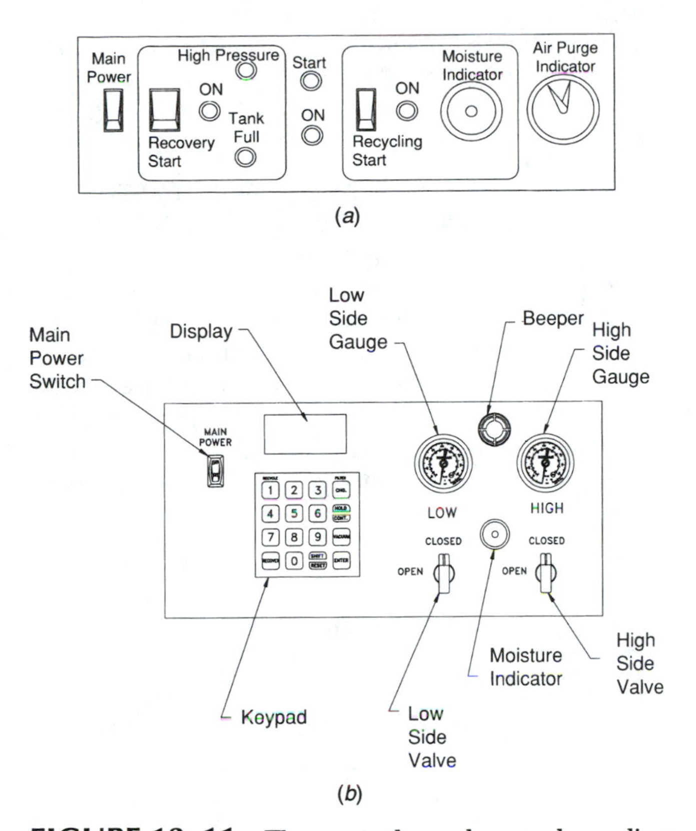

FIGURE 13–11 The control panels vary depending on the capabilities of a particular recovery-only, recovery and recycling (*a*), or full-service (*b*) machine. (*Courtesy of Robinair, SPX Corporation*)

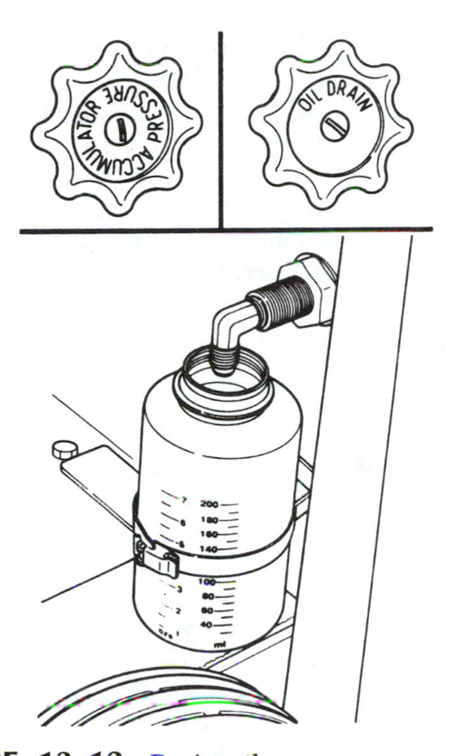

FIGURE 13–12 During the recovery process, oil from the system is separated into a container so the technician will know how much oil was removed. (*Reprinted with permission of General Motors Corporation*)

The system can now be repaired. Present standards and good work habits require that recovered refrigerant be recycled before reuse, even if it is to be returned to the system from which it was recovered.

13.3.3.1 Recovering Contaminated Refrigerant Special procedures should be followed when you need to remove a contaminated or a blend refrigerant from a system. You should not use an R-12 or R-134a recovery machine to remove a contaminated or a blend refrigerant. As with a gauge set, an R-12 machine can be converted to recover blends, but this machine should be used only for blends unless it is decontaminated after each usage. If the mixture contains more than 4% hydrocarbon, it should be considered explosive, and this mixture should not be recovered using an electric-powered machine. Air-powered recovery machines are available and can be safely used for explosive mixtures.

Hydrocarbon refrigerants, blends, and unknown mixtures should be recovered into containers clearly labeled CONTAMINATED REFRIGERANT. The proper color for these containers is gray with a yellow top. Recovery of contaminated refrigerant using a recovery machine is essentially the same as recovering uncontaminated refrigerant except that the contaminated material must be sent off for disposal or off-site recycling.

Refrigerant recovery without a machine using the following procedure is *not* a truly effective process.

To recover contaminated refrigerant without using a recovery machine, you should:

1. Connect a gauge set to the system.
2. Connect the container for the recovered mixture to the center hose and place the container in a larger, tublike container with ice to chill it as much as possible.
3. Open the high and low side gauge valves and allow the mixture to leave the system.
4. Start the engine, disconnect the compressor clutch wire, and turn on the system to high heat and high blower to warm it up as much as possible.
5. When recovery is completed as much as possible, close all valves, shut the system and engine off, and ship the recovered mixture to an approved refrigerant reclaimer.

13.3.4 Recycling Refrigerant

Most recycling units pump the recovered refrigerant through a very fine filter to remove foreign particles, past a desiccant to remove water, and through an oil separator to remove any excess oil. Air is removed by venting it, using the noncondensable purge, from the top of the liquid refrigerant. Remember that recycled refrigerant must meet the same purity standards as new (virgin) refrigerant: less than 15 ppm moisture, less than 4,000 ppm oil, and less than 330 ppm air. Some machines have a sight glass equipped with a moisture indicator so that the operator can tell when the moisture has been removed. Some are designed to stop operation if a filter or desiccant change is needed. Some can perform the recycling process in a single pass from the storage container through the cleaning process and back to the storage container; these machines often complete the recycling process while the system is being evacuated. Others require several passes, and the recovery process continues to circulate the refrigerant as long as necessary (Figure 13–13).

The recycling machine is dedicated to a particular type of refrigerant, and a recycled blend can only be recharged back into the vehicle it came from or another vehicle from the same fleet.

To recycle refrigerant, you should:

1. Open the valves or perform the programming steps required by the machine manufacturer and turn on the machine.
2. The machine operates until excess foreign particles and water have been removed or for a programmed length of time and then shuts off. Check the moisture indicator to ensure that the refrigerant is dry. If the machine does not shut off in the proper amount of time, its internal filters or desiccant probably require service (Figure 13–14).

SERVICE TIP

Sometimes all of the air will not be removed in one recycling pass. Repeat the recycling as needed to remove all of the excess air.

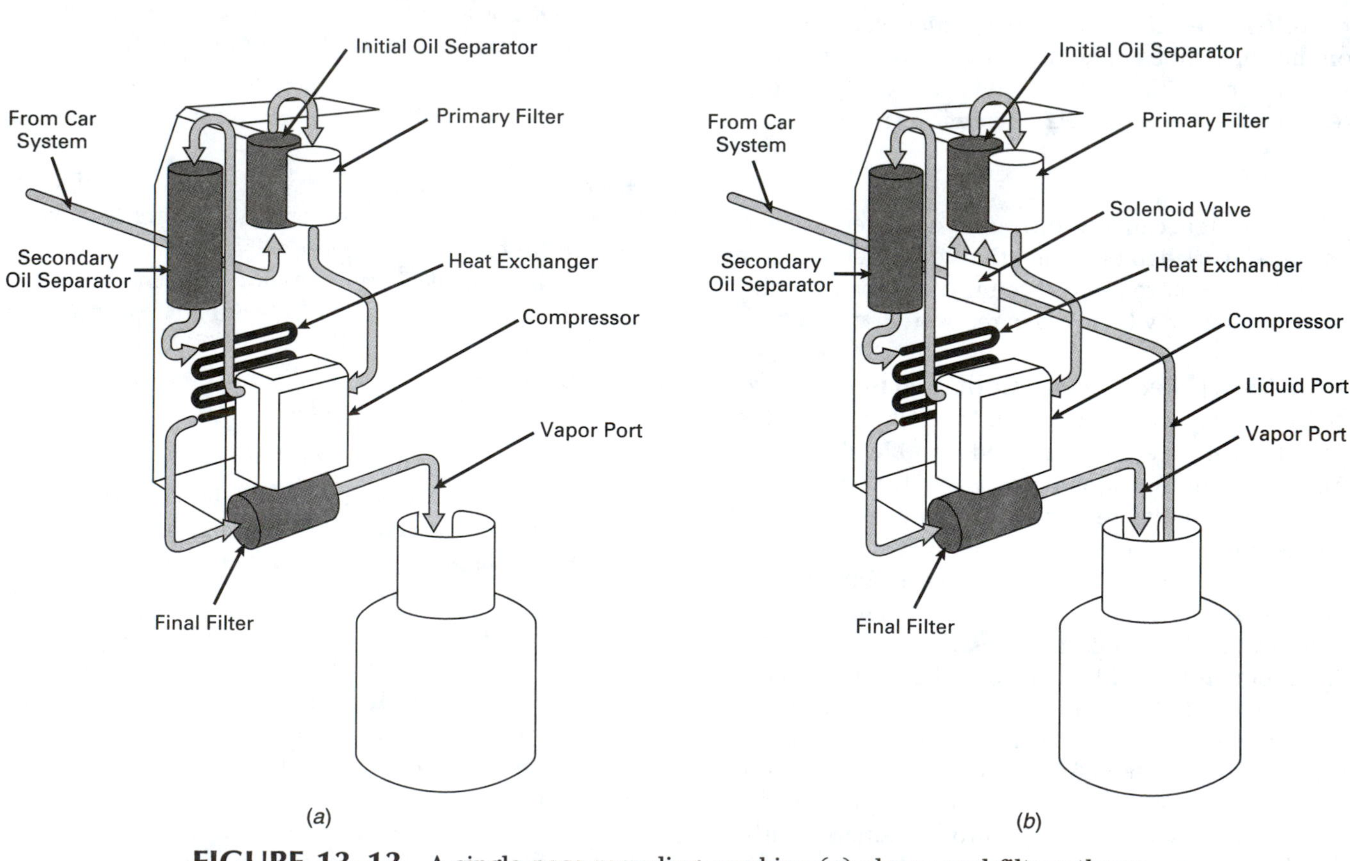

FIGURE 13–13 A single pass recycling machine (*a*) cleans and filters the refrigerant as it is being recovered. A multipass machine (*b*) recovers the refrigerant in one operation and then cycles the refrigerant through filters and separators in another operation.

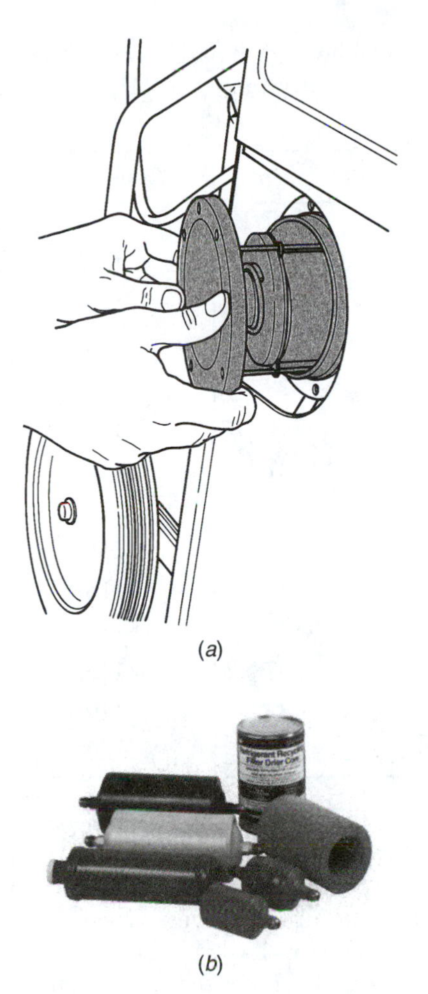

(a)

(b)

FIGURE 13–14 Recycling machines have a filter and desiccant that must be replaced after a certain amount of use (*a*). These parts are available from various sources (*b*). (a *is reprinted with permission of General Motors Corporation;* b *is courtesy of Robinair, SPX Corporation*)

13.3.4.1 Air Contamination Checks After recovery is completed, you can check for excess air in the refrigerant by evaluating the pressure temperature relationship (PT). This is best done after the temperature has stabilized through the liquid refrigerant; the start of the workday is an ideal time.

To check the PT relationship, you should:

1. Keep the storage container at a temperature above 65°F (21°C) and away from direct sunlight for 12 hours.
2. Read the pressure in the container using a calibrated pressure gauge with 1-psi increments.
3. Read the temperature of the air next to the container.
4. Compare the pressure and temperature readings with Table 13–1 for R-12 or Table 13–2 for R-134a.

Note: *All of the pressures are referenced for Sea Level. These pressures will be lower at higher elevations.*

If the pressure for a particular temperature is less than that given in the table, the refrigerant does not contain an excess amount of air and is uncontaminated. If the pressure is greater than that given in the table, slowly vent or purge gas from the top of the container (red valve) until the pressure drops below that given in the table. If the pressure does not drop, the refrigerant must be recycled or sent off for disposal or recovery. A recent development is a two-gauge unit that attaches onto the recovery container. One gauge is a pressure gauge that shows the actual refrigerant pressure; the other gauge is a thermometer calibrated to show the ideal pressure. Excess air is indicated if the actual pressure is higher (Figure 13–15).

TABLE 13–1 If the pressure of a container of R-12 exceeds that shown here for a particular temperature, the refrigerant is contaminated, probably with air (SAE J1989). (*Reprinted with permission from SAE Document M-106, © 1992, Society of Automotive Engineers, Inc.*)

Temp. (°F)	*psig*	*Temp. (°F)*	*psig*	*Temp. (°F)*	*psig*	*Temp. (°F)*	*psig*	*Temp. (°F)*	*psig*
65	74	75	87	85	102	95	118	105	136
66	75	76	88	86	103	96	120	106	138
67	76	77	90	87	105	97	122	107	140
68	78	78	92	88	107	98	124	108	142
69	79	79	94	89	108	99	125	109	144
70	80	80	96	90	110	100	127	110	146
71	82	81	98	91	111	101	129	111	148
72	83	82	99	92	113	102	130	112	150
73	84	83	100	93	115	103	132	113	152
74	86	84	101	94	116	104	134	114	154

TABLE 13–2 If the pressure of a container of R-134a exceeds that shown here for a particular temperature, the refrigerant is contaminated, probably with air (SAE J2211). *(Reprinted with permission from SAE Document M-106, © 1992, Society of Automotive Engineers, Inc.)*

Temp. (°F)	*psig*	*Temp. (°F)*	*psig*	*Temp. (°F)*	*psig*	*Temp. (°F)*	*psig*
65	69	79	90	93	115	107	144
66	70	80	91	94	117	108	146
67	71	81	93	95	118	109	149
68	73	82	95	96	120	110	151
69	74	83	96	97	122	111	153
70	76	84	98	98	125	112	156
71	77	85	100	99	127	113	158
72	79	86	102	100	129	114	160
73	80	87	103	101	131	115	163
74	82	88	105	102	133	116	165
75	83	89	107	103	135	117	168
76	85	90	109	104	137	118	171
77	86	91	111	105	139	119	173
78	88	92	113	106	142	120	176

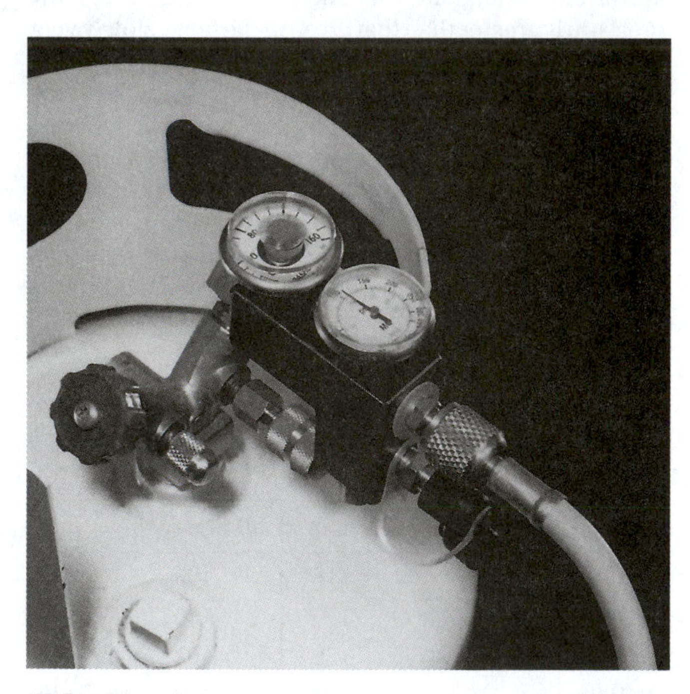

FIGURE 13–15 This dual-gauge set has a pressure gauge and a thermometer calibrated to read the pressure that should correspond to that temperature. A comparison of the two indicates refrigerant purity. *(Courtesy of Mastercool)*

An air purge is a manual or automatic operation that can take place during the recycle process of some recycling machines (Figure 13–16). Air purge is an automatic operation in some machines that occurs during the evacuation cycle; shortcutting the evacuation time can cause an incomplete air purge. If recycling is attempted with contaminated refrigerants or some blends, the entire mix can be dumped as the machine attempts to purge air.

When the air purge operates, some refrigerant boils off as the pressure drops, and this boiling absorbs heat and lowers the liquid temperature, which in turn lowers the pressure. This drop in pressure can make you falsely believe that you have purged all of the air. With refrigerant that is severely contaminated with air, it is thought that some of the excess air is contained in the liquid, not just sitting on top of it, making it much more difficult to purge. If you encounter this, you should make a partial air purge and then let the PT relationship stabilize over a period of time. Then you can recheck the PT and purge again if necessary.

A new unit that monitors refrigerant optically automatically bleeds out excess air (Figure 13–17). One unit is designed to be used directly on an A/C system, a recovery–recycling machine, or a refrigerant container; some units are intended to be used only on containers. Air purge units bleed off or "burp" the air in short cycles and then monitor the remaining gas, repeating the purge cycles as necessary.

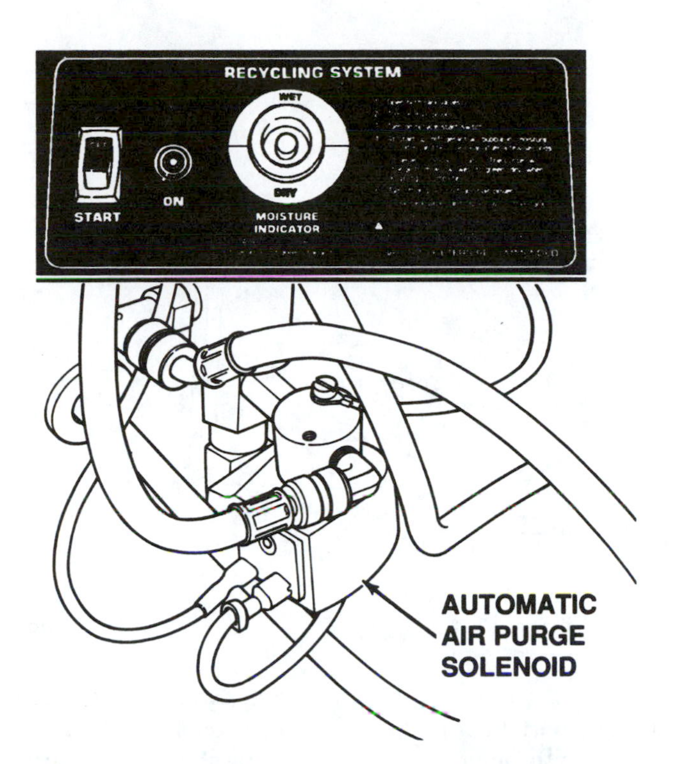

FIGURE 13–16 Some recycling machines have an automatic air purge that vents air from the storage container if the pressure is too high for the temperature. Note the moisture indicator on this unit. *(Reprinted with permission of General Motors Corporation)*

FIGURE 13–17 This automatic air purge tool can be connected to an A/C system or bottle of refrigerant. It checks for air contamination, automatically purges a small amount of gas, and repeats this process. *(Reprinted by permission of Neutronics Inc.)*

13.3.5 Flushing an A/C System

When a compressor fails, it usually sends solid compressor particles into the high side, and this material can plug the condenser passages and OT. It is always a good practice to check the OT or TXV filter screen for debris when there is a compressor failure and to flush the system, install a filter, or both before completing the job.

✓SERVICE TIP

Debris from a failed compressor is often a fine, black material called **goo** or **black death** by many technicians. This mixture of small particles of aluminum and Teflon mixed with refrigerant oil will be trapped between the compressor discharge port and the orifice tube. It is very difficult to flush out, especially from a modern condenser.

Flushing is done by pumping a liquid material through the passages in a reverse or normal flow direction. A gas such as shop air or nitrogen is not really effective in removing solid material; remember that many condensers have several small passages that can be easily clogged and that air can bypass a clogged passage. Flushing agents can be a commercial flushing solution, liquid refrigerant, or very lightweight ester oil; liquid refrigerant and ester oil have very little or no solvent action, and remaining ester oil might be difficult to remove from the system.

A compressor cannot be flushed. An accumulator or receiver–drier is usually not flushed; it is simply replaced. There is some question about whether an evaporator should be flushed; most experts believe that the OT or TXV screen will filter out larger debris. Most flushing is done on the high side, from the compressor connection of the discharge line to the OT or receiver–drier fitting of the liquid line or vice versa (Figure 13–18).

✓SERVICE TIP

It is very difficult or impossible to flush a modern flat tube condenser with its multiple, small passages. One or more of the small passages can be plugged so the flush material will pass through the open passages.

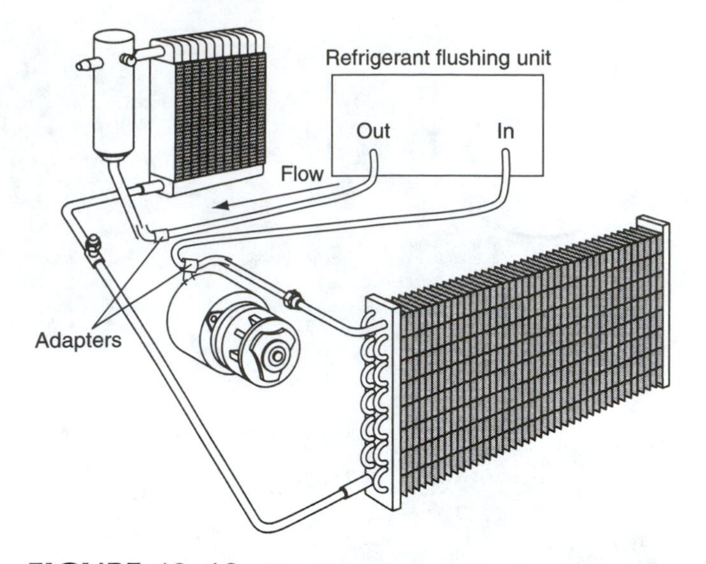

FIGURE 13–18 Portions of an A/C system can be flushed to remove debris and excess oil. Adapters are used to connect a flushing unit, which pumps the flushing material through the components.

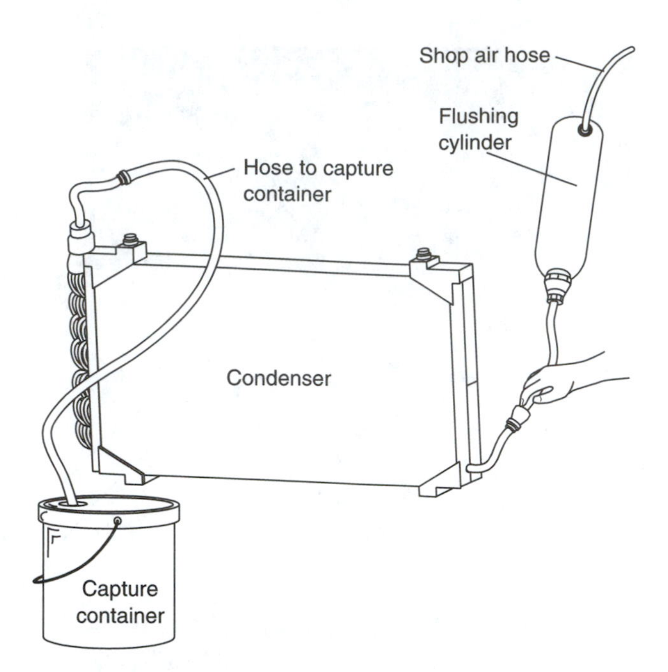

FIGURE 13–19 A/C components can also be flushed using a portable tool kit and special adapters. The flush is forced through the condenser using shop air pressure.

✓SERVICE TIP

If flushing a system using air pressure, the air must be clean. Shop air often contains oil and metal debris from the air lines, and these can contaminate the system being cleaned. Experienced technicians recommend using bottled nitrogen gas, which is clean and inert.

At this time, there are two schools of thought on how to flush. One is to use a solvent to clean the discharge line, condenser, and liquid line up to the OT or receiver–drier (Figure 13–19). A solvent is effective in dissolving foreign material, but contamination and later problems may result if all of the solvent is not removed from the system. The other procedure uses liquid refrigerant to flush the entire system (except for the compressor), from the suction line at the compressor inlet to the discharge line at the compressor outlet. A refrigerant flush removes most solid particles and oil, and it is fairly easy to remove all of the refrigerant from the system. Most recovery–recycling machines can be fitted with a flushing kit that allows them to pump liquid refrigerant into one line and recover that refrigerant from the other line (Figure 13–20).

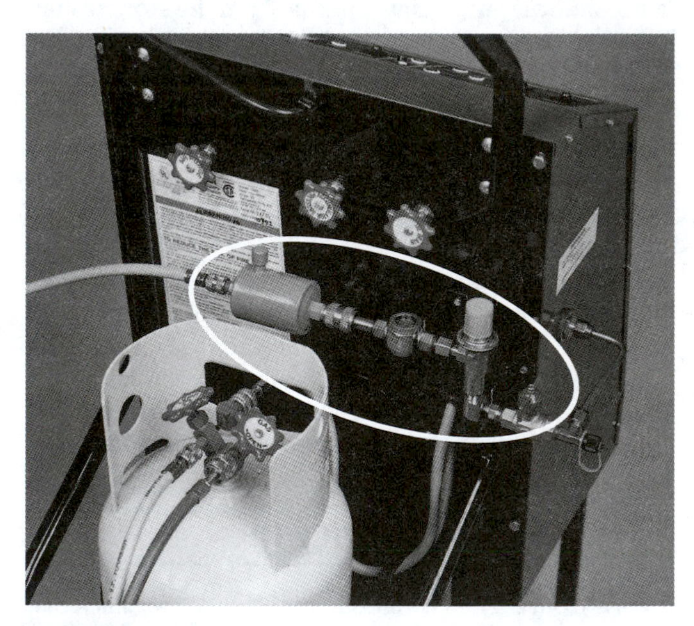

FIGURE 13–20 This recovery and recycling system has an optional flushing kit that uses recovered refrigerant to flush a system clean. The refrigerant is then recycled for reuse. *(Courtesy of Robinair, SPX Corporation)*

The flushing process requires an adapter fitting at each of the connections; these fittings can be made in the shop or purchased (Figure 13–21). It also requires a container for the flushing agent and a way to force it into the line; some systems use a pump, and others use shop air pressure (Figure 13–22). Because the flushing agent is either expensive or has ozone-depleting or greenhouse-effect gas restrictions, a catch container is connected to the outlet end to capture it. The liquid flushing agent is then forced through the condenser and lines until as much of the debris as possible is removed.

FIGURE 13–21 Adapters to connect the flushing unit into the system can be commercial fittings (*a*) or fittings made in the shop (*b*). (a *is courtesy of MACS;* b *is courtesy of Robinair, SPX Corporation)*

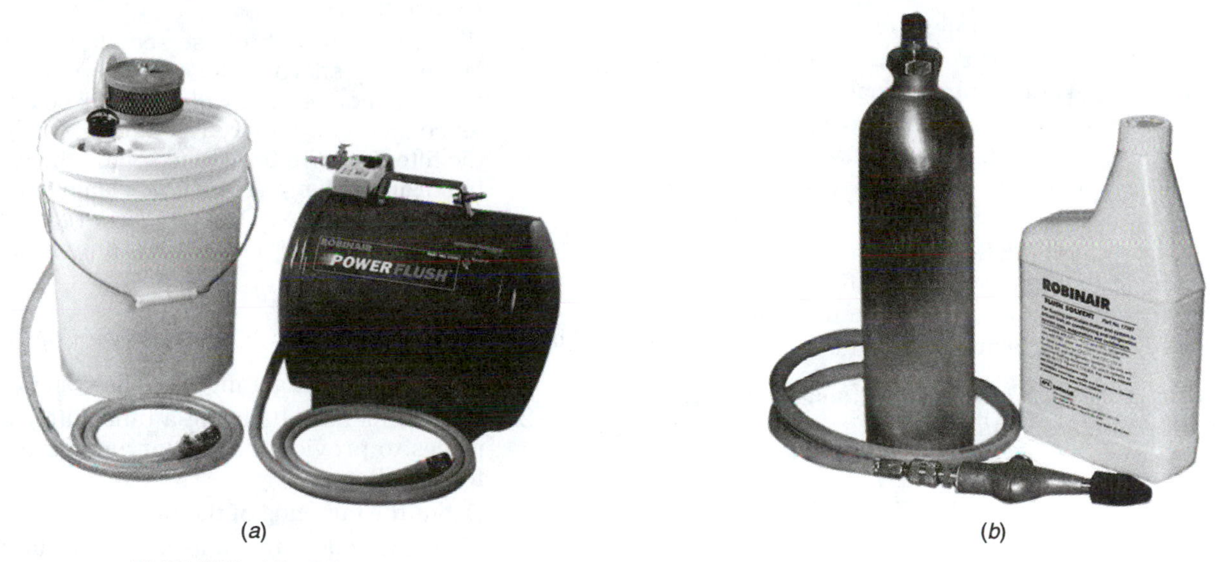

FIGURE 13–22 The power flush unit (with catch container at left) forces the flushing solvent through the lines in a pulsating manner (*a*); the operator of the flush gun (*b*) controls the flow at the nozzle. *(Courtesy of Robinair, SPX Corporation)*

✓REAL WORLD FIX

The 1990 Infiniti (87,000 miles) came in with no A/C because of a leaking evaporator. The system was recharged, but it still did not cool. The pressures were 25 psi on the low side and 110 psi on the high side. The compressor and TXV were replaced using OEM parts, but the pressures stayed the same and there was still no cooling. Checking the parts that were just installed showed the new TXV plugged with desiccant so the TXV and receiver–drier were replaced. After recharging again, the same problem was found.

Fix: A check of the TXV showed that it was plugged again. Removal of the liquid line allowed a thorough cleaning, removed all of the desiccant particles, and fixed this problem.

Author's Note: The liquid line could have been flushed or a filter installed earlier in this repair to prevent some of the problems from occurring. I'm not sure why the new TXV was replaced instead of cleaning it.

✓SERVICE TIP

If you suspect that a system has too much oil in it, flush the system using liquid refrigerant, and add the specified amount of oil for the system.

✓SERVICE TIP

Some technicians place a paper coffee filter to catch the debris in the flush material going to the catch container. This allows them to inspect the material being flushed from the system and to determine the effectiveness of the flushing process.

✓REAL WORLD FIX

The 1993 Jaguar (170,000 miles) came in with no A/C. The high side pressure was slightly low, and the low side pressure would drop into a vacuum. The pressures would equalize when the system was shut off. The refrigerant was recovered, and a check of the TXV showed desiccant pellets. The receiver–drier and the TXV were replaced, air was blown through the liquid line, and the system was evacuated and recharged. A retest of the system shows almost the same problem: The system cools well for a few minutes, and then the low side pressure drops into a vacuum.

Fix: Flushing the liquid line and installation of a filter along with evacuation and recharging the system fixed this problem.

✓SERVICE TIP

On vehicles with dual evaporators, remember to check the liquid line going to the rear evaporator and TXV for contaminants.

13.3.6 Installing an In-Line Filter

If a compressor fails, debris in the form of metal particles often travels downstream through the condenser and is trapped in the receiver–drier or OT, where it can cause plugging. Some compressor failures send debris into the suction line (see Section 14.2). A similar problem can occur because of scale from a poorly cleaned condenser or evaporator; evaporator scale can damage the compressor. The technician can install a supplementary in-line filter in either the high side or low side. This filter is sometimes changed after an hour of operation and is called **live flushing**. The high side filter is usually placed in the liquid line between the condenser and the receiver–drier or OT, and the low side filter is installed in the suction line to the compressor. In-line filters are available with push-on, barb-type connections for rubber or nylon hoses or compression-type connections for metal tubing. They are available with connections to fit different line sizes. Some filters include a new OT (Figure 13–23).

To install an in-line filter, you should:

1. Recover the refrigerant from the system.
2. Select the location for the filter. The flexibility of hose allows a great deal of freedom. With metal tubing, you need to locate a straight section of tubing slightly longer than the filter's connections.
3. *With hose,* cut out a hose section with a hose cutter, very sharp knife, or sharp utility knife, so a neat connection will be made; coat the filter connections with refrigerant oil; firmly push the filter into the hose; and secure the connection using either a screw-type clamp or a metal ferrule ring. (*Note: Push-on, barb-type fittings and screw-type clamps are not approved for R-134a systems.*)

To install an in-line filter in a metal line, you should:

1. Recover the refrigerant from the system.
2. Cut the metal tube using a tube cutter in two places to provide the proper amount of room for the filter.
3. Debur the cut ends of the tube.
4. Clean the ends of the tube inside and out; the inside can be cleaned using a cotton swab dipped in mineral oil.
5. Place the filter, ferrules, and nut (without the O-rings) in position, and tighten the nuts to the correct torque to "set" the ferrules (Figure 13–24).
6. Remove the nuts, wet the O-rings with mineral oil, and slide them into place. Retighten the nuts to the correct torque to complete the installation.

Installation of a compressor suction screen is described in Section 14.2.1.

Hose and tubing connections are described in more detail in Section 14.3.

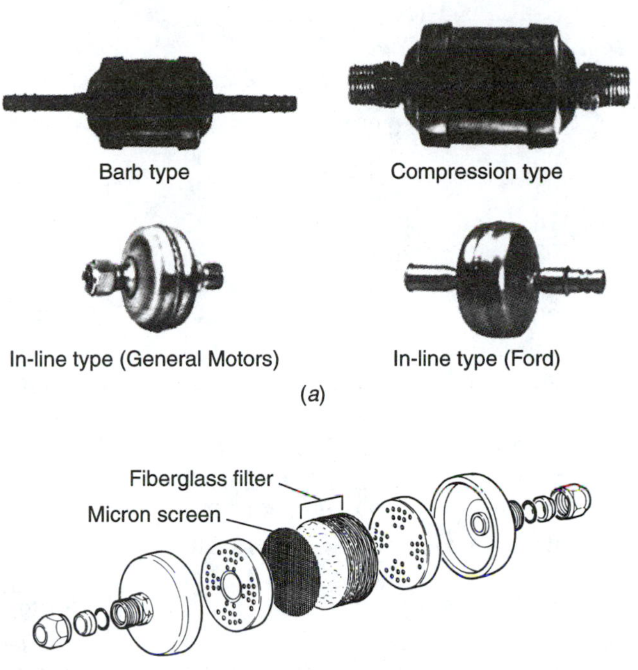

FIGURE 13–23 In-line filters are available in different forms, depending on how they will be connected into the system (*a*). An exploded view of an in-line filter is shown in *b*. (a *is courtesy of Four Seasons)*

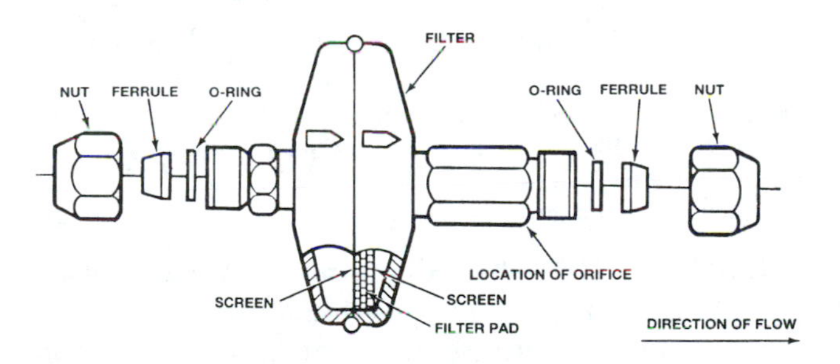

FIGURE 13–24 This in-line filter is installed using two ferrules that are compressed as the nuts are tightened. A short line section must be removed to make room for the filter. Note that this unit includes an OT. *(Reprinted with permission of General Motors Corporation)*

13.3.7 Checking and Correcting Oil Levels

Too little oil in a system can cause compressor wear and damage. Too much oil can cause excessive low and high side pressures. There is a tendency for some technicians in this industry to add oil when it is not needed, causing an overfill.

Early A/C systems used compressors with oil sumps, and it was relatively easy to check the oil level using a dipstick or the level plug in an A-6 compressor. Most modern compressors do not have an oil sump. The only way to check the amount of oil is to remove the compressor and drain it or to recover the refrigerant and note how much oil comes out. You should also note that it is almost impossible to drain all of the oil out of a compressor or evaporator because of the internal shape or out of an accumulator or receiver–drier because of the oil absorbed in the desiccant bag.

✓SERVICE TIP

An observant technician gets a good idea of the oil charge when he or she opens a line connection, like an orifice tube, and looks inside. It should always be damp with oil. A dry OT indicates low oil charge. One that is dripping with oil indicates too much oil.

✓SERVICE TIP

Some technicians drain the accumulator or receiver–drier and measure the amount of oil. If 6 or 8 ounces of oil come out when 3/4 ounce was expected to drain, the system has too much oil. If very little oil comes out, the system is oil starved.

The proper amount of oil can be kept in a system by adding oil to replace the amount of oil removed when a component is replaced. This is usually only a small amount as shown in Figure 14–58. Compressor replacement requires a slightly larger amount as described in Section 14.2.6.

There is a procedure to check the volume of oil charge in the refrigerant, and one manufacturer, Seltec, recommends a charge of 3 to 6% oil to properly lubricate its compressors. This procedure uses the weight of the refrigerant with its oil charge, and it requires a service port close to and upstream of the TXV or OT, an empty, generic oil charge container, and a scale that is accurate to a gram.

To determine oil charge, you should:

1. Attach the empty container and valve to a vacuum pump, and pull a deep vacuum. Close the valve, remove from the pump, and weigh the container and valve. Record this weight.
2. Run the A/C system and allow it to stabilize. Attach the container to the liquid line service port, and open the valve to allow the internal vacuum

to pull in refrigerant from the system. Close the container, and disconnect it from the system.

3. Weigh the container of refrigerant with the oil it contains, and record this weight.
4. Connect the container to a recovery unit, and recover the refrigerant. This should remove the refrigerant but leave the oil.
5. Weigh the container and remaining oil, and record this weight.
6. Subtract the weight of the container (step 1) from the weight of the container and refrigerant (step 3). The result is the weight of the refrigerant with oil. Record this weight.
7. Subtract the weight of the container (step 1) from the weight of the container with oil (step 5). The result is the weight of the oil that was contained in the refrigerant. Record this weight.
8. Divide the weight of the oil (step 7) by the weight of refrigerant (step 6). Multiply this by 100 to obtain the percentage of oil.

✓SERVICE TIP

Do not let PAG oil contact the vehicle's painted surfaces. PAG can damage paint.

✓SERVICE TIP

You can test the oil in a system by placing a few drops into water. Mineral and ester oils will react like normal oils and form floating beads. Some PAGS will dissolve into the water and disappear; other PAGS will form a film layer beneath the surface of the water.

13.3.8 Evacuating a System

Evacuating is also called **pumping the system down**. After a system has been repaired, all of the air and moisture that might have entered must be removed. Removing air is fairly easy; it is simply pumped out by the vacuum pump. Water removal is more difficult because it must be boiled and the water vapor pumped out with the vacuum pump. The water boils because of the reduced pressure in a vacuum; Figure 13–25 shows the boiling points. Note that it takes a vacuum of 28″ Hg or lower to boil water at room temperature.

Temperature (°F)	*Vacuum (inches)*	*Microns**	*Pressure (psi)*
212	0.00	759,968	14,696
205	4.92	535,000	12.279
194	9.23	525,526	10.162
176	15.94	355,092	6.866
158	20.72	233,680	4.519
140	24.04	149,352	2.888
122	26.28	92,456	1.788
104	27.75	55,118	1.066
86	28.67	31,750	0.614
80	28.92	25,400	0.491
76	29.02	22,860	0.442
72	29.12	20,320	0.393
69	29.22	17,780	0.344
64	29.32	15,240	0.295
59	29.42	12,700	0.246
53	29.52	10,160	0.196
45	29.62	7,620	0.147
32	29.74	4,572	0.088
21	29.82	2,540	0.049
6	29.87	1,270	0.0245
−24	29.91	254	0.0049
−35	29.915	127	0.00245
−60	29.919	25.4	0.00049
−70	29.9195	12.7	0.00024
−90	29.9199	2.54	0.000049

*Remaining pressure in systems in microns:
1.000 inch = 25,400 microns = 2.540 cm = 25.40 mm
0.100 inch = 2,540 microns = 0.254 cm = 2.54 mm
0.039 inch = 1,000 microns = 0.100 cm = 1.00 mm

FIGURE 13–25 The boiling point of water is reduced as a system is put into a vacuum. Microns provide a much more accurate way of measuring vacuum. *(Courtesy of Robinair, SPX Corporation)*

Vacuum pumps are rated by both **cubic feet per minute (cfm)** and **micron** ratings. The cfm rating tells us the volume it can pump; the micron rating tells us how deep a vacuum it can pull. An automotive A/C system requires about a 1.2- to 1.5-cfm rating, whereas a larger system used in a bus or truck needs about a 5- to 6-cfm vacuum pump. Using a vacuum pump that is too small requires a much longer evacuation time, which causes excess wear on the vacuum pump. A perfect vacuum is 29.92″ Hg, or 0 microns (a micron is equal to one-millionth of a meter). The micron rating of a vacuum pump tells us how deep a vacuum it can pull under field conditions—the lower the better. For example, a vacuum pump that pulls down to only 27″ Hg (685,800 microns) is only effective on temperatures above 110°F (43°C). A good vacuum pump will pull a

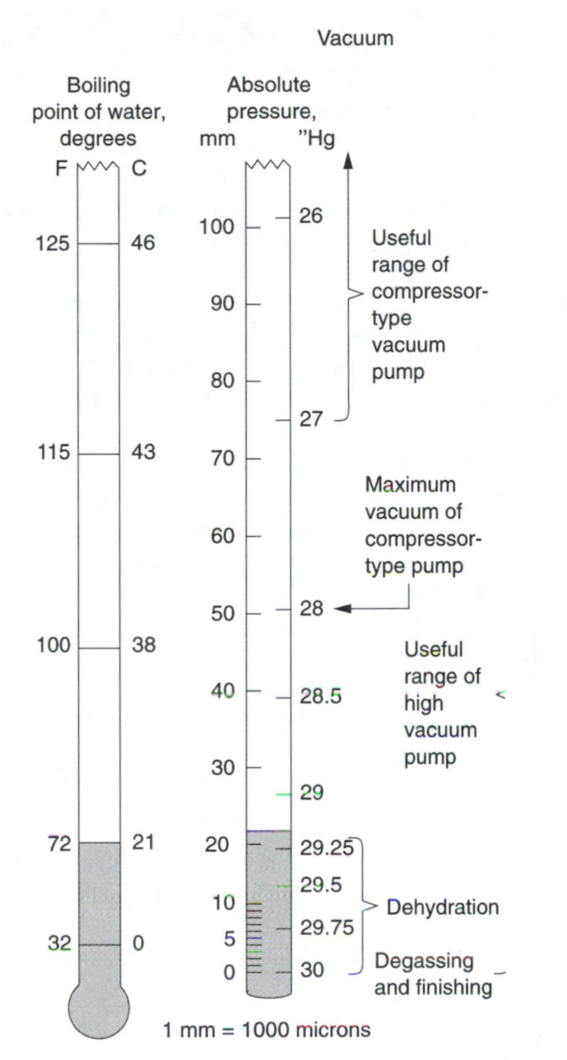

FIGURE 13–26 Vacuum pumps differ in their ability to remove all of the refrigerant and water from a system.

system down to under 1,500 microns (29.85″ Hg); this drops the boiling point of water to around 10°F (−6°C) (Figure 13–26).

✓SERVICE TIP

Some sources recommend using an electronic, thermistic vacuum gauge that is capable of reading micron levels in a deep vacuum. If the system drops below 1,500 microns, all of the water should have boiled out. One technician recommends comparing the reading in microns to the atmospheric pressure for that day. If the barometric atmospheric pressure is 29.99, a vacuum of 28.49 microns (95%) is good enough. At this point, there is no need to evacuate any longer and wear out the vacuum pump.

✓SERVICE TIP

A vacuum pump can be tested to see if it will boil water. Connect a service port to the lid of a glass jar (many 16 to 32 oz food jars will work), and connect this to the vacuum pump inlet. Add about an inch of water into the container, and connect the lid tightly. Start the vacuum pump, and observe the water. If it boils, the pump is good. It is recommended to change the oil because of the water that entered the pump. If the water does not boil, try changing the oil in the vacuum pump, and repeat the test to see if that fixes it.

A vacuum pump must be maintained with the proper oil level and periodic oil changes to maintain proper operation. If service is neglected, it will not pump to its design capabilities. Some modern electronic units flash a warning if the oil has not been changed at the proper interval (about 10 hours) (Figure 13–27).

New refrigerant service machines are equipped with oil-less vacuum pumps. These units eliminate the need to service the vacuum pump. One particular pump is a two-stage pump that is rated at 95% of barometric pressure, it will pull a 29″ vacuum in a 29.2 atmosphere. To run without lubricating oil, the pump uses two high-temperature plastic pistons operating in anodized aluminum cylinders with a Teflon-type cup seal. The piston and connecting rod are one piece. The crankshaft and connecting rod bearings use a lubricated and sealed ball bearing.

It is accepted practice to evacuate an automotive system for 20 to 30 minutes. This time can be shortened on a dry system and should be lengthened on a wet system. The longer evacuation helps ensure proper long-term system operation.

✓SERVICE TIP

Some compressors have the service ports mounted very low on the side of the compressor. This position allows the vacuum pump to draw the oil out of the compressor. When evacuating a system with this service port location, check to see if you have pulled out a significant amount of oil; if so, replace it with new oil.

If the system is contaminated with moisture, it is accepted trade practice to replace the accumulator or receiver–drier and, if there is metal or debris, to install an in-line filter.

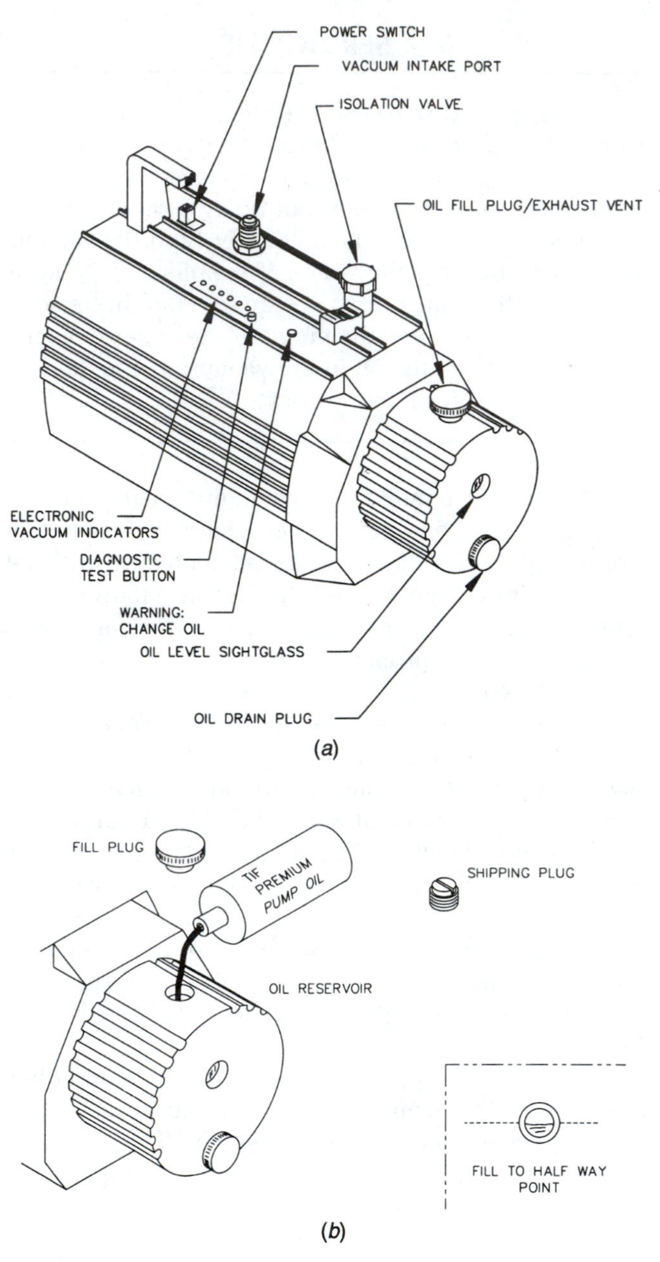

FIGURE 13–27 This vacuum pump includes an oil change warning light (*a*). Old oil is drained out, and new oil is added to the correct level (*b*). (*Courtesy of TIF Instruments*)

✓SERVICE TIP

At the end of your selected time period for evacuation, shut off the vacuum pump, close the manifold valves, and after a few minutes check the low side gauge for a pressure rise. An increase in pressure indicates either something—refrigerant or water—is changing state to become a gas or that there is a leak.

✓SERVICE TIP

Experience can teach you how to tell a wet system from a dry system by watching the low side pressure drop during evacuation. This pressure drop depends on the rating of the vacuum pump, size of the system, and any water (or oil-absorbed refrigerant) contained in the system. As water or refrigerant boils, it produces a much greater volume of gas to be pumped out. The more gas there is to pump, the longer it will take to drop the pressure. A system that takes a short time to drop down to 28″ Hg is fairly dry. A system that pumps right down to 20 to 26 inches, holds at this pressure for a while, and then drops to 28 inches probably contained a liquid that boiled out while the vacuum dropped from 20 to 28 inches. It is good practice to continue evacuation for at least 15 to 20 minutes after the pressure has dropped to its lowest reading.

✓SERVICE TIP

If you suspect a low side restriction, evacuate through the high side service port with the low side manifold valve closed. If it takes a long time period for the low side to drop, there is a problem.

To evacuate a system using a manifold gauge set and portable vacuum pump, you should:

1. Connect the manifold service hoses to the system service ports if necessary; normally these are still connected from the recovery process. There should be no or very little pressure in the system. Connect the center service hose to the vacuum pump inlet (Figure 13–28).
2. Open both manifold valves completely (and the vacuum pump valve if there is one) and start the vacuum pump. You should notice an air discharge from the vacuum pump and a drop in gauge pressures.
3. Check the gauge pressures periodically. After about 5 minutes, the pressure should be lower than 20″ Hg. A leak is usually the cause if the pressure has not dropped this low. You can confirm a leak by closing all valves, shutting off the vacuum pump, and watching the pressure. If it increases, there is a leak that must be located and repaired before continuing.

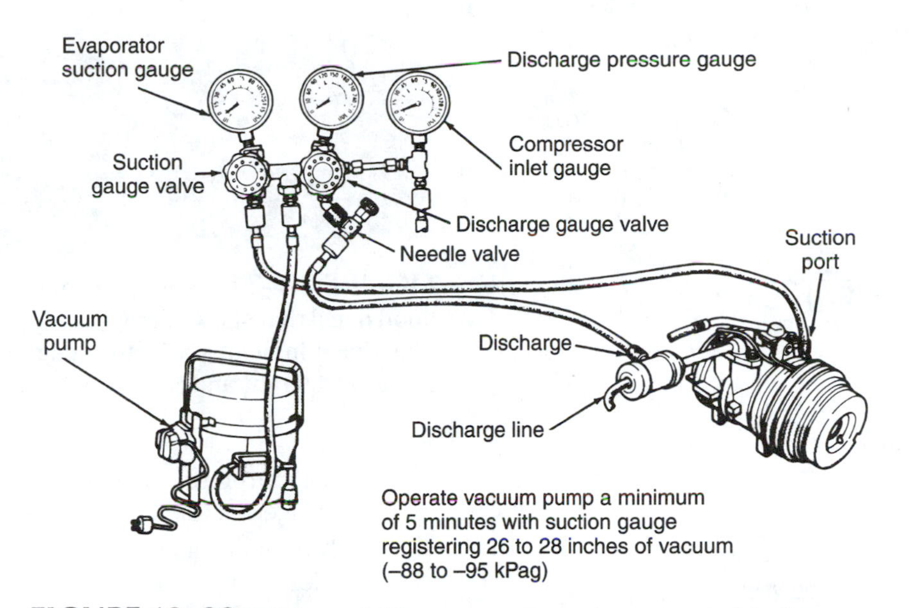

FIGURE 13–28 This portable vacuum pump is connected to the center service hose. Both gauge valves are open during the evacuation process. *(Courtesy of DaimlerChrysler Corporation)*

4. Continue evacuation for the desired length of time, close all valves, shut off the vacuum pump, and note the low side pressure (Figure 13–29).
5. After 5 minutes, recheck the low side pressure. If the vacuum held steady, the system is good and ready to be recharged. If the low side pressure increased, a possible leak is indicated; note the comments in Section 10.3.1.

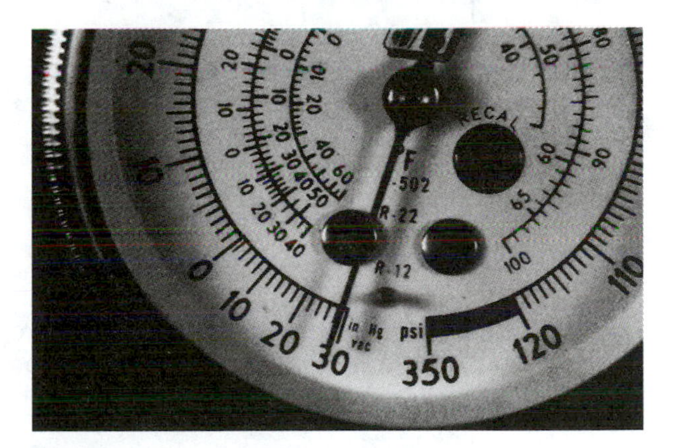

FIGURE 13–29 After the proper amount of time at 28″ Hg of vacuum, the vacuum pump is shut off and the manifold valves are closed. A tight system holds this vacuum for at least 5 minutes. *(Courtesy of Nissan Motor Corporation in USA)*

When using an A/C service unit, evacuation is often simply a matter of flipping a switch, assuming the service hoses are still connected from the recovery operation. Some modern stations use electric solenoids to control the flow inside the machine, and starting the evacuation process opens the solenoids needed for this process. Some also include a microprocessor that can be programmed to run the vacuum pump for the desired length of time. Some older charging stations are purely mechanical; on these units, the proper valves must be opened as the vacuum pump is started (Figure 13–30).

13.3.9 Recharging a System

After the system has been evacuated, it can be recharged with the correct amount of new or recycled refrigerant. Charge level is system specific, and as the volume of HVAC systems has been downsized, the amount of refrigerant has become critical. The charge level is normally found on a specification decal fastened to some location under the hood; on many modern vehicles, this decal tells both the type and volume of refrigerant used. The decal is attached to the compressor on some older systems. If the decal is missing or illegible, charge specifications are also given in service manuals published by the vehicle manufacturer and aftermarket A/C component suppliers (Figure 13–31).

The best method to charge a system is to start with an empty system and charge in the specified amount as measured by accurate scales. When systems were larger, refrigerant was cheaper, and environmental concerns

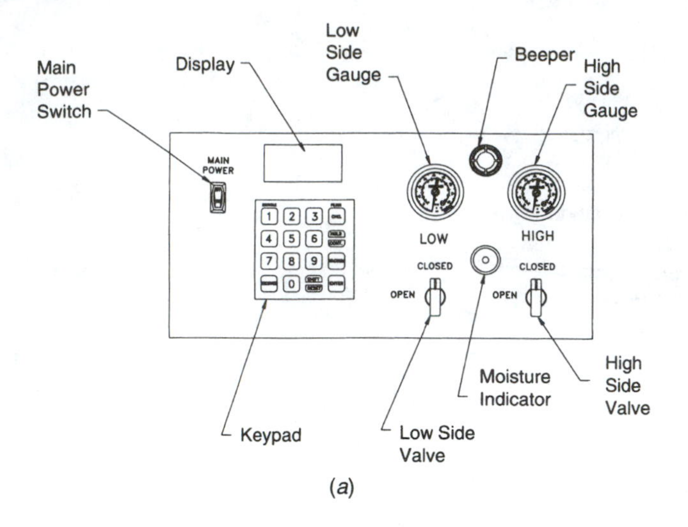

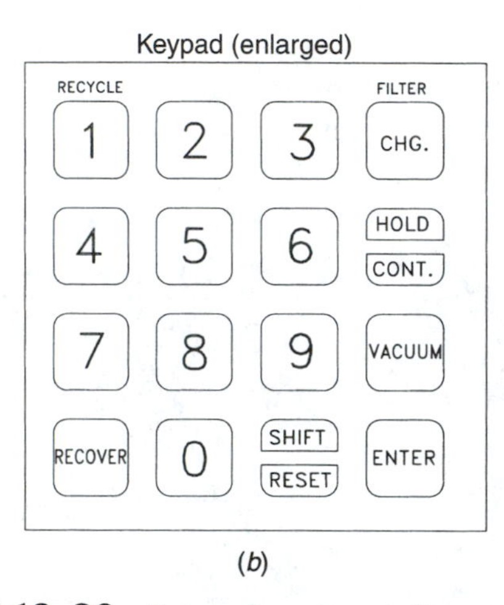

FIGURE 13–30 Some charging stations can be programmed to vacuum a system for a certain period. *(Courtesy of Robinair, SPX Corporation)*

were not as great, **partial charging or topping off** was a common practice: Refrigerant was added to a system with a low charge level.

Partial charging is no longer a recommended practice for several reasons:

- The leak in the system should be located and repaired to stop refrigerant loss into the atmosphere.
- It is very easy to overcharge a system as the partial charge is added. Older systems with large receiver–driers and accumulators had a tolerance for slight overcharges that newer systems do not have. An overcharge causes excessively high pressures that can damage the system and cause refrigerant release through the pressure relief valve.

FIGURE 13–31 Modern vehicles include an underhood decal that shows refrigerant type and charge level. This decal indicates a factory charge of 0.907 kg (2.00 lb) of refrigerant and SP-20 PAG oil.

- A nonprofessional might have added the wrong refrigerant and contaminated the system.
- Topping off is illegal in some areas and countries.

Now that recovery and recycling equipment is common, it is relatively easy to remove and recycle the system's refrigerant, repair any leaks, and recharge the system with the specified amount of new or recycled refrigerant. This gives a much more reliable repair and removes the guesswork of partial charging.

✓REAL WORLD FIX

The front TXV (front + rear system) was replaced on the 1993 Dodge Caravan (95,000 miles) and the system was recharged, but the A/C does not cool. The high side pressure is excessive, over 450 psi, the low side pressure is normal. The technician questions if the charge specification is correct at 61 oz.

Fix: Following advice, the technician rechecked his math. He had interpreted the sticker reading of 3.13 to be 3 lbs and 13 oz, which equals 61 oz. It really means 3.13 lbs, or 3.13 x 16 or 50 oz. Recovering the refrigerant and recharging to the proper level of 50 oz fixed this problem.

13.3.9.1 Partial Charging Occasionally partial charging or topping off still has a place with R-12 vehicles if the vehicle owner cannot afford to retrofit the vehicle or the value of the vehicle is not worth the cost of a retrofit. Partial charging adds refrigerant into a partially filled system until it has a full charge. The technician usually determines the charge level by watching the sight glass or watching the high side pressure. Remember from Chapter 10 that neither of these methods is completely reliable in determining charge level. Several things can cause bubbles in the sight glass or affect the high side pressure, and there are two better methods of determining charge level (as described in Section 10.3.5). Also remember that the reason for the low charge level is often a leak, and experienced technicians are very thorough in their attempts to prevent refrigerant leaks, especially of R-12, into the atmosphere.

When performing a partial charge, the following practices are recommended:

- Check the system for leaks; if a leak is found and the cause is located, repair the leak.
- Identify the refrigerant remaining in the system. If the refrigerant is contaminated with air or a foreign refrigerant, fix the system rather than charge it.
- Inject fluorescent dye during the partial charge, and call the vehicle back in a week or so to check for a leak. When the leak is located, you can recover the refrigerant, repair the leak, and evacuate and recharge the system.

Partial charging is a continuation of the performance test described in Section 10.3. At step 6, on a system with a low charge level, you will note bubbles or foam in the sight glass or an accumulator that is warmer than normal, system high side pressure that is lower than normal, and poor cooling at the A/C ducts. From this point, with the system operating, the technician slowly charges refrigerant into the low side until the sight glass clears up or the accumulator begins to feel cold at the bottom; then 1/4 lb of additional refrigerant is added to complete the charge while carefully watching the high-pressure gauge to make sure that the pressure does not increase as the additional refrigerant is added. If the high side pressure or the level of cold air output does not become normal, something else is wrong with the system.

✓SERVICE TIP

Professional service technicians prefer not to top off a system because the system probably has a leak that should be repaired. Also, the system contains an unknown amount of refrigerant, oil, and possible contaminants. It is a much better practice to recover the refrigerant and see how much refrigerant and oil come out. The refrigerant can now be recycled to remove any air that might have been present and then returned to the system along with additional refrigerant to fill the system with the specified amount of clean oil.

✓REAL WORLD FIX

The 1996 Explorer (95,000 miles) was brought in with a request to add more refrigerant because of poor cooling. The technician informed the customer that he felt it was already overcharged.

Fix: The refrigerant in the system was recovered, and there were 3.25 lbs. The system was recharged to the 1.5-lbs specification, and this fixed the poor cooling problem.

13.3.9.2 Charging from Large Containers

Most shops use larger refrigerant containers, in the 20- to 30-lb size, so the amount to be charged into the system must be measured. The units most commonly used are the **charging cylinder** and **electronic scales**. These can be either individual portable units or parts of a charging station.

The charging cylinder unit, dial-a-charge, uses a clear sight column where the volume of refrigerant is shown. A calibrated shroud around the outside of the unit is dialed or rotated until the graduated numbers for that refrigerant at that pressure (gauge at the top) are next to the sight column. The various sets of graduations compensate for volume changes due to temperature. The technician normally adds refrigerant to the charging cylinder until it contains the amount needed for a system; this amount is then charged into the system (Figure 13–32).

Modern scale units use electronic scales that can be programmed for the desired charge level. The refrigerant container is placed on the scale, and a hose is used to connect its valve to the scale. The operator then programs in the charge volume desired and starts the charge process. When the proper amount of refrigerant has left the container, an electric solenoid in the unit shuts off the refrigerant flow. These units can also be operated manually, with the operator holding down a button or switch until the desired amount of refrigerant has left the container (Figure 13–33).

Moving the refrigerant into the system requires that the charging container pressure be greater than the system pressure. Because the process begins with the system in a vacuum, the first portion goes rather quickly, but the first 1/2 lb or so fills the internal volume and starts generating pressure. As refrigerant boils and leaves the container, it cools the remaining refrigerant and causes a pressure drop. Many charging stations include heaters for the refrigerant container to raise its internal pressure to help force refrigerant into the system. When heaters are used, the system does not need to be operated. The

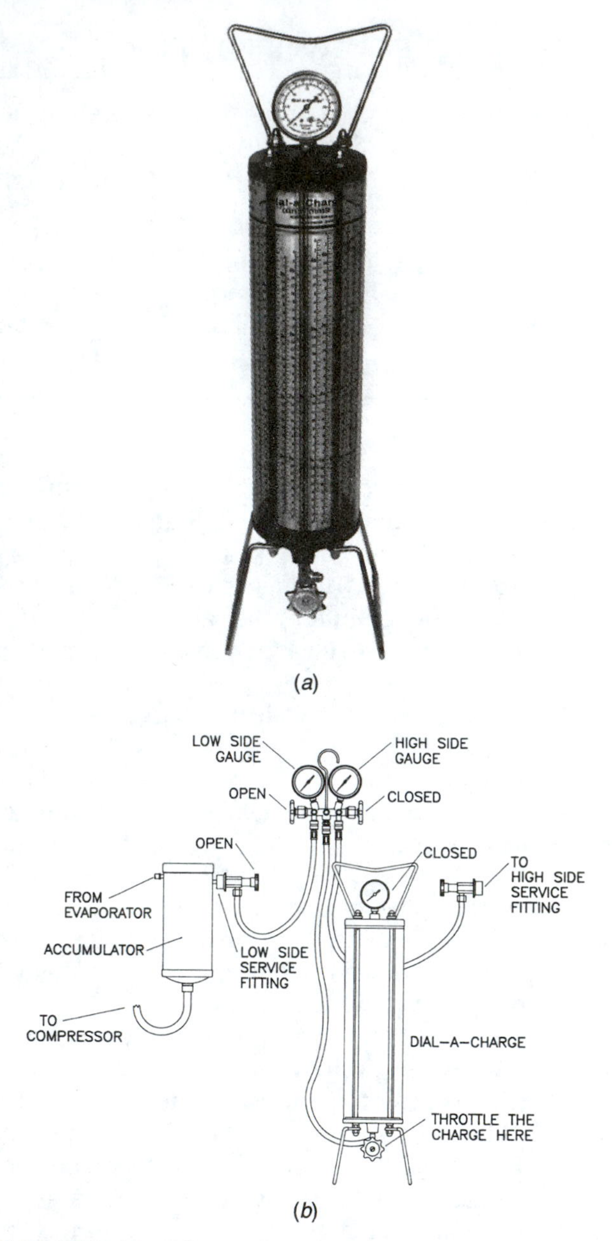

FIGURE 13–32 This charging cylinder has a calibrated shroud marked with the refrigerant volumes for different temperatures (*a*). Refrigerant is added to the cylinder from a large container; this amount of refrigerant is then charged into the system (*b*). (*Courtesy of Robinair, SPX Corporation)*

pressure difference can also be increased by starting the A/C system so the low side pressure drops and then charging only into the low side. *Remember that the high side manifold valve should never be open while the system is operating.* Systems with low-pressure cutout will not operate, and older GM cars with thermal limiter fuses will burn out the fuse, when the charge level is very low. Jumper devices can be used to temporarily bypass these low-pressure protection devices (Figure 13–34).

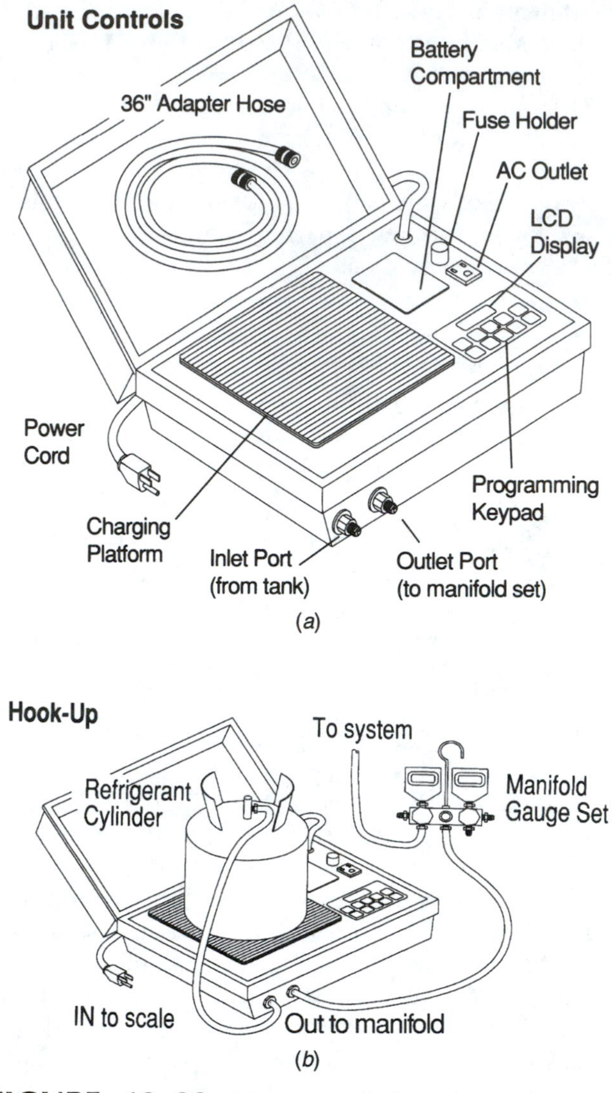

FIGURE 13–33 This charging scale includes a weighing platform and a shutoff solenoid valve (*a*). The container of refrigerant is placed on the scale, and the unit is programmed to allow the correct amount of refrigerant to enter the system (*b*). (*Courtesy of TIF Instruments)*

✓SERVICE TIP

Some technicians prefer to begin the charging process into the high side (system off). The high side pressure should increase immediately as the valve is opened. There should be a slight delay and a slower low side pressure increase as refrigerant bleeds through the TXV or OT. A pressure increase that is too rapid indicates a possible compressor internal leak; too slow an increase indicates a possible restriction of the TXV or OT (Figure 13–35).

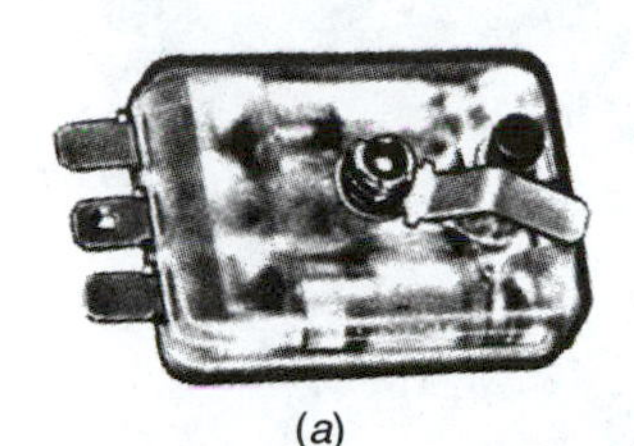

(a)

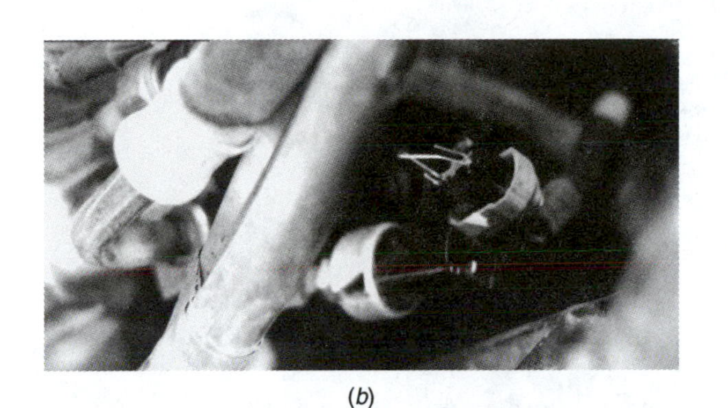

(b)

FIGURE 13–34 This tester and adapter is designed to test GM thermal fuse circuits and to allow bypass of the thermal fuse during charging operations (*a*). A bent paper clip can often be used as a jumper to bypass pressure switches (*b*). (a *is courtesy of Four Seasons)*

✓SERVICE TIP

Some technicians prefer to complete the charging process with the system operating so they can watch the gauge pressures and temperatures normalize, even though this is not necessary with most modern charging stations. Normal A/C system operation should occur about 1/4 to 1/2 lb before the system is completely charged. This last volume becomes reserve to be stored in the receiver–drier or accumulator. At this point, the sight glass should clear up or the bottom of the accumulator should become cold. The high side pressure should not increase as the remaining refrigerant is charged into the system.

With the system running, caution should be exercised if charging liquid into it to avoid **slugging** the compressor; letting liquid refrigerant enter a running compressor can cause severe damage. If the container is upright, gas will leave it; if the container is upside down, liquid will leave it. With approved recovery containers, either liquid or gas can be removed from an upright container, depending on which valve is opened. Putting 1 lb of liquid into a system is much faster than putting in 1 lb of gas because of a liquid's much smaller volume (Figure 13–36).

LOW SIDE VALVE Closed
HIGH SIDE VALVE Open

After low side gauge shows pressure, open low side valve.

REFRIGERANT SOURCE

FIGURE 13–35 Some technicians prefer to begin the charging procedure by adding pressure to the high side; a pressure increase in the low side should begin a few seconds later as the refrigerant passes through the expansion device.

To charge a system using a charging station, you should:

1. Enter the specified amount of refrigerant into the charging scale or into the charging cylinder unit. This charge process begins with the system still under a vacuum, with the manifold valves closed (Figure 13–37).
2. Depending on the equipment being used, open the necessary valves and push the correct buttons to program the unit or fill the charging cylinder with the correct amount of refrigerant. Once the gauges show a pressure increase in the system, turn on the refrigerant heater (if so equipped).
3. When the charge volume has entered the system, close the necessary valves.
4. Start the A/C system, let it run until the pressures stabilize, and note system pressures as described in Section 10.2.4. They should be normal.

✓SERVICE TIP

If liquid has been charged into a system, it is possible that liquid refrigerant has entered the compressor because of the location of the compressor and service ports. If you suspect this, rotate the compressor drive plate by hand at least one revolution to ensure that this has not occurred.

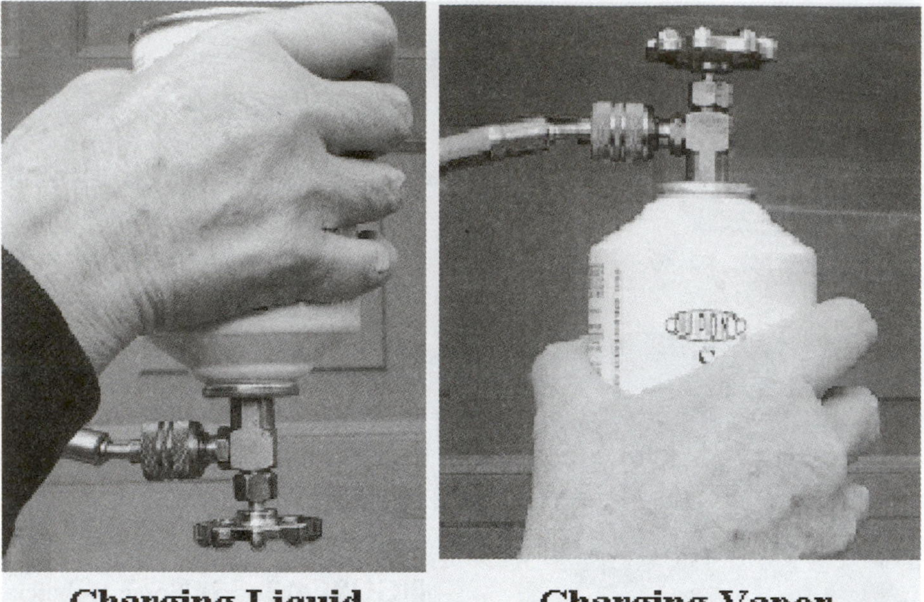

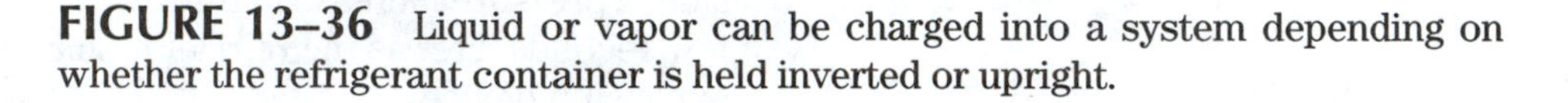

FIGURE 13–36 Liquid or vapor can be charged into a system depending on whether the refrigerant container is held inverted or upright.

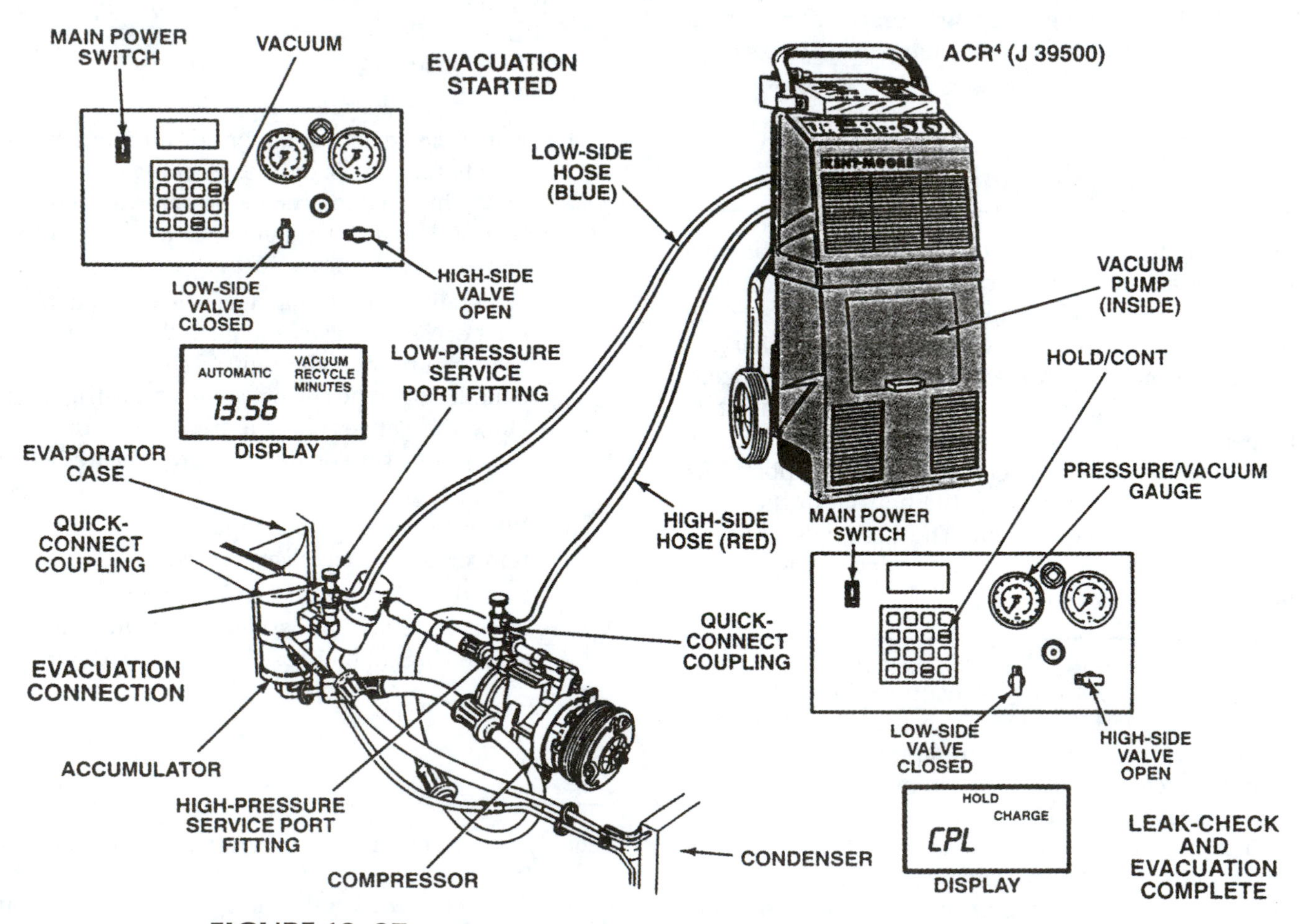

FIGURE 13–37 Many A/C service units can be programmed to operate through a leak check and evacuation procedure. If the system passes the leak check, the service unit will be programmed to charge the proper amount of refrigerant into the system. *(Reprinted with permission of General Motors Corporation)*

13.3.9.3 Charging from Small Cans When a system is to be charged from small cans, the technician needs to determine how many cans are needed. This is done by dividing the system's specification by the net weight of the cans. For example, the specification for a system might be 2¼ lb, 2.25 lb, or 2 lb 4 oz (all of these are the same); this amount needs to be converted so that the total weight is in ounces.

- There are 16 ounces in a pound. Multiply 2 × 16 to get 32, and 32 + 4 = 36. This system holds 36 oz of refrigerant.
- Next, divide 36 by the weight in a small can, which in this case is 14 oz (some are 12 oz); 36 ÷ 14 = 2 with 8 left over.
- To charge this 36-oz system using 14-oz cans, add two full cans plus 8 oz from a third can, leaving 6 oz in the third can.

Measuring the amount removed from the third can is done most accurately by weighing the can, tapper, and hose as the refrigerant is pulled out. For example, if this weight is 24 oz when we start, we stop when the weight has dropped to 16 oz, 8 oz less. The remaining refrigerant can be approximated by feeling the can temperature (the liquid level will be colder than the gas portion at the top) or by tipping the can (suction pressure will increase when liquid enters the hose) (Figure 13–38).

Small cans are sealed and must be tapped to get the refrigerant out. Older cans had a seam at the top, and the can tapper gripped this seam. Some can tappers fit around the can and pierce the side. These side tappers are faster to use: A quick squeezing action is all that is needed to attach the tapper and pierce the can, and it is removed using one quick motion. Can tappers for R-134a and some blends thread onto the top of the can (Figure 13–39).

To charge a system using small containers, you should:

1. Locate the specifications and determine the number of small cans needed. The charge process should begin with the system still under a vacuum, with the manifold valves closed.
2. Attach the can tapper to the center service hose, connect the can to the tapper, and pierce the first can of refrigerant.
3. Open the high side service valve, watch the low side pressure increase, and then open the low side service valve. The refrigerant can becomes cold as the refrigerant boils and transfers to the system. If the manifold is equipped with a sight

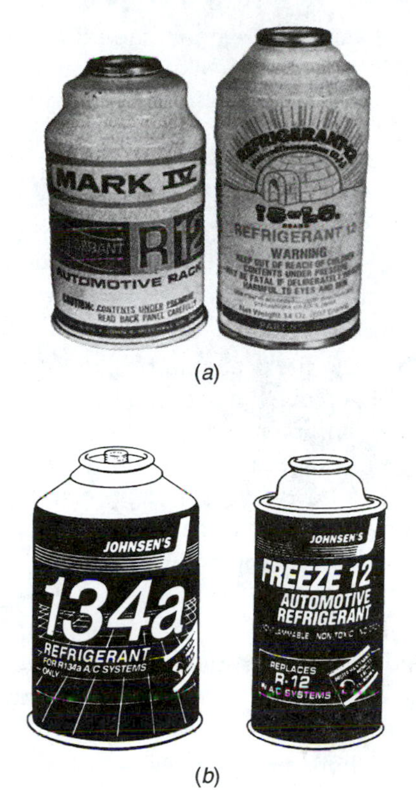

FIGURE 13–38 If using small cans, be sure to note the net weight of the can. The cans of R-12 (*a*) hold 14 and 15 oz of refrigerant; the cans of Freeze 12 and R-134a (*b*) hold 12 oz. (b *is courtesy of Technical Chemical Company [TCC]*)

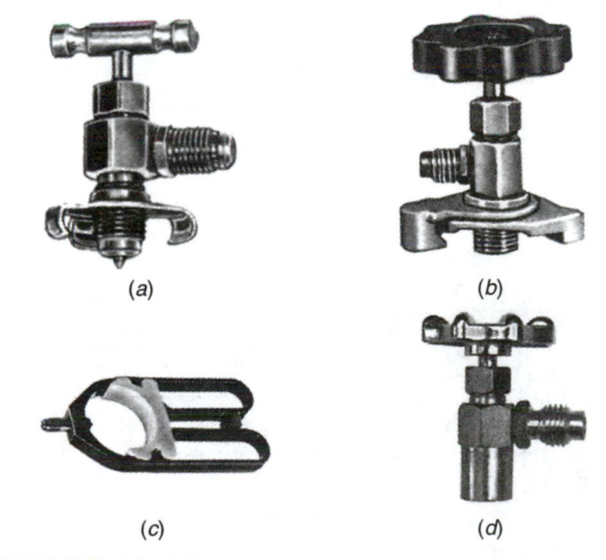

FIGURE 13–39 Three different can tappers can be used with small R-12 cans (*a–c*). Different adapters that thread onto the can are required for R-134a (*d*) and blends. (*Courtesy of Four Seasons*)

glass, liquid refrigerant should be seen moving into the system. Shaking the can of refrigerant allows you to note the remaining liquid volume and also allows the whole can area to warm the remaining refrigerant.

4. When the charging action slows down, close the high side valve, make sure the hoses are clear of the soon-to-be-rotating engine parts, start the engine, and turn on the A/C system and blower to high speed.
5. When the first can empties, close the low side valve, remove the first can from the tapper, and install and pierce the next can. With some can tappers, air can enter, and the center hose will have to be evacuated or purged. If the charge level is less than 28 oz, stop charging before the second can empties.
6. Depending on system capacity, repeat step 5 until the last can is reached. Measure the amount of refrigerant you remove from the final can to ensure that the right amount of refrigerant enters the system.
7. Close both manifold valves. With the system operating, note the low and high side pressures; they both should be normal.

✓SERVICE TIP

When a can empties, allow the low side pressure to drop as low as it will go to pull as much refrigerant as possible from the can. It is easy to leave as much as an ounce in the can, and this can cause an undercharged system or one without any refrigerant reserve.

✓SERVICE TIP

Tampering has become a major concern with many A/C service technicians; once a system has been serviced and put in proper operating condition, they want it to stay that way. A set of seals has been developed so it is easy to tell if a system has been opened. These seals are plastic sleeves that are placed over the service fittings and heated to shrink them to fit the fitting and cap (Figure 13–40). The seal must be cut in order to remove the service cap.

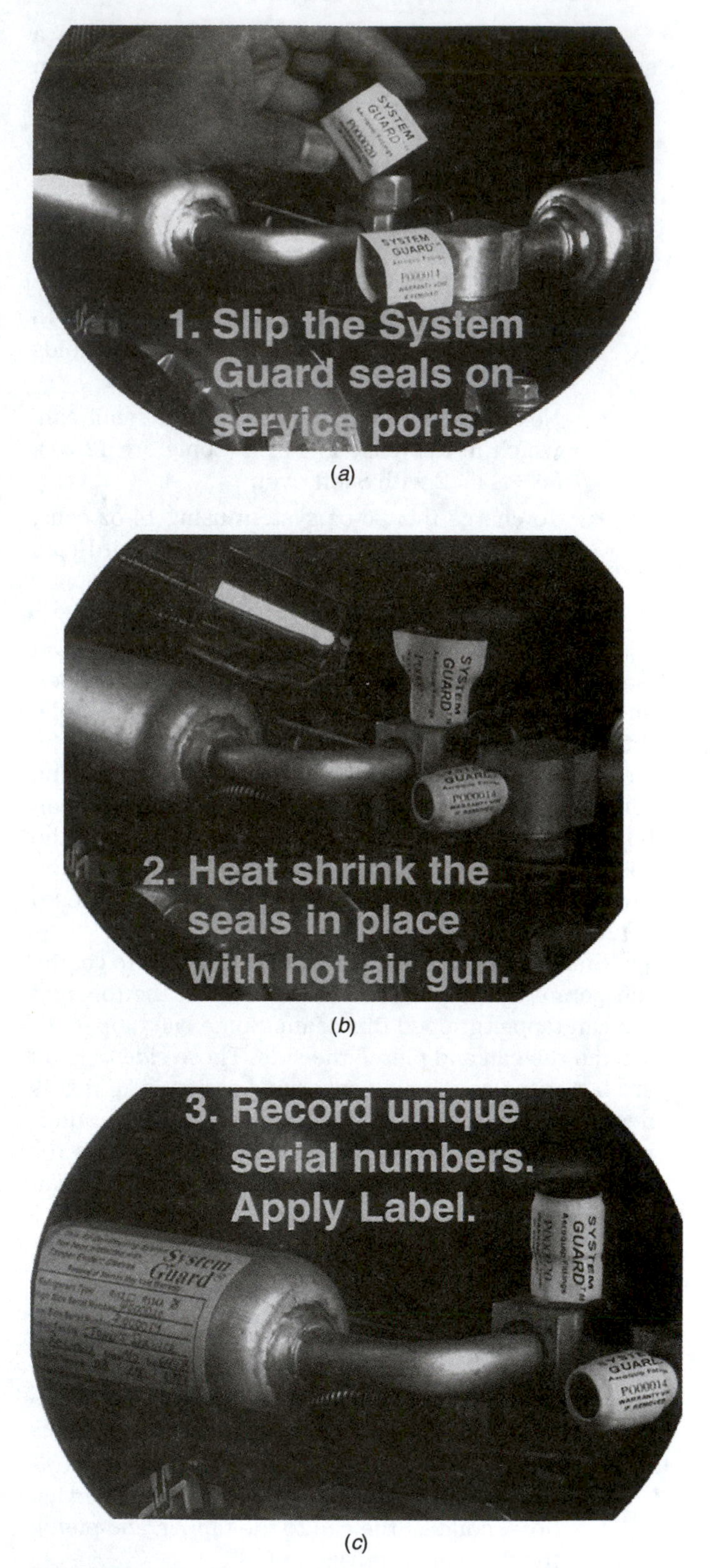

FIGURE 13–40 System Guard Seals are placed over the service ports and locked in place by shrinking them using a hot air gun. The system cannot be opened without cutting the seal. *(Courtesy of System Guard)*

13.4 RETROFITTING R-134a INTO AN R-12 SYSTEM

Retrofitting is normally a **repair-driven operation**; it is not done until absolutely necessary. All experts agree that if a system was designed for R-12, R-12 should be used in it when it requires service, even though some systems cool better after changing to R-134a. We do know that new (virgin) R-12 will only be available for a limited time, so if a system requires service, consideration should be given to converting it to R-134a. This is especially true if major repairs, such as a compressor replacement, are needed. When R-12 is no longer available, the choice will be between retrofitting and not having A/C at all.

13.4.1 Refrigerant Choice for Retrofit

Any EPA SNAP-approved refrigerant, listed in Appendix D, can be used to replace R-12. Some of these are called "drop-in refrigerants"—but remember that there is no such thing. All of them require a similar retrofit procedure. Some of the things you should consider when choosing which refrigerant to use are the following:

- *It must be EPA approved and have unique service fittings and label.*
- R-134a is the refrigerant used in every new vehicle.
- Unique service, recovery, and charging equipment are required for each refrigerant; most shops have R-12 and R-134a equipment.
- Most major compressor manufacturers and builders will not warranty a compressor that failed if a blend refrigerant was used in it.
- Some of the blends cost three or four times as much as R-134a and must be sent off for disposal or recycling.
- Blends that contain R-134a (some people call them dirty R-134a) share the same compressor problems as R-134a.
- Blends should be charged into a system as a liquid to reduce fractionation.
- Blends that contain R-22 must use barrier hoses and compatible inner seals (see Figure 3–15).
- Blends that contain HCFC (R-22) will be phased out when global warming limits are enacted.
- More than 2% hydrocarbon content can produce a flammable mixture.
- The vehicle might travel to an area where the blend and its service equipment are not available.
- Blends are outlawed in some areas, including at least one Canadian province.
- Some blends do not work as well with leak detectors and fluorescent dyes.

All vehicle manufacturers, both international mobile air conditioning organizations (IMACA and MACS), and most service shops favor using R-134a for retrofits. It is relatively inexpensive, used in all new vehicles (so the service equipment is fairly common), readily available, and, because it is a single-compound refrigerant, can be recycled in service shops. Many shops are recovering any R-12 in systems to be retrofitted and saving that refrigerant for use in vehicles that are difficult or expensive to retrofit.

13.4.2 Possible Retrofit Problems

R-134a refrigerant works quite well in most R-12 systems. However, some areas of concern include:

- Oil degradation from chemical contamination, resulting in possible compressor damage.
- Incompatibility of compressor seal material, resulting in leakage. A compressor that contains Viton seals cannot be used with R-134a.
- Incompatibility of desiccant type, causing further contamination.
- Permeability of hoses, resulting in leakage.
- Increased high side pressure, causing pressure release through relief valve, increased loss through hoses, and compressor damage.
- Systems with a pressure relief valve must have a *high-pressure cutoff switch*, also called a *refrigerant containment switch*, installed.
- Inability of compressor type or clutch to work reliably at higher high side pressures.

✓REAL WORLD FIX

The 1989 Honda (100,000 miles) has a bad compressor, and the vehicle owner would like to repair the system and retrofit it to use R-134a. The technician is concerned that a new or repaired version of this compressor will not stand up to the pressures of an R-134a system. The compressors that were received were longer than the original and would not fit.

Fix: Research found a supplier with a kit containing a new compressor with brackets, an accumulator, and hose sections that allowed a more durable compressor to be fitted to this system.

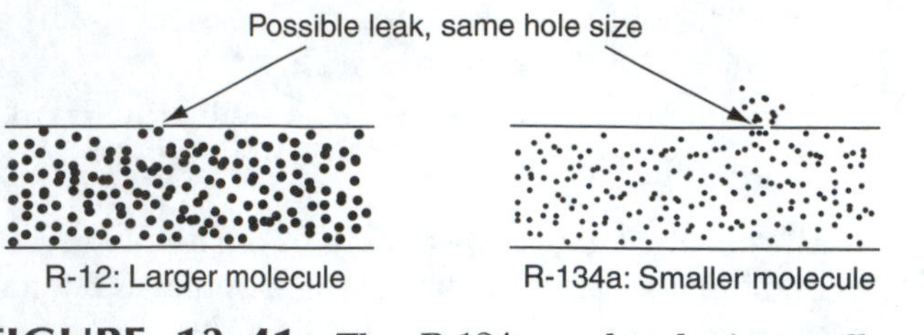

FIGURE 13–41 The R-134a molecule is smaller than an R-12 molecule, so the leak rate through the same hole is greater.

✓SERVICE TIP

The Ford FX-15 compressor might not handle the higher pressure when a system is retrofitted to R-134a. It was replaced by the FS-10 compressor in mid 1993; the FS-10 is a much stronger version of the FX-15. Two ways to identify an FX-15 are (1) note the date on the back side of the rear mounting flange; if the date is 94 or later, it is an FS-10; and (2) pull the clutch plate, and count the number of splines; the FX-15 has 16 splines and the FS-10 has 21.

An R-134a molecule is smaller than an R-12 molecule. Consequently, a system with a very small R-12 leak will have a bigger, more rapid leak with R-134a. All leaks in a system to be retrofitted should be repaired (Figure 13–41).

Some of the compressors that have or might have problems with an R-134a retrofit are listed in Appendix E at the back of this book.

Some RWD vehicles are proving difficult to retrofit; this difficulty is related to condenser capacity and airflow through the condenser.

✓SERVICE TIP

IMACA has recommended the following guidelines: Check the high side pressure of the system operating with R-12 and compare the pressure to ambient temperature (AT) plus 40. If the high side pressure is less than AT + 40, the system should operate quite well with R-134a. If the pressure is above AT + 40, the pressure might become excessive with R-134a, and the best time to make improvements is during the retrofit process, after the R-12 has been recovered. In some cases, a more efficient or larger condenser is needed. In most cases, however, increased airflow from a higher-capacity fan, better fan shrouding, seals around the condenser to control the airflow, or recalibrated fan controls will solve this problem.

13.4.3 Retrofit Procedure

There are two types of retrofit procedures:

- *High performance*, also called *Type I*, is the most effective and follows the recommendations of the vehicle manufacturer and ensures that the performance will be equal to the original operation.
- *Economy*, also called *Type II*, is the least expensive and replaces only those parts necessary to comply with federal requirements. This procedure is the one used by most shops and is described in this section.

The exact procedure used to retrofit R-134a varies depending on the design and materials in a particular system. For example, many late model R-12 systems are equipped with barrier-type hoses along with compressors and accumulators or receiver–driers that are completely compatible with R-134a, so there is no need to replace these parts. Older cars probably require replacement of both the compressor and hoses, especially if they show a lot of wear.

Some odd things are turning up as a result of retrofit tests. Apparently, chlorine from R-12 seasons a compressor so that a used, good R-12 compressor would probably operate satisfactorily with R-134a. Some new R-12 compressors have a higher failure rate with R-134a than used R-12 compressors. Rebuilt compressors, compatible with R-134a, are available. Also, mineral oil apparently coats the inside of the hoses, so the leak rate through them is much lower than expected when retrofitting a system to R-134a.

When retrofitting, at least 99% of the R-12 must be removed from the system and as much of the mineral oil as practical. Remember that more than 2% R-12 will cause excessive high side pressures and other problems and that the oil has absorbed R-12. A recovery unit removes the R-12 vapor and some of the oil and measures the amount of oil removed; some units measure the amount of R-12 also. Most recovery units cannot pull the system down into a deep vacuum, making it necessary to connect a vacuum pump for the final R-12 removal. Complete R-12 removal can be difficult because of the amount absorbed by any remaining oil. Remember that thorough oil removal also helps eliminate R-12 from the system.

Oil removal is more difficult because it must be either drained out, flushed out, or pulled out with the R-12. Complete draining of some components requires removal of the component, which greatly increases the labor costs. As described in Section 13.3.5, liquid R-12 can flush oil out (Figure 13–42). Any remaining mineral oil will probably gel and settle in the bottom of the accumulator or evaporator. It is known that flushing the system with R-11 causes future problems.

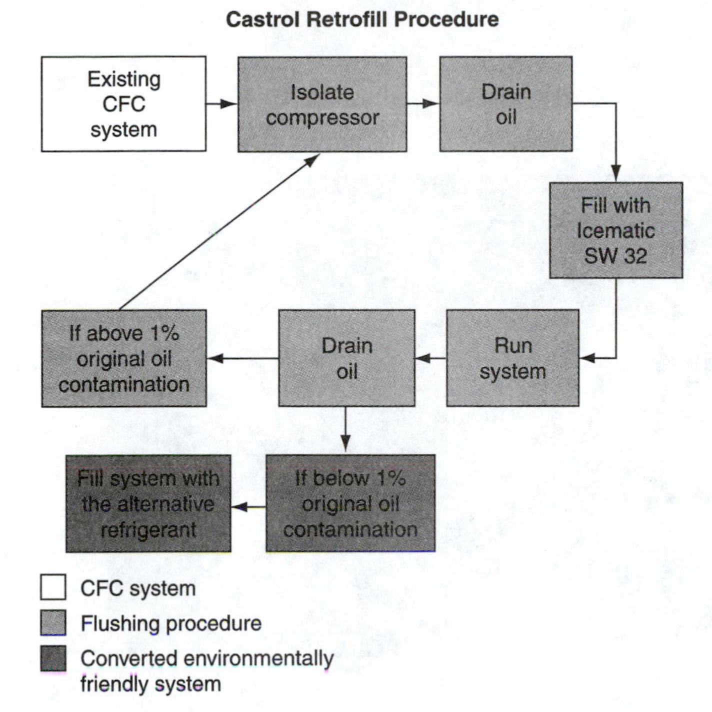

FIGURE 13–42 A procedure used to remove oil from an R-12 system before an R-134a retrofit. *(Courtesy of Castrol North America)*

✓SERVICE TIP

One way to partially identify the oil type in a system is to put a few drops of the oil into a container of water. PAG is very hygroscopic and will be dissolved into the water. An end-capped PAG will form an oil film. Ester and mineral oil will float on top of water.

✓SERVICE TIP

Most vehicle manufacturers recommend using R-12 in an R-12 system as long as this refrigerant is available, but most of them have also developed a TSB that describes the recommended retrofit procedure for those vehicles that can be retrofitted. These procedures should be used when R-12 is not easily available. A binder containing these OEM retrofit procedures is available from IMACA.

✓SERVICE TIP

PAG oil is proving more durable, and there are fewer compressor failures in systems where it is used in the retrofit, especially in cases with borderline lubrication.

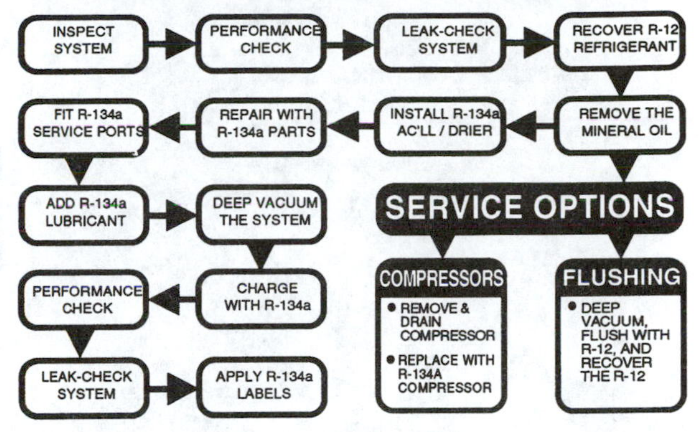

FIGURE 13–43 This flowchart shows the procedure to follow when retrofitting an R-12 system with R-134a. *(Courtesy of the International Mobile Air Conditioning Association [IMACA])*

In actual practice, shops are quite successful with Type II retrofits of both FWD and RWD vehicles. Most retrofits of FWD vehicles are quite easy (Figure 13–43).

R-134a is lighter than R-12, so a system should be charged to about 80 to 90% the amount of the R-12 capacity. Some aftermarket sources have made R-134a retrofit charge capacities available. If you need to figure the new charge level, simply multiply the R-12 charge level by 0.8. For example, if the R-12 charge is 2 lb (32 oz), $32 \times 0.8 = 27$ oz. General Motors guidelines for retrofitting its vehicles recommend multiplying the R-12 charge level by 0.9 and subtracting 0.25 lb (4 oz). Using this procedure, $32 \times 0.9 = 29 - 4 = 25$, slightly less than the 27 oz (80% level). Refer to Section 10.2.5.2 for methods to get the proper charge in the system.

✓REAL WORLD FIX

The 1990 Plymouth Voyager (88,000 miles) had a leak in the high side and two faulty TXVs (rear air), so it was decided to retrofit to R-134a. The system was flushed, and both TXVs, the compressor, receiver–drier, discharge line, and all seals were replaced. Eight oz of PAG oil were added, and the system was evacuated for 90 minutes before charging it with 34.5 oz of R-134a. When tested, the system did not cool well, and the pressures were excessive: 75 psi low side and 225 psi high side. The heater hoses were clamped off, but this did not help.

Fix: The refrigerant was recovered, and the system was recharged with 30 oz of R-134a for a 70% charge level. This smaller refrigerant charge fixed this problem.

Author's Note: The original charge specification for this vehicle is 43 oz, and DaimlerChrysler Corporation recommends a retrofit charge level of 32 oz of R-134a.

FIGURE 13–44 A retrofit kit can be simple and inexpensive (*a*) or very complete, designed for a particular vehicle (*b*). (*a*) includes oil, R-134a service fittings, decals and new O-rings. (*b*) contains all of these items plus a replacement condenser, control switch, hoses, receiver–drier, and TXV. (a *is courtesy of Castrol North America;* b *is courtesy of Wynn Oil Company*)

Both vehicle and aftermarket manufacturers produce kits that include the parts needed to convert particular systems (Figure 13–44). Vehicle manufacturer kits are for Type I retrofit, and most aftermarket sources provide Type II kits. Depending on the particular make and model, a kit can include the following:

- A sticker to identify that it is an R-134a system, along with the charge level (Figure 13–45)
- R-134a-type service fittings to be permanently installed over the existing service fittings
- Replacement hoses with the R-134a service fittings
- Replacement O-rings
- Ester or PAG oil
- Replacement receiver–drier or accumulator with XH-7 or XH-9 desiccant
- Replacement system switches calibrated for R-134a pressure
- Replacement TXV or OT calibrated for R-134a pressure
- High-pressure cutoff switch

Simpler kits cost less than $50.

The R-12 service fittings must either be converted to R-134a service fittings or permanently capped to make them unusable. Normally, they are converted using a **conversion fitting** (Figure 13–46). A conversion fitting is threaded over the R-12 fitting, using the same Schrader valve, and a thread-lock adhesive locks it in place. It is a good practice to replace the old Schrader valves with ones that are compatible with R-134a. If an R-12 fitting is in a bad location or is unusable, it can be capped and a saddle clamp with the R-134a fitting can

NOTICE: RETROFITTED TO R-134a

⚠ CAUTION: System to be serviced by qualified personnel. R-134a refrigerant under high pressure.

Retrofit procedure performed to SAE J1661.

134a Charge Amount:

134a Lubricant Type: Amount:

Retrofit Performed by:

Name:

Address:

City: State: Zip:

Date of Retrofit:

FIGURE 13–45 A label that shows retrofit information must be placed over the old R-12 information decal. (*Courtesy of Four Seasons*)

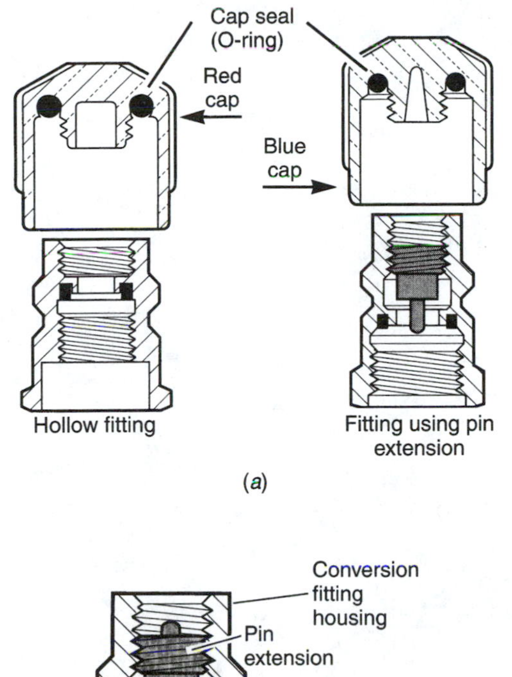

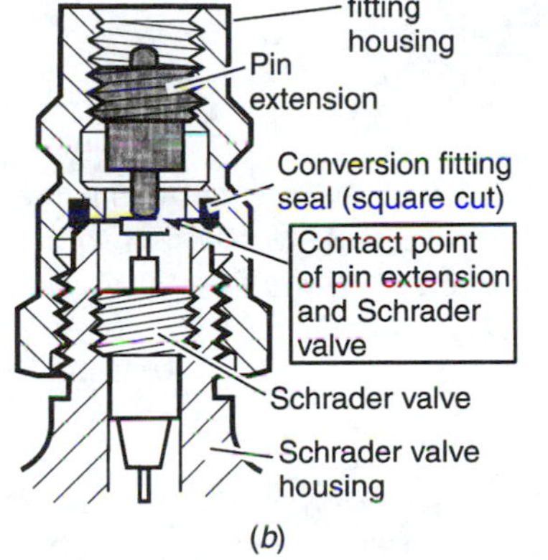

FIGURE 13–46 The two most popular conversion fittings thread directly onto the R-12 service ports (*a*). Fittings that use a pin extension should be adjusted after installation so there is slight clearance between the pin extension and the Schrader valve (*b*). (*Courtesy of Four Seasons*)

be installed. A saddle clamp is placed over a straight section of metal tubing and locked in place, and the tubing is pierced to provide an opening (Figure 13–47).

✓SERVICE TIP

The most popular conversion fittings are either hollow or contain a pin extension to contact the valve stem. The hollow style is quite short and is simply threaded over the service port. Fittings with a pin extension ensure better valve opening as the service hose is connected, but the pin extension must be adjusted after installation. Turn the pin extension inward until it just contacts the valve stem and then turn it outward one half-turn.

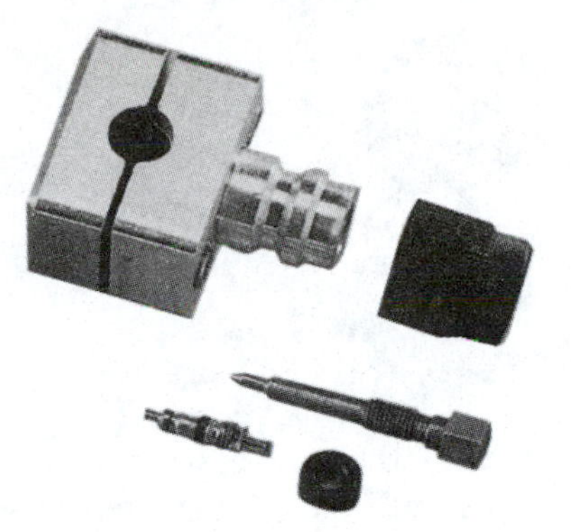

FIGURE 13–47 This port is part of a saddle clamp that can be placed over a metal line; the line is then pierced and a Schrader valve is installed. (*Courtesy of ACDelco*)

The retrofit process will probably consist of the following steps:

1. Visually inspect the system to ensure good condition, install a gauge set, and operate the system to bring it up to operating temperatures. Check for proper operation and note any needed repairs. Record the high side pressure for later comparison.
2. Recover the R-12 from the system, restarting the recovery procedure to remove as much oil-dissolved R-12 as possible.
3. Make any repairs to the system to cure problems that were found in step 1.
4. If the compressor failed, remove the failed compressor, flush the system and/or install a high side filter, and install the replacement compressor along with a new accumulator or receiver–drier.
5. Check the system to determine whether a high-pressure relief valve is used. If it has one, a high-pressure cutoff switch must be installed to stop the compressor before pressure relief valve release pressures occur (Figure 13–48). The switch is installed so it senses high side pressure and is wired into the clutch wire or relay so it can interrupt clutch operation.
6. If directed, R&R the receiver–drier or accumulator.
7. R&R any hoses that were leaking, or as directed. If a blend refrigerant that contains R-22 is to be used, check the hoses to make sure that they are barrier hoses. You should be able to find the identification name and number on the hose.

FIGURE 13–48 This high-pressure cutoff switch can be installed in a high side port and is connected into the clutch wiring so the compressor will be shut off before it reaches pressures that must be released through the relief valve. *(Courtesy of ACDelco)*

8. R&R any line-fitting O-rings on connections that were disturbed, or as directed.
9. R&R any switches and valves as directed.
10. Add the proper type and amount of oil—ester or PAG—into the oil fill port or suction port.
11. Install the R-134a service fittings. Any old Schrader valves that remain in service should be replaced with new R-134a valves.
12. Fill out and install the identifying decal to properly identify the system. The old label must be rendered unreadable.
13. Connect a vacuum pump to the system and pull a minimum vacuum of 25″ Hg for at least 30 minutes to evacuate the system.
14. Recharge the system using R-134a. Charge the system with 80 to 90% of the specified amount of R-12. Proper charge level can be verified by any of the methods described in Section 10.2.5.2.
15. Operate the system and check for proper operation, paying careful attention to the high side pressure.
16. Test for leaks.

✓SERVICE TIP

If a system that you retrofitted does not cool sufficiently and has excess high side pressure, an auxiliary fan may solve your problem. These fans are available as pusher or puller fans, for mounting at either the front or the rear of the condenser, and in sizes from 9 to 16 inches across. It is recommended that you use the largest fan that you can fit into place. Several aftermarket sources supply the fans along with mounting kits, wiring harnesses, and installation directions.

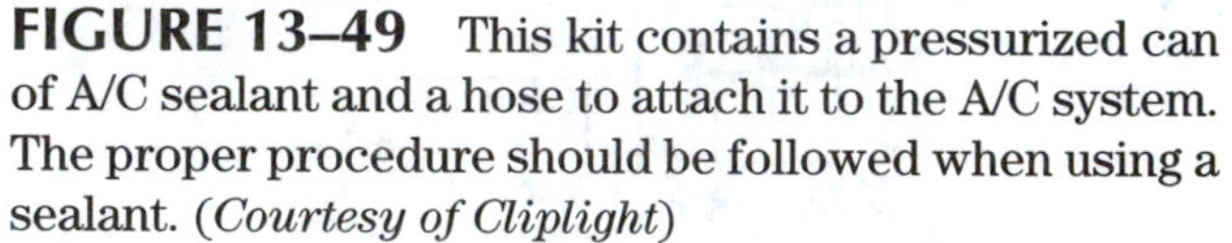

FIGURE 13–49 This kit contains a pressurized can of A/C sealant and a hose to attach it to the A/C system. The proper procedure should be followed when using a sealant. *(Courtesy of Cliplight)*

13.5 USING AND INSTALLING SEALANTS AND STOP LEAKS

Many technicians dislike stop leaks as they are considered an inadequate or temporary repair method and the best repair is to actually fix the leak. Stop leaks also have a reputation of plugging up small orifices that might reduce the proper flow of refrigerant or oil; plus there is a question of whether the stop leak material will contaminate the rather delicate chemical balance within an A/C system. In spite of these misgivings, sealants are becoming more popular; some very good technicians use them as a less expensive way to either fix or reduce the leak rate of systems in which a very small leak cannot be found or is too expensive to repair by normal methods. Sealants are not considered a cure-all or first-resort repair, but as another way to cure particular leaks. Large leaks will not seal; the cause of the leak must be repaired using normal methods.

Modern sealants are single or two-part chemicals that are moisture activated; such a chemical remains fluid until it contacts moisture. The sealant is purchased as a kit that includes the sealant with a valve and short adapter hose (Figure 13–49). The absence of moisture inside the A/C system along with the heat and pressure prevent the sealant from hardening inside the system.

To use a sealant, you should:

1. Recover the refrigerant from the system.
2. Evacuate the system for 5 minutes. The vacuum should drop to at least 25 inches, one source recommends 29 inches, and it should hold a 23 inch vacuum, one source says 25 inches, for at least 5 minutes.

   Note: If the system will not pass this test, the leak is probably too large to be closed effectively using sealant.
3. Evacuate the system with a deep vacuum for at least 20 minutes to remove all traces of moisture from the system.

   Tip: Heat the accumulator/receiver–drier to ensure that all moisture is driven out
4. Attach the adapter hose to the sealant container and the low side service port. Pierce the sealant container if necessary.
5. With the sealant container inverted, open the valve and let the system vacuum pull the contents into the system. If used, immediately repeat this step with the second part of the sealant.
6. Remove the sealant adapter, reattach the service hoses and gauge set, and recharge the system. One source recommends waiting 5 minutes before starting the system.
7. Start the system and run it for at least 15 to 20 minutes to circulate the sealant.

CAUTION

Refrigerant service machines have experienced problems of lines plugging up and solenoids sticking after recovering refrigerant form systems containing sealants. One compressor manufacturer will not warranty compressors if sealant residue is found in them.

CHAPTER QUIZ

These questions help you study this chapter. Enter the proper word(s) in the blanks to complete each statement.

1. A dirty condenser with debris at the front can be cleaned using ________ ________, ________ ________, or a(n) ________.
2. The first step in any refrigerant service procedure should be to ________ what type of ________ is in the system.
3. More than ________ ________ air in the refrigerant of an A/C system is considered contaminated.
4. Refrigerant contaminated with air should be ________ and then ________.
5. A shop that does full refrigerant service on both older and newer A/C systems must have ________ sets of service ________.
6. A recycling machine has the ability to remove ________, ________, ________, and solid ________ from the refrigerant.
7. Refrigerant that is contaminated with air will have a(n) ________ that is ________ than normal for a particular temperature.
8. A(n) ________ or ________ should never be flushed, and it is doubtful that a flat-tube ________ can be cleaned by flushing.
9. A(n) ________ ________ can be installed in the liquid line to trap debris that can plug the OT.
10. Water is removed from inside an A/C system by ________ the system to a(n) ________ lower than the boiling point of water.
11. The ________ in most vacuum pumps must be ________ periodically in order to keep the pump operating properly.
12. Partial charging of a system is not recommended because of the difficulty of adding the ________ ________ of refrigerant.
13. The proper charge amount for a vehicle can be found on the ________ under the vehicle's ________.
14. It will take ________ small (12 oz) ________ of refrigerant to charge a system that holds $2^{1}/_{4}$ lbs of refrigerant.
15. If retrofitting an R-12 system to R-134a, a small refrigerant leak will become ________, and high side pressures will probably ________.
16. If retrofitting a system that contains a high-pressure relief valve, a(n) ________ ________ ________ ________ must be installed.
17. When retrofitting a system, new ________ ________ must be permanently installed over the existing ones, and a(n) ________ ________ must be filled out and placed over the old one.

REVIEW QUESTIONS

These questions allow you to check what you have learned. Select the answer that correctly completes each statement.

1. Technician A says that eye protection should be worn when working with refrigerants. Technician B says to avoid skin contact with refrigerants and their oil. Who is correct?
   a. A only **c.** Both A and B
   b. B only **d.** Neither A nor B
2. Two technicians are discussing how to remove refrigerant from a system. Technician A says that you can vent it into the atmosphere as long as it has moisture in it. Technician B says that all refrigerant must be captured. Who is correct?
   a. A only **c.** Both A and B
   b. B only **d.** Neither A nor B
3. Technician A says that the same recovery unit can be used for R-12 and R-134a. Technician B says that the service hoses from an R-134a recovery unit must have shutoff valves within 18 inches of the end. Who is correct?
   a. A only **c.** Both A and B
   b. B only **d.** Neither A nor B
4. Technician A says that high pressure in recycled refrigerant is only caused by air contamination. Technician B says that recycled refrigerant should have the same purity standards as new refrigerant. Who is correct?
   a. A only **c.** Both A and B
   b. B only **d.** Neither A nor B
5. Technician A says that compressor failure sends metal and debris through the system so it should be flushed using R-11. Technician B says that a good method to solve this problem is to replace the compressor and install an in-line filter. Who is correct?
   a. A only **c.** Both A and B
   b. B only **d.** Neither A nor B
6. A system contaminated with moisture should be evacuated for a minimum of 15 minutes.
   a. True **b.** False
7. Technician A says that checking the pressure–temperature relationship is a reliable way to check for contaminated refrigerant. Technician B says that refrigerant with more than 3% of a foreign refrigerant normally should be sent off for reclaiming or disposal. Who is correct?
   a. A only **c.** Both A and B
   b. B only **d.** Neither A nor B
8. Refrigerant can be contaminated with NCG if the
   a. recovery unit hose is not tight.
   b. system is not thoroughly evacuated before recharging.
   c. technician uses sloppy service procedures.
   d. All of these
9. Technician A says that a system should be evacuated twice as long if the vacuum pump will only pull it down to 20″ Hg. Technician B says that the system might have a leak in it if it will not pull down into a deep vacuum. Who is correct?
   a. A only **c.** Both A and B
   b. B only **d.** Neither A nor B
10. Technician A says that refrigerant charge levels are given in service manuals. Technician B says that charge levels are printed on the under-hood decal. Who is correct?
    a. A only **c.** Both A and B
    b. B only **d.** Neither A nor B
11. Technician A says that an overcharge will cause excessively high system pressures. Technician B says that a small can of refrigerant holds 1 lb. Who is correct?
    a. A only **c.** Both A and B
    b. B only **d.** Neither A nor B
12. We normally charge refrigerant into the high side with the engine running.
    a. True **b.** False
13. Technician A says that retrofitting a system is simply a matter of recovering the R-12 and recharging it with a drop-in refrigerant. Technician B says that the charge level specification is the same with R-134a as with R-12. Who is correct?
    a. A only **c.** Both A and B
    b. B only **d.** Neither A nor B
14. Technician A says that SNAP approval indicates a good refrigerant that will have no problems in a system. Technician B says that using any refrigerant that contains R-22 requires the system to have barrier hoses. Who is correct?
    a. A only **c.** Both A and B
    b. B only **d.** Neither A nor B

A/C System Repair

LEARNING OBJECTIVES

After completing this chapter, you should be able to:

- R&R an A/C compressor.
- Perform the standard bench repairs on a compressor and clutch, given the proper service manual.
- R&R an A/C hose and make any necessary repairs to that hose.
- R&R any A/C component, given the proper service manual.
- Complete the A/C system repairs listed in the ASE Task List, Section B.

TERMS TO LEARN

barb-type fitting
beadlock fitting
bubble-style crimp
compressor bench checks
compressor leak checks
ferrule
finger-style crimp
gut pack
lip-type seal
Lokring fitting
seal protector
spring lock coupling
two-part seal

14.1 INTRODUCTION

Much of A/C system repair consists of removing and replacing faulty components. Most repair operations recover the refrigerant, make the actual repair, and then evacuate and recharge the system. Most specific repair operations are described in vehicle manufacturers' service manuals; in printed and computer service information published by Alldata, Chilton, Motors, and National Service Data; and in bulletins published by IMACA and MACS. These resources usually describe the fastest, easiest way to make a repair on a particular make and model of car. When a technician performs a repair for the first time on a specific car make and model, he or she should locate the repair in the service information and follow that procedure. *The descriptions given in this text are very general.*

✓SERVICE TIP

Before beginning any refrigerant service operations, protect your eyes and skin.

In a survey of its members, MACS determined that the most frequent A/C repairs involve (in order) the compressor (more than 20%), hoses, control devices,

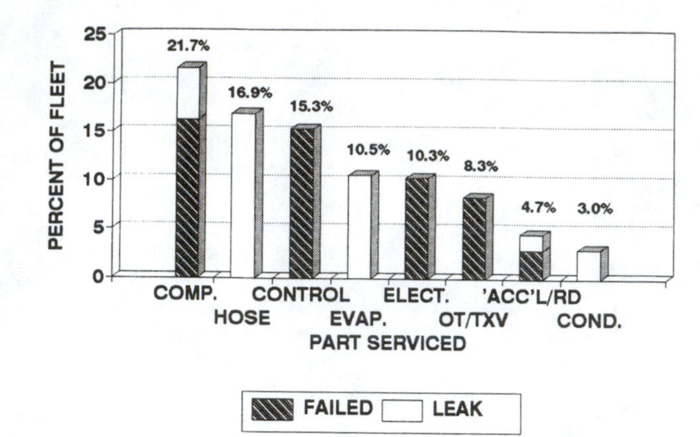

FIGURE 14–1 The most common reasons for servicing automotive A/C systems are shown in this graph, the outgrowth of a survey by MACS of its members in the A/C service business. (*Courtesy of MACS*)

the evaporator, electrical circuits, the TXV or OT, the receiver–drier or accumulator, and the condenser (3%). These repairs were either to repair or replace a failed part or to correct a leak (Figure 14–1).

14.2 COMPRESSOR REPAIR

Normal compressor repair includes replacement of the entire unit or of the clutch and clutch coil, shaft seal, gaskets and O-rings, and, occasionally, the reed valve plate. The extent of repair depends on the particular compressor, the availability of repair parts, the special tools required to make the repair, and the skill of the technician. Many technicians prefer to repair the original compressor, if practical, to eliminate problems in obtaining the correct replacement. On many compressors, it is very easy to remove and replace the head, so the head, reed valve plate, and O-ring or gasket are often replaced. Compressors with severe internal damage or for which the repairs are difficult or time consuming will probably be replaced with new or rebuilt units. This is another case in which the repair description in a service manual can help the technician determine what course to take.

✓SERVICE TIP

Compressor failure rate is increasing on modern vehicles. This increase has been attributed to:

- reduced refrigerant charge levels with less tolerance for leakage,
- smaller compressors with tighter tolerances, and
- smaller, tighter engine compartments with much higher under-hood temperatures.

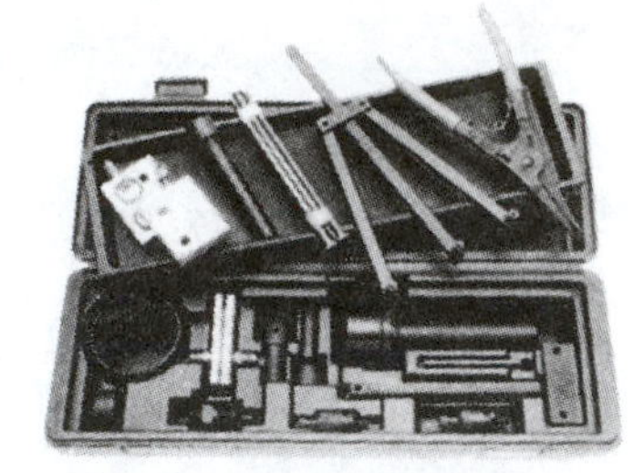

FIGURE 14–2 This set contains the special tools needed to service the clutch and shaft seal of Ford FS-6 and Chrysler C-171 compressors. (*Courtesy of Kent-Moore*)

When a compressor is replaced, some of the variables that should be checked to verify their correct replacement are the following:

- Clutch pulley (see Section 14.2.2)
- Compressor make
- Mounting lug type, location, and spacing
- Switch location and types
- Clutch coil connector location
- Port location, size, and type
- Refrigerant type

Occasionally, the repair operation can be performed with the compressor still in place on the engine, but it is usually removed. Bench repairs usually allow for better cleanup and access to all portions of the compressor.

Special service tools are available through the vehicle manufacturer and from several aftermarket sources, some of which supply the manufacturer. Special tools are usually required for clutch removal and replacement, seal removal and replacement, and internal compressor repairs. Special tool kits contain the specific tools for different compressor models and often include pressure test plates, which allow the compressor to be checked for leaks while still on the bench (Figure 14–2).

✓SERVICE TIP

Occasionally a vane compressor will appear to have failed if the vanes stick in the innermost position. Sometimes they can be freed up by adding oil through the low side port, running the engine at about 3,000 rpm, and cycling the clutch in and out.

✓REAL WORLD FIX

The 1990 Taurus came from the body shop after damage repairs from an accident had been repaired. The condenser had been replaced and the system recharged, but it did not cool. The technician recovered the refrigerant, evacuated the system, and tried to recharge it, but it would not take the last 8 oz of the 2 lb, 12 oz charge specification. The compressor clutch operates, but both gauges show a pressure of 60 psi.

Fix: A more thorough check showed that the compressor shaft was broken. Replacement of the compressor along with a new accumulator fixed this problem.

14.2.1 Compressor R&R

Depending on access to the compressor and its mounting bolts, compressor R&R can be either fairly easy or very difficult. A compressor is often removed from the car to repair the clutch or to replace the seal, gaskets, or O-rings. Like most other operations, the exact procedure to follow on a particular car is given in the service manual. If a failed compressor is replaced, it is recommended that the receiver–drier or accumulator be replaced at the same time and the amount of oil in the system be checked and adjusted.

✓SERVICE TIP

If the failed compressor sent debris into the discharge line, the line and the condenser should be replaced or flushed to remove the debris or a filter should be installed to keep it from traveling downstream. Many technicians install one or two in-line filters to ensure the system operates properly or to protect the new compressor.

✓SERVICE TIP

When compressors fail internally, it will often send metal particles into the suction line. The compressor must be replaced with a new or rebuilt unit, but this material can enter the new compressor and cause early failure. A suction filter screen has been developed to trap the debris in the suction line (Figure 14–3).

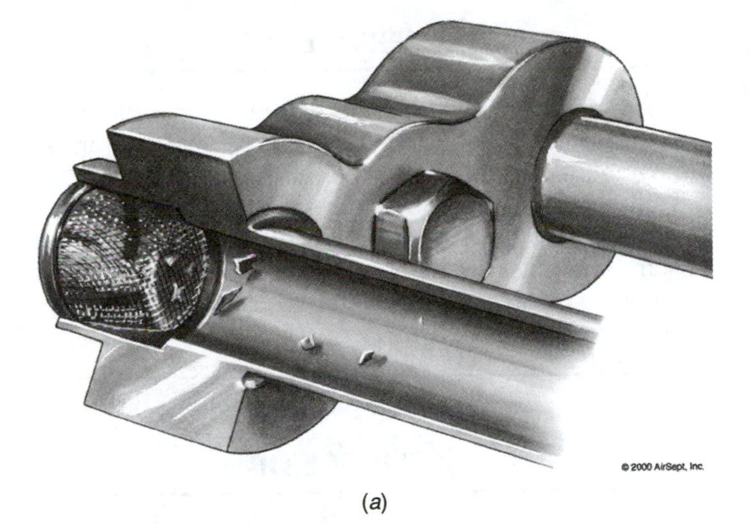

(*a*)

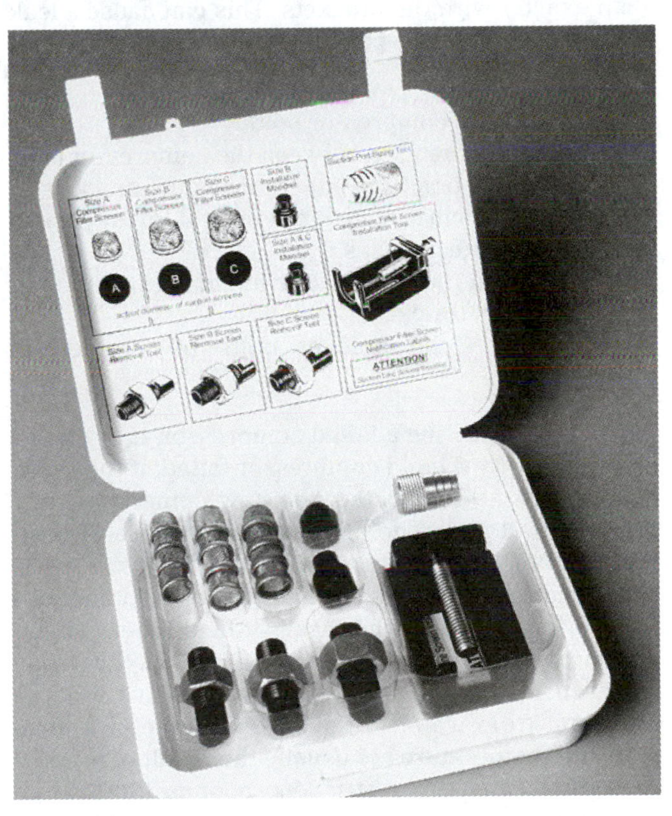

(*b*)

FIGURE 14–3 Failure of some compressors will send debris into the suction line. A filter screen (*a*) can be installed into the suction using a special tool set (*b*). *(Courtesy of Air Sept)*

✓SERVICE TIP

Many new and rebuilt compressors come with an oil charge for the whole system; some come with an oil charge for just the compressor. Some replacement R-12 compressors come with a mineral oil charge, an R-134a compressor comes with a PAG or POE oil charge, and some compressors intended for either refrigerant come empty of oil. Be sure to read the information that comes with the compressor.

✓SERVICE TIP

The compressor housing can become distorted as it is tightened onto the mounts if the mounting pads do not align exactly with the brackets. This can cause a leak between the housing parts. To ensure that this does not occur:

- Snug the compressor bolts.
- If there are gaps between the compressor pads and the brackets, either loosen the brackets to allow them to align properly or shim the pads to eliminate the gaps.
- Tighten the compressor mounting bolts to the correct torque.
- Tighten the bracket bolts to the correct torque.

Before replacing a failed compressor, it is wise to determine why the old compressor failed; if the cause of failure is not corrected, the new compressor will probably fail also. The most common cause of compressor failure is poor lubrication, excessive pressures and temperatures, or a burned-out clutch. Poor lubrication is usually the result of insufficient oil, insufficient refrigerant, the wrong type of oil, the wrong oil viscosity, oil breakdown from excessive heat, or acidic oil from water in the system. Excessive pressure and temperature are usually the result of internal or external condenser restrictions, poor airflow through the condenser, an overcharge, or contaminated refrigerant. Clutch burnout is usually the result of insufficient voltage at the clutch or excessive high side pressure. During the replacement procedure, a wise technician will remove as much of the old oil from the system as practical and replace the accumulator or receiver–drier. This allows installation of a complete system charge of new, clean oil of the proper type and viscosity.

✓SERVICE TIP

When installing a new or rebuilt/remanufactured compressor, follow these guidelines:

- Return the system to as close to like-new condition as possible.
- Replace the accumulator or receiver–drier.
- If the compressor failure sent metal debris into the high or low side lines,
  - Flush the lines and/or install liquid line filter or a suction-line filter screen.
  - Replace the OT or clean the TXV filter screen.
  - With modern flat-tube condensers, be aware of probable internal condenser plugging.
- Add the correct amount of the proper oil, and split the oil, half to the compressor and half to the accumulator/receiver–drier.
- Check the clutch air gap, and correct it if necessary.
- Check the clutch operating voltage; it should be 12 V or greater with the engine running.
- Tighten the compressor mounting bolts evenly to the correct torque.
- Evacuate the system a minimum of 30 minutes (some recommend 45 to 60 minutes).
- Charge the system with the proper amount of clean refrigerant.
- Rotate the compressor shaft by hand a few revolutions (some recommend ten revolutions) to ensure that refrigerant oil has not filled a cylinder.

To remove a compressor, you should:

1. Recover the refrigerant from the system.
2. Disconnect the compressor clutch and any switch or sensor wires.
3. Disconnect the drive belt.
4. Disconnect the discharge and suction hoses and cap the hoses to prevent dirt and moisture from entering. On some vehicles, reverse steps 4 and 5 for different access to the hose connections.
5. Remove the compressor mounting bolts and remove the compressor (Figure 14–4).
6. With most newer systems, drain the oil from the compressor by placing it over a drain pan with the discharge and suction ports downward; some compressors have a drain plug that must be removed. Measure the amount of oil drained

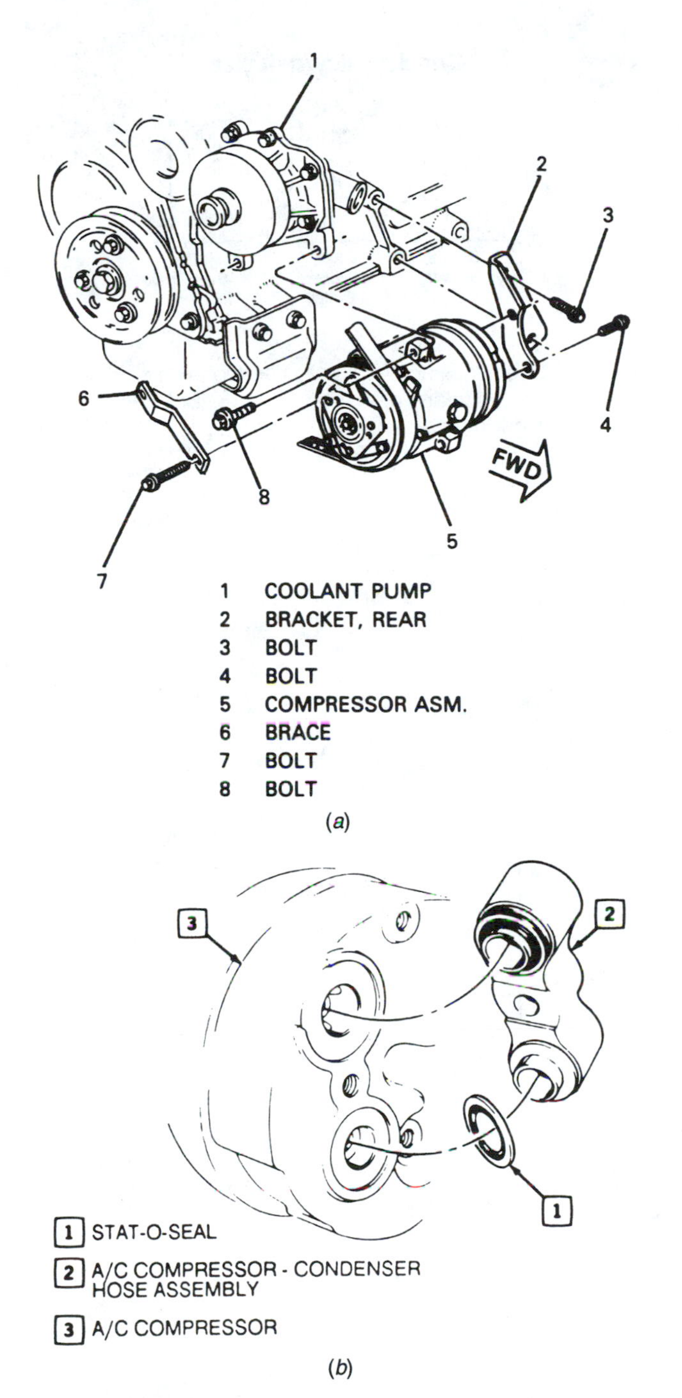

FIGURE 14–4 A compressor and its mounting bracket (*a*). New seals should be used at the line connections when a compressor is replaced (*b*). (*Reprinted with permission of General Motors Corporation*)

out. Note the condition of this oil and discard it in the approved manner (Figure 14–5). Oil with metal particles indicates a failed compressor; brown or dark oil indicates overheating from excessive high side pressures.

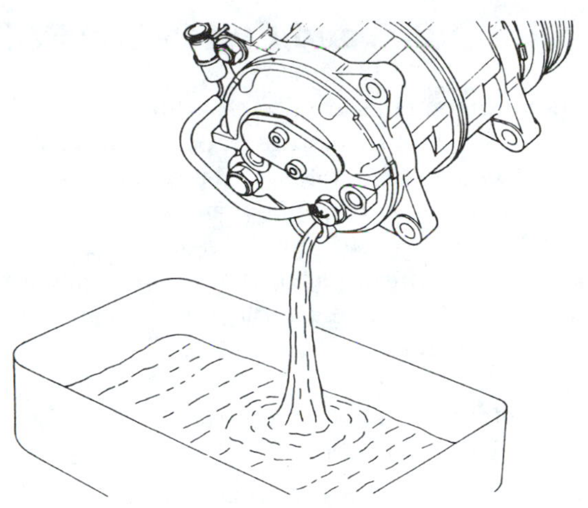

FIGURE 14–5 Oil is drained by pouring it out of the suction and discharge ports; it can also be drained out of the oil drain or oil level–checking plug opening, if so equipped. (*Courtesy of Zexel USA Corporation*)

To replace a compressor, you should:

1. Adjust the oil level in the compressor as instructed by the vehicle or compressor manufacturer.
2. Install the compressor on the engine and replace the mounting bolts.
3. Install the drive belt, adjust the belt tension, and tighten all mounting bolts to the correct torque.
4. Using new gaskets or O-rings, connect the discharge and suction lines.
5. Connect the clutch and switch or sensor wires.
6. Evacuate and recharge the system and check for leaks.

✓ REAL WORLD FIX

The 1992 Nissan Pathfinder (110,000 kilometers) came in with a bad compressor. The clutch had overheated and welded itself together. The compressor and clutch were replaced along with the receiver–drier. The system was recharged, but it did not cool properly. The system pressures at idle were 85 psi on the low side and 105 psi on the high side; at 2,000 rpm, the pressures changed to 55 psi on the low side and 150 psi on the high side. The TXV was replaced twice, but this did not help.

Fix: It was determined that the rebuilt compressor was faulty. A second compressor replacement fixed this problem.

14.2.2 Clutch and Clutch Coil R&R

Most modern clutches are three-part assemblies with a separate drive hub (armature), rotor pulley, and coil. Many earlier clutches were two-part units that combined the hub and rotor pulley, and a few were single units with all three parts combined in a single assembly. When replacing these parts, the technician must use care to ensure the correct replacement. This checking procedure should include the following:

1. Determine shaft type (tapered or straight) (Figure 14–6).
2. Determine pulley and belt size and type.
3. On one- or two-part units, determine whether a single- or double-row ball bearing is used.
4. Determine compressor make and model.
5. Measure pulley diameter.
6. Measure pulley groove placement.
7. With some units, determine clutch manufacturer (Figure 14–7).

✓SERVICE TIP

A slipping clutch will have a blue, overheated appearance. It can generate a temperature of 700 to 800°F in three minutes or less. The clutch bearing lubricant melts at about 300°F and can run out of the bearing. The bearing will fail shortly after this, possibly causing damage to the compressor snout.

✓SERVICE TIP

Some late model clutch assemblies have undergone changes to improve clutch holding power. Early and late versions of these units should not be combined because this can cause rubbing between the parts or clutch slippage. If one of these parts has to be replaced, all three portions should be replaced as a unit. Another point to remember on three-part assemblies is that the clearance, or air gap, between the hub and rotor pulley must be adjusted during replacement and that magnetic action pulls the coil into the rotor pulley as the clutch operates. These parts must be properly positioned as they are assembled (Figure 14–8).

Check pulley/belt size

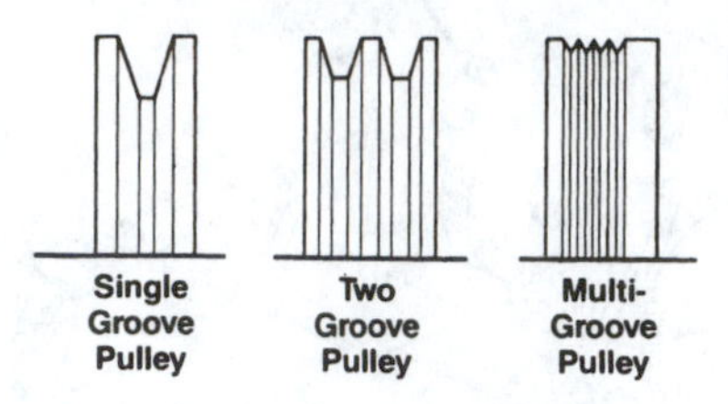

Count the number of belts needed to drive the clutch and determine their size. Belt size is given by dimensions "C" and "D" in the dimensional information section of this catalog.

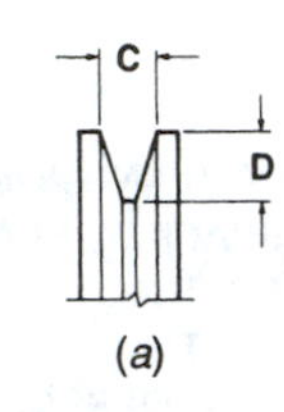

(*a*)

Measure diameter of pulley

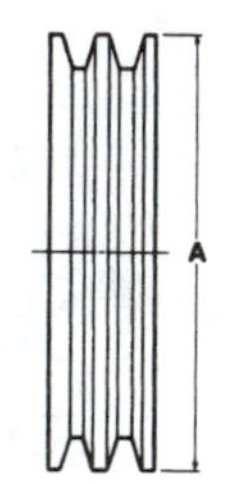

Measure the pulley from edge to edge across its face in the middle of the clutch. This diameter is listed in the dimension charts as "A."

(*b*)

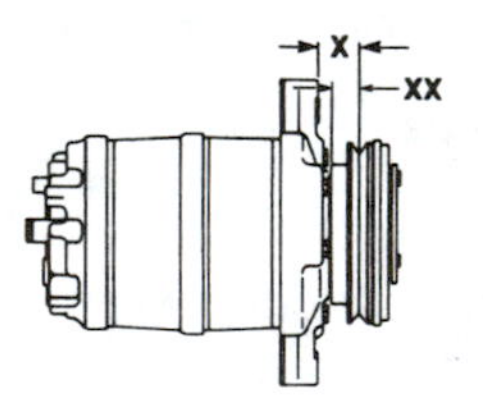

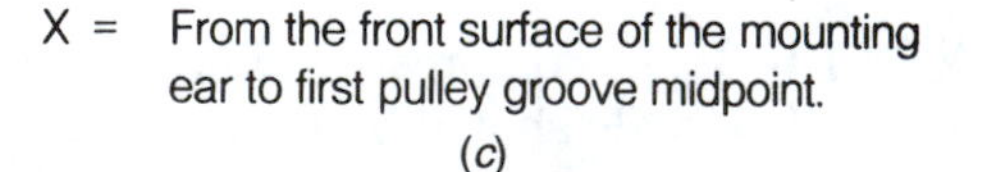

(*c*)

FIGURE 14–6 When replacing a clutch or pulley, check the type and size of the pulley belt grooves (*a*), the pulley diameter (*b*), and groove spacing (*c*) to make sure the replacement part is correct. (*Courtesy of Warner Electric*)

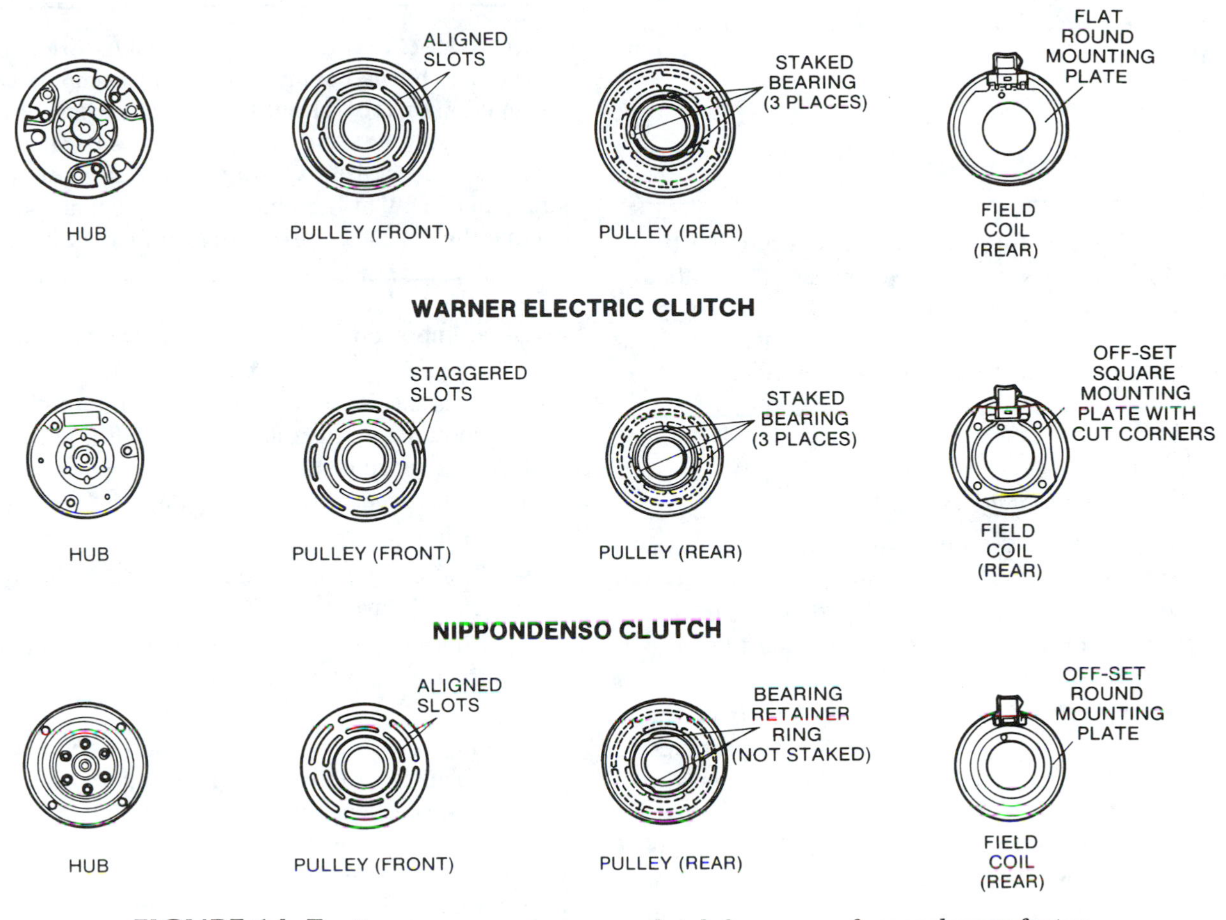

FIGURE 14–7 Some compressors use a clutch from one of several manufacturers, but parts from one manufacturer should not be used with parts from another. (*Courtesy of Warner Electric*)

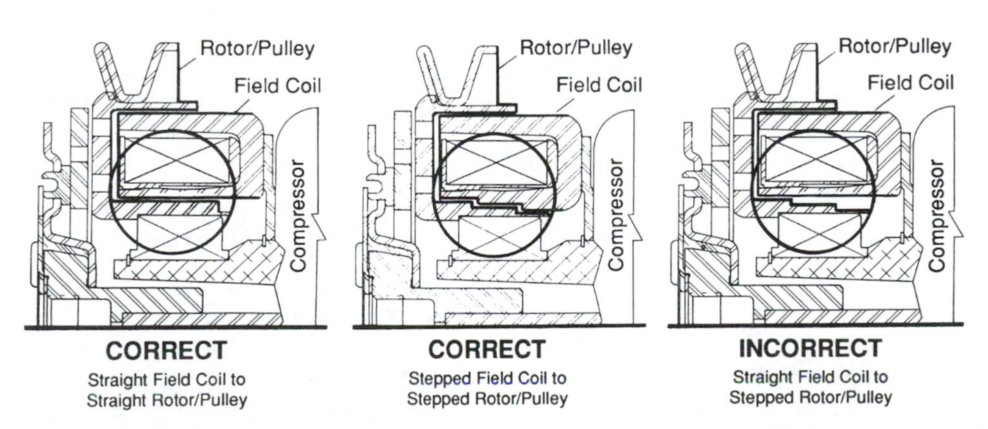

FIGURE 14–8 The correct pulley being used with the correct field coil (left and center). A weak mismatch with possible rubbing occurs if the parts are switched (right). (*Courtesy of Warner Electric*)

In some cases, the clutch repairs can be made with the compressor in place on the engine. If compressor removal is necessary, follow the procedure described in Section 14.2.1.

✓SERVICE TIP

If a slipping clutch is found, check to determine if the clutch surfaces were wet or dry. Wet surfaces indicate a seal leak caused the problem. Dry surfaces indicate excessive clearance, weak clutch coil magnetism, or excessive high side pressure caused the problem.

To remove a clutch assembly, you should:

1. Remove the locknut or bolt from the compressor shaft. A clutch hub wrench is often required to keep the hub from turning. Some compressors do not use a locknut (Figure 14–9).
2. Use the correct tool to pull the hub from the compressor shaft. On some compressors, a group of adjustment shims are under the clutch hub.
3. Remove the rotor pulley retaining ring and the rotor pulley. A special puller is required on most compressors; the rotor pulley can be slid off some compressors, such as the Nippondenso styles.
4. Mark the location of the clutch coil wire connector, remove the retainer ring, and remove the coil. Some compressors use a press fit to secure the coil in place. These coils are removed using a special puller.

If one or more clutch parts fail, inspect the parts to determine the cause of failure. If the parts are removed to gain access to the compressor seal, check them to make sure they are serviceable.

✓SERVICE TIP

If the compressor is mounted with the drive belt in place, engage the clutch. The magnetic force will lock the hub and hold it from turning while you loosen the nut.

✓SERVICE TIP

A burned-out clutch with blued, heavily scored contact surfaces and burned paint was obviously slipping and is not functioning. You should determine what may have caused the slippage and failure. In this case, make sure there are at least 10.8 V (some sources recommend a minimum of 12 V at the clutch connector) at the clutch and that the discharge pressure is not excessively high.

Other important points to check are the following:

- Contact faces for the hub and rotor pulley should be flat and fairly smooth, without excess scoring (Figure 14–10).
- Hub drive springs or rubber should not be broken.
- The pulley should rotate smoothly and quietly on its bearing.
- There should be no rubbing contact between the rotor pulley and coil.

A faulty rotor pulley bearing can be removed and replaced. This is usually a simple process of driving or pressing the old bearing out and installing a new one. When installing a new bearing, it is important not to exert a driving or pressing force across the inner bearing parts; press or drive only on the outer bearing race, where it fits tightly with the rotor. On some units, the bearing is retained by a snap ring; on many newer units, the bearing is staked. Staking involves upsetting or deforming metal so it overlaps the bearing and bore. As a staked bearing is driven out, the staked metal is bent out of the way. The rotor pulley must be restaked to lock the new bearing in place (Figure 14–11).

✓REAL WORLD FIX

The compressor clutch on the 1991 Lincoln (87,000 miles) operates intermittently. There is B+ voltage through the A/C relay to the compressor, and the wiring looks good.

Fix: With the engine running, the technician carefully tapped the front of the clutch hub with a broom handle, and the clutch engaged. With the engine off, the clutch air gap was measured and found to be too wide, over 0.060″. Adjusting the clutch air gap to the proper width, 0.020″, fixed this problem.

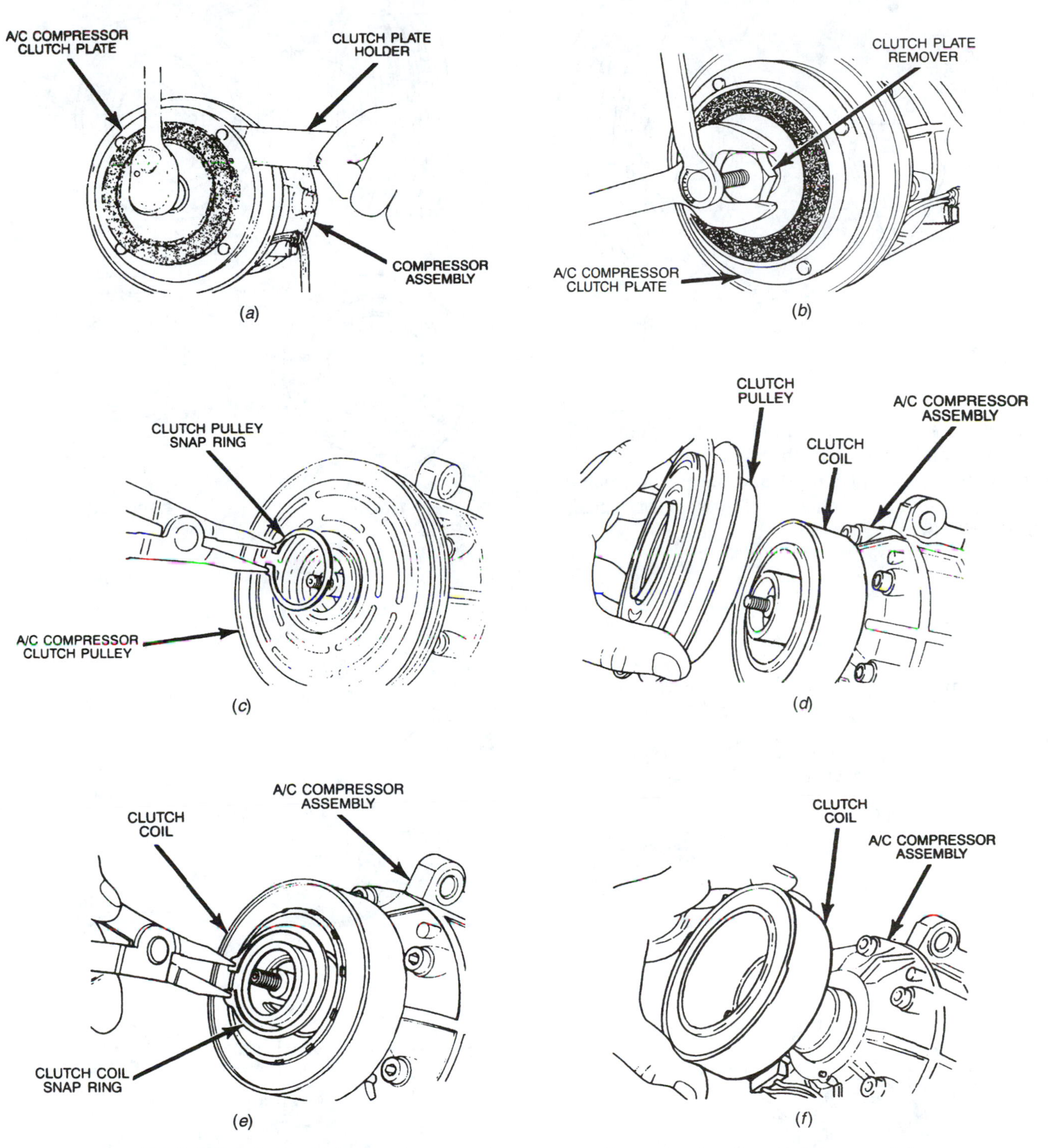

FIGURE 14–9 This clutch assembly is removed by removing the retaining nut (*a*), pulling the clutch plate (*b*), removing the pulley retaining ring (*c*), removing the pulley (*d*), and removing the coil retaining ring and the coil (*e*) and (*f*). Special pullers are often required to remove the plate, pulley, or coil. (*Courtesy of DaimlerChrysler Corporation*)

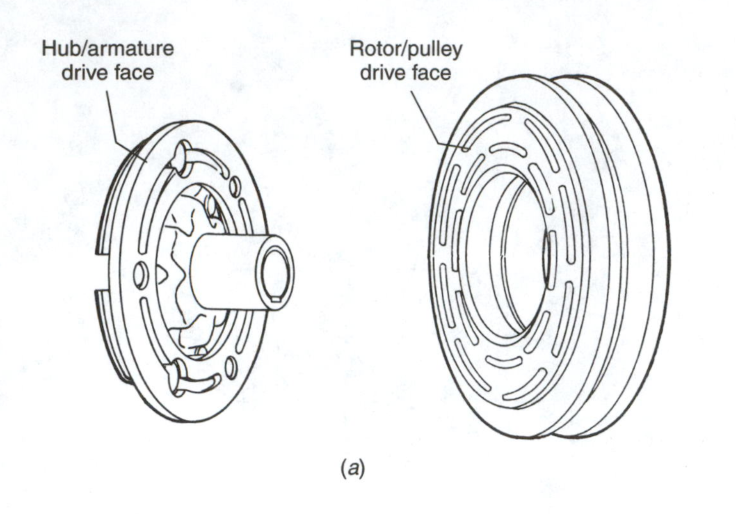

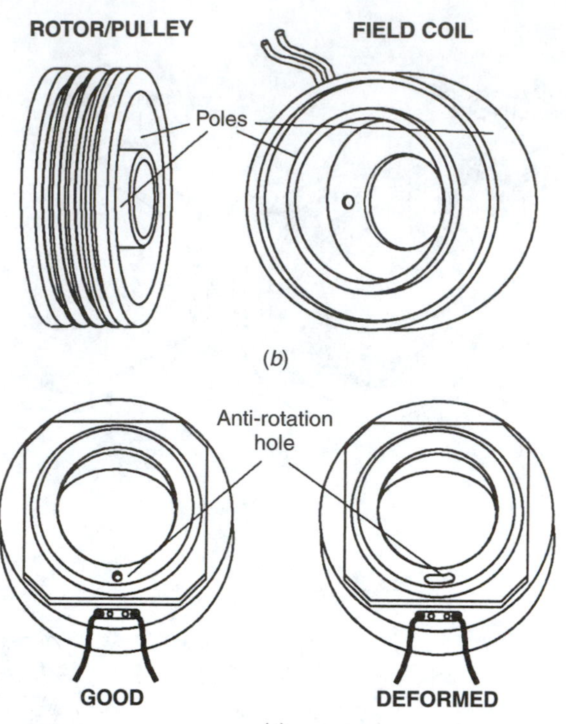

FIGURE 14–10 As the clutch is removed, the face of the hub and pulley (*a*), the pulley bearing and the poles of the pulley and coil (*b*), and the coil antirotation hole(s) (*c*) should be checked for wear or damage. (*Courtesy of Warner Electric*)

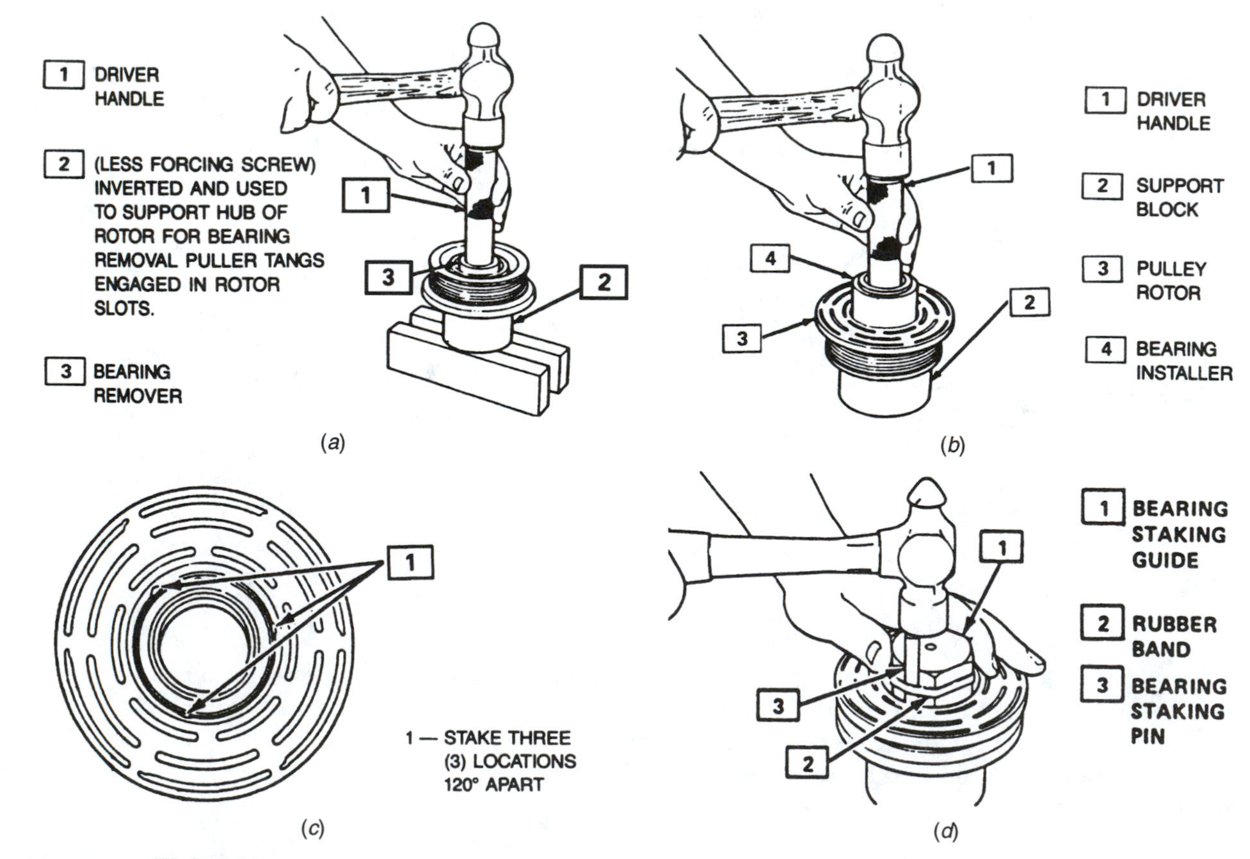

FIGURE 14–11 A faulty bearing being removed (*a*) and a new bearing being installed (*b*) into a pulley. Special tools are required to ensure that the pulley and bearing are not damaged. Another tool is required to stake the pulley (*c*) to lock the bearing in place (*d*). (*Reprinted with permission of General Motors Corporation*)

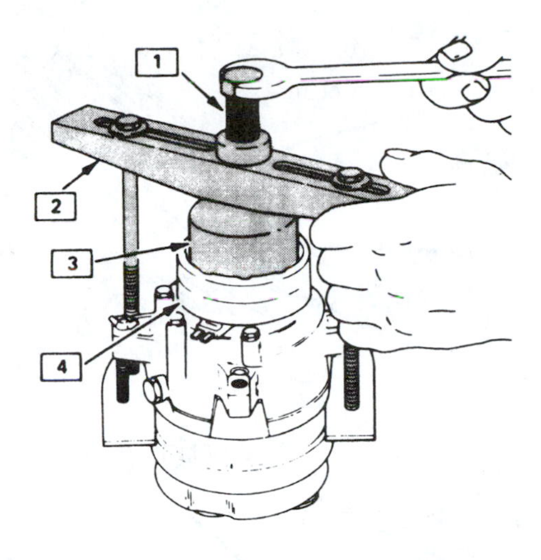

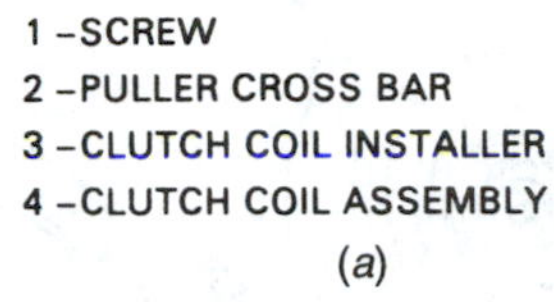

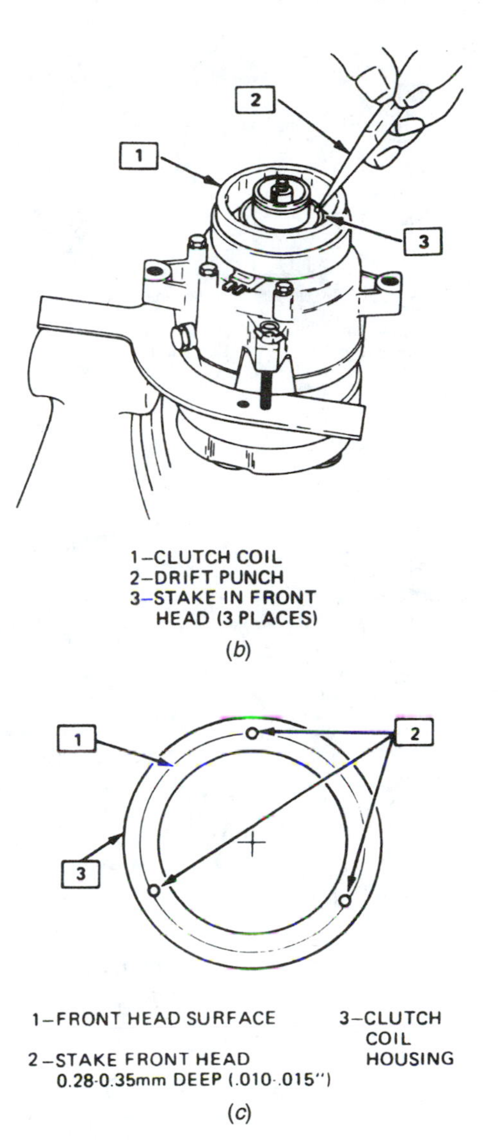

FIGURE 14–12 On this compressor, the clutch oil must be pressed into place (*a*) and locked with three stake marks from a punch (*b*). The staking (*c*) upsets metal to firmly hold the coil in position. (*Reprinted with permission of General Motors Corporation*)

To replace a clutch assembly, you should:

1. Replace the coil, making sure that the antirotation pins and holes are aligned and the wire connector is in the correct position. The coil must be pressed in place on some compressors (Figure 14–12).
2. Replace the coil retaining ring. On some compressors, the retaining ring has an inner bevel; this bevel must face away from the coil. Be sure that the retaining ring enters its groove and fits tightly (Figure 14–13).
3. Install the rotor pulley and replace the retainer ring. Some retainer rings have a beveled face; this side must face away from the pulley. Test this installation by rotating the rotor pulley; it must rotate freely, with no interference.
4. On some compressors, install the adjusting shims onto the shaft, install the drive key, and align and install the hub. On many GM compressors, the hub must be pulled onto the shaft. It should be pulled on just far enough to get the correct air gap.
5. Install the shaft locknut or bolt and tighten it to the correct torque.
6. Check the clutch air gap at three locations around the clutch. The clearance should be within specifications at all three points. If it is too wide or too narrow, readjust the air gap. If it is too wide at one point and too narrow at another, replace the hub (Figure 14–14).

✓SERVICE TIP

A clutch used in an R-134a system will not tolerate an air gap that is too wide. It will slip when high pressures are encountered.

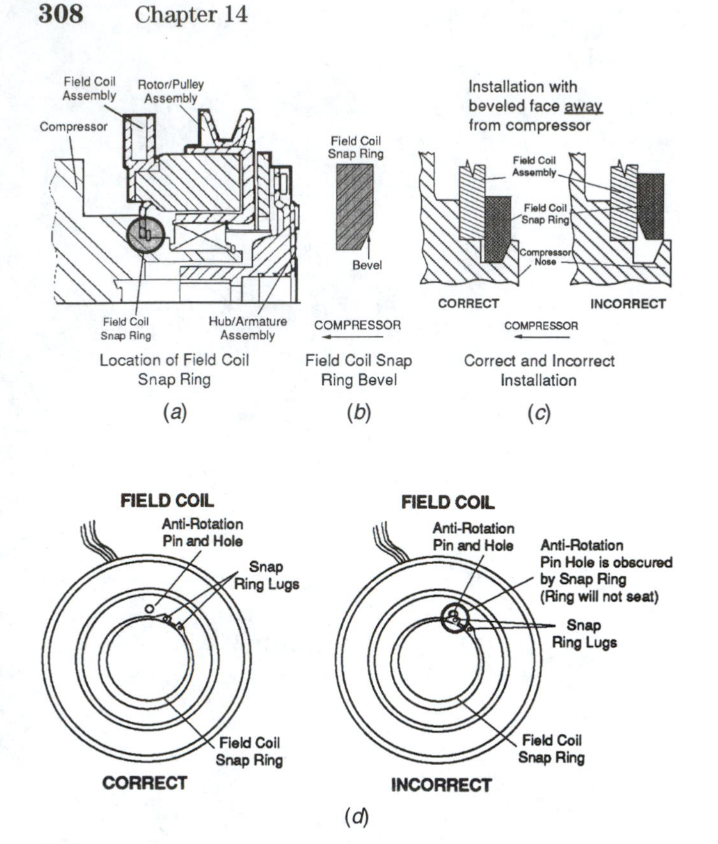

FIGURE 14–13 When parts are held in place using a snap ring (*a*), the bevel portion of the ring (*b*) should be positioned so it enters the groove to tighten against the part being retained (*c*). Placing the lugs of the snap ring so they are on the antirotation pin can keep the ring from properly entering its groove (*d*). (*Courtesy of Warner Electric*)

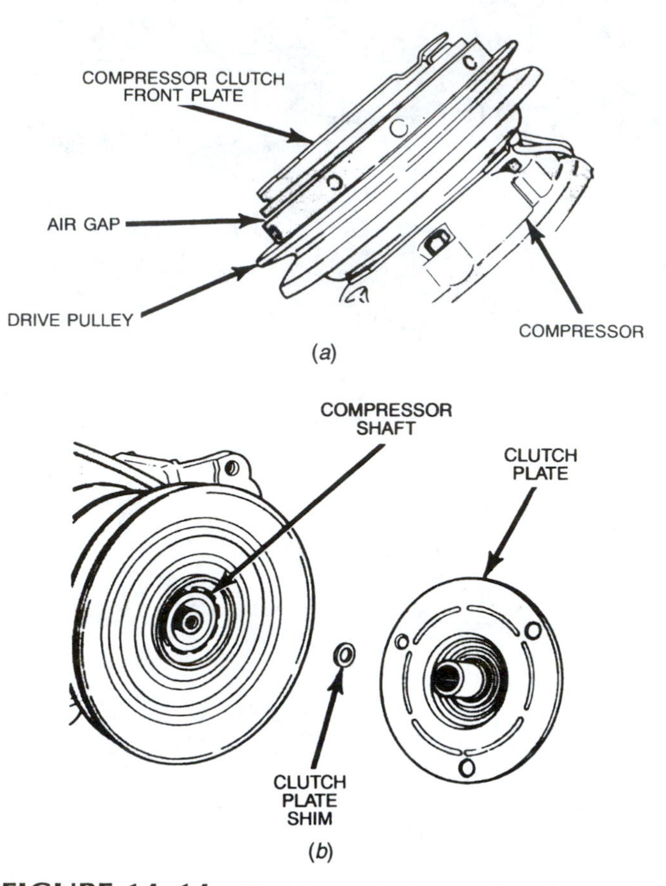

FIGURE 14–14 The air gap between the plate and pulley should be correct (*a*); some compressors use a shim pack under the plate hub to adjust this air gap (*b*). (*Courtesy of DaimlerChrysler Corporation*)

✓REAL WORLD FIX

The 1986 Mercedes (141,000 miles) came in with a complaint of no A/C. A bad A/C relay was found and replaced. The A/C would operate, but at 1,800 rpm the compressor would cut out. The compressor rpm sensor was replaced. This helped, but the compressor would still shut off intermittently.

Fix: The clutch air gap was checked and found to be too wide, 0.040 in. with a specification of 0.015 in. Removal of a few shims to get the correct gap fixed this problem.

✓SERVICE TIP

Many clutch and seal snap rings have a slight bevel on one side, and the snap ring groove is also beveled. The beveled edge is placed away from the clutch coil or pulley bearing, which causes the snap ring to move toward the part it retains, holding it tighter.

✓SERVICE TIP

When installing new clutch components, it is good practice to burnish the contact surfaces before putting the clutch in service. Do this by cycling the clutch on and off at a rate of 10 to 15 times a minute for a total of 10 to 15 cycles with the engine speed at 2,500 to 3,000 rpm. One manufacturer recommends 50 cycles. Burnishing wears the surfaces to perfectly match each other, increasing the holding power, before the clutch has to engage at full head pressures. Recheck the air gap after burnishing.

14.2.3 Compressor Shaft Seal R&R

With most compressors, the shaft seal is removed from the front of the compressor after the clutch assembly has been removed. On some compressors, the seal is positioned inside the front head, and partial disassembly of the compressor is required to get to it. Two styles of seals are used: two-part seal assemblies and lip-type seals.

The two-part seal is currently most common. It consists of a seal seat and a seal cartridge. The seal seat is stationary and is secured to the front of the compressor. An O-ring or gasket is used to prevent refrigerant from leaking between the compressor and the seal seat. The seal seat is made from either cast iron (gray) or ceramic (white) material and has a polished sealing face. The seal cartridge fits over the compressor shaft and rotates with the shaft. Many compressors have flat areas on the shaft where the cartridge fits for a positive drive connection. The actual sealing member is a disk of carbon with a highly polished face; this face is pushed against the seal seat by a spring. A rubber bellows or O-ring is also included in the cartridge to seal the seat to the shaft and prevent a refrigerant leak in this area. A film of oil between the sealing faces of the seat and cartridge forms the final seal and prevents wear (Figure 14–15).

A lip-type seal uses a Teflon, TFE material seal lip, which seals against the compressor shaft. The seal lip is bonded to a steel shell mounted in the front of the compressor. An O-ring is used to seal the shell to the compressor. Internal refrigerant pressure helps keep the sealing lip against the shaft to stop refrigerant from escaping. Compressor oil provides the final seal and also prevents seal lip and shaft wear (Figure 14–16).

✓SERVICE TIP

Some seal leaks are caused by pits in the compressor shaft in the area of the seal lip. A replacement seal is available that has two seal lips, and each one is offset from the original seal lip position.

Seals should not be reused; new seal parts should always be installed. These parts should be thoroughly lubricated with refrigerant oil as they are installed.

✓SERVICE TIP

When handling the new seal parts, avoid touching the sealing surfaces with your fingers. These sealing surfaces can be damaged by acids from your skin.

Special tools are required for most seal replacements. These tools are designed to grip the seal cartridge or seat so they can be quickly pulled out or slid back into the proper position.

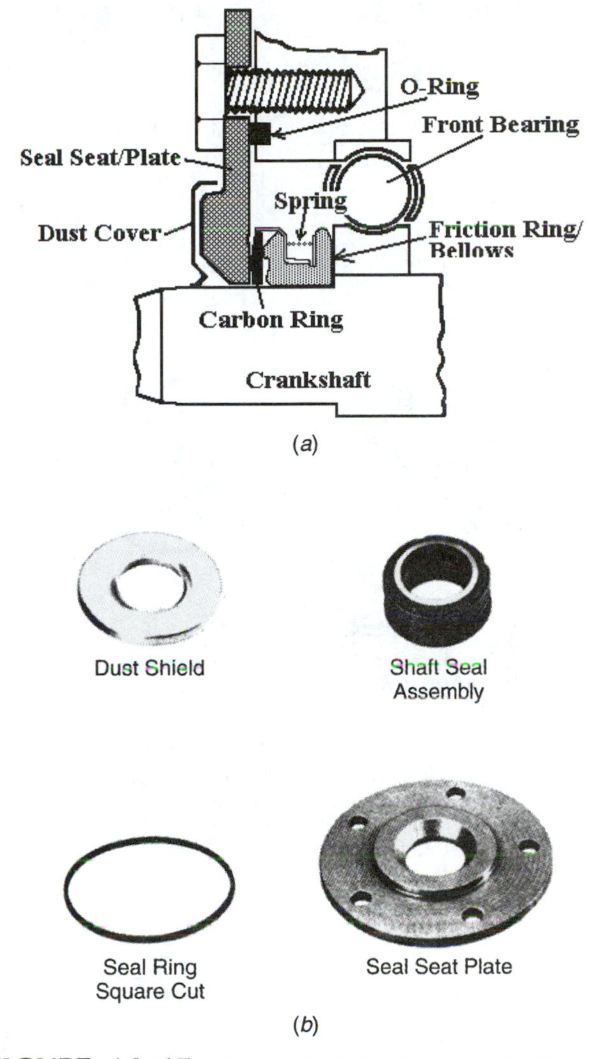

FIGURE 14–15 Many shaft seals consist of a cartridge with a carbon ring, that fits tightly onto the compressor shaft (a), and the carbon ring seals against the seal seat. A seal kit with these parts is shown (b). (b *is reprinted with permission of General Motors Corporation*)

To remove a lip-type seal, you should:

1. Recover the refrigerant.
2. Remove the compressor clutch assembly.
3. Remove the clutch hub drive key if it is still in the shaft.
4. Remove the seal retaining ring (Figure 14–17).
5. Attach the seal puller to the seal and, using a rotating motion, pull the seal from the compressor.
6. Thoroughly clean the area inside the compressor neck where the seal is located.
7. Use an O-ring pick tool to hook and remove the O-ring. Discard the O-ring and seal.
8. Clean the O-ring groove and compressor shaft.

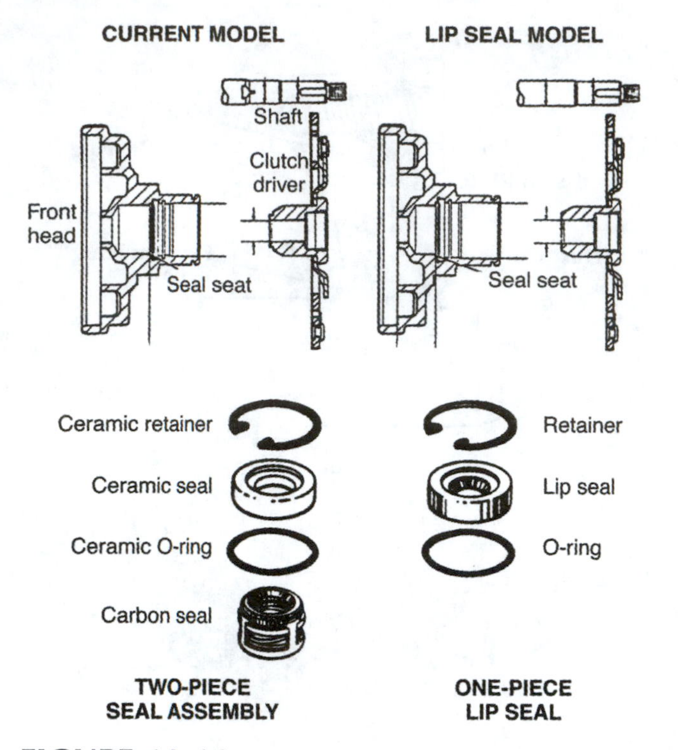

FIGURE 14–16 Early GM compressors use a two-piece seal (left), whereas newer ones use a one-piece lip seal (right). The seal lip seals directly against the smooth shaft (top). (*Reprinted with permission of General Motors Corporation*)

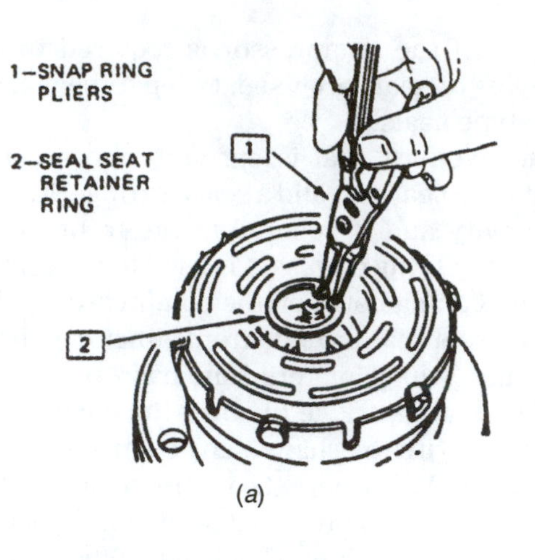

(*a*)

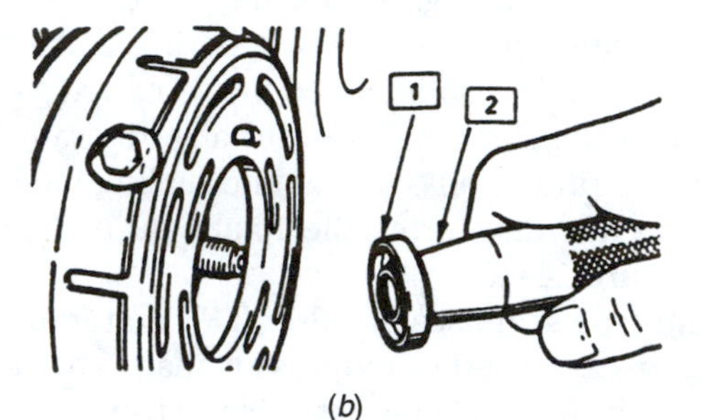

(*b*)

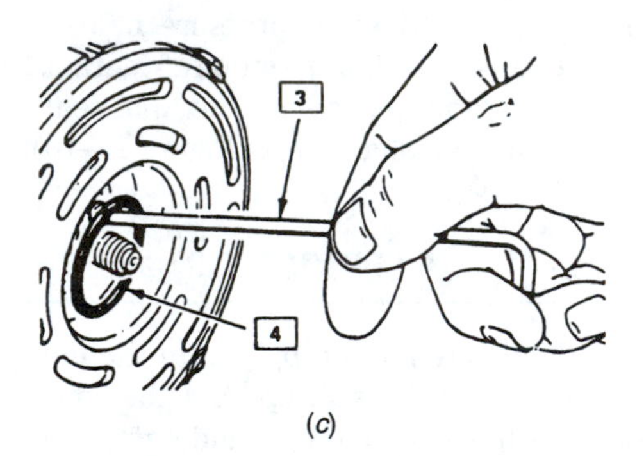

(*c*)

FIGURE 14–17 A lip-type seal is removed by removing the retaining ring (*a*), removing the seal (*b*), and removing the O-ring (*c*). (*Reprinted with permission of General Motors Corporation*)

✓SERVICE TIP

Use mineral refrigerant oil to lubricate the parts as they are assembled. It is less hygroscopic and safer for you to use, and the small amount that is used will not affect an R-134a system. It is currently available in aerosol containers, making its use easier and faster (Figure 14–18).

To replace a lip-type seal, you should:

1. Check the sealing surface on the compressor shaft to make sure that it is smooth and clean. Oil the new O-ring, position it on the O-ring installer, and install the O-ring into its groove. Make sure that it is seated completely in its groove.
2. Place the seal lip protector over the compressor shaft to prevent damage to the seal as it is installed (Figure 14–19).
3. Place the new seal onto the installer, oil the seal lip, and install the seal using a rotating motion.
4. Install the new retaining ring with the flat side of the ring against the seal and the beveled side outward. Make sure the ring enters its groove all the way around.
5. Rotate the compressor shaft a few turns, pressurize the compressor, and check for leaks. (Testing a compressor for leaks is described in Section 14.2.5.)
6. Evacuate and recharge the system and check for leaks.

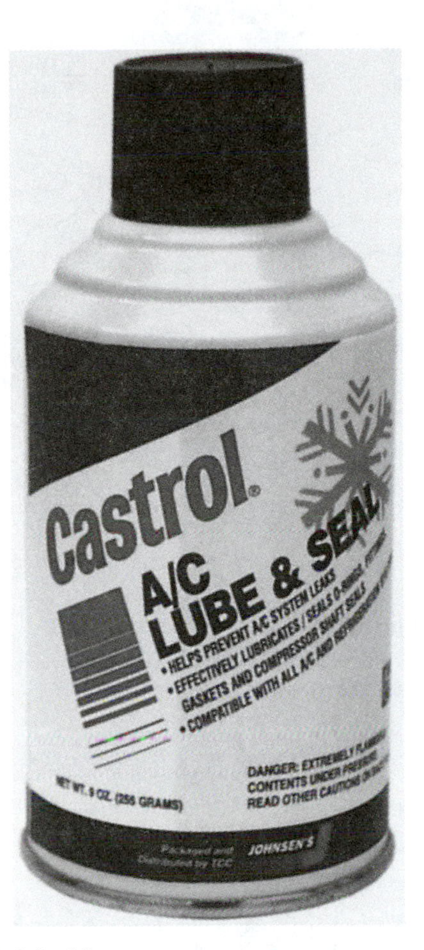

FIGURE 14–18 A/C lube & Seal is a premium mineral lubricant in an aerosol container. It is a convenient method of lubricating parts, O-rings, and fittings as they are assembled. *(Courtesy of Technical Chemical Company, TCC)*

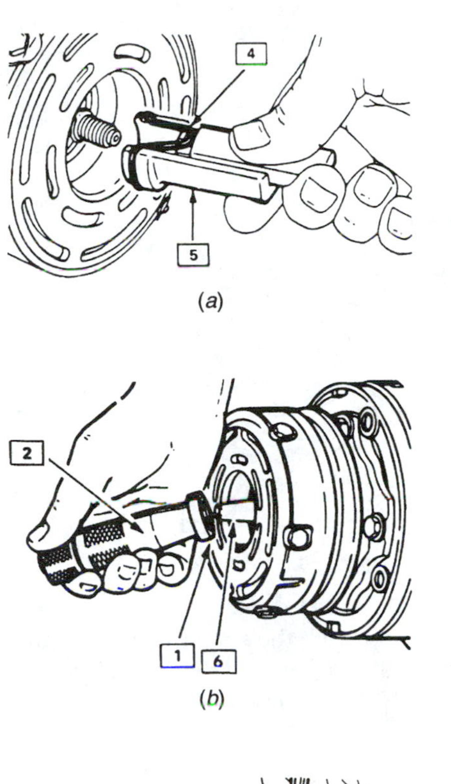

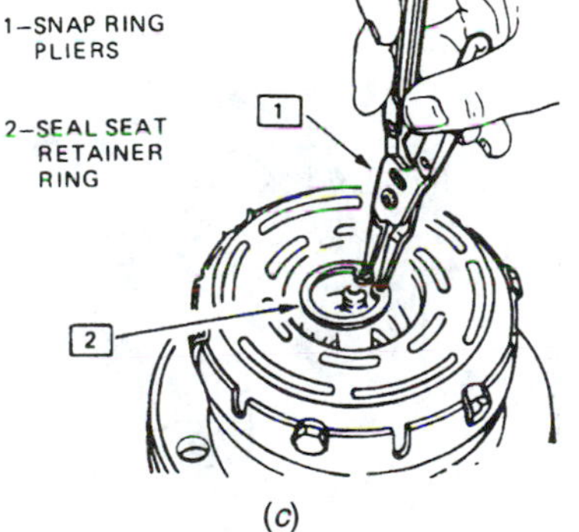

FIGURE 14–19 A lip-type seal is replaced by installing the O-ring (*a*), installing the seal (*b*), and replacing the retaining ring (*c*). The O-ring and seal lip should be wetted with refrigerant oil. Note the seal protector (6), which prevents damage to the seal lip during installation. *(Reprinted with permission of General Motors Corporation)*

To remove a two-part seal, you should:

1. Recover the refrigerant.
2. Remove the compressor clutch assembly.
3. Remove the clutch hub drive key from the shaft.
4. Remove the seal seat. On some compressors:
   a. The seal seat is retained by a group of bolts at the front of the compressor. Remove these bolts, and remove the seat.
   b. The seal seat is held in the compressor neck by a retaining ring. Remove the retaining ring, attach a puller to the seal seat, and pull the seat out using a rotating motion (Figure 14–20).
   c. The front head must be removed; the center cylinder sections should be clamped together to prevent disturbing the center O-ring before this is done. Install the clamp, remove the bolts that hold the compressor together, remove the front head, remove the reed plate from the front head, and drive the seal seat out the rear of the front head (Figure 14–21).
5. Remove the seal cartridge from the shaft. A special tool is required on most compressors.
6. Remove the seal seat O-ring if it was not removed with the seal seat.
7. Thoroughly clean the seal and seal seat areas.

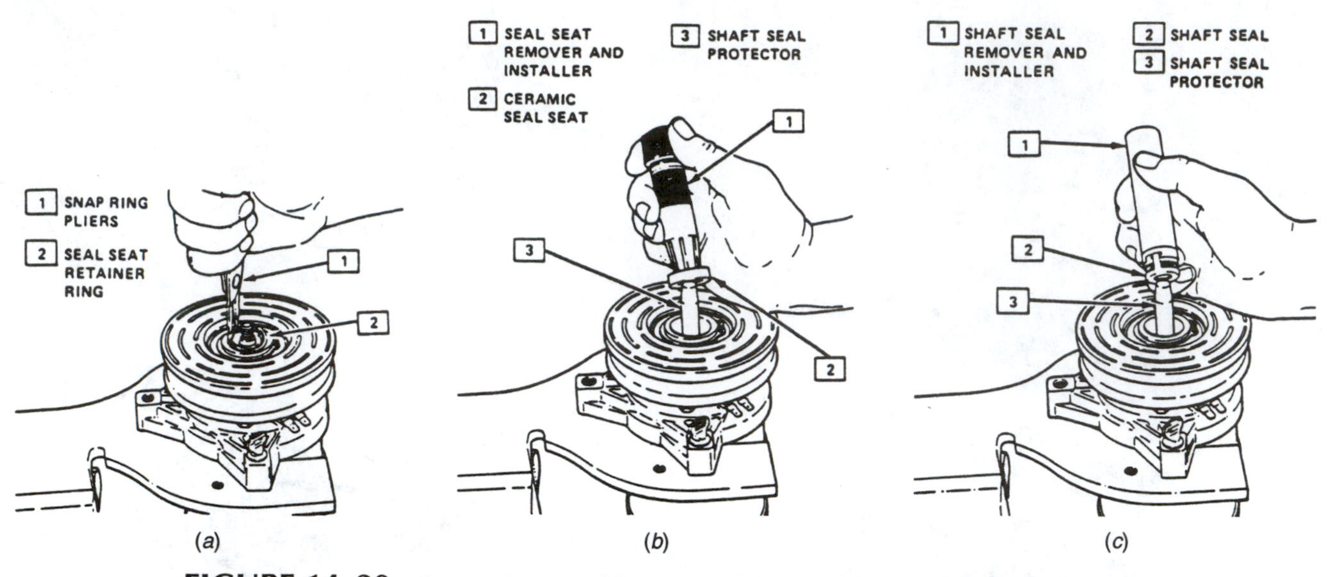

FIGURE 14–20 A two-piece seal is removed from a GM compressor by removing the snap ring (*a*), removing the seal seat (*b*), removing the seal cartridge (*c*), and removing the O-ring (see Figure 14–17*c*). (*Reprinted with permission of General Motors Corporation*)

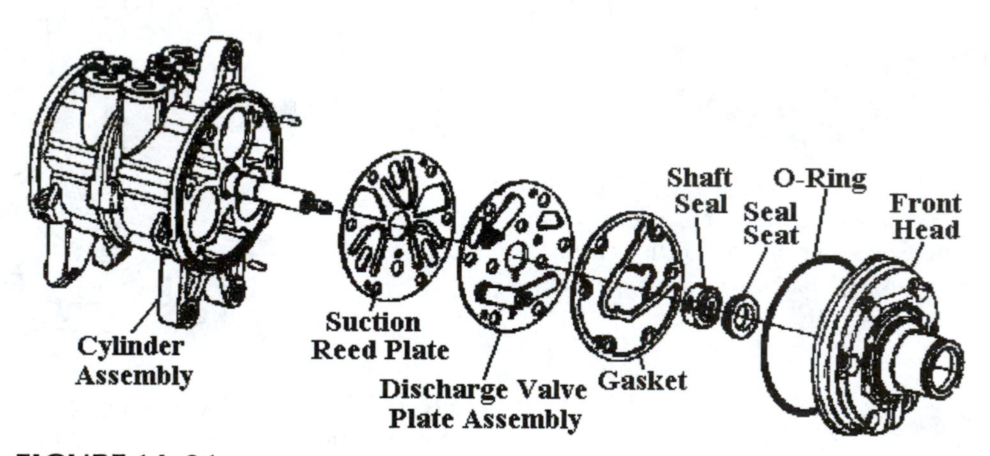

FIGURE 14–21 The shaft seal of many Nippondenso-type compressors is inside the front head. The compressor must be partially disassembled in order to remove and replace the seal.

To replace a two-part seal, you should:

1. Install a seal protector over the compressor shaft.
2. On some compressors, oil the new O-ring and position it in its groove; a special tool is sometimes needed.
3. Oil the new seal and slide it over the compressor shaft. On some compressors, the seal must be rotated to engage the flat areas. You should feel the seal move in farther and the compressor shaft rotate as these flat areas are engaged. Some compressors require a special installer for the seal (Figure 14–22).
4. Lubricate the seal seat and slide it into position. If a special tool was required to pull the seat, use the same tool to install it to ensure that it is positioned correctly. On compressors with the seal inside the front head, drive the seal seat and O-ring into the front head using the special tool, install a new O-ring in the front head, position the reed valve over the dowel pins, position the front head over the dowel pins, and install the bolts and tighten them to the correct torque.

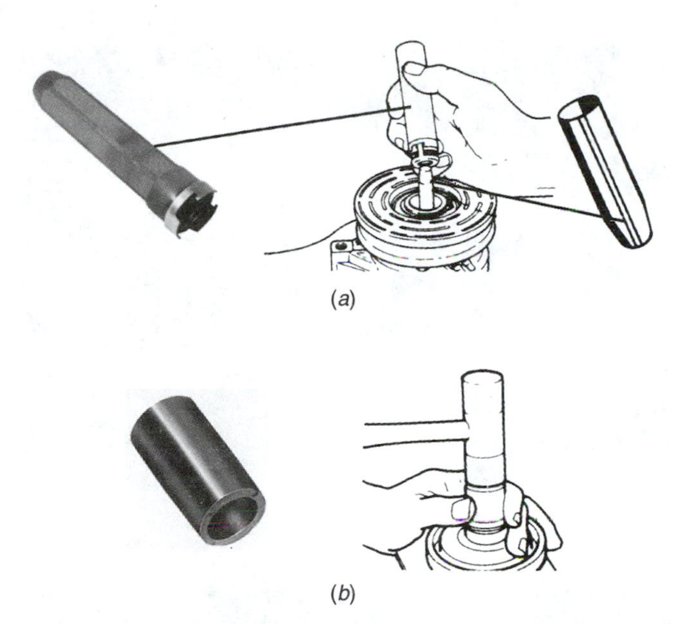

FIGURE 14–22 Installation of a GM two-piece seal is essentially the reverse of removal. A seal protector should be used as the seal is installed (*a*), and another special tool is required to tap the seal seat downward so the snap ring can enter its groove (*b*). (*Courtesy of Kent-Moore*)

5. Install the seal seat retainer. Remember that the seal cartridge should exert a forward pressure against the seat. *If bolts are used*, center the seal seat to the shaft and tighten the bolts to the correct torque. *If a retaining ring is used*, position the flat side of the retaining ring against the seat and, using a special tool, push inward on the ring until it enters its groove.
6. Rotate the compressor shaft a few turns, pressurize the compressor, and check for leaks.
7. Evacuate and recharge the system and check for leaks.

14.2.4 Compressor Repair

Compressor internal repair is limited by the availability of repair parts, service information, and the skill level of the technician. A noisy or roughly operating compressor that passes metal or plastic particles and shavings out the discharge is normally replaced with a new or rebuilt unit. If a system's high side pressure is too low and low side pressure too high, with these pressures within 30 to 50 psi of each other, the compressor is probably bad and should be either repaired or replaced. (This problem can be caused by faulty reed valves or an internal leak.) Another indication of a malfunctioning compressor is metal or plastic debris at the OT or TXV screen. When replacing a compressor, be sure that the replacement part is the right one and is compatible with the refrigerant used. Remember that if the compressor failed, the system is probably contaminated; the receiver–drier or accumulator is normally replaced also, and, in some cases, the condenser should be flushed or an in-line filter installed.

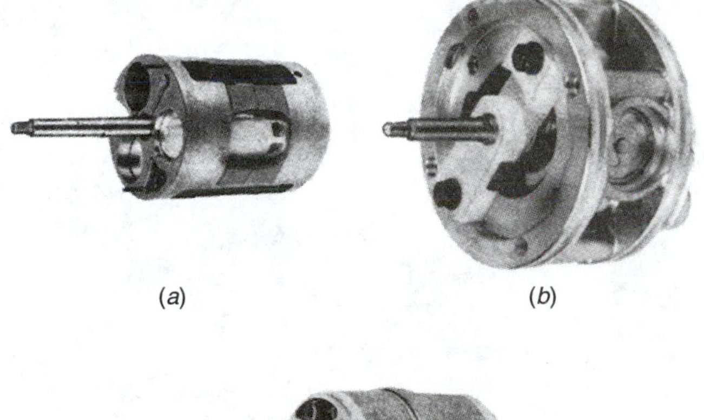

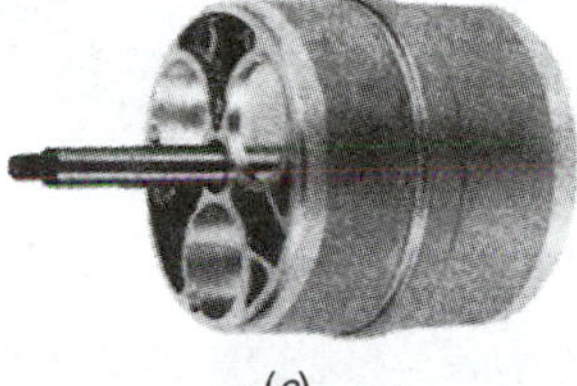

FIGURE 14–23 New cylinder assemblies or "gut packs" are available for some compressors. Replacements for an A-6 (*a*), R-4 (*b*), and DA-6 (*c*) are shown here. (*Reprinted with permission of General Motors Corporation*)

✓SERVICE TIP

At least one A/C system on a fairly popular vehicle has two low side service ports, and an inexperienced technician installed both service hoses to these low side ports (an adapter was needed to connect the high side hose). The gauges read the same pressure, indicating a bad compressor, but intuition and other indications suggested that the compressor was operating. Seeking advice from an experienced A/C technician produced a few laughs and advice about where to connect the high side service hose.

Internal gasket and O-ring, reed valve sets, and replacement heads are available for some compressors, and replacing a cylinder head and reed valve plate is usually a relatively simple process. An internal assembly or "gut pack" is available for some compressors: This is an assembly that contains all of the wearing parts (Figure 14–23). Installing a gut pack allows the compressor shell, heads, and clutch to be reused. As

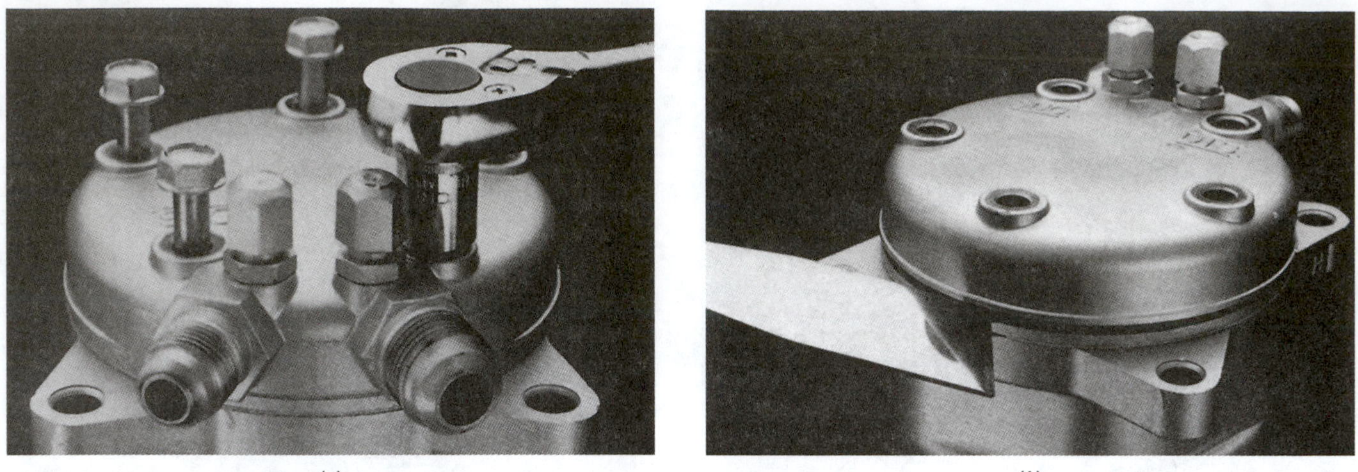

(a) (b)

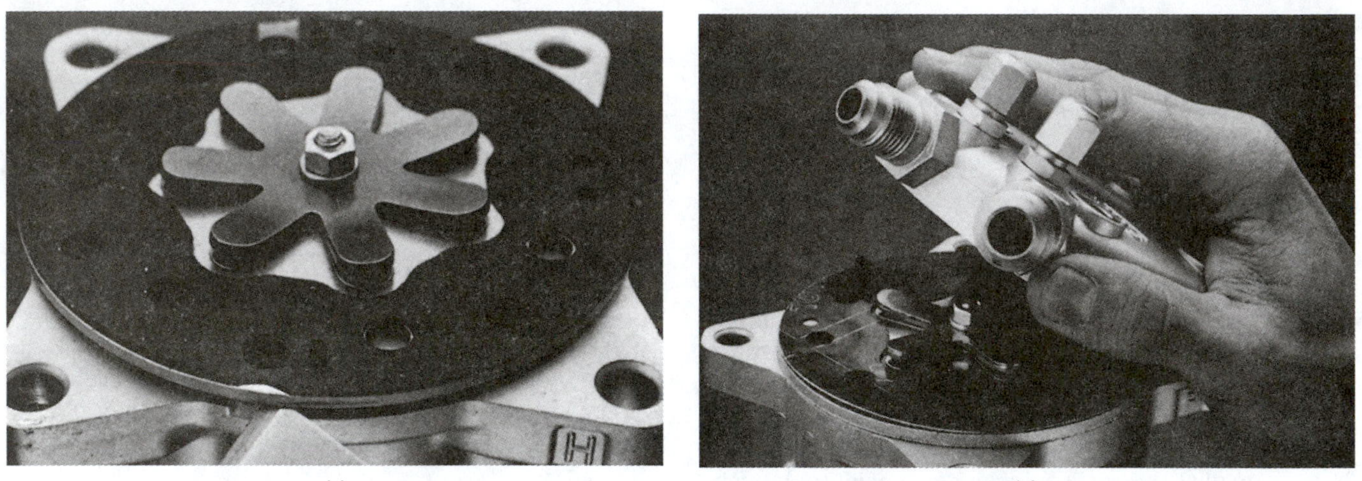

(c) (d)

FIGURE 14–24 The valve plate of a Sankyo compressor can be replaced by removing the bolts (*a*) and the head (*b*), removing the valve plate (*c*) and old gasket, and installing a new valve plate with gaskets (*d*). Note that (*c*) is a 7-cylinder compressor while (*a*), (*b*), and (*d*) show a 5-cylinder compressor. (*Courtesy of Sanden International*)

mentioned earlier, the actual service procedure given in the manufacturer's service manual should be followed.

Many shops prefer to install a new or rebuilt compressor rather than make internal repairs on an old unit.

To remove a reed valve assembly, you should:

1. Recover the refrigerant from the system. If necessary, remove the compressor as described in Section 14.2.1.
2. Drain the oil out of the compressor and mount it in a holding fixture. Inspect the oil for debris and contamination.
3. Remove the head bolts and separate the head from the compressor.
4. Remove the reed valve assembly from the cylinder block or head (Figure 14–24).
5. Locate and remove the alignment dowel pins used on most compressors.
6. Remove the upper and lower gaskets or O-rings; there is usually one on each side of the reed valve plate.
7. Wash the cylinder head and top of the cylinder with clean solvent and air dry.

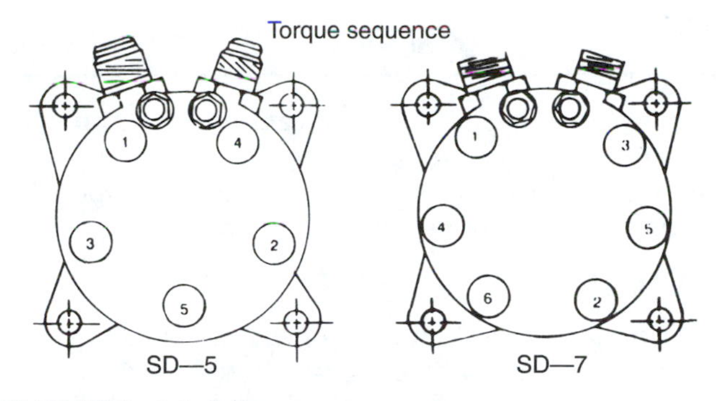

FIGURE 14–25 When a compressor head is replaced, the bolts should be tightened to the correct torque and in the correct sequence. (*Courtesy of Sanden International*)

To replace a reed valve assembly, you should:

1. Install the alignment dowels into the holes in the cylinder assembly.
2. If a gasket is used between the reed valves and cylinders, wet the gasket with refrigerant oil and place it in the correct position.
3. Lubricate the reed valves with refrigerant oil and place them in the proper position, being sure to align the dowel pins and passages.
4. Wet the upper gasket or O-ring with refrigerant oil and position it.
5. Position the head, being sure to align the dowel pins.
6. Install the head bolts and tighten them to the specified torque using a diagonally opposite tightening sequence (Figure 14–25).
7. Rotate the compressor shaft to test the assembly, pressurize the compressor, and test for leaks.
8. Replace the compressor as described in Section 14.2.1.

14.2.5 Compressor Bench Checks

Compressor bench checks consist of rotation checks of the compressor and clutch and leak checks. The rotation checks are for excessive drag, roughness, and excessive runout. These checks should always be made on a used compressor before installing it (Figure 14–26).

To make a rotating compressor check, you should:

1. Rotate the compressor drive hub and observe the air gap or clearance between the drive plate and rotor pulley. On single-unit and two-part clutches, hold the rotor pulley stationary during this check.

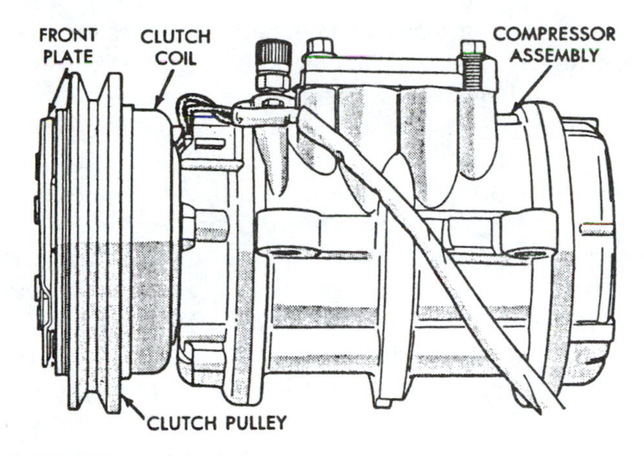

FIGURE 14–26 A compressor can be bench checked to ensure that the pulley turns smoothly without excessive runout and with no interference. The front plate (and compressor shaft) should also rotate smoothly without excessive runout. Excessive drag, noise, or clearance inside the compressor indicates a faulty compressor. (*Courtesy of DaimlerChrysler Corporation*)

2. After rotating the shaft several turns, feel for roughness and excessive drag. One manufacturer gives a rotating torque specification of 7 lb-ft maximum.
3. Rotate the rotor pulley while holding the shaft stationary and observe the air gap. The pulley should rotate smoothly and quietly.

✓SERVICE TIP

A bent plate or compressor shaft is indicated if the runout is greater than the air gap specification range.

Stop the rotation and quickly turn the shaft backward as you feel for excessive internal clearance. The shaft should rotate smoothly and fairly easily, with no internal free play.

✓SERVICE TIP

A bent rotor pulley is indicated if runout is greater than the air gap specification range.

Compressor leak checks require adapters that connect the discharge and suction ports to the service hoses of a manifold gauge set. Commercial adapters are available for most compressors, or they can be made in the shop by adapting the appropriate line fitting or block-off plate to the proper service valve fitting. Clean, oil-free, and water-free shop air can be used for these checks. An internal leak check looks for leaks from the discharge to the suction side in the compressor; an external leak check looks for leaks to atmosphere. While making an internal leak check, remember that reed valves do not make a perfect seal, and some leakage can be expected.

To make compressor leak checks, you should:

1. Install the pressure test adapters to the compressor ports.
2. Connect the service hoses of a manifold gauge set to the service valves and a 60-psi or greater pressure source (see Figure 10–66).
3. Open the high side hand valve and watch the pressure gauges. The high side pressure gauge should increase to source pressure immediately as the valve is opened. The low side pressure gauge should show a delayed and much slower pressure increase. An internal compressor leak is indicated if the low side pressure increases too rapidly.
4. Open both valves until both gauges show source pressure and then close both valves. Observe the gauge pressures—they both should hold steady. An external leak is indicated if they show a pressure loss.
5. Apply a soap solution to probable leak locations: the shaft seal, cylinder head, any other compressor body joint, any bolts, any switches or pressure relief valves, and the drain plug. A leak will produce bubbles.

14.2.6 Compressor Oil Level Checks

At one time, most compressors had an oil sump; the level was checked relatively quickly, using a dipstick. With most modern compressors, there is no oil sump or way to use a dipstick. Checking the oil level is further complicated by the fact that the oil travels through the system with the refrigerant and can migrate while the system is shut off. To get a truly accurate check, the system should be operated for 10 to 20 minutes before checking levels. Too little oil in a system contributes to compressor wear or damage; too much can cause slugging in the compressor or excessive discharge pressures.

✓SERVICE TIP

Most systems have two specifications. One is for the amount in the system; on average, this is about 10 oz. The other is for the compressor; this is a specified amount somewhere between 3 and 6 oz.

Normally, an A/C system does not lose oil or contaminate that oil. Oil loss is usually the result of a refrigerant leak. Whenever a major system component is replaced, a certain amount of oil should be added to the new part to compensate for the oil removed. Oil contamination is usually caused by compressor failure, water or other chemicals, or system overheating. Contaminated oil points to a problem inside the system. If the proper equipment is available, the system should be flushed with liquid refrigerant, an in-line filter installed, the receiver–drier or accumulator replaced, and the system thoroughly evacuated. System oil level checking procedure is described in Section 13.3.7.

✓SERVICE TIP

Remember that refrigerant oil, especially PAG oil, is hygroscopic and will absorb water from the atmosphere if possible. Always keep the oil container capped; replace the cap immediately after pouring oil out of it. If you are working with PAG oils, also remember that you should wear nonpermeable gloves so the oil will not come into contact with your skin.

✓SERVICE TIP

If you add oil to a system into the suction port or line, the oil may become trapped on top of the pistons when it starts operation, and this can damage the compressor. Rotating the compressor shaft at least 10 turns by hand should pump any trapped oil out of the cylinders.

To check the oil level in a compressor with dipstick provision, you should:

1. Operate the system (if it can be operated) for 10 to 20 minutes to normalize the oil levels.
2. *If the system is equipped with service valves,* isolate the compressor from the system (Figure 14–27). *If the system is equipped with Schrader valves,* recover the refrigerant from the system.

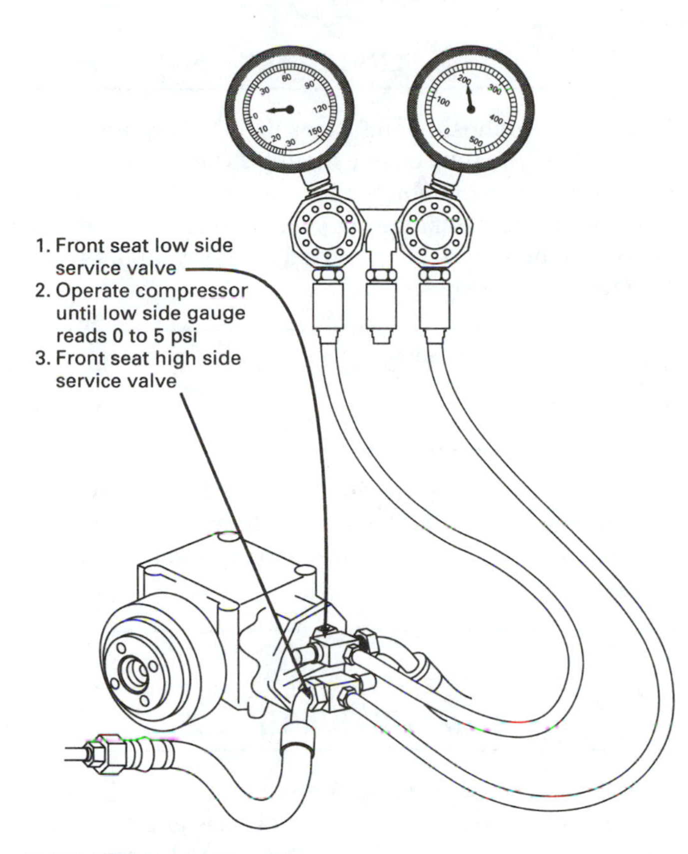

FIGURE 14–27 If the compressor is equipped with service valves, it can be isolated from the system following this procedure.

3. Remove the checking plug and insert the dipstick into the bottom of the compressor. The compressor shaft may have to be rotated to allow the dipstick to enter (Figure 14–28).
4. Remove the dipstick and measure the oil level.
5. Compare the oil level with the specification and adjust the level if necessary.
6. Replace the checking plug and tighten it to the correct torque.
7. Evacuate and recharge the system and test for leaks to return it to service. For service valve systems, back-seat the service valves to return it to service.

✓SERVICE TIP

Some compressor designs make it virtually impossible to drain all of the oil out of them.

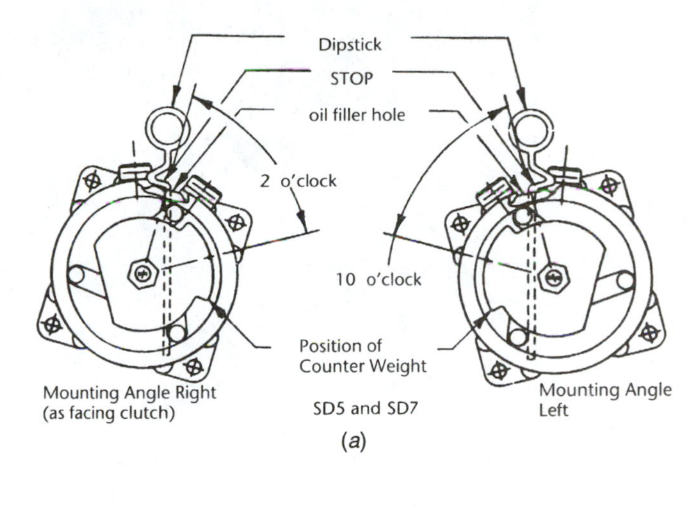

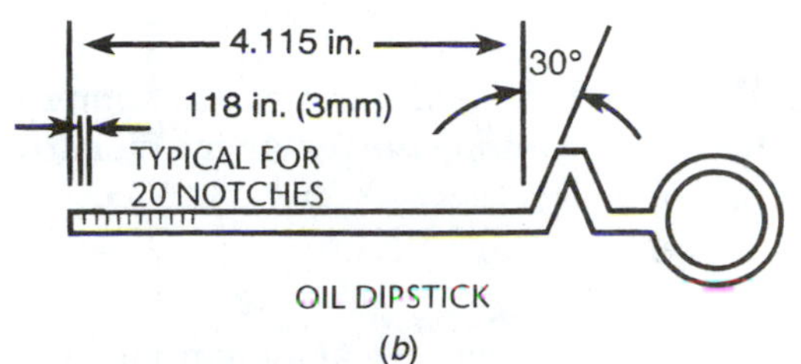

Mounting Angle (Degrees)	Acceptable Oil Level In Increments					
	505	507	508	510	708	709
0	4-6	3-5	4-6	2-4	4-6	3-5
10	6-8	5-7	6-8	4-5	5-7	4-6
20	8-10	6-8	7-9	5-6	6-8	5-7
30	10-11	7-9	8-10	6-7	7-9	6-8
40	11-12	8-10	9-11	7-9	8-10	7-9
50	12-13	8-10	9-11	9-10	9-11	8-10
60	12-13	9-11	9-12	10-12	10-12	9-11
90	15-16	9-11	9-12	12-13	11-13	10-13

(c)

FIGURE 14–28 A Sankyo compressor has an oil sump and provision to use a dipstick to check the oil level (*a*). The checking procedure and the acceptable levels are shown here (*b* and *c*). (*Courtesy of Sanden International*)

To check the oil level in a compressor without dipstick provision, you should:

1. Operate the system (if it can be operated) for 10 to 20 minutes to normalize the oil levels.
2. Recover the refrigerant.
3. Remove the compressor as described in Section 14.2.1.
4. Drain the oil from either the drain opening or the service ports, measure the quantity drained

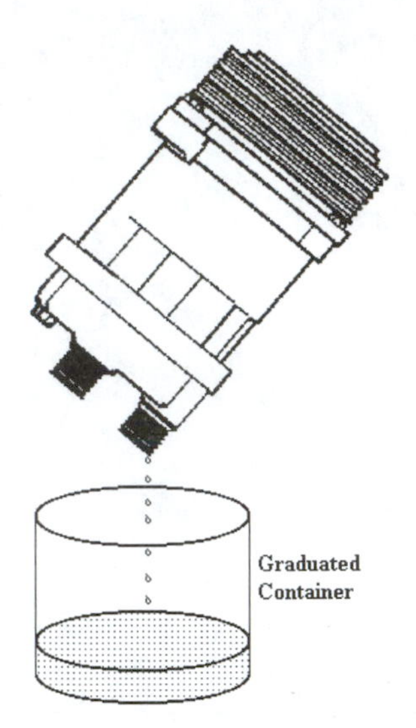

FIGURE 14–29 The oil level in this compressor is being checked by pouring the remaining oil into a graduated container.

out, and compare the amount drained out with the specifications. Check the condition of this oil and dispose of it in an approved manner (Figure 14–29).

5. Pour the specified amount of new oil of the proper type and viscosity into the compressor inlet and replace the compressor as described in Section 14.2.1.
6. Evacuate and recharge the system and check for leaks.

✓SERVICE TIP

If the amount of oil drained from the compressor is low, the compressor design is trapping oil or the system is low on oil; in the latter case, extra oil should be added. If the amount of oil is high, the system is wet with too much oil, and an undercharge of new oil should be added.

14.3 HOSE AND FITTING REPAIR

Most refrigerant leaks are through a faulty hose or hose connection, and, depending on the nature of the leak, several repair methods are possible. These can be as simple as tightening a loose fitting or as complex as making up a refrigerant hose.

✓SERVICE TIP

Early failure of hoses and fittings in a FWD vehicle can be the result of faulty engine mounts. The top of the engine in this type of vehicle rocks back and forth, from the front to the back, during vehicle operation; an excessive amount of rocking caused by faulty mounts can lead to premature hose failure.

✓SERVICE TIP

Early failure of hoses or fittings near the compressor can be caused by loose compressor mounting bolts. Compressor vibration can cause cracks and breakup of metal parts.

✓SERVICE TIP

Dual O-ring fittings can be difficult to separate; the preferred method is to use a special tool designed for that purpose. An alternate method if there is still pressure in the system is to loosen the fitting nut and operate the system; the internal pressure can loosen the fitting.

14.3.1 Fitting Repair

Leaky line fittings cannot be made to seal by overtightening them, and the size of the fitting nuts and the wrenches used with them make overtightening easy. If the fitting is tight and still leaks, it must be taken apart and inspected for damage, and a new O-ring must be installed. Torque specifications for the various line fittings are provided by the vehicle manufacturer (Figure 14–30).

Most line fittings are sealed by one or two O-rings. At this time there are at least 5 different O-ring materials and 37 different sizes used in automotive A/C systems. Some of the sizes are very similar, the only difference being a slight change in thickness. O-rings are sized by internal diameter and cross section; there are both fractional-inch and metric sizes. Most O-rings have a round cross-section; some are oval.

The chemical names for O-rings can be very confusing because of the seemingly conflicting use of chemical manufacturers' special names. O-rings must be chemically compatible with the refrigerant and oil used in the system (see Figure 3–15). The most com-

Metal Tube Outside Diameter (in inches)	Thread and Fitting Size (in inches)	Steel Tubing Torque		Aluminum or Copper Tubing Torque		Nominal Torque Wrench Span (in inches)
		N • m	*lb-ft*	*N • m*	*lb-ft*	
1/4	7/16	14–20	10–15	7/8	5–7	5/8
3/8	5/8	41–48	30–36	15–18	11–13	3/4
1/2	3/4	41–48	30–36	20–27	15–20	7/8
5/8	7/8	41–48	30–36	28–37	21–27	11/16
3/4	1 1/16	41–48	30–36	38–44	28–33	1 1/4

FIGURE 14–30 This chart shows the tightening torque for the various sizes of hose and line fittings. It is easy to overtighten most of these because of the large wrenches. (*Reprinted with permission of General Motors Corporation*)

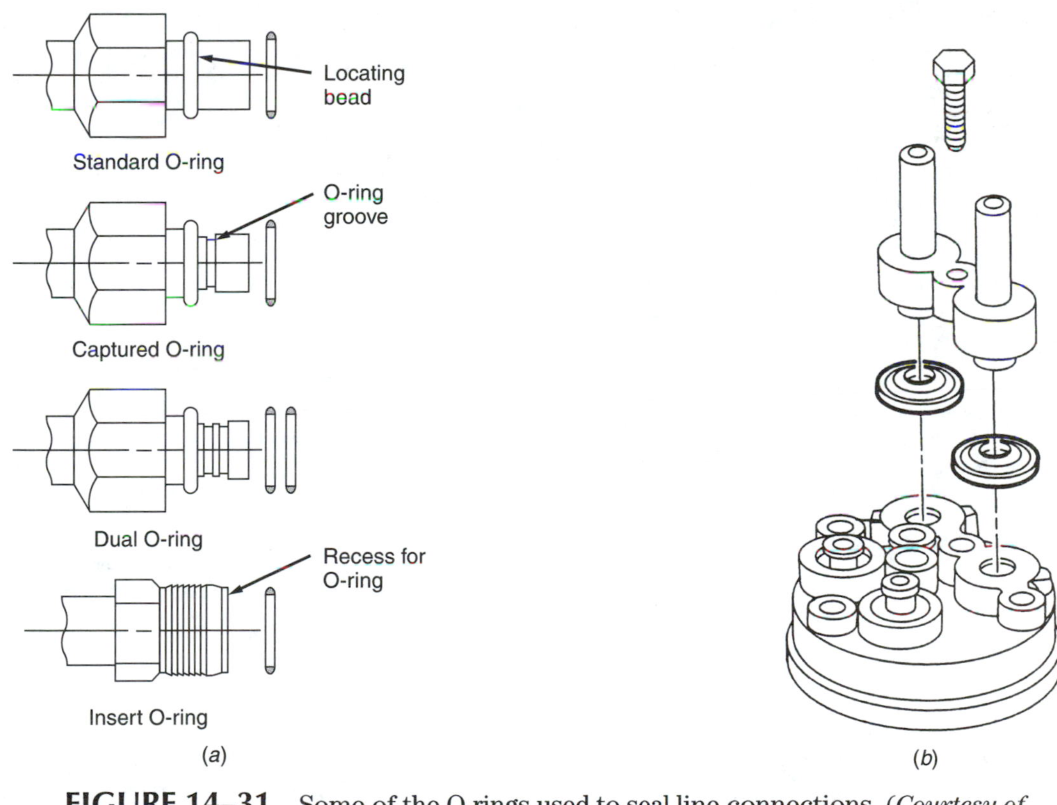

FIGURE 14–31 Some of the O-rings used to seal line connections. (*Courtesy of ACDelco*)

monly used references are *nitrile butadiene rubber* (*NBR*) and *hydrogenated nitrile butadiene rubber* (*HNBR*). NBR, nitrile, or *Buna N* rubber is the standard O-ring material; this is a black material used in R-12 systems. Buna N is commonly used for *static installations* in which both parts are stationary, relative to each other. A different, more resilient material must be used in *dynamic installations* with movement (as in a spring lock coupler), and this is also a black material. HNBR is commonly used with R-134a, and neoprene O-rings are used in OEM installations. HNBR is also a black material, but it has a blue tint. Some O-rings used with R-134a are made from *highly saturated nitrile (HSN)* and have a green color. Teflon O-rings are used with the service fittings on some early compressors; this material is white (Figure 14–31).

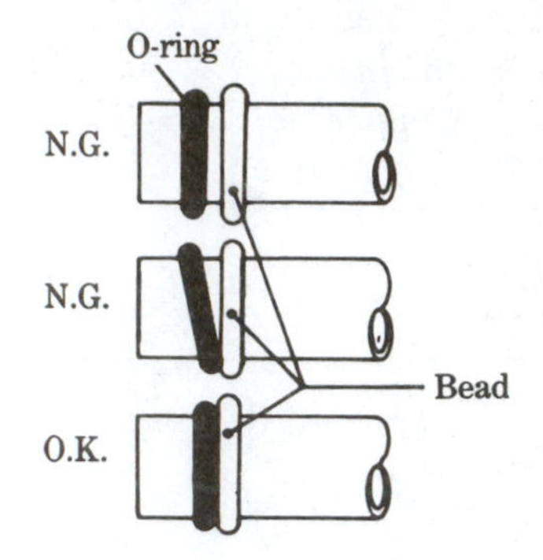

FIGURE 14–32 When an O-ring is installed, it should be placed in its proper position next to the locating bead or in its groove. (*Courtesy of Nissan Motor Corporation in USA*)

✓SERVICE TIP

Do not trust the color of the O-ring to tell you the material it was made from. Purchase O-rings from a reputable source and specify their intended use.

✓SERVICE TIP

HNBR and HSN O-rings work for both R-12 and R-134a installations. These O-rings should be used for all R-12 systems to help make such systems ready for future retrofitting and also to reduce the need to stock NBR O-rings. All O-rings should be lubricated with standard mineral refrigerant oil as they are installed unless specified differently by the manufacturer.

To repair a leaky fitting with threaded connectors, you should:

1. Recover the refrigerant from the system.
2. Clean the fitting and disassemble it. Note that after the initial loosening, the nut should turn freely on the threads.
3. Remove the old O-ring and inspect both O-ring mating surfaces for damage. Nicked or deformed sealing surfaces require replacement.
4. Obtain the correct O-ring replacement and position it right next to the locating bead or, if a captive O-ring, in its groove (Figure 14–32).
5. Wet the O-ring with mineral type refrigerant oil (Figure 14–33). Wet the threads of the fitting with the same oil.
6. Carefully align the assembly and finger tighten the nut.

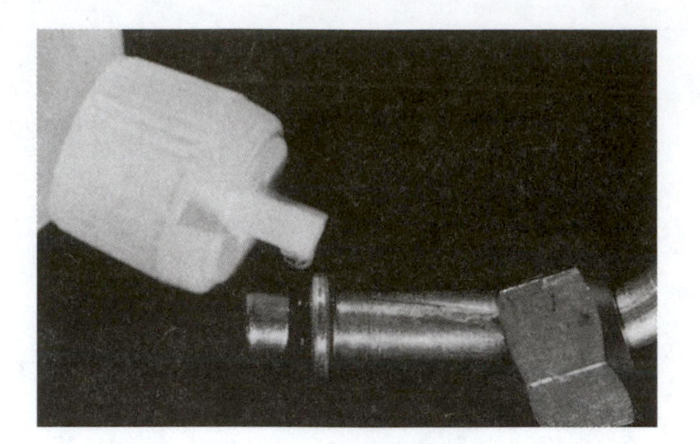

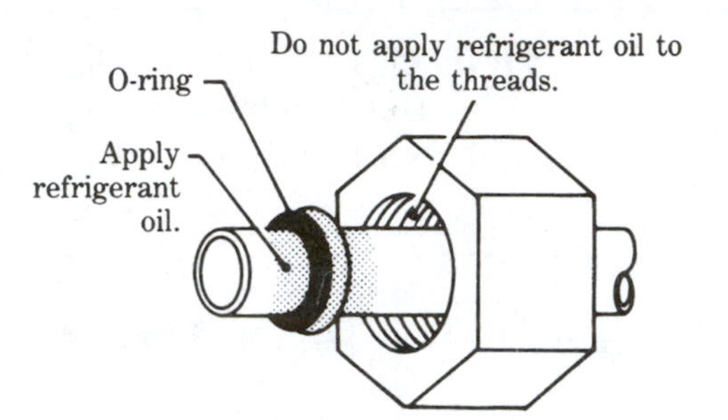

FIGURE 14–33 The O-ring must be wetted with ordinary refrigerant oil as it is installed. (*Courtesy of Nissan Motor Corporation in USA*)

7. Use one wrench to keep the fitting from turning and, using a torque wrench, tighten the nut to the correct torque (Figure 14–34).
8. Evacuate and recharge the system and test for leaks.

✓SERVICE TIP

An iron fitting nut tends to seize onto the aluminum, male fitting of some components. If this occurs, heat the fitting nut using a hot air gun, and carefully work the nut back and forth until it is free. To prevent this problem from occurring, some technicians put never-seize compound or mineral refrigerant oil on these threads when assembling a component.

To repair a leaky fitting with a spring lock connector, you should:

1. Recover the refrigerant from the system.
2. Fit the spring lock coupling tool of the correct size over the line, close the tool, and push the tool into the cage opening to release the garter spring (Figure 14–35).

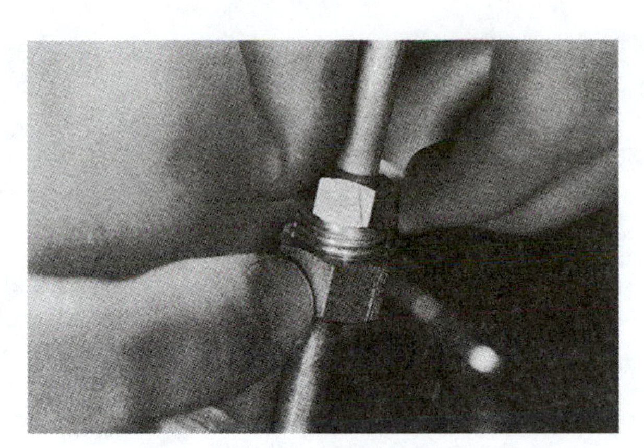

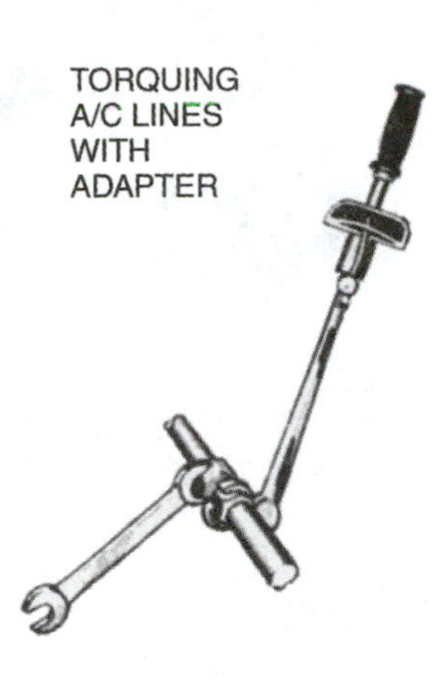

(a)

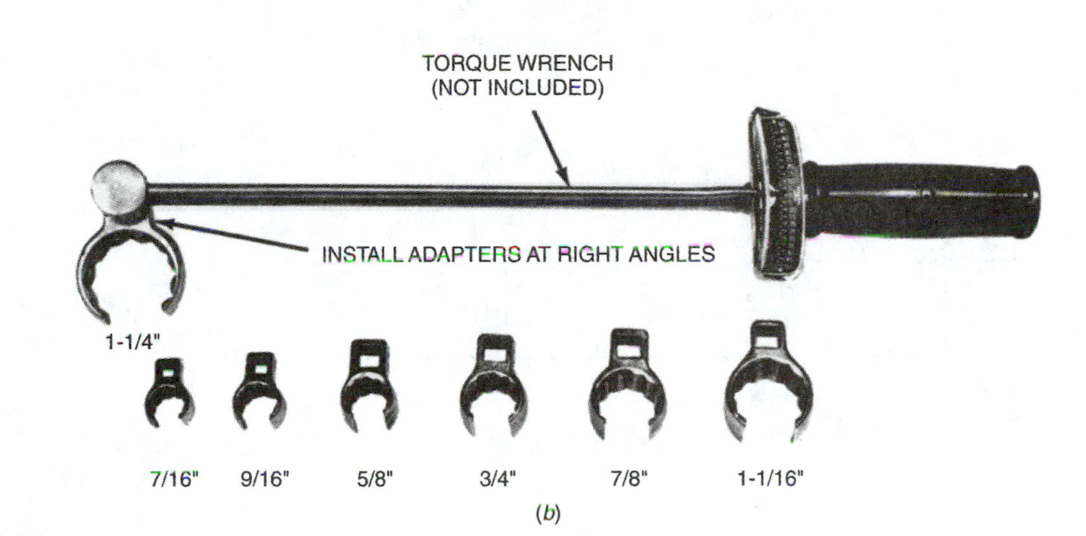

(b)

FIGURE 14–34 The fitting should be assembled finger tight (*a*) and then tightened using a torque wrench with adapters to fit the nut (*b*). Note the use of a second wrench to keep the line from twisting as the nut is tightened. (a *is courtesy of Nissan Motor Corporation in USA;* b *is reprinted with permission of General Motors Corporation*)

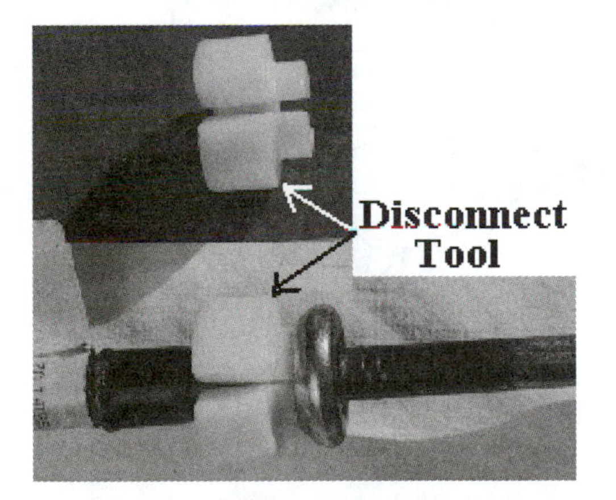

FIGURE 14–35 A spring lock coupling is disconnected by placing a special tool over the line and pushing the tool into the coupling to release the spring so the coupling can be pulled apart.

3. Pull the male and female fittings apart and remove the tool.
4. Clean the fitting parts, remove the old O-rings using a wood or plastic pick, and inspect the fittings for damaged sealing surfaces or garter spring. A damaged or bad garter spring should be replaced. Soft, silicon-impregnated brushes are available to clean the female fitting.
5. Place the new O-rings in position and wet them with mineral type refrigerant oil.
6. Assemble the fitting by simply pushing the two parts together with a slight twisting motion. Visually check your assembly; the flared end of the female fitting should have disappeared behind the garter spring (Figure 14–36).
7. Evacuate and recharge the system and test for leaks.

✓SERVICE TIP

Special clamps are available from aftermarket sources to help cure spring lock coupling leakage. These clamps secure the two coupling parts to eliminate the relative motion that causes the leak (Figure 14–37).

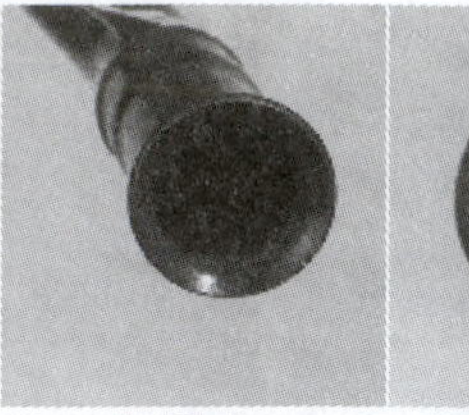

FIGURE 14–36 A leaking spring lock coupling can often be repaired by cleaning the O-ring grooves and the male and female ends. After installing and lubricating new O-rings, the fitting is assembled, making sure the spring traps the flared end of the male fitting.

14.3.2 Line Replacement

A faulty hose or metal line is often repaired by removing and replacing it with a new or repaired line. This normally involves disconnecting each end of the line at a fitting. Some hoses, like the suction and discharge, are combined with another line, making a double connection at the compressor, condenser, or TXV and evaporator. A manifold-type connector sealed with a pair of O-rings is commonly used at these connections (Figure 14–38).

To remove and replace a hose or metal line, you should:

1. Recover the refrigerant from the system.
2. Clean the line connections and disconnect each end of the line.

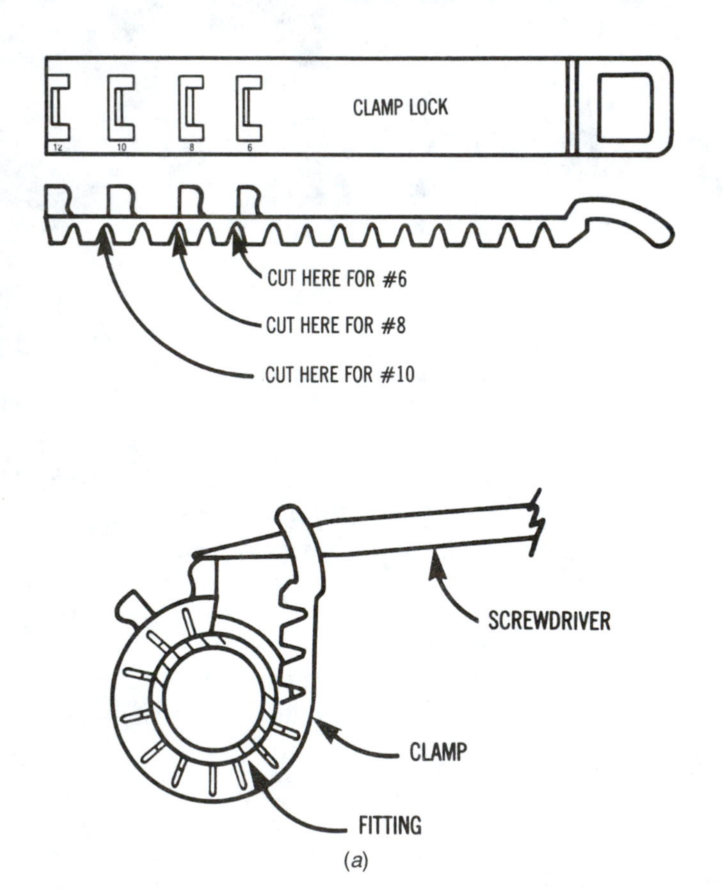

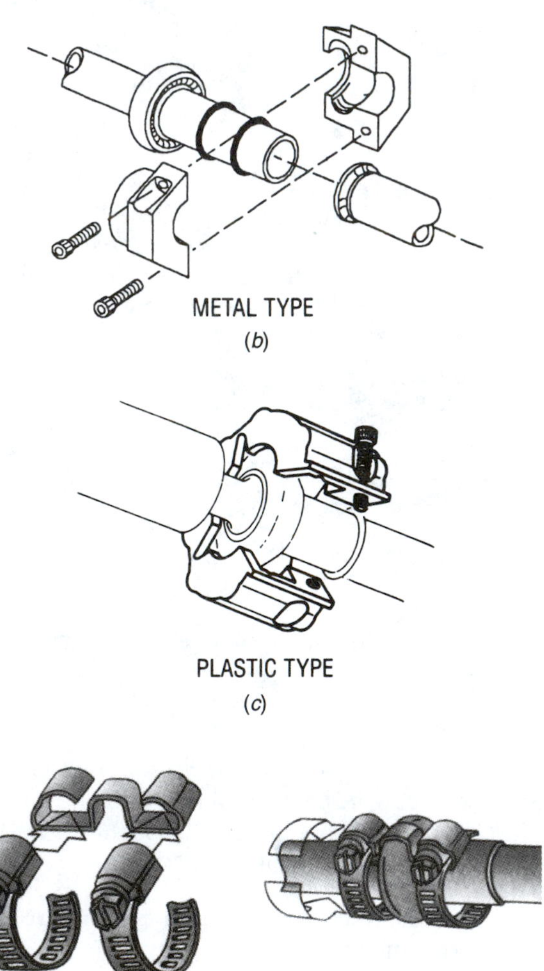

FIGURE 14–37 Four different clamping systems are available to lock a spring lock coupling together to prevent future leakage. (a, b, *and* c *are courtesy of Four Seasons;* d *is courtesy of T.D.R. Stabilizer Clamp Company*)

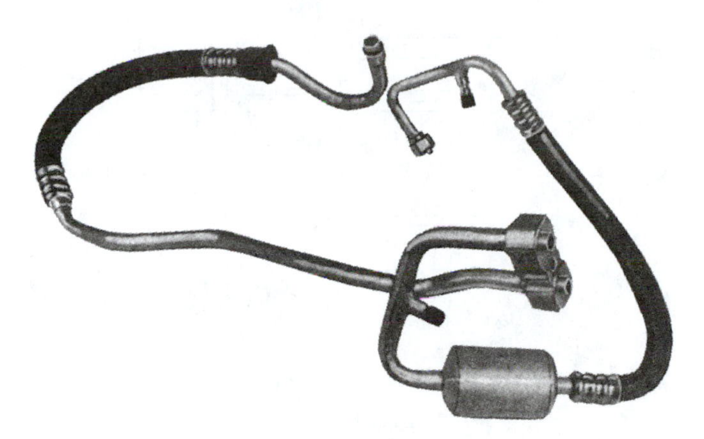

FIGURE 14–38 This OEM suction and discharge line assembly uses one manifold-type fitting at the compressor; this connection is usually sealed using a pair of O-rings. (*Courtesy of Four Seasons*)

3. Note how the line is routed, remove any line holding brackets and hose ties, and remove the line.
4. Clean and inspect the sealing surfaces for damage at each line connector.
5. Install new O-rings at each sealing connection, wetting the O-rings with refrigerant oil.
6. Assemble each connection and tighten each fitting nut or retaining bolt to the correct torque.
7. Replace all line brackets and hose ties.
8. Evacuate and recharge the system and test for leaks.

14.3.3 Hose Repair and Makeup

A hose can sometimes be repaired by replacing one of the ends or by cutting out a damaged section and splicing it back together. Many early systems used rather long hoses. Refrigerant hose is expensive; repairing a damaged hose can reduce repair costs. The fittings described here allow the technician to make up a replacement hose if a new hose is not available or is too expensive.

Most OEM hoses are connected to the line fitting with a captive, beadlock ferrule so the hose is gripped by both the line connector and the ferrule (Figure 14–39). These line connectors have grooves or very small barbs to grip the hose; they cannot be reused with the hose secured by a separate ferrule or clamp. Aftermarket R-12 repair fittings use three large, raised rings or barbs to grip the line securely. These fittings are connected to the hose with a metal shell.

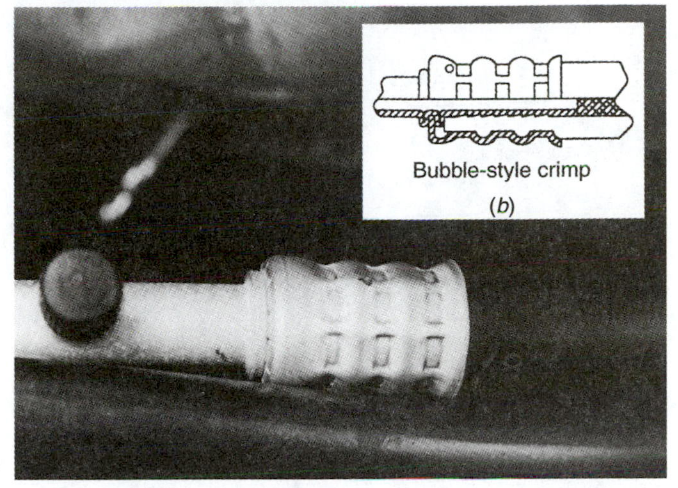

(a)

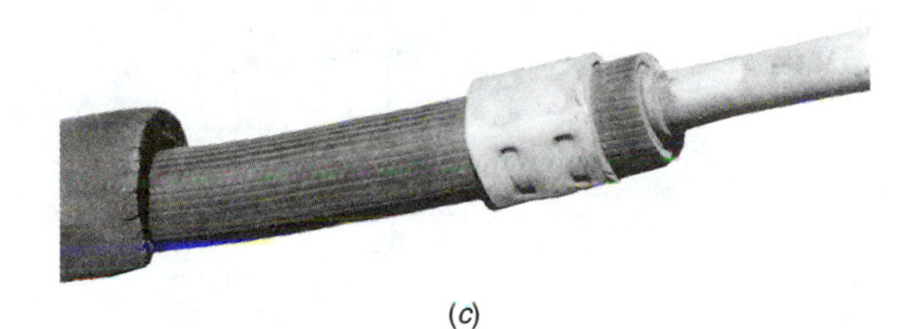

(c)

FIGURE 14–39 An OEM hose clamped in place using a captive, beadlock ferrule (*a*); a beadlock fitting makes a very secure connection (*b*). A fitting using a ring-type ferrule is shown in (*c*). (c *is courtesy of ACDelco*)

✓SERVICE TIP

Some A/C repair shops have tubing repair centers that allow them to duplicate almost any OEM A/C line. They can bend new tube to the desired shape; form the proper beads at the end of the tube; attach a barb fitting at the end of the tube; secure a hose to the fitting; and pressure test the assembly to make sure there are no leaks.

✓SERVICE TIP

In the past, a screw or worm gear clamp was commonly used, but these clamps are no longer recommended because they will not pass the coupling integrity performance requirements. Some screw clamps are fitted with a locator extension that places the clamp directly over the barbs (Figure 14–40).

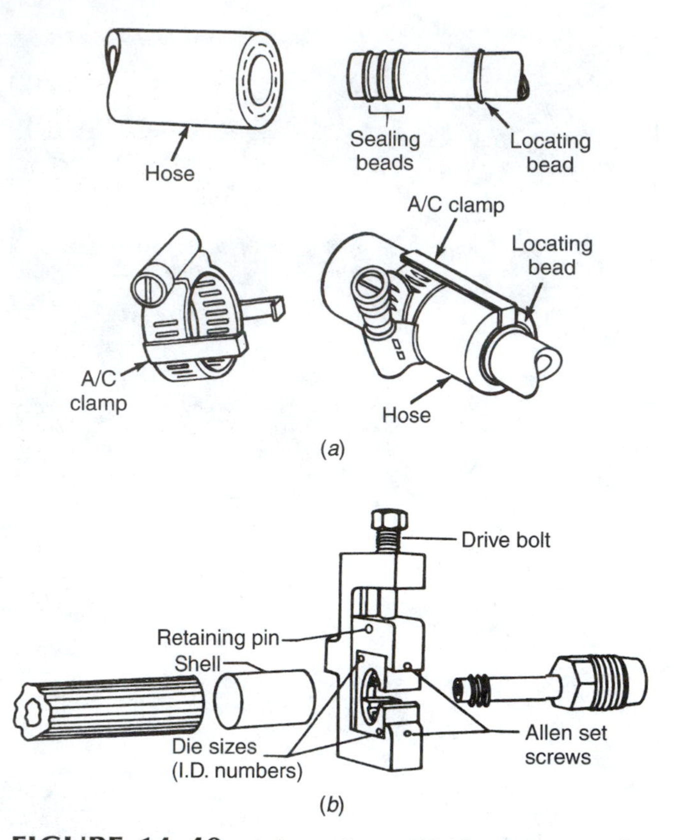

FIGURE 14–40 A hose for an R-12 system can be fastened to a fitting using a screw clamp (*a*) or a metal shell (*b*), which is crimped using a special tool. A screw-type A/C hose clamp is no longer a recommended service procedure. (a *is reprinted with permission of General Motors Corporation;* b *is courtesy of Four Seasons*)

It should be noted that these barb fittings may cut the inner liner of barrier hoses and that neither a screw clamp nor a noncaptive ferrule meets the new SAE leakage specifications for R-134a systems. New service fittings with a captive metal ferrule and a new clamping method have been developed to allow field repair and makeup of barrier hoses. These fittings are commonly called **bubble crimp** or **one-piece beadlock** fittings. Beadlock fittings can be used to repair all types of hoses (Figure 14–41).

A recent innovation is the E-Z Clip system developed by Aeroquip. This system requires the use of a particular hose, but it allows the fitting to be installed quickly and simply. After the hose is slid onto the fitting, the cage is snapped into place, and then the clamps are positioned and secured using the special pliers.

Service fittings of both the barb and beadlock types are available in standard and jump sizes, with various styles of fitting types and bends. The standard size fitting matches the thread diameter of the connector with the diameter of the hose (Figure 14–42).

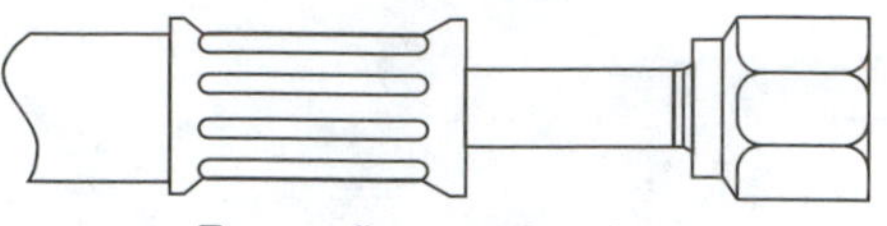

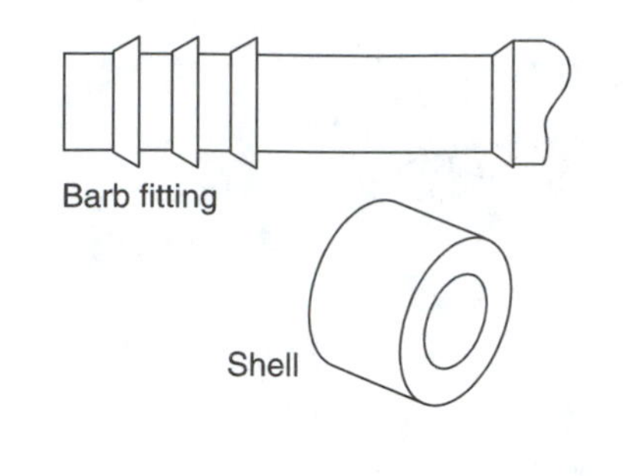

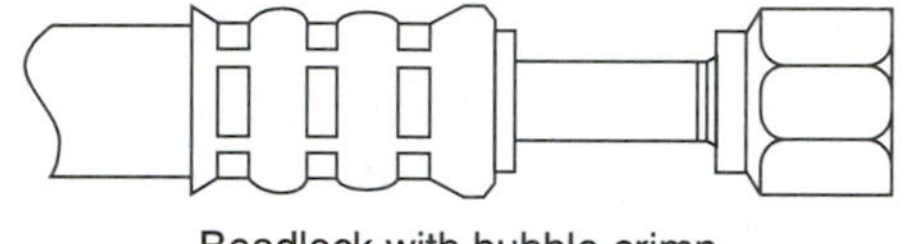

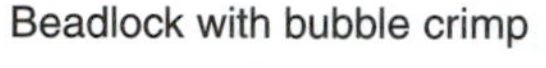

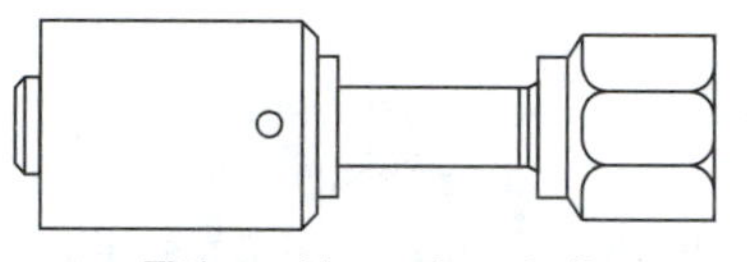

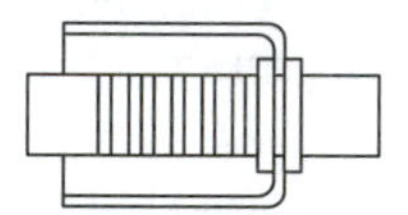

FIGURE 14–41 Barb-type service fittings are of two parts; the fitting and a shell that uses a finger-style crimp (top). A beadlock fitting has a captive shell that is locked in place using a bubble crimp (bottom). (*Courtesy of ACDelco*)

FITTING SIZE (IN INCHES)

Size	*Tube OD*	*Thread Diameter*	*Hose ID*
6	3/8 (0.375)	5/8	5/16
8	1/2 (0.500)	3/4	13/32
10	5/8 (0.625)	7/8	1/2
12	3/4 (0.750)	1 1/16	5/8

FIGURE 14–42 Most replacement hose fittings match the hose ID to a particular tube and nut diameter.

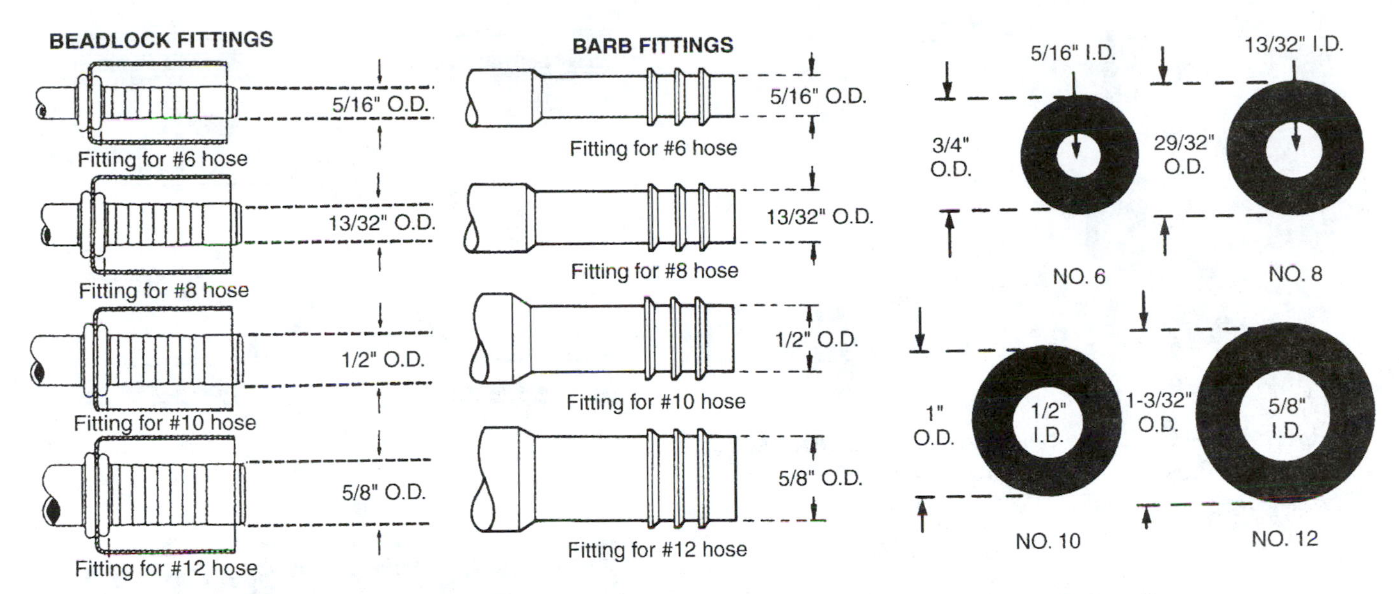

FIGURE 14–43 The stem section of beadlock and barb fittings matches the hose ID. (*Courtesy of Four Seasons*)

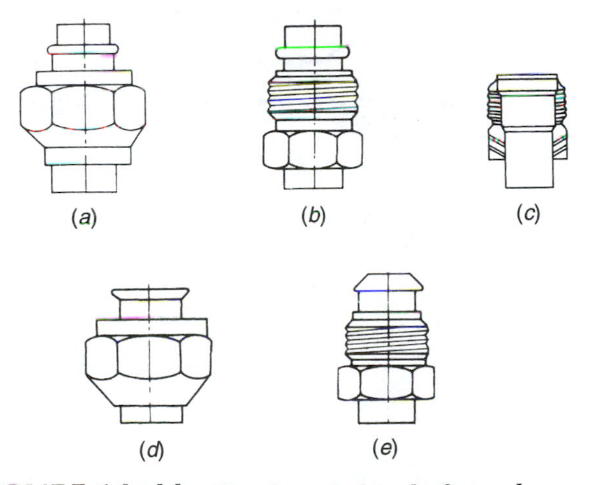

FIGURE 14–44 The threaded end of a replacement line fitting is available in female O-ring (*a*), male O-ring (*b*), male insert O-ring (*c*), female flare (*d*), and male flare (*e*) types. (*Courtesy of Four Seasons*)

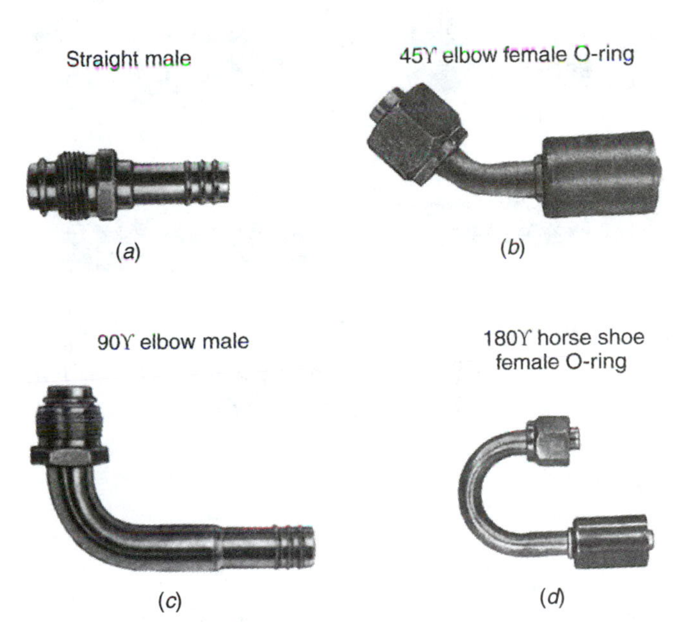

FIGURE 14–45 Replacement line fittings are available in straight (*a*), and 45° (*b*) and 90° (*c*) bends and sometimes in a 180° (*d*) horseshoe bend; they can be either beadlock or barb fittings. (*Courtesy of Four Seasons*)

✓SERVICE TIP

Remember that the fitting sizes are identified by the diameter where the hose is connected (the diameter is roughly in 1/16): 6 = 6/16 (3/8), 8 = 8/16 (1/2), 10 = 10/16 (5/8), and 12 = 12/16 (3/4) (Figure 14–43). For example, the connector nut for a #8 hose uses a 3/4-inch, 16 thread, and probably a 7/8-inch wrench. Jump sizes, also called step-up or step-down sizes, allow a different size hose to be fitted to a particular connector (e.g., a 3/4-inch fitting to a #8 or #12 hose).

The most common fitting styles are female and male O-rings (determined by the nut threads), female and male flares, male insert fittings, and spring lock styles (Figure 14–44). These fittings are usually available for the four common hose sizes and in straight, 45° bend and 90° bend shapes (Figure 14–45).

Special fittings are available in sizes and configurations to fit special installations and can combine several features, including a service valve, muffler, compressor

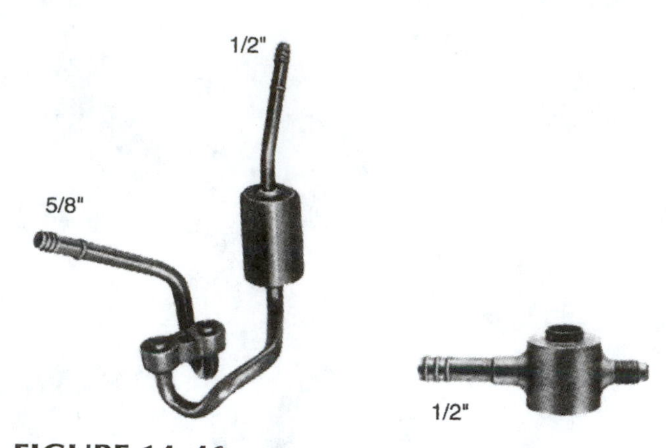

FIGURE 14–46 These two special fittings are used to connect a hose to a GM compressor. (*Courtesy of Four Seasons*)

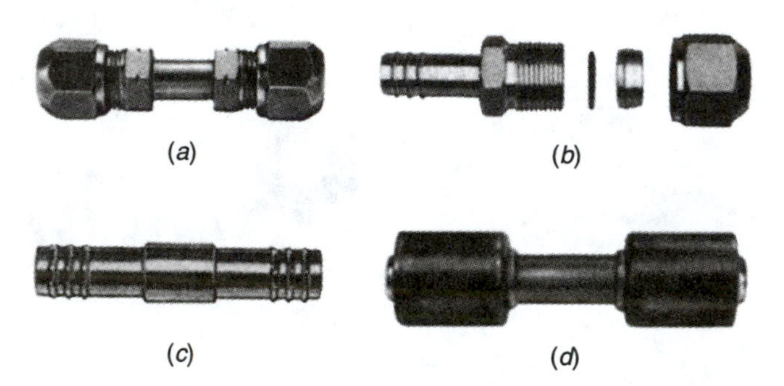

FIGURE 14–47 The fittings to splice two metal lines together (*a*), a metal line to a hose (*b*), or two hoses together (*c* and *d*). (*Courtesy of Four Seasons*)

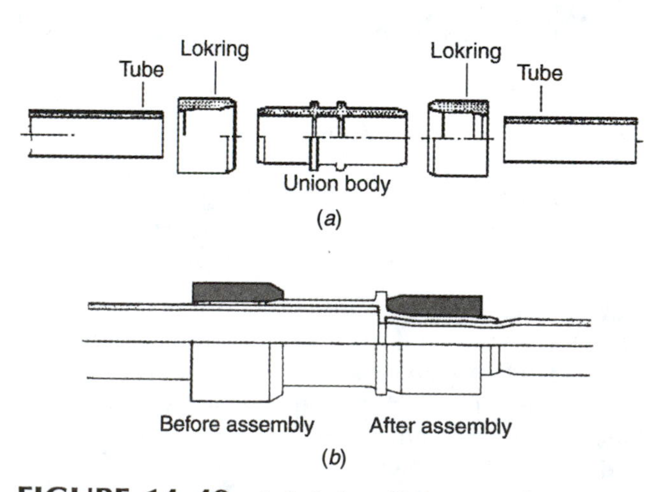

FIGURE 14–48 A Lokring fitting can be used to splice two metal lines or a fitting to a line. (*Courtesy of American Lokring*)

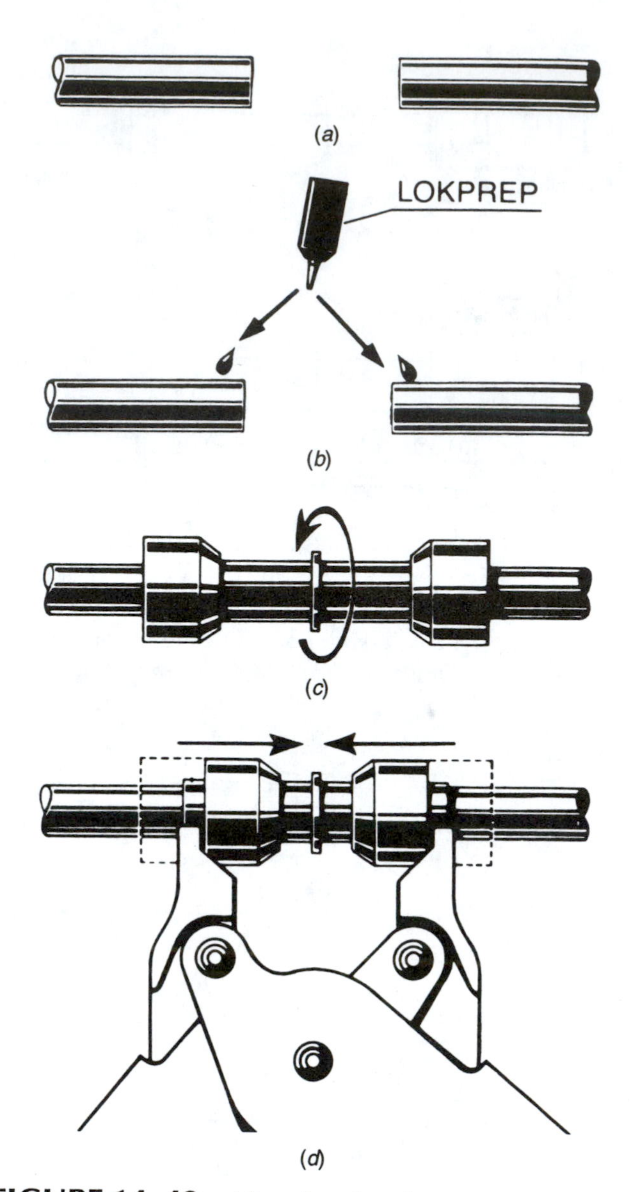

FIGURE 14–49 After cleaning the metal lines (*a*), a few drops of Lokprep are applied to the tubes (*b*), the tube ends are inserted into the union body (*c*), and Lokrings are squeezed over the union body using special pliers (*d*). (*Courtesy of American Lokring*)

connection, or tee connection (Figure 14–46). Another important fitting splices hose, metal tubing, or metal tubing to a hose. This fitting can combine a compression fitting with a barb or beadlock connector (Figure 14–47). Another splicing method is the Lokring that allows two metal lines to be joined using a rather simple and quick procedure (Figure 14–48). It allows you to join a fitting to a metal line or to cut out a faulty part of a metal line and rejoin the ends. This system requires a special handheld tool to compress the Lokrings (Figure 14–49).

It is also possible to reuse the original metal ends when replacing the hose portion. This is done by care-

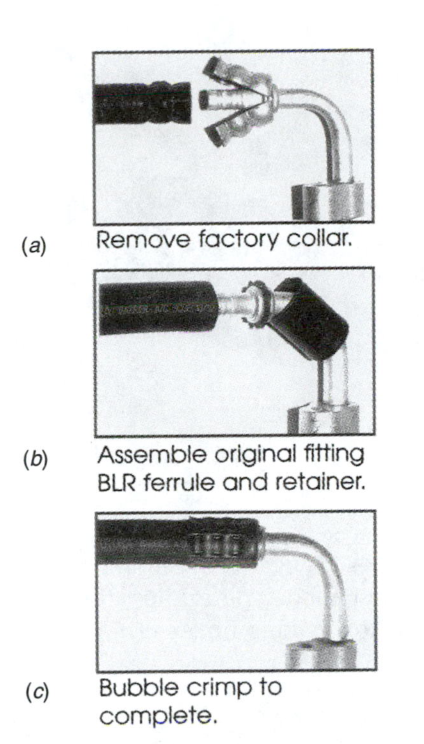

FIGURE 14–50 A damaged hose can be removed from the fitting by cutting the ferrule and prying it open (*a*). Then a special ferrule and retaining ring can be positioned on the fitting (*b*). The hose is then installed and the new ferrule is crimped to lock it in place (*c*). *(Courtesy of BLR Enterprises)*

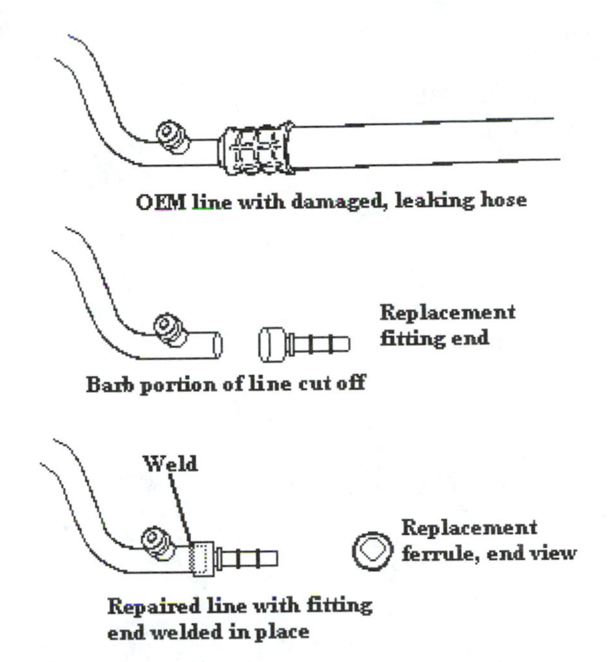

FIGURE 14–51 A line can be repaired by cutting the hose portion off and welding a new end fitting onto it. The replacement ferrule has an oversize, offset ferrule that can be slid into position to secure the new hose onto the fitting.

fully slicing the metal shell on two sides, being very careful not to cut into the inner metal fitting. This allows spreading the metal shell and sliding the old hose off the fitting. Two special replacement shell designs are available. One design passes over the crimp for the original shell and a retainer ring fits into the groove in the crimp to retain the shell (Figure 14–50). The other design uses a large, offset hole in the shell; it is positioned into the fitting groove so the smaller portion of the hole fits into the groove. After placing either type of shell into position, the replacement hose is slid in place and the shell is crimped to lock the hose in place.

✓SERVICE TIP

A special repair fitting that is designed to save an OEM fitting is available (Figure 14–51). This fitting is the hose portion with a groove for a special ferrule design. The hose end of the OEM fitting is cut off, and the repair fitting is welded onto it. A new hose can then be attached to the original fitting.

Refrigerant hose is available in bulk lengths and in both double-braid and barrier-type hoses. Barrier hose has a thin, nonpermeable nylon layer to reduce leakage.

Fittings are secured onto a hose by crimping the metal shell with a special hose crimper; a different crimper is required for each crimp style.

To attach a barb fitting to an R-12 nonbarrier hose using a finger-style crimp, you should:

1. Cut the hose to the proper length with either a hose cutter or a very sharp knife (Figure 14–52).
2. Wet the inside of the hose and the fitting barbs with refrigerant oil.
3. Place the shell of the proper size over the end of the hose and slide the fitting into the hose until the shell is snug between the hose end and the locating bead.
4. Make sure the dies are the proper size and place the fitting in the crimping tool with the shell centered in the dies (Figure 14–53).
5. Tighten the crimp tool bolt until the dies almost touch each other so the shell is completely crimped.

(a) (b)

FIGURE 14–52 Refrigerant hose can be cut with a special cutter (*a*) or a sharp knife (*b*). (a *is courtesy of Four Seasons*)

(a)

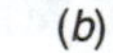

(b)

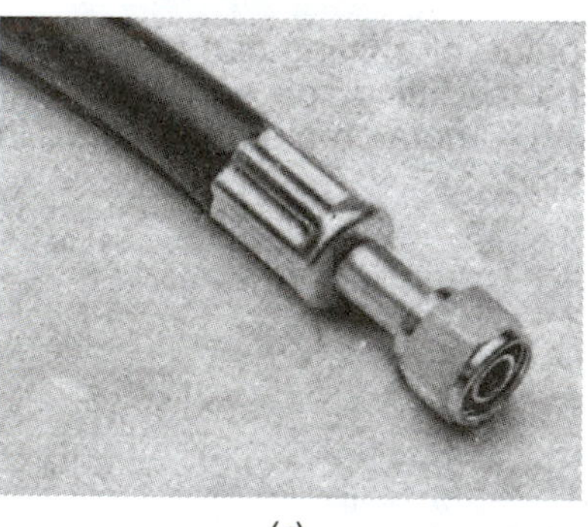

(c)

FIGURE 14–53 When making a finger-style crimp on the metal shell, dies of the correct size are installed in the tool (*a*) and the shell is crimped by tightening the tool drive bolt (*b*). A finished installation is shown in (*c*). (*Courtesy of Four Seasons*)

FIGURE 14–54 A beadlock crimping tool and a set of dies to make bubble-type crimps; the procedure is similar to that for making finger crimps. (*Courtesy of Four Seasons*)

To attach a beadlock fitting to a hose using a bubble-style crimp, you should:

1. Cut the hose to the proper length with either a hose cutter or a very sharp knife.
2. Wet the inside of the hose and the fitting with refrigerant oil.
3. Insert the fitting into the hose until the cut end is visible in the hose-locating hole.
4. Make sure the dies are the proper size and place the fitting in the crimping tool with the shell centered in the dies (Figure 14–54).
5. Tighten the crimp tool bolt until the dies almost touch each other so the shell is completely crimped.

Refrigerant hose repair varies greatly depending on the affected hose, the damage to it, and the parts available. The following describes the replacement of a hose section between an accumulator and a compressor connection. Remember that the liner in a barrier hose can be damaged if used with a barb fitting. The procedures used to attach a fitting to a hose using either a finger- or bubble-style crimp are almost the same; the major difference is the crimping tool.

To repair this refrigerant hose, you should:

1. Recover the refrigerant and remove the hose as described in Section 14.3.2.
2. Determine the fitting type for each end. In this case, we will use a 90° female O-ring fitting at

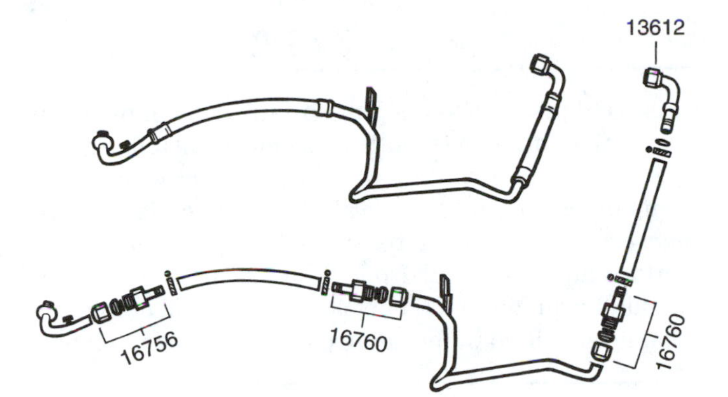

FIGURE 14–55 An OEM hose with three metal portions (top) can be replaced using two new hose sections, one line fitting (13612), and three connectors (16756 and 16760). *(Courtesy of Four Seasons)*

one end and a compression-style fitting at the other (Figure 14–55). Two more compression-style fittings and another hose section are used at the other end of this line.

3. Determine the size of the fitting by measuring the tube OD size right next to the nut (Figure 14–56).
4. Determine the size of the hose by measuring the tube OD size right next to the hose. Check the size charts to see if a jump size (step-up or step-down size) is needed.
5. Determine the best fitting angle.
6. Determine the length of hose needed; cut the hose with a hose cutter or sharp knife.
7. Determine the best location for the compression fitting. Cut the metal tube with a tubing cutter. If you use a hacksaw, be sure to file the rough end so it is smooth and clean out the resulting metal shavings. A compression fitting requires a straight section of metal tube about 1 inch long.
8. Clean the cut end and slide the nut and compression rings over the tube.
9. Wet the compression rings and tube end with refrigerant oil and slide the fitting in place.
10. Finger tighten the nut onto the compression fitting and, using two wrenches, tighten the nut to the correct torque.
11. Crimp the shell, as previously described, to secure the hose to the fitting.
12. Repeat steps 10 and 11 to make the other hose connections.
13. Replace the hose using new O-rings.
14. Evacuate and recharge the system and check for leaks.

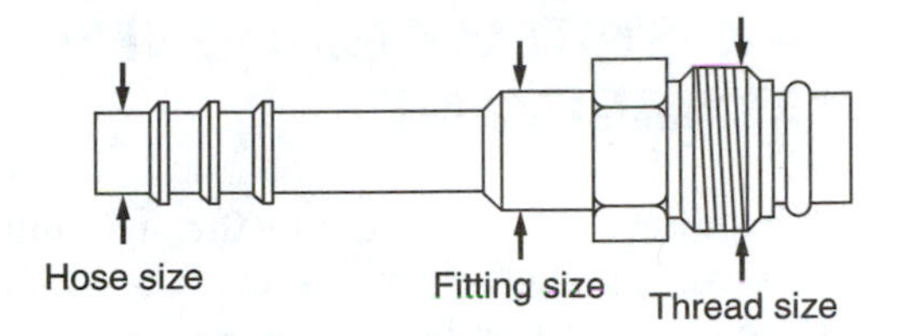

FIGURE 14–56 A fitting size is determined by measuring the tubing diameter next to the nut the hose size is determined by measuring the diameter of the hose pilot (not the raised barbs), and the nut size is determined by measuring the diameter of the threads.

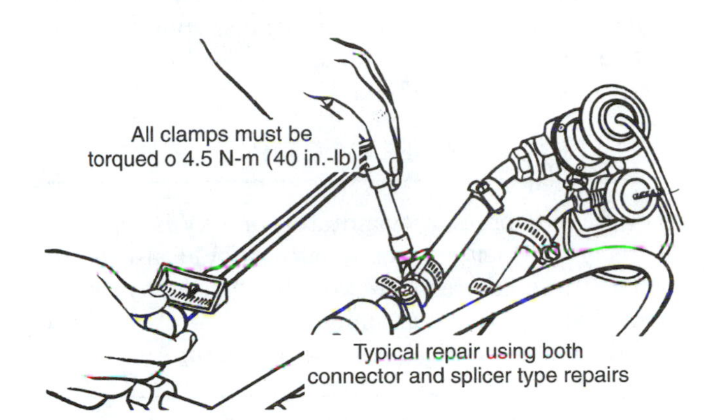

FIGURE 14–57 If making an R-12 hose repair secured by a screw clamp, the clamp should be tightened to the correct torque. As the clamp is tightened, the rubber hose will squeeze up to fill the slots in the clamp. *(Reprinted with permission of General Motors Corporation)*

✓SERVICE TIP

As mentioned, screw-type clamps are no longer a recommended hose-clamping method. If you must use them, place them over the hose and tighten the clamp to the proper torque. At this tightness, the outer rubber layer of the hose will be squeezed outward into the slots of the clamp (Figure 14–57).

✓SERVICE TIP

Many A/C repair shops have the ability to pressure test hose assemblies that they have made up. This allows them to check the integrity of the new fittings.

14.4 A/C SYSTEM COMPONENT REPLACEMENT

If faulty, a major A/C component (the accumulator, condenser, evaporator, OT, receiver–drier, or TXV) is repaired by removing and replacing it with a new one. Except for an OT, which requires a puller and some TXVs, the R&R operations are very similar. Replacement with most of these components is relatively easy because there is rather good access, except when working with evaporators. In most cases, getting the evaporator case out of the car is tedious and time consuming, sometimes requiring the vehicle or evaporator case to be cut. Many technicians will not change an evaporator without consulting a service manual.

✓SERVICE TIP

When the compressor, evaporator, or TXV is replaced, the receiver–drier or accumulator should also be replaced. These failures were possibly caused by system contamination. This contamination will probably have loaded the desiccant capacity with moisture.

✓SERVICE TIP

A possible cause for early evaporator failure can be the moisture and acids generated by decaying leaves in the bottom of the evaporator case. If this type of debris is found while replacing an evaporator, it is recommended that a screen be added to the fresh air intake to reduce chances of future problems.

✓SERVICE TIP

It is good practice to keep the plastic caps in place on the new components until just before installation to keep as much moisture out of the system as possible. Remember, too, that removal of an accumulator, condenser, evaporator, or receiver–drier also removes a certain amount of oil from the system, and new oil should be added to the new part. The actual amount is usually specified in the service manual (Figure 14–58).

✓SERVICE TIP

Be aware that it is possible to cause a vehicle's airbags to deploy if the wrong electrical wires are connected. This deployment can possibly cause injury to the technician or bystanders; it will cause an expensive repair. It is recommended that the air bags be deactivated using the manufacturer's recommended procedure before doing any work under the instrument panel.

To remove and replace a major A/C component, you should:

1. Locate and review the repair procedure in a service manual.
2. Recover the refrigerant from the system.
3. Disconnect and cap the refrigerant lines to the component.
4. Disconnect any mounting brackets and wires connected to the component and remove it (Figure 14–59).

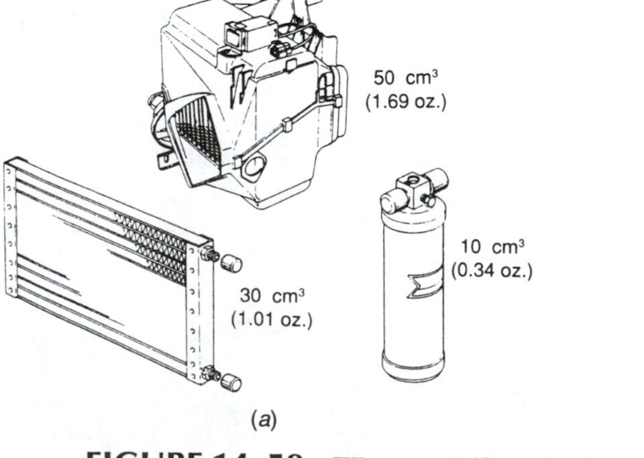

(a)

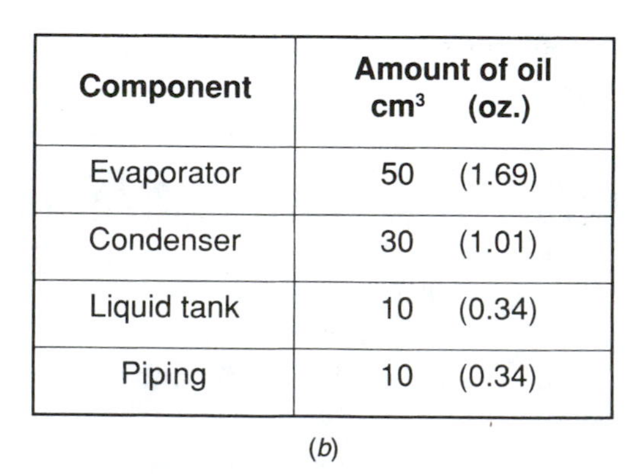

Component	Amount of oil cm³ (oz.)
Evaporator	50 (1.69)
Condenser	30 (1.01)
Liquid tank	10 (0.34)
Piping	10 (0.34)

(b)

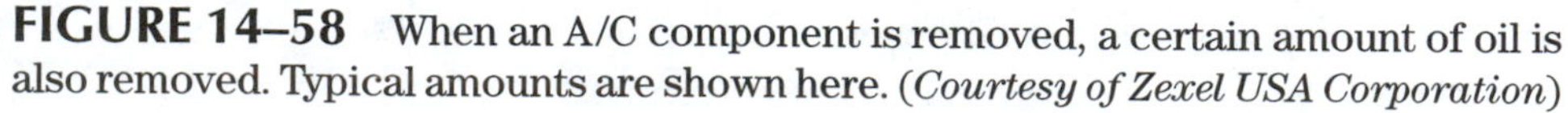

FIGURE 14–58 When an A/C component is removed, a certain amount of oil is also removed. Typical amounts are shown here. *(Courtesy of Zexel USA Corporation)*

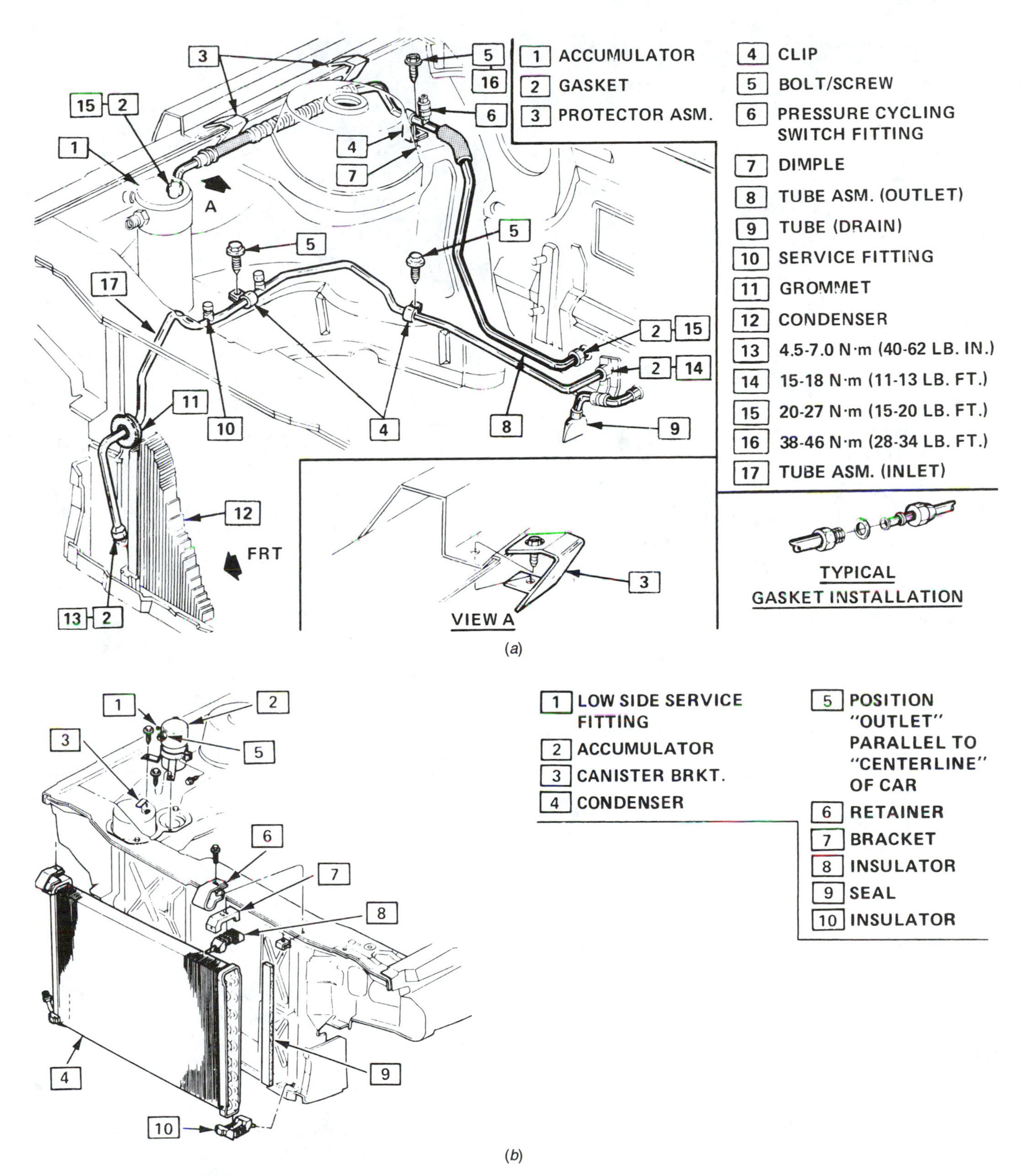

FIGURE 14–59 When an A/C component is removed and replaced, the lines (*a*), fittings, clamps, and brackets (*b*) should be replaced in the proper manner. *(Reprinted with the permission of General Motors Company)*

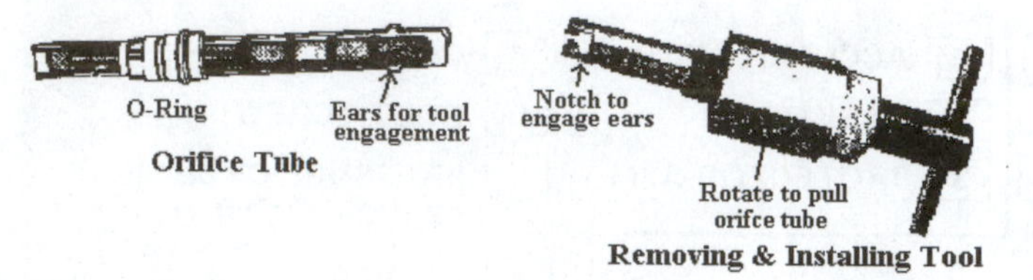

FIGURE 14–60 The removing and installing tool uses a pair of notches to engage the two ears on the orifice tube. Rotating the large nut will pull the orifice tube from the evaporator inlet.

5. Install the new part and attach any mounting brackets and wires.
6. Remove the line caps, pour the proper amount of the correct refrigerant oil into the component or line, and connect the refrigerant lines.
7. Evacuate and recharge the system and test for leaks.

A special puller that attaches to the OT is normally used to pull the OT out of the evaporator inlet tube and to install the new one. This helps ensure it is installed in the right direction (Figure 14–60).

✓SERVICE TIP

Occasionally an OT sticks and breaks apart; special tools to pull broken OTs are available. A stuck OT can sometimes be loosened by heating (preferably with a heat gun) the evaporator inlet tube, but this requires great care. Another method that sometimes works is applying air pressure to the evaporator outlet in an attempt to blow the OT out of the inlet. (Before trying this, a shop cloth should be wrapped around the inlet tube to catch the flying OT.)

✓SERVICE TIP

If the end of the OT breaks off, thread a long #8 or #10 drywall screw into the OT, heat the outer tube, and, using pliers or vise grips, pull outward on the screw to remove the OT.

If these methods fail, a service kit to replace the evaporator inlet tube is available. This kit is used by cutting off the evaporator inlet with the stuck OT still inside and installing a replacement tube and a new OT (Figure 14–61). A similar kit is used to replace an in-line OT. In this case, locate the indentations in the liquid line, which show where the OT is, and cut out that line section. The kit includes a new line section and OT and two compression fittings to connect it to the liquid line (Figure 14–62).

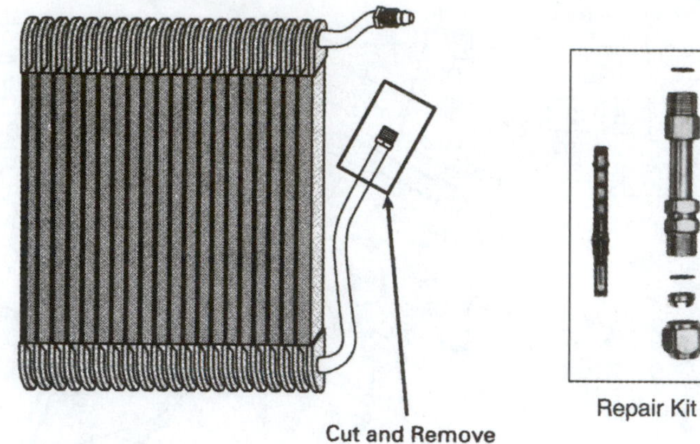

FIGURE 14–61 If an OT cannot be removed or the evaporator inlet is damaged, the inlet tube can be cut off and a replacement kit can be installed. (*Courtesy of Four Seasons*)

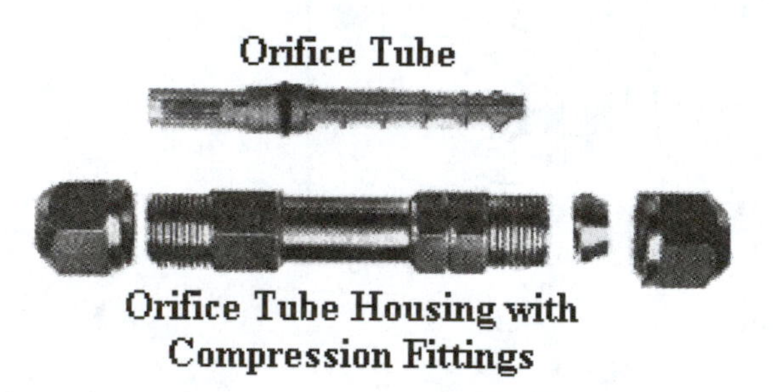

FIGURE 14–62 Some orifice tubes are mounted into a section of the liquid line. This section must be cut out of the line, and a replacement housing section with the orifice tube is installed into the line.

When replacing a TXV (other than a block-type TXV), the thermal bulb must be securely attached to the evaporator outlet tube. This area must be clean to ensure good heat transfer. After attaching the thermal bulb, it must be wrapped with insulating refrigerant tape. This thick, pliable tape is used to keep outside heat from reaching the thermal bulb (Figure 14–63).

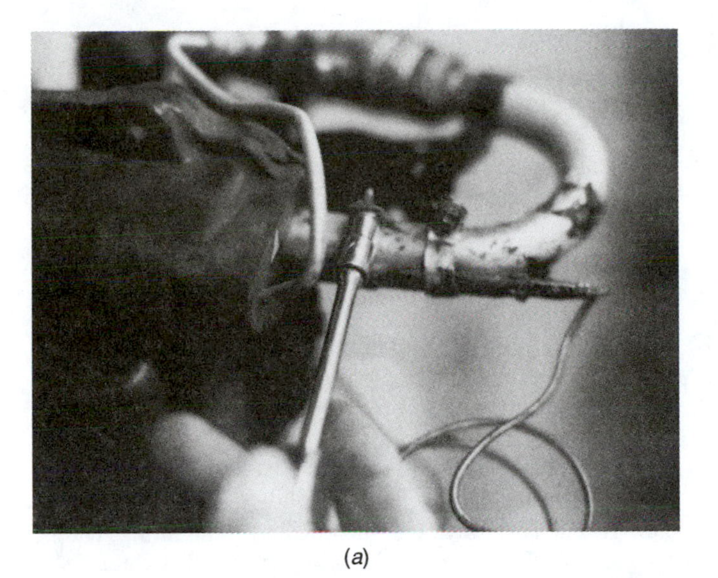

(a)

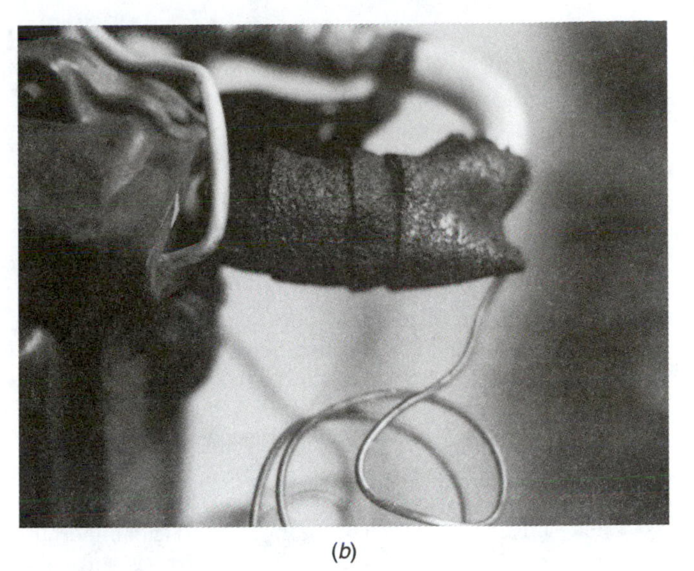

(b)

FIGURE 14–63 When a TXV is installed, the thermal bulb should be clamped securely to the evaporator outlet (*a*) and then wrapped with insulating tape (*b*).

✓REAL WORLD FIX

The 1994 Jeep Grand Cherokee (103,000 miles) came in with no refrigerant because of a faulty hose. The hose and receiver–drier were replaced and the system recharged. It worked well but came back in a few days with a complaint of high engine temperature and poor A/C. The fan clutch was replaced, and this fixed the hot engine problem. The A/C was okay during normal driving but did not cool very well during stop and go conditions. The high side pressure was good, but the low side pressure was excessive.

Fix: The technician found out that the evaporator had been replaced a year ago, and an inspection of the evaporator revealed a poor quality unit that was too small for the case. Replacement of the evaporator core with an OEM comparable unit fixed this problem.

CHAPTER QUIZ

These questions help you study this chapter. Enter the proper word(s) in the blanks to complete each statement.

1. When installing a new or rebuilt compressor, it is a good practice to read the ________ ________ to determine the ________ and ________ of oil to put into the replacement compressor.
2. Always check the ________ ________ whenever a clutch is installed, and ________ it to the proper clearance if necessary.
3. When a clutch slips, it will ________ in a short period of time and probably ________ the bearing.
4. A lip-type compressor shaft seal must have a shaft surface that is ________ and not ________.
5. As you install a shaft seal cartridge, you should ________ ________ that the driving flats of the seal engage the ________ on the ________.
6. It is a good practice to test a used replacement compressor to determine there are no ________ or ________ leaks and the shaft ________ smoothly with no excessive ________.
7. A replacement O-ring must have the correct ________ and ________ and be made from the proper ________ of ________.
8. When an A/C fitting is replaced, it must be tightened to the correct ________ of ________.
9. Leaks at a spring lock coupling can often be stopped by using a(n) ________ ________.
10. A replacement fitting for an R-134a hose should be of the ________ style and crimped using a(n) ________ style of crimp.
11. When a new component like an accumulator or evaporator is replaced, the proper amount of ________ should be ________ to that component.
12. When replacing an older-style TXV, be sure to ________ the thermal bulb to the evaporator tailpipe and wrap it ________ tape.

REVIEW QUESTIONS

These questions allow you to check what you have learned. Select the answer that correctly completes each statement.

1. Normal compressor repairs include replacement of the
   a. shaft seal.
   b. clutch coil.
   c. reed valves.
   d. All of these
2. Technician A says that when the compressor is replaced, the oil level should be checked. Technician B says that the oil from the old compressor can be reused. Who is correct?
   a. A only
   b. B only
   c. Both A and B
   d. Neither A nor B
3. Technician A says that as the clutch drive hub is replaced, the air gap should be checked at three different places. Technician B says that the air gap is not important as long as the clutch ohms are correct. Who is correct?
   a. A only
   b. B only
   c. Both A and B
   d. Neither A nor B
4. Technician A says that it is important to position snap rings correctly when replacing clutch parts on a compressor. Technician B says that the whole clutch assembly should be replaced if the pulley bearing fails. Who is correct?
   a. A only
   b. B only
   c. Both A and B
   d. Neither A nor B
5. The clutch assembly must be removed and replaced to
   a. repair or replace one of the clutch parts.
   b. get access to the compressor shaft seal.
   c. remove and replace the compressor.
   d. Both A and B
6. Technician A says that shaft seal parts should be well lubricated as they are installed. Technician B says that special tools are often required when removing or replacing seal parts. Who is correct?
   a. A only
   b. B only
   c. Both A and B
   d. Neither A nor B
7. Oil level in an A/C compressor can be checked at any time.
   a. True
   b. False
8. When a compressor is replaced, it is a good practice to also replace the
   a. TXV.
   b. OT.
   c. Both A and B
   d. receiver–drier or accumulator.
9. Technician A says that you must install the compressor and charge the system to check for seal leaks. Technician B says that if you apply pressure to a compressor discharge port, pressure will show up immediately at the suction port. Who is correct?
   a. A only
   b. B only
   c. Both A and B
   d. Neither A nor B
10. Technician A says that most leaks at line fittings can be stopped by tightening the fitting nut a little tighter. Technician B says that line fitting O-rings should be properly positioned and lubricated with a drop of refrigerant oil as they are installed. Who is correct?
    a. A only
    b. B only
    c. Both A and B
    d. Neither A nor B
11. A new hose can be made up in the shop by installing the correct ends onto the right size, type, and length of hose.
    a. True
    b. False
12. Technician A says that new refrigerant oil should be added to new components such as accumulators or condensers as they are installed. Technician B says that two wrenches should be used when tightening most line fittings. Who is correct?
    a. A only
    b. B only
    c. Both A and B
    d. Neither A nor B

CHAPTER

Cooling System Theory

LEARNING OBJECTIVES

After completing this chapter, you should:

- Understand the cooling system's role in maintaining proper engine temperature.
- Realize the effects that overcooling and overheating have on an engine.
- Understand what the parts of the cooling system do and how they operate.
- Understand what coolant is and what can happen if it is not properly maintained.

TERMS TO LEARN

coolant
coolant recovery reservoir (CRR)
core hole plug
crossflow
downflow
ethylene glycol (EG)
extended life
fan clutch
freeze plug
hose
organic acid (OA)
pH buffer
pressure cap
propylene glycol (PG)
reverse flow
serpentine belt
soft plug
thermostat
V-ribbed belt
water jacket
water pump

15.1 INTRODUCTION

Automotive engines use a cooling system to remove excess combustion heat. Modern cooling systems are designed to maintain an even temperature of about 180 to 230°F (82 to 113°C); this is at or above the boiling point of water. A coolant mixture of water and antifreeze is used and, because it is under pressure, the coolant has a still higher boiling point.

An engine will be damaged if the operating temperature gets too high, above the coolant boiling point, which is about 260°F (125°C). This damage can show as preignition and detonation, a warped or cracked cylinder head or block, or piston-to-cylinder-wall scuffing.

An engine performs poorly if it operates at too low a temperature, below the specified thermostat setting. Cold engine problems include poor oil flow, sludge formation in the oil, and poor fuel vaporization. Also, computer-controlled engines and transmissions will not go into normal operating modes if they are too cold; this causes poor performance and fuel mileage. Some vehicles set a diagnostic trouble code (DTC), and turn on the malfunction indicator lamp (MIL) if it takes too long to warm up to operating temperature.

Most modern vehicles are liquid cooled: Liquid coolant is used to transfer heat to the airstream. Many small gas engines and motorcycles are air cooled. Air is

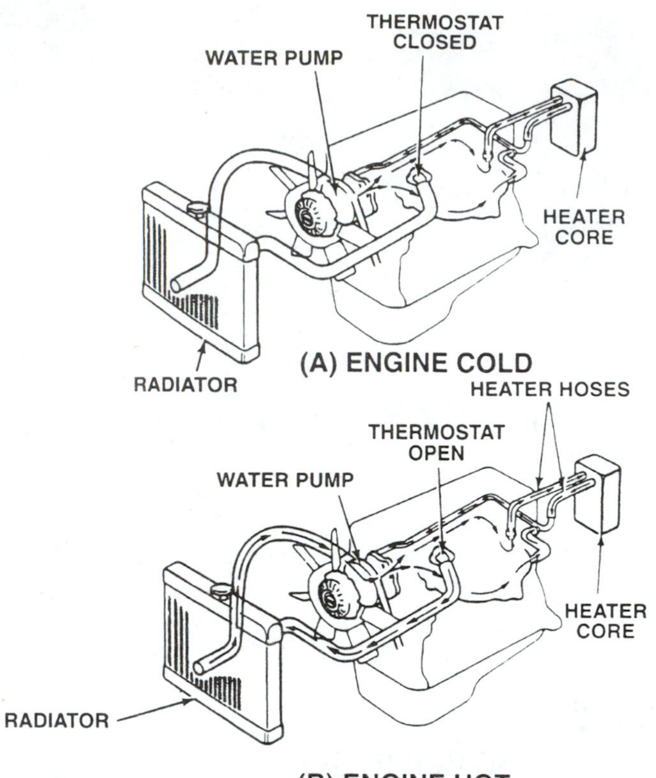

FIGURE 15–1 The major cooling system components; the water pump and water jackets are not shown. *(Reprinted with permission of General Motors Corporation)*

FIGURE 15–2 Water jackets are chambers that surround the cylinders and head of the combustion chamber.

blown across fins attached to the cylinders and heads to remove excess heat. Air-cooled engines are generally simpler and lighter weight, but liquid cooling provides much better temperature control over the cylinders and combustion chambers.

An automotive cooling system consists of the engine's water jackets, thermostat, water pump, radiator, fan, radiator hoses, and coolant. When an engine is at operating temperature on a warm day, the coolant circulates through the water jackets and radiator. The excess heat absorbed by the coolant in the water jackets is passed on to the air flowing through the radiator. When the engine is below operating temperature, the thermostat blocks the flow to the radiator, and the coolant circulates through the water jackets to raise the temperature evenly to an efficient temperature. On cool days when the engine is at operating temperature, the thermostat modulates the flow and allows just enough circulation through the radiator to remove excess heat and maintain an efficient temperature in the engine. On warm days, the thermostat is probably wide open (Figure 15–1).

✓PROBLEM SOLVING

The engine in your friend's vehicle would not run so he pulled the cylinder head to see what was wrong. Two exhaust valves were badly burned, and the head was warped so coolant leaked into the combustion chambers. He has decided to install a rebuilt head. What do you think caused this problem? What checks should he make before he completes this job?

15.2 WATER JACKETS

When an engine cylinder block and head are cast, cavities called **water jackets** are formed around the cylinder walls and combustion chambers. These water jackets allow coolant to circulate around the very hot areas, including the exhaust valve seats, as well as the relatively cooler areas of the lower cylinders. The coolant absorbs heat from the hot areas and transfers this heat to the colder areas in the engine or radiator (Figure 15–2).

The engine block is cast by pouring molten iron or aluminum into a mold or form that is the shape of the outside and inside of the engine. The area for water jackets is filled with a sand core; projections or risers from the core extend to the outer mold to hold it in position. These risers leave a hole in the finished casting called a **core hole.** Core holes are normally filled with a cup-shaped metal plug or a threaded pipe plug. These cup plugs are commonly called **freeze plugs** or **soft plugs**. (Early motorists and mechanics thought that these plugs were put in the engine as safety plugs and that the plugs would pop out if the engine froze to protect the block from cracking. This idea is wrong; **soft plug** and **core hole plug** are better names.) The side of the engine block also includes a coolant drain plug that enters into the bottom of the water jacket (Figure 15–3).

15.3 THERMOSTAT

The thermostat is a temperature-controlled coolant valve. In most engines, it is located at the upper radiator hose connector, which forms the thermostat housing. In a few engines the thermostat is located at the lower radiator hose or inlet connection (Figure 15–4).

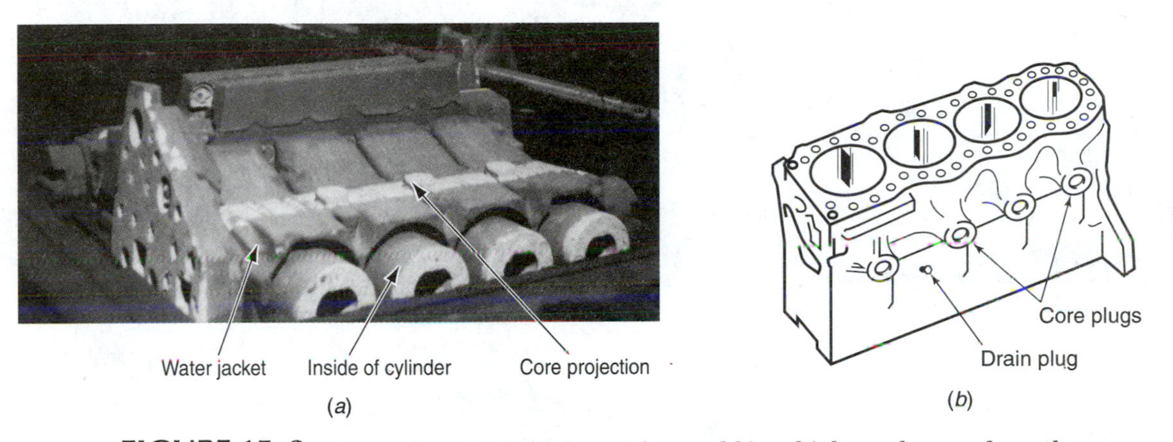

FIGURE 15–3 A cylinder block (*a*) is cast in a mold in which sand cores form the cylinders and water jackets (*b*). Molten cast iron fills the voids in the mold to form the outside of the block and the cylinder walls. (a *is Courtesy of Ashland Chemical*)

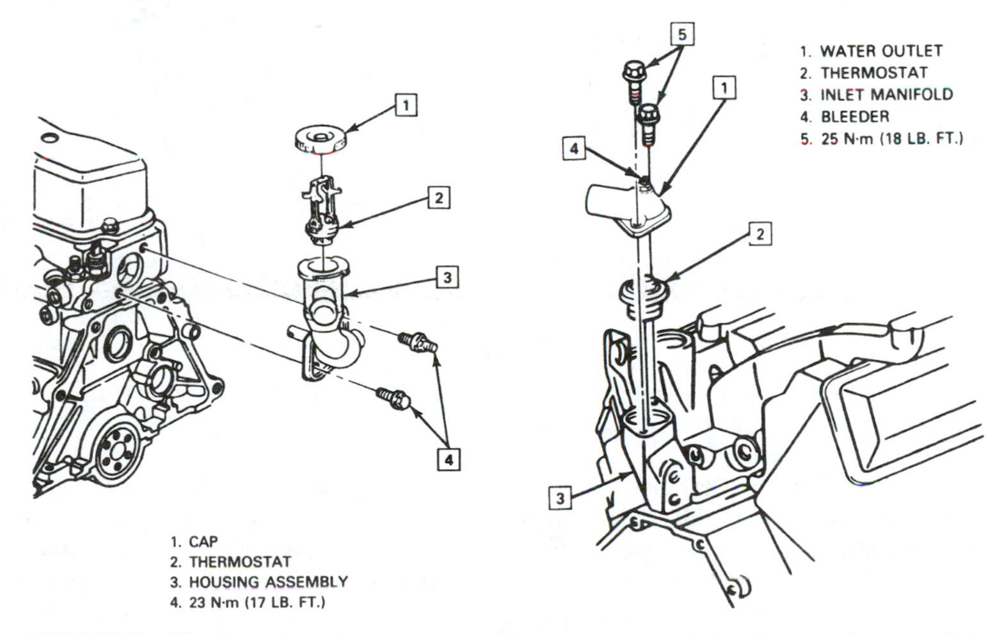

FIGURE 15–4 Most thermostats (2) are mounted in the coolant outlet. (*Reprinted with permission of General Motor Corporation*)

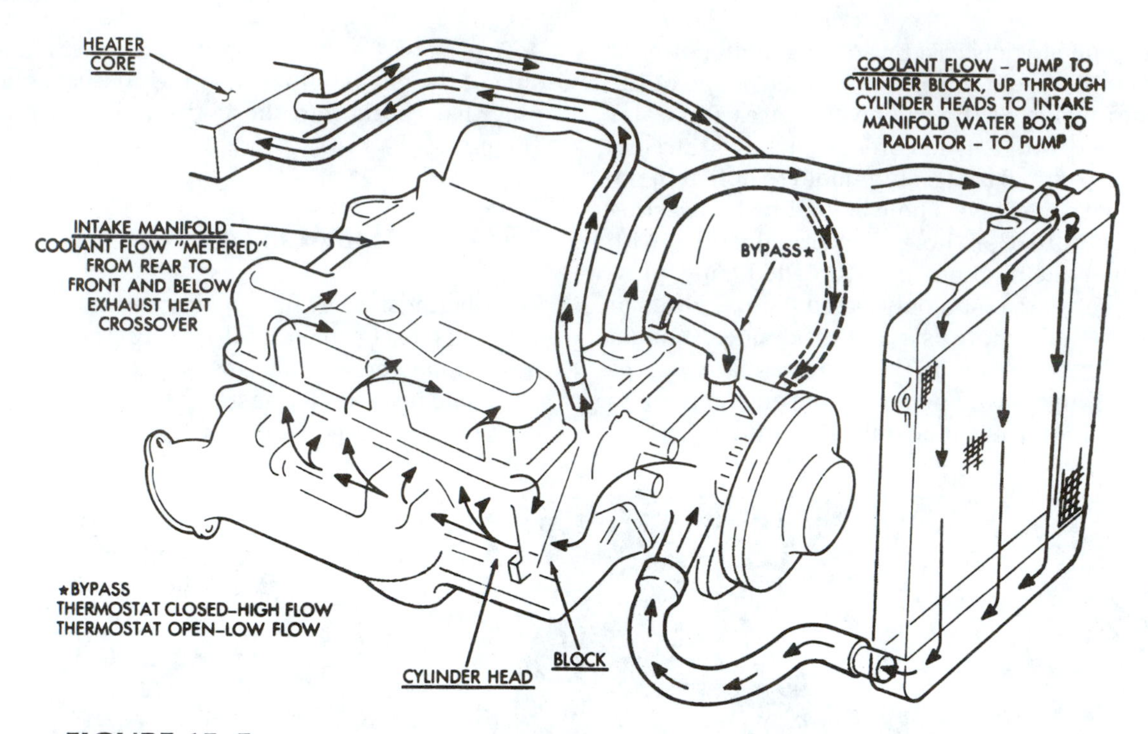

FIGURE 15–5 In most engines, a bypass hose allows coolant to circulate back to the water pump when the thermostat is closed. *(Courtesy of DaimlerChrysler Corporation)*

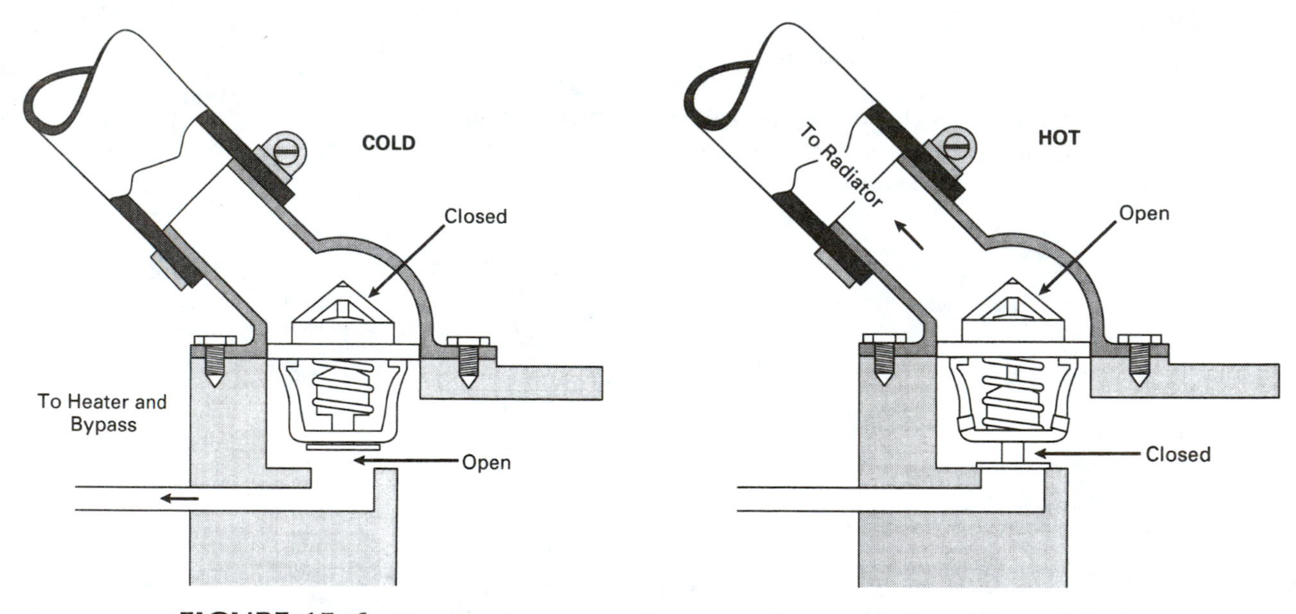

FIGURE 15–6 Some systems use a three-way thermostat that closes off the bypass when the thermostat opens.

The two major functions of a thermostat are to speed up engine warm-up and to regulate operating temperature. The thermostat is closed when the coolant is cold; it opens when the coolant warms up to a specific temperature. Most modern vehicles use a 180, 190, or 195°F (82, 88, or 91°C) thermostat. A thermostat begins to open within a few degrees of the rating and should be fully open about 20°F (11°C) higher. The operating temperature point is often stamped on the thermostat.

A bypass is located right next to the thermostat. During cold engine operation while the thermostat is closed, coolant circulates through this bypass to the water pump and back to the water jackets (Figure 15–5). This circulation warms up the engine parts evenly and also ensures that the thermostat is warmed up by the coolant. Some engines use a three-way thermostat; this design shuts off the bypass when the thermostat opens, and it directs all of the warm coolant to the radiator (Figure 15–6).

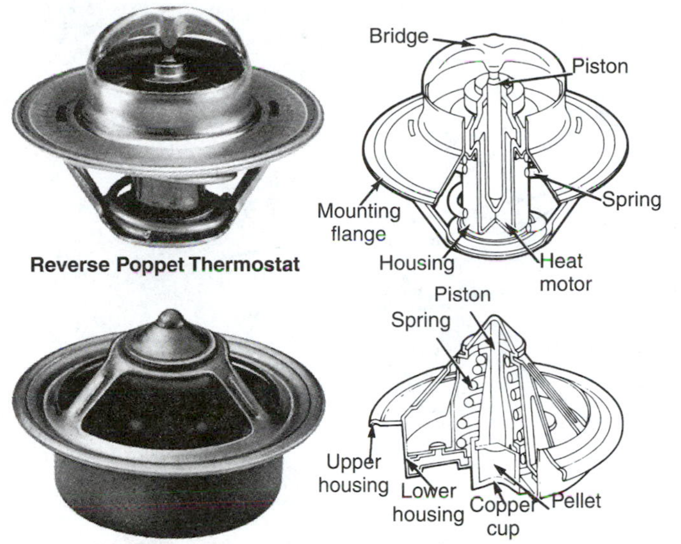

FIGURE 15–7 When the heat motor of the thermostat reaches the correct temperature, it pushes the piston outward and opens the thermostat. (*Courtesy of Stant Manufacturing*)

Prior to the 1950s, the two common thermostat heat ranges were 160 and 180°F. The 160°F thermostat was called a **winter thermostat** because 160°F was below the boiling point of the alcohol-based antifreeze used in those days. Alcohol antifreeze was called "temporary" because it was used only for winter operation; the alcohol would boil away if used in the summer. In the spring, when the days and nights warmed up, the motorist drained the antifreeze, changed to a 180°F **summer thermostat**, and refilled the system with water. Those days are long gone, but for some reason, the 160°F thermostat is still quite popular. A 160°F thermostat is too cold for modern engines.

The valve in most thermostats is closed by a spring and is opened by a wax motor or pellet. When the wax warms up and melts, it expands and compresses the rubber diaphragm to push away from the stemlike piston to open the valve. The motor operates a valve of either reverse-poppet or balanced-sleeve design. This action is quite reliable and trouble free. However, a thermostat can fail, and if it fails, it will stick in either the open or closed position. If it sticks open, the engine will have a very long, slow warm-up period with delayed heater operation. If it sticks closed, the engine will overheat (Figure 15–7). On computer-controlled engines, replacement thermostats must have the same temperature rating as the original.

Many thermostats have a bleed notch or hole to prevent air lock as the system is filled. A **jiggle pin** is often placed in the hole to prevent plugging.

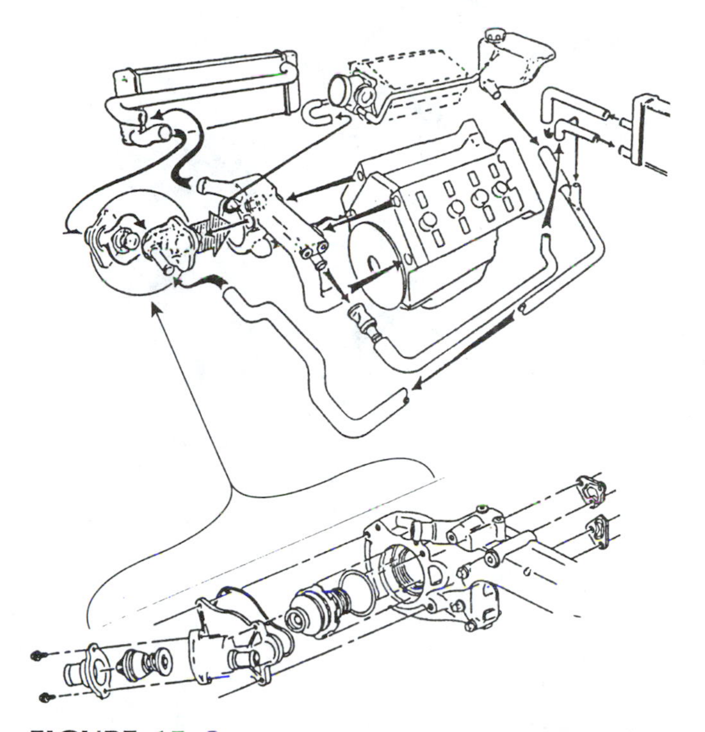

FIGURE 15–8 An inlet-side thermostat. This thermostat controls the coolant flow into the water pump from either the radiator or the engine outlet. (*Reprinted with permission of General Motors Corporation*)

In a few engines, the thermostat is located at the engine inlet to reduce thermal shock. When an engine operates on a cold day, the thermostat can open to allow cold coolant to enter and hot coolant to leave the engine. The thermostat recloses when the cold coolant reaches it but only after partial engine cooldown has occurred. Inlet thermostat systems prevent thermal shock that can cause warpage, especially with aluminum engine parts (Figure 15–8).

✓ PROBLEM SOLVING

The engine in a friend's vehicle overheats, and she has asked your advice. In trying to cure the problem, her father has removed the thermostat, but the car still overheats. Was this a wise move?

15.4 WATER PUMP

In most vehicles, the water pump is driven by the accessory drive belt, commonly called the **drive belt** or **fan belt**, at the front of the engine. In some modern engines, the water pump is gear driven (Figure 15–9). The water pump is a nonpositive displacement centrifugal pump. Coolant enters at the center of the pump, is

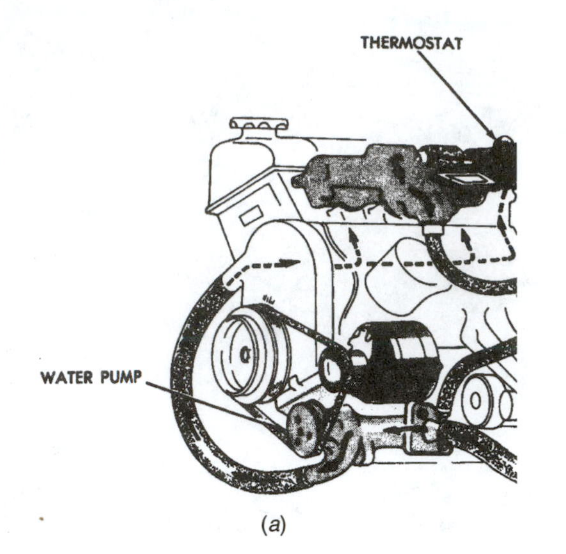

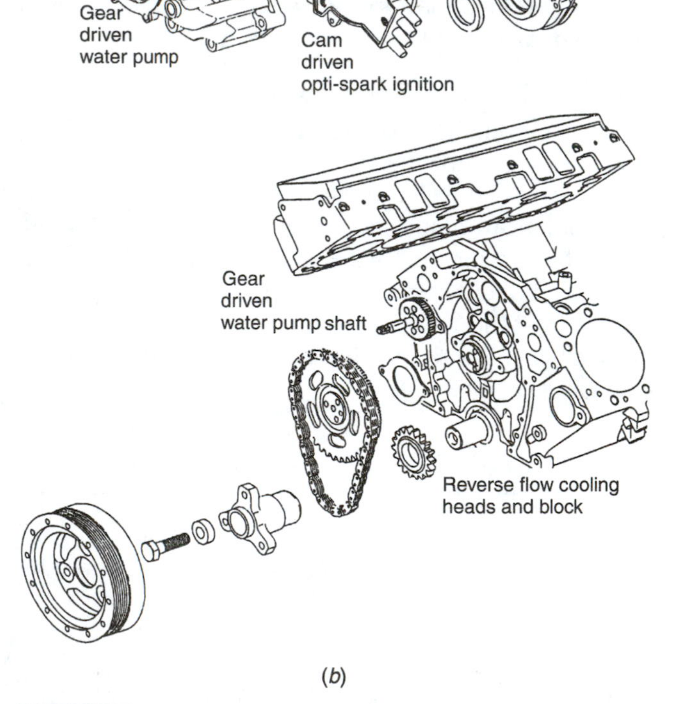

FIGURE 15–9 In most vehicles, the water pump is driven by a drive belt from the crankshaft (*a*). Some engines drive the water pump using a shaft from the camshaft (*b*). *(a is courtesy of DaimlerChrysler Corporation;* b *is reprinted with permission of General Motors Corporation)*

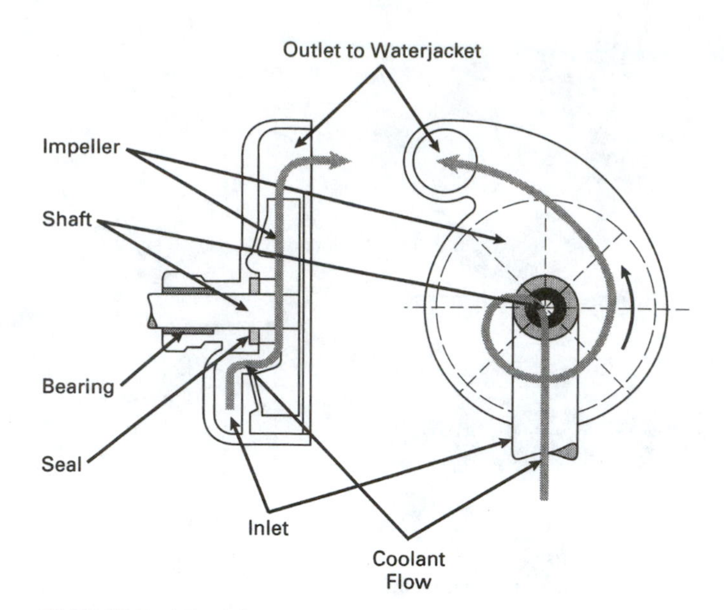

FIGURE 15–10 Coolant enters the water pump at the center of the impeller. From here, the impeller spins the coolant, and centrifugal force sends it through the outlet to the water jackets.

caught by the impeller, and is spun outward to the outlet. The term **coolant pump** is used by some vehicle manufacturers (Figure 15–10).

In most engines the coolant is pumped into the lower water jackets, around the cylinders. It then flows toward the rear of the engine, upward to the cylinder head, and then to the outlet at the front of the head or intake manifold (Figure 15–11). On many transverse-mounted engines, the coolant flows into the engine at the water pump at the right end and outward at the thermostat housing at the other end. The placement and sizes of the passages in the head gasket control the flow to ensure adequate flow to all areas, especially the hottest.

A few engines use a **reverse coolant flow** so the coolant flows in at the cylinder head and out at the bottom of the block. Reverse flow brings the coolest coolant to the exhaust valve area, which is the hottest part of the engine. The cooler combustion chamber temperatures allowed Chevrolet engineers to increase the compression ratio of the LT1 engine, improving both horsepower and fuel mileage. Reverse flow also brings the heated coolant to the lower cylinder walls, which are the coolest. This results in more even temperatures throughout the engine and improved piston and ring wear (Figure 15–12).

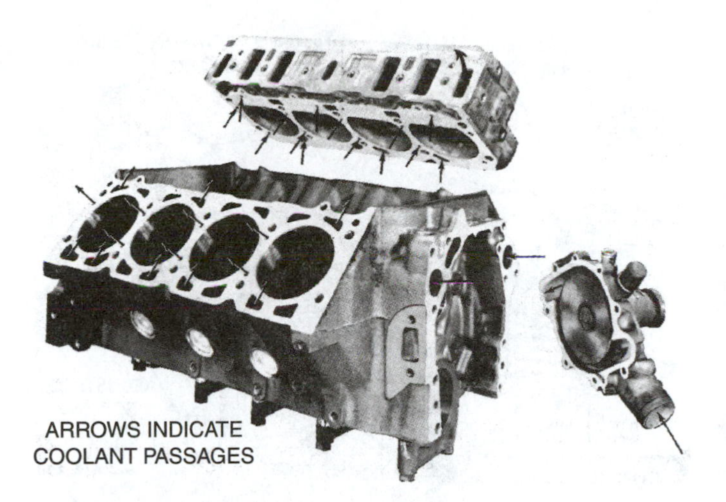

FIGURE 15–11 In most engines, the coolant flows into the block's water jackets, upward to the head, and out the passages at the front of the head. *(Reprinted with permission of General Motors Corporation)*

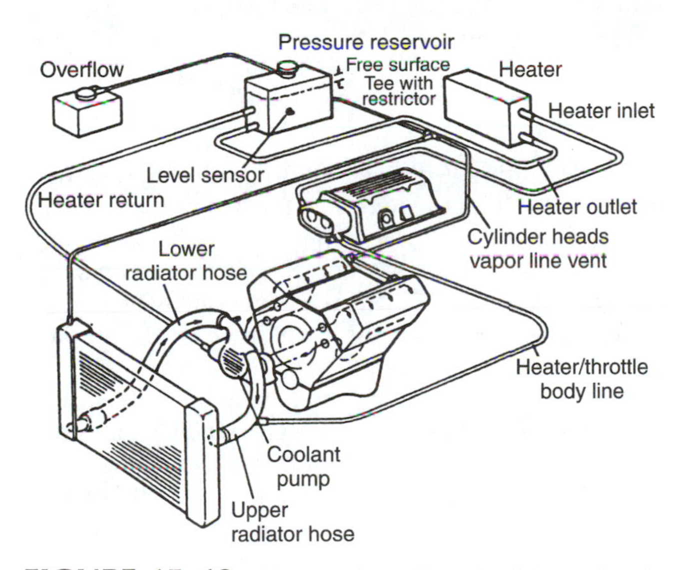

FIGURE 15–12 The coolant flow in this engine is from the coolant pump, upward to the cylinder heads, downward to the cylinder block, and out to the coolant pump. *(Reprinted with permission of General Motors Corporation)*

Water pumps use a metal or plastic impeller with the curved pumping fins; the impeller is mounted on the end of the pump drive shaft. The shaft is usually supported by a permanently lubricated, double-row ball bearing. A carbon face seal is mounted where the shaft enters the pumping chamber, and a weep hole is placed between the seal and the bearing. The weep hole allows any coolant that has seeped through the seal to escape rather than pass into the bearing (Figure 15–13).

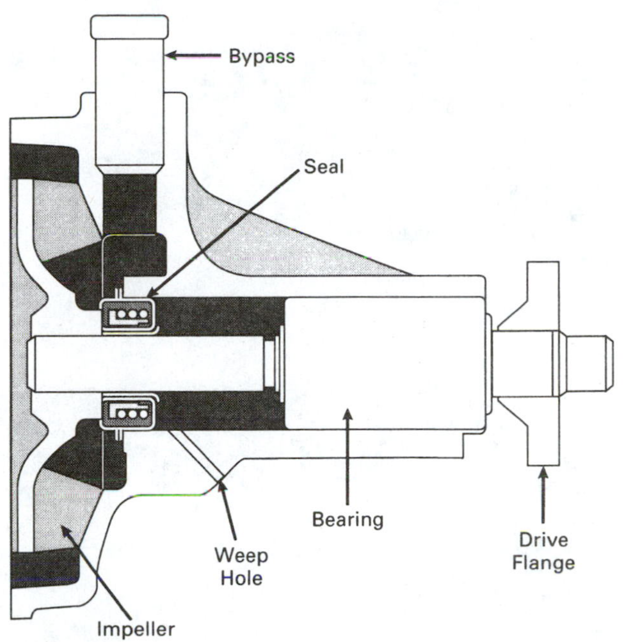

FIGURE 15–13 A cutaway view of a water pump. The inlet brings coolant in the same area as the bypass; note the weep hole to allow any seepage past the seal to escape. *(Reprinted with permission of General Motors Corporation)*

15.4.1 Drive Belt

In the past, most vehicles used one or more **V belts** to drive the water pump and other accessories. Many modern vehicles use a single, wide, **serpentine V-ribbed belt.** A V-ribbed belt is also called a **multi-V** or **poly-V belt** (Figure 15–14).

A V belt gets its name from the V shape of the belt and pulley groove. This belt is designed with a tensile member just under the top cover. This tensile member is a strong layer of cording, which prevents stretching (Figure 15–15). When the belt turns a corner around a pulley, the lower section must shorten, and the sides of the belt grip the pulley because of a slight thickening of the sides. A V belt can only transmit a certain amount of power, so in many cases three or four belts were needed to drive the A/C compressor, air pump, alternator,

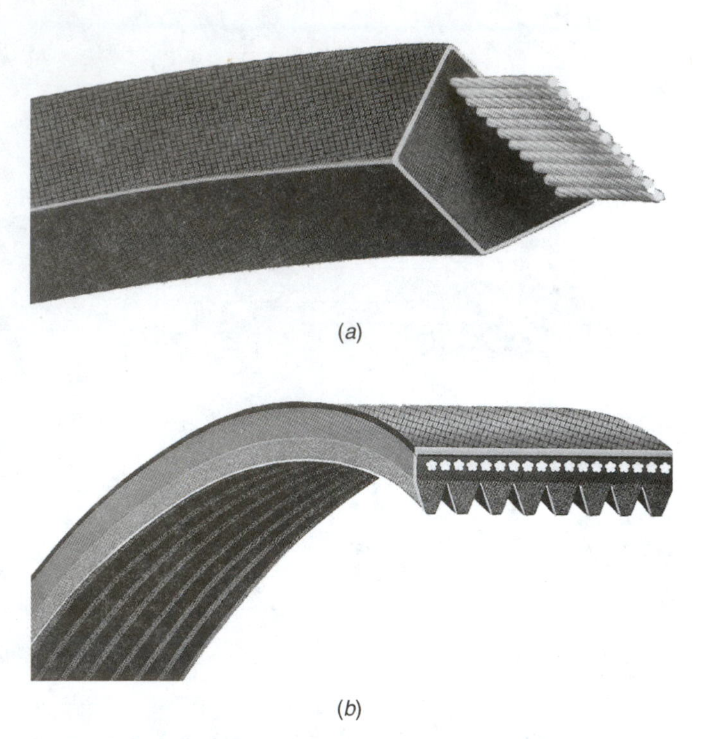

FIGURE 15–14 A V (*a*) and a V-ribbed belt (*b*). The V-ribbed belt has smaller V sections. (*Courtesy of Goodyear Tire & Rubber Company*)

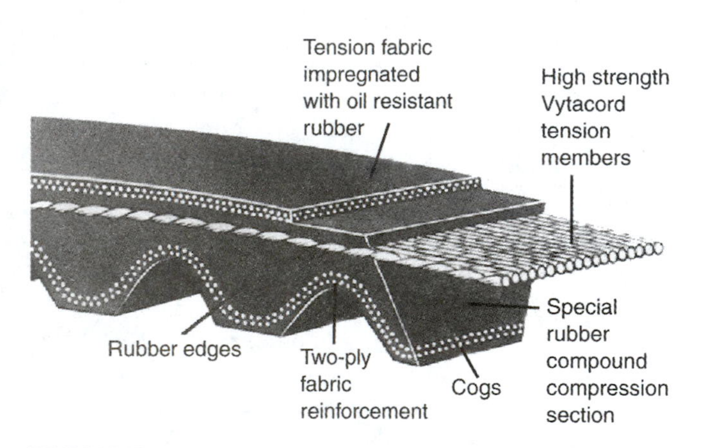

FIGURE 15–15 A V belt is made from different compounds, each having a special purpose. (*Courtesy of Goodyear Tire & Rubber Company*)

power steering pump, and water pump and fan. A loose belt will slip; it usually makes a squealing sound as it slips. This slippage generates heat, which can cause the belt to harden and glaze. One method of adjusting belt tension is to pivot one of the driven members or an idler pulley (Figure 15–16).

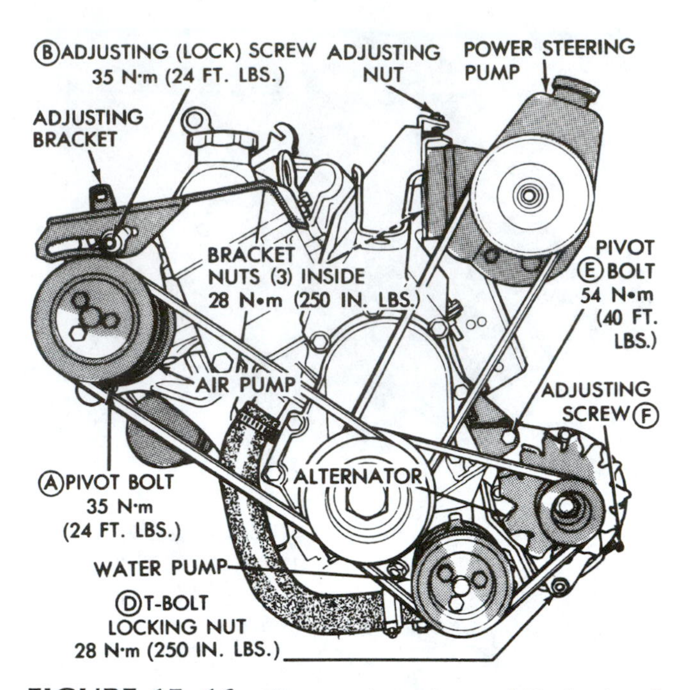

FIGURE 15–16 The tension of most V belts is adjusted by pivoting the driven component. (*Courtesy of Chrysler Corporation*)

A serpentine V-ribbed belt drives through the friction of the belt on the pulley so the belt must wrap further around the pulley. A serpentine belt bends in both directions so that the back of the belt is used on some pulleys (Figure 15–17). The tensioner pulley or an idler pulley is positioned so that it will increase the amount of wrap around certain pulleys. The amount of wrap and its width allow the belt to handle the horsepower requirement. Proper belt tension is usually maintained by a spring-loaded tensioner pulley (Figure 15–18). Some tensioners include markings to show that a belt has stretched too far and should be replaced . It should be noted that the pulleys driven by the back of this belt, usually the water pump and fan, rotate in the opposite or counterclockwise direction; these parts are redesigned to operate properly in this direction (Figure 15–19). Some engines use two V-ribbed belts or a V-ribbed belt and a V belt.

15.5 RADIATOR

The radiator is a heat exchanger that gets rid of excess engine heat. Most radiators are of **fin-and-tube** design.

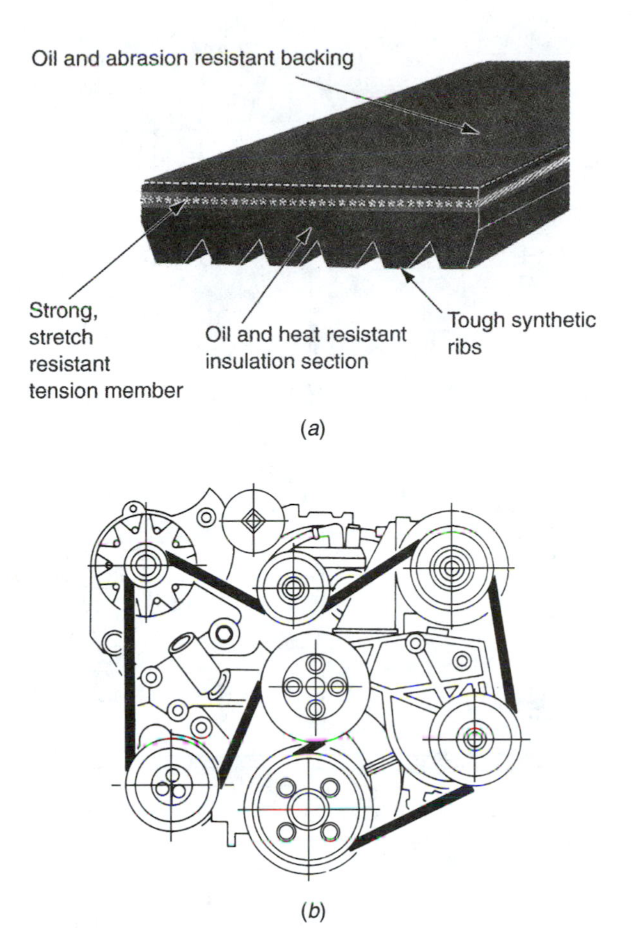

FIGURE 15–17 A cutaway view of a V-ribbed belt showing its internal construction (*a*). (*b*) shows a typical belt routing and demonstrates why this belt is called a serpentine belt. Note that the water pump rotates in an opposite direction because it is driven by the back of the belt. *(Courtesy of Goodyear Tire & Rubber Company)*

FIGURE 15–18 This idler pulley includes an automatic tensioner to maintain the proper tension on the V-ribbed belt.

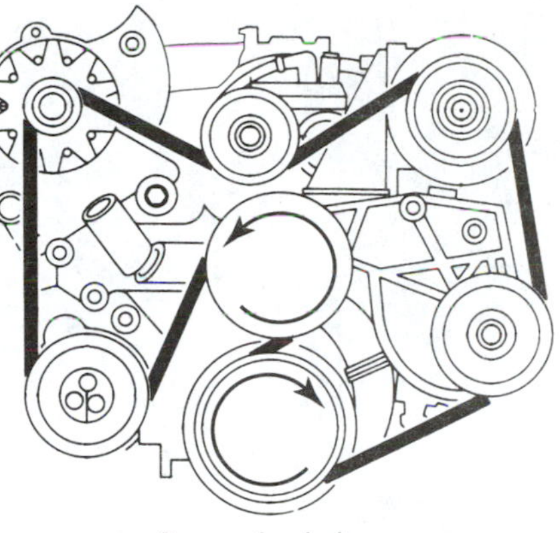

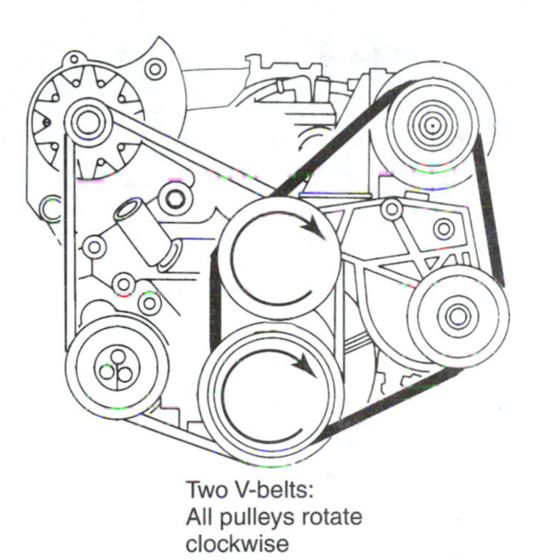

FIGURE 15–19 When a V belt is used, all the driven components rotate in a clockwise direction. With a serpentine belt, those components driven by the back of the belt rotate counterclockwise.

outlet tank. The fins are attached to the tubes to provide the air contact area. The fin-and-tube area is commonly called the **core** (Figure 15–20).

Most older vehicles use **downflow** radiators: The flow direction is from top to bottom. Most newer vehicles use **crossflow** designs in which the flow is from one side to the other. A downflow radiator uses the natural flow direction (cooled water drops while heated water rises). A few engines use this principle in **thermosyphon** cooling systems, which have no water pump. A crossflow radiator produces better cooling in modern vehicles with low, wide hood lines. The tubes in

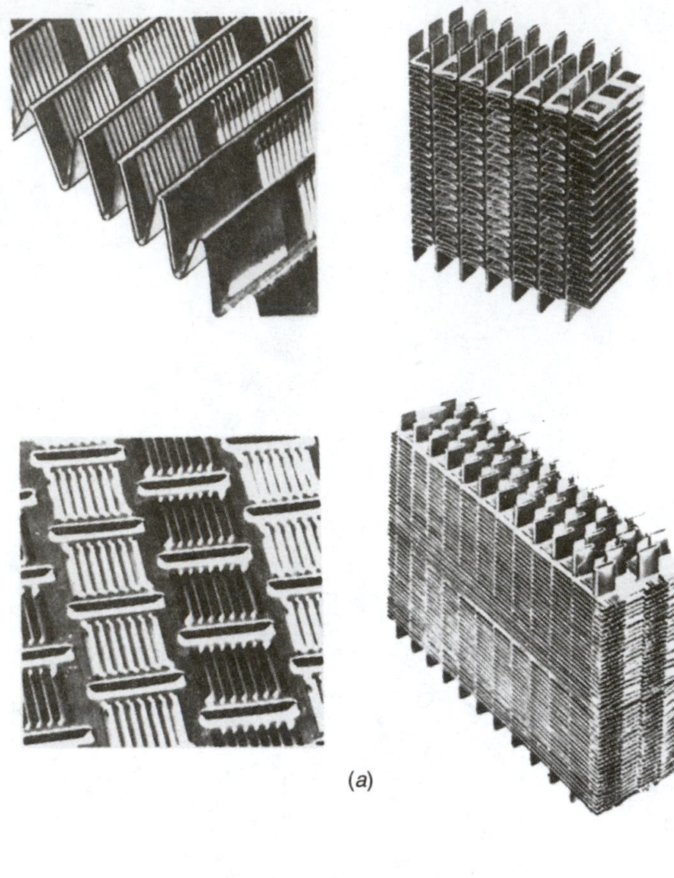

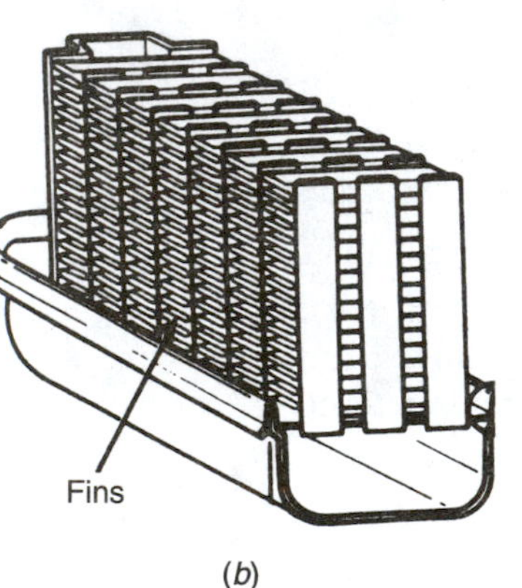

FIGURE 15–20 A radiator core is made up of tubes and fins (*a*) that join to a header and tank (*b*). (*Courtesy of Modine Manufacturing*)

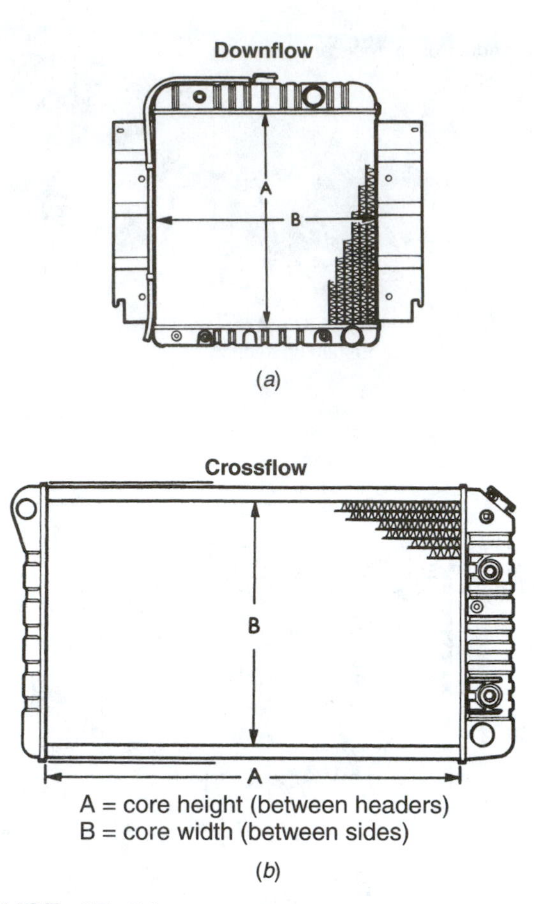

FIGURE 15–21 Most older vehicles use a downflow radiator design (*a*) in which the coolant flows downward. Most newer vehicles use a crossflow design (*b*) in which the coolant flows across the radiator. (*Courtesy of Modine Manufacturing*)

a crossflow radiator can be longer; this provides more time for cooling (Figure 15–21).

Most older radiators are made of copper or brass. They were assembled by soldering the tubes to the headers and soldering the headers to the tanks. Many newer radiators are made of aluminum. The core is vacuum brazed together, and on many, plastic tanks are held onto the core by crimped metal tabs. A specially shaped O-ring is used to seal the tanks to the core headers (Figure 15–22).

The radiator cap is usually mounted in the top tank of a downflow design and in the colder, lower-pressure outlet tank of a crossflow radiator. On vehicles with an automatic transmission, a transmission oil cooler is mounted in the bottom or outlet tank. All newer vehicles use a **pressure cap** and a **coolant recovery reservoir (CRR)**.

15.5.1 Pressure Cap

The pressure cap seals the cooling system so it will hold pressure. Pressure is used to raise the coolant boiling point so the engine can operate at a higher, more efficient temperature. A spring in the cap pushes

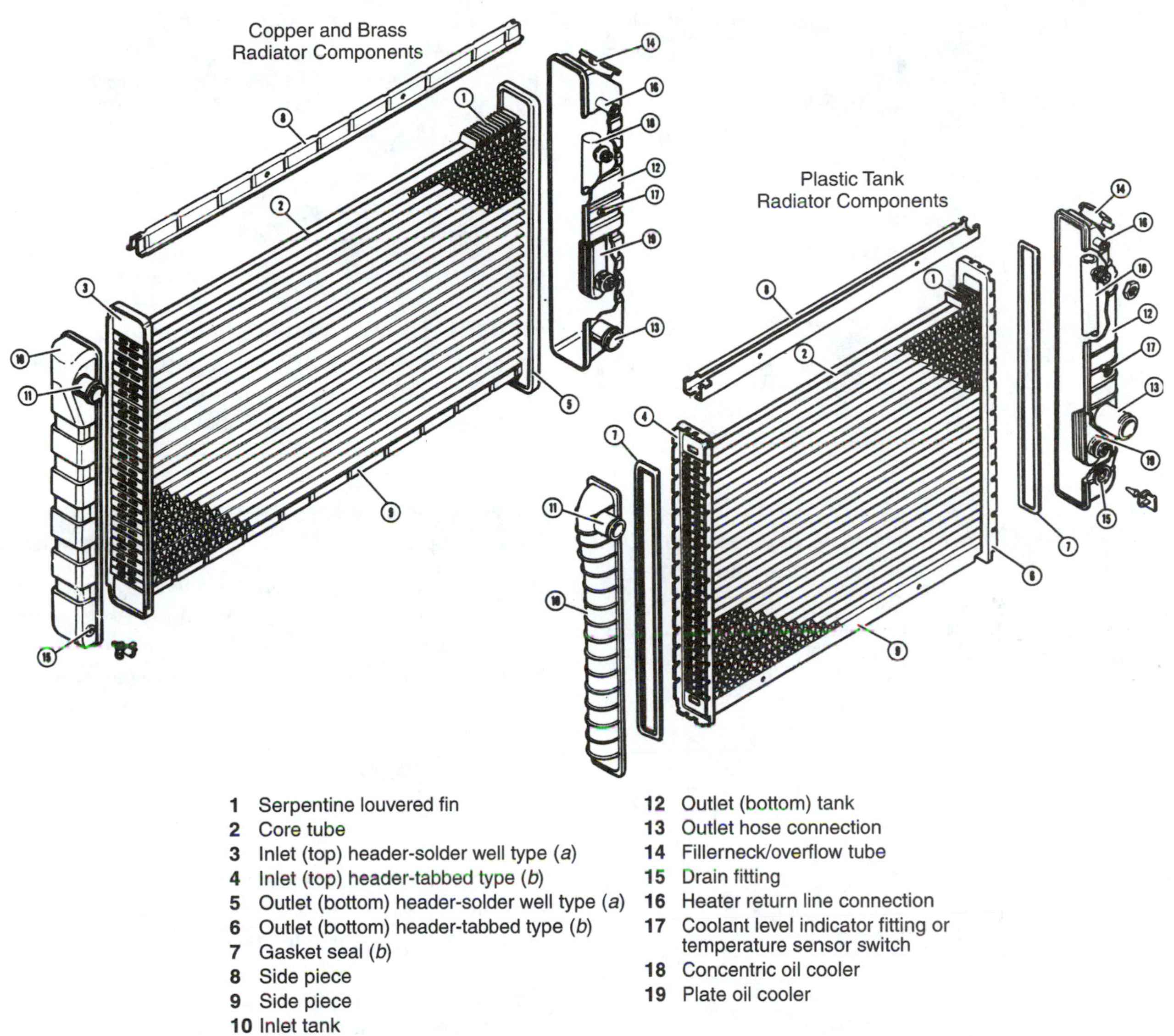

FIGURE 15–22 Most older radiators used a brass or copper core with the tanks soldered in place (*a*). Many newer radiators use an aluminum or brass core with plastic tanks (*b*). (*Courtesy of Modine Manufacturing*)

the cap's lower gasket against a seat in the filler neck to form a seal. When coolant pressure gets high enough, it pushes the lower sealing gasket off the seat, and some coolant escapes to the CRR or out the overflow hose. Most modern vehicles use a 15-lb (103-kPa) rated cap (Figure 15–23).

Coolant pressure comes from the expansion of the coolant as it is heated by the engine. Coolant volume increases about 10% as it heats from 70 to 180°F (21 to 82°C).

As described in Section 2.5, raising the pressure of a liquid increases the liquid's boiling point. Pressure of 15 lb increases the boiling point of water to 250°F (121°C). With a coolant mix of 50% antifreeze, the boiling point at 15 psi is increased to 262°F (125°C) (Figure 15-24).

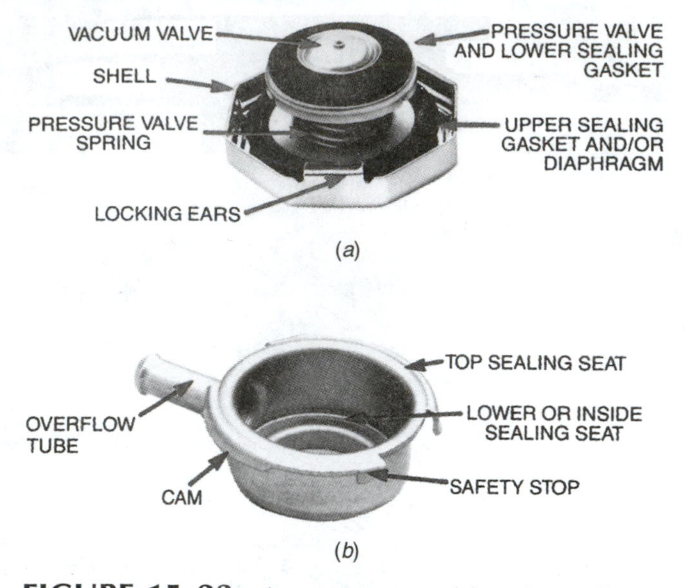

FIGURE 15–23 A pressure cap (*a*) and a radiator filler neck (*b*). As the cap is installed, the cam portion of the filler neck pulls the cap downward to seat the lower sealing gasket firmly on the sealing seat. (*Courtesy of Stant Manufacturing*)

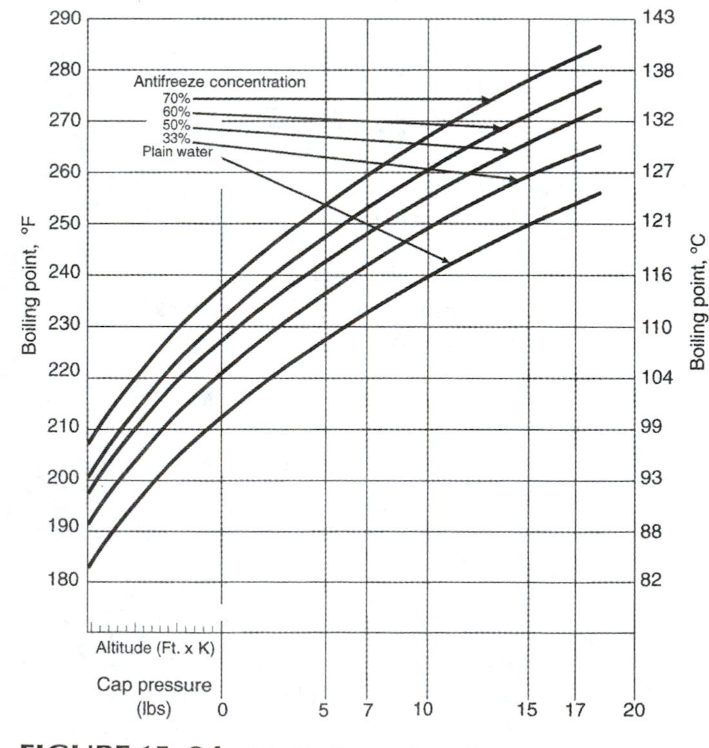

FIGURE 15–24 The boiling point of water increases with pressure and the addition of antifreeze.

FIGURE 15–25 Do not remove the cap from a hot system unless absolutely necessary. If you have to remove it, follow these precautions. Release of cooling system pressure can produce instant and severe boiling.

CAUTION

Remember that many vehicles have coolant temperatures above the boiling point. Many sources recommend that the pressure cap never be removed from a hot engine. Releasing system pressure can cause immediate and violent boiling. Numerous injuries and at least one fatality have occurred from burns caused by boiling coolant (Figure 15–25).

As a cap is installed, the cam on the filler neck or the threads on some caps pull the cap downward to seat the cap's lower gasket against the seat in the filler neck. During a normal warm-up cycle, coolant expands to form pressure, and, as mentioned, a certain amount of coolant pushes past the gasket and leaves the radiator, either to the CRR or out the overflow tube. During cooldown, the coolant contracts, and because of the pressure drop, the vacuum relief valve in the cap opens and allows coolant to return from the CRR. In older vehicles, air reenters through the overflow tube (Figure 15–26).

15.5.2 Coolant Recovery Reservoir

In older vehicles, a certain volume of air was kept in the radiator to allow for expansion. Air adds oxygen to the system; this leads to oxidation and corrosion. Air also mixes with coolant as it passes through the radiator, which reduces efficiency. An air or steam pocket in a water jacket can allow the temperature of some portions to reach critical points (Figure 15–27).

The CRR, also called an **expansion tank**, allows the cooling system to purge all air and be 100% full of coolant.

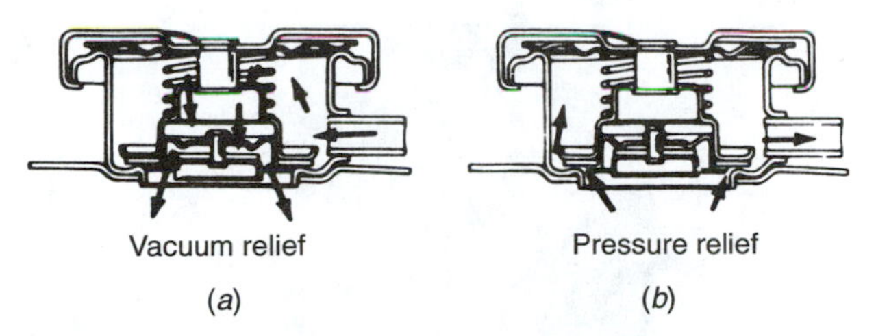

FIGURE 15–26 When a cooling system cools down and the coolant contracts, the vacuum relief valve in the cap opens to allow coolant to return from the recovery reservoir (*a*). When the cooling system heats and the coolant expands, system pressure forces the pressure valve off the cap seat and allows a coolant to the recovery reservoir (*b*). (*Reprinted with permission of General Motors Corporation*)

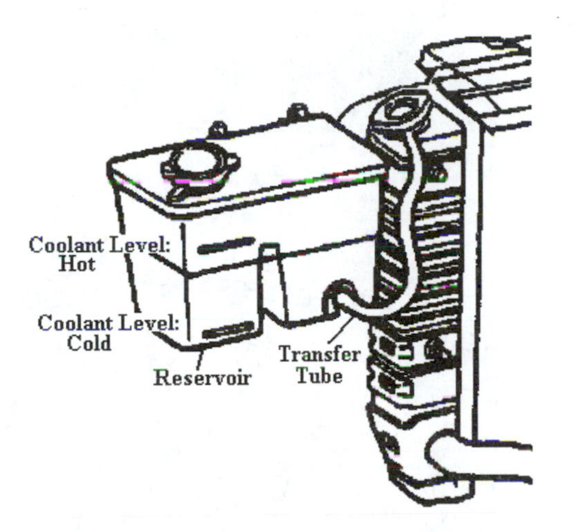

FIGURE 15–27 When the cap relieves pressure, coolant travels through the transfer tube to the reservoir. The coolant returns to the radiator during cooldown.

In most vehicles, the CRR is simply a semitransparent molded plastic container. Coolant level can be checked by looking at the coolant level; graduations indicate the correct level. The expansion hose from the radiator connects to the bottom of the CRR, and the CRR is vented to atmosphere. When the engine warms, expanding coolant flows into the CRR, and the level rises. When the engine cools, coolant returns to the radiator and the level drops.

15.5.3 Cooling Modules

Modern vehicles require more than one system be air cooled; these systems include the cooling system, A/C system, automatic transmission fluid, engine oil, power steering fluid, and incoming air charge from a supercharger. If these units can be combined, assembly of the

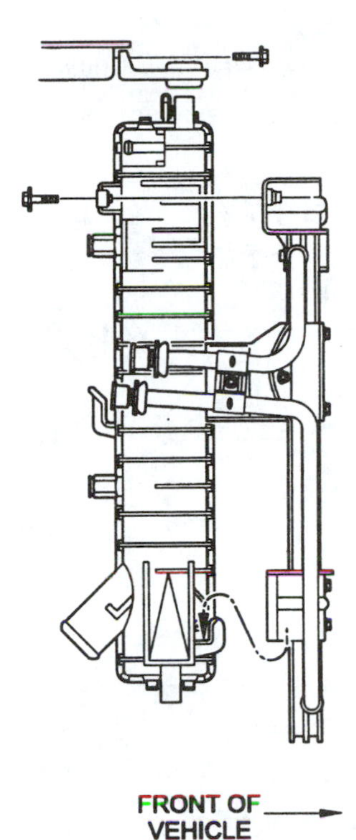

FIGURE 15–28 This A/C condenser is attached to the radiator to form a cooling module. Some modules include an oil cooler for power steering or engine oil. (*Courtesy of Visteon*)

vehicle during manufacture is simplified and made easier. A cooling module combines some or all of these heat exchangers into a single unit that often includes the electric cooling fan assembly with shroud (Figure 15–28).

A cooling module also allows one component to give structural support to another, and this allows the overall weight and size to be reduced. Most modules are designed with connections of the quick-connect style.

15.6 FAN

The **fan** ensures adequate airflow through the radiator while the vehicle is stopped or moving at low speeds. At cruising speeds, the fan is not needed because ram air through the grill supplies ample air. This airflow is improved by an air dam under the front bumper of some vehicles (Figure 15–29).

A fan requires power to drive it (about 2 to 6 hp). This power requirement varies with diameter, blade pitch, and number of blades. Larger diameter, higher blade pitch, or more blades move more air but also require more power and can be very noisy. Much design work has been done

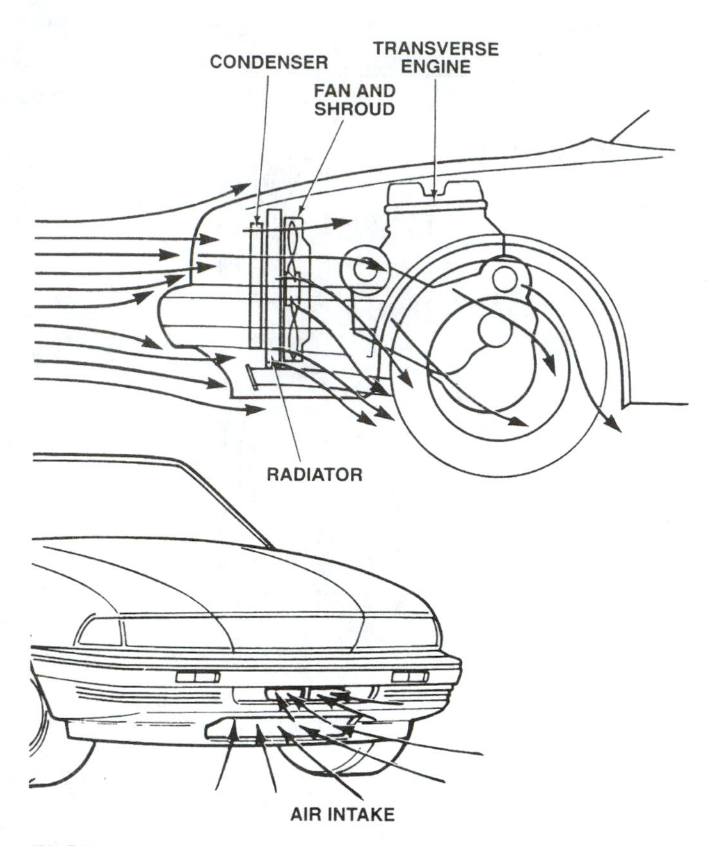

FIGURE 15–29 Air enters the radiator from the grill area above or below the front bumper. It exits through the engine compartment to the front wheel wells and below the vehicle. (*Reprinted with permission of General Motors Corporation*)

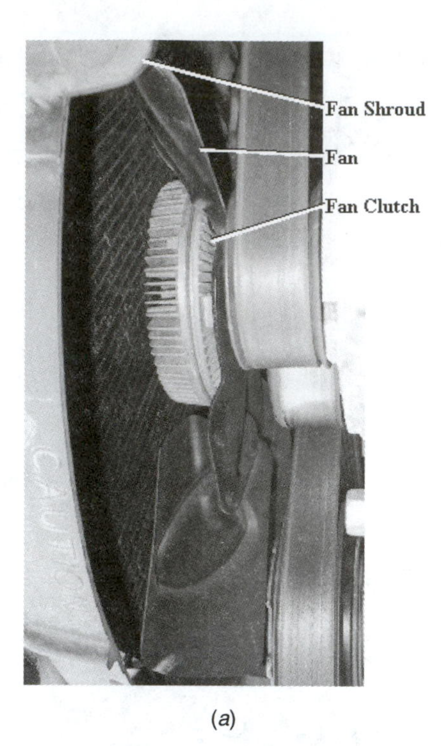

(*a*)

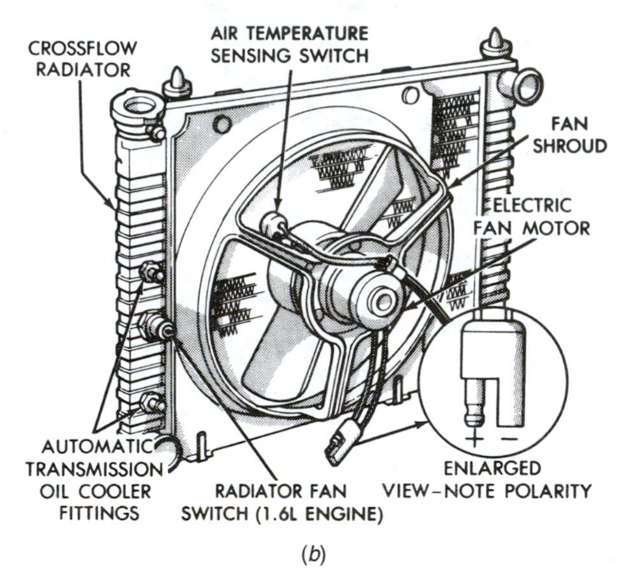

(*b*)

FIGURE 15–30 Most RWD vehicles drive the fan through a clutch that is mounted on the water pump shaft (*a*). Most FWD vehicles use an electric fan (*b*). (b *is courtesy of DaimlerChrysler Corporation*)

in shaping the blades to reduce both power draw and noise. Many fans have unequal blade spacing and blades with curved tips to reduce noise. Most fans also operate in a **shroud** to improve airflow (Figure 15–30).

On most RWD cars, the fan is mounted on the water pump shaft and driven through a **fan clutch**. A few RWD vehicles use **variable pitch fans** with flexible blades. Most FWD cars use one or two **electric fans.**

CAUTION

When working on a vehicle with the hood open, exercise caution around the fan. Fan blades can break and be thrown off with great force; at least one person has been killed when struck by a flying fan blade. Check a fan for cracks at the blades that might lead to breaking, and never stand in line with the fan of a running engine (Figure 15–31).

15.6.1 Fan Clutch

Most fan clutches are temperature-controlled fluid couplings and are called **thermal** or **thermostatic** fan clutches. They are designed to slip when cold and can only drive the fan to certain speeds when hot. There is

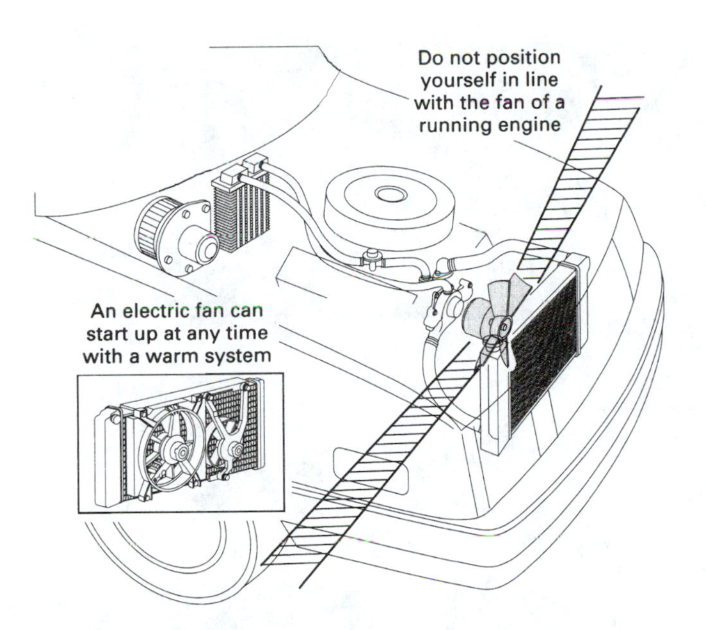

FIGURE 15–31 Use caution around a running engine. Things can get caught in the fan or belts, or a fan blade can break off and fly outward with great force. An electric fan can start up at any time on many vehicles. (*Courtesy of Everco Industries*)

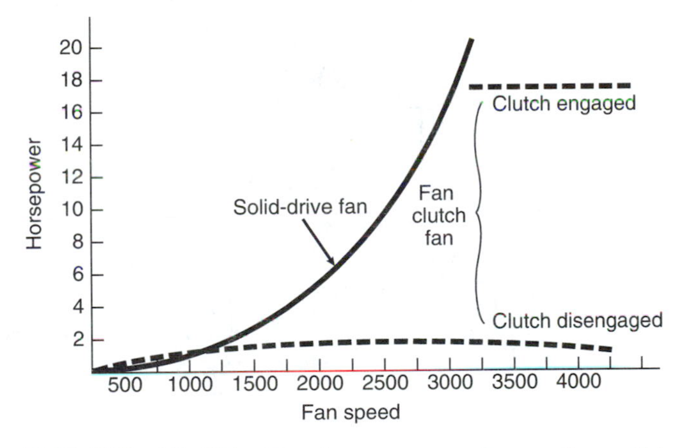

FIGURE 15–32 The horsepower draw of a solid drive fan increases significantly as engine speed increases. With the clutch disengaged, draw is limited to about 2 horsepower (hp). With the clutch engaged, draw is limited to about 18 hp.

also a nonthermal design that can only drive the fan up to a speed of about 1,200 to 2,200 rpm. A fan clutch improves fuel mileage because it reduces the fan's horsepower draw, and it also greatly reduces the noise level. When a fan clutch is engaged, the increased airflow and noise level from cold to hot are quite noticeable (Figure 15–32).

The fan is bolted to the body of the fan clutch. The clutch plate inside the body is connected to the water

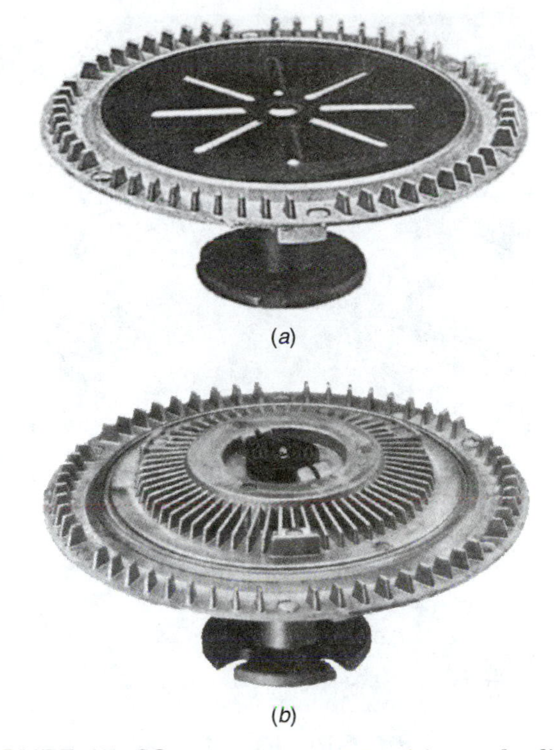

FIGURE 15–33 A nonthermal fan (*a*) merely slips as speed increases; a thermal fan (*b*) disengages almost completely when it is cool. (*Courtesy of Stant Manufacturing*)

pump shaft. A bearing on the shaft supports the body. The body also contains a reservoir in which silicone fluid can be stored. The fluid stays in the clutch chamber of a nonthermal or torque-limited fan clutch and can only transmit a small amount of torque, so it simply slips when the fan load reaches its torque rating.

When a thermal design unit is cold, a small portion of fluid is in the working chamber so that the clutch can only transfer a small amount of power. The fan can be driven to speeds of about 1,000 rpm. When the unit is hot, a thermostatic bimetal or coil spring moves to open a passage between the working chamber and reservoir, and silicone fluid fills the working chamber. This allows the clutch to transfer more power, and the fan speed increases to about 1,600 to 2,000 rpm (Figure 15–33).

15.6.2 Electric Fans

The fan blade is attached directly to the shaft of the electric motor; the motor mounting bracket often includes the shroud for the fan. Fan motors and fans of different sizes are used by vehicle manufacturers and are used as single units or in pairs. They can also be mounted as **pusher** fans in front of the radiator or

FIGURE 15–34 This vehicle uses two electric fans (arrows) mounted in a fan shroud that helps improve the airflow.

puller fans behind it. Larger fans and fans with more blades can move more air, but they are noisier and require larger motors, which in turn require more electrical current (Figure 15–34).

Fan motors are usually controlled by a **relay**. In some vehicles, the relay is controlled by the ECM or BCM. Some motor relays are controlled by a coolant temperature switch. These switches turn on when the coolant temperature goes above a certain point, say 200°F (93°C), and turn off when the coolant temperature drops a few degrees below the on point. Some dual fan units turn the second fan on if the coolant temperature continues to rise; the on setting is about 10 to 20°F above the first fan. Some vehicles use two-speed fans in a similar manner. Fan motors are also controlled by A/C operation. A fan can be turned on when the A/C is turned on or when the A/C high side pressure reaches a certain point. A dual fan unit can use both of these.

Another fan switch variation is that most systems use a simple high side pressure switch, and some units use a pressure transducer. The pressure transducer allows the control module to monitor high side pressure. When the pressure reaches a certain point relative to engine temperature and other factors, the control module turns a fan on or off (Figure 15–35).

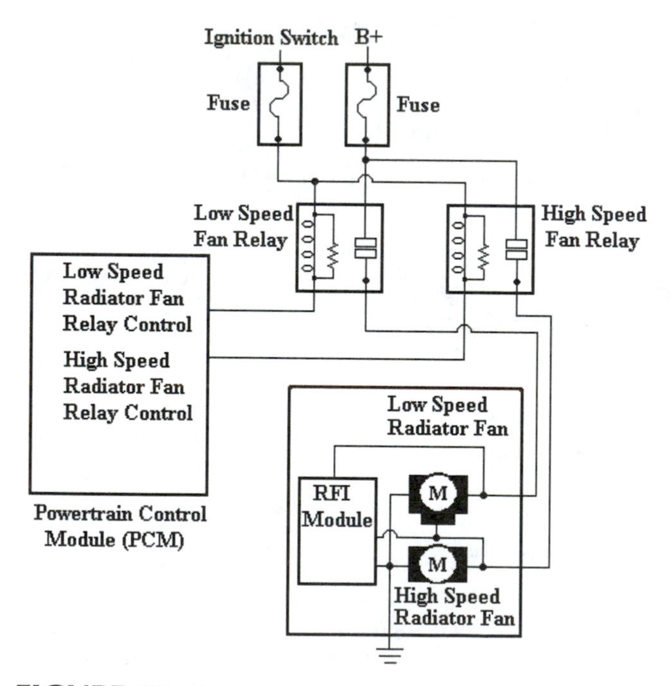

FIGURE 15–35 These two fan motors are controlled by two relays that are switched on and off by the PCM.

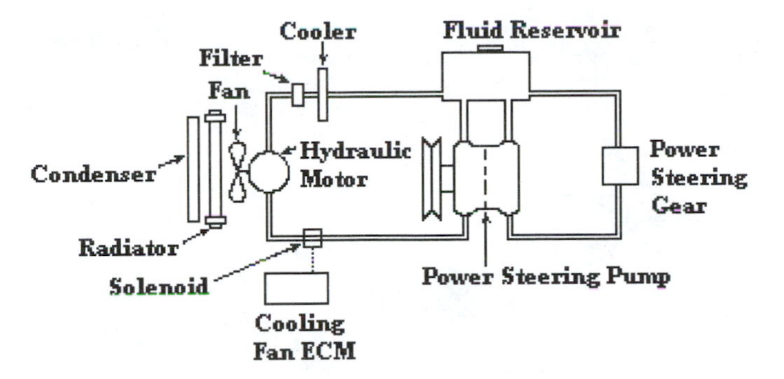

FIGURE 15–36 This cooling fan is driven by hydraulic pressure from the power steering pump. The hydraulic fluid flow is controlled by an electric solenoid that is controlled by an electronic module.

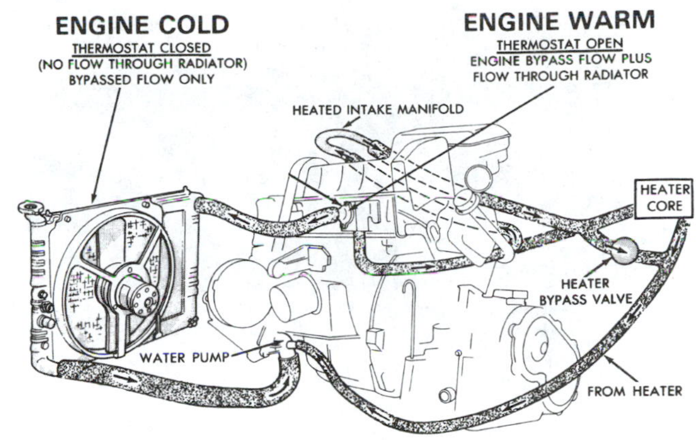

FIGURE 15–37 The upper and lower radiator hoses connect the radiator to the water pump and thermostat housing. Hoses also route coolant to the heater and intake manifold. (*Courtesy of Chrysler Corporation*)

Many fan relays are connected directly to B+ through a fuse link or circuit breaker; the circuit is not controlled by the ignition switch. These fan motors can turn on anytime the conditions are right; do not put yourself in a position where you can be injured by the unexpected start-up of a fan.

15.6.3 Hydraulic Cooling Fan

Some modern vehicles use a cooling fan that is driven by hydraulic pressure from the power steering system (Figure 15–36). An electric solenoid valve controls the hydraulic flow to the hydraulic fan motor, and the solenoid valve is controlled by an electronic control module (ECM). This allows control of fan speed, as well as when it operates.

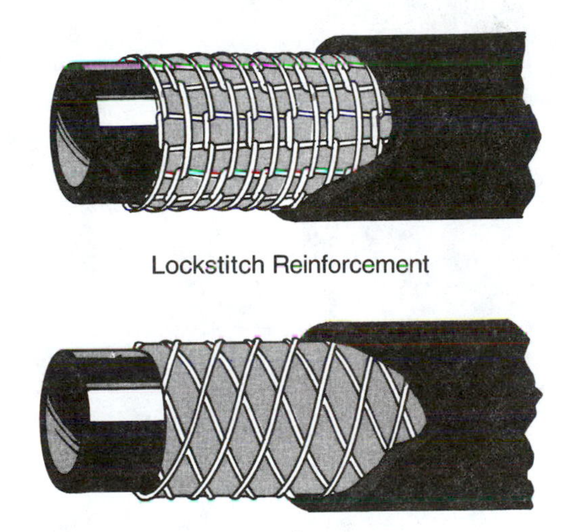

FIGURE 15–38 A cutaway radiator hose showing internal reinforcement and a heater hose showing spiral reinforcement. (*Courtesy of The Gates Rubber Company*)

15.7 HOSES

Two large **hoses** connect the engine's water jackets to the radiator. The upper hose transfers hot coolant to the radiator; the lower hose returns the coolant, which is now cooler, to the engine. These hoses must be flexible to allow the engine to move on the motor mounts. Hose flexibility also removes the need for close alignment of the connections between the radiator and engine. Hose flexibility also removes engine vibrations, which can shake a radiator apart (Figure 15–37).

Hoses are constructed around an inside tube of synthetic rubber. One or two layers (or plies) of woven textile are placed around this inner tube; then an outer layer of rubber is placed around the textile plies (Figure 15–38). To increase the life of the hose, better hoses use rubber materials that are resistant to oil, heat, and electrochemical degradation. Electrochemical degradation is caused by electrical current flow between the radiator and engine that causes small cracks in the inner hose liner; chemical action in these cracks breaks down the hose materials. Oil tends to swell and soften rubber, and heat tends to make it hard and brittle. Silicone rubber materials are used for severe-duty, high-heat conditions. Lower radiator hoses should always contain a springlike coil of wire or reinforcement

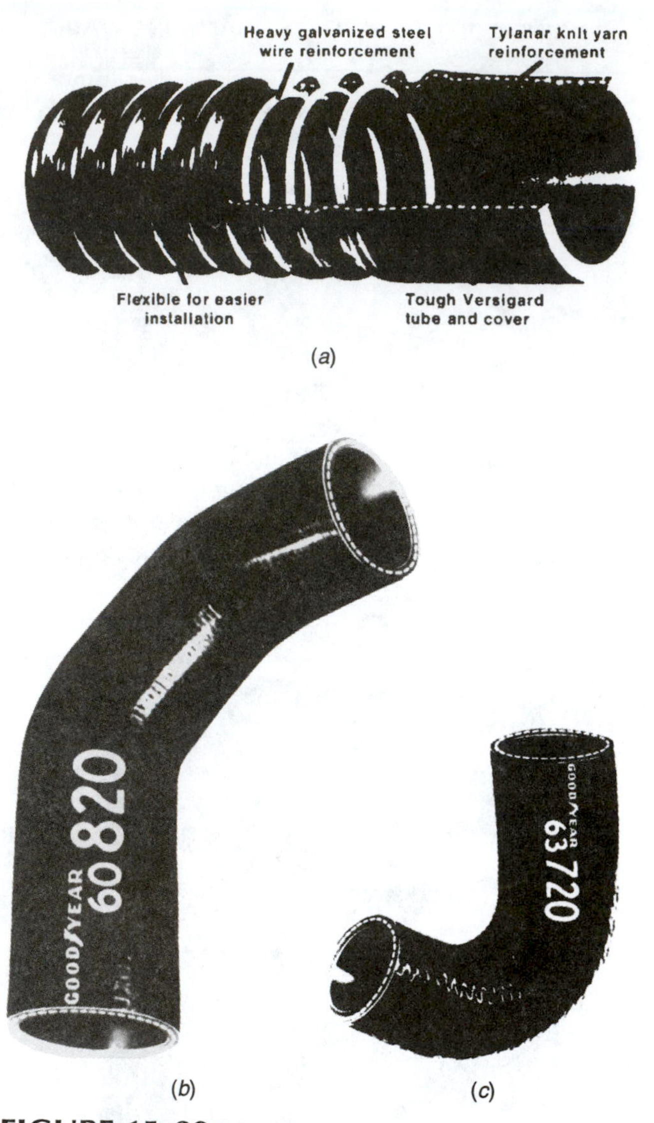

FIGURE 15–39 Besides the straight form, radiator hoses are available in flexible forms (*a*) and molded shapes (*b* and *c*). (*Courtesy of Goodyear Tire & Rubber Company*)

to prevent them from collapsing from water pump suction. If the lower hose collapses, it can shut off the flow of coolant into the engine.

Some hoses are molded to a particular shape for a particular installation. Some hoses even have alignment marks to ensure they are installed so they will not be deformed. Hoses are also available in a straight form and in a flexible, corrugated form (Figure 15–39). One or more of the hoses used in modern vehicles can be branched or T-shaped; for example, the lower radiator hose of one passenger car has a branch that connects to the CRR (Figure 15–40).

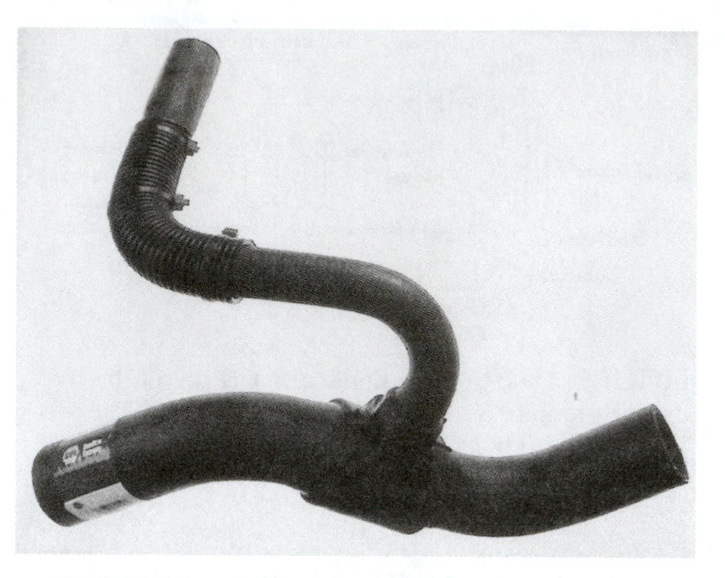

FIGURE 15–40 This lower radiator hose connects to the radiator outlet connection, the water pump inlet, and the coolant recovery reservoir.

Several types of clamps are available to make leak-proof connections. The better clamps can exert a high amount of clamping pressure evenly around the hose in a manner that does not cut the hose (Figure 15–41). A recently developed clamp uses a thermal plastic ring that is placed over the hose end and heated using a heat gun to shrink it and tighten it in place. This band has the ability to compensate for diameter changes as the hose and connector heat and cool. It is cut off to remove it. A spring-style clamp also allows for expansion and contraction.

15.8 GAUGES

The gauge indicates the engine's coolant temperature. Remember that the normal cooling system operating range is about 180 to 245°F (84 to 120°C). The lower end is the thermostat opening temperature, and the upper end is just below the coolant boiling point with the specified pressure cap and antifreeze mixture (Figure 15–42).

Most older vehicles used mechanical gauges that use a sending unit connected to the gauge by a thin capillary tube. The sending unit contains a gas that expands as it is heated. This gas pressure acts on a Bourdon tube to give a gauge reading (Figure 15–43).

Most vehicles today use an electric gauge that is essentially an ammeter. This gauge is connected to

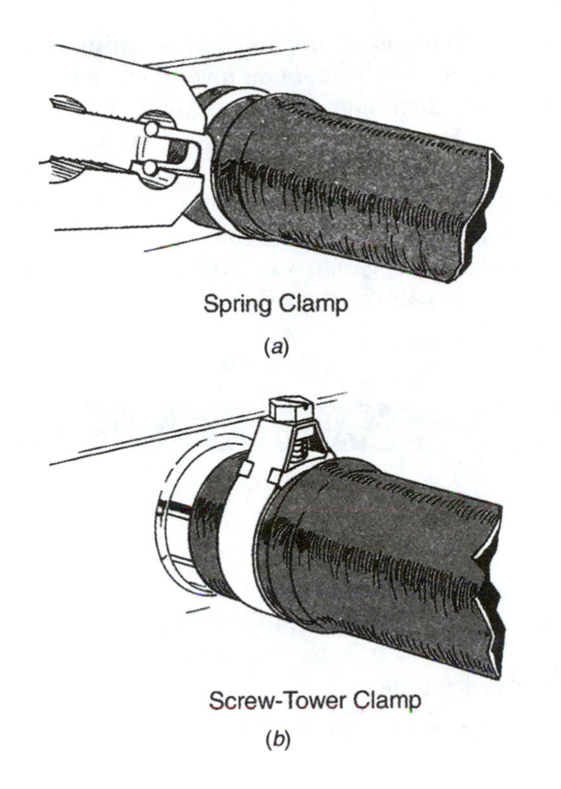

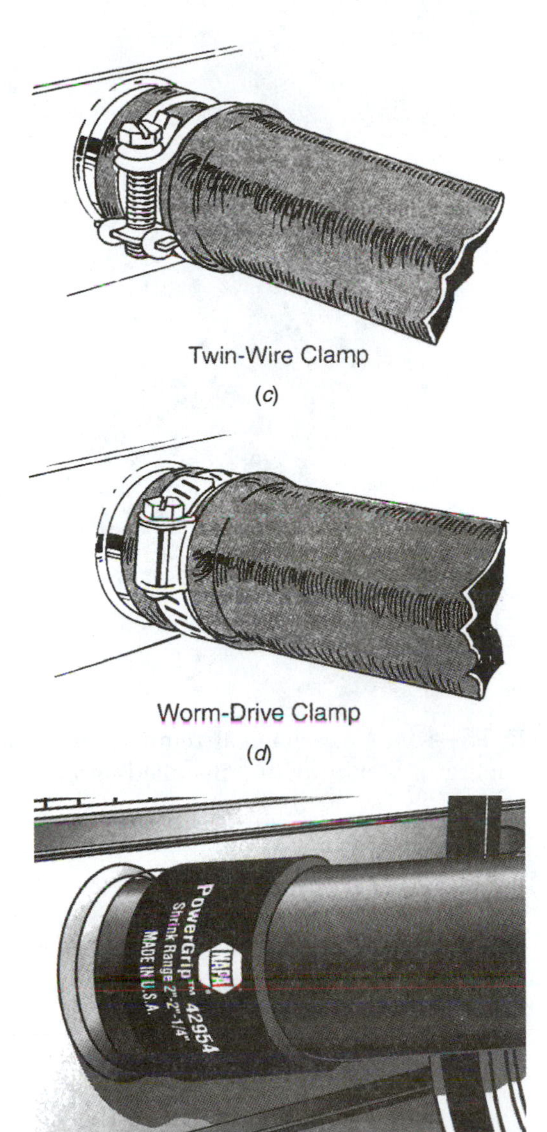

FIGURE 15–41 Common hose clamp types. Spring clamps (*a*) are common on O.E. installations; screw-tower clamps (*b*) and twin-wire clamps (*c*) were popular in the past; and worm-drive clamps (*d*) are popular replacements. PowerGrip clamps (*e*) are heat-sensitive material shrunk to fit using a heat gun. (*Courtesy of The Gates Rubber Company*)

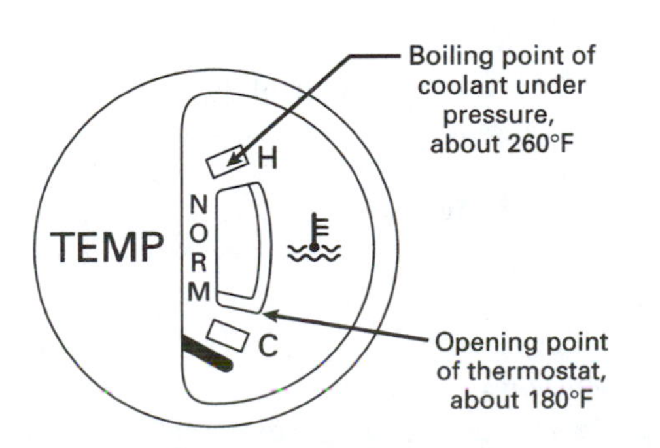

FIGURE 15–42 Most temperature gauges do not show the actual temperature; "normal" is 150 to 240°F.

the sending unit in the engine by an electric wire (Figure 15–44). The sending unit changes resistance value as the engine warms up. This causes the gauge to respond to the increased current flow that results. Electric gauges require a voltage source that is more constant than that produced by an alternator; a constant voltage regulator or instrument voltage regulator is used in many gauge circuits.

Many modern vehicles use electronic displays to show instrument panel functions. These panels consist of electronic modules with an information sensor for each function.

Some older vehicles use one or two coolant temperature lights. A few vehicles have a green light that

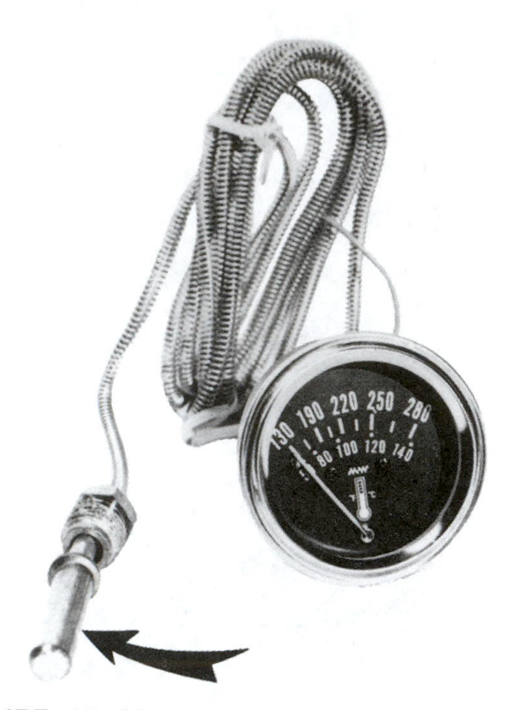

FIGURE 15–43 A mechanical temperature gauge uses a temperature bulb (arrow) installed into the water jackets.

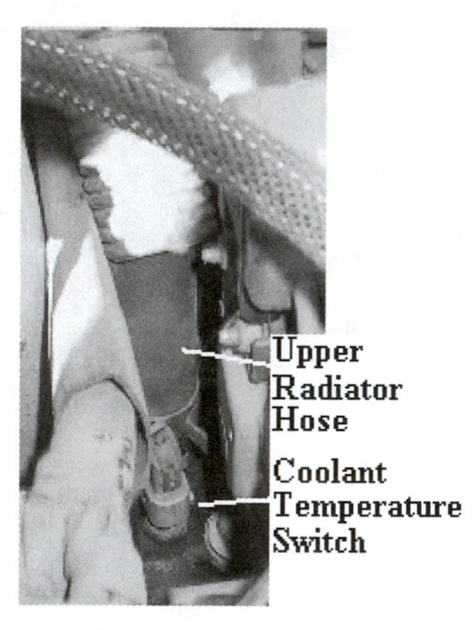

FIGURE 15–44 This coolant temperature switch (CTS) is mounted close to the thermostat and the outlet to the upper radiator hose.

comes on when the engine is at operating temperature, and many vehicles have a red light that comes on when the temperature is close to the boiling point. These lights use a simple circuit that goes from the light to the sending unit in the engine's water jackets. The sending unit is a temperature switch that closes at temperatures above a certain value.

Some vehicles use a **low coolant warning light** to warn the driver of a low coolant level. This light uses a sensor or switch mounted in one of the radiator tanks. When the coolant level drops below the switch, its contacts close, and this turns the light on. The other three styles of temperature indicators require coolant to bring the temperature to the sending unit. If there is a sudden loss of coolant, the gauge will probably not respond, and the engine can overheat while showing a normal gauge reading.

CAUTION

Coolants, especially ethylene glycol but also propylene glycol, are toxic and also have a sweet taste so they are attractive to children and pets. In one year, 1990, 585 people ingested antifreeze; 26 of these victims went to intensive care, and 4 of them died. In Europe, several manufacturers add a bittering agent to antifreeze to make it less desirable.

15.9 COOLANT

Coolant is the working fluid that transfers heat in a cooling system, much like the refrigerant in an A/C system. Water is the base coolant because it is plentiful, inexpensive, flows easily, and has an excellent ability to absorb and release heat. Water has drawbacks, though: It freezes at too high a temperature, boils at too low a temperature, and can cause metal corrosion (rust on steel and oxide on aluminum). The chemicals in high-quality antifreeze improve water to make it an excellent coolant. These chemicals are designed to lower its freezing point, raise its boiling point, reduce foaming, reduce cavitation, and prevent rust and corrosion. Some of the major chemicals used are the following:

- *Corrosion inhibitors:* Silicates, phosphates, and borates to help protect aluminum from corrosion, but silicates are abrasive to water-pump seals, leading to early failure. Extended-life coolants use carboxylate technology (a mixture of organic acids) for corrosion protection.
- *pH buffers:* To maintain proper acid–alkaline balance and prevent electrolytic corrosion.
- *Ethylene glycol (EG)* or *propylene glycol (PG):* To reduce freezing point and increase boiling point.
- *Dye:* To provide color (Figure 15–45).

FIGURE 15–45 This section of an antifreeze container gives important information about the capabilities of the product.

EG is the base for most domestic antifreeze brands. PG is the base for some domestic and European antifreezes; it is advertised as the safe antifreeze. Most manufacturers recommend a 50% coolant mixture of antifreeze and water (half and half) to get adequate corrosion protection. In very cold areas, the mixture can be more concentrated, but the limit is 67% (two-thirds antifreeze and one-third water). Higher antifreeze concentrations have a higher freezing point along with poor heat transfer. Antifreeze has a higher viscosity than water, so it does not flow as well; it also does not pick up and release heat as well (Figure 15–46).

EG and PG come from the base chemical ethylene oxide (EO), and both have a problem that if they are exposed to higher than normal temperatures, they become blackened, corrosive, and foul smelling in a relatively short period of time. Future emission requirements, especially with diesel engines, will probably increase coolant temperatures. Another coolant, Propanediol (PDO), is currently being developed and tested for high temperature use. PDO is also derived from EO, and will probably be supplied with extended-life chemistries.

Both EG and PG are good antifreeze base chemicals. The biggest fault is toxicity, and PG is less toxic and does not have the sweet taste that EG has. These two antifreeze types should not be mixed. They have different specific gravities; the specific gravity of EG is higher. The antifreeze concentration of a mixture cannot be accurately tested with either a hydrometer or refractometer. It is recommended that a system using PG be labeled to avoid future mix-ups.

Coolant life in an engine is determined by the life of the corrosion-protection package. If allowed to remain in an engine too long, rust, corrosion, and erosion begin.

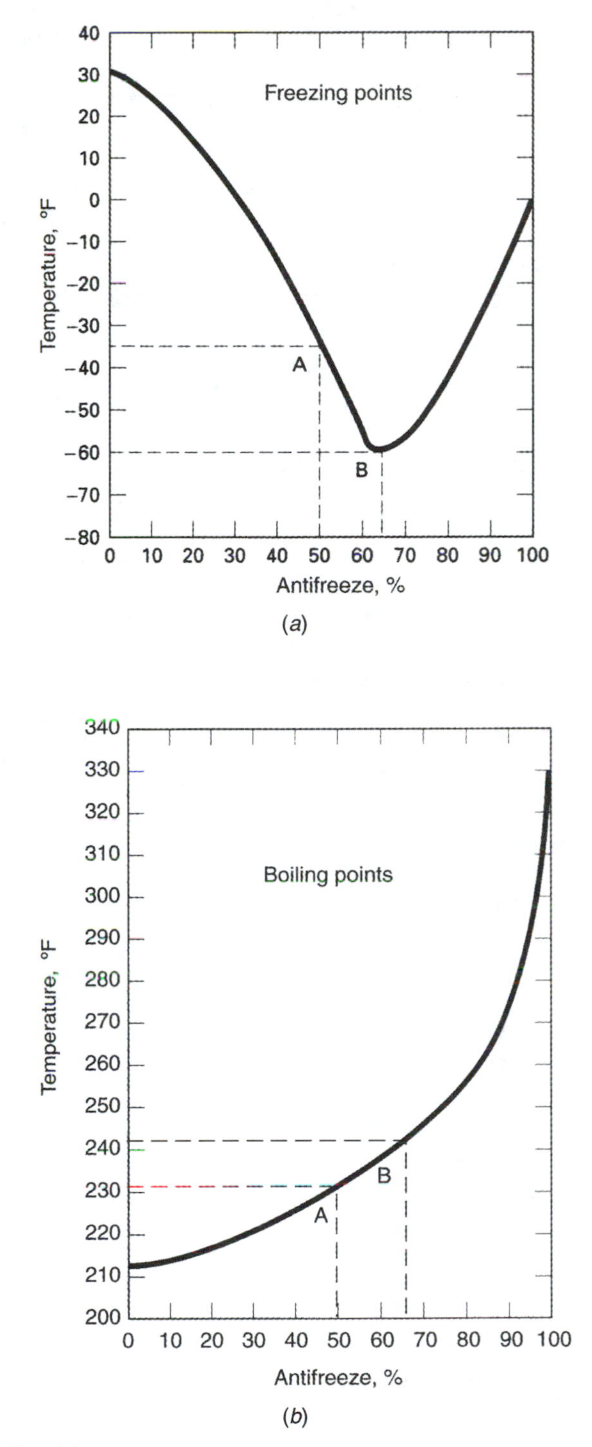

FIGURE 15–46 Coolant boiling point increases as antifreeze is added. The coolant freezing point drops as antifreeze is added until it reaches about two-thirds antifreeze. Recommended concentration is between 50% (*a*) and 66% (*b*).

Most antifreeze (dyed green) should be changed every 1 or 2 years. Extended-life antifreeze (dyed orange) has a recommended life of 5 years or 150,000 miles. Extended-life antifreeze uses an inhibitor package based on organic acids (OA) instead of amines, borates, nitrates, nitrites, phosphates, or silicates that can be used in conventional antifreeze. Extended-life and conventional antifreeze should not be mixed because it can reduce the service life to that of conventional antifreeze.

✓SERVICE TIP

Most domestic original equipment (OE) and aftermarket antifreeze use the following color coding: Light green indicates a material using silicone–silicate inhibitors; orange indicates organic acid, extended-life inhibitors. Antifreeze can be either EG or PG based. There is no universal color coding, so foreign OE antifreeze can use red, blue, green, or orange dyes for any of the antifreeze types.

✓SERVICE TIP

It is recommended to replace coolant using a mixture of the same type and concentration of antifreeze as originally used. If replaced with an extended-life antifreeze, the duration is the same as the original coolant. Extended-life coolant must be used in the initial or first fill of the vehicle in order to obtain extended-life performance.

Aluminum radiator and heater cores are subject to erosion; the coolant flow wears into the metal until it becomes too thin to contain the internal pressure. The erosion rate increases as the amount of dirt, rust, hard water deposits, and abrasive material dropping out of antifreeze increases. Silicates with other antifreeze chemicals can form abrasives, which in turn wear anything they pass by, including soft aluminum and water-pump seals.

The old antifreeze that is drained out must be recycled or disposed of properly. Although most of the ingredients will biodegrade, coolant absorbs heavy metals—copper, lead, and zinc—from the engine and radiator, and these are toxic. As mentioned earlier, coolant should not be allowed to stand in an open container because children or animals may drink it.

15.10 ENGINE BLOCK HEATER

Engine block heaters are small, electrical-resistance heater units that can be mounted in the block. These heaters are plugged into ordinary 110-volt AC house current and are used in very cold areas to warm an engine while it is shut off. Block heaters provide easier engine starting and faster warm-up (Figure 15–47).

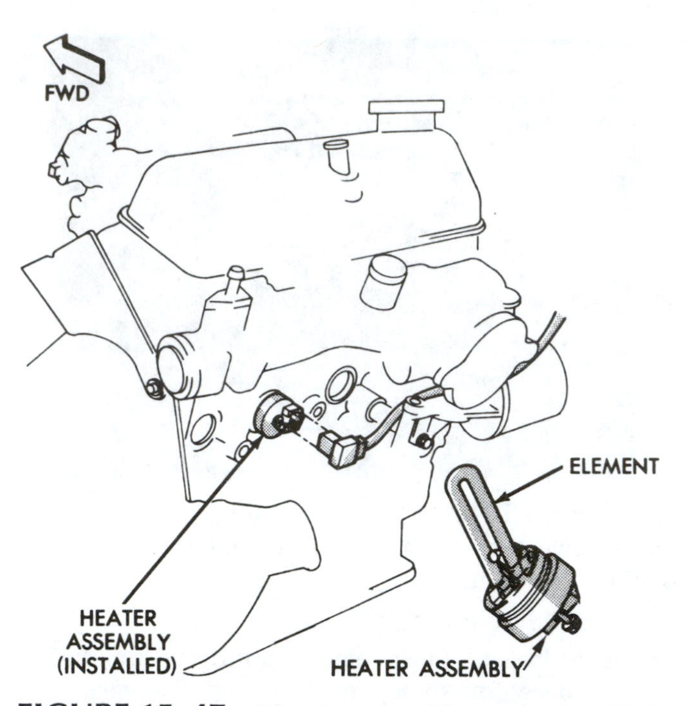

FIGURE 15–47 A heater assembly can be installed in place of a core plug. The cord for this heater is plugged into common house current to warm the coolant and prevent freeze-up. (*Courtesy of DaimlerChrysler Corporation*)

✓PROBLEM SOLVING

A friend asks you why his pickup may be overheating. The radiator core, hoses, and fan appear normal. When you look at the coolant in the radiator, you find it has a brown color, and there is a brown, slimy film on the valve of the radiator cap. What might be wrong with this system? What recommendations should you make to your friend?

Another friend has a similar complaint of an overheating car; it boils out coolant under a moderate load. When you look at the coolant level, you find that it is low and the coolant has a slightly brownish color. You fill the radiator, but you can find no sign of leaks with the engine either off or running. There are some streaks around the CRR that look like they might have come from hot coolant. What could be wrong with this system? What should you recommend your friend do?

CHAPTER QUIZ

These questions help you study this chapter. Enter the proper word(s) in the blanks to complete each statement.

1. The vehicle's cooling system is designed to keep the engine at the __________ __________ __________.
2. The plugs used to seal the large holes in the sides of the engine block are properly called __________ __________ __________.
3. The __________ opens at a temperature of about 200°F to allow coolant to circulate between the engine and __________.
4. While the thermostat is closed, __________ will circulate between the engine and the __________ __________.
5. Many thermostats use a(n) __________ __________ to open the valve at the correct __________.
6. The coolant flow is from the __________ to the lower hose, and then it is __________ through the water jackets.
7. The coolant flow through most radiators is from the __________ to the __________ or from one __________ to the other.
8. Most modern vehicles use a(n) __________ - __________ drive belt that is also called a(n) __________ belt from the way that it is routed around the pulleys.
9. Cooling systems use a pressure cap in order to raise the __________ __________ of the __________.
10. As an engine warms up to operating temperature, coolant will flow past the pressure cap to the __________ __________ __________.
11. Do not remove the pressure cap of a hot system because the coolant will probably __________ __________.
12. A fan clutch helps reduce fan __________ __________ and __________ when a high airflow is not needed.
13. An electric fan can be mounted in front of the radiator and __________ the air through it or behind the radiator and __________ the air through.
14. The ideal coolant mix for most vehicles is __________ antifreeze and __________ water.
15. Dirty coolant is a sign that the __________ should be __________.

REVIEW QUESTIONS

These questions allow you to check what you have learned. Select the answer that correctly completes each statement.

1. Modern cooling systems are designed to maintain an engine temperature of
   a. 150 to 270°F. **c.** 180 to 230°F.
   b. 100 to 160°F. **d.** 170 to 300°F.
2. Two students are discussing thermostats and coolant circulation. Student A says that the coolant does not circulate while the thermostat is closed. Student B says that the heater doesn't blow warm air while the engine is cold because of this. Who is correct?
   a. A only **c.** Both A and B
   b. B only **d.** Neither A nor B
3. While discussing cooling system operation, Technician A says that the engine will overheat if the thermostat sticks in the open position. Technician B says that heater operation will be affected if the thermostat sticks open. Who is correct?
   a. A only **c.** Both A and B
   b. B only **d.** Neither A nor B
4. On most engine blocks, the coolant flow is into the
   a. top and out the bottom.
   b. top and out the top.
   c. bottom and out the top.
   d. bottom and out the bottom.
5. Two students are discussing coolant circulation. Student A says that some modern vehicles use thermosyphon coolant circulation. Student B says that the coolant flows from one side of the radiator to the other in a crossflow design. Who is correct?
   a. A only **c.** Both A and B
   b. B only **d.** Neither A nor B
6. The boiling point of water with a 15-psi pressure cap is about
   a. 212°F **c.** 257°F
   b. 290°F **d.** 300°F
7. The amount of power required to drive a fan is affected by the
   a. diameter. **c.** number of blades.
   b. blade pitch. **d.** All of these
8. Two students are discussing fan clutches. Student A says that some fan clutches have a thermostatic valve that allows the clutch to drive the fan when it gets hot. Student B says that a fan clutch allows a large fan to be used with a lower noise level and power loss. Who is correct?
   a. A only **c.** Both A and B
   b. B only **d.** Neither A nor B
9. An electric cooling system fan is controlled by
   a. cooling system temperature.
   b. A/C system operation.
   Which is correct?
   a. A only **c.** Both A and B
   b. B only **d.** Neither A nor B
10. The best amount of ethylene glycol to mix into the coolant is
    a. 25%. **c.** 75%.
    b. 50%. **d.** 100%.

Cooling System Inspection, Trouble Diagnosis, and Service

LEARNING OBJECTIVES

After completing this chapter, you should:

- Be familiar with the preventive maintenance operations and adjustments necessary to keep a cooling system operating properly.
- Be able to inspect a cooling system to determine whether it is operating correctly.
- Be able to perform the standard cooling system tests to locate the cause of a problem.
- Be able to R&R a faulty cooling system component.
- Be able to complete the engine cooling system diagnosis and repair operations in the ASE Task List, Section C (Appendix A).

TERMS TO LEARN

back flush
chemical flush
coolant voltage
finger check
flush
hydrometer
pH balance
pressure test
refractometer
reserve alkalinity (RA)
test strip

16.1 INTRODUCTION

Engine cooling systems require periodic maintenance to keep them operating properly. These checks range from checking coolant level and condition, inspecting belts and hoses, and replacing coolant. Properly maintained cooling systems normally give lifetime trouble-free operation. Coolant should be replaced before the inhibitor package breaks down so problems caused by corrosion, electrolysis, and erosion can be prevented.

✓SERVICE TIP

Electrolysis can cause pinholes in the radiator or heater core. An aluminum component will be discolored, black and pitted; copper and brass components will probably develop blue-green corrosion. The pinholes tend to occur in the center of the heater core and an inch or so away from the radiator tanks or in the vicinity of the cooling-fan mounts.

Cooling system problems do occur, and these problems usually show up as leaks, overcooling, or overheating. Overcooling is first noticed by most drivers

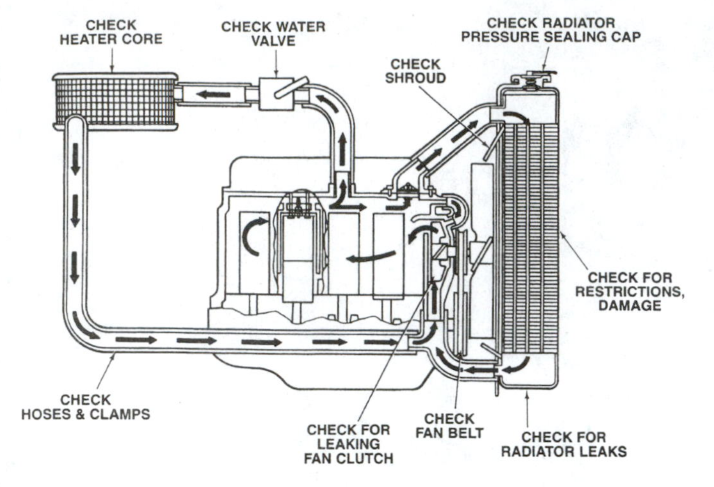

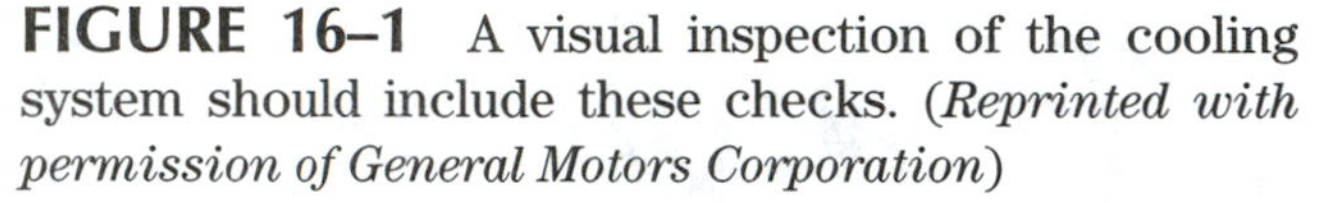

FIGURE 16–1 A visual inspection of the cooling system should include these checks. (*Reprinted with permission of General Motors Corporation*)

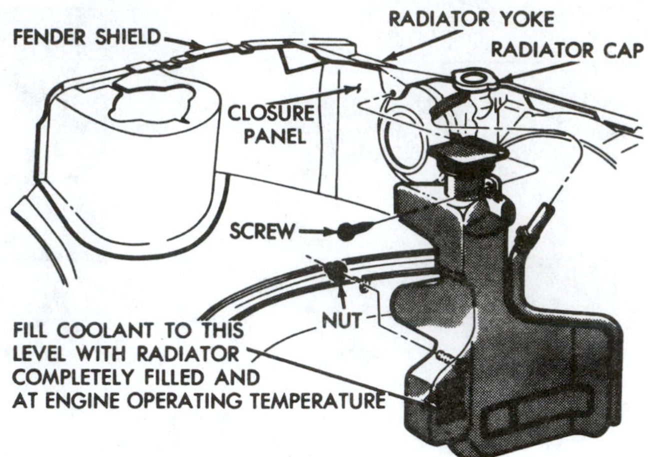

FIGURE 16–2 Coolant level in a modern system, using a coolant recovery reservoir, is checked at the reservoir. (*Courtesy of DaimlerChrysler Corporation*)

as poor heater operation in the winter; it also causes poor engine performance, reduced fuel mileage, and excess engine wear. Overheating is usually more serious. More than one engine has been ruined by the effects of overheating, which include warped and cracked cylinder heads and burned and badly scored pistons, cylinders, and valves. Several tests are used to locate cooling system problems, including visual and hand-feel checks, temperature tests, and pressure tests (Figure 16–1).

Cooling system service includes radiator and engine water jacket cleaning and flushing and the repair and replacement of faulty components.

✓SERVICE TIP

Erosion is wear from dirty, abrasive coolant. Erosion normally occurs at the radiator or heater core inlet and tube ends.

16.2 PREVENTIVE MAINTENANCE OPERATIONS

Preventive maintenance operations are usually listed in the vehicle owner's manual and in the maintenance and lubrication section of the service manual (along with the maintenance schedule). These schedules usually call for regular inspection of the coolant level (some recommend this at each gasoline refill), annual inspection of the belts and hoses, and a change of the coolant at two-or three-year intervals.

16.2.1 Coolant Level

In a modern semiclosed system with a pressure radiator cap and coolant recovery reservoir (CRR), coolant level is checked by looking at the level in the semitransparent CRR. The cap does not need to be removed except to check coolant condition and cap performance (Figure 16–2).

CAUTION

Extreme care should be exercised when removing the pressure cap from a system that is at operating temperature or above. Remember that coolant temperature can be above 212°F (100°C) and boil immediately if the pressure is removed.

✓SERVICE TIP

Some technicians squeeze the upper hose. If it is hard from internal pressure, **do not remove the cap**.

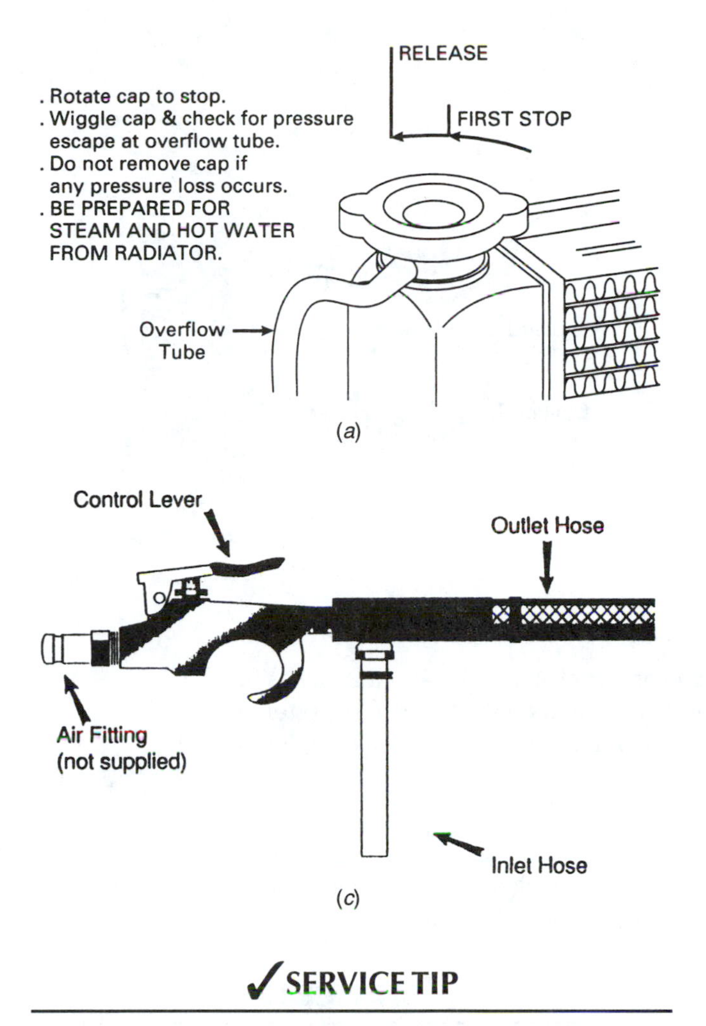

(b)

FIGURE 16–3 If necessary, follow these steps to remove the cap from a hot system (*a*). A safety cap has a lever that is raised to release any pressure from the system (*b*). The inlet hose of this coolant pressure reliever is connected to the radiator overflow hose, and the control lever is operated to safely remove any pressure from a system (*c*). (b *is courtesy of Stant Manufacturing;* c *is courtesy of Wynn Oil Company*)

✓SERVICE TIP

When removing the pressure cap from a warm system, carefully follow these steps:

- Place a shop rag over the cap.
- Stay at arm's length from the cap.
- Carefully rotate the cap to its safety stop (Figure 16–3).
- Wiggle the cap as you check for system pressure.
- If there is escaping pressure, either step away until the pressure is released or cool the system down until the pressure release stops.
- If there is no escaping pressure, finish removing the cap.

The radiator of a CRR system should be full to the level of the sealing surface at the filler neck. A lower coolant level indicates a system leak, faulty cap, or leak in the CRR hose. The coolant level in a non-CRR system should be about 2 inches below the filler neck on a cold system and almost to the filler neck seat on a warm system. This lower level allows for coolant expansion as the system warms up (Figure 16–4).

16.2.2 Coolant Mixtures

As mentioned earlier, coolant is normally a mixture of antifreeze and water. In some areas, there is a legal definition for four types of coolant:

- *Engine coolant concentrate:* Essentially pure antifreeze
- *Prediluted engine coolant:* A 50–50 mixture of antifreeze and water
- *Recycled engine coolant concentrate:* Recycled EG that meets new coolant specifications
- *Recycled coolant:* A 50–50 mixture of recycled antifreeze and water

Antifreeze is commonly available as pure 100% antifreeze or in a 50% mixture as a new or recycled product.

Water can absorb more heat than antifreeze, but it freezes at 32°F (0°C), boils at 212°F (100°C), and allows corrosion and other contamination to occur. Antifreeze lowers the freezing point, raises the boiling point, and contains corrosion inhibitors, but it cannot transfer as much heat as water. The lowest freezing point occurs in a mixture of one-third water and two-thirds (66.6%)

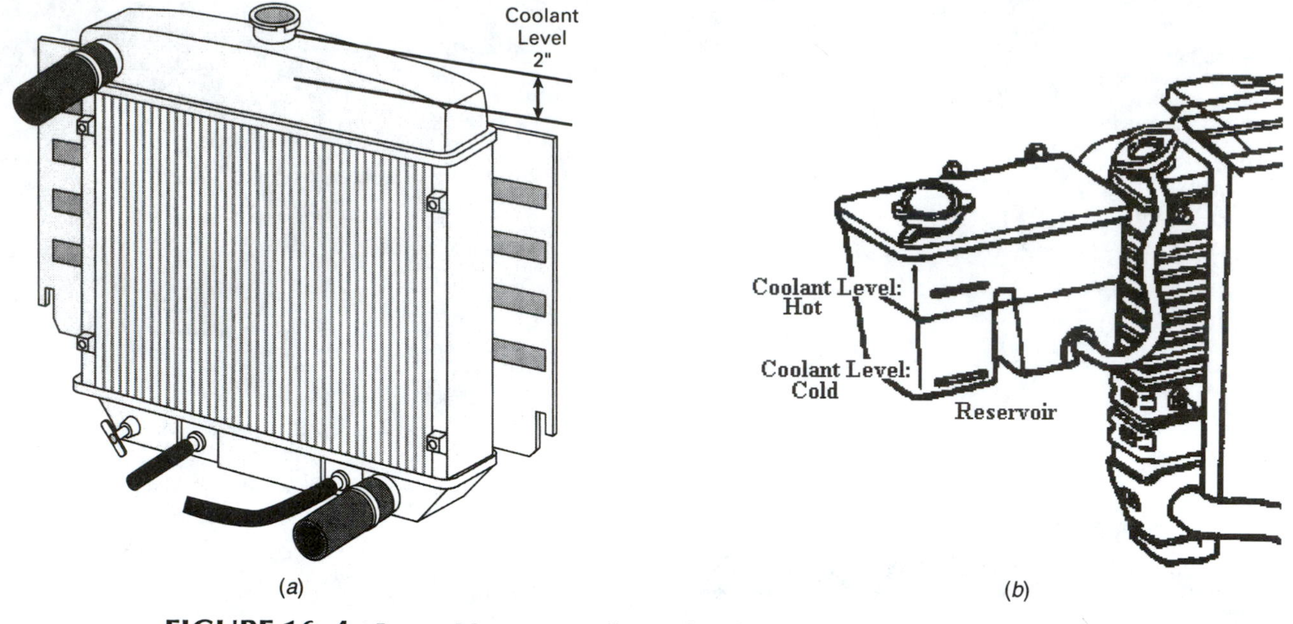

FIGURE 16–4 In an older system, the coolant level should be about 2 inches below the seat in the filler neck (*a*). The coolant level in a coolant recovery system is indicated on the recovery reservoir (*b*), and the radiator should be filled completely. (a *is courtesy of Everco Industries)*

antifreeze. Also, corrosion inhibitors do not dissolve properly without water. With 100% antifreeze, the inhibitors can form deposits in the system and become insulation, raising the temperature inside the system.

Most sources recommend a 50% mixture, one-half water and one-half antifreeze; this ratio provides a freeze protection to –35°F (–37°C). In very cold areas, the antifreeze portion can be increased to 67%; this mixture prevents freezing down to about –61°F (–52°C). The antifreeze concentration should be at least 40% but never greater than 67%.

✓ SERVICE TIP

EG can burn, and a mist of pressurized EG coolant can produce an explosion if it contacts an open flame.

16.2.3 Coolant Condition

Good coolant should have a bright yellow–green or orange color and a very slight slippery feel. Coolant normally deteriorates to a brownish, rusty color. As it loses its corrosion- and rust-protection abilities, rust forms on the iron parts and corrosion forms on the aluminum portions of the water jackets. Dirt and rust are also deposited on the horizontal surfaces and at the bottom of the water jackets. These are indications that the coolant has gone past the point where it should be changed.

✓ SERVICE TIP

A simple **finger check** (rubbing a finger around the inside of the filler neck) provides the technician a quick way to check for faulty coolant and a dirty system (Figure 16–5).

✓ SERVICE TIP

Many technicians will lower the coolant level to expose some of the radiator tubes. This allows a limited, visual inspection of the radiator tubes.

✓ SERVICE TIP

A brown, muddy sludge has been found in systems using Dexcool, an extended-life antifreeze developed by Havoline. The cause of this sludge has been attributed to incomplete bleeding that left air in the system.

A slightly dirty system should be flushed with plain water as the coolant is changed. A dirty or rusty system

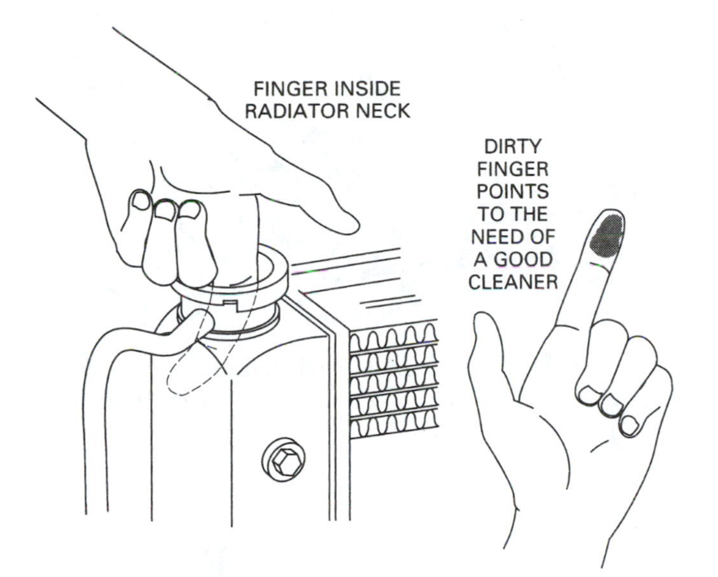

FIGURE 16–5 A simple check for cleanliness is to rub a finger around the inside of the filler neck. Any dirt on your finger indicates a dirty system that should be cleaned.

can usually be cleaned using a **mechanical reverse flush** or a **chemical flush**; these procedures are described in Section 16.4.1.

Another coolant check is for the antifreeze concentration to determine the freeze point. The freeze point can be measured using a hydrometer, refractometer, or test strips (Figure 16–6). A hydrometer measures the specific gravity of the coolant, and a refractometer measures its light-refracting qualities. The **hydrometer** is the least accurate, especially if propylene glycol or a mixture of EG and PG is used. Most hydrometers sold in the United States are calibrated for EG. Some **refractometers** are calibrated for both EG and PG, but a mixture produces inaccurate readings. A refractometer is more accurate than a hydrometer, but it is also much more expensive. A **test strip** is dipped into the coolant, and this produces a color change. Comparing the color with a chart determines the freeze point. Test strips are relatively inexpensive and are relatively accurate with EG, PG, and mixtures of the two. They also offer an advantage in that test strips check reserve alkalinity at the same time. Test strips can be saved or attached to the inspection sheet or work order to provide a record. Normally a 50–50 coolant mix with a freezing point of –34°F (–37°C) is recommended.

Reserve alkalinity (RA) is a check of the **coolant pH balance**. A coolant can become acidic and produce severe corrosion and electrolysis when the buffers for reserve alkalinity are used up. Some shops use a pH meter to check the level. Fresh EG coolant mixes should measure between pH 9 and 11; fresh OA antifreeze mixtures should be pH 8.3. As a general rule, pH should

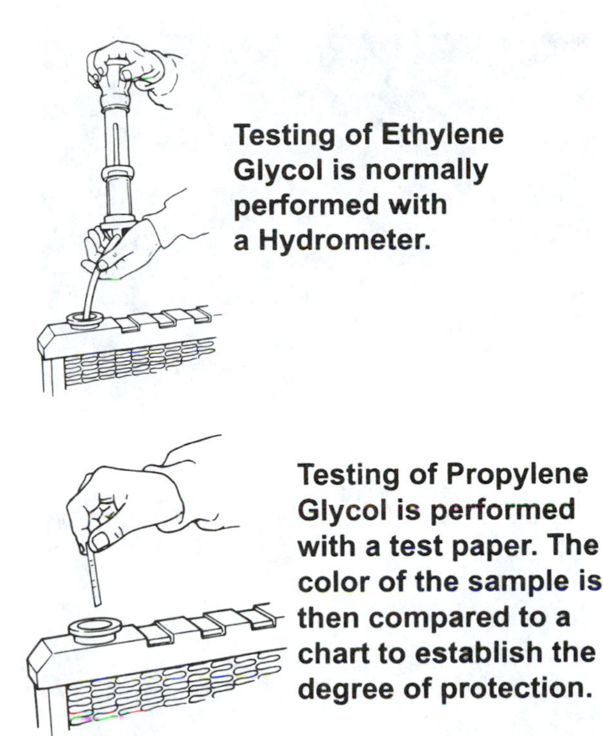

FIGURE 16–6 Ethylene glycol–based coolant mixtures are commonly tested using a hydrometer or refractometer; test strips are used for propylene glycol mixtures. (*Courtesy of Four Seasons*)

be between 7.5 and 10. A lower level indicates that the coolant is becoming acidic and corrosive. If the pH is above 11, buildup will occur.

Another check is for coolant electrolysis, also called **coolant voltage**. This quick and simple check measures the voltage in the coolant; less than 0.3 V is acceptable. Some sources recommend 0.1 V or less. A higher voltage reading indicates acidic coolant or the possibility of stray current passing through the coolant, possibly from an improperly grounded cooling fan motor. This problem has increased with radiators that have plastic tanks because they do not provide a good electrical ground path.

To measure coolant concentration and condition using test strips, you should:

1. Run the engine to mix the coolant thoroughly and to bring it to room temperature.
2. Dip the strip into the coolant, remove it immediately, and shake off the excess coolant.
3. Determine the freezing and boiling points by immediately comparing the color of the end pad with the color chart (Figure 16–7).
4. After 15 seconds, compare the second pad with the color chart to determine acid corrosion protection.

(a)

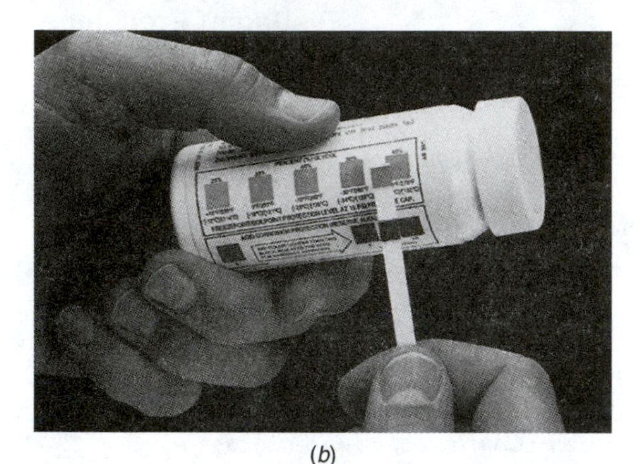

(b)

FIGURE 16–7 A test strip is dipped into the coolant (*a*) and the color change is compared with a chart (*b*) to determine freeze point, boil point, and acid corrosion protection. *(Courtesy of Environmental Test Systems, Inc., Elkhart, IN)*

To measure coolant concentration using a refractometer, you should:

1. Run the engine to mix the coolant thoroughly.
2. Clean the window and the plastic cover so they are clean and dry. Close the cover (Figure 16–8).
3. Transfer a few drops of coolant so the drops will run between the window and the cover (Figure 16–9).
4. Point the unit toward light and look into the eyepiece. The coolant protection is at the point between the light and dark portions of the scale (Figure 16–10).

To measure coolant concentration using a hydrometer, you should:

1. Run the engine to mix the coolant thoroughly.
2. Draw a sample of coolant into the hydrometer so the float is floating. Tap the hydrometer

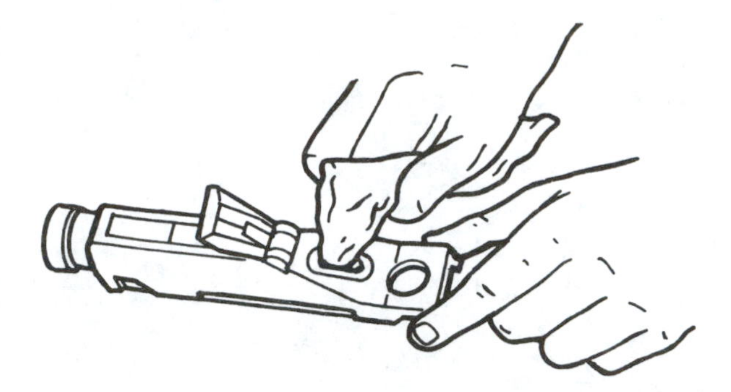

FIGURE 16–8 The first step in using a refractometer is to clean the measuring window and cover. *(Reprinted with permission of General Motors Corporation)*

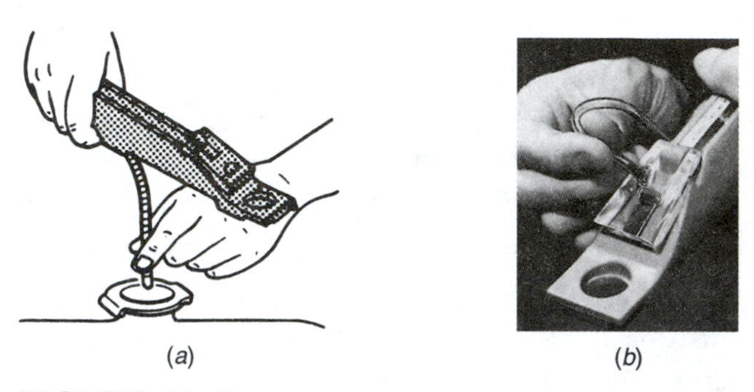

(a) (b)

FIGURE 16–9 The plastic pump is used to draw a few drops of coolant (*a*) and transfer it onto the measuring window (*b*). The window is then closed (a *is reprinted with permission of General Motors Corporation;* b *is courtesy of Wynn Oil Company)*

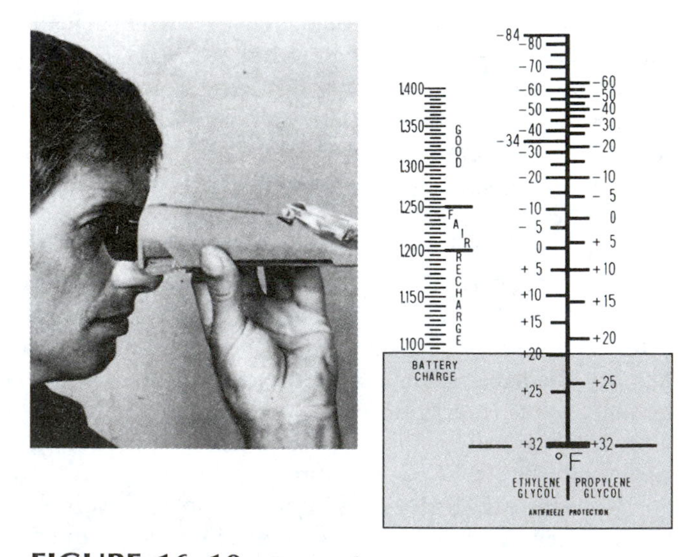

FIGURE 16–10 Point the refractometer toward a light source and look through the eyepiece. The freeze point is indicated at the dividing line between dark and light (the edge of the shadow). *(Courtesy of Wynn Oil Company)*

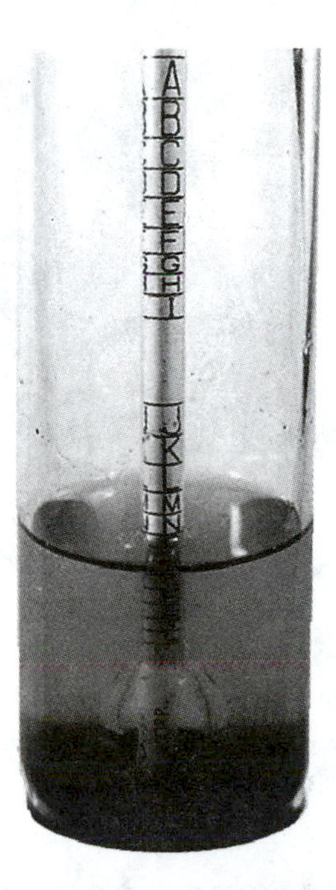

FIGURE 16–11 A hydrometer is read by drawing enough coolant into the unit to lift the float and noting the graduation at the liquid level.

to ensure that the float is not stuck (Figure 16–11).

3. Check the thermometer; adjust the temperature-calibrating sleeve if so equipped.
4. Read the scale at the level of the coolant on the float. On some units, a chart is used to read the freezing-point temperature by the letter indicated on the float (Figure 16–12).

To measure cooling system voltage, you should:

1. Run the engine to mix the coolant thoroughly.
2. Shut off the engine, and connect the negative (−) lead of a digital voltmeter to the metal radiator filler neck, metal part close to the filler neck, battery ground, or engine ground.
3. Insert the positive (+) lead into the coolant, making sure it does not touch the filler neck or core, and read the meter (Figure 16–13). A reading of more than 0.1 V indicates a problem. *If electrolysis is suspected,* watch the meter while turning on or off the electrical devices that might be causing the problem; a change in the meter reading indicates a problem.
4. Start the engine as you watch the voltmeter reading. A higher reading during this dynamic check indicates an electrolysis caused by an engine electrical component.

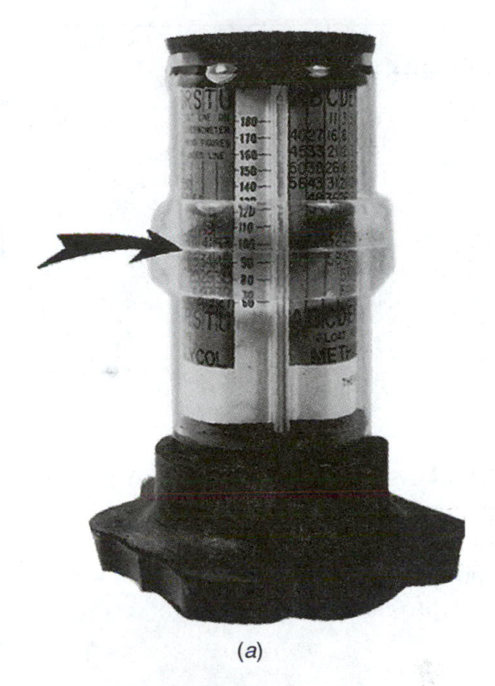

(*a*)

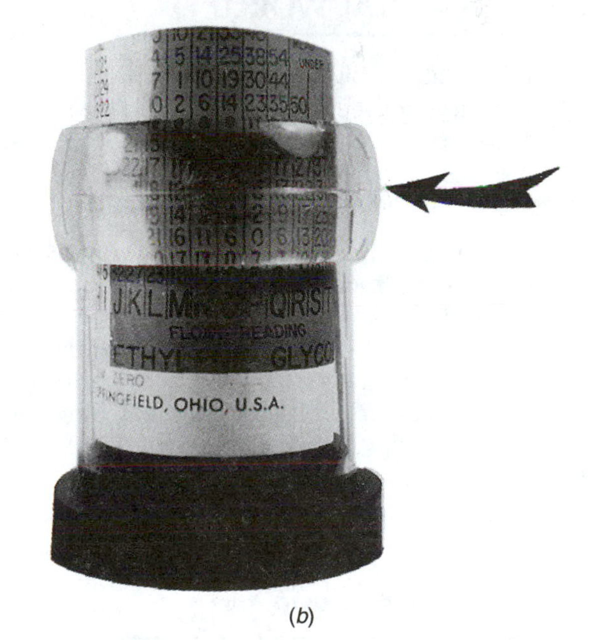

(*b*)

FIGURE 16–12 Many antifreeze hydrometers include a thermometer and slide. Set the slide to the thermometer reading (*a*) and read the freezepoint on the proper scale (*b*).

✓ SERVICE TIP

To find the cause of a dynamic electrolysis problem, watch the voltmeter as a helper switches the vehicle's electrical devices off. A drop in cooling system voltage indicates the problem source.

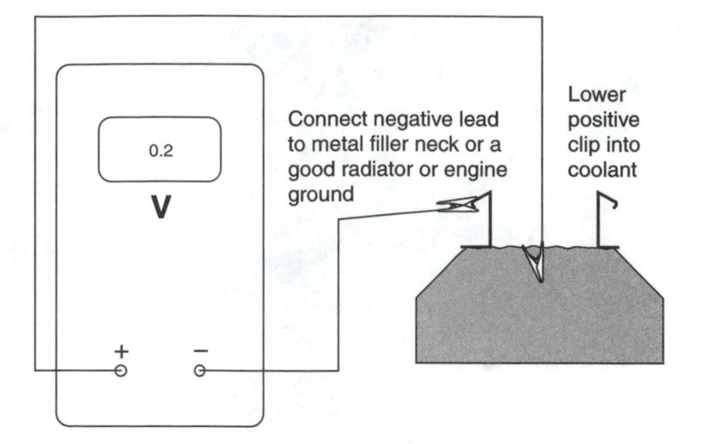

FIGURE 16–13 Radiator voltage is measured by clipping one voltmeter lead to a good ground or the filler neck of all-metal radiators. Dip the clip on the other lead into the coolant; a reading of 0.3 V or greater indicates a problem.

✓ REAL WORLD FIX

The 1993 Jeep had a severe leak at the lower radiator hose connection. Inspection showed that the lower radiator was severely eaten away, leaving a lot of blue-white corrosion deposits. Talking with the customer revealed that this radiator was only six weeks old.

Fix: Checking with an ohmmeter showed a lot of resistance between the radiator and the vehicle's frame. Inspection of the radiator and A/C condenser showed that one of the ground straps was missing and the other one was installed by a very loose bolt. Proper replacement of the missing ground strap and tightening of the other bolts along with repair of the radiator fixed this problem.

✓ SERVICE TIP

Cooling system voltage can result from faulty component grounds; in fact, most electrolysis problems are caused by a faulty starter or cooling fan ground. To check for these, observe the voltmeter reading as the engine is cranking and when the cooling fan operates. If the radiator voltage reading increases during these events, the ground circuit is probably inadequate. Another way to check this is to have a helper connect a jumper to provide a ground as the component operates. One end of this jumper should be connected to the battery ground.

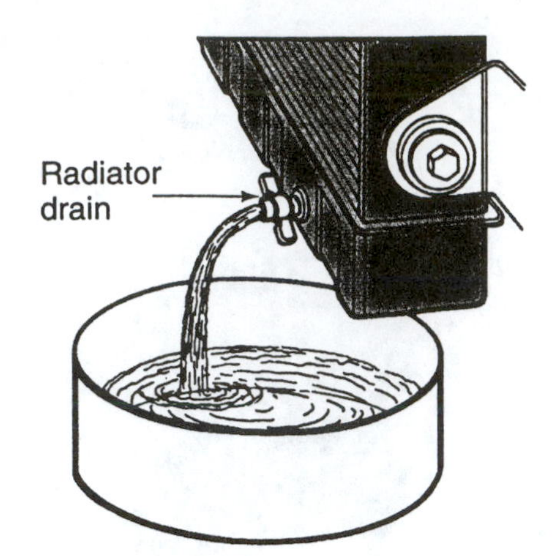

FIGURE 16–14 Most radiators include a plug or petcock to allow draining the coolant. Be sure to catch the coolant for proper disposal. *(Courtesy of Chrysler Corporation)*

16.2.4 Changing Coolant

Dirty coolant should be changed and either recycled or disposed of. Many sources recommend changing coolant every two years.

To change coolant, you should:

1. Run the engine to bring the coolant to operating temperature and shut the engine off.
2. Position a container to catch the draining coolant and open the radiator drain (Figure 16–14).
3. Remove the radiator cap after the CRR has emptied.
4. Remove the cylinder block drain plug (V engines have two) and drain the coolant into a container (Figure 16–15).
5. Drain the CRR if it did not empty.
6. If the system is slightly dirty, close and replace the drain plugs, fill the system with clean water, and repeat steps 1–5. If the system is very dirty, backflush the system as described in Section 16.4.1.
7. Replace the drain plug(s) and close the radiator drain.
8. Refill the system with a coolant mixture of 50% new or recycled antifreeze and 50% water. The water jacket portion of most engines refills quite slowly because the air has to bleed out through the small hole in the thermostat. Some systems have an air bleed valve that should be opened to allow the air to leave the water jackets (Figure 16–16).

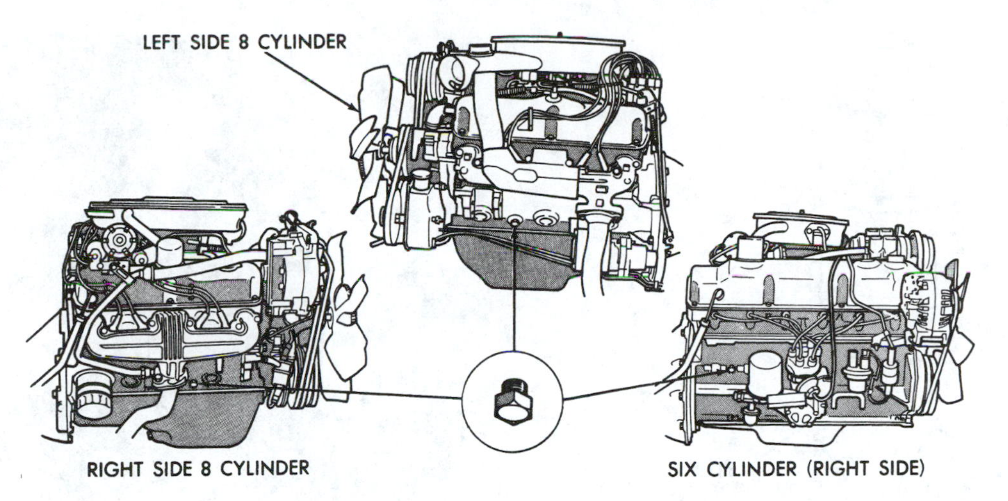

FIGURE 16–15 An in-line engine usually has one drain plug (left); a V-type engine usually has a drain plug in each cylinder bank. (*Courtesy of Chrysler Corporation*)

9. Replace the radiator cap. Run the engine until the thermostat opens and the upper radiator hose becomes hot. On some vehicles, you should turn the heater on to open the valve and allow coolant to flow through the heater core.
10. Stop the engine. Fill the radiator to the filler neck seat and the CRR to the proper level.
11. Replace the radiator cap.

✓SERVICE TIP

Leave the radiator cap in place as you begin draining the system; the coolant will be pulled out of the CRR. When the CRR is empty, remove the radiator cap to speed up draining.

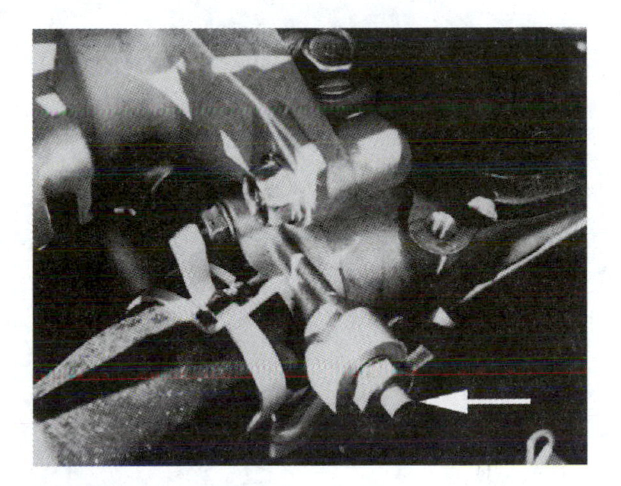

FIGURE 16–16 Some cooling systems include a bleed valve to allow easy removal of air as the water jackets are filled.

✓SERVICE TIP

If using pure antifreeze and water, locate the cooling-system capacity specification and pour one-half this amount of new or recycled antifreeze into the system. Fill the radiator with water.

✓SERVICE TIP

If using extended-life antifreeze, distilled water is recommended. Chlorine from tap water can react with this antifreeze to reduce corrosion protection qualities.

✓SERVICE TIP

Used antifreeze should be properly handled for several reasons. It is poisonous, but animals such as dogs and cats will drink it because it has a sweet taste.

✓SERVICE TIP

Many areas consider antifreeze to be hazardous waste, and shops and individuals who pour antifreeze into the ground or down the drain may be subject to severe fines.

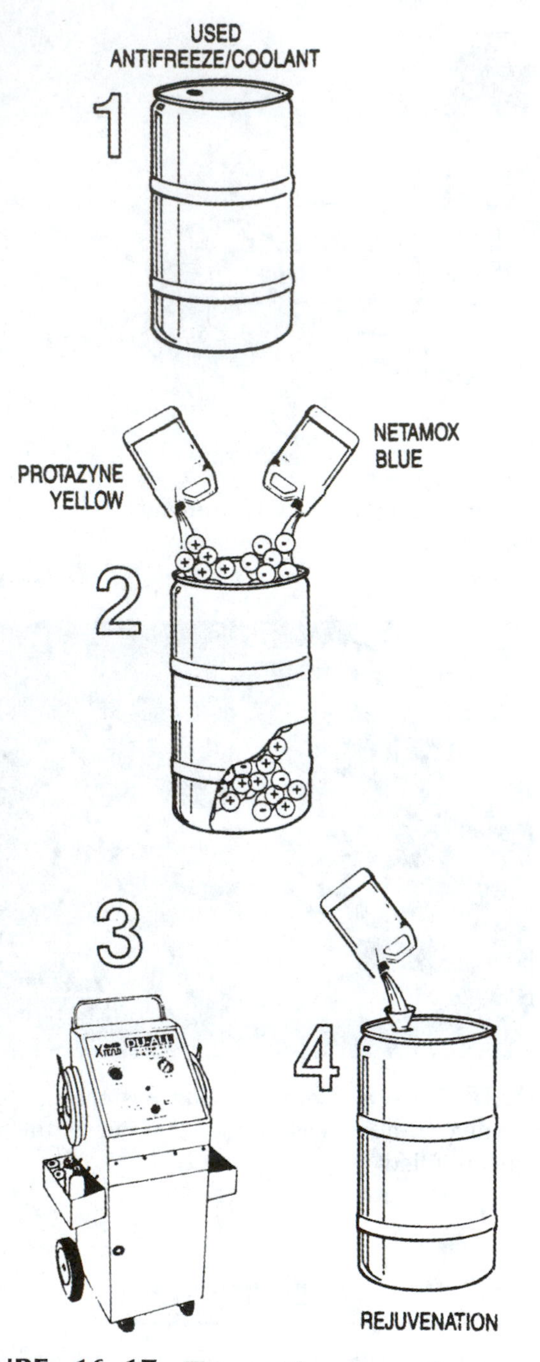

FIGURE 16–17 This antifreeze-recycling process adds two chemicals to the used coolant before it is run through the recycling unit. After reclaiming, new antifreeze is added to adjust the coolant concentration. (*Courtesy of Wynn Oil Company*)

FIGURE 16–18 A coolant exchange unit is used to pull out the old coolant and replace it with a fresh water and antifreeze mixture. (*Courtesy of Robinair, SPX Corporation*)

Antifreeze-recycling machines allow shops to reclaim antifreeze for reuse. These machines use several different processes to filter out impurities; chemicals are added to restore the additives. Most recycling units start with dirty, used coolant and end up with a clean 50–50-mix coolant ready for reuse. Companies in many metropolitan areas provide recovery and reclaiming services (Figure 16–17).

At this time, many shops use an exchange unit to allow them to replace the coolant easier and faster as well as doing a better job by reducing the chance of air bubbles trapped in the system (Figure 16–18). These machines pump fresh coolant into the system as the old coolant is removed and stored in the machine. Some units include a vacuum fill procedure to remove any air pockets in the cooling system. Some coolant exchange units also are designed to recycle the coolant removed from the vehicle.

To change coolant using a coolant exchange unit:

1. Connect the machine to the vehicle. This is normally done by removing the upper radiator hose and installing a pair of adapters.
2. Follow the machine's operating procedure to complete the coolant exchange.

16.2.4.1 Air Locks. Air locked in the water jackets, heater core, or hoses will prevent complete filling of the cooling system. The air is trapped by the closed thermostat or by a system design that has portions of the system above the filler cap. Systems with air bleeds reduce this problem by making it easier to remove the air at the air bleeds, but air bleeds are not used in all systems, and they can be difficult to locate when they are used.

A recent innovation is a tool that evacuates the cooling system through the cooling system filler opening

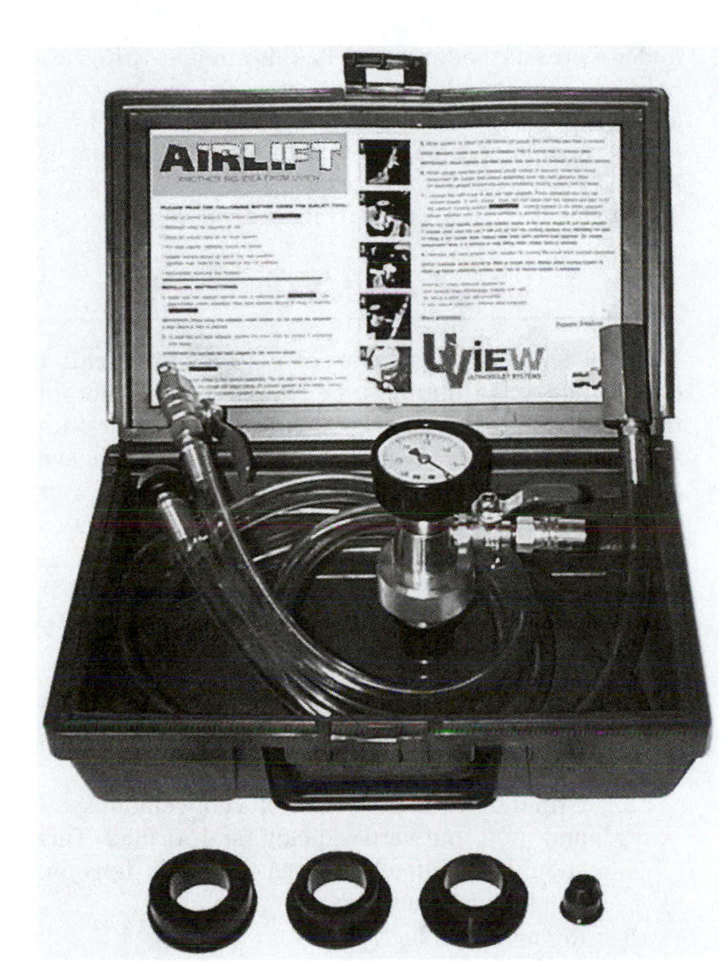

FIGURE 16–19 An airlift unit is designed to evacuate the cooling system of all air, check for vacuum leaks, and quickly refill the system completely with new coolant. It attaches to the radiator filler neck with another connection to a shop air hose and then to the coolant container. *(Courtesy of UView UltravioletSystems)*

(Figure 16–19). After pulling the system into a vacuum, a shut-off valve is closed to trap the vacuum. A leak in the system is indicated if it does not hold the vacuum. If the system is tight and holds the vacuum, a supply of coolant is connected to the device, and the valve is opened. The coolant will be pulled into the system by the vacuum, completely filling the system in a short period of time.

✓ REAL WORLD FIX

The 1996 Chevrolet Blazer (33,000 miles) has a noise problem; it sounds like water is sloshing around under the glovebox. The technician suspects that the evaporator drain is plugged. The drain was located and checked and found to be okay.

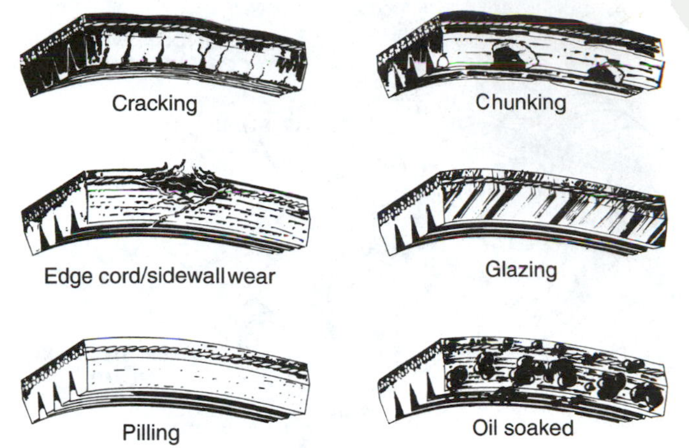

FIGURE 16–20 A drive belt should be replaced if it shows any of these faults. *(Courtesy of The Gates Rubber Company)*

Fix: Following advice, the technician checked for a heater core air lock. A sludgy buildup was found at the heater core connections, so the system was flushed and the coolant was replaced. Making sure that all of the air was bled out of the heater core fixed this noise problem.

16.2.5 Drive Belt Inspection and Replacement

Engine drive belts should be checked periodically for damage and proper tension. If a belt shows excessive wear, severe glazing, rubber breakdown, or frayed cords, it should be replaced. Small cracks on the ribs of a V-ribbed belt are acceptable, but if chunks of ribs are missing, the belt should be replaced. As mentioned earlier, the failure of many belts cannot always be foreseen, so, to be on the safe side, many sources recommend belt replacement at four- or five-year intervals (Figure 16–20). The automatic tensioner on some vehicles includes a belt stretch indicator; if belt stretch of more than 1% is indicated, the belt should be replaced.

A belt that is too loose will slip. Slippage causes an annoying, screeching noise as the engine accelerates. It also causes glazing of the belt, which can cause more slipping. A belt that is too tight causes an excessively high load on the bearings of the components driven by the belt. Traditionally, belt tension is checked by pushing the center of the belt inward and then pulling it outward. The belt should deflect about 1/8 to 1/4 inch

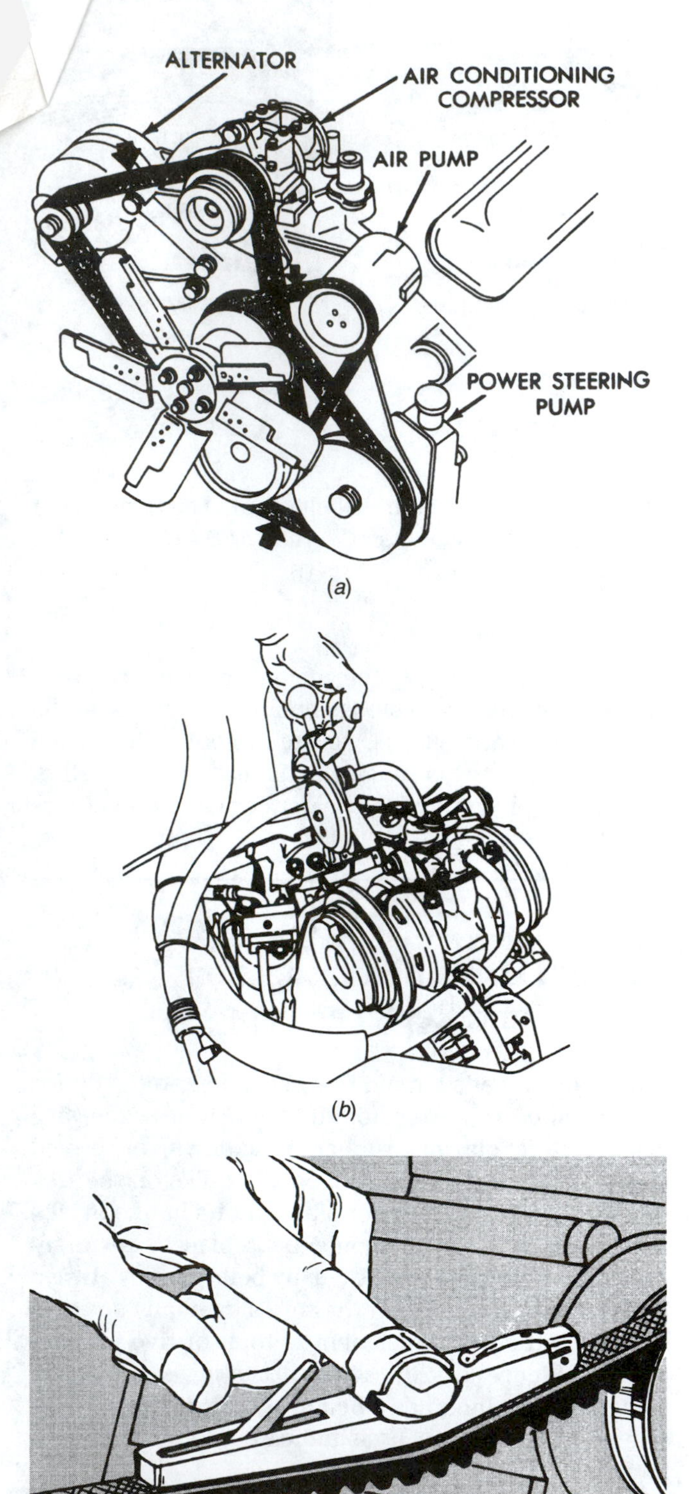

(c)

FIGURE 16–21 Drive belt tension can be checked by measuring how far the belt deflects under a light pressure (*a*) or by using a tension gauge, which is more accurate (*b* and *c*). (a *and* b *are courtesy of DaimlerChrysler Corporation;* c *is courtesy of The Gates Rubber Company*)

under a pressure of around 5 lb; this amount varies according to the length of the belt span. This is not a very accurate method of checking tension: Most manufacturers recommend using a belt tension gauge that is hooked onto the belt and uses a scale to show tension. This gauge is most accurate (Figure 16–21).

✓SERVICE TIP

If the belt shows signs of failing early, check for **parallel** or **angular misalignment** (Figure 16–22). Either of these conditions can cause excessive belt temperature and shortened life (Figure 16–23). Pulley and sheave alignment is checked by sighting along the pulleys or placing a straightedge or round bar along the pulleys.

To replace a drive belt that uses a set tensioner, you should:

1. Locate the device that is used to adjust belt tension and loosen both the adjuster bolt and the pivot bolt (Figure 16–24).
2. Grip the belt at the center of a convenient span and pull outward quickly and firmly. This should completely loosen the belt tension, making installation easy.
3. Remove the belt.
4. Compare the old and new belts to ensure you have the correct replacement.
5. Place the new belt in position, making sure that it is seated in each pulley groove.
6. Swing the belt adjuster outward, tightening the belt to the correct tension, and tighten the adjuster and pivot bolts. Some units include a provision for a wrench or pry bar to make this procedure easier. A belt pulley jack can also be used. Do not use excessive prying force on fragile components (Figure 16–25).
7. Connect a tension gauge and check for correct tension. Readjust the belt if necessary (Figure 16–26).
8. Start the engine and check for proper belt operation. Excessive belt slap indicates a need for readjustment.

✓SERVICE TIP

Some technicians test belt tension by trying to rotate the alternator pulley (with the engine off) by pushing or pulling on a fan blade. If the pulley can be rotated, the belt is too loose.

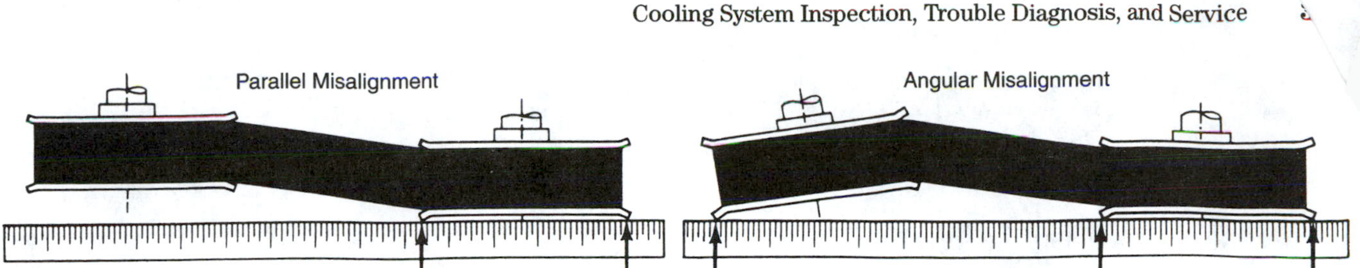

FIGURE 16–22 Misalignment of the pulleys can cause belt noise and excessive wear of the belt and pulleys. (*Courtesy of The Gates Rubber Company*)

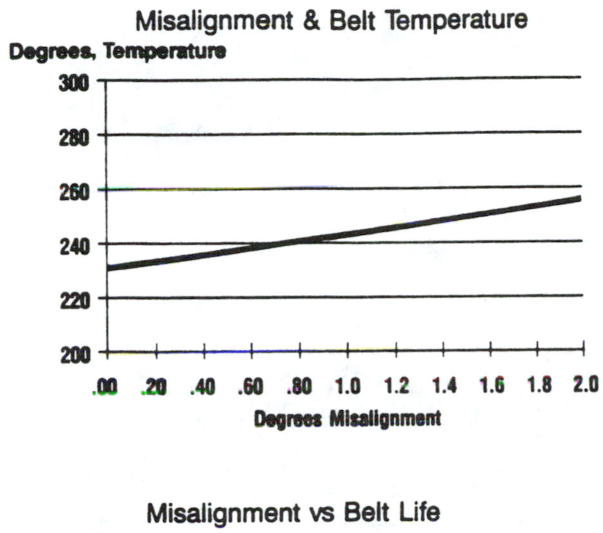

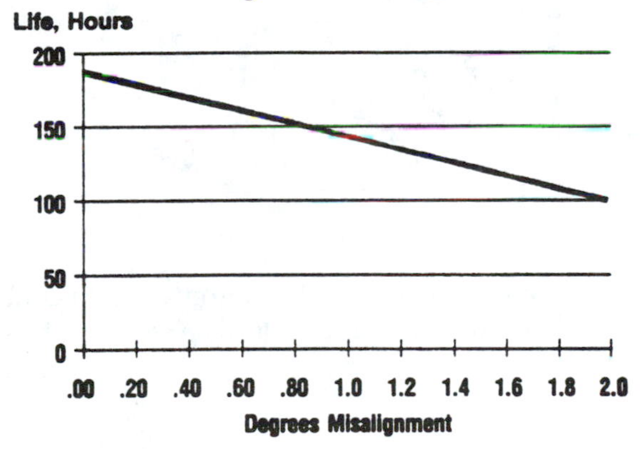

FIGURE 16–23 As shown here, a small amount of misalignment can increase drive belt temperature and wear. (*Courtesy of The Gates Rubber Company*)

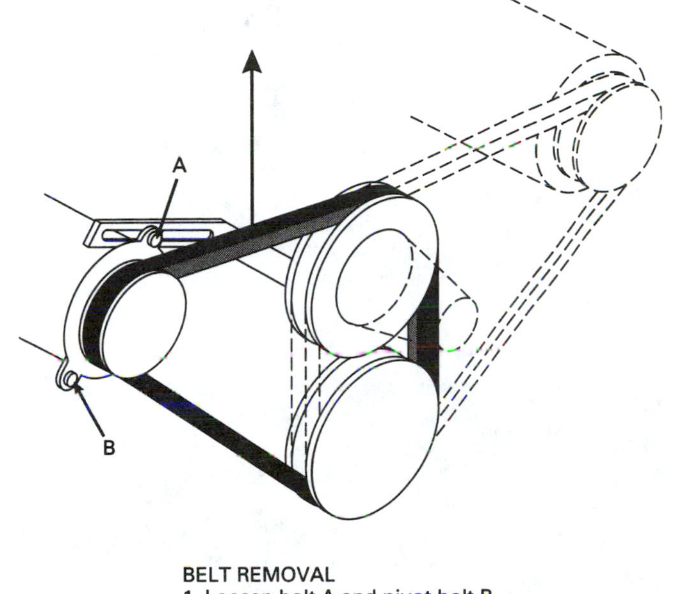

FIGURE 16–24 To remove a drive belt, loosen the adjuster and pivot bolts and pull sharply at the center of the belt. This loosens the adjustment, making installation of the new belt easier.

To replace a drive belt using an automatic tensioner, you should:

1. Note the routing of the belt.
2. Relieve the belt tension and slip the belt off the pulleys. A wrench can be used for this procedure on some tensioners; others provide for a pry bar (Figure 16–27).
3. Install the belt on some pulleys, rotate the tensioner, slide the belt into the proper position, and release the tensioner.
4. Check to ensure proper belt placement on each pulley (Figure 16–28).
5. Start the engine and check for proper belt operation.

16.2.6 Hose Inspection and Replacement

Radiator and heater hoses should be checked periodically. As with drive belts, potential problems can be difficult to see. Signs of failure include cracks, cuts, swelling, and hardening of the hose material. Some hoses become hard, stiff, and brittle, and others become soft and swollen. A hose can appear to be good on the outside, but the inner layer can collapse and

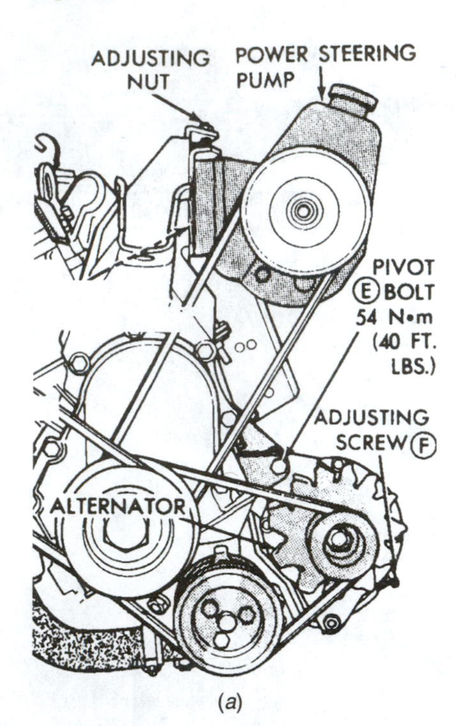

(*a*)

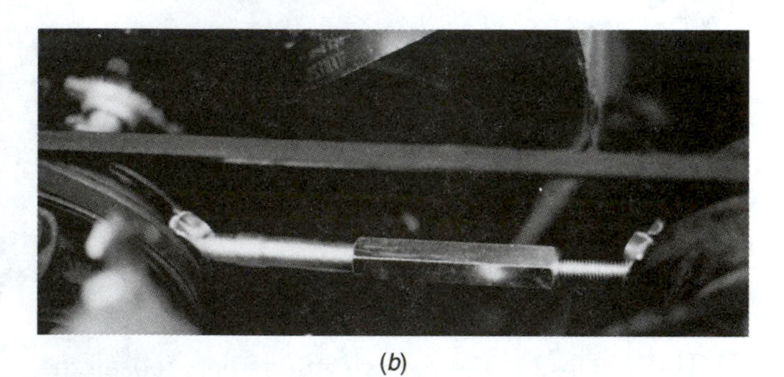

(*b*)

FIGURE 16–25 Depending on the installation, you can pry on the adjustable component using a screwdriver or bar, swing it using a wrench, move it using a threaded adjuster (*a*), or use a pulley jack (*b*). (a *is courtesy of DaimlerChrysler Corporation*)

FIGURE 16–26 This type of belt tension gauge is hooked onto the belt and quickly released to measure belt tension.

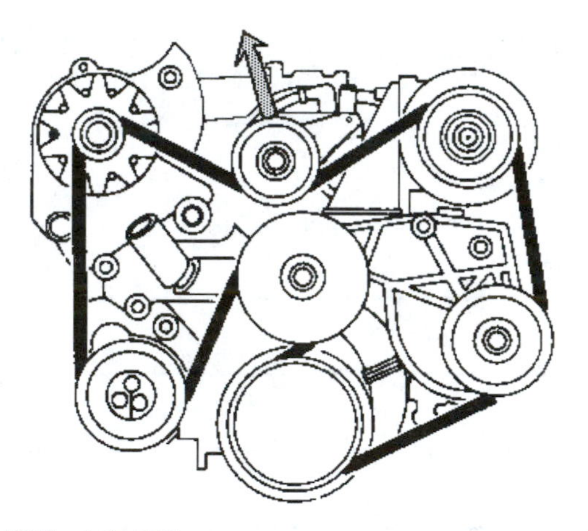

FIGURE 16–27 Moving the automatic tensioner outward allows the serpentine belt to be removed from the pulleys.

partially block coolant flow. Squeeze and bend the hose as you check it for signs of these problems. As with a belt, some sources recommend replacing the hoses at four- or five-year intervals to prevent failure (Figure 16–29).

✓SERVICE TIP

Squeeze the hose near the connectors and compare the feel with the center. Electrochemical degradation usually occurs within 2 inches of the ends of a hose, and soft spots, gaps, or channels indicate a weak hose.

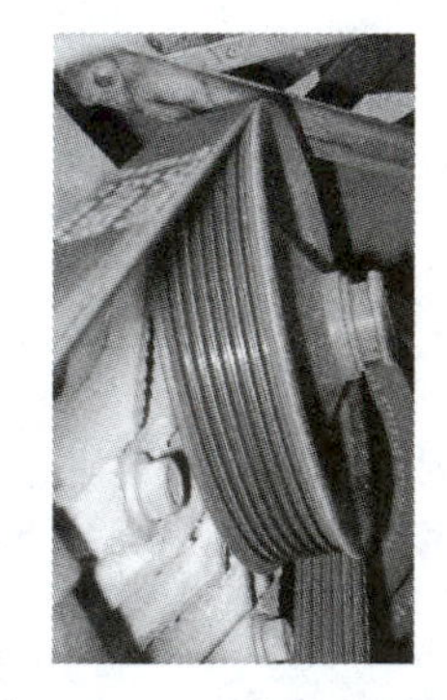

FIGURE 16–28 When a V-ribbed belt is installed, make sure the belt is positioned correctly in the pulley grooves.

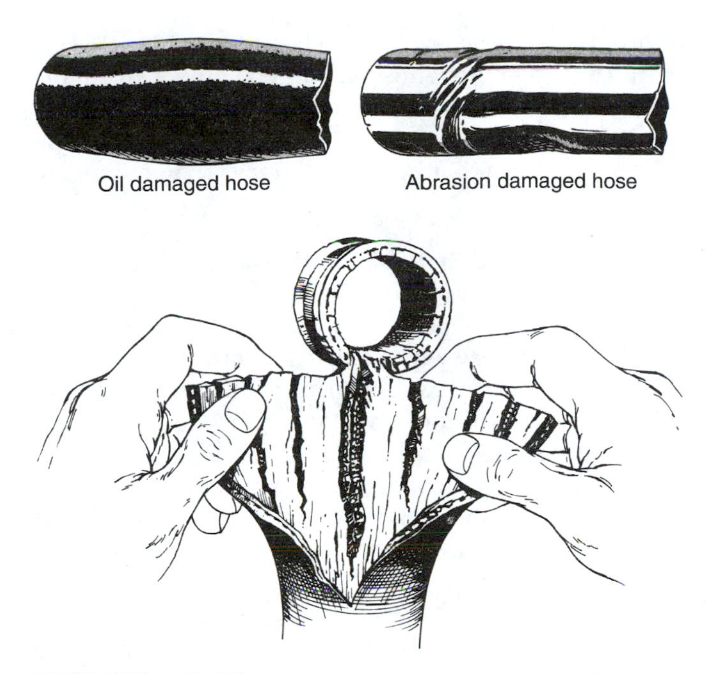

FIGURE 16–29 Oil damage and abrasion damage from rubbing are seen during a visual inspection. Internal electrochemical degradation. (*Courtesy of The Gates Rubber Company*)

✓SERVICE TIP

The hose is usually stuck in place; to prevent damage to the connector, many technicians slice the end of the hose with a sharp knife. Then the hose end can be peeled loose (Figure 16–30).

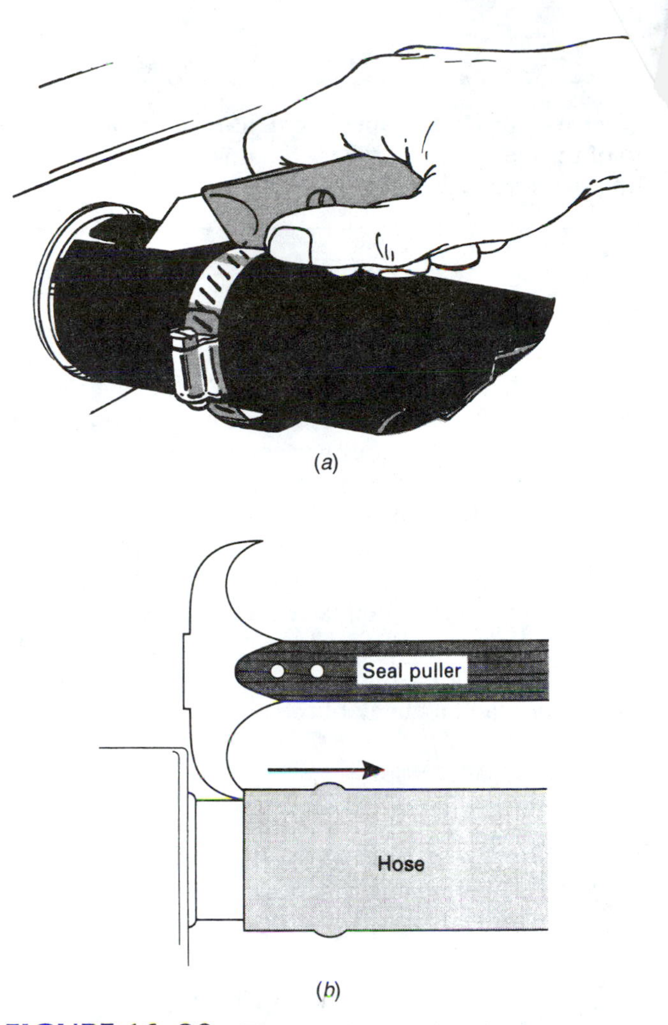

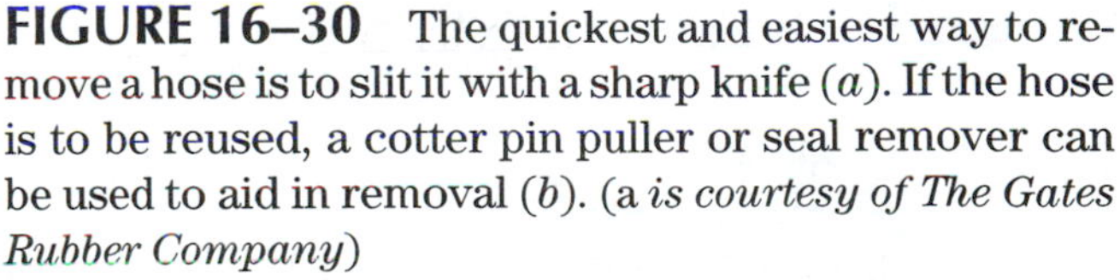

FIGURE 16–30 The quickest and easiest way to remove a hose is to slit it with a sharp knife (*a*). If the hose is to be reused, a cotter pin puller or seal remover can be used to aid in removal (*b*). (a *is courtesy of The Gates Rubber Company*)

To replace a hose, you should:

1. Partially drain the coolant so the level is below the hose connections.
2. Loosen the clamp at each end of the hose.
3. Slide the hose off the connection.
4. Clean any rust or corrosion from the connection.
5. Slide the new hose with clamps (loose) over the connectors, engine end first, and rotate the hose on the connections until it is properly positioned and free of kinks. Some manufacturers recommend the use of a water-resistant sealant at these connections.
6. Position the clamps so they are right next to the bead on the connector and tighten them (Figure 16–31).
7. Fill the cooling system, run the engine until it reaches operating temperature, and check for leaks as you retighten the clamps.

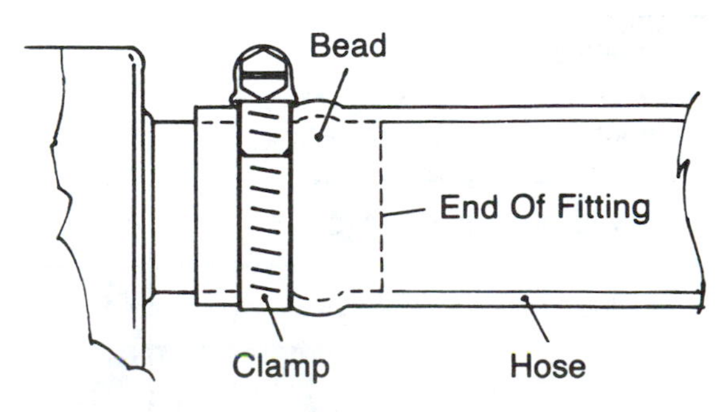

FIGURE 16–31 The hose clamp should be located next to the bead on the connector, and the clamp should be tightened to the correct torque.

✓SERVICE TIP

A screw-type clamp should be tightened to 30 to 40 in.-lb of torque (the point where the hose rubber squeezes into the slots of the clamp).

16.3 TROUBLE DIAGNOSIS

As mentioned, the three major cooling system problems are leaks, overcooling, and overheating. Additional problems are improper gauge operation, which indicates that a normally operating system is overcooled or overheated, and excessive pressure caused by a faulty cap. A technician normally performs a visual inspection; a check of the coolant antifreeze concentration, condition, and level; a pressure test on the cap and system for leaks; a test for combustion leakage; a test of thermostat operation (if warranted by the inspection); and a check of the actual system temperature. The tests do not have to be performed in this order: The order depends on the equipment available and the nature of the problem. A troubleshooting chart such as the one shown in Figure 16–32 is often followed to ensure that important items are not skipped.

16.3.1 Cooling System Inspection

Like system inspection for A/C diagnosis, cooling system inspection is used to locate any obvious problems that need correction. It is also a test that can involve all of your senses. Besides a visual inspection, you feel for bearing looseness, hard or soft hoses, temperatures of various items, and flow or pressure of coolant in a hose; you listen for malfunctioning water-pump bearings and gurgles or thumps inside the radiator and water jackets; and you notice the smell of hot antifreeze leaking out of a system. Many shops use an inspection checklist to provide inspection information for the customer (Figure 16–33). The cooling system inspection consists of a number of steps.

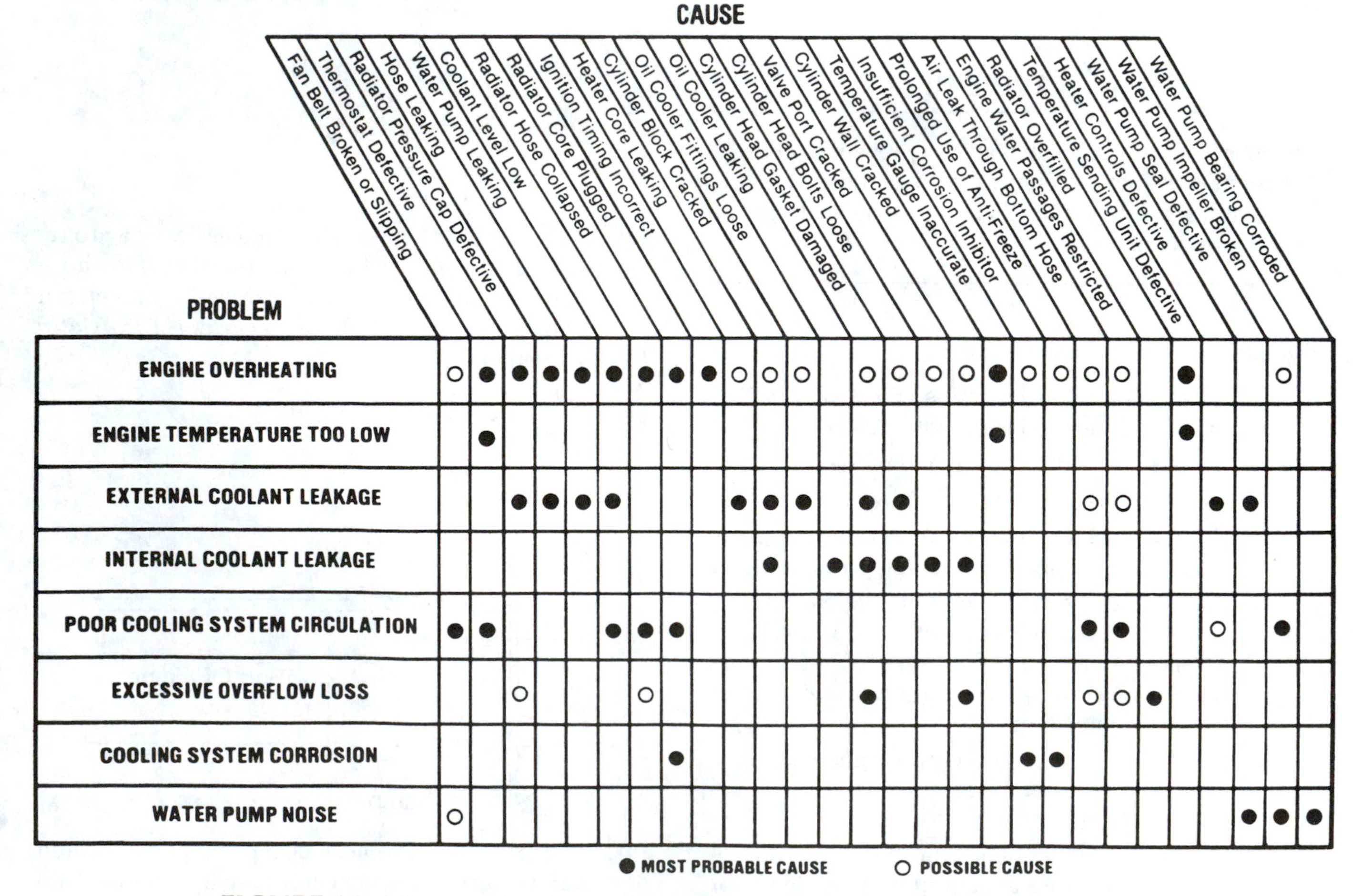

CAUSE

PROBLEM	Fan Belt Broken or Slipping	Thermostat Defective	Radiator Pressure Cap Defective	Hose Leaking	Water Pump Leaking	Coolant Level Low	Radiator Hose Collapsed	Radiator Core Plugged	Ignition Timing Incorrect	Heater Core Leaking	Cylinder Block Cracked	Oil Cooler Fittings Loose	Oil Cooler Leaking	Cylinder Head Gasket Damaged	Cylinder Head Bolts Loose	Valve Port Cracked	Cylinder Wall Cracked	Temperature Gauge Inaccurate	Insufficient Corrosion Inhibitor	Prolonged Use of Anti-Freeze	Air Leak Through Bottom Hose	Engine Water Passages Restricted	Radiator Overfilled	Temperature Sending Unit Defective	Heater Controls Defective	Water Pump Seal Defective	Water Pump Impeller Broken	Water Pump Bearing Corroded
ENGINE OVERHEATING	○	●	●	●	●	●	●	●	●	○	○	○		○	○	○	○	●	○	○	○	○		●			○	
ENGINE TEMPERATURE TOO LOW		●																●						●				
EXTERNAL COOLANT LEAKAGE			●	●	●	●				●	●	●		●	●						○	○			●	●		
INTERNAL COOLANT LEAKAGE											●		●	●	●	●	●											
POOR COOLING SYSTEM CIRCULATION	●	●				●	●	●													●	●			○		●	
EXCESSIVE OVERFLOW LOSS			○				○							●			●				○	○	●					
COOLING SYSTEM CORROSION								●											●	●								
WATER PUMP NOISE	○																									●	●	●

● MOST PROBABLE CAUSE ○ POSSIBLE CAUSE

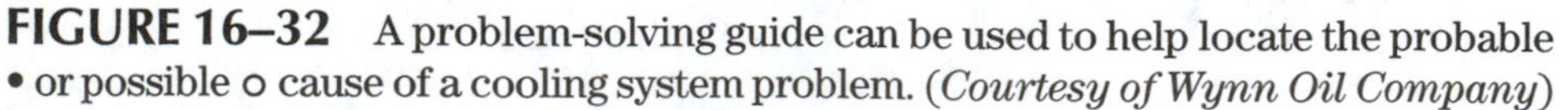

FIGURE 16–32 A problem-solving guide can be used to help locate the probable • or possible o cause of a cooling system problem. (*Courtesy of Wynn Oil Company*)

CUSTOMER ______________ PHONE ______________ DATE __________

VEHICLE/MAKE ______________ YEAR __________ LICENSE __________ MILEAGE __________

INSPECTION AREA	INSPECTION RESULTS						RECOMMENDATION
HOSES:	OK	HARD/ BRITTLE	SPLIT/ CRACKED	SOFT/ SPONGY	OIL SOAKED	REPLACE	
UPPER RADIATOR							
LOWER RADIATOR							
BY PASS							
HEATER							
FANBELTS:	OK	FRAYED	SPLIT/ CRACKED	GLAZED	LOOSE	REPLACE	
ALT./GEN.							
POWER STEERING							
A-C COMPRESSOR							
WATER PUMP:	OK	LOOSE SHAFT BEARING	BLEED HOLE LEAKS	GASKET LEAKS	FAN CLUTCH LOOSE	REPLACE/ POWER FLUSH	
COOLANT RECOVERY TANK:	OK	DIRTY	EMPTY	REFILL	MISSING	INSTALL	
RADIATOR CAP PRESSURE CHECK:	OK	SWOLLEN GASKET	BROKEN GASKET	CORRODED CAP	WEAK CAP	REPLACE	
RADIATOR PRESSURE CHECK:	OK	FINS PLUGGED	RUSTY/ OILY	LEAKS	BLOCKED CORES	ROD OUT/ POWER-FLUSH	
COOLANT:	OK	DIRTY	RUSTY	OILY	FOAMY	REPLACE/ POWER-FLUSH	
THERMOSTAT:	OK	REPLACE/ OPEN-MISSING	STUCK/ CLOSED	FLUSH TEST	GASKET LEAKS	REPLACE	

FIGURE 16–33 Many technicians use a checklist during cooling system inspection. The checklist provides a professional display of any problems for the customer and helps ensure that important checks are not skipped. (*Courtesy of Wynn Oil Company*)

With the system cold and the engine off, the checks are as follows:

1. Remove the radiator cap and check the level, condition, and antifreeze concentration of the coolant.
2. Using a light, look for corrosion, dirt, and rust inside the radiator (Figure 16–34). Insert a finger into the filler neck and rub it around the surrounding area. Visible dirt or rust in the radiator or on your finger indicates a need for flushing.
3. Check the exterior of the radiator for stains, which might indicate a leak, and debris on the front or bent fins, which can interfere with airflow.
4. Check the drive belt(s) for possible damage and proper tension.

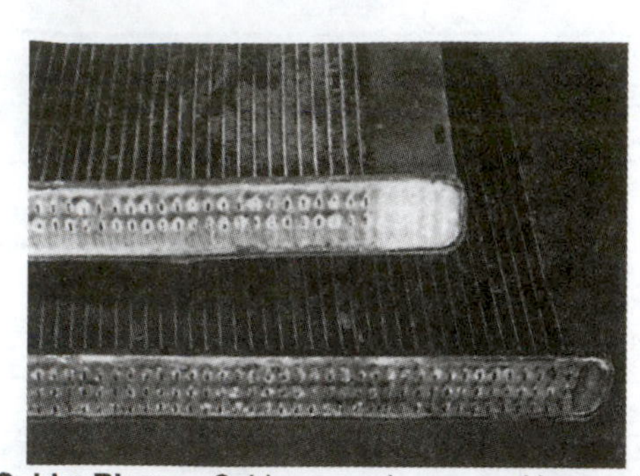

Solder Bloom—Solder corrosion caused by poorly inhibited antifreeze. Tube-to-header joints are weakened, and corrosion can restrict coolant flow.

Internal Deposits—Rust and leak inhibitors can form solids that collect in the cooling system and restrict flow.

Fin Deterioration—A chemical deterioration of the fins most often caused by road salt or sea water.

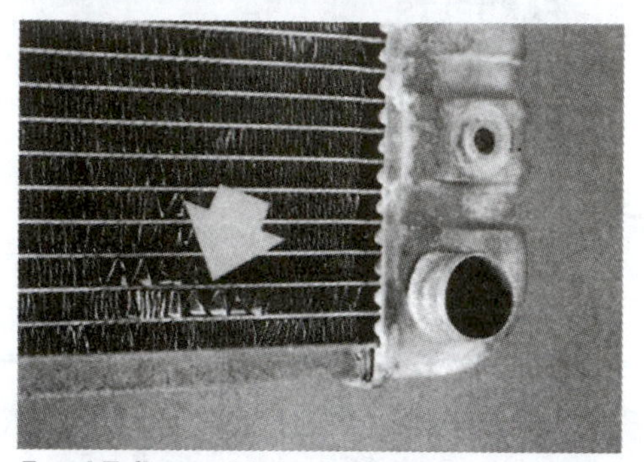

Fin Bond Failure—A loss of solder bond between fins and tubes. Fins are loose in core. This causes loss of heat transfer and reduces the strength of the radiator.

Tube-to-Header Leaks—Failure of the solder joint results in coolant loss.

Leaky Tank-to-Header Seam—Failure of the solder joint or cracked headers is generally the result of pressure.

Blown Tank-to-Header Seam—Usually an indication that the radiator has seen extreme pressures.

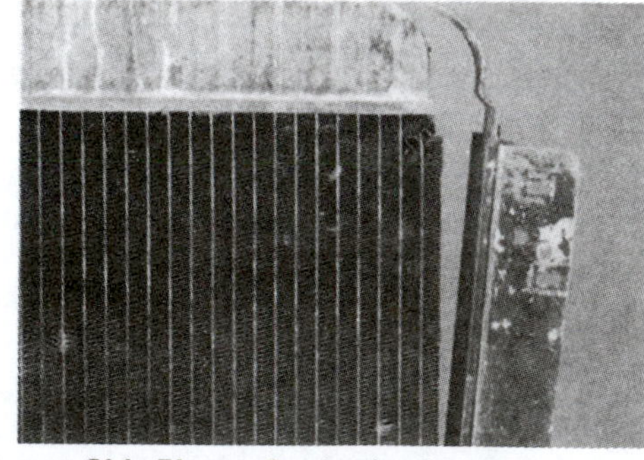

Loose Side Piece—Can lead to flexing of the core and radiator tube failure.

FIGURE 16–34 The most commonly encountered radiator faults. Most of these problems require replacement of the radiator or tanks. (*Courtesy of Modine Manufacturing*)

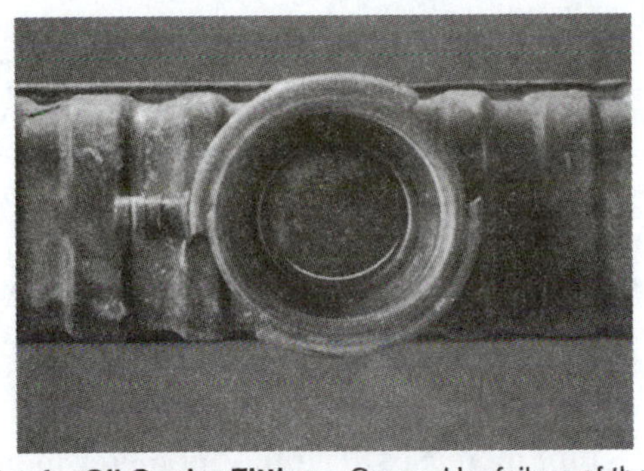

Leaky Oil Cooler Fitting—Caused by failure of the solder joint between the radiator tank and oil cooler.

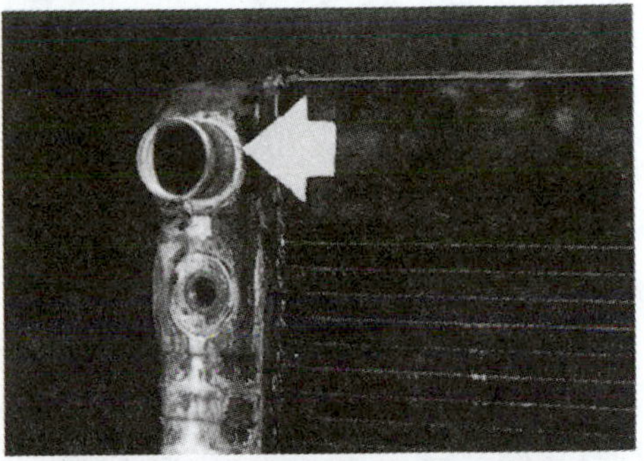

Leaky Inlet-Outlet Fitting—Leaks in this area of the radiator can be caused by fatigue or by corrosion of the solder joint.

Fan Damage—Minor collisions of failed water pump can result in damage to the radiator.

Over Pressurization—Excessive pressure in the radiator by a defective pressure cap or engine exhaust gas leak can destroy the radiator.

Electrolysis—Stray electric current can cause excessive corrosion of metal components.

Electrolysis—Stray electric current can cause an electrochemical reaction that will produce voids in tubes.

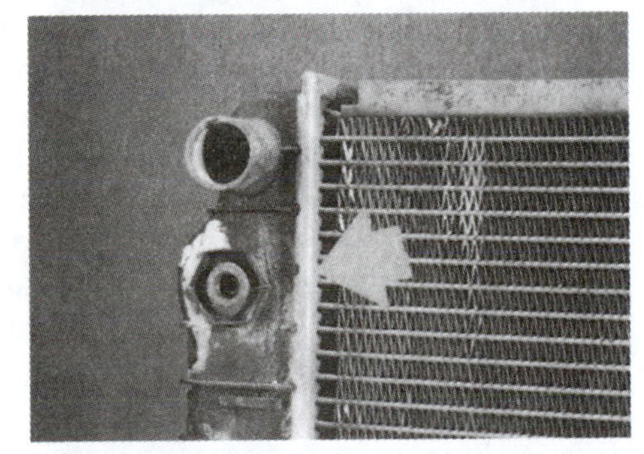

Cracked Plastic Tanks—High stress in radiator can cause premature failure of the plastic tanks.

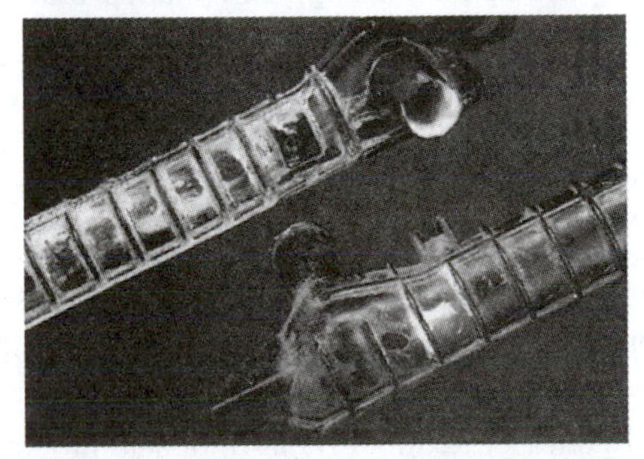

Steam Erosion—Steam can break down the plastic tank which will produce thinning and ultimately holes in the plastic tank. Frequently, white deposits are also found.

FIGURE 16–34 (Continued)

FIGURE 16–35 If the fan is rotated, you should notice a definite drag of the fan clutch. Also check for excessive play in either a front-to-back or sideways direction (arrows).

5. Check the hoses for visible damage and signs of leaking and bend and squeeze the hoses as you check them.
6. Check the water pump for signs of leakage. Try moving the pump shaft from side to side; there should be no perceptible movement.
7. *On RWD cars,* check the fan, fan clutch, and fan shroud. Fan blades should be secure and straight. The fan should rotate at the clutch, but there should be definite resistance. There should be no oil leaks from the fan clutch body. Grip the tips of two opposite blades, and try to move the fan from side to side and from front to back. More than 1/8 inch of movement indicates a faulty fan clutch (Figure 16–35). *On FWD cars,* check the fan, motor, and shroud-mounting bracket for damage, proper mounting, and free rotation of the fan.

With the engine running, the checks are as follows:

1. Place a thermometer into the radiator filler neck and start the engine.

   OR: check the upper radiator hose temperature using an infrared thermometer.
2. Squeeze the upper radiator hose. With the engine cold, the thermostat should be closed and there should be little or no flow through the hose.
3. Check the drive belts. They should be running true, with a small amount of whipping at the longest span or the last span before the crankshaft pulley. All pulleys and shafts should be running smoothly, with very little or no runout.
4. After 5 to 10 minutes, the engine should reach operating temperature, and the thermostat should open. At this time, coolant flow can be felt in the upper hose, and the hose temperature increases. The thermometer indicates the coolant temperature and the opening point of the thermostat (Figure 16–36).
5. Increase the engine speed as you squeeze the upper radiator hose. The flow inside the hose should increase as the water pump speeds up.
6. The electric cooling fan of a FWD car should start operation when the engine warms up, about 20 to 40°F above the thermostat opening point.

✓SERVICE TIP

Infrared thermometers make temperature checks much easier; merely point the unit at the heat source to be checked. The temperature is then read on the unit's display (Figure 16–37).

With the system warm to hot and the engine off, the checks are as follows:

1. Compare the temperatures indicated on the thermometer and on the vehicle temperature gauge; they should be equal.
2. Listen to the cooling system. Light snapping, popping noises are normal heat expansion and contraction noises. A slight pressure release at the radiator cap (when in place) is also normal. Heavy thumping sounds from inside the engine indicate overheating in localized areas; the noise is probably caused by an internal restriction.
3. On RWD cars, spin the fan. With the engine warm, a faulty fan clutch is indicated if the fan spins more than one revolution.

16.3.2 Pressure Tests

Pressure tests are used to check for leaks and proper operation of a pressure cap. A pressure test can also be used to check for an internal combustion leak.

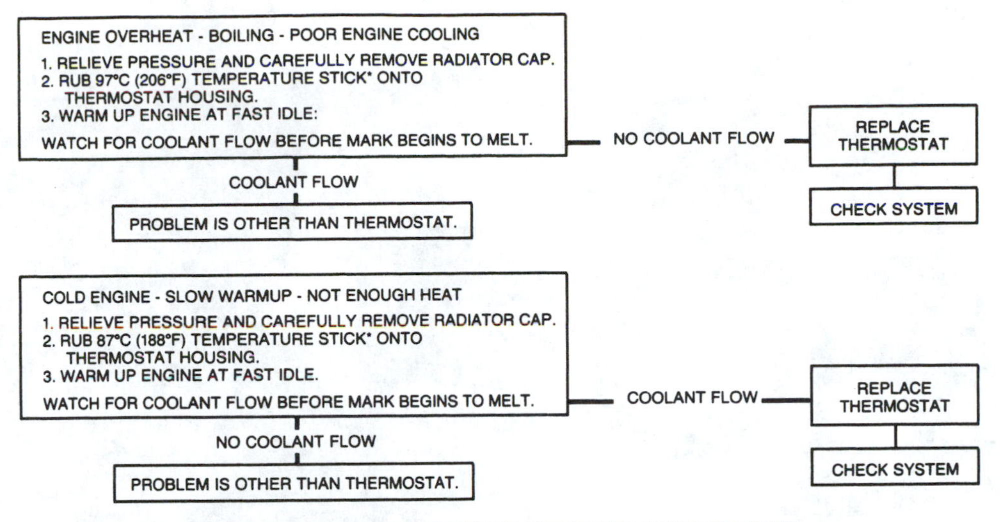

* NOTE: THE TEMPERATURE STICK IS A PENCIL LIKE DEVICE WHICH HAS A WAX MATERIAL CONTAINING CERTAIN CHEMICALS WHICH MELT AT A GIVEN TEMPERATURE. TEMPERATURE STICKS CAN BE USED TO DETERMINE A THERMOSTAT'S OPERATING TEMPERATURE BY RUBBING 87°C (188°F) AND 97°C (206°F) STICKS ON THE THERMOSTAT HOUSING. THE MARKS MADE BY THE STICKS SHOULD MELT WHEN COOLANT TEMPERATURES OF 87°C (188°F) AND 97°C (206°F) ARE REACHED, RESPECTIVELY. THESE TEMPERATURES ARE THE NORMAL OPERATING RANGE OF THE THERMOSTAT.

(*a*)

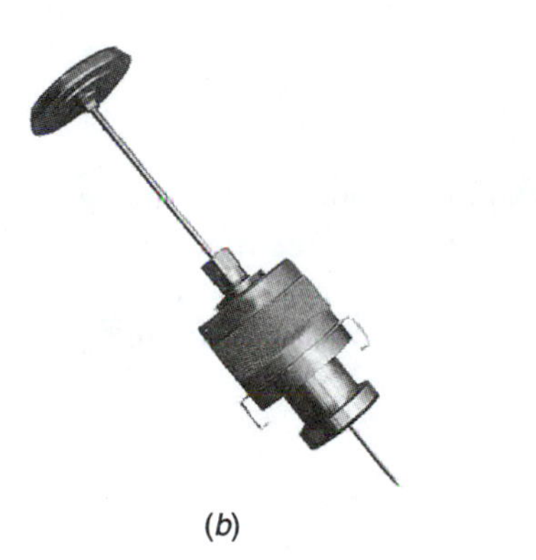

(*b*)

(*c*)

FIGURE 16–36 A faulty thermostat causes the coolant flow to occur at too high or too low a temperature (*a*). A thermometer installed in the radiator filler neck can be used to check the temperature of the coolant. The adapter is also used for pressure checks (*b* and *c*). (a *is reprinted with permission of General Motors Corporation;* b *and* c *are courtesy of Waekon Industries*)

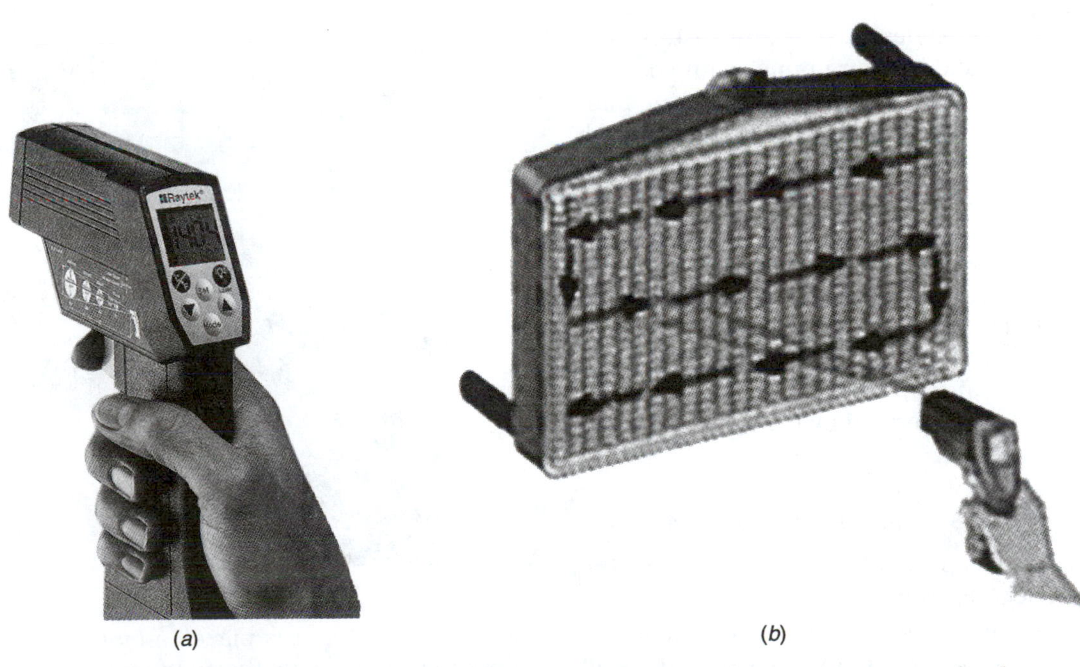

(*a*) (*b*)

FIGURE 16–37 An infrared thermometer allows fast temperature checks. Merely point the unit at the spot to be checked, pull the trigger, and read the temperature. (*Courtesy of Raytek Corporation*)

FIGURE 16–38 Cooling system pressure testers include a style using a hand pump (*a*) and ones that use shop air pressure or cooling system pressure (*b*). (a *is courtesy of Stant Manufacturing;* b *is courtesy of Waekon Industries*)

There are several styles of pressure testers. The most common is a hand pressure pump with adapters that connect to the pressure cap and radiator filler neck. An adapter that connects into the system through the temperature sending unit port is also available for some hand pumps. Another style of pressure tester connects to the system through the coolant overflow hose. This unit is essentially a pressure regulator connected to shop air. A third style of tester replaces the radiator cap with an adapter that allows a pressure or temperature probe to be inserted. This unit can use a pressure regulator and shop air as a pressure source or measure the pressure generated by the system (Figure 16–38).

To test a cap using a hand pump tester, you should:

1. Remove the cap from the system and clean the cap gasket with water.
2. Select the proper adapter for the cap, immerse the cap in water to wet the gasket, and install the cap on the adapter and the adapter on the tester (Figure 16–39).

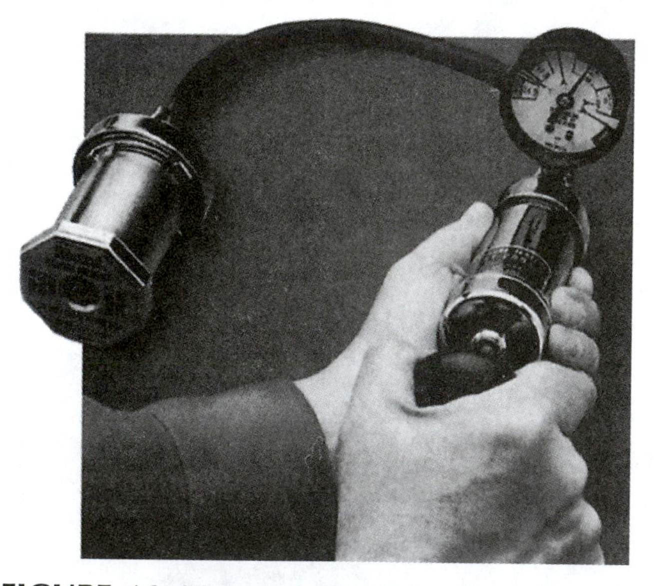

FIGURE 16–39 After wetting the cap gasket with coolant or water, it is placed on the adapter. Then the tester is pumped to the point where pressure is released. (This pressure should match the cap's rating.) (*Courtesy of Stant Manufacturing*)

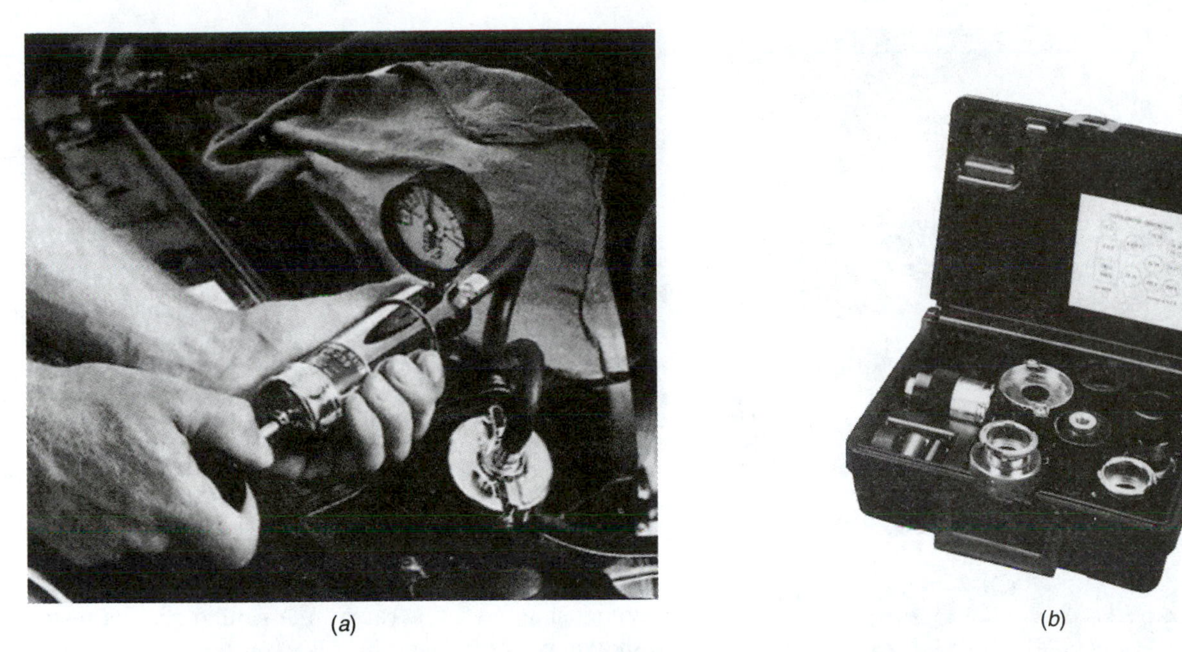

(a) (b)

FIGURE 16–40 A system's pressure is checked by connecting the tester to the radiator and bringing it up to the proper pressure (*a*). Adapters are available to fit the various filler neck styles (*b*). (*Courtesy of Stant Manufacturing*)

3. Slowly operate the tester pump as you watch the gauge pressure. The pressure should increase to a point and then stop increasing as the cap releases pressure; this point should be the pressure value of the cap. The cap should hold this pressure for 1 minute.
4. Release the pressure and repeat step 3 to make sure the test is accurate. If the cap releases pressure early or late or will not hold pressure, it should be replaced.

To pressure test a system using a hand pump tester, you should:

1. Fill the radiator.
2. Connect the pressure tester to the radiator filler neck (Figure 16–40). Adapters are required to connect the tester to some radiators. *Or* remove the temperature sending unit, install the adapter into this opening, and connect the pressure tester to this adapter. Note that with this connection you are testing the cap also.
3. Slowly pump the tester to the pressure rating of the system or the pressure range indicated on the cap. If using the adapter so that the radiator cap is in place and the cap releases early, replace the cap.
4. Observe the pressure gauge: It should hold steady for at least 2 minutes. If the pressure drops, coolant is probably leaking out; check for leaks in the system.
5. If using the adapter so that the radiator cap is in place, increase the pressure using the pump and observe the coolant overflow hose. The cap should release pressure and cause flow through the hose at its rated pressure.

To pressure test a system using a pressure regulator and shop air pressure, you should:

1. With the system cool, remove the radiator cap, install the adapter, and install the pressure probe into the adapter.

   Or remove the overflow hose from the filler neck and connect the pressure regulator unit to the filler neck connection (Figure 16–41).
2. Connect a shop air hose to the test unit and increase the pressure regulator setting to the rated pressure of the system.
3. After the system pressure is reached, shut off the air supply and watch the pressure gauge. The system should hold pressure for 2 minutes. A leak is indicated if the pressure drops.

(a)

(b)

FIGURE 16–41 A pressure probe with regulator has been installed in the radiator neck using the same adapter shown in Figure 14–31*b* (*a*). Shop air can be put into the system using the fitting (arrow). A similar style of tester is shown in *b*. (a *is courtesy of Waekon Industries*)

CAUTION

Be careful while using either type of pressure tester to check for combustion leaks. With the tester replacing the cap, there is a closed system with nothing to release excess pressure. You must be ready to relieve excess pressure in a harmless manner. Never permit the pressure in the system to exceed 20 psi. In a normal system, the pressure gradually increases as the coolant warms and expands. In a system with a combustion leak, the pressure can increase very rapidly.

To test for a combustion leak, you should:

1. Connect either the hand pump unit or the pressure regulator unit to the radiator filler neck.
2. Start the engine and observe the pressure gauge. A rapid pressure increase indicates a leak from the combustion chamber to the water jacket. Pulsations of the gauge needle also indicate combustion gases being forced into the water jacket. If this happens, you can short out the spark plugs until the pulsations stop to find the bad cylinder. Be ready to release excess pressure (Figure 16–42).

✓SERVICE TIP

Some technicians completely fill the radiator and crank over the engine using the starter. A combustion leak is indicated by pressure pulses on the gauge.

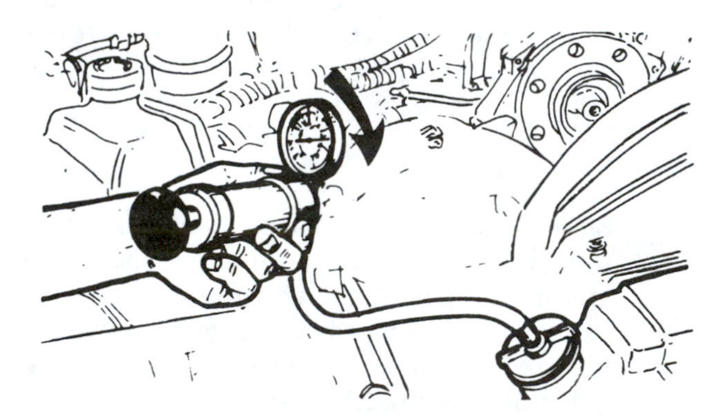

FIGURE 16–42 A combustion leak will cause a rise in system pressure if the engine is started. Be ready to relieve any excess pressure.

16.3.3 Combustion Leak Checks

A combustion leak can be caused by a blown head gasket, warped head or cylinder block, or cracked head or block. This type of leak allows hot, high-pressure gases to pass from the combustion chamber to the water jacket. The result is usually rapid overheating and deterioration of the coolant. Sometimes the leak allows coolant to flow into the combustion chamber; this can cause excess steaming from the exhaust or a hydrostatic lock in a cylinder during cranking (Figure 16–43).

Occasionally, an internal leak allows coolant to enter the oil. Minor leaks cause excess water droplets in the oil; major leaks can cause gray, repulsive-looking oil.

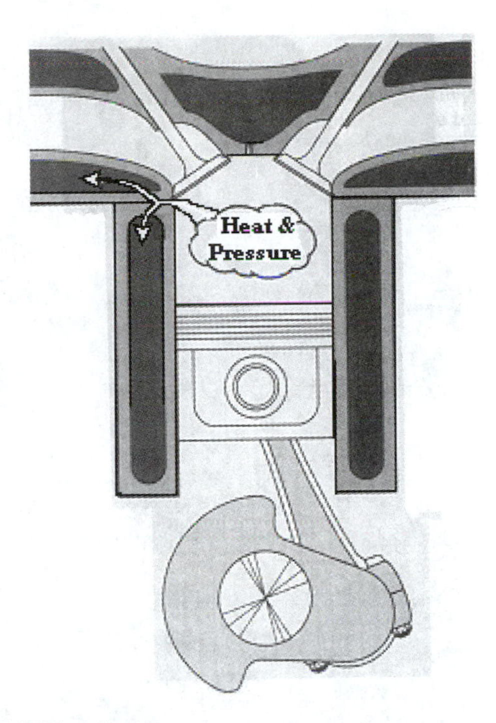

FIGURE 16–43 A combustion leak is caused by a crack or a faulty head gasket, which allows combustion pressure to enter the water jackets.

✓ SERVICE TIP

A simple test to determine whether there is water in the oil is to transfer a few drops of oil from the dipstick to a hot exhaust manifold. Plain oil will spread out and burn off as light-colored smoke. Any water in the oil will sizzle and spatter.

Checks for combustion leaks look for gas bubbles in the coolant or the presence of carbon dioxide (CO_2). CO_2 is indicated by a color change in the test fluid. Exhaust gas analyzers can also be used to test for CO or CO_2 in coolant, but great care should be exercised to prevent coolant from being drawn into the tester.

✓ SERVICE TIP

When using an exhaust gas analyzer, some technicians place a close fitting tube, about 6″ long, into the filler neck. This is used to direct any escaping gases to the tester probe. Many technicians insert the analyzer probe into the CRR because there is usually ample room above the coolant, and the CRR tends to trap any gases escaping from the cooling systems.

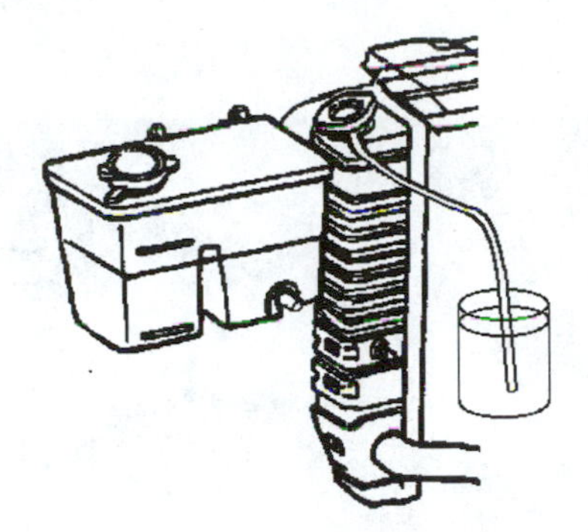

FIGURE 16–44 A simple check for a combustion leak is to immerse the overflow/transfer tube into a container of water and run the engine. Bubbles from the hose indicate a combustion leak.

✓ SERVICE TIP

Some technicians check for HC in the cooling system as an indication of a combustion leak, but this can be misleading because many chemicals can cause false HC readings. It is recommended, if using a modern combustion analyzer, to check for carbon monoxide (CO); a CO reading of greater than 0.02 parts per million (ppm) indicates there is a probable combustion leak into the cooling system.

To check for gas bubbles, you should:

1. Disconnect the overflow hose from the CRR and place it in a container of water (Figure 16–44).
2. Loosen the radiator cap so it will release pressure.
3. Start the engine and watch the end of the overflow hose. A stream of bubbles indicates a combustion leak or a suction leak at the water pump or lower radiator hose. The combustion leak rate increases if the load on the engine is increased.

To check for CO_2 in the coolant, you should:

1. Remove the radiator cap and partially drain the coolant so the level is 2 or 3 inches below the filler neck.
2. Start the engine and let it idle.
3. Partially fill the test unit to the correct level with the test fluid (Figure 16–45).
4. Place the test unit into the filler neck and draw air from the radiator through the test fluid using the aspirator bulb. Continue this step for up to 2 minutes.

(a)

(b)

FIGURE 16–45 A combustion leak tester is partially filled with test fluid (*a*). Working the test bulb draws gases from the system through the fluid (*b*). There is a combustion leak if the fluid changes color.

5. Observe the test fluid. If it stays blue, there is no combustion leak. If the color changes from blue to yellow, there is a combustion leak.

✓SERVICE TIP

The exact cylinder that is leaking can usually be determined through a compression test. Another method is to leave the system connected to the regulated pressure tester for a period of time with the engine shut off. Pressure from the tester can force coolant to fill the faulty cylinder. If the spark plugs are removed and the engine cranked, coolant will be pumped out of the bad cylinder.

16.3.4 Thermostat Check

For some reason, many people believe that an overheating engine can be "cured" by removing the thermostat completely or by replacing it with one with a cooler rating. This is wrong. The cooling system is designed to operate with the restriction provided by a normally operating thermostat and cannot operate properly without it. With all modern, computer-controlled engines, you must use the thermostat specified by the manufacturer to ensure fast warm-up and proper emission-free engine operation. A faulty cooling system overheats and boils the coolant; all a thermostat does in a faulty system is reduce the time to boil.

A thermostat that is stuck closed causes the engine to overheat; one that is stuck open causes overcooling, with slow, prolonged warm-up. Either of these conditions should show up during a visual inspection. The test described here is used to either confirm a suspicion or evaluate an unknown thermostat.

To test a thermostat, you should:

1. Force the valve of the thermostat partially open, slide a thin feeler gauge through the opening, and allow the thermostat to close and catch the feeler gauge.

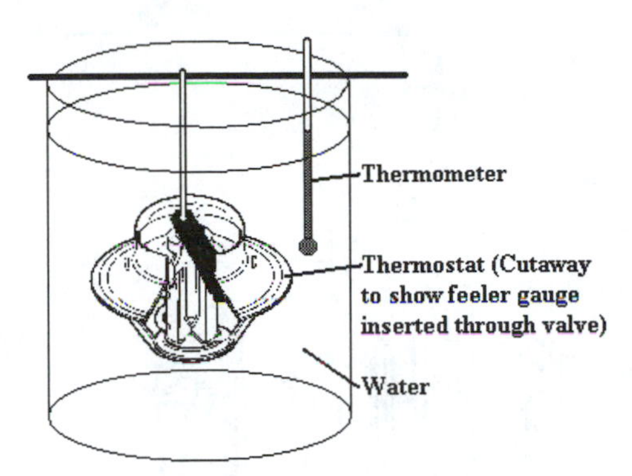

FIGURE 16–46 Place a thin feeler gauge into the valve portion of a thermostat and suspend it by a string in a container of water. Heat the water, and the valve should open to let the thermostat drop. This should occur at the thermostat's rated temperature.

2. Place a container of water on a stove or hot plate.
3. Tie a string to the feeler gauge and suspend the thermostat in the water so it does not touch the sides or bottom of the container (Figure 16–46).
4. Suspend a thermometer in the water near the thermostat.
5. Begin heating the water and observe the thermostat. It should open and fall off the feeler gauge when it reaches operating temperature.
6. When the thermostat opens, note the water temperature.
   a. If the thermostat opens at a temperature within 10°F (18°C) of the rating stamped on it and opens completely, it is functioning properly and can be reused.
   b. If the thermostat opens too early (cooler) or does not close completely, it should be replaced.
   c. If the thermostat opens too late or does not open, it should be replaced.

Always replace a faulty thermostat with the one specified by the vehicle manufacturer. If the engine has a properly functioning thermostat and still overheats, something else is wrong.

✓ REAL WORLD FIX

The 1996 Mazda MPV van (65,000 miles) overheats after a 10- to 20-minute drive. The thermostat has been replaced twice and the radiator once, but this did not help.

Fix: The technician removed the heater hoses to check the coolant flow and discovered that there was no flow. Checking further, it was found that the plastic water-pump impeller had cracked so that when it got hot, it would slip on the shaft.

Author's Note: Feeling for coolant flow by squeezing the upper hose could have indicated this problem.

16.3.5 Fan Clutch Test

In addition to the checks made during the inspection, the operation of a fan clutch can be tested using a fan clutch tester, strobe light tachometer, or timing light. When cold, the fan speed should be limited; when warmed up, the fan speed should be at or not above a specified rpm. The flashes of the strobe light make the fan and fan clutch appear to stand still when the flashes occur at the same frequency as the fan speed.

To quick check a fan clutch, you should:

1. Start the engine, and turn on the A/C to add heat to the radiator; place cardboard or shop cloths in front of the radiator to block the airflow.
2. A roaring, rumbling noise with an increase of airflow indicates that the fan clutch has engaged.

To test a fan clutch, you should:

1. Using a suitable marker, mark one of the fan blades and the water pump pulley; some technicians put a different number on each of the blades.
2. Start the engine and direct the strobe light toward the fan and pulley. Adjust the engine speed to 1,000 rpm and the strobe light speed so the fan and pulley appear to stand still. The fan and pulley should be at close to the same speed.
3. Increase the engine to 1,500 rpm. If cold, the clutch should be slipping, and there should be a speed difference between the fan and pulley.
4. Block off the front of the radiator to increase the temperature so the fan clutch engages, increase the engine speed to 2,000 rpm, and recheck the fan speed. The fan speed should be limited to about 1,500 to 2,000 rpm.

16.3.6 Electric Fan Tests

Electric fans are powered through a relay and one or more switches or the ECM as described in Chapter 12. A faulty fan motor is removed and replaced. Other

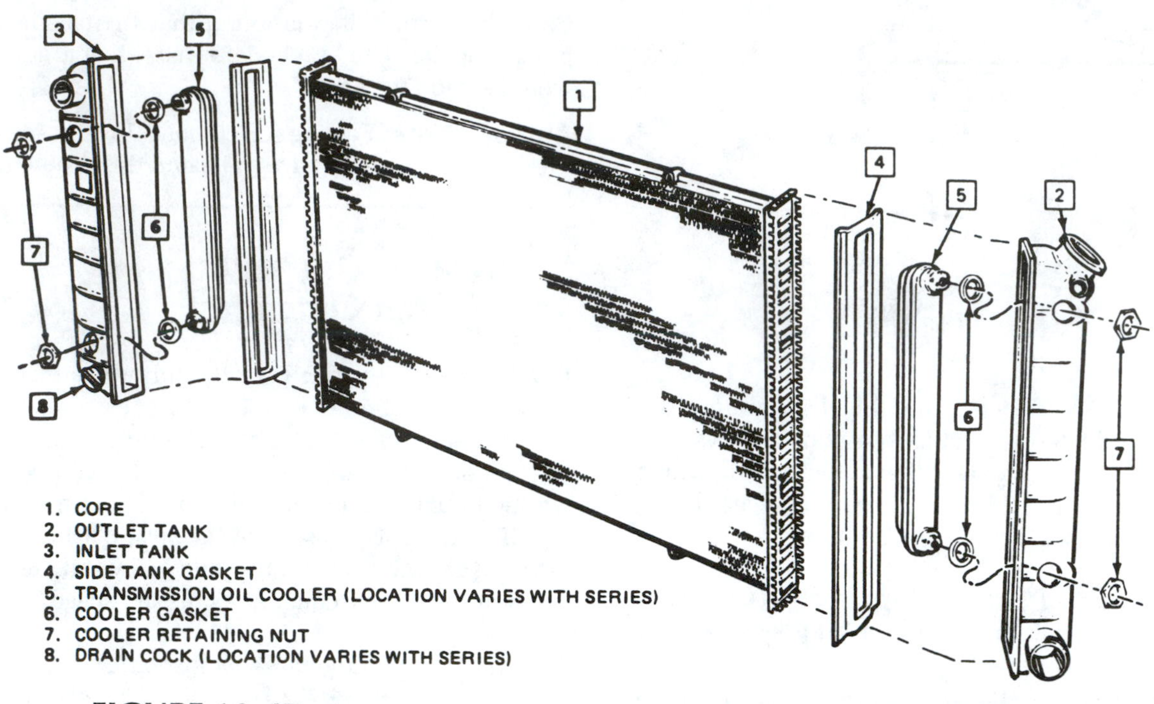

FIGURE 16–47 The tanks on some radiators can be removed for service or replacement of some of the components. (*Reprinted with the permission of General Motors Corporation*)

problems such as vibration or poor airflow can be caused by damaged or broken fan blades or a damaged shroud/housing. Reversed airflow can be caused by reversed connections at the fan motor.

Cavitation is a disruption of the airflow that also causes poor airflow. You can test for cavitation by measuring the fan motor current draw.

To test for cavitation, you should:

1. Remove the fan so it can operate in the open with free airflow in and out. Be careful so you or your clothing does not get caught by the blades.
2. Connect the fan motor to a battery using the proper polarity and with an ammeter in series with one of the leads. While the motor is running, the amount of current draw should be indicated on the ammeter.
3. You can determine the effect of cavitation by laying the fan motor so the inlet side is flat on a bench top or floor and repeating step 2. You should be able to hear the motor run at a higher speed with a lower current draw; this tells you that the fan is moving less air.
4. Replace the fan, and remeasure the current draw. If the amount is less than what was measured in step 2, the fan is cavitating because of an airflow problem.

16.4 COOLING SYSTEM SERVICE AND REPAIR

The major cooling system service and repair operations are flushing to clean out a dirty system and replacement of faulty components. Overheating can be caused by defective engine parts (e.g., a blown head gasket or warped head), but this text does not cover engine teardown. Many good information sources describe engine disassembly, inspection, and repair.

Radiator repair is a specialized skill that is normally performed by specialty shops. Some modern radiators use an aluminum core with tanks that are secured by bent or crimped connections. The tanks are removed and replaced to allow replacement of faulty items in many vehicle dealerships; they normally have the tools and information required for these operations (Figure 16–47).

16.4.1 Flushing a System

A dirty, rusty cooling system can often be cleaned using a chemical cleaning agent and a pressure backflush. Heavy rust or scale buildup in the water jackets probably requires engine teardown and a thorough boil-out or bake-oven cleaning. Rust and dirt are insulators that slow heat flow to the coolant in the water jackets and from the coolant to the radiator. A 1/16-inch layer of rust slows heat flow as much as a 1/4-inch layer of iron.

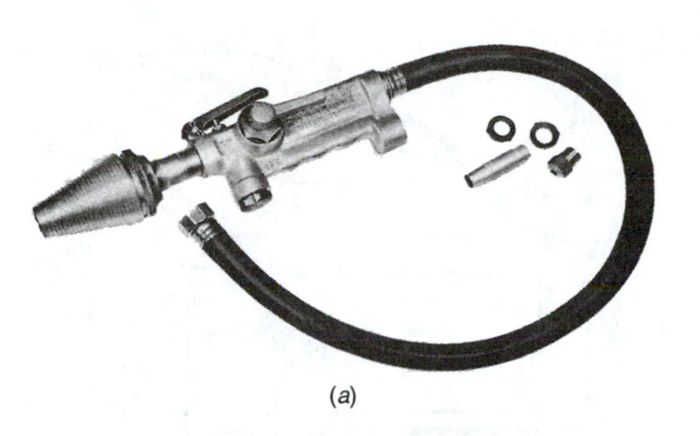

(a)

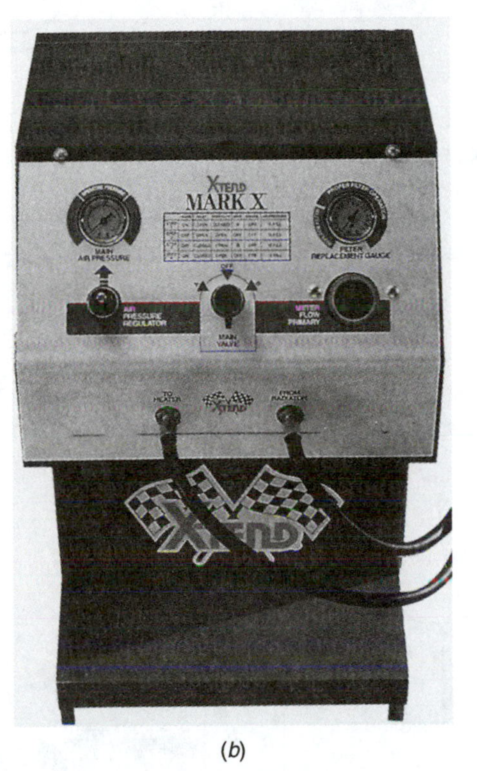

(b)

FIGURE 16–48 A flushing gun (*a*) and a flushing machine (*b*) can be used to flush a cooling system. This particular flushing machine will also recycle coolant. (a *is courtesy of DaimlerChrysler Corporation;* b *is courtesy of Wynn Oil Company*)

Chemical flushing uses strong chemicals designed to loosen or dissolve oil film, rust, scale, and other debris in the water jacket or radiator. These chemicals are either poured into the cooling system and circulated by the water pump or pumped through the system by a flushing unit. If using either of these, be sure to follow the directions on the container or the flushing unit.

Pressure flushing is often called **backflushing** because flow is the reverse of the normal direction. Reverse flow helps loosen debris. This flow is often pulsed, using air pressure to break more material loose (Figure 16–48). The handheld backflush gun is an effective flushing tool but is becoming less popular because of the amount of operator time it requires. Machine flushing units can be attached quickly to a system; after the flushing operation begins, the operator can leave and perform other tasks.

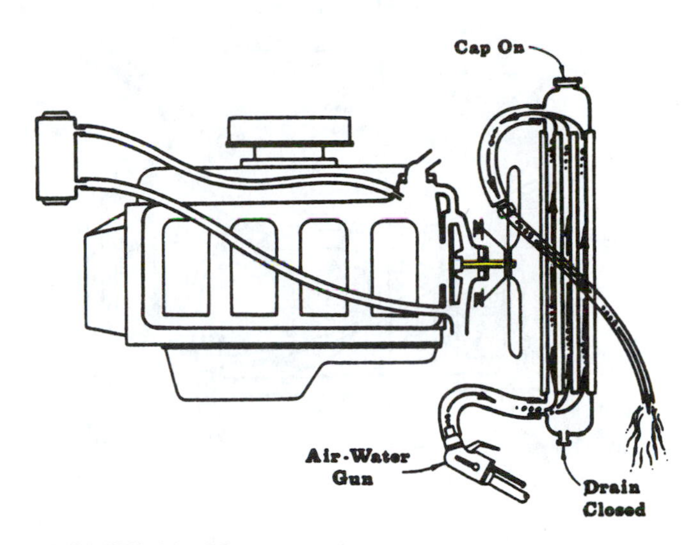

FIGURE 16–49 A radiator is backflushed by running water with added air bursts backward through the core.

To flush a system using a handheld flushing gun, you should:

1. Add a flushing chemical to the radiator if the system is dirty enough to warrant it. Follow the directions on the container, which usually involve running the engine a prescribed length of time after it has warmed up. Drain the system and dispose of the spent cleaning solution in the proper manner.
2. Disconnect the upper radiator hose at the radiator and the lower hose at the engine.
3. Attach the flushing gun to the lower radiator hose; connect a hose to the upper radiator connector to direct water away from the engine (Figure 16–49).
4. Connect the flushing gun to a water hose and an air hose to a regulated air supply. Adjust the air pressure regulator to 20 psi.
5. Turn on the water and allow it to fill the radiator.
6. Operate the air valve in short bursts so the radiator stays full of water.
7. Watch the water flow from the radiator; repeat step 6 until the water runs clear.
8. Remove the thermostat from the engine and replace the thermostat cover.
9. Attach the flushing gun to the upper radiator hose (leading to the engine water jacket) and repeat steps 5 and 6 until the water running out of the engine is clear. When flushing the engine,

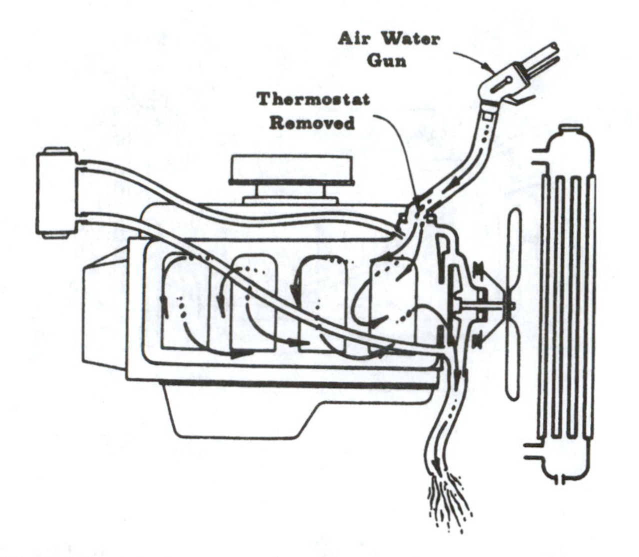

FIGURE 16–50 Water jackets are backflushed by running water with added air bursts backward through them. Be sure to remove the thermostat before backflushing.

the air pressure can be increased to 40 psi (Figure 16–50).

10. The heater core can be flushed by connecting the flushing gun to the heater hoses and repeating steps 5, 6, and 7. Air pressure should be adjusted to 15 psi when flushing a heater core (Figure 16–51).
11. Replace the thermostat and radiator hoses, fill the system with a 50–50 mix of antifreeze and water, and check for leaks.

As for the procedure for coolant exchange, most shops today use a flushing machine. This machine is quickly attached to the vehicle's cooling system at the upper radiator hose or one or both of the heater hoses. Then the machine is used to circulate a chemical cleaning agent through the cooling system until it is clean.

If the radiator still shows scale deposits, its flow can be tested to determine if it is restricted. A radiator shop usually has the equipment to boil out, rod, or replace the core in a very dirty radiator. The rodding operation involves removing the radiator tanks and running a metal rod through each of the tubes.

16.4.2 Component Replacement

Replacement of most cooling system components essentially follows a procedure of draining the coolant, removing the old part, installing the new part, replacing the coolant, and checking for leaks. The exact procedure to follow for each vehicle is described in the service manual. Descriptions given here are to point out service operations performed by some technicians to ensure proper operation.

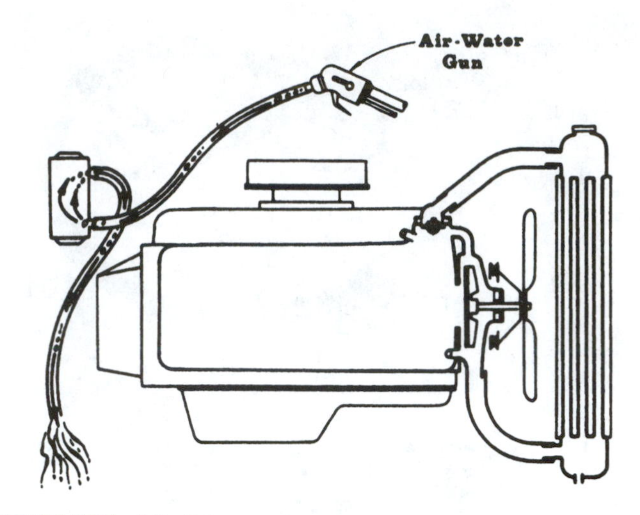

FIGURE 16–51 A heater core can also be backflushed, but be sure to keep the pressures low so as not to rupture the core.

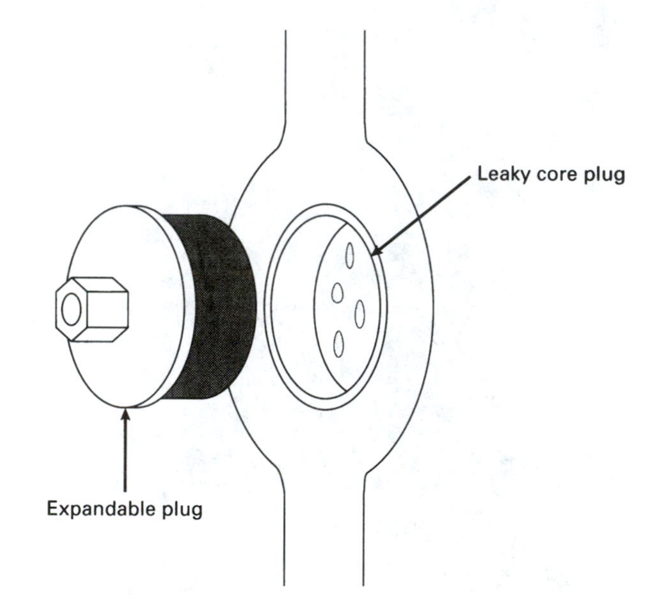

FIGURE 16–52 A leaky core plug can often be repaired by installing an expandable plug into the core plug.

16.4.2.1 R&R Engine Core or Soft Plugs Core plugs are normally replaced when an engine is overhauled or when they begin leaking. The biggest problem typically encountered is getting access to the plug: It is often behind an exhaust pipe, motor mount, or starter. Normally the old plug is removed and a new one driven in place.

✓SERVICE TIP

If a cup plug is in a hard-to-reach location, the old plug can be cleaned and left in place with an expandable rubber plug inserted into it (Figure 16–52).

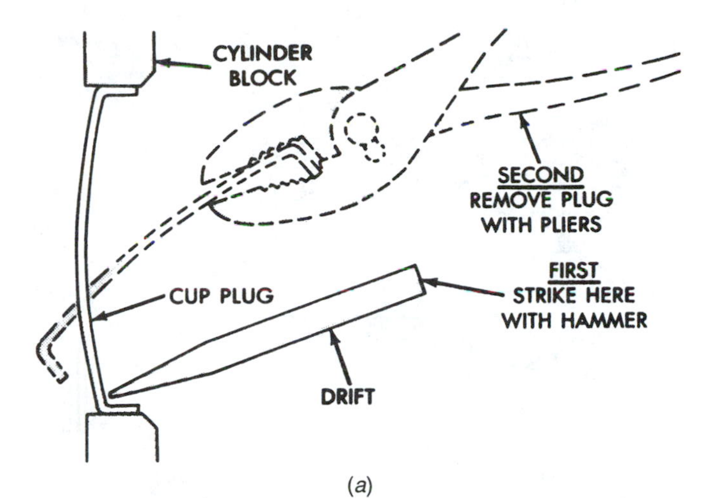

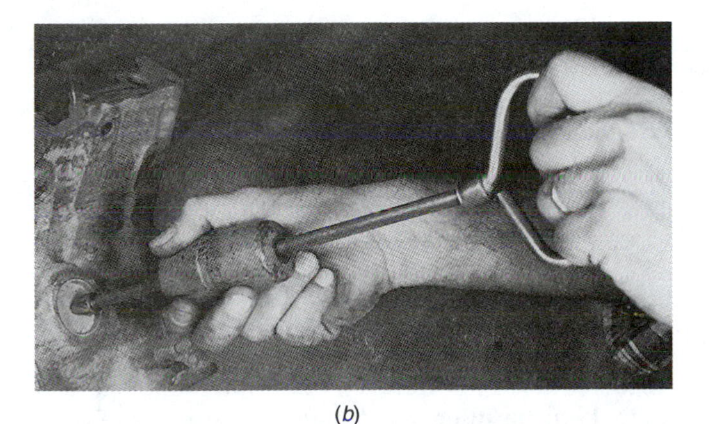

(b)

FIGURE 16–53 A cup plug can be removed by rotating the plug using a punch and hammer and pulling it out with pliers (*a*). Another way to do this is to use a slide hammer with a puller screw (*b*). (a *is courtesy of DaimlerChrysler Corporation*)

✓SERVICE TIP

Many technicians coat the inner side of a metal plug with nonhardening gasket sealer or room temperature volcanizing (RTV) sealant to serve as a corrosion barrier and the side of the plug to serve as a sealant. Rubber expandable plugs should be installed dry.

To R&R a leaky core plug, you should:

1. Drain the coolant.
2. If practical, remove the old plug. This is normally done using a slide hammer or punch and pliers, as shown in Figure 16–53.
3. Clean any corrosion from the opening.
4. Coat the new plug with sealant and drive it into the opening. The driving tool should loosely fit the inner diameter of the plug. Try to position the plug so its inner edge is flush with the inside of the water jacket to eliminate a debris buildup pocket.
5. Refill the system with coolant and check for leaks.

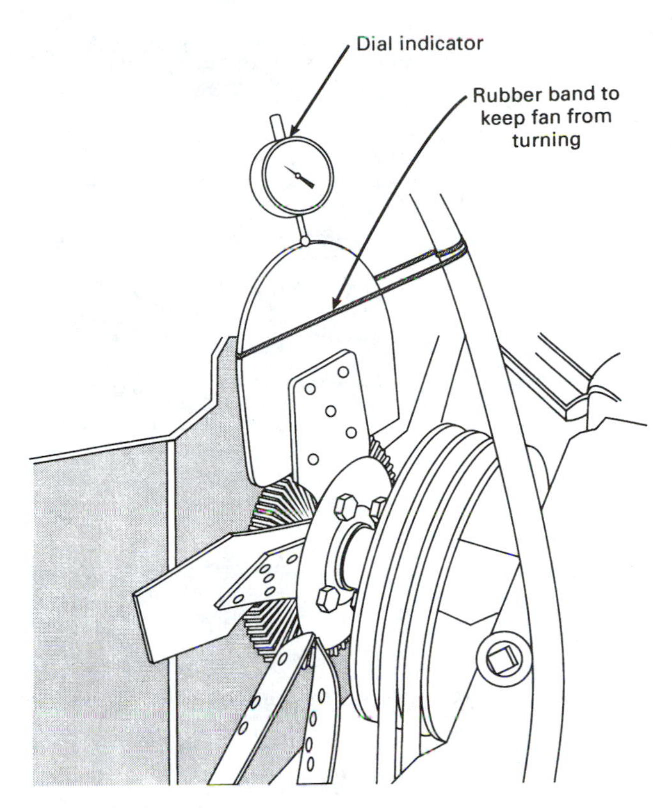

FIGURE 16–54 Fan runout is measured by holding the fan stationary using a rubber band, mounting a dial indicator at one of the blades, and cranking the engine. Excessive runout indicates faulty fan or fan clutch mounting.

16.4.2.2 R&R Fan and Fan Clutch Except for the tight fit between the radiator and some of the bolts, R&R of the fan and fan clutch is a fairly easy task. Be careful to tighten the mounting bolts onto the water pump shaft evenly so that a good, straight fit is accomplished. Improper mounting can make the fan and fan clutch run off center. This can cause vibration, which in turn can damage the water-pump bearings or shaft. If improper mounting is suspected, a fairly simple fan runout check can be made.

To check fan runout, you should:

1. Disable the ignition system so the engine will not start.
2. Keep a fan blade from rotating by looping a strong rubber band over the blade and an engine part (Figure 16–54).

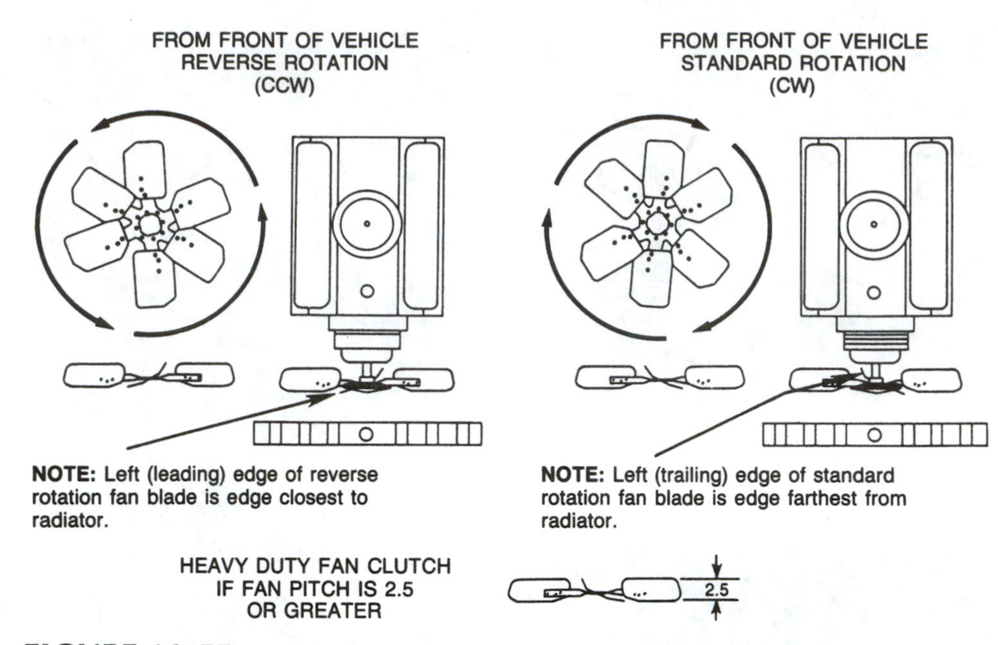

FIGURE 16–55 If replacing a fan, make sure that the replacement is designed to rotate in the correct direction. (*Courtesy of Stant Manufacturing*)

3. Crank the engine as you observe for side-to-side motion of the fan. It should stay still.

If the fan moves more than 0.010 inch, check the mounting to ensure that it is straight. If necessary, shim the mounting point to remove the runout. If the fan is damaged and requires replacement, be sure to replace it with a fan of the correct size and type.

✓SERVICE TIP

If replacing a fan or fan clutch, pay attention to the direction of rotation. Most modern engines that use a serpentine, V-ribbed belt drive the water pump and fan off the back side of the belt so it turns in a counterclockwise direction. Engines that use a plain V belt drive the fan and water pump from the inside of the belt, and the fan turns in a clockwise direction. Both the fan and fan clutch must match the direction of rotation (Figure 16–55).

✓SERVICE TIP

Some fan clutches thread onto the drive shaft from the water pump. They are removed by unscrewing them in a direction opposite to normal rotation. On some vehicles, a special wrench is required for this operation.

16.4.2.3 R&R Radiator Probably the most difficult part of radiator removal is the disassembly of the oil cooler lines for automatic transmissions. On older vehicles, a flare fitting is used, and two wrenches are required for disconnection: one wrench to turn the fitting nut, and one to keep the fitting from turning. Some newer vehicles use a quick-connect coupling that requires a special tool.

To R&R a radiator, you should:

1. Drain the coolant.
2. Disconnect the upper and lower radiator hoses and the overflow hose from the radiator. On vehicles with automatic transmissions, disconnect the oil cooler lines.
3. Disconnect the shroud, if necessary, and remove the bolts that attach the radiator (Figure 16–56).
4. Lift out the radiator.
5. Slide the new or repaired radiator in place, being careful not to bend any of the fins.
6. Replace the radiator mounting bolts and the shroud.
7. Reconnect the hoses and lines to the radiator.
8. Refill the system with coolant and check for leaks.

16.4.2.4 R&R Thermostat The thermostat in most vehicles is located under its cover at the upper outlet of the engine. In some vehicles, it is located at the lower coolant inlet.

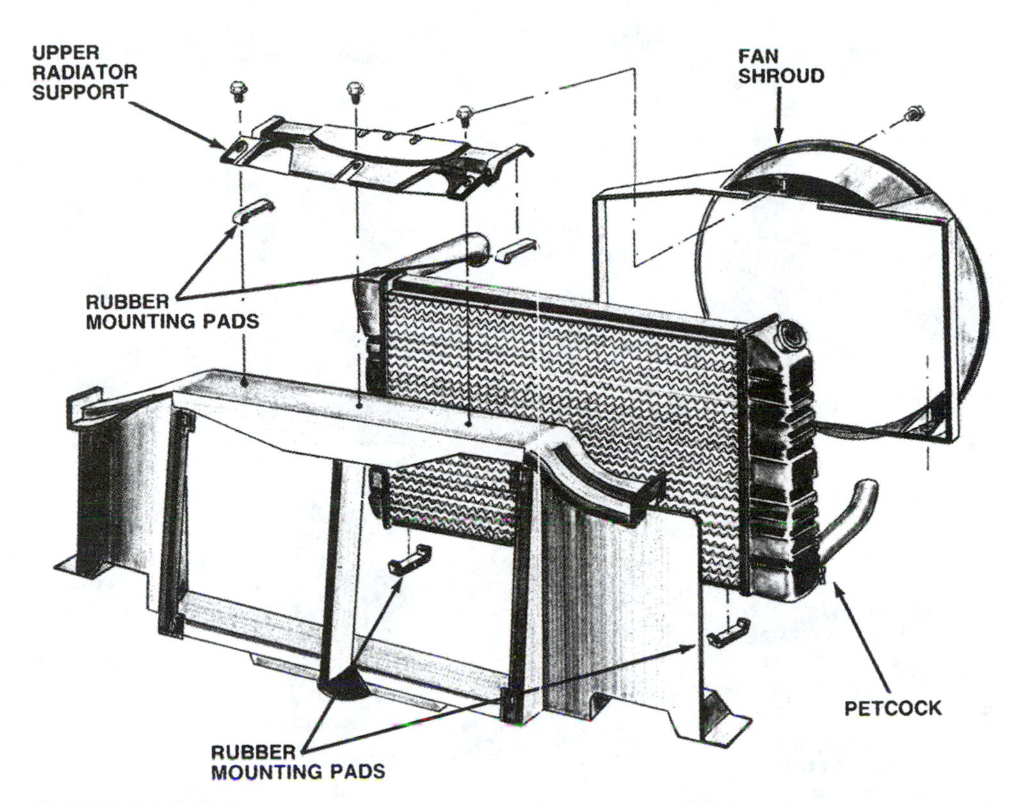

FIGURE 16–56 An exploded view of radiator assembly showing parts removed during replacement. (*Courtesy of Everco Industries*)

✓SERVICE TIP

A thermostat must always be installed so that the heat motor or sensing portion is toward the hot water. A close inspection of their mountings shows that most thermostats fit into a recess (Figure 16–57). The gasket not only seals the cover but also prevents coolant flow past the thermostat. Some thermostats must be aligned as they are installed into their housing.

✓SERVICE TIP

Many thermostats have an air bleed hole or notch in them. On some, the bleed hole has a jiggle pin in it to keep it open. On vertically mounted thermostats, the air bleed should be positioned in the uppermost position to bleed as much air as possible.

To R&R a thermostat, you should:

1. Partially drain the coolant so the level is below the thermostat cover.

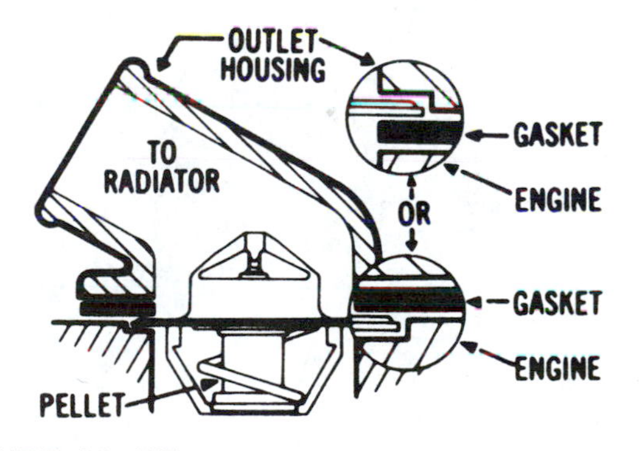

FIGURE 16–57 A thermostat is normally set in a recess in the engine or outlet housing; the gasket often prevents flow around the thermostat. (*Courtesy of Stant Manufacturing*)

2. Disconnect the upper radiator hose at the thermostat cover and disconnect the bypass hose (if so equipped). On some vehicles, just remove the radiator cap–like thermostat cover (Figure 16–58).
3. Remove the thermostat housing bolts and the thermostat housing.
4. Remove the thermostat and gasket (Figure 16–59).

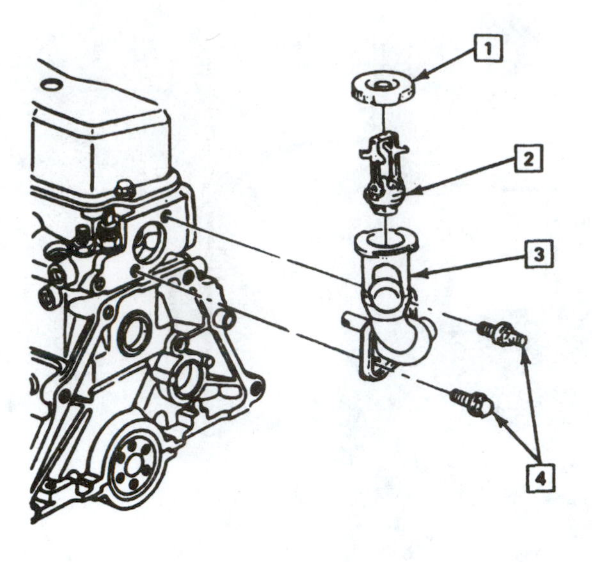

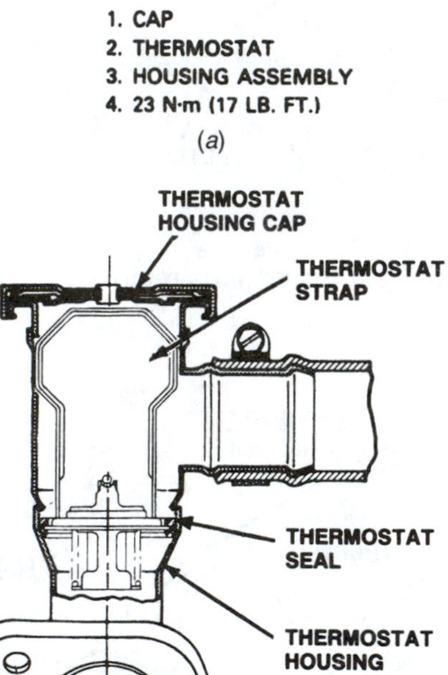

FIGURE 16–58 An exploded and cutaway view of a cap-mounted thermostat. (a *is reprinted with permission of General Motors Corporation;* b *is courtesy of Stant Manufacturing*)

5. Clean the gasket surfaces and the thermostat recess.
6. Install the thermostat, new gasket, and cover. A sealant is usually recommended for the gasket.
7. Replace the bolts and tighten them to the correct torque.
8. Refill the system with coolant and check for leaks.

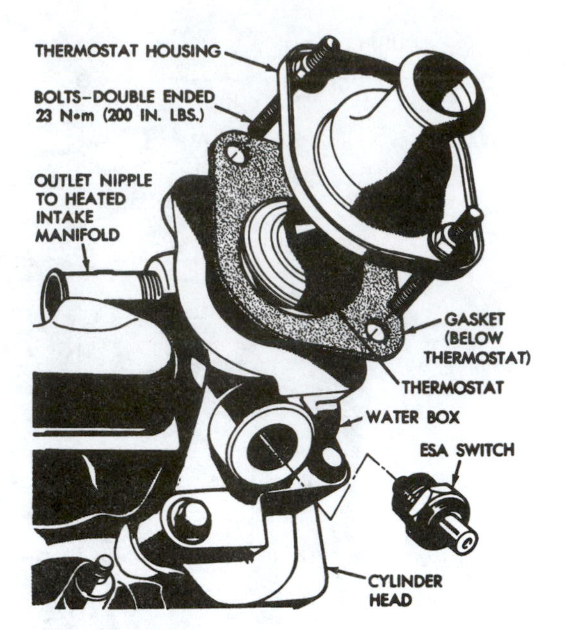

FIGURE 16–59 The thermostat is normally mounted near the heater hose outlet and temperature sensing (ESA) switch. (*Courtesy of DaimlerChrysler Corporation*)

✓SERVICE TIP

When replacing a thermostat, some technicians force the valve open and place an aspirin tablet in the valve to keep it from closing completely. The partially open valve allows the air from the water jackets to escape so the coolant can enter quickly. The tablet will dissolve shortly after the system is filled.

✓SERVICE TIP

Vertically mounted thermostats can be difficult to hold in position while the cover is replaced. A tip is to loop a rubber band through the thermostat and then slip it out after the cover is in place (Figure 16–60). Gaskets with an adhesive coating that will stick onto the housing can also be used to hold the thermostat in the correct position.

16.4.2.5 R&R Water Pump As with core plugs, the biggest problem faced when replacing a water pump is often access to the bolts. On some engines, the water pump bolts also secure the timing chain cover, and replacement of the water pump can break the oil seal be-

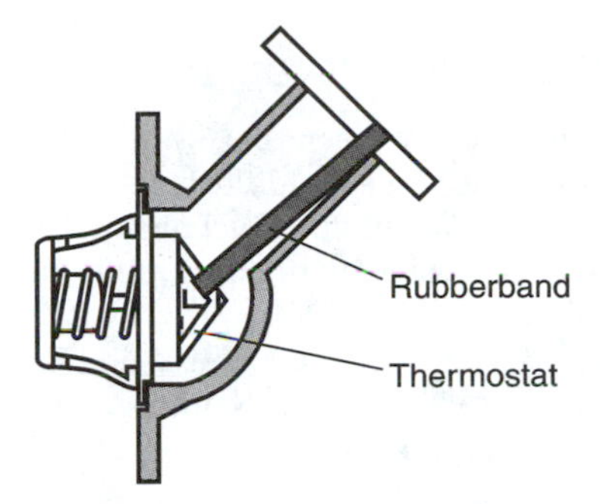

FIGURE 16–60 The rubber band is holding the thermostat in place in the housing. After the housing is fastened in place, the rubber band will be removed.

tween this cover and the block. It is best to follow the manufacturer's replacement procedure.

To R&R a water pump, you should:

1. Drain the coolant and remove the drive belt. On RWD vehicles, remove the fan and fan clutch.
2. Remove the water pump mounting bolts and remove the pump. If necessary, use a hammer to tap the side of the pump to break the gasket loose.
3. Clean off the old gasket and sealant; clean out the pump impeller cavity in the block if it is exposed.
4. Install two guide pins (headless bolts) to hold the gasket in place, slide the gasket into place, and slide the water pump into place. A coating of gasket sealant or RTV sealant is often recommended for the gasket (Figure 16–61).
5. Install the bolts and tighten them to the correct torque.
6. Replace the fan and fan clutch.
7. Install and adjust the drive belt.
8. Refill the system with coolant and check for leaks.

FIGURE 16–61 A pair of guide pins is being used to hold the gasket in position as this water pump is installed.

CHAPTER QUIZ

These questions help you study this chapter. Enter the proper word(s) in the blanks to complete each statement.

1. The coolant level, in a modern vehicle, should be at the ________ ________ on the CRR, and if the radiator cap is removed, it should be even with the sealing surface in the ________ ________.
2. If removing the radiator cap from a hot cooling system, be very careful because the ________ might suddenly ________.
3. Dirty coolant indicates that the coolant should be ________ and probably that the cooling system be ________.
4. Coolant freeze protection can be checked using a(n) ________, ________, or ________ ________.
5. The most accurate checking method is the ________.
6. A reading of more than 0.3 V at the radiator filler neck indicates a potential problem of ________.
7. The normal coolant is ________ antifreeze and ____ water; in very cold climates, the percentage of antifreeze can be increased to ________.
8. Used antifreeze should be ________ or disposed of in the ________ ________.
9. Early drive belt failure can be caused by ________ or improper ________.
10. When a hose is replaced, the ________ should be positioned right next to the connector bead, and it is a good practice to recheck clamp ________ after the engine has been warmed up and cooled down.
11. A radiator cap is tested using a(n) ________ ________.
12. Bubbles coming up through the coolant is a radiator is a sign of a(n) ________ ________.
13. A faulty thermostat usually sticks ________ or ________, and it should be ________ with one that meets the manufacturer's specifications.
14. A dirty cooling system can be cleaned using a(n) ________ flush, ________ flush, or ________ flush.

REVIEW QUESTIONS

These questions allow you to check what you have learned. Select the answer that correctly completes each statement.

1. In a modern cooling system, the coolant level is checked at the
   a. radiator.
   b. thermostat cover.
   c. coolant recovery reservoir.
   d. None of these
2. While discussing coolant changes, Technician A says that the old coolant can simply be poured down the drain. Technician B says that used coolant must be recycled. Who is correct?
   a. A only c. Both A and B
   b. B only d. Neither A nor B
3. Two technicians are discussing coolant flow. Technician A says that you should not be able to feel a coolant flow through the upper radiator hose while the engine is cold. Technician B says that when the thermostat opens, the upper hose should start getting hot and you should feel a flow through it. Who is correct?
   a. A only c. Both A and B
   b. B only d. Neither A nor B
4. Technician A says that a cooling system should hold pressure for at least 2 minutes when its pressure is checked. Technician B says that the pressure tester can also be used to check for combustion leaks. Who is correct?
   a. A only c. Both A and B
   b. B only d. Neither A nor B
5. A thermostat that is stuck open causes the engine to overheat.
   a. True b. False
6. Technician A says that a leaky heater core shows up as a drip onto the carpet. Technician B says that you can test a heater core for leaks using a pressure tester. Who is correct?
   a. A only c. Both A and B
   b. B only d. Neither A nor B
7. Technician A says that fan clutch operation can be tested using a strobe light. Technician B says that the fan clutch should be locked up when the engine is cold. Who is correct?
   a. A only c. Both A and B
   b. B only d. Neither A nor B
8. Two technicians are discussing system flushing. Technician A says that you should flush a system in the direction of normal coolant flow. Technician B says that a system should be flushed until clean water flows out of the system. Who is correct?
   a. A only c. Both A and B
   b. B only d. Neither A nor B
9. Technician A says that if a core plug is leaking and is in a difficult location, an expandable rubber plug can be inserted into it. Technician B says that rubber plugs should always be installed dry. Who is correct?
   a. A only c. Both A and B
   b. B only d. Neither A nor B
10. Two technicians are discussing radiator removal. Technician A says that you should use two wrenches as you loosen the cooler lines to the automatic transmission. Technician B says that cooler line on modern vehicles might require special tools to disconnect the quick-connect–type coupler system. Who is correct?
    a. A only c. Both A and B
    b. B only d. Neither A nor B
11. Technician A says that the major cause of engine overheating is a faulty thermostat that opens too late. Technician B says that the bleed notch or jiggle pin should be at the top when installing vertically mounted thermostats. Who is correct?
    a. A only c. Both A and B
    b. B only d. Neither A nor B
12. Technician A says that after service, it is good practice to retighten hose clamps after the engine has been run and then cooled down. Technician B says that when coolant is replaced, the coolant level should be checked after the engine has been warmed up. Who is correct?
    a. A only c. Both A and B
    b. B only d. Neither A nor B
13. Two technicians are discussing antifreeze. Technician A says that the antifreeze should be changed every year or two to keep a cooling system clean. Technician B says that an overheating system can be kept from boiling by using straight antifreeze for a coolant. Who is correct?
    a. A only c. Both A and B
    b. B only d. Neither A nor B
14. Technician A says that the fan used with modern RWD vehicles always rotates in a clockwise direction, the same as the crankshaft. Technician B says that the fan clutch used on FWD vehicles can cause fan rotation in either direction. Who is correct?
    a. A only c. Both A and B
    b. B only d. Neither A nor B

APPENDIX

ASE Certification and Task List

Mechanics have the opportunity to voluntarily take ASE certification tests to become ASE-certified technicians. ASE is short for National Institute for Automotive Service Excellence. Certification helps technicians prove their abilities to themselves, their employers, and their customers, many of whom are suspicious of the automotive repair profession.

ASE certification requires that you pass one or more tests and have at least two years of automotive repair work experience. School training can be used to substitute for part of the work experience requirement, and you may take the test or tests before completing the work experience requirement. Initially you will receive the test score report; when the experience requirement is completed, you will receive certification.

There are eight automotive service tests, and one is A7, Heating and Air Conditioning.* The A7 test has 50 questions that are taken from the following content areas:

Content Area	*Questions*
A. A/C System Diagnosis and Repair	12
B. Refrigeration System Component Diagnosis and Repair	11
1. Compressor and Clutch	(5)
2. Evaporator, Receiver–Drier, Condenser, and so forth	(6)
C. Heating and Engine Cooling Systems Diagnosis and Repair	6
D. Operating Systems and Related Controls Diagnosis and Repair	16
1. Electrical	(7)
2. Vacuum or Mechanical	(4)
3. Automatic and Semi-automatic Temperature Controls	(5)
E. Refrigerant Recovery, Recycling, and Handling	5

*The content areas and task list are provided courtesy of the National Institute for Automotive Service Excellence.

If you intend to take the A7 test and feel a need to study for it, all content areas are divided into groups of tasks. The tasks focus on the things that a heating and air conditioning technician should be able to do. For an up-to-date task list, call ASE at 703-713-3800 and request the Automobile Preparation Guide.

HEATING AND AIR CONDITIONING TEST TASK LIST

A. A/C System Diagnosis and Repair

1. Diagnose the cause of unusual operating noises of the A/C system; determine needed repairs.
2. Identify system type and conduct performance test on the A/C system; determine needed repairs.

3. Diagnose A/C system problems indicated by refrigerant flow past the sight glass (for systems using a sight glass); determine needed repairs.
4. Diagnose A/C system problems indicated by pressure gauge readings; determine needed repairs.
5. Diagnose A/C system problems indicated by sight, sound, smell, and touch procedures; determine needed repairs.
6. Leak test the A/C system for leaks; determine needed repairs.
7. Identify and recover A/C system refrigerant.
8. Evacuate A/C system.
9. Clean A/C system components and hoses.
10. Charge A/C system with refrigerant (liquid or vapor)
11. Identify lubricant type; inspect level in A/C system.

B. Refrigeration System Component Diagnosis and Repair

1. Compressor and Clutch

1. Diagnose A/C system problems that cause the protection devices (pressure, thermal, and control modules) to interrupt system operation; determine needed repairs.
2. Inspect, test, and replace A/C system pressure and thermal protection devices.
3. Inspect, adjust, and replace A/C compressor drive belts and pulleys.
4. Inspect, test, service, and replace A/C compressor clutch components or assembly.
5. Identify required lubricant type; inspect and correct level in A/C compressor.
6. Inspect, test, service, or replace A/C compressor.
7. Inspect, repair, and replace A/C compressor mountings.

2. Evaporator, Condenser, and Related Components

1. Inspect, repair, or replace A/C system mufflers, hoses, lines, filters, fittings, and seals.
2. Inspect A/C condenser for airflow restrictions.
3. Inspect, test, and replace A/C system condenser and mountings.
4. Inspect and replace receiver–drier or accumulator–drier.
5. Inspect, test, and replace expansion valve.
6. Inspect and replace orifice tube.
7. Inspect, test, or replace evaporator.
8. Inspect, clean, and repair evaporator, housing, and water drain.
9. Inspect, test, and replace evaporator pressure–temperature control systems and devices.
10. Identify, inspect, and replace A/C system service valves (gauge connections).
11. Inspect and replace A/C system high-pressure relief device.

C. Heating and Engine Cooling Systems Diagnosis and Repair

1. Diagnose the cause of temperature control problems in the heater/ventilation system; determine needed repairs.
2. Diagnose window fogging problems; determine needed repairs.
3. Perform cooling system tests; determine needed repairs.
4. Inspect and replace engine cooling and heater system hoses.
5. Inspect, test, and replace radiator, pressure cap, coolant recovery system, and water pump.
6. Inspect, test, and replace thermostat, bypass, and housing.
7. Identify, inspect, and recover coolant; flush and refill system with proper coolant.
8. Inspect, test, and replace fan (both electrical and mechanical), fan clutch, fan belts, fan shroud, and air dams.
9. Inspect, test, and replace heater coolant control valve (manual, vacuum, and electrical types).
10. Inspect, flush, and replace heater core.

D. Operating Systems and Related Controls Diagnosis and Repair

1. Electrical

1. Diagnose the cause of failures in the electrical control system of heating, ventilating, and A/C systems; determine needed repairs.

2. Inspect, test, repair, and replace A/C-heater blower motors, resistors, switches, relay/modules, wiring, and protection devices.
3. Inspect, test, repair, and replace A/C compressor clutch, relays/modules, wiring, sensors, switches, diodes, and protection devices.
4. Inspect, test, repair, replace, and adjust A/C-related engine control systems.
5. Inspect, test, repair, replace, and adjust load-sensitive A/C compressor cutoff systems.
6. Inspect, test, repair, and replace engine cooling/condenser fan motors, relays/modules, switches, sensors, wiring, and protection devices.
7. Inspect, test, adjust, repair, and replace electric actuator motors, relays/modules, switches, sensors, wiring, and protection devices.
8. Inspect, test, service, or replace heating, ventilating, and A/C control-panel assemblies.

2. *Vacuum/Mechanical*

1. Diagnose the cause of failures in the vacuum and mechanical switches and controls of the heating, ventilating, and A/C systems; determine needed repairs.
2. Inspect, test, service, or replace heating, ventilating, and A/C control-panel assemblies.
3. Inspect, test, adjust, and replace heating, ventilating, and A/C control cables and linkages.
4. Inspect, test, and replace heating, ventilating, and A/C vacuum actuators (diaphragms/motors) and hoses.
5. Identify, inspect, test, and replace heating, ventilating, and A/C vacuum reservoir check valve, and restrictors.
6. Inspect, test, adjust, repair, or replace heating, ventilating, and A/C ducts, doors, and outlets.

3. *Automatic and Semi-Automatic Heating, Ventilating, and A/C Systems*

1. Diagnose temperature control system problems; determine needed repairs.
2. Diagnose blower system problems; determine needed repairs.
3. Diagnose air distribution system problems; determine needed repairs.
4. Diagnose compressor clutch control system; determine needed repairs.
5. Inspect, test, adjust, or replace climate control temperature and sunload sensors.
6. Inspect, test, adjust, and replace temperature blend door actuator.
7. Inspect, test, and replace low engine coolant temperature blower control system.
8. Inspect, test, and replace heater water valve and controls.
9. Inspect, test, and replace electric and vacuum motors, solenoids, and switches.
10. Inspect, test, and replace ATC control panel.
11. Inspect, test, adjust, or replace ATC microprocessor (climate control computer/programmer).
12. Check and adjust calibration of ATC system.

E. Refrigerant Recovery, Recycling, and Handling

1. Maintain and verify correct operation of certified equipment.
2. Identify and recover A/C system refrigerant.
3. Recycle or properly dispose of refrigerant.
4. Label and store refrigerant.
5. Test recycled refrigerant for noncondensable gases.

APPENDIX

Temperature–Pressure Charts

R-134a				*R-12*			
Temperature °C (°F)	*Pressure kPa (psi)*	*Temperature °C (°F)*	*Pressure kPa (psi)*	*Temperature °C (°F)*	*Pressure kPa (psi)*	*Temperature °C (°F)*	*Pressure kPa (psi)*
–9 (16)	106 (15)	38 (100)	857 (124)	–9 (16)	127 (18)	38 (100)	808 (117)
–8 (18)	115 (17)	39 (102)	887 (129)	–8 (18)	136 (20)	39 (102)	893 (121)
–7 (20)	124 (18)	40 (104)	917 (133)	–7 (20)	145 (21)	40 (104)	859 (125)
–6 (22) —	134 (19)	41 (106)	948 (137)	–6 (22) —	155 (22)	41 (106)	893 (129)
–4 (24)	144 (21)	42 (108)	980 (142)	–4 (24)	165 (24)	42 (108)	917 (133)
–3 (26)	155 (22)	43 (110) —	1,012 (147)	–3 (26)	175 (25)	43 (110) —	940 (136)
–2 (28) Evaporator Range	166 (24)	44 (112)	1,045 (152)	–2 (28) Evaporator Range	185 (27)	44 (112)	969 (140)
–1 (30)	177 (26)	46 (114)	1,079 (157)	–1 (30)	196 (28)	46 (114)	997 (145)
0 (32)	188 (27)	47 (116)	1,114 (162)	0 (32)	207 (30)	47 (116)	1,027 (149)
1 (34)	200 (29)	48 (118)	1,149 (167)	1 (34)	219 (32)	48 (118)	1,057 (153)
2 (36)	212 (31)	49 (120)	1,185 (172)	2 (36)	230 (33)	49 (120)	1,087 (158)
3 (38)	225 (33)	50 (122) Condenser Range	1,222 (177)	3 (38)	249 (36)	50 (122) Condenser Range	1,118 (162)
4 (40)	238 (35)	51 (124)	1,260 (183)	4 (40)	255 (37)	51 (124)	1,150 (167)
7 (45)	272 (40)	52 (126)	1,298 (188)	7 (45)	287 (42)	52 (126)	1,182 (171)
10 (50) —	310 (45)	53 (128)	1,337 (194)	10 (50) —	322 (47)	53 (128)	1,215 (176)
13 (55)	350 (51)	54 (130)	1,377 (200)	13 (55)	359 (52)	54 (130)	1,248 (181)
16 (60)	392 (57)	57 (135)	1,481 (215)	16 (60)	398 (58)	57 (135)	1,334 (194)
18 (65)	438 (64)	60 (140)	1,590 (231)	18 (65)	440 (64)	60 (140)	1,425 (207)
21 (70)	487 (71)	63 (145)	1,704 (247)	21 (70)	484 (70)	63 (145)	1,519 (220)
24 (75)	540 (78)	66 (150)	1,823 (264)	24 (75)	531 (77)	66 (150)	1,618 (235)
27 (80)	609 (88)	68 (155)	1,948 (283)	27 (80)	580 (84)	68 (155)	1,721 (250)
30 (85)	655 (95)	71 (160)	2,079 (301)	30 (85)	633 (92)	71 (160)	1,828 (265)
32 (90)	718 (104)	74 (165) —	2,215 (321)	32 (90)	688 (100)	74 (165) —	1,940 (281)
35 (95)	786 (114)	77 (170)	2,358 (342)	35 (95)	746 (108)	77 (170)	2,057 (298)

Note: Evaporator pressures represent gas temperatures inside the coil and not at the coil surfaces. Add to temperature for coil and air-off temperatures (4 to 6°C or 8 to 10°F). Condenser temperatures are not ambient temperatures. Add to ambient (19 to 22°C or 35 to 40°F) for proper heat transfer; then refer to chart.

Example: 32°C + 22°C = 54°C Condenser temperature = 1,377 kPa (R-134) or 1,248 kPa (R-12), based on 30-mph airflow.

Conditions vary for different system configurations. Refer to the manufacturer's specifications.

(Reprinted with permission of General Motors Corporation)

APPENDIX

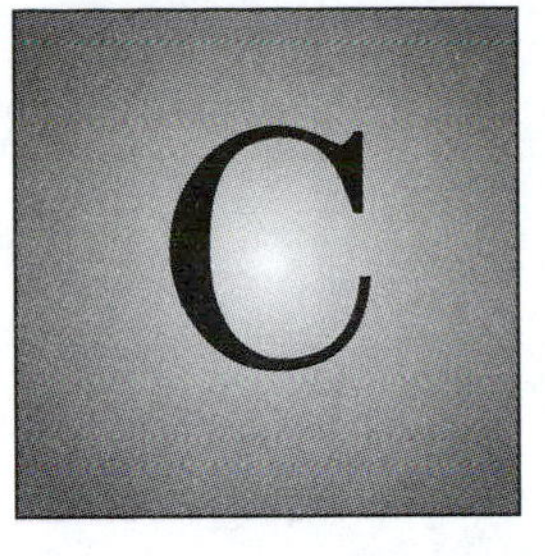

A/C Service Procedure Documents

Proper A/C service procedures are described in documents developed by the federal government, the Society of Automotive Engineers (SAE), and the American Refrigeration Institute (ARI). Following is a listing of those that have a definite impact on modern A/C operations:

ARI 700-88 Purity Standard for Reclaimed R-12 Intended for Resale or Mobile Use.

SAE J639 Safety and Containment of Refrigerant for Mechanical Vapor Compression Systems Used for Mobile Air Conditioning Systems.

SAE J658 Alternate Refrigerant Consistency Criteria for Use in Mobile Air Conditioning Systems.

SAE J659 Vehicle Testing Requirements for Replacement Refrigerants for CFC-12 (R-12) Mobile Air Conditioning Systems.

SAE J1627 Rating Criteria for Electronic Leak Detectors.

SAE J1628 Technician Procedure for Using Electronic Refrigerant Leak Detectors for Service of Mobile Air Conditioning Systems.

SAE 1629 Cautionary Statements for Handling HFC-134a During Mobile Air Conditioning Service.

SAE J1657 Selection Criteria for Retrofit Refrigerants to Replace CFC-12 (R-12) in Mobile Air Conditioning Systems.

SAE J1658 Alternate Refrigerant Consistency Criteria for Use in Mobile Air Conditioning Systems.

SAE J1660 Fittings and Labels for Retrofit of R-12 Mobile Air Conditioning Systems to R-134a.

SAE J1661 Procedure for Retrofitting R-12 Mobile Air Conditioning Systems to R-134a.

SAE J1662 Compatibility of Retrofit Refrigerants with Air Conditioning System Materials.

SAE J1732 HFC-134a (Refrigerant Recovery Equipment for Mobile Automotive Air Conditioning Systems).

SAE J1989 Recommended Procedure for the Containment of R-12.

SAE J1990 Extraction and Recycle Equipment for Mobile Automotive Air Conditioning Systems.

SAE J1991 Standard of Purity for Use in Mobile Air Conditioning Systems.

SAE J2099 Standard of Purity for Recycled HFC-134a for Use in Mobile Air-Conditioning Systems.

SAE J2196 Service Hose for Automotive Air Conditioning.

SAE J2197 HFC-134a Service Hose Fittings for Automotive Air-Conditioning Service Equipment.

SAE J2209 CFC-12 Extraction Equipment for Mobile Automotive Air-Conditioning Systems.

SAE J2210 HFC-134a Recycling Equipment for Mobile Air Conditioning Systems.

SAE J2211 Recommended Service Procedure for the Containment of HFC-134a.

SAE J2219 Mobile Air Conditioning Industry Criteria and Guidelines.

SAE J2297 Ultraviolet Leak Detection: Stability and Compatibility Criteria of Fluorescent Leak Detection Dyes for Mobile Air Conditioning Systems.

SAE J2298 Ultraviolet Leak Detection: Procedure for use of Refrigerant Leak Detection Dyes for Service of Mobile Air Conditioning Systems.

SAE J2299 Ultraviolet Leak Detection: Performance Requirements for Fluorescent Refrigerant Leak Detection Dye Injection Equipment for Aftermarket Service of Mobile Air Conditioning Systems.

APPENDIX

EPA SNAP-Approved Refrigerants as Substitutes for CFC-12

Refrigerant	*Supplier*	*Chemical Composition*	*Lubricant Recommended*	*Desiccant Recommended*	*Label Color Background*	*Label Color Foreground*
R-134a	many	100% HFC-134a	PAG POE	XH-7 or XH-9	Sky blue	Black
Freeze 12	Technical Chemical	80% HFC-134a 20% HCFC-142b	MO POE	XH-7 or XH-9	Yellow	Black
Free Zone RB-276	Refrigerant Gases	79% HFC-134a 19% HCFC-142b 2% lubricant	MO	XH-7 or XH-9	Light green	White
FRIGC FR-12	Intercool	59% HFC-134a 39% HCFC-124 2% R-600a	POE	XH-7 or XH-9	Gray	Black
FX-56, 409A*		60% HCFC-22 25% HCFC-124 15% HCFC-142b				
GHG 406A* McCool	People's Welding	55% HCFC-22 41% HCFC-142b 4% R-600a	MO Alkylbenzene	XH-9	Black	
GHG-X4, 414A* Autofrost Chill-it	People's Welding	51% HCFC-22 28.5% HCFC-124 16.5% HCFC-142b 4% R-600a	MO Alkylbenzene	XH-9	Red	White
GHG-X5	People's Welding	41% HCFC-22 15% HCFC-142b 40% HFC-227ea 4% R-600a			Orange	

Refrigerant	*Supplier*	*Chemical Composition*	*Lubricant Recommended*	*Desiccant Recommended*	*Label Color Background*	*Label Color Foreground*
GHG-HP	People's Welding	65% HCFC-22 31% HCFC-142b 4% R-600a	MO Alkylbenzene		Not yet developed	
Hot Shot, 414B* **Kar Kool**	Icor	50% HCFC-22 39% HCFC-124 9.5% HCFC-142b 1.5% R-600a	MO	XH-9	Medium blue	Black
Ikon		25% HFC-152a 75% FIC-1311**	MO			
MP-39, 401A*		53% HCFC-22 34% HCFC-124 13% HFC-152a				
MP-66, 401B*		61% HCFC-22 28% HCFC-124 11% HFC-152a				
MP-52, 401C*		33% HCFC-22 52% HCFC-124 15% HFC-152a				
SP34E	Solpower	Composition claimed as confidential business information.				

*American Society of Heating, Refrigerating, and Air-Conditioning Engineers (ASHRAE) number.
**FIC, triodide fluorodocarbon.
Oils: MO=Mineral oil; PAG=Polyalkaline glycol; POE=polyolester.
All refrigerants subject to use conditions: Unique fittings and label required; barriers hoses required for refrigerants that contain HCFC-22.
Blank spaces indicate information is not available.

UNACCEPTABLE REFRIGERANTS FOR AUTOMOTIVE USAGE

OZ-12/HC-12a, Duracool 12a, R-176 (Artic Chill), R-405A (G2015)

SERVICE FITTING SIZES

	High Side Service Port			*Low Side Service Port*		
Refrigerant	*Diameter*	*Threads per inch (TPI)*	*Direction*	*Diameter*	*Threads per inch (TPI)*	*Direction*
R-12, before 1987	7/16 inch	20	Right	7/16 inch	20	Right
R-12, after 1987	6/16 inch (3/8 inch)	24	Right	7/16 inch	20	Right
R-134a	16 mm	Quick connect		13 mm	Quick connect	
Freeze 12	7/16 inch	14	Left	8/16 inch (1/2 inch)	18	Right
Free Zone	8/16 inch (1/2 inch)	13	Right	9/16 inch	18	Right
FRIGC	17 mm	Quick connect		15 mm	Quick connect	
GHG 406A	0.305 inch	32	Left	0.368 inch	26	Left
GHG X-4*, Autofrost	0.305 inch	32	Right	0.368 inch	26	Right
Chill-it	6/16 inch (3/8 inch)	24	Left	7/16 inch	20	Left
GHG X-5	8/16 inch (1/2 inch)	20	Left	9/16 inch	18	Left
GHG HP	NYT			NYT		
Hot Shot	10/16 inch (5/8 inch)	18	Left	10/16 inch (5/8 inch)	18	Right

*GHG X-4 is sold by two different suppliers, and there is a unique fitting for each.
NYT, not yet developed.

APPENDIX

Potential Compressor Problems with R-134a

Some compressors have problems and experience failure in systems retrofitted with R-134a; others might experience problems that usually result from the retrofit. They are listed by the vehicle they were installed on and by the manufacturer or remanufacturer.

Audi Nippondenso units after date codes 1605 and up, 2893 and up, 3901 and up are okay; earlier ones should have the seal replaced.

Chrysler Dayton A590/C171 might not handle the pressure. A Sanden clone with casting numbers 709CA or 709 CC is okay. Remanufactured A590/C171 are probably okay. Mitsubishi scroll compressor (some models between 1988 and 1993) require control valve change.

Ford Tecumseh HR-980 and Panasonic/Matsushita (on Probes) use viton seals; not okay. FX15 might not handle the pressure; check the orifice tube for metal debris.

GM The following compressors will not work: V5 with blue labels and date codes from 1/88 to 8/89, R4 compressors with date codes from 1/1/90 to 6/18/93, all DA-6 (if the date code is after 1988, it is an HR6 and is okay), Nippondenso 10PA20 used on Corvette from 1988 to 1993, and Sanden 7-cylinder used on medium-duty trucks from 1990 to 1993.

Honda Keihin compressor will probably not handle the pressure. Conversion kits for a different compressor are available.

Moog Automotive, Temperature Control Division Rebuilts Remanufactured compressors since 1993 are compatible with both R-12 and R-134a. However, they do not recommend that an R-12 system using a DA-6 or FX15 compressor be converted. Replace the DA-6 with an HR6 or HD6 and replace the FX15 with an FS10.

Nissan V5 compressor used on Q45 requires replacement with new V6.

Saab Some Sanden 5-cylinder compressors use viton seals; not okay.

Toyota Units without the ink label *RBR* (rubber both refrigerants) require O-ring change at compressor block fitting.

APPENDIX F

English–Metric Conversion Table

Multiply	*By*	*To Get/Multiply*	*By*	*To Get*
Length				
inch (″)	25.4	millimeter (mm)	0.3939	inch
mile	1.609	kilometer	0.621	
Area				
$inch^2$	645.2	$millimeter^2$	0.0015	psi
Pressure				
psi	6.895	kilopascals (kPa)	0.145	$inch^2$
psi	0.0703	kilogram/centimeter (obsolete system)	14.22	psi
psi	0.0689	bar	14.5	psi
psi	0.006895	megapascal (mPa)	0.000145	psi
Volume				
$inch^3$	16,387	$millimeter^3$	0.00006	$inch^3$
$inch^3$	6.45	$centimeter^3$	0.061	$inch^3$
$inch^3$	0.016	liter	61.024	$inch^3$
quart	0.946	liter	1.057	quart
gallon	3.785	liter	0.264	gallon
Weight				
ounce	28.35	gram (g)	0.035	ounce
pound	0.453	kilogram (kg)	2.205	pound
Torque				
inch-pound	0.113	Newton-meter (N-m)	8.851	inch-pound
foot-pound	1.356	Newton-meter	0.738	foot-pound
Velocity				
miles/hour	1.609	kilometer/hour	0.6214	miles/hour
Temperature				
(degrees Fahrenheit − 32) × 0.556 = degrees Celsius				
(degrees Celsius × 1.8) + 32 × degrees Fahrenheit				

APPENDIX

Bolt Torque Tightening Chart

The tightening value for a particular nut or bolt is normally determined by the diameter and grade of the bolt, the material into which the bolt is being threaded, and whether the threads are lubricated. Torque value specifications are normally printed in the service manuals, but if none are available, the following chart can be used as a guide.

SAE STANDARD

Grade Size	*1 and 2*	*5*	*8*	*Wrench Size*	
				Bolt	*Nut*
1/4	5	7	10	3/8	7/16
5/16	9	14	22	1/2	9/16
3/8	15	25	37	9/16	5/8
7/16	24	40	60	5/8	3/4
1/2	37	60	92	3/4	13/16

The following values are given in foot-pounds. They can be converted to inch-pounds by multiplying them by 12 or to Newton-meters by multiplying them by 1.356. These values are given for clean, lubricated bolts.

METRIC STANDARD

Grade Size	*5*	*8*	*10*	*12*	*Wrench Size Bolt*
6 mm	5	9	11	12	10 mm
8 mm	12	21	26	32	14 mm
10 mm	23	40	50	60	17 mm
12 mm	40	70	87	105	19 mm
14 mm	65	110	135	160	22 mm

Full-Size Illustrations of Fittings

The following drawings are reprinted with permission of General Motors Corporation.

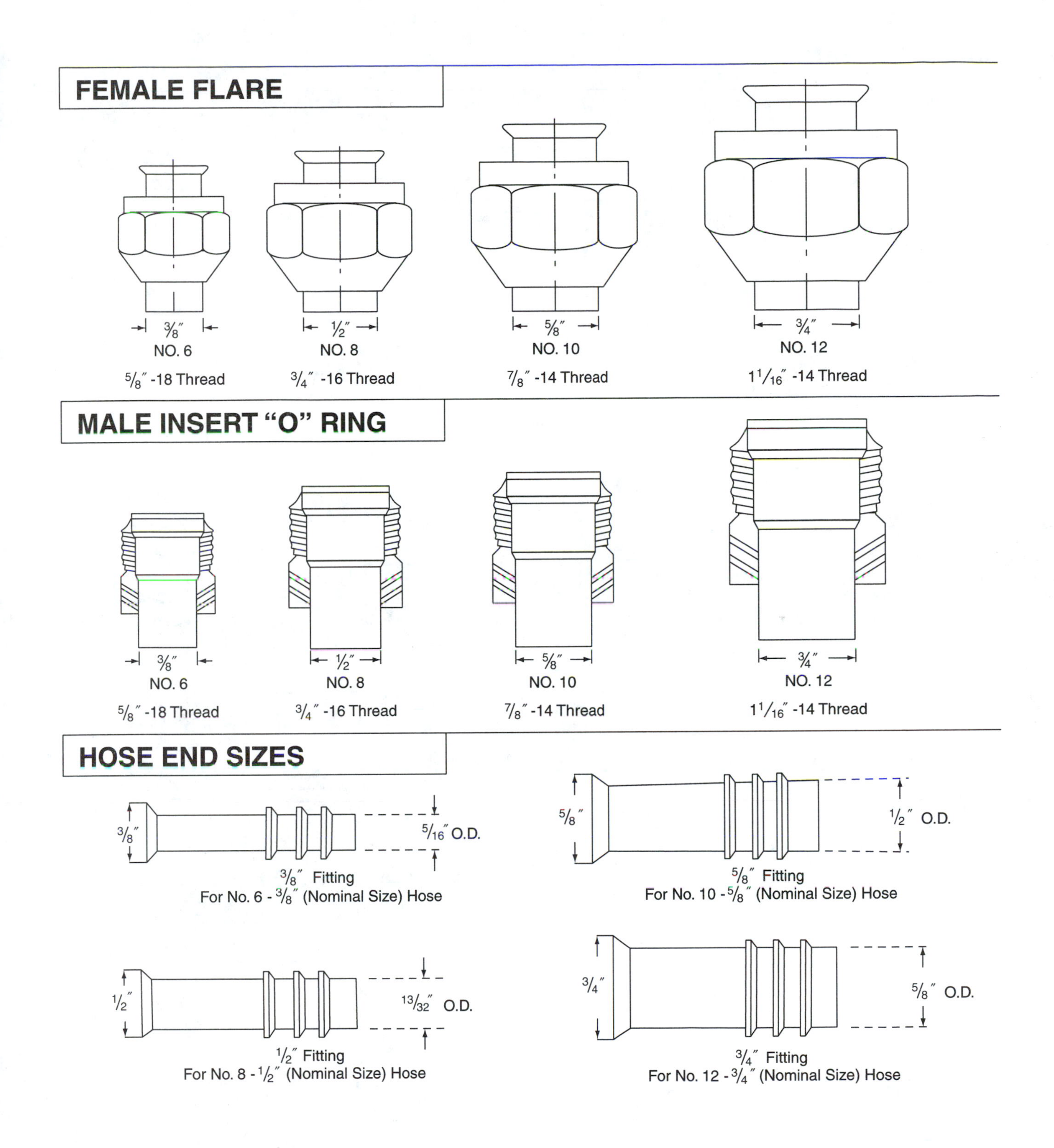
FEMALE FLARE
3/8"
NO. 6
5/8" -18 Thread
1/2"
NO. 8
3/4" -16 Thread
5/8"
NO. 10
7/8" -14 Thread
3/4"
NO. 12
1 1/16" -14 Thread
MALE INSERT "O" RING
3/8"
NO. 6
5/8" -18 Thread
1/2"
NO. 8
3/4" -16 Thread
5/8"
NO. 10
7/8" -14 Thread
3/4"
NO. 12
1 1/16" -14 Thread
HOSE END SIZES
3/8"
5/16" O.D.
3/8" Fitting
For No. 6 - 3/8" (Nominal Size) Hose
5/8"
1/2" O.D.
5/8" Fitting
For No. 10 - 5/8" (Nominal Size) Hose
1/2"
13/32" O.D.
1/2" Fitting
For No. 8 - 1/2" (Nominal Size) Hose
3/4"
5/8" O.D.
3/4" Fitting
For No. 12 - 3/4" (Nominal Size) Hose

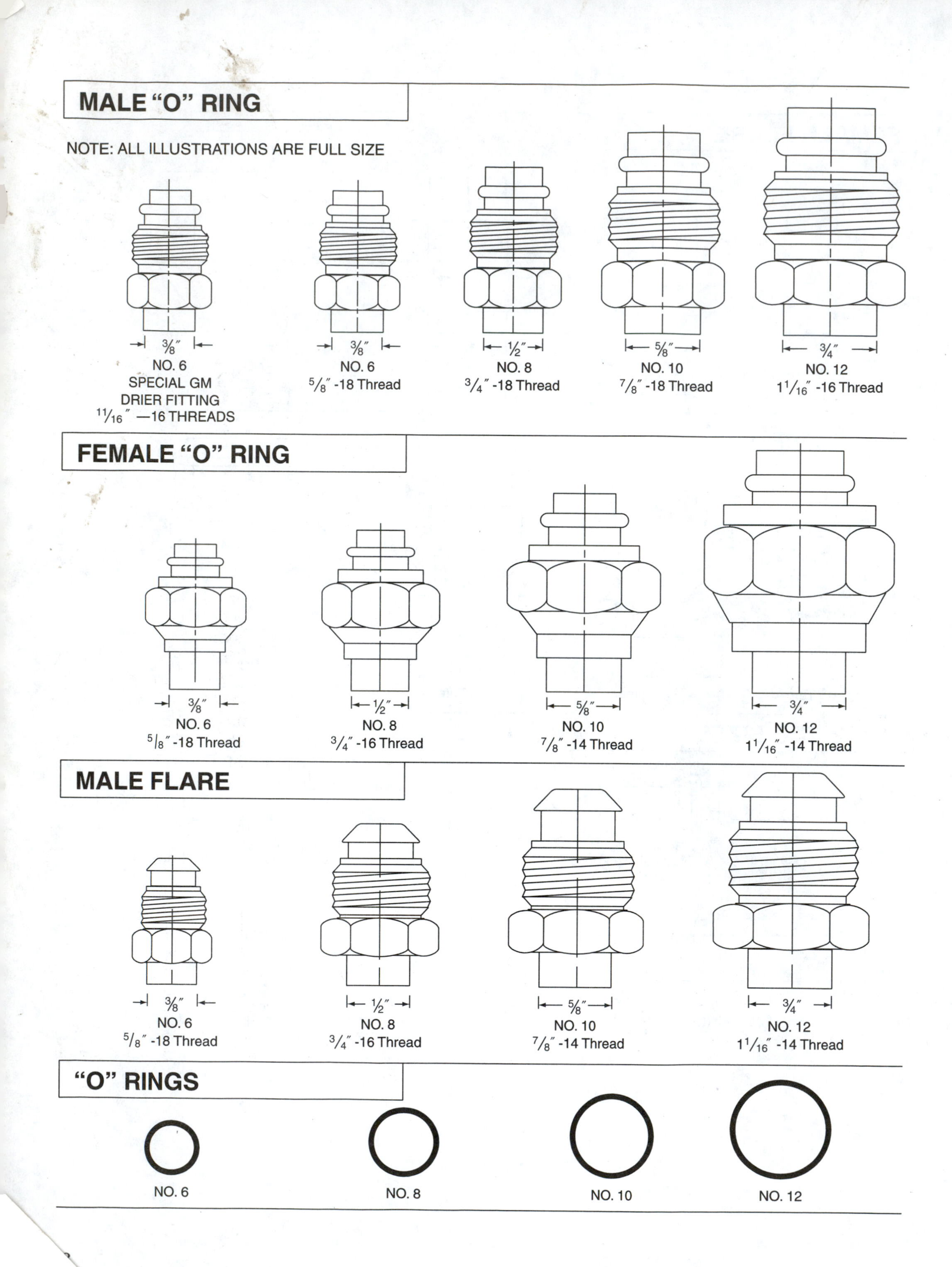
MALE "O" RING
NOTE: ALL ILLUSTRATIONS ARE FULL SIZE
3/8"
NO. 6
SPECIAL GM
DRIER FITTING
11/16" —16 THREADS
3/8"
NO. 6
5/8" -18 Thread
1/2"
NO. 8
3/4" -18 Thread
5/8"
NO. 10
7/8" -18 Thread
3/4"
NO. 12
1 1/16" -16 Thread
FEMALE "O" RING
3/8"
NO. 6
5/8" -18 Thread
1/2"
NO. 8
3/4" -16 Thread
5/8"
NO. 10
7/8" -14 Thread
3/4"
NO. 12
1 1/16" -14 Thread
MALE FLARE
3/8"
NO. 6
5/8" -18 Thread
1/2"
NO. 8
3/4" -16 Thread
5/8"
NO. 10
7/8" -14 Thread
3/4"
NO. 12
1 1/16" -14 Thread
"O" RINGS
NO. 6
NO. 8
NO. 10
NO. 12

APPENDIX

Refrigerant Numbering System

The procedure for determining the refrigerant number is rather tedious and is of no value to the refrigerant technician. It is included in this text to reduce some misconceptions and wild stories that have passed through the industry.

The system is based on the number of carbon, hydrogen, and fluorine atoms in the refrigerant molecule. It is a three-digit system with the letter "a" added to indicate a nonsymmetrical or asymmetrical molecule. Two methods can be used to detemine the number:

- C (minus 1), H (plus 1), and F. R-12 has 1 C (1–1=0); O H (0+1=1); and 2 F (2), to get 012, or 12.
- Add 90 to the number; the sequence is carbon, hydrogen, and fluorine. With R-12, 12+90=102, so this tells us there is 1 C, 0 H, and 2 F.

Look at the R-12 and R-134a molecules in Figure 2–9. You will see that the R-12 molecule is symmetrical; both sides are the same. The R-134a molecule is asymmetrical; the left and right sides are different so an *a* is added to it. There is an HFC-134 molecule that is symmetrical, but it is not used in automotive refrigerant systems.

Technically, we should use the name CFC-12, HFC-134a, or HCFC-22 to indicate that the refrigerant is a chlorofluorocarbon, hydrofluorocarbon, or hydrochlorofluorocarbon, but with an industry in which many people like to call everything "freon," using "R" for refrigerant is a good start. Those of you with better habits than this author can use the proper prefix.

The chemical name for CFC-12, dichlorodifluoromethane, tells us that the molecule has two (di) chlorine atoms and two fluorine atoms, and the suffix methane tells us there is one carbon atom. The prefix, di = 2; tri = 3, tetra = 4, penta = 5, and so on tell us the number of atoms. With carbon-based molecules, single C molecules are methanes, double C molecules are ethanes, three C molecules are propanes, and so on. HFC-134a, tetrafluoroethane, has four (tetra) F atoms, two H (not mentioned in the name), and the suffix ethane, telling us there are two carbon atoms.

This explanation is not complete; chemists use other designations to show bonds within the molecule and different number combinations for blend refrigerants. This numbering system is based on the carbon chain that the molecule is built around. In this system, HFC-134a is 1, 1, 1, 2 tetrafluoroethane.

GLOSSARY

Absolute pressure: Pressure measured above absolute zero pressure.

Absolute zero: The total absence of pressure or temperature.

A/C: Air conditioning; a device that controls the temperature and humidity of the air inside a vehicle.

Accumulator: The component in some A/C systems that contains the desiccant and provides a place to store reserve refrigerant. It is found in the low side.

Actuator: A device that moves another item, such as an air control door.

Aftermarket: An item or part installed after the vehicle was made; not made by the vehicle manufacturer.

Air conditioning: Control of cleanliness, humidity, movement, and temperature of air.

Air inlet door: The door that controls the source of air entering the A/C and heating unit.

Air lock: An incomplete filling problem encountered when filling a cooling system if the air inside the system cannot escape as the coolant is added.

Alternate refrigerant: An approved refrigerant type used to retrofit an R-12 system.

Ambient temperature: The temperature of air surrounding an object; the temperature of air outside a vehicle.

Ambient temperature sensor: A sensor that measures the temperature of outside air entering the vehicle.

Antifreeze: A substance mixed with water to lower the freezing point of the coolant.

Aspirator: A component that pulls air past the in-vehicle temperature sensor; it uses a Venturi principle.

Atmospheric pressure: The pressure of air around us; 14.7 psi at sea level.

Atom: The smallest particle of matter.

Automatic temperature control (ATC): A system that automatically adjusts the output of the heat and A/C system to provide the desired in-vehicle temperature.

Auxiliary gauge: The third gauge used to measure pressure on the outlet side of an EPR valve or STV.

Axial compressor: A compressor with pistons arranged around the center shaft.

Backflush: An operation that forces a liquid through a system in the reverse direction to clean the system.

Back seat: Turning a service valve all the way outward to allow flow from the compressor to the suction or discharge line.

Barrier hose: A hose with a nonpermeable inner liner.

Belt lock system: A strategy that will shut off the A/C compressor if it is operating slower that it should be to prevent a seized compressor from destroying the accessory drive belt.

Black death: A term referring to the black, gooey mess resulting from a catastrophic compressor failure.

Bleed: To loosen or open a fitting or hose to allow air to escape from a cooling system or heater.

Blend: A mixture of refrigerants to form a non-CFC refrigerant usable in an R-12 system.

Blend air: A system that mixes cold and warm air to get the desired temperature.

Blower fan: The fan that forces air through the evaporator and heater core and into the passenger compartment.

Blower speed controller: A solid-state control unit that regulates the blower speed from signals from the control module.

Boiling: Conversion of a liquid to a vapor that is accompanied by bubbles as vapor rises through the liquid.

Boiling point: The temperature at which a liquid boils.

Bowden cable: A cable or wire inside a housing used to control the position of a remote door, valve, or other device.

British thermal unit (Btu): A measurement of heat quantity; the amount of heat needed to increase the temperature of 1 lb of water 1°F.

Calorie: A measurement of heat quantity; the amount of heat required to increase the temperature of 1 g of water 1°C.

Capillary tube: A very thin, gas-filled tube used to sense temperature.
Celsius: Temperature scale based on 0° as the freezing point of water and 100° as the boiling point.
Centigrade: Same as Celsius.
Charge: The specific amount of refrigerant needed to properly fill a system; the act of putting the correct amount of refrigerant into a system.
Charging station: A piece of equipment that includes the things needed for normal A/C system service: gauges, hoses, valves, a vacuum pump, a means of measuring refrigerant, and a storage container for refrigerant.
Chlorofluorocarbon (CFC): A human-made compound that contains chlorine, fluorine, and carbon; R-12.
Climate control: The act of controlling the temperature, humidity, and purity of air.
Clutch: The device used to connect and disconnect the compressor pulley with the compressor shaft.
Clutch cycling switch: The temperature or pressure switch that turns the compressor off and on to prevent the evaporator from freezing.
Cold: The absence of heat.
Cold engine lockout: A device to prevent blower motor operation while the engine coolant is cold.
Compound gauge: A gauge that measures both a pressure and a vacuum; the low side manifold gauge.
Compressor: The component in an A/C system that compresses refrigerant vapor and causes that vapor to move through the system.
Condensation: Conversion of a vapor into a liquid.
Condenser: The component in an A/C system in which heat is removed and vapor is changed to liquid.
Conduction: Heat transfer directly through a material.
Contamination: R-12 or R-134a that contains 3% or more of noncondensable gas or a foreign refrigerant.
Control head: The panel where the controls for A/C and heat are located.
Convection: Heat transfer by circulation of a liquid or vapor.
Conversion fitting: A fitting installed over an R-12 service fitting to allow use of an alternate refrigerant.
Cool: The act of removing heat.
Coolant: A mixture of antifreeze and water used in the cooling system.
Corrosion: Chemical action that eats away a metal substance.
Cycling: A repeated off–on operation at regular intervals.
Cycling clutch: A method of controlling A/C temperature by cycling the compressor off and on.
Cycling clutch orifice tube (CCOT): A system that uses a fixed orifice tube and that cycles the clutch to prevent icing of the evaporator.

Dehumidify: Remove water vapor from air.
Delta T: An engineering term that refers to the difference in temperature between two points.
Desiccant: A drying agent found in the receiver–drier or accumulator that removes moisture.
Diagnostic trouble code (DTC): Numeric code used to indicate the nature of trouble or problems.
Dichlorodifluoromethane: Cl_2Fl_2C, CFC-12, R-12.
Discharge: The act of removing refrigerant from a system (*see* Venting); the outlet side of the compressor.
Discharge line: The line that connects the compressor discharge port to the condenser.
Discharge pressure: High side pressure.
Drier: *See* Receiver–drier.
Drive belt: A belt driven by the crankshaft that drives the A/C compressor, water pump, and other engine accessories.
Dryer: *See* Receiver–drier.

Engine coolant temperature (ECT) sensor: A device to monitor engine coolant temperature and convert it into an electrical signal.
Environmentally friendly: Procedures and products designed to prevent harm to Earth or its atmosphere; also called environmentally aware, environmentally conscious, environmentally safe.
Evacuate: To pump a vacuum into an A/C system to remove water.
Evaporation: The change of state from a liquid to a vapor.
Evaporator: The component in an A/C system in which heat is absorbed and liquid is changed to a vapor.
Evaporator pressure regulator (EPR) valve: A valve used to maintain evaporator pressure to prevent icing.
Expansion tube: *See* Orifice tube (OT).
Expansion valve: *See* Thermal expansion valve (TXV).

Fahrenheit: Temperature scale based on 32° as the freezing point of water and 212° as the boiling point.
Fan clutch: A device used to control the speed of the radiator cooling fan.

Filter: A device used to remove solid foreign particles from the refrigerant or coolant.
Fixed orifice tube: *See* Orifice tube (OT).
Flooding: A term that refers to overcharging a system so liquid refrigerant fills an area.
Fluorocarbon: A compound that contains fluorine and carbon.
Flush: To clean a system by pumping a liquid through it.
Freezing point: The temperature at which a liquid freezes.
Freon: Brand name for some of the refrigerants sold by Du Pont.
Fresh air: Air that enters the A/C and heat system from the outside.
Fresh air filter: A filter that removes dust and pollen from fresh air.
Front seat: Turning a service valve all the way inward to shut off the flow from the compressor to the suction or discharge line.
Fuse: An electrical device designed to open a circuit if there is excess current flow.

Gas: A vapor without any droplets of liquid.
Gauge pressure: Referring to a pressure scale in which 0 is equal to atmospheric pressure.
Gauge set: *See* Manifold gauge set.

Head pressure: The same as high side pressure.
Heat: A form of energy that raises temperature.
Heater core: The component in a heater system that transfers heat from the coolant to the air.
Heat exchanger: A device used to move heat from one item to another; a vehicle's condenser, evaporator, and radiator are heat exchangers.
Heat pump: A combination A/C and heater system in which the line connections at the compressor can be reversed to provide both A/C and heat.
Hg: The chemical symbol for mercury; vacuum is measured in inches of mercury (″ Hg).
High-pressure relief valve: A valve in the high side of a system designed to open and release excessive pressure.
High side: Part of the A/C system that has high pressure and contains liquid refrigerant.
Humidity: Water vapor in the air.
HVAC: Heating, ventilation, and air conditioning.
HVAC air filter: A filter contained in the air distribution system that is designed to remove dust and pollen particles from the air stream.
Hydraulic fan: A fan-drive system that operates by hydraulic flow. It usually incorporates electronic controls.
Hydrochlorofluorocarbon (HCFC): A human-made compound that contains hydrogen, chlorine, fluorine, and carbon.
Hydrofluorocarbon (HFC): A human-made compound that contains hydrogen, fluorine, and carbon; R-134a.
Hygroscopic: The ability to readily take in and retain moisture.

Identifier: A piece of equipment used to determine the purity of a refrigerant.
In-car sensor: A temperature sensor that measures the temperature of the air inside the vehicle.
Insulate: Add a barrier to isolate or seal with a nonconductor.

Latent heat: Heat that causes a change in state without a change in temperature.
Leak detector: Equipment used to locate refrigerant leaks.
Liquid: A state of matter in which the atoms move freely but do not separate as in a vapor.
Liquid line: The line that connects the condenser outlet to the receiver–drier inlet or orifice tube and the receiver–drier outlet to the TXV.
Low side: Part of the A/C system that has low pressure and contains refrigerant vapor.

Magnetic clutch: *See* Clutch.
Malfunction indicator light (MIL): A light on the instrument panel to inform the driver that a malfunction or problem has occurred.
Manifold gauge set: Service equipment that consists of two pressure gauges, three or four service hoses, and two valves; used for pressure testing and servicing an A/C system.
Matter: The substance created when atoms join together.
Mode door: A flap valve that moves to divert airflow to the floor, instrument panel, or windshield outlets.
Muffler: A device that quiets compressor pumping sounds.
Multiplexing: A system of sending two or more signals through a common circuit; microprocessors are required at each end of the circuit.

Noncondensable gases (NCG): Gases, such as air in an A/C system, that do not readily change to a liquid state as the refrigerant does.

OEM: Original equipment manufacturer.
Oil bleed hole: An orifice at the bottom of the accumulator that allows oil and some liquid

refrigerant to flow to the compressor for lubrication.

Organic acid technology (OAT): The inhibitor package used with extended-life antifreeze.

Orifice tube (OT): The component in some A/C systems that provides a restriction to meter the refrigerant flow into the evaporator. It causes a pressure buildup in the high side and a pressure drop in the evaporator and low side.

Overcharge: Too much refrigerant in a system.

Ozone: A chemical link between three oxygen atoms (O_3).

Ozone layer: Layer that acts as a shield against ultraviolet radiation in the upper atmosphere.

Panel registers: Outlets in the instrument panel used to direct air into the passenger compartment.

Performance test: The standard diagnostic test of system pressures and temperatures to determine whether a system is operating properly.

Pilot-operated absolute (POA) valve: A valve used to maintain evaporator pressure to prevent icing.

Plenum: The chamber in the A/C and heat unit in which the cool and warm air are blended for the desired temperature.

Polyalkylene glycol (PAG): A synthetic lubricant used with R-134a refrigerant.

Polyolester (POE): A synthetic lubricant used with R-134a refrigerant.

Pressure: A force per unit of area. *See also* psi.

Pressure cutoff switch: A switch that stops compressor operation when high side pressures get too high.

Programmer: An electronic device that controls ATC system operation.

psi: Pounds per square inch.

psig: *See* Gauge pressure.

Pulse width–modulated (PWM) blower speed: A method of controlling blower speed by changing the duty cycle.

Pump down: *See* Evacuate.

Purge: To flush a system with liquid refrigerant or dry nitrogen to remove all air or moisture; to thoroughly evacuate a system.

Radiation: Heat transfer from the sun's rays.

Radiator: The component in a cooling system that transfers heat from the coolant to air.

Ram air: Air forced through the condenser and radiator by the forward movement of the vehicle.

Receiver–drier: The component in some A/C systems that contains the desiccant and provides a place to store liquid refrigerant. It is found in the high side.

Recirculated air: Airflow into the A/C and heat system from the passenger compartment.

Reclaim: The act of removing all impurities from refrigerant so it is essentially identical to new, unused refrigerant. Reclamation is not performed in automotive service shops.

Recover: The act of removing refrigerant from a system so that it can be recycled for reuse.

Recycle: The act of using portable equipment to clean debris, oil, water, and noncondensable gases from refrigerant so it can be reused in a system.

Refrigerant: The agent in a liquid and vapor form in the A/C system used to absorb, transfer, and release heat.

Refrigeration: The use of mechanical means to remove heat.

Relative humidity: The amount of water vapor contained in air relative to the maximum amount it can contain at a given temperature.

Resistor: A device used in electrical circuits to reduce current flow or drop voltage.

Retrofit: The act of replacing the R-12 in a system with an EPA-approved refrigerant.

Saddle clamp: A fitting that is clamped onto a metal line to allow attachment of a service port.

Saturated vapor: A vapor that is in contact with its liquid in an enclosed space.

Schrader valve: A spring-closed valve used at A/C service ports.

Sealant: A one- or two-part chemical that is charged into an A/C or cooling system that is designed to seal small leaks.

Secondary loop system: An A/C system design that uses one cooling system contained completely under the hood and another loop to transfer heat from the passenger compartment to the primary cooling loop.

Semiautomatic temperature control (SATC): A climate control system that can automatically maintain the air temperature and blower speed; the operator controls where the air is discharged.

Sensible heat: Heat that causes a temperature change without a change of state.

Service hoses: Hoses used to connect a manifold gauge set or charging station to a vehicle's A/C system.

Service ports: A fitting to which service hoses are attached.

Servomotor: A motor that is capable of positioning an air control door or other item to a particular position.

Short cycling: A condition in which the compressor cycles off and on too frequently.

Sight glass: A window in the liquid line or receiver–drier that allows observation of refrigerant flow.

Solid: A state of matter that has a fixed shape in which the atoms have restricted movement.

Static pressure: Pressure in a system that is not operating.

Stop leak: *See* Sealant.

Subcooling: Liquid that is cooled below the point of condensation; the decrease in temperature of the liquid leaving the condenser.

Suction line: The line that connects the evaporator outlet to the compressor suction port.

Suction pressure: Same as low side pressure.

Suction side: *See* Low side.

Suction throttling valve (STV): A valve used to maintain evaporator pressure to prevent icing.

Sunload sensor: A sensor that detects the amount of sunlight and radiant heat entering the vehicle.

Superheat: Vapor that is heated above the boiling point; the increase in temperature of the evaporator outlet above the inlet.

Temperature: Heat intensity measured in degrees.

Temperature-blend door: A door in the A/C and heat system that mixes cold and hot air to get the desired temperature.

Temperature sensor: A device used to determine a temperature and convert it into an electrical signal.

Tetrafluoroethane: HFC-134a, R-134a.

Thermal: Refers to heat.

Thermal expansion valve (TXV): The component in some A/C systems that meters the refrigerant flow into the evaporator. It causes a pressure buildup in the high side and a pressure drop in the evaporator and low side.

Thermal limiter: A device that opens a circuit when the temperature gets too high.

Thermistor: A device used in electrical circuits that changes resistance relative to temperature.

Thermostat: An automatic device for regulating temperatures; used in a cooling system to control the circulation between the engine water jackets and the radiator.

Thermostatic switch: A control switch used in some A/C systems that monitors evaporator air temperature and opens the compressor clutch circuit to prevent icing.

Topping off: The act of adding refrigerant to a system without making any repairs.

Transducer: A device used to change an incoming signal of one type to an outgoing signal of another type.

Undercharge: A system that is low on refrigerant.

Under-dash unit: An industry name that refers to hang-on A/C systems; aftermarket systems.

Vacuum: A pressure below atmospheric; usually read in inches of mercury.

Vacuum actuator: A device used to move air control doors.

Vacuum motor: *See* Vacuum actuator.

Vacuum pump: A device used to evacuate a refrigeration system.

Vapor: A gas; the gaseous state of a refrigerant or water.

Variable orifice valve (VOV): An orifice tube with a valve that changes orifice size in response to refrigerant flow and pressure.

Venting: The act of releasing refrigerant or a gas into the atmosphere. Venting a refrigerant is illegal.

Viscosity: The relative thickness of a liquid.

Water valve: A device for controlling the flow through a heater core.

INDEX

A

B

C

S